MEDICAL CELL BIOLOGY

CHARLES J. FLICKINGER, M.D.
Professor, Department of Anatomy

JAY C. BROWN, Ph.D.
Associate Professor of Microbiology

HOWARD C. KUTCHAI, Ph.D.
Associate Professor of Physiology

JAMES W. OGILVIE, Ph.D.
Associate Professor of Biochemistry

All of the University of Virginia School of Medicine, Charlottesville, Virginia

with a contribution by
MICHAEL J. PEACH, Ph.D.
Professor of Pharmacology,
University of Virginia School of Medicine

1979

W. B. SAUNDERS COMPANY • Philadelphia • London • Toronto

W. B. Saunders Company: West Washington Square
Philadelphia, PA 19105

1 St. Anne's Road
Eastbourne, East Sussex BN21 3UN, England

1 Goldthorne Avenue
Toronto, Ontario M8Z 5T9, Canada

Medical Cell Biology ISBN 0-7216-3721-3

Last digit is the print number: 9 8 7 6 5 4 3 2 1

PREFACE

This book is intended to serve as the text for a cell biology course for first-year medical students, other health professionals, or advanced undergraduates as well as a source for review and update of cell biology for residents and practicing physicians. In writing it, the most difficult question was what material to include from the broad field of cell biology. This difficulty was compounded by the fact that the subject matter of cell biology courses varies considerably. Our approach to this problem has been to attempt to make the coverage of the book broad enough to be adequate for a wide range of courses. We hope that this will permit individual course directors to choose sections that are most pertinent to their needs and that they will find it useful to have additional related material close at hand for reference.

The text presents an integrated approach to cell biology, encompassing information on cell morphology, physiology, molecular biology, biochemistry, and pharmacology. The rationale for this approach is our belief that an accurate and comprehensive picture of current knowledge of the cell can be attained only by utilizing information from several disciplines.

To emphasize the relevance of the book to medical practice, examples of applications of cell biology in clinical medicine and pathology are interspersed throughout, and Chapter 20 illustrates the use of principles of cell biology in pharmacology. The mention of clinical applications is intended to stimulate the reader's interest in basic information. Clearly it is not possible to be comprehensive in this effort. We hope that students will realize that the examples included represent only a few of the many interesting situations in which knowledge of cell biology can enhance their understanding of the human organism.

The authors have participated in a course given for several years to first-year medical students at the University of Virginia School of Medicine. The selection of material and the manner of interrelating subject matter is based on our teaching experience. We were also fortunate to be able to enlist the aid of Michael J. Peach, who wrote Chapter 20, "Interactions of Drugs with Cells." Because of the integrated nature of the approach, authors' names are not assigned to chapters. The book is a thoroughly collaborative effort, and the entire manuscript was edited and rewritten by Charles J. Flickinger, with the aim of achieving continuity and uniformity of style.

We wish to thank all those who have contributed to this book and assisted in its preparation. We appreciate very much the micrographs, tables, and graphs contributed by the many individuals who are identified in figure legends. The assistance of John J. Hanley and Roberta Kangilaski and their associates at W.

B. Saunders Company was invaluable at virtually all stages of preparation of the book and is gratefully acknowledged. Thanks are owed to our colleagues Gary Ackers, Virginia Huxley, Robert Kadner, Richard Murphy, Lionel Rebhun, Stephen Vogel, and George Zavoico, who read selected parts of the text. We are indebted to Mary Loving for typing much of the manuscript, to Mary Stuart for printing micrographs, and to Bette Flickinger and Bev Ogilvie for typing drafts of the text. Finally, we are indebted to many University of Virginia medical and graduate students for their assistance in helping us to refine and clarify the material presented here.

Charles J. Flickinger
Jay C. Brown
Howard C. Kutchai
James W. Ogilvie

CONTENTS

1

INTRODUCTION TO THE STUDY OF CELL STRUCTURE AND CHEMISTRY

The study of cells was begun in 1665 by the Englishman Robert Hooke, who built a primitive microscope and with its aid discovered the cellulose walls of plant cells in hand-cut slices of cork. In 1674, Leeuwenhoek in Holland made a remarkable series of observations on "free" cells, such as protozoa and bacteria, from diverse sources, including pond water and plant infusions. Understanding of the biological importance of cellular organization, however, remained in a primitive state until there was a renewed interest in cells in the early part of the 19th century. This activity led to formulation of the cell theory of Schleiden and Schwann, which stated simply that all plants and animals are composed of cells and their products. Although it is so obvious today as to seem trivial, in its time this articulation of the ubiquity and fundamental importance of cells was a significant advance and served as a foundation for subsequent work. In the late 19th and early 20th centuries there was a period of great activity in cytology. Many of our fundamental concepts about cells were established, including the ideas that cells reproduce themselves, contain a nucleus and a cytoplasm, and exhibit chromosomes that have a complicated and consistent behavior during mitosis. Cellular structure was described in great detail at the light microscope level in a wide variety of different cell types. The work of Rudolf Virchow during this period is of particular medical interest, because it was he who applied the new knowledge about cells to pathology and suggested that the pathological changes seen in disease originate from alterations in the activities of cells.

The past 25 years have been another period of great productivity in cell biology. Like the earlier surge in activity around the turn of the century, which may be attributed to technical improvements in the optical microscope, the current increase in knowledge of cells is due, at least in part, to technical innovations. These include the development of the electron microscope, the ultracentrifuge, and a large number of other instruments and methods. During this time there also has been a convergence of cytology, which is the study of cell structure, and other disciplines, such as biochemistry and genetics. These various approaches reveal different aspects of cellular structure and function. Combined, they can provide a comprehensive picture of how cells perform.

BASIC CYTOLOGICAL METHODS

It would be desirable to base a description of cell structure on the appearance of living rather than killed cells, but for technical reasons the study of living cells is often impractical. Most tissues are too thick for microscopic study and must be cut in sections. Furthermore, morphological study of living cells is at present limited to light microscopy because of the need for very thin specimens to be placed in a vacuum for examination in the electron microscope. Experiments on the use of high-voltage electron microscopes to examine hydrated specimens offer some hope of applying electron microscopy to living cells in the future.

Some special light microscope techniques have proved especially useful for the study of living cells. An example is the phase microscope, which makes use of differences in refractive index to produce differences in contrast between parts of cells. With the phase microscope, the nucleus, nucleolus, and some cytoplasmic components, such as mitochondria, can be distinguished in living cells (Fig. 1–1). The use of cell and tissue culture methods has produced a large amount of information about the behavior and biochemistry of living cells. Study of living cells in culture by cinemicrography has proved useful in analyzing cell motility and movements of organelles within cells.

In most cases, however, cell structure is studied in killed cells through the use of fixed preparations (Fig. 1–2). The principles of specimen preparation are similar for light microscopy and transmission electron microscopy (Fig. 1–3), although the materials differ. Cells are killed, and their parts are precipitated in as lifelike a pattern as possible by a chemical fixative. Various aldehydes, acids, and other agents are used for light microscopy. The quality of preservation during fixation required for electron microscopy is higher than that for light microscopy, and solutions of glutaraldehyde and other aldehydes are commonly used. This step must be followed by postfixation in buffered osmium tetroxide, which results in the attachment of atoms of heavy metal to parts of the cell, increasing their visibility in the electron microscope.

After fixation the specimens are dehydrated and embedded in a firm substance, so that sections can be cut with a microtome. Paraffin is the most commonly used embedding medium for light microscopy, and sections 5 to 15 μm

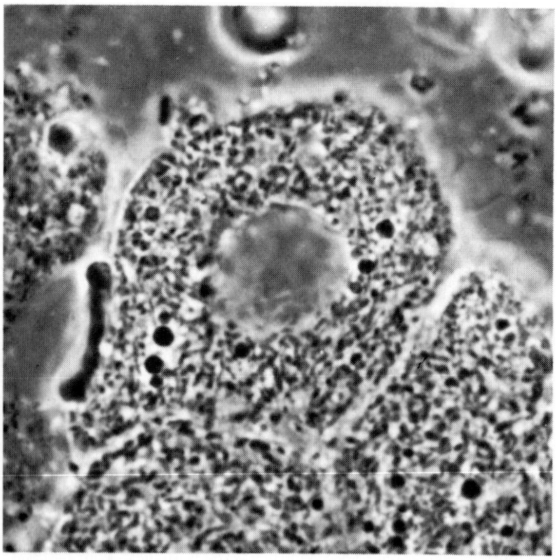

FIGURE 1–1 Phase contrast micrograph of living isolated rat liver cells. The spherical nucleus is visible in the center of the cell, surrounded by a cytoplasm rich in organelles and inclusions. Many of the filamentous and globular structures in the cytoplasm are mitochondria. × 2400. (Micrograph courtesy of P. Drochmans.)

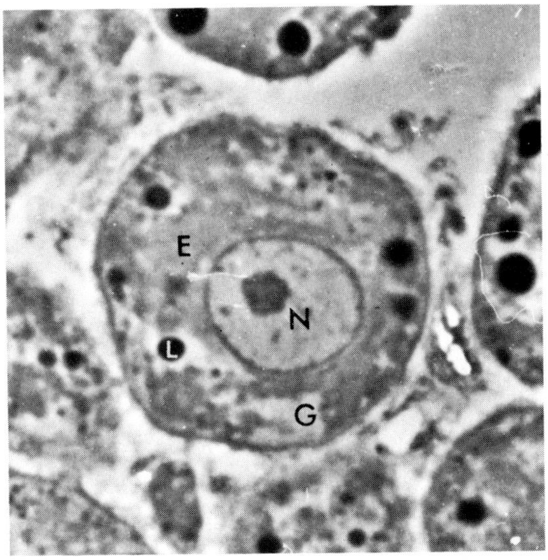

FIGURE 1–2 Phase contrast micrograph of fixed and sectioned isolated rat liver cells. The nucleus (N) contains a large dense nucleolus. Several areas are distinguishable in the cytoplasm, including dark regions (E), occupied by rough endoplasmic reticulum and mitochondria, and pale glycogen-rich regions (G). L, lipid droplet. × 2400. Compare with Figure 1–1. Micrograph courtesy of Drochmans, P., Wanson, J.–C., and Mosselmans, R.: J. Cell Biol., 66: 1, 1975. Reproduced by permission of the Rockefeller University Press.)

thick are cut with steel knives. An alternate method of specimen preparation is the freezing process. Samples are hardened by freezing them rapidly, thus eliminating the need for fixation and embedding. Because of the speed of the process it is used in pathology for the rapid examination of surgical specimens. Since chemical fixatives and solvents are avoided, it also finds application in histochemical studies. The necessity for thinner sections for electron microscopy requires the use of harder embedding media. These are usually plastics, such as the epoxy resins, which are infiltrated into the tissue in a fluid, monomeric form

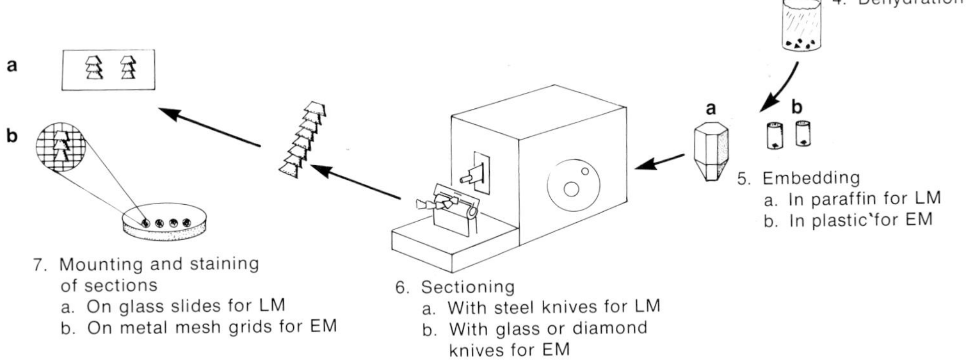

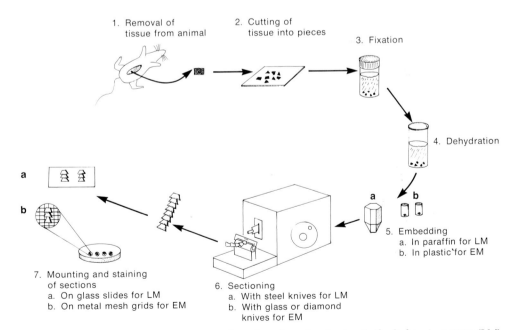

FIGURE 1–3 The main steps in preparation of sections for viewing in the light microscope (LM) or the transmission electron microscope (EM). The principles are similar in both cases, but the chemicals and other materials used differ because of the need for better preservation and thinner sections for electron microscopy.

and then polymerized to a hard polymer. Sectioning of this material at 50 to 100 nm is accomplished with glass or diamond knives.

Sections for light microscopy are usually mounted on glass slides and stained with any of a large number of dyes that reveal different features of cell structure (Fig. 1–4). A colored image is produced by differences in the absorption of light of varying wavelengths by different parts of the specimen. Alternatively, unstained sections can be viewed with the phase contrast microscope (Fig. 1–2).

Sections for electron microscopy are usually mounted on circular metal mesh grids and "stained" with a solution of a heavy metal, such as lead citrate or uranyl acetate (Fig. 1–5). In the transmission electron microscope, an image is produced on a fluorescent screen (or photographic film) by electrons passing through the section, similar to the way in which light passing through sections forms the image viewed in the light microscope (Fig. 1–6). However, in the case of the electron microscope, contrast in the image is produced by differences in the electron-scattering ability of different parts of the specimen. This electron-scattering ability is directly related to the atomic weight of the atoms present, so that regions containing more elements of high atomic weight scatter electrons more effectively and appear darker than surrounding areas. The fact that most biological specimens contain mainly light atoms, such as C, H, O, and N, accounts for the use of heavy metals, such as Os, Pb, and U, in fixatives and stains in electron microscopy, since their addition enhances the electron-scattering ability of the specimen.

An alternate method of specimen preparation for transmission electron microscopy is the freeze-cleave process. This is particularly applicable to the study of membranes and cell contact specializations and will be described in Chapter 14.

In recent years, increasing use has been made of the scanning electron microscope, which furnishes a view of the surfaces of cells and tissues. In this case, the specimen is fixed, dried, and coated with a thin layer of metal. In the microscope, the surface of the specimen is scanned by a pencil of electrons. The secondary electrons that are emitted are recorded and furnish a picture of the contours of the surface of the specimen (Fig. 1–7).

Recall the following measurements:

1 centimeter (cm)	=	10 millimeters (mm)
1 millimeter (mm)	=	1000 micrometers (μm)
1 micrometer (μm)	=	1000 nanometers (nm)
1 nanometer (nm)	=	10 angstroms (Å)

Approximate dimensions of the components of the hierarchy of organization in cells are given here.

Cells	μm to cm
Organelles	nm to μm
Macromolecules (proteins)	5 to 500 nm
Small molecules (amino acids)	1 nm
Atoms	Å

Most of the light micrographs reproduced in this book provide views of cells at magnifications of a few hundred up to one or two thousand times. The electron micrographs include much higher magnifications, ranging from a few thousand to

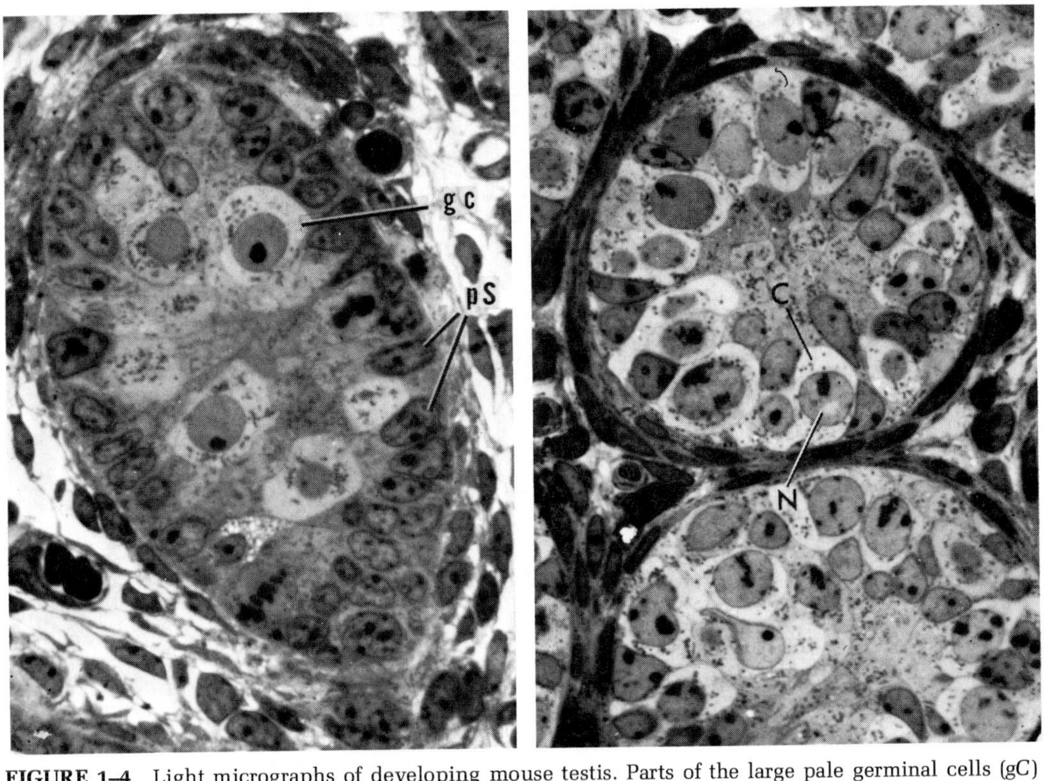

FIGURE 1–4 Light micrographs of developing mouse testis. Parts of the large pale germinal cells (gC) visible in light microscope sections include the nucleus (N) with its dense nucleolus, and the cytoplasm (C), which contains granules representing mitochondria and other organelles. Similar germinal cells are shown in electron micrographs in Figure 1–5 and 1–11. The smaller, denser cells (pS) are immature Sertoli cells, the supporting cells of the seminiferous tubules. Toluidine blue stain. (From Flickinger, C.: Z. Zellforsch., *78*:92, 1967. Reproduced by permission of Springer-Verlag.)

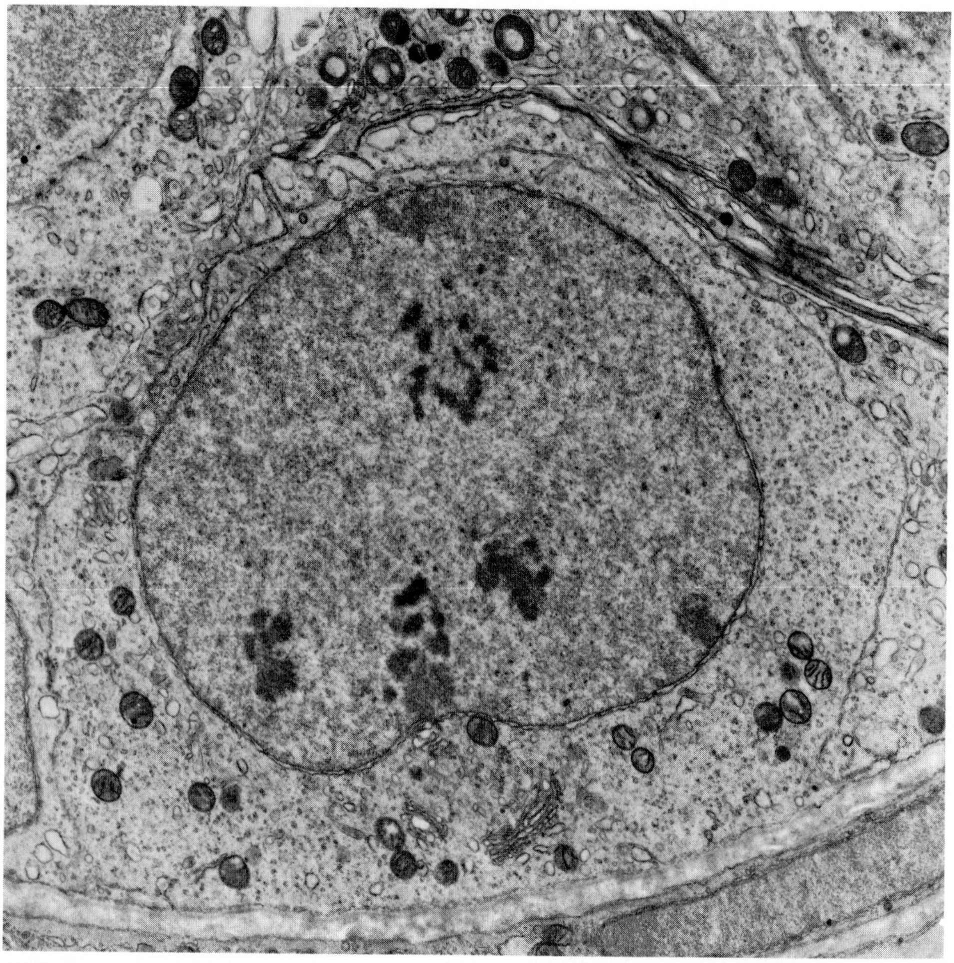

FIGURE 1–5 Electron micrograph of a germ cell in the developing mouse testis. This cell is similar to the one that is labeled in Figure 1–4. In comparison with the light micrograph, further details of the organization of the nucleus and cytoplasm are made visible by the greater resolution, thinner sections, and more refined preparative procedures employed in electron microscopy. × 12,000.

hundreds of thousands of times. In viewing the illustrations it is necessary to appreciate the relative sizes of the structures depicted and to bear in mind which structures are visible with the light microscope and which are visible only with the increased magnification of the electron microscope. The increase in effective magnification of electron micrographs is due to the greater resolution, or resolving power, of the electron microscope compared with the light microscope.

Resolution may be defined as the minimum distance that two points in a specimen can be separated and still be distinguishable as two individual points. For example, if the resolution of a microscope were 1 μm, then two points 1 μm or more from each another could be distinguished. However, if they were closer together than 1 μm, they would appear as a blurred single image. The limit of resolution due to diffraction is given by the following equation:

$$\text{Resolution} = \frac{0.61\lambda}{\text{NA}} = \frac{0.61\lambda}{\text{n}\ \sin\alpha}$$

where λ is the wavelength of the radiation (light or electrons), NA is the numerical aperture of the objective lens, n is the refractive index of the medium, and α

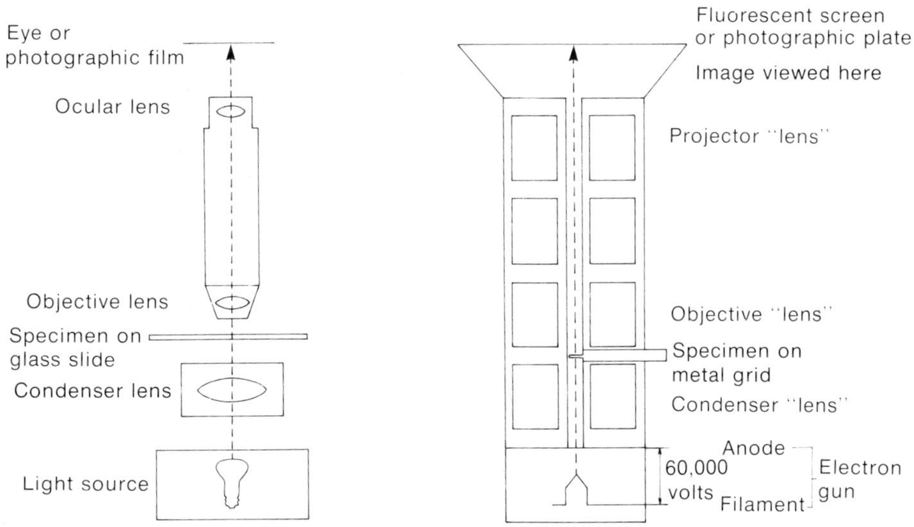

FIGURE 1–6 Diagram of the light and the transmission electron microscopes. The analogy between components of the two instruments is illustrated. The "lenses" in the electron microscope are electromagnetic coils, which are capable of deflecting the path of electrons. For comparison with the light microscope, the column of the electron microscope has been inverted from its usual orientation. In most transmission electron microscopes the electrons pass vertically from top to bottom rather than as shown here.

equals one half the angle of the cone of radiation entering the objective lens. In the case of the electron microscope, the wavelength of the electrons is related to the accelerating voltage and is much less (~ 0.05 Å) than the wavelength of light (~ 5000 Å, green light); consequently much smaller structures can be resolved with the electron microscope. A further discussion of these and other factors, such as aberrations in lenses, which determine the resolution of microscopes must be left to more specialized books, but it should be noted that the resolution of light microscopes is about 0.2 μm, while that of current transmission electron microscopes is in the range of 2 to 10 Å.

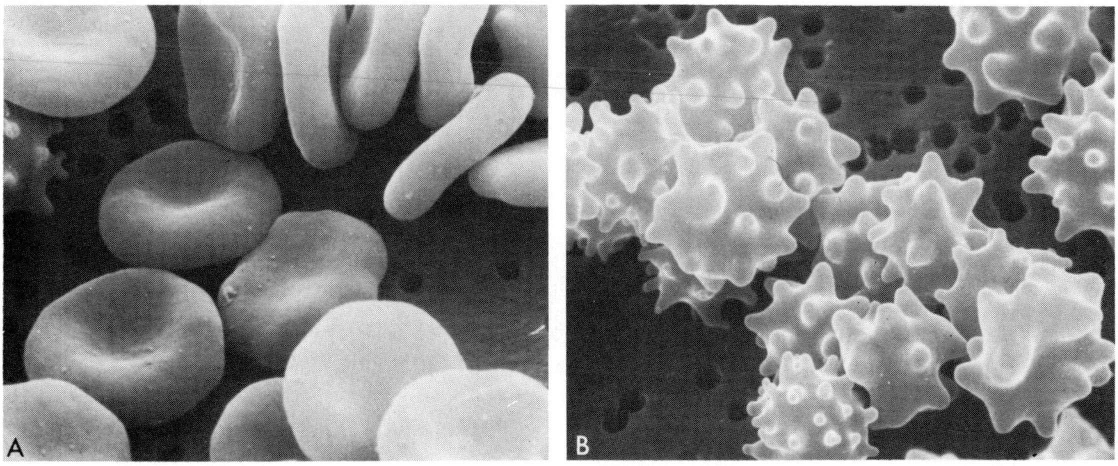

FIGURE 1–7 Scanning electron micrographs of human red blood cells. The scanning electron microscope is very useful in studying the shape of cells and the configuration of their surfaces. The cells in *A* were freshly obtained from serum and placed in a buffer. *B* illustrates the crenated shape of the cells after 24 hours in buffer, which leads to depletion of intracellular ATP. (Micrograph courtesy of Sheetz, M. P., Painter, R. G., and Singer, S. J.: J. Cell Biol., *70*: 193, 1976. Reproduced by permission of the Rockefeller University Press.)

With this background on how images of cell structure are usually obtained, the structural features of prokaryotic and eukaryotic cells will now be outlined. This brief overview of cell structure is intended to serve as a framework for the more detailed study of cell structure, chemistry, and physiology that follows in subsequent chapters.

STRUCTURAL FEATURES OF PROKARYOTIC AND EUKARYOTIC CELLS

Two main kinds of cells, the prokaryotes and the eukaryotes, are distinguished on the basis of certain fundamental structural and biochemical differences (Table 1–1). The simpler prokaryotes include bacteria, blue-green algae, and PPLO's (pleuropneumonia-like organisms, or *Mycoplasma*). The eukaryotes comprise all other "higher" cells, including those of fungi, plants, and animals. In subsequent chapters attention will be directed toward eukaryotic cells, because emphasis in medicine is of course placed on understanding normal and pathological states of human cells. Prokaryotes, however, are also important in medicine because some of them are disease agents. In addition, much information about how cells function, especially at the molecular level, has been obtained from prokaryotes.

Prokaryotes

Most bacteria (Figs. 1–8, 1–9) are small cells approximately 1 μm long. They are surrounded by a rigid cell wall, a structure not present in animal cells. Underlying the cell wall is the plasma membrane, which is a feature of all cells. The plasma membrane delimits the external margin of the cell's cytoplasm and usually appears trilaminar in electron micrographs. The plasma membrane serves as a barrier between the cell and its external environment and is important in maintaining the internal composition of the cell.

The interior of the bacterial cell is divisible into two main regions, the nucleoid and the cytoplasm. However, there is no membrane or nuclear envelope separating the two regions. The nucleoid has a fibrillar or filamentous ultrastructure. It contains the bacterial chromosome, which is a single circular molecule of DNA. Prokaryotic nucleoids do not contain a nucleolus, and the DNA is not complexed with histones, as in eukaryotes.

Table 1–1 Some Differences Between Prokaryotic and Eukaryotic Cells*

Prokaryotes	Eukaryotes
No nuclear envelope	Nuclear envelope present
No nucleolus	Nucleolus present
No histones	DNA complexed with histones
Few intracellular membranes	Many membranous organelles
60S ribosomes	80S ribosomes

*Differences in the size of ribosomes and in the protein synthesizing systems in prokaryotes and eukaryotes will be discussed in Chapter 7.

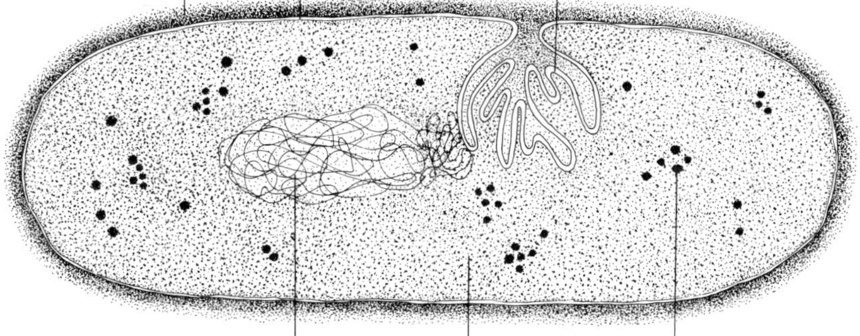

FIGURE 1–8 Diagram of the main structural components of a prokaryotic cell, a bacterium.

The cytoplasm is chemically complex, but virtually the only visible structures are the bacterial ribosomes, which contain RNA and protein and appear as dense granules about 150 Å in diameter. Aggregates of ribosomes, the polyribosomes, are active in synthesis of proteins. The mesosome, an infolding of the plasma membrane, may also be visible in the cytoplasm. It contains respiratory enzymes, and the bacterial chromosome is attached to it. It has been proposed that the mesosome plays a role in the separation of daughter chromosomes during cell division, and it is thought that DNA synthesis is initiated at the point of attachment of the chromosome to the membrane.

Eukaryotes

Besides the greater size of eukaryotic cells, which range from 4 μm to many centimeters in their dimensions, the most obvious structural difference between prokaryotes and eukaryotes is the compartmentalization of the eukaryotic cell (Figs. 1–10 and 1–11). Eukaryotes contain a large number of intracellular membranes which separate parts of the cell into different organelles. Some of the organelles, such as mitochondria and endoplasmic reticulum, are visible with the light microscope, but the electron microscope is required for definition of the details of their organization.

The most easily recognizable division in eukaryotic cells is the separation of the nucleus from the cytoplasm by the nuclear envelope (Figs. 1–11 and 1–12). The nuclear envelope is composed of two nuclear membranes that enclose a space, the perinuclear cisterna. The two membranes of the nuclear envelope are fused at intervals to form nuclear pores, channels through which the karyoplasm inside the nucleus is in apparent communication with the cytoplasm. Two main structures are discernible in the interior of the nucleus (Fig. 1–11). The chromatin consists of fine fibrils about 100 Å thick which fill much of the interior of the nucleus. Chromatin is composed of DNA and protein and represents the genetic material. The nucleolus is a compact, often spherical, body with different regions composed of granules and fibrils. It is rich in RNA and protein and is the source of ribosomes.

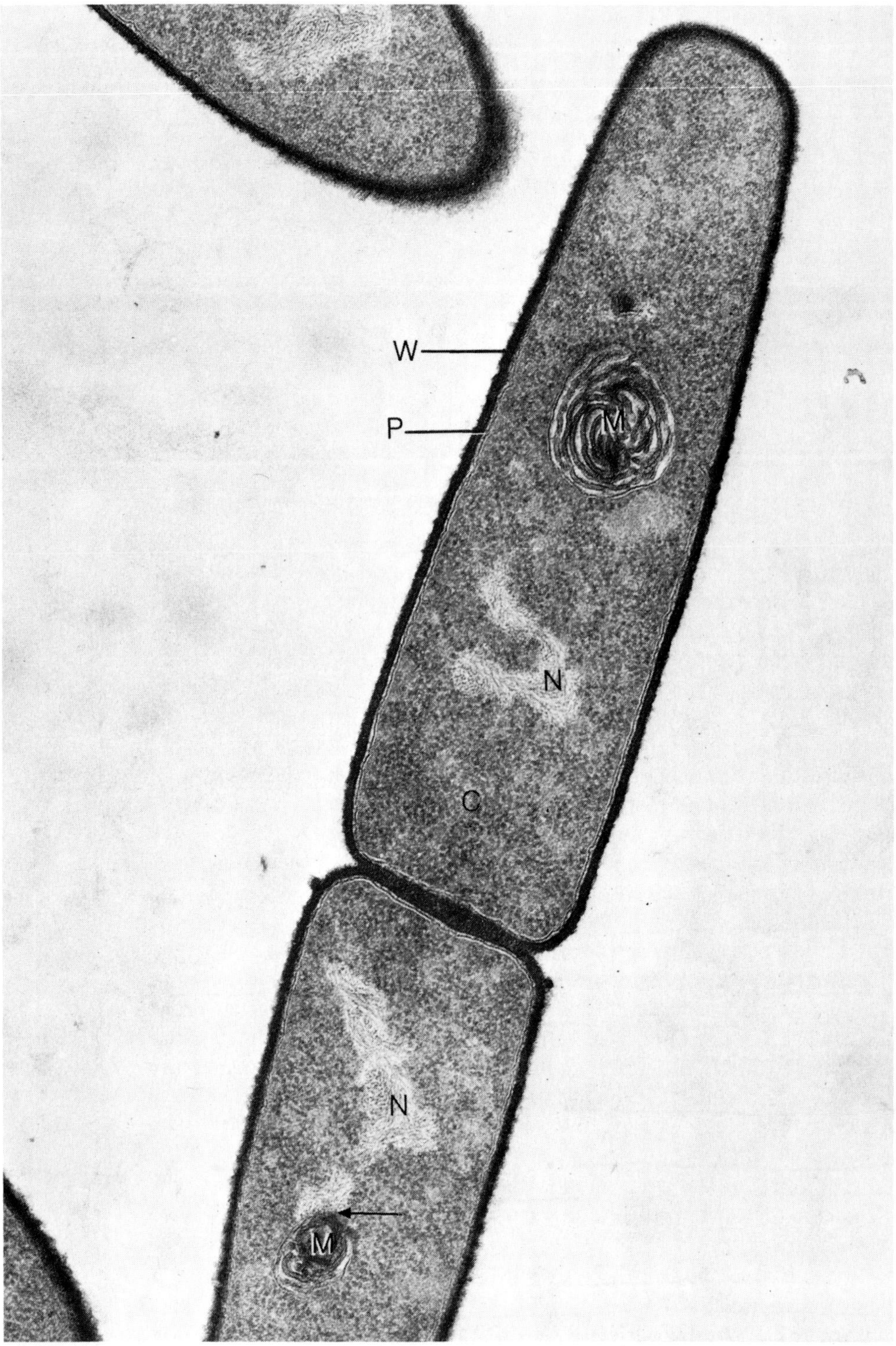

FIGURE 1–9 Electron micrograph of cells of the bacterium *Bacillus subtilis*. Each cell is surrounded by a densely staining cell wall (W). Immediately underlying the wall is the trilaminar plasma membrane (P). Ribosomes are visible as dense granules within the cytoplasm (C). The nucleoid (N) is relatively electron-lucent and has a fibrillar texture. The mesosome (M) is a membranous invagination of the plasma membrane. In the cell in the lower portion of the micrograph, a region of attachment of the genetic material in the nucleoid to the mesosome can be seen at the arrow. (Micrograph courtesy of A. Ryter.)

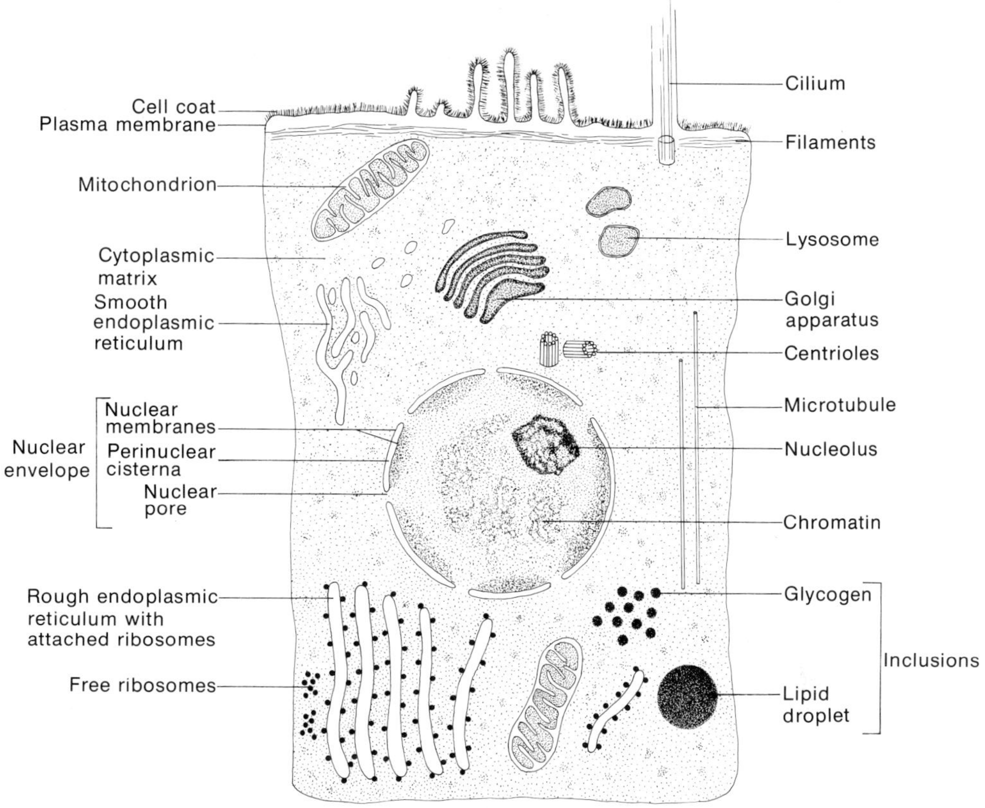

FIGURE 1–10 Diagram of the main structural components of a eukaryotic cell.

The endoplasmic reticulum is a system of membrane-limited cavities, prominent in the cytoplasm of most eukaryotic cells (Fig. 1–13). Two forms, rough and smooth, are distinguished. The rough endoplasmic reticulum is characterized by the presence of ribosomes studding the surface of the membranes. It is often found in the form of flat sacs, or cisternae. In gland cells the attached ribosomes are the site of synthesis of secretory proteins. Cytoplasmic regions occupied by rough endoplasmic reticulum are often detectable as "basophilic bodies" in the light microscope because of the tendency of RNA in the ribosomes to bind basic dyes. The smooth endoplasmic reticulum (Fig. 1–14) lacks ribosomes and often occurs in the form of membranous tubules. It has many functions, including a role in the biosynthesis of steroid hormones and the metabolism of drugs.

The Golgi complex is a system of smooth-surfaced membranes (Fig. 1–15). Its distinctive components are cisternae, or saccules, which are arranged in parallel to form stacks. The cisternae are accompanied by variable numbers of smooth-surfaced vesicles and vacuoles. The Golgi apparatus is important in the "packaging" of secretory materials into secretory vacuoles in gland cells.

Mitochondria (Figs. 1–14 and 1–16) are bounded by two membranes, the inner of which is folded to form the mitochondrial cristae. Mitochondria are the main site of production of the high-energy compound, ATP. Thus they are particularly abundant in cells with high energy demands, such as certain muscle cells. The plastids are membranous organelles found in the cytoplasm of plant cells. They will not be considered further since the emphasis of this volume is on medical aspects of cell biology, but it is worth recalling that chloroplasts have

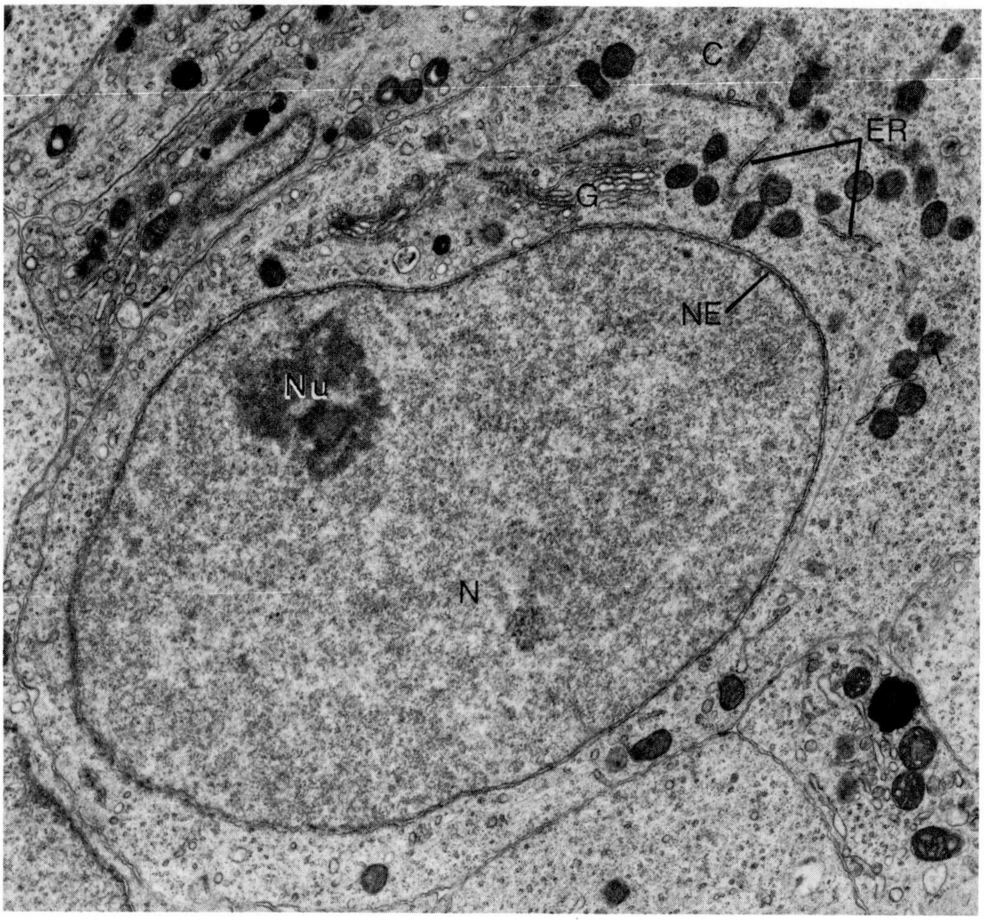

FIGURE 1–11 Many fundamental features of eukaryotic cells in general are visible in this low-power electron micrograph of a germ cell in the mouse testis. The separation of the nucleus (N) and cytoplasm (C) by the nuclear envelope (NE) is clearly defined. Other structures in the field are the nucleolus (Nu), mitochondria, Golgi apparatus (G), and a small amount of endoplasmic reticulum (ER). × 14,000.

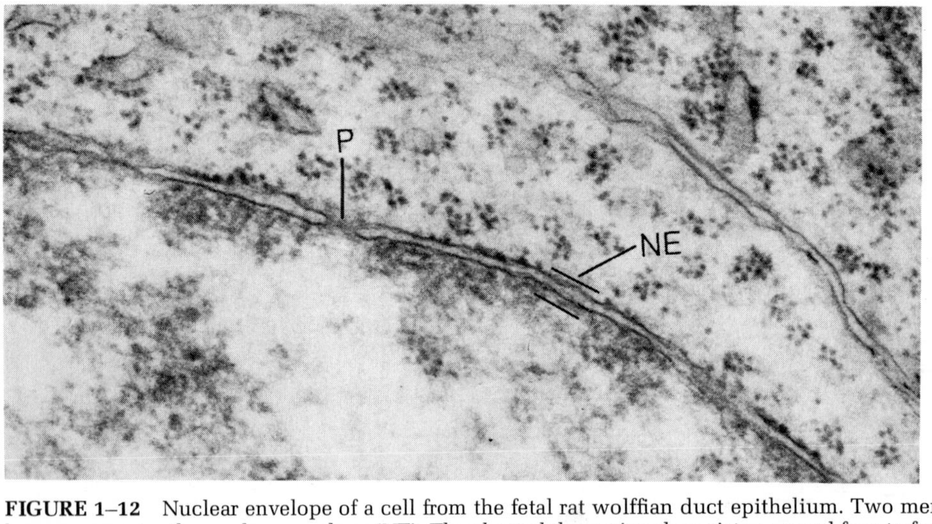

FIGURE 1–12 Nuclear envelope of a cell from the fetal rat wolffian duct epithelium. Two membranes comprise the nuclear envelope (NE). They bound the perinuclear cisternae and fuse to form nuclear pores (P). × 60,000.

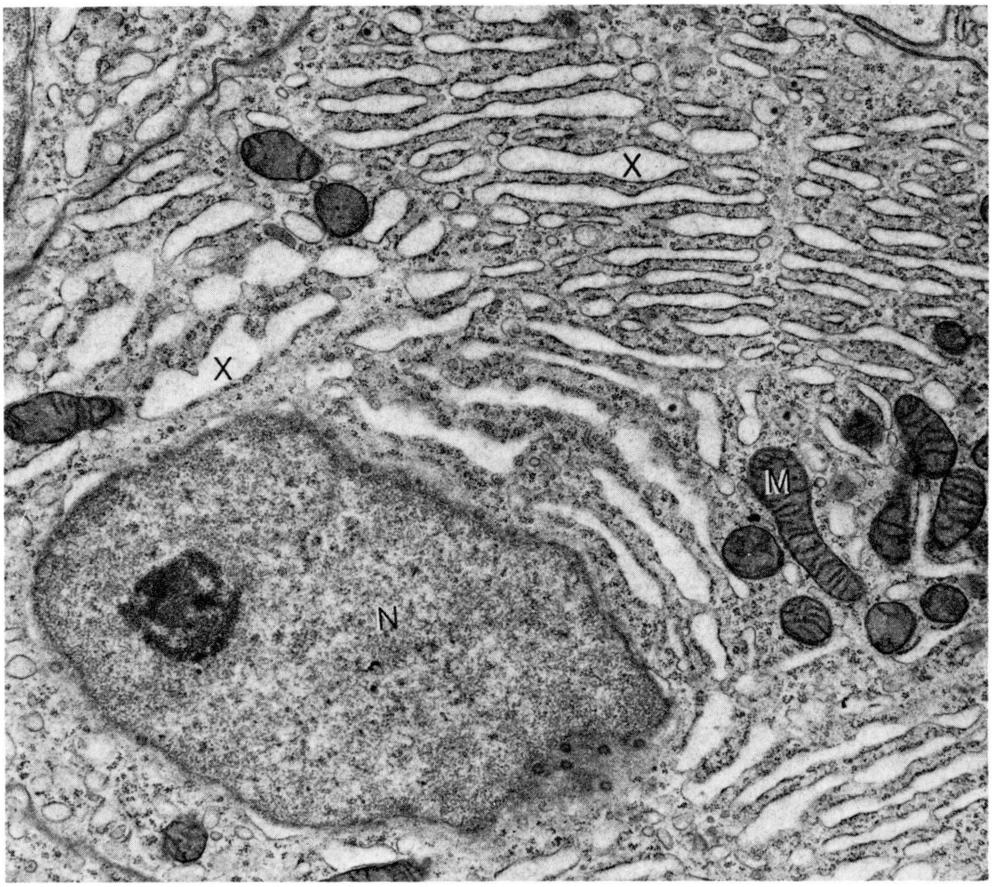

FIGURE 1–13 Ventral prostate gland of a rat. The extent of the rough endoplasmic reticulum can be appreciated readily in this preparation because the interior of the reticulum (X) is lighter than the surrounding cytoplasmic matrix. At this low magnification, ribosomes appear as small dense granules on the surface of the endoplasmic reticulum membranes. N, nucleus; M, mitochondria. × 15,000.

great biological importance because they are the site of photosynthesis. Furthermore, chloroplasts are of current interest in cell biology because, like mitochondria, they contain DNA, RNA, and a protein-synthesizing system, and their relation to other parts of the cell is a complicated and interesting one.

Lysosomes (Fig. 1–17) are structures that are bounded by a single membrane and contain a variety of acid hydrolases. They function in intracellular digestion and thus may be seen to contain a variety of materials in different stages of disintegration.

In addition to ribosomes attached to membranes of the rough endoplasmic reticulum, eukaryotic cells possess free or unattached ribosomes, as do prokaryotes (Fig. 1–16). Unattached polyribosomes are thought to function mainly in the synthesis of proteins used within the cell.

Microtubules are long hollow cylinders about 250 Å in diameter (Fig. 1–18). In some cells they appear to have a skeletal function in determining cell shape. They are also an important component of the mitotic spindle. Microtubules course longitudinally in the core of cilia and flagella. In this location they frequently have a characteristic "9 + 2" distribution when viewed in cross section (Fig. 1–19), with a central pair of single tubules being surrounded by nine doublet tubules. Fine filaments of different types are widely distributed in the cytoplasm of eu-

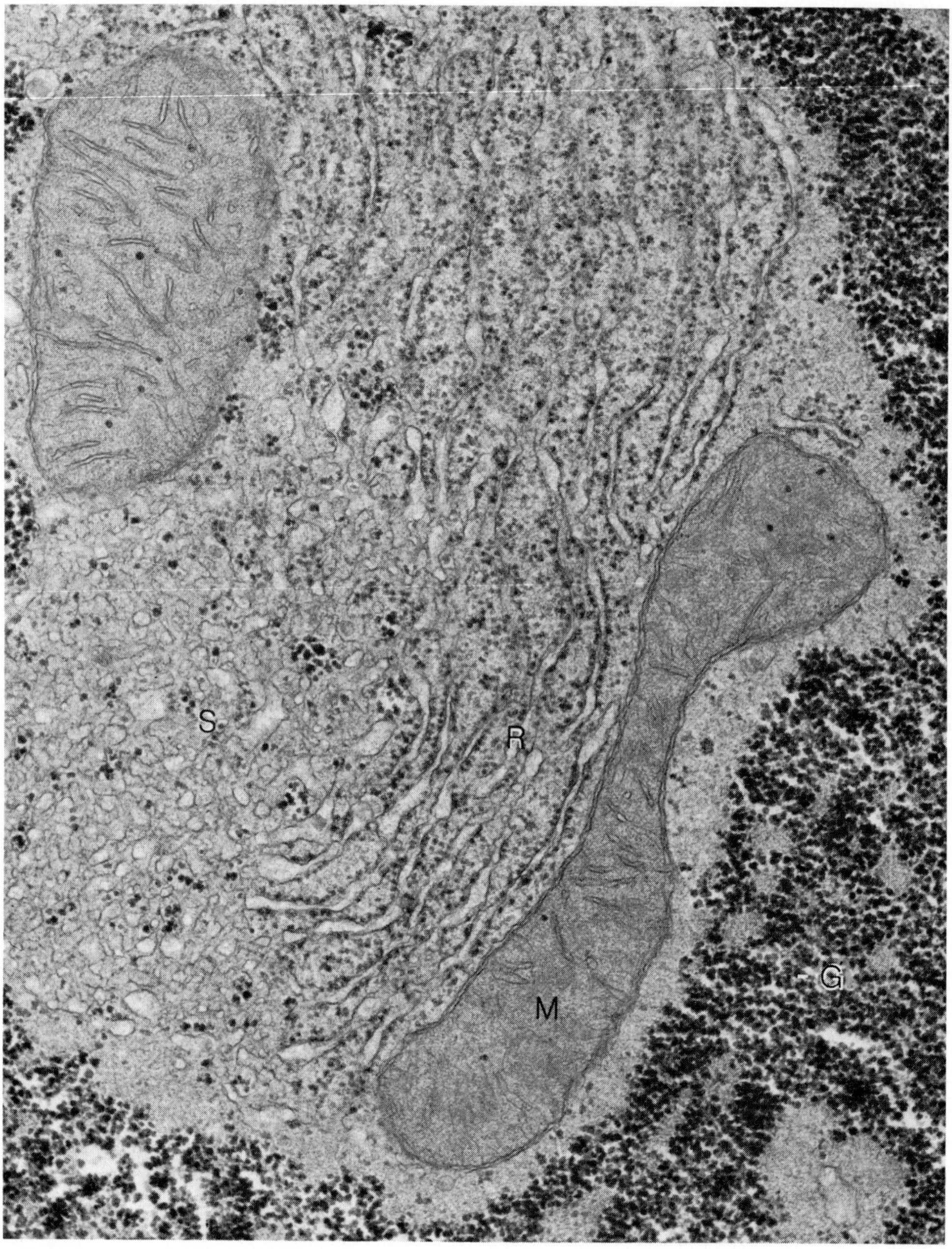

FIGURE 1–14 Cytoplasm of a rat liver cell, showing densely staining particles of glycogen (G) and tubules of smooth endoplasmic reticulum (S). R, rough endoplasmic reticulum; M, mitochondrion. × 36,000. (Micrograph courtesy of R. R. Cardell.)

karyotic cells, and some are thought to play a role in cell motility. Filaments of the proteins actin and myosin are very abundant in muscle cells, which are specialized for motility.

The cellular organelles described here are the most evident components of the cell in electron micrographs. It should be remembered, however, that these organelles are suspended in the cytoplasmic matrix, which pervades the entire cytoplasm between the organelles. It is a solution of many important proteins and

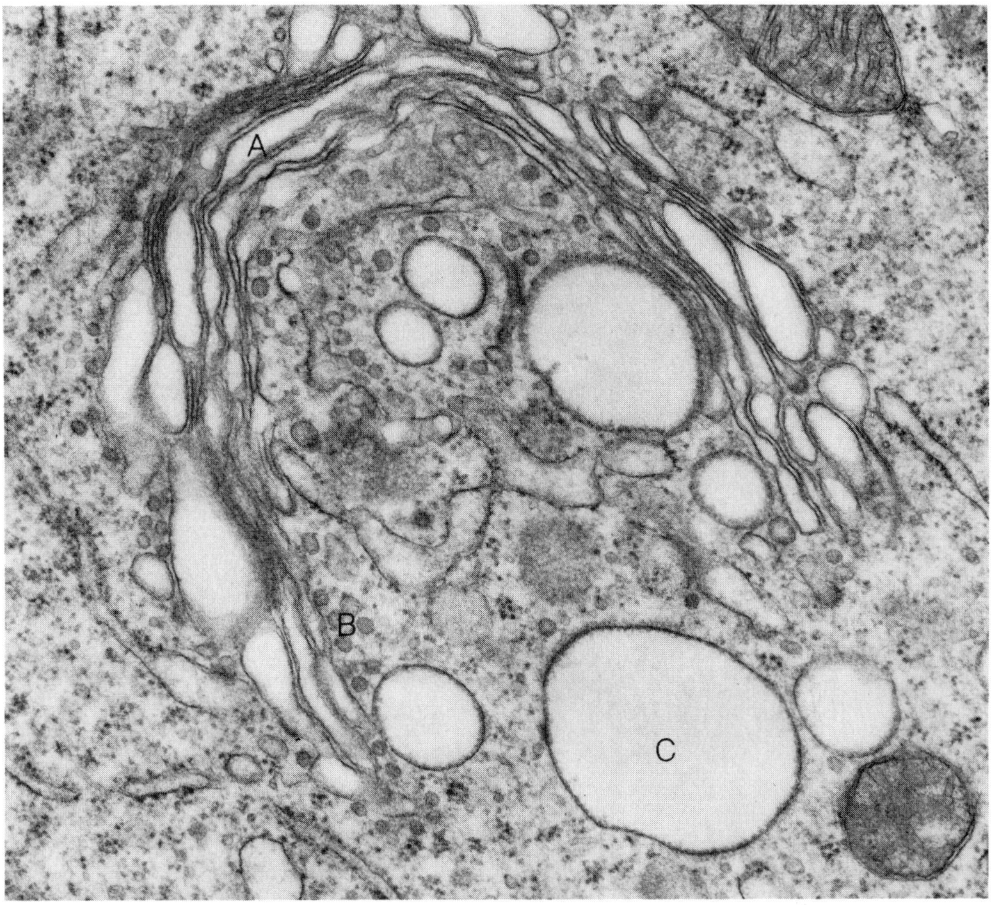

FIGURE 1–15 Rat ventral prostate. The three basic membranous components of the Golgi apparatus are cisternae (A), small vesicles (B), and vacuoles (C). × 40,000.

other molecules, including many of the enzymes of intermediary metabolism to be studied in subsequent chapters. In addition to organelles, various inclusions may be suspended in the cytoplasmic matrix. These include substances such as glycogen granules (Fig. 1–14) and lipid droplets (Fig. 1–20) which are neither membrane-bound nor constant components of cells.

Surrounding the cytoplasm is the lipoprotein plasma membrane (Fig. 1–21), which is important not only in determination of the internal composition of the cell but also in relations between cells in tissues of multicellular organisms. In eukaryotic cells a variety of structures are found external to the plasma membrane (Fig. 1–21). The term glycocalyx has been applied to encompass them because they are rich in carbohydrates. The cell wall of plant cells is a prominent example. Animal cells lack this rigid wall, but they do have an external polysaccharide-rich coat which is firmly attached to the external surface of the plasma membrane and is thought to be responsible for many of the properties of cell surfaces. This cell coat is very highly developed on the luminal surface of the absorptive cells of the intestine, but it is present in a less conspicuous form in most other animal cells.

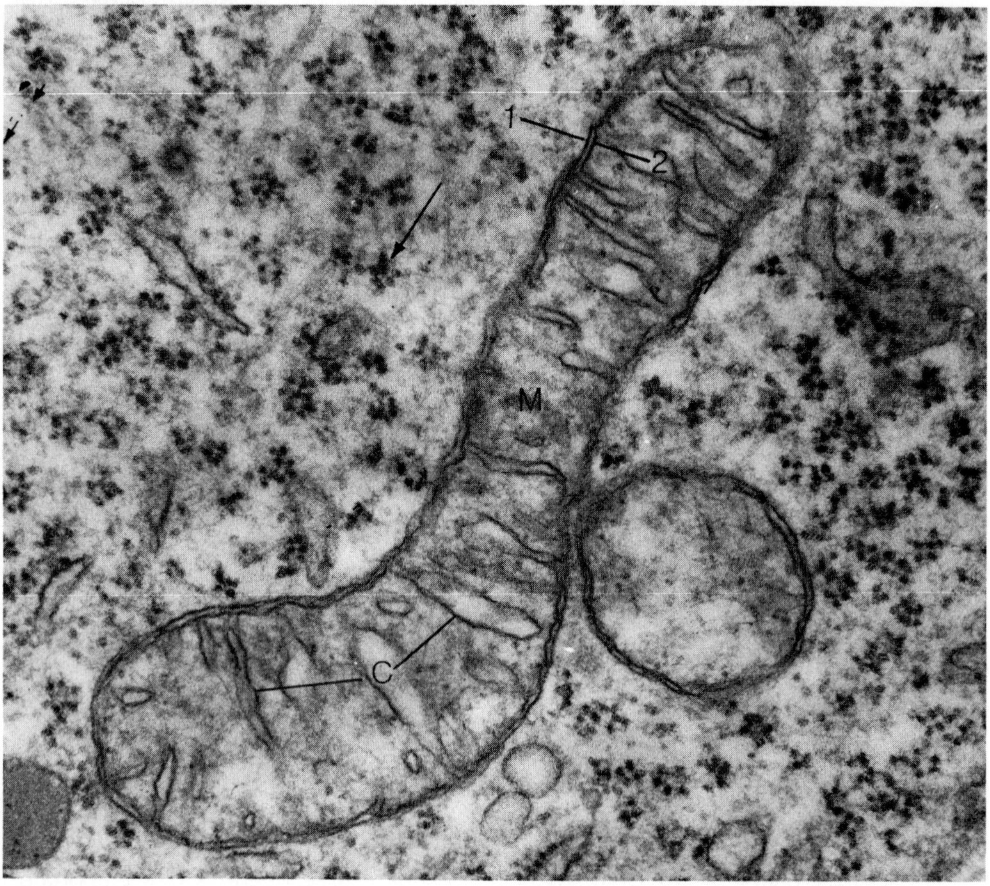

FIGURE 1–16 Cytoplasm of fetal rat seminal vesicle epithelium. Mitochondria (M) are bounded by two membranes (1 and 2), the inner of which is folded to form mitochondrial cristae (C). Unattached ribosomes in the cytoplasmic matrix are found in groups or rosettes (*arrow*) representing polysomes, which are active in protein synthesis. × 60,000.

INTRODUCTION TO SELECTED TECHNIQUES

In general, methodology is beyond the scope of this book. It will be considered in subsequent chapters only when it is pertinent to an understanding of the material under discussion. Several techniques will be described at this point, however, because their use will be noted repeatedly in the following sections, and familiarity with their basic principles is needed to understand the succeeding discussions. These methods are selective staining, cytochemical staining, radioautography, and cell fractionation. Another reason for emphasizing these techniques is that they are frequently used in conjunction with structural methods to furnish information on the localization of chemical compounds in cells. These methods are of particular importance in cell biology because they help to relate morphological observations to biochemical studies.

Selective Staining

Staining of sections for routine light microscopy is necessary simply to make most structures visible. Even commonly used dyes, however, are capable of

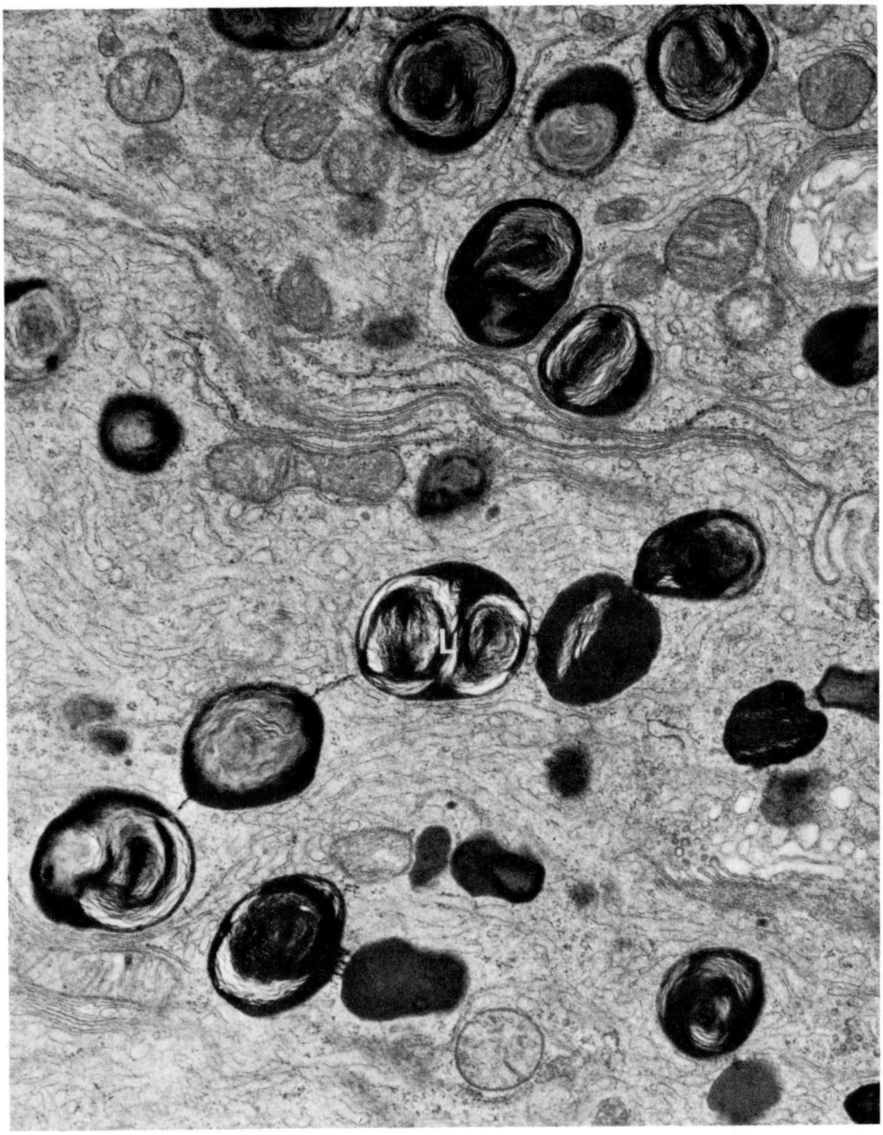

FIGURE 1–17 Lysosomes in the rat epididymal epithelium. These lysosomes (L) have dense interiors which include whorls of membranous material. × 19,000.

providing information about the chemical composition of cellular structures. Basic dyes, such as toluidine blue (Fig. 1–4), are charged positively at the pH of the staining solution. Thus dye molecules combine with acidic substances in cells, which have a negative charge. These acidic compounds are therefore referred to as basophilic substances because of their affinity for basic dyes. Examples of basophilic substances in cells are the nucleic acids DNA and RNA.

Acidic dyes, such as acid fuchsin, are charged negatively at the pH of the staining solution. They combine with basic substances in cells and with many cellular proteins, which have a net positive charge at the pH of the staining solution. Thus cellular components that combine with acid dyes are referred to as acidophilic substances. Examples of acidophilic substances include proteins of the cytoplasmic matrix, mitochondria, and smooth endoplasmic reticulum. Hematoxylin and eosin are commonly used stains for routine histological and

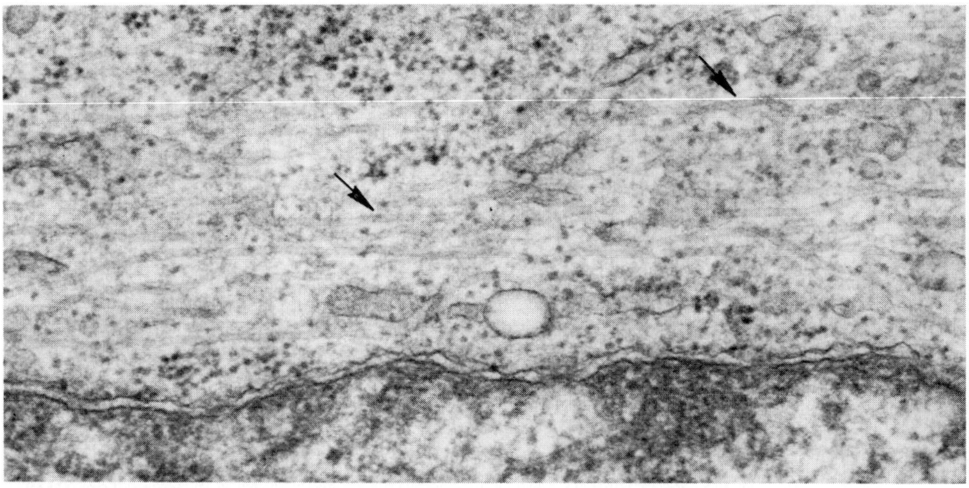

FIGURE 1–18 Epithelial cell of the epididymis of the newborn rat. Several long cylindrical microtubules (*arrows*) course through the cytoplasm in the field. × 52,000.

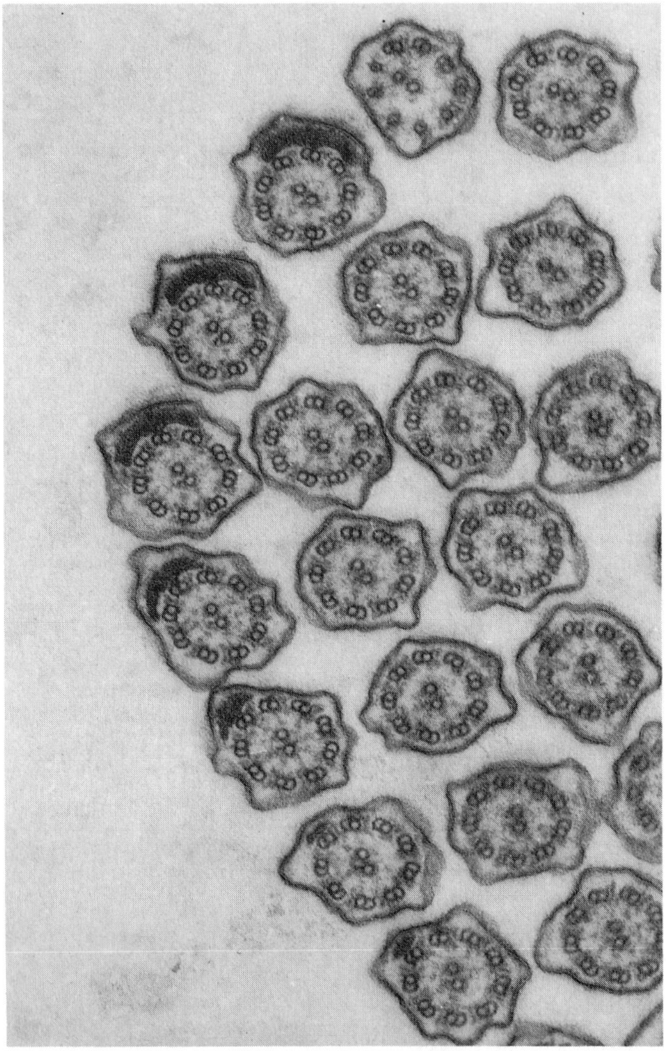

FIGURE 1–19 Cross sections of cilia of a protozoan. The "9 + 2" arrangement of microtubules in the axoneme consists of a central pair of single tubules surrounded by a ring of nine double tubules. × 73,000. (Micrograph courtesy of Sattler, C. A., and Staehelin, L. A., J. Cell Biol., *62*: 473, 1974. Reproduced by permission of the Rockefeller University Press.)

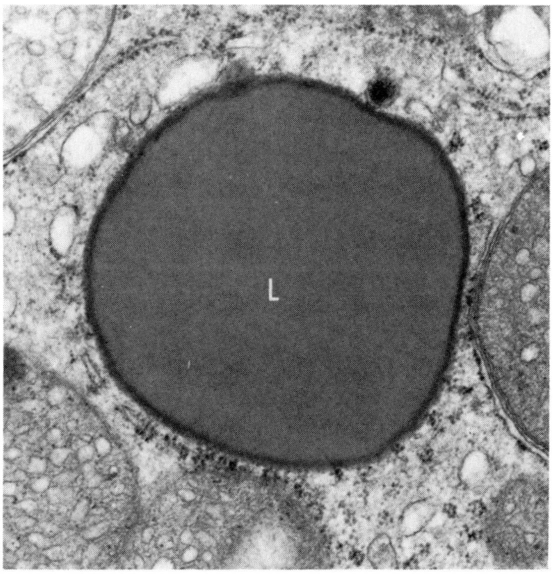

FIGURE 1–20 Sertoli cell of the rat testis. Lipid droplets (L) are dense, homogeneous, spherical structures. They lie in the cytoplasmic matrix and usually are not bounded by a membrane. × 28,000.

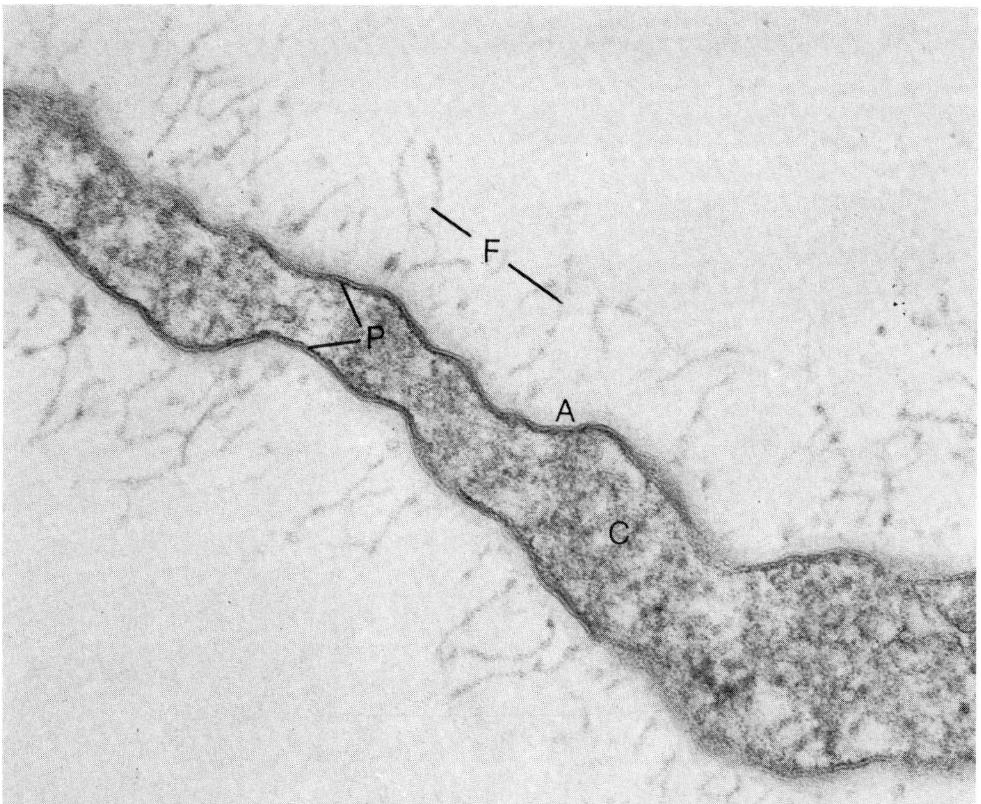

FIGURE 1–21 Cell surface of an amoeba. The carbohydrate-rich cell coat is highly developed in these cells. It comprises an amorphous layer (*A*) next to the plasma membrane (*P*) and long filaments (*F*) that extend outward from the plasma membrane. × 101,000.

pathological studies of paraffin sections. As a general rule, hematoxylin stains most basophilic substances, while eosin stains acidophilic substances. It may be noted, however, that hematoxylin does not react with acidic substances through a salt linkage or an acid-base reaction, and the exact mechanism of hematoxylin staining is poorly understood.

Histochemistry and Cytochemistry

Other stains are more specific and provide further information on the chemical composition of parts of tissues (histochemistry) and parts of cells (cytochemistry). The Feulgen reaction, for example, is specific for DNA. The periodic acid-Schiff method (PAS) stains complex carbohydrates, such as glycogen and glycoproteins (Fig. 1–22). The use of a particular stain can be combined with enzymatic digestion to furnish still more specific information about the chemical composition of cells. For example, if a portion of the cytoplasm of liver cells stains with PAS, the presence of glycogen may be suspected but is not proved, because PAS is capable of staining other carbohydrate-rich substances besides glycogen.

The presence of glycogen can be confirmed in the following way. Some sections are exposed to the enzyme diastase, which breaks down glycogen, while control sections are not exposed to the enzyme. Both digested and control sections are stained with PAS. If staining is abolished in the sections digested with diastase, it can be attributed to the presence of glycogen.

Methods are available by which the presence and localization of an enzyme within cells can be detected by incubating cells in a special medium. The enzyme,

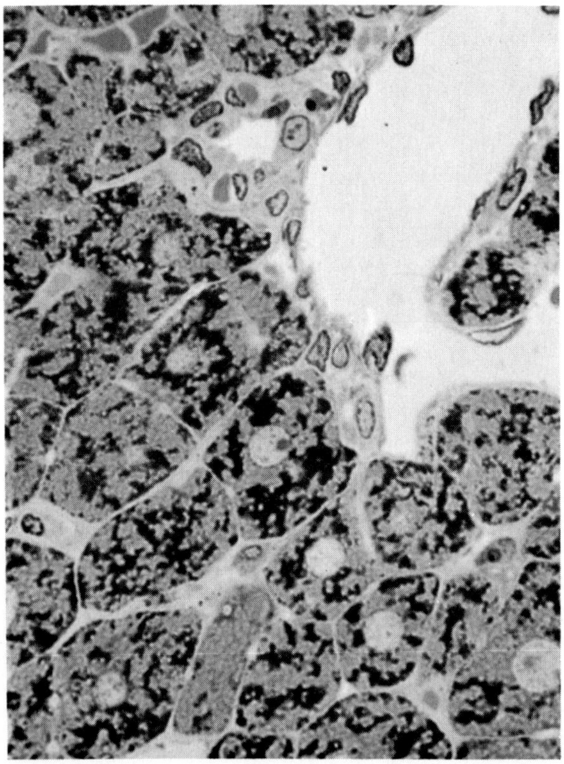

FIGURE 1–22 Light micrograph of rat liver cells stained with the periodic acid-Schiff (PAS) and toluidine blue technique. Glycogen in the liver cells appears as dense masses of PAS-positive material. × 800. (Micrograph courtesy of Babcock, M. B., and Cardell, R. R.: Am. J. Anat., *143*:399, 1975. Reproduced by permission of the Wistar Institute Press.)

which is within the cell, is used to catalyze a reaction that converts a substrate, which is furnished in the medium, to a product that is visible in the light microscope or the electron microscope. An example of this is the cytochemical method for the enzyme glucose-6-phosphatase. Cells are briefly fixed in an aldehyde. They are then exposed at 37°C to an incubation medium containing glucose-6-phosphate (the substrate for the enzyme), lead nitrate, and a buffer. The enzyme catalyzes the following reaction:

$$\text{Glucose-6-P} \rightarrow \text{glucose} + \text{phosphate}$$

The phosphate ions produced are then precipitated by the lead ions in the medium to form lead phosphate, which is electron-opaque.

$$\text{phosphate} + Pb^{+2} \rightarrow Pb_3\,(PO_4)_2 \downarrow \text{ (electron opaque)}$$

The cells are then dehydrated, embedded, and sectioned as usual. When they are examined in the electron microscope, a dense precipitate of the reaction product, lead phosphate, marks the location of the enzyme. If this method is applied to liver cells (Fig. 1–23), the reaction product is found within the endoplasmic reticulum, indicating that the endoplasmic reticulum of liver cells contains the enzyme glucose-6-phosphatase. Another widely used cytochemical technique is the Gomori method for acid phosphatase, which is performed in a similar way with β-glycerol phosphate as substrate. It is often used to identify lysosomes.

Immunocytochemistry is used to locate chemical compounds within cells with a high degree of specificity. Sections of cells are reacted with specific antibodies to the substance in question. The antibody combines with its antigen, and the location of the complex can be visualized if the antibodies are labeled with the fluorescent dye fluorescein for light microscopy, or with ferritin or peroxidase for electron microscopy. An alternate approach is the use of unlabeled antibody in the first step, followed by reaction of the specimens with labeled antibodies to proteins of the species in which the first antibody was made.

Radioautography

Radioautography is a technique for localizing radioactive compounds within cells. It can be carried out using either the light microscope or the electron microscope. Living cells are exposed to a radioactive precursor of some cellular component. Usually the labeled precursor is a molecule in which one or more of the hydrogen atoms (^1H) have been replaced by atoms of the radioactive isotope tritium (^3H). For example, thymidine -^3H is used as a precursor of DNA, uridine -^3H as a precursor of RNA, and various tritiated amino acids as precursors of proteins. The cells take up the labeled precursor, and if they are active in synthesis they incorporate it into the appropriate macromolecule along with unlabeled precursor molecules. The cells are then fixed, and sections are prepared in the usual way. The presence of a labeled component within the cells is detected as follows.

In a darkroom, a liquid photographic emulsion of silver halide crystals is applied to the surface of the sections. After the emulsion dries, the preparations are stored in light-tight boxes to permit the radioactivity in the sections to "expose" the overlying emulsion. The time required ranges from days to several weeks for light microscope radioautographs up to several months for electron

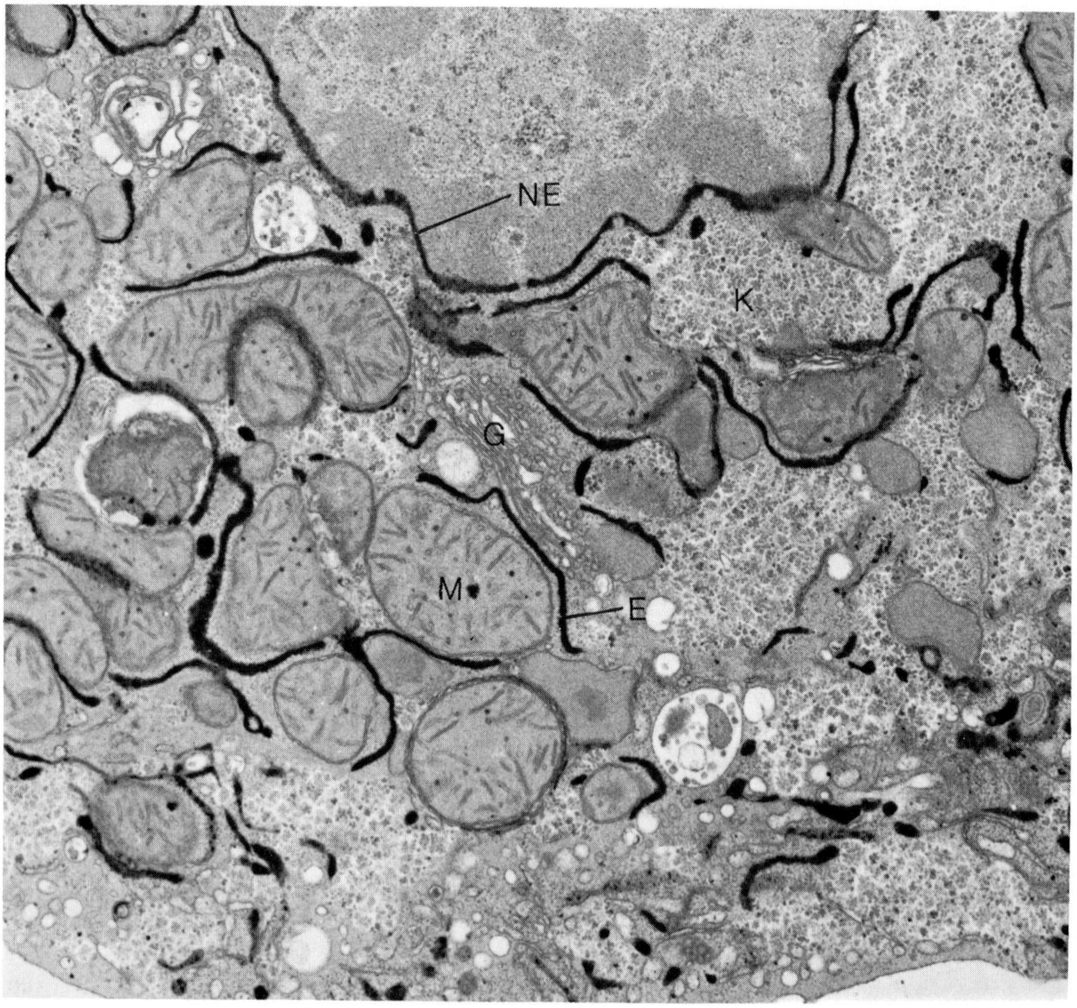

FIGURE 1–23 Electron microscope cytochemical staining for glucose-6-phosphatase in a rat liver cell. Dense deposits of the reaction product, lead phosphate, are present in the endoplasmic reticulum (E) and nuclear envelope (NE). No reaction product is visible in the Golgi apparatus (G), mitochondria (M), or glycogen (K). × 21,000. (Micrograph courtesy of Drochmans, P., Wanson, J.–C., and Mosselmans, R.: J. Cell Biol., *66*:1, 1975. Reproduced by permission of the Rockefeller University Press.)

microscope radioautographs. The preparations are then developed and fixed in a manner similar to that used to process ordinary photographic film. Reduced silver grains appear wherever the emulsion was exposed by radioactivity. Thus the silver grains overlie those parts of the cell that contain radioactive molecules. The silver grains appear as small black dots in light microscope preparations (Fig. 1–24). Their shape in electron microscope radioautographs varies with the type of development, but commonly they appear as twisted black threads (Fig. 1–25).

As an example of the use of radioautography, the method can be used to follow the synthesis and transport of secretory proteins. In the experiment shown in Figure 1–25, rats were injected with leucine-^{3}H. At intervals thereafter animals were killed, and radioautographs were prepared from sections of the prostate gland. Four minutes after injection, grains appeared overlying the rough endoplasmic reticulum, indicating that leucine-^{3}H was taken from the blood by the prostatic epithelium and incorporated into protein by the polyribosomes attached

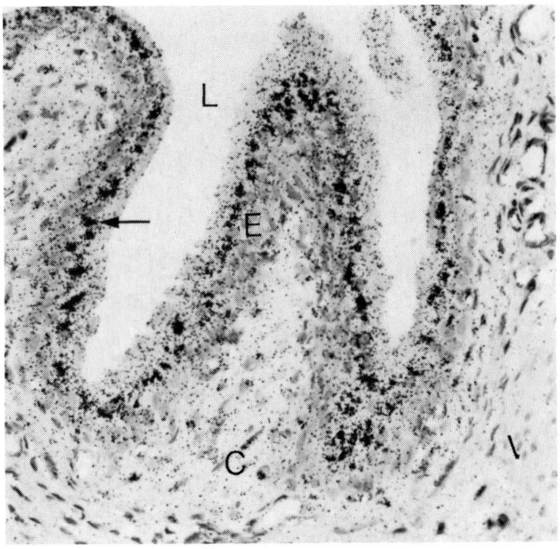

FIGURE 1–24 Light microscope radioautograph of rat bladder 10 minutes after injection of ^{3}H-fucose. Silver grains appear as black dots and indicate the presence of radioactive material in the underlying section. The heaviest incorporation of radioactivity (*arrow*) is into parts of the epithelial cells near the nucleus, corresponding to the location of the Golgi apparatus. This reflects the role of the Golgi apparatus in the assembly of fucose-containing glycoproteins. L, lumen; E. epithelium; C, connective tissue. × 250. (Micrograph courtesy of Bennett, G., Leblond, C. P., and Haddad, A.: J. Cell Biol., *60*:258, 1974. Reproduced by permission of the Rockefeller University Press.)

to the rough endoplasmic reticulum. By 30 minutes after the injection (Fig. 1–25), grains overlay the Golgi apparatus and secretory vacuoles, reflecting the intracellular transport of labeled secretory proteins from the endoplasmic reticulum to these other organelles. At later times after the injections, radioactive proteins were released from the cells.

Numerous other examples of information obtained from radioautography will be seen in the following chapters. The following points should be kept in mind to understand these examples. First, in most radioautographic studies it is necessary that the radioactive precursor be incorporated into a *macromolecule*, which is insoluble in the solutions used to process the tissue. Small molecules that are soluble in water, including *unincorporated precursor,* will be washed out and lost in the processing solutions. Complicated special methods must be used for radioautography of water-soluble substances. Second, incorporation of the precursor will occur only if the cell is active in the appropriate synthesis when the precursor is available. For example, cells in culture may be exposed to a five-minute pulse of thymidine-^{3}H (the thymidine-^{3}H is available to the cells for only five minutes). Only those cells that were synthesizing DNA during this five-minute period will exhibit labeled nuclei upon preparation of radioautographs. Cells that completed DNA synthesis prior to the pulse or began it subsequently will not have labeled nuclei.

Cell Culture

Many different kinds of cells from humans and other mammals have been grown in culture. Cells are obtained for culture by removing an organ, cutting it into small pieces, and dissociating the cells by treating them with the proteolytic enzyme trypsin. The cells are plated on the surface of a sterile vessel containing one of the many culture media that have been developed. After growth and proliferation of the cells has occurred, subculturing is performed by treating the cultures with trypsin to obtain the cells in suspension and then replating them with fresh medium in a new vessel. Such a subculturing is known as a "passage" in culture. Methods are now available for cloning, the culture of progeny of a single mammalian cell.

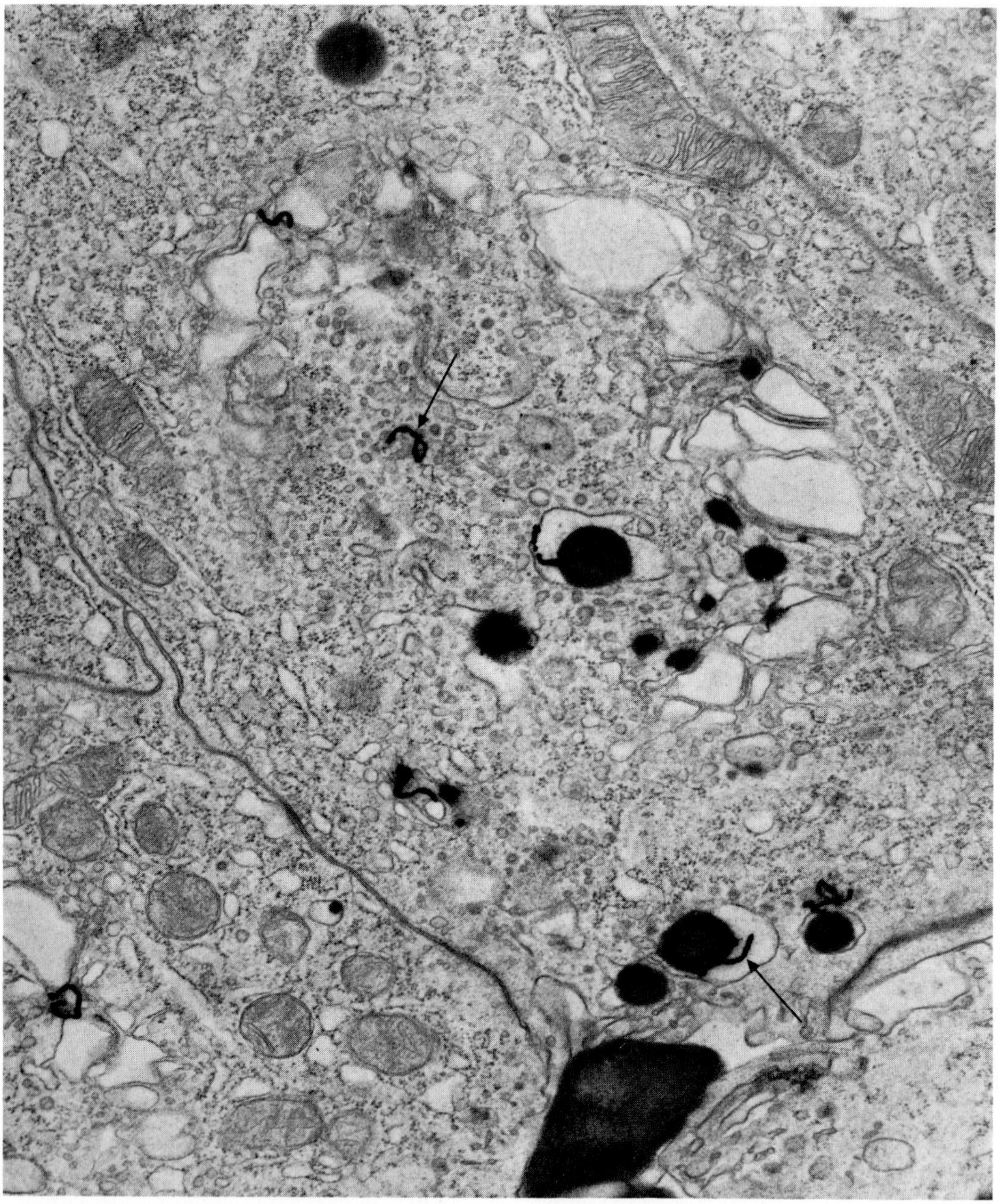

FIGURE 1–25 Electron microscope radioautograph of rat seminal vesicle 30 minutes after an intraperitoneal injection of leucine-^{3}H. Silver grains (*arrows*) overlie secretory vacuoles and elements of the Golgi complex, reflecting the presence of radioactive secretory proteins. × 23,000.

Once the cells obtained from an animal have been passaged in culture, they are known as a *primary cell line*. Primary cell lines are diploid and retain many characteristics of the tissue from which they were derived. Normal cells are not capable of indefinite propagation in culture; although they may be subcultured up to 70 to 100 times, after a certain number of passages their growth rate declines and the cultures eventually die out. However, sometimes the primary cell line may develop into an *established cell line*, which is capable of indefinite propagation *in vitro*. This change may occur gradually or it may appear rapidly by *transformation,* which designates a striking change in the behavior and properties of the cells. Such change includes increase in growth rate, loss of the properties of the cell of origin, and chromosomal alterations from a diploid to an aneuploid state.

Transformation may occur spontaneously with different frequencies in various primary cell lines, and it can be induced by carcinogenic chemicals or tumor-producing viruses. Cells transformed by tumor viruses, such as the polyoma virus, are malignant, as can be demonstrated by their ability to produce tumors upon introduction into animals. Cell lines can also be derived directly from tumors, and these behave from the outset as if they were transformed. An example of an established cell line that has been widely studied is the HeLa cell, which was originally obtained from a human cervical carcinoma. The properties of normal and transformed cell lines in culture have been widely studied in relation to changes that occur in cancer cells.

Another application of cell culture that has become of great importance in recent years is the study of somatic cell genetics. Cells in culture can be induced to fuse with one another by treating them with certain viruses, such as the Sendai virus. If cells from two different species are fused, at first a heterokaryon with two different nuclei is formed. When this cell divides, the two sets of chromosomes form a single metaphase figure, and the progeny are hybrids which have chromosomes of the two species in a single nucleus. The chromosomal complement of these hybrids is frequently unstable. In human-rodent fusions, e.g., human-mouse hybrids multiplying in culture, the human chromosomes tend selectively to be lost. This has made it possible to map the human chromosomes, because the loss of a particular human trait can often be correlated with the loss of a certain chromosome.

Cell Fractionation

An important objective in cell biology is to gain an understanding of the structure and function of cells and their organelles at the molecular level. Many advances in this area have been made possible by the development of procedures for cell fractionation. By isolating parts of cells, the details of their activities and the characteristics of their chemical components can be more readily studied than in intact cells. Thus, although some information on the composition and function of various cell organelles can be obtained by microscopic methods such as cytochemical staining and radioautography, cell fractionation provides a more general and powerful approach to answering important questions, such as the following:

1. What are the properties and functions of the various organelles that make up a cell, and how do these give rise to the characteristics of the intact cell?
2. What is the chemical composition of each of the organelles?
3. What are the properties of the chemical components of the organelles, and how do these give rise to the features of the organelles of the cell?

Cell fractionation requires disruption of the plasma membrane and isolation of the various types of organelles from the disrupted cell. There are three basic steps in a successful cell fractionation. First, the cell must be disrupted, but without total destruction of the organelles of interest. Second, each of the various types of organelles or organelle fragments should be isolated in a relatively pure form, free from significant contamination by other types of organelles or fragments. Third, it is necessary to identify the organelle or fragmented organelle in each fraction; that is, the relationship between the material present in each fraction and the structures or organelles observed in the intact cell must be established. Each of these three steps in cell fractionation will be considered briefly.

CELL DISRUPTION

Some cells can be disrupted by osmotic lysis. The plasma membrane is permeable to water, but it is impermeable to a large fraction of the molecules and ions contained within the cell. Thus if cells are suspended in water or a dilute buffer solution, there will be a net movement of water across the membrane from the solution outside of the cell to the inside of the cell because the activity or "effective concentration" of water is greater outside the cell. As a result of the movement of water into the cell, the volume of fluid and the pressure within the cell increase, and this eventually leads to disruption of the plasma membrane and release of the cell's contents into the suspending medium.

While osmotic lysis is a particularly efficient method for disrupting certain cells such as erythrocytes, which exist as individual cells unattached to other cells, the more general methods for disrupting cells involve the use of shear forces generated by mechanical means. A common procedure for generating shear forces is through the use of a homogenizer, such as the one depicted schematically in Figure 1–26. As the pestle is turned, usually by a motor, the barrel containing the bits of tissue is alternately raised and lowered, forcing the suspension to pass through the narrow space between the barrel and pestle at a very high velocity. This leads to the development of large shear forces which disrupt the cells.

Other methods commonly used to disrupt cells depend upon shear forces developed by irradiation of cell suspensions with ultrasonic energy or by repeated freezing and thawing of the cell suspension. Thus the cell disruption step is relatively simple, and the method of choice depends upon the cell type or tissue to be disrupted as well as upon the types of organelles that are to be isolated.

FRACTIONATION

Microscopic study shows that the various organelles differ in their size, shape, and staining properties. The differences in staining characteristics are due to differences in the chemical composition of the organelles. Differences in chemical composition are often reflected in differences in the densities of the organelles. Thus subcellular organelles may differ in three physical properties—size, shape, and density. These differences form the basis for the separation of organelles in an ultracentrifuge into nuclear, mitochondrial, microsomal, and other fractions.

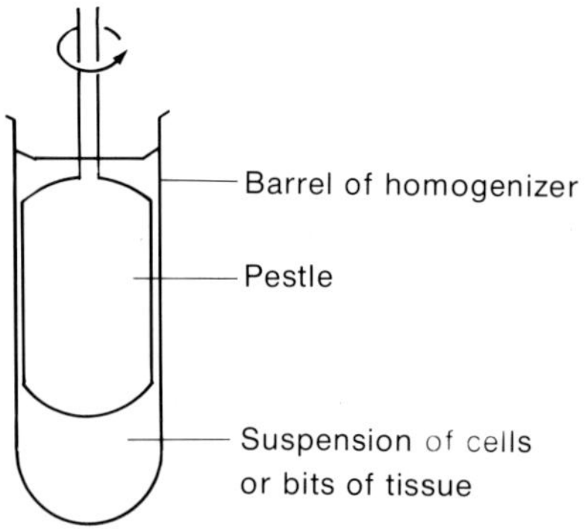

FIGURE 1–26 Diagram of a homogenizer.

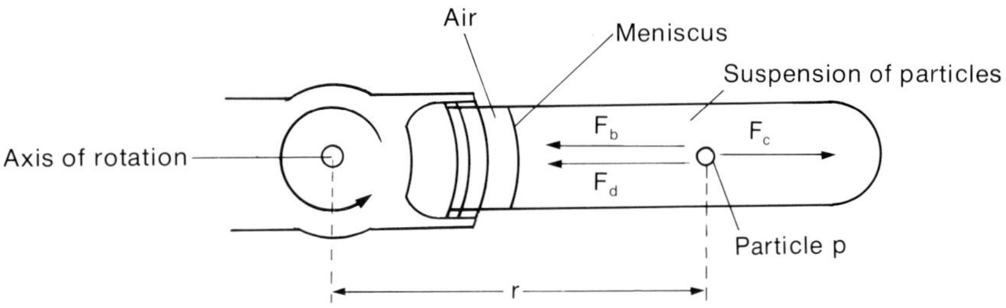

FIGURE 1–27 Diagram of a centrifuge as seen from above.

An ultracentrifuge is capable of generating centrifugal fields in excess of 100,000 × gravity. In order to understand the two basic ways in which an ultracentrifuge is used to achieve fractionation, some principles of centrifugation will be considered briefly. A diagram of a centrifuge, as seen from above, is shown in Figure 1–27. The centrifuge tubes are rotating about an axis at an angular velocity of ω radians/second. A particle, p, suspended in the solution in the tube, will be subjected to two forces: (1) a centrifugal force, F_c, directed radially toward the bottom of the tube, and (2) a buoyant force, F_b, also directed radially but in the opposite direction to F_c, as indicated in the diagram. The magnitude of these forces can be calculated from the following equations:

(1) $$F_c = \omega^2 r m_p$$

(2) $$F_b = -\omega^2 r m_s$$

where r is the distance of the particle from the axis of rotation, m_p is the mass of the particle p, and m_s is the mass of the solution displaced by particle p. If F_c is greater than F_b, (i.e., if the mass of the particle p is greater than the mass of the solution it displaces), the particle will be accelerated toward the bottom of the tube. However, owing to the motion of the particle through the solution, a new force, F_d, a frictional or drag force, develops. The magnitude of this frictional force may be represented by equation 3:

(3) $$F_d = -fv$$

where f is the frictional coefficient of the particle and is a function of the size and shape of the particle, and v is the velocity at which the particle is sedimenting. The negative sign for F_d and F_b indicates that these forces oppose F_c.

Particle p will continue to accelerate radially toward the bottom of the tube until the net force acting upon it is zero, whereupon it will sediment toward the bottom of the tube at constant velocity. This constant velocity is achieved very rapidly by a particle in an ultracentrifuge, and an expression for the magnitude of the constant velocity attained by the particle is readily derived, as follows:

$$F_c + F_b + F_d = 0 \text{ when constant velocity is attained}$$
$$\omega^2 r m_p - \omega^2 r m_s - fv = 0$$

or

(4) $$v = \frac{\omega^2 r (m_p - m_s)}{f}$$

From equation 4 it is apparent that the constant velocity attained by a particle is proportional to the centrifugal field, $\omega^2 r$, and to the effective mass of the particle, $m_p - m_s$, but is inversely proportional to the frictional coefficient, f. Since m_p is proportional to the density and volume of particle p, and f is a function of the size and shape of particle p, we see that the velocity at which a particle will sediment is a function of its size, shape, and density, the very properties in which the various organelles of the cell may differ.

One way in which the centrifuge is employed to separate various organelles is *differential sedimentation*. This method is based on the fact that the *velocities* at which the various organelles sediment are different. If one type of organelle sediments much more rapidly than the other types, centrifugation will result in the formation of a pellet at the bottom of the tube which is enriched in the faster-sedimenting organelle (Fig. 1–28). After decanting the supernatant fraction, the pellet can be resuspended in buffer and recentrifuged, thereby attaining an additional enrichment in the more rapidly sedimenting organelle. Relatively homogeneous preparations of some organelles can be isolated by this procedure.

Isopycnic centrifugation is a second type of fractionation procedure which is based solely on the differing *densities* of the organelles to be separated. Since mass = density × volume, m_s will equal m_p when the density of the solution is equal to the density of the displacing particle. Therefore, when the densities of the solution and the displacing particle are equal, $(m_p - m_s)$ will equal zero, and from equation 4 it can be seen that the velocity of sedimentation of the particle will be zero. In isopycnic centrifugation the centrifuge tube is filled with a solution in such a way as to create a gradient in density between the top and bottom of the tube, with the density being greatest at the bottom of the tube. This gradient in density is obtained by generating a gradient in the concentration of a solute, such as sucrose, which can be readily prepared with a simple gradient-mixing device. A suspension of the cellular components to be fractionated is layered on top of the density gradient in the centrifuge tube (Fig. 1–29), and the tube is centrifuged. Each type of organelle will sediment through the gradient until it reaches the position in the gradient at which the density of the solution is equal to the density of the organelle. The zones or bands of separated organelles thus generated can then be collected as discrete fractions by carefully draining the contents of the tube. A typical fractionation scheme employing sedimentation velocity and isopycnic centrifugation is depicted in Figure 1–30. By such fractionation schemes, relatively pure fractions of functionally intact subcellular organelles can be isolated.

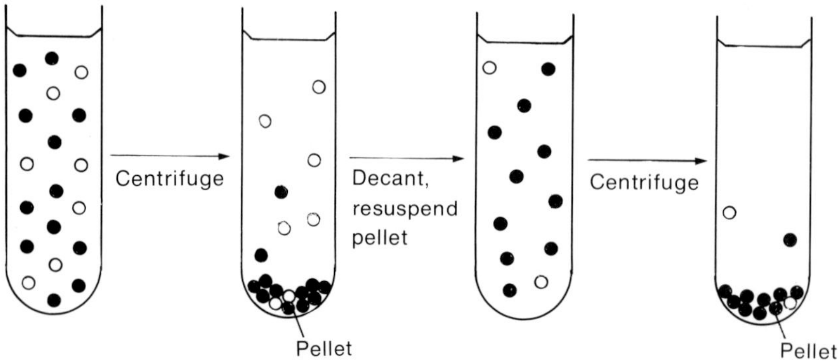

FIGURE 1–28 Schematic representation of fractionation by differential centrifugation.

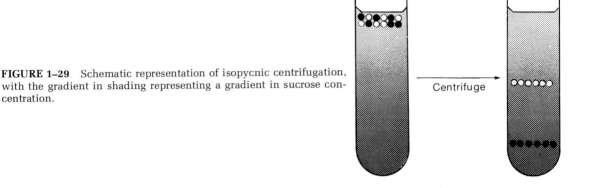

FIGURE 1–29 Schematic representation of isopycnic centrifugation, with the gradient in shading representing a gradient in sucrose concentration.

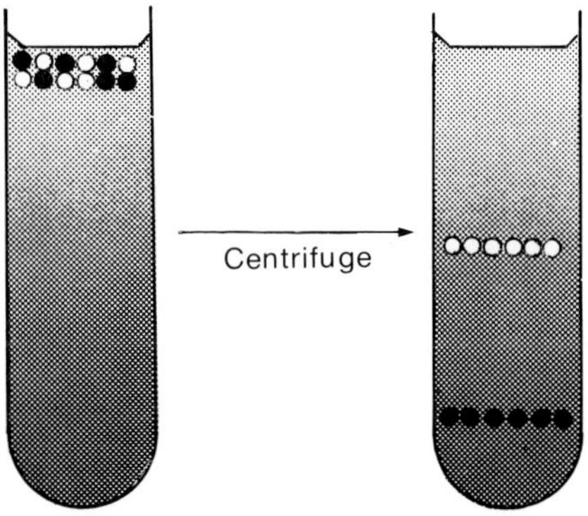

Mixtures of macromolecules, such as proteins and nucleic acids, can also be fractionated by differential sedimentation and isopycnic centrifugation. Furthermore, sedimentation properties of a macromolecule can be used to determine its molecular weight. In sedimentation-velocity studies of macromolecules, the velocity of sedimentation of the macromolecule, $\frac{dr}{dt}$, is measured in the ultracentrifuge. The sedimentation coefficient, s, is determined by dividing the velocity of sedimentation by $\omega^2 r$, the centrifugal field strength.

$$s = \frac{\frac{dr}{dt}}{\omega^2 r}$$

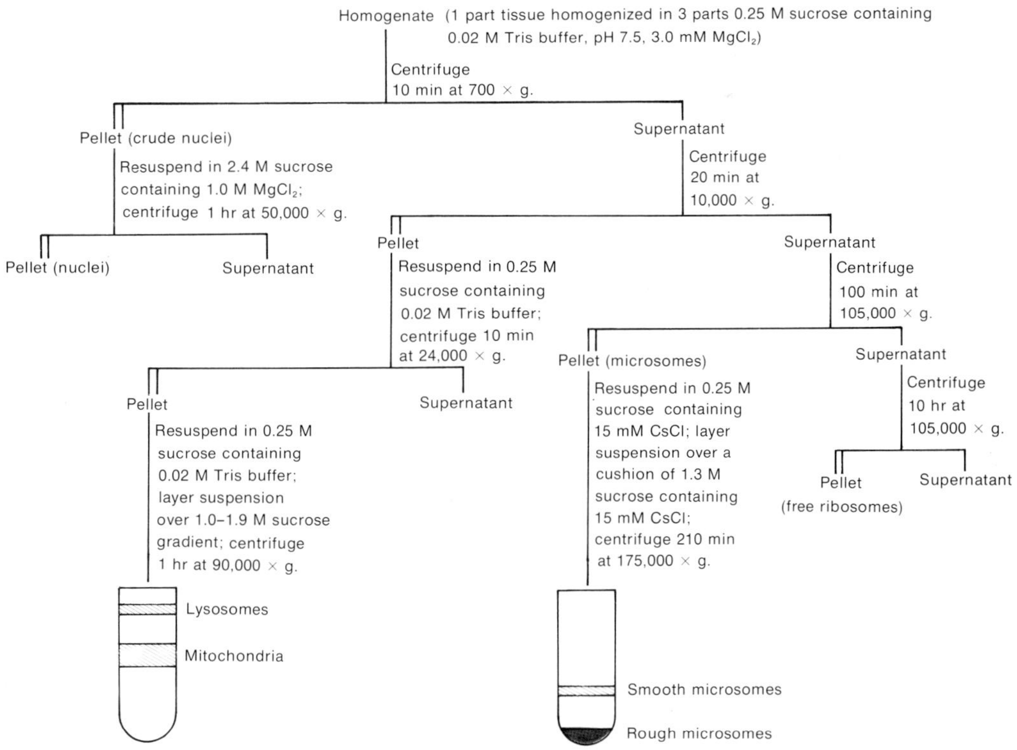

FIGURE 1–30 Diagram of a cell fractionation procedure.

The units of s are seconds. Since the sedimentation coefficients of macromolecules are generally in the range of 1×10^{-13} sec to 200×10^{-13} sec, the sedimentation coefficient is expressed in Svedberg units, S, where the Svedberg unit is 1×10^{-13} sec. If the sedimentation and diffusion coefficients of a macromolecule are known, the molecular weight of the macromolecule can be calculated by the Svedberg equation

$$M = \frac{RTs}{D(1 - \bar{v}\rho)}$$

where M is the molecular weight, R the gas constant, T the temperature in degrees Kelvin, s the sedimentation coefficient, D the diffusion coefficient, $\bar{v}$ the partial specific volume of the macromolecule, and ρ the density of the solution.

IDENTIFICATION

In order to relate information from cell fractionation to the structure and function of the intact cell, it is important to determine the composition of the cell fractions obtained by ultracentrifugation. In particular, it is necessary to determine the degree of purity of a given fraction, that is, the extent to which the fraction is composed of a single organelle or is contaminated with some other cellular constituent. The difficulty in identifying the isolated fraction varies from one organelle to another. Isolated fractions of nuclei or mitochondria are easily identified by electron microscopy, since their morphology after isolation is very similar to that in the intact cell. On the other hand, none of the isolated fractions contains membranous elements resembling the endoplasmic reticulum of the intact cell. The microsomal fraction, which sediments at 100,000 × gravity, contains small membrane-bound vesicles. It can be further fractionated, by the procedure outlined in Figure 1–30, into a pellet containing "rough microsomes" and a fraction containing "smooth microsomes." The "rough microsomes" consist of small vesicles to which are attached particles approximately 150 to 200 Å in diameter, which resemble the ribosomes observed in the rough endoplasmic reticulum. The "smooth microsomes" are small vesicles that appear to be very similar to the "rough microsomes," except that they do not have the small particles attached. Cells that are seen by electron microscopy to contain large amounts of rough endoplasmic reticulum yield a large "rough microsomal" fraction when fractionated; cells shown to contain a large amount of smooth endoplasmic reticulum by electron microscopy yield a large "smooth microsomal" fraction. These observations and others indicate that the endoplasmic reticulum is easily broken in the process of disrupting the cell, and that the resulting fragments vesiculate and constitute the microsomal fraction (see also Chapter 9).

Fortunately, isolated organelles as well as fragmented organelles often retain some of their functional activities. Thus, with relatively homogenous fractions of the various organelles in hand, it is possible to determine the chemical composition of each type of organelle and to correlate the functional properties of the organelles with their chemical composition. Chemical analyses show that in addition to water, salts, and small molecules, all organelles are composed of very large molecules with molecular weights of 10^4 to 10^9 or greater. These large molecules are known as macromolecules, and they include the following types. Proteins are present in all organelles. Deoxyribonucleic acids (DNA) are found mainly in the nucleus, with small amounts appearing in the mitochondria. Ribonucleic acids (RNA) are present mainly in the ribosomes, with lesser amounts in the nucleus, mitochondria, and cytosol. Polysaccharides of different types are present in some cell fractions. Organelles with membranous elements always contain lipids, a heterogeneous class of compounds which in general have a very low solubility in water and show a tendency to associate via noncovalent interactions to form large and often highly organized structures in an aqueous environment.

Since the key to understanding the structure and function of organelles at the molecular level lies in the understanding of the structure and properties of the macromolecules that make up the organelles, the next three chapters will be devoted to the structures and properties of these macromolecules. Then each of the main cellular organelles will be considered in turn.

FURTHER READING

Beck, F., and Lloyd, J. B. (eds.): The Cell in Medical Science. Vols. 1–4. New York, Academic Press, 1974.
Birnie, G. D. (ed.): Subcellular Components: Preparation and Fractionation. Baltimore, University Park Press, 1972.
Bittar, E. E. (ed.): Cell Biology in Medicine. New York, John Wiley and Sons, 1973.
Brachet, J., and Mirsky, A. E. (eds.): The Cell: Biochemistry, Physiology, Morphology. Vols. I–VI. New York, Academic Press, 1959–1964.
Dawes, C. J. (ed.): Biological Techniques in Electron Microscopy. New York, Barnes & Noble, 1971.
De Duve, C.: The separation and characterization of subcellular particles. Harvey Lect., *59*:49, 1965.
De Robertis, E. D., Nowinski, W. W., and Saez, F. A. (eds.): Cell Biology. 6th ed. Philadelphia, W. B. Saunders, 1975.
Fawcett, D. W.: The Cell: Its Organelles and Inclusions. Philadelphia, W. B. Saunders, 1966.
Grant, J. K. (ed.): Methods of separation of subcellular structural components. Biochem. Soc. Symp., *23*:1963.
Grossman, L., and Moldave, K. (eds.): Preparation of tissue organelles. *In*: Methods in Enzymology. Vol. XII A, Section III. New York, Academic Press, 1967, pp. 416–529.
Hayat, M. A. (ed.): Electron Microscopy of Enzymes. Vols. 1–3. New York, Van Nostrand Reinhold, 1973–1974.
Hayat, M. A. (ed.): Principles and Techniques of Electron Microscopy. Vols. 1–7. New York, Van Nostrand Reinhold, 1972.
Hayflick, L.: The cell biology of human aging. N. Engl. J. Med., *295*:1302, 1976.
Koehler, J. K. (ed.): Advanced Techniques in Electron Microscopy. New York, Springer-Verlag, 1973.
Paul, J. (ed.): Cell and Tissue Culture. 4th ed. Baltimore, Williams & Wilkins, 1972.
Pearse, A. G. E. (ed.): Histochemistry. 3rd ed. Boston, Little, Brown, 1968.
Pease, D. C. (ed): Histological Techniques for Electron Microscopy. New York, Academic Press, 1960.
Porter, K. R., and Bonneville, M. A. (eds.): An Introduction to the Fine Structure of Cells and Tissues. 2nd ed. Philadelphia, Lea & Febiger, 1964.
Prescott, D. M. (ed.): Methods in Cell Physiology; Methods in Cell Biology, A series. New York, Academic Press, 1966.
Ruthmann, A. (ed.): Methods in Cell Research. Ithaca, Cornell University Press, 1966.
Salpeter, M. M., and Bachmann, L.: Autoradiography. *In*: Hayat, M. A. (ed.): Principles and Techniques of Electron Microscopy. Vol. 2. New York, Van Nostrand Reinhold, 1972, p. 219.

2

BIOCHEMISTRY OF MACROMOLECULES

It is apparent from even a brief outline of cell structure that a cell is an extremely complex entity, consisting of a number of highly organized subcellular organelles, such as the nucleus, mitochondria, Golgi apparatus, lysosomes, rough and smooth endoplasmic reticulum, and plasma membrane. Most of these cell organelles can be separated from one another by cell fractionation techniques similar to those outlined in Chapter 1. In this way, individual fractions are obtained, each of which contains primarily a single type of cell organelle. In many cases, the isolated organelles retain a number of the functional properties they possessed in the intact cell. Furthermore, it is possible to disrupt the isolated organelles and to separate and isolate by chromatographic, electrophoretic, and other techniques the different kinds of molecules that make up the organelles. These component molecules are water, small organic and inorganic molecules and salts, and the very large organic macromolecules. On the basis of their chemical and physical properties the macromolecules can be divided into three broad classes: the proteins, the nucleic acids, and the polysaccharides.

Even after being subjected to the conditions employed in their isolation, many of the isolated macromolecules retain specific functional properties that can be identified with functions of the organelle from which they were derived. A list of some of the functions of macromolecules is presented in Table 2–1. Extensive chemical degradation of these macromolecules to form smaller molecules generally results in a loss of function, suggesting that intact macromolecules are responsible for the structural and functional characteristics of the cell. Thus, an understanding of the structure and properties of the macromolecules is essential to an understanding of the structure and function of the cell and cell organelles. In this chapter we shall begin a consideration of the proteins, polysaccharides, and nucleic acids, all of which are biopolymers, by exploring three basic questions:

1. What are the structures of these macromolecules?
2. What are the chemical and physical properties of the macromolecules?
3. How do these macromolecules generate the structures and system properties of the cell?

Consideration of these questions is not limited to this chapter but will be continued as an integral part of all subsequent chapters.

SOME FUNCTIONS OF MACROMOLECULES **TABLE 2–1**

Function	Macromolecules
Catalysts	The proteins known as enzymes
Information storage	DNA
Information read-out	RNA
Contractile elements	The proteins actin and myosin
Structural elements	The proteins such as α-keratin (in hair, nails, and feathers), elastin (in tendons, arteries, and elastic connective tissue), and collagen (in connective tissue, such as tendon, cartilage, and bone)
Transport	Proteins, such as the carrier proteins in membranes and hemoglobin in the erythrocyte
Protective role	Proteins, such as antibodies and fibrinogen in the blood of vertebrates
Energy storage	Polysaccharides, such as glycogen

MONOMERIC UNITS AND CHEMICAL BONDS OF MACROMOLECULES

The macromolecules are all polymers that are formed from much smaller molecules designated monomeric units, or monomers. The nature of the monomers and the way they are linked together to form the proteins, polysaccharides, and nucleic acids are considered in the following section.

Proteins—Linear Polymers of Amino Acids

Proteins are the most abundant organic molecules in the cell, constituting 50 per cent or more of the dry weight of the cell. They are also functionally the most diverse of the macromolecules, serving as catalytic, contractile, structural, transporting, and protective elements in the cell. Proteins differ not only in their functional properties but also in such physical properties as solubility, size, and shape. The simplest of cells contains hundreds of different kinds of proteins. Thus far, the proteins of the *E. coli* cell have been resolved into approximately 1100 different components by two-dimensional gel electrophoresis. For convenience, proteins may be divided into two major classes, simple proteins and conjugated proteins. Simple proteins yield only a mixture of α-amino acids when hydrolyzed, whereas conjugated proteins yield a mixture of α-amino acids plus one or more prosthetic groups consisting of other organic or inorganic compounds. Regardless of the source of the protein, whether from humans, microorganisms, plants, or other organisms, only 20 different α-amino acids are commonly found in the hydrolysates of proteins.

STRUCTURE AND PROPERTIES OF α-AMINO ACIDS

An α-amino acid may be represented by the general structure

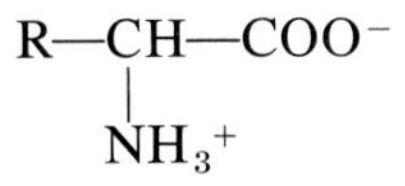

where R is a substituent group attached to the α-carbon. The 20 α-amino acids commonly isolated from the hydrolysates of proteins are presented in Table 2–2. All of the amino acids isolated from proteins are L-amino acids except for glycine, which is not asymmetric because the two hydrogens attached to its α-carbon are identical groups. (A brief discussion of the structure and configuration of α-amino acids is presented in Appendix A.) The three-letter code used to designate each amino acid is also given in Table 2–2, following the name of the amino acid.

Amino acids can be divided into subclasses, such as aliphatic amino acids, acidic amino acids, and so forth, on the basis of the chemical nature of the R group attached to the α-carbon. Some aspects of the structure of the different subclasses of amino acids that are important in determining protein structure will now be considered. Since all amino acids have a carboxyl group, an amino group, and a hydrogen atom attached to the α-carbon, the present discussion will be confined to the remaining group attached to the α-carbon, the R group.

Aliphatic Amino Acids. The aliphatic amino acids are glycine, alanine, valine, leucine, and isoleucine. The R groups of aliphatic amino acids are acyclic and contain only carbon and hydrogen. These R groups are nonpolar, i.e., they do not have a permanent dipole.

Aromatic Amino Acids. The aromatic amino acids are phenylalanine, tyrosine, and tryptophan. These three amino acids each have an aromatic R group, and as a result each exhibits an ultraviolet absorption spectrum with an absorption maximum between 259 nm and 280 nm. The presence of these amino acids in proteins leads to the absorption maxima near 278 nm that is characteristic of most proteins. The R group of phenylalanine is nonpolar, while the R groups of tyrosine and tryptophan are somewhat polar, owing to the presence of the —OH and $>$N—H groups.

Acidic Amino Acids. Aspartic acid and glutamic acid are classified as acidic amino acids because each possesses an R group containing a carboxyl group. At the pH existing in a cell, these carboxyl groups are largely dissociated to carboxylate anions, as indicated in Table 2–2. Hence, the R groups of aspartic and glutamic acids are polar and negatively charged at the pH of the cell.

Amide-Containing Amino Acids. Asparagine and glutamine are the two amino acids whose R groups contain carboxamide groups ($\text{—}\overset{\overset{\displaystyle O}{\|}}{C}\text{—}NH_2$). These R groups are not charged at cellular pH, but they are very polar, owing to the presence of the amide group.

Basic Amino Acids. Lysine, arginine, and histidine all possess an R group containing a basic nitrogenous group and are classified as basic amino acids. In the physiological pH range, the ϵ-amino group of lysine and the guanidino group ($H_2N\text{—}\overset{\overset{\displaystyle NH}{\|}}{C}\text{—}NH\text{—}$) of arginine each exists in the protonated form and bears a positive charge, as indicated in Table 2–2. Under these same conditions, approximately one half of the histidine molecules contain a protonated and positively charged imidazole group, 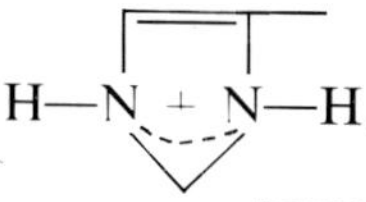, while the other half bear an unprotonated and neutral imidazole group, 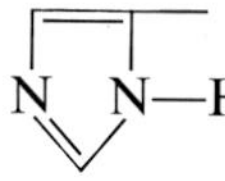. Thus the R groups of the basic amino acids are polar and are positively charged at physiological pH.

Hydroxy Amino Acids. This subclass of amino acids consists of serine and threonine, each of which has an R group bearing a hydroxyl group (—OH). Owing to the presence of the hydroxyl group, these R groups are polar.

Mercapto Amino Acids. Only cysteine has an R group with a sulfhydryl or mercapto group (—SH) as a substituent. This group plays an important role in determining the structure of some proteins and makes the R group of cysteine polar.

Thioether-Containing Amino Acid. The R group of methionine contains a thioether group (CH_3—S—CH_2—CH_2—) and is nonpolar.

Heterocyclic Amino Acid. Proline is unique among these 20 amino acids in that the α-amino group is a secondary amine as a result of its involvement in a five-membered heterocyclic ring. As we shall see, proline plays an important role in determining protein structure. The R group of proline is —CH_2—CH_2—CH_2—, which is nonpolar.

The polarity and charge of each R group on amino acids has been emphasized because this is an important determinant of protein structure. It is energetically unfavorable to place large nonpolar R groups, such as those of valine, leucine, isoleucine, and phenylalanine, into an aqueous environment because they disrupt the normal structure of water. Thus the large nonpolar groups of valine, leucine, isoleucine, and phenylalanine are known as hydrophobic groups. When present in a protein, they are usually found in the interior of the molecule where they are isolated from the aqueous environment that surrounds a protein molecule in solution. Smaller nonpolar groups, such as the R groups of glycine and alanine, cause less change in the normal structure of water. As a result, these R groups may be found on the exterior of a protein molecule, where they are exposed to the aqueous solution, as well as in the interior of the protein molecule, where they are isolated from the aqueous environment.

By contrast, the charged and polar R groups of aspartate, glutamate, lysine, histidine, and arginine are hydrophilic groups and are usually found on the exterior of the protein molecule, where they are exposed to the aqueous environment. The neutral but polar R groups, such as those of tyrosine, tryptophan, serine, threonine, glutamine, asparagine, and cysteine, are usually found on the exterior of the protein molecule, but they are also found, albeit less frequently, in the interior of the molecule.

COVALENT BONDS INVOLVED IN FORMING PROTEINS FROM α-AMINO ACIDS

The Peptide Bond. Proteins are linear polymers of amino acids in which the amino acids are joined together by peptide bonds. The peptide bond is an amide bond formed between the α-carboxyl group of one amino acid molecule and the α-amino group of another amino acid molecule. Diagrammatically, peptide bond formation can be represented as the result of the elimination of the elements of H_2O between two amino acids.

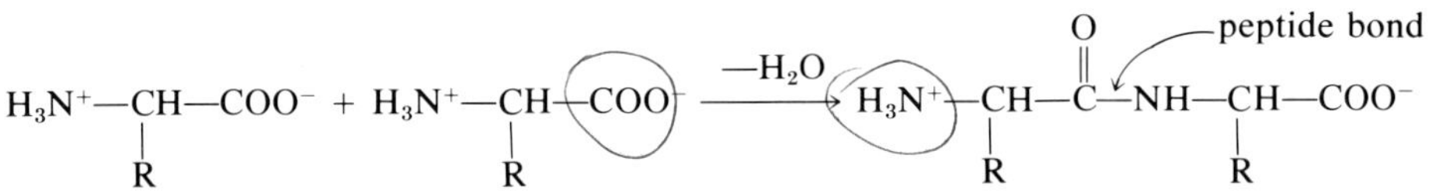

a dipeptide

A molecule containing amino acids linked by peptide bonds is known as a peptide. The number of amino acid molecules or residues contained in the peptide

TABLE 2–2 STRUCTURES OF THE L-AMINO ACIDS

A. Aliphatic Amino Acids

acyclic
-only C + H
nonpolar

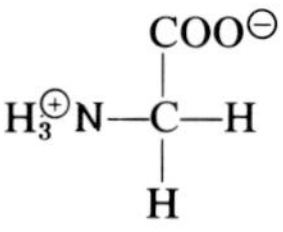

Glycine (Gly)

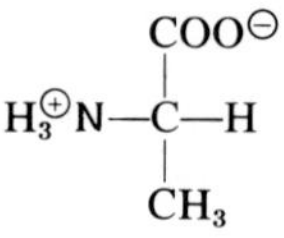

L-Alanine (Ala)

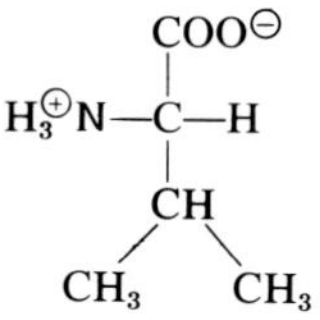

L-Valine (Val)

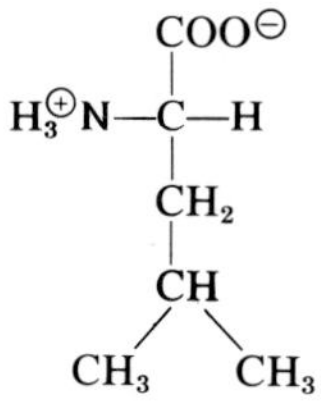

L-Leucine (Leu)

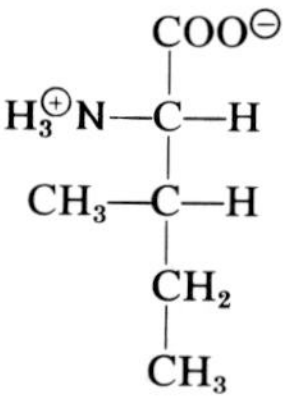

L-Isoleucine (Ile)

B. Aromatic Amino Acids

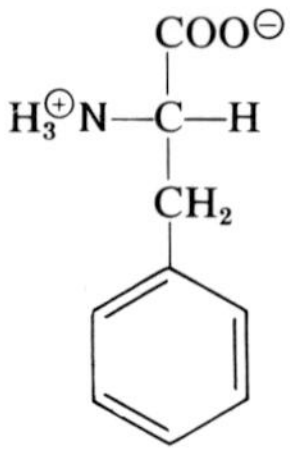

non polar

L-Phenylalanine (Phe)

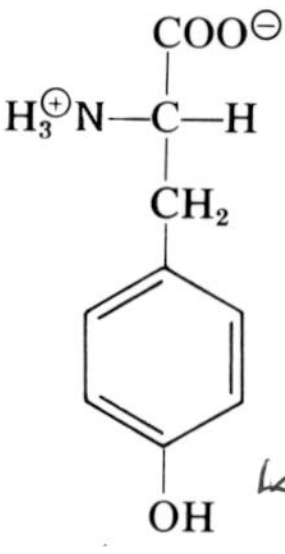

L-Tyrosine (Tyr)

somewhat polar

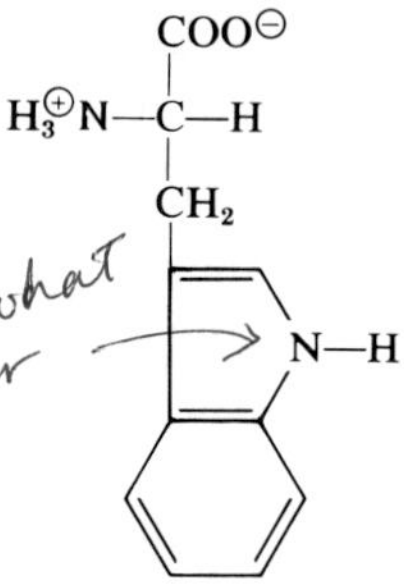

L-Tryptophan (Trp)

C. Acidic Amino Acids

COO- dissociate to COO- ions

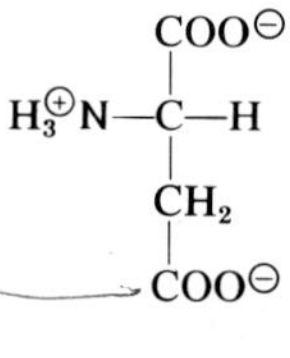

polar (−) charged

L-Aspartic Acid (Asp)

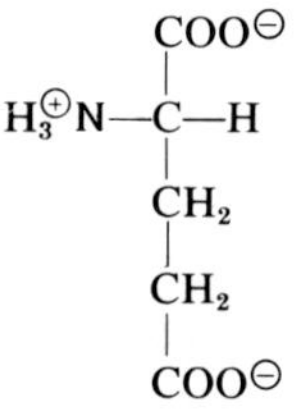

L-Glutamic Acid (Glu)

D. Amide-containing Amino Acids

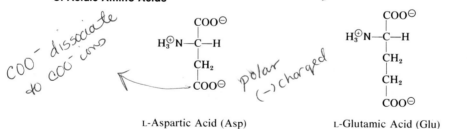

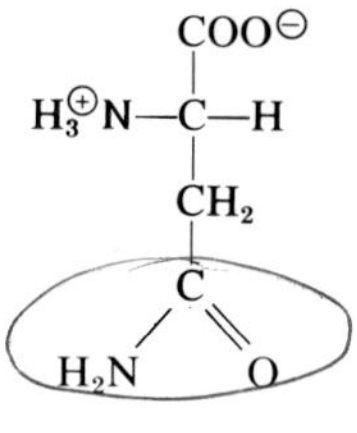

amide

L-Asparagine (Asn)

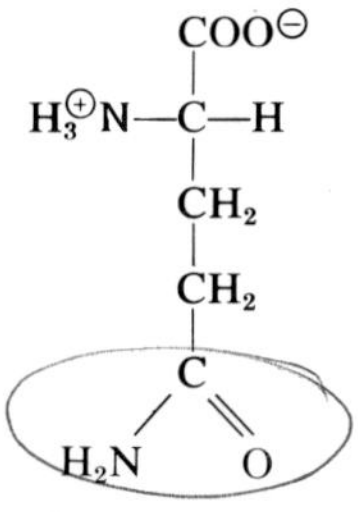

L-Glutamine (Gln)

TABLE 2–2 ***Continued***

E. Basic Amino Acids

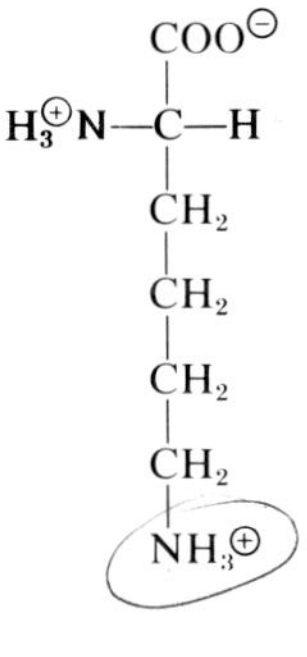

L-Lysine (Lys)

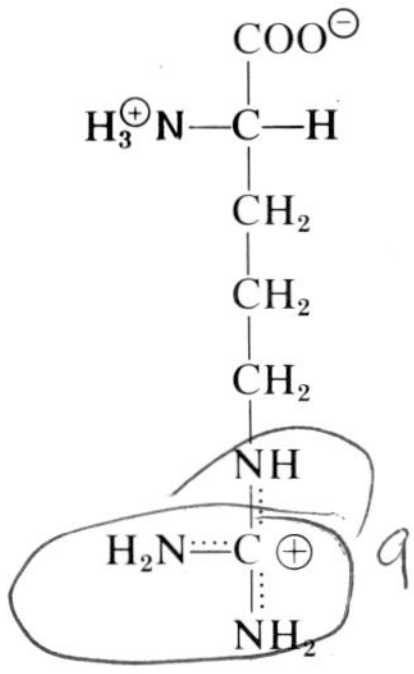

L-Arginine (Arg)

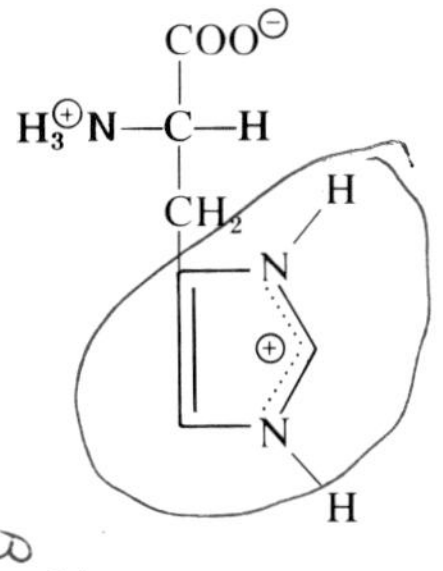

L-Histidine (His)

F. Hydroxy Amino Acids

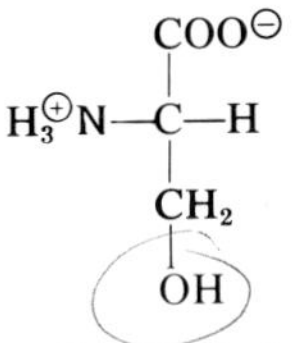

L-Serine (Ser)

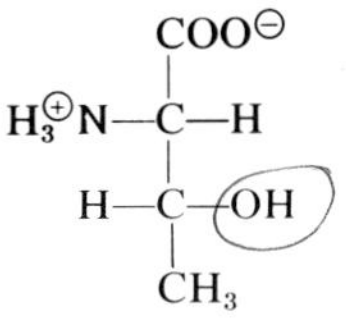

L-Threonine (Thr)

G. Mercapto Amino Acid

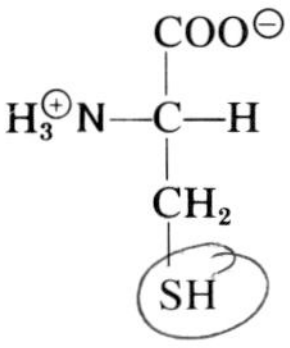

L-Cysteine (Cys)

H. Thioether-Containing Amino Acid

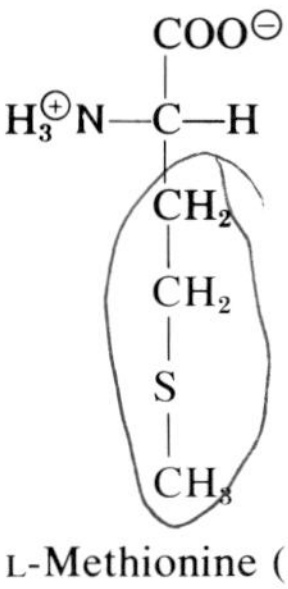

L-Methionine (Met)

I. Heterocyclic Amino Acid

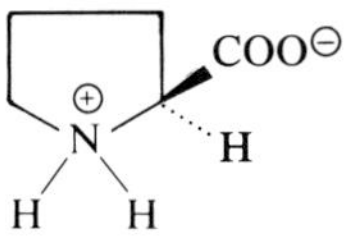

L-Proline (Pro)

is designated by a prefix, i.e., dipeptide, tripeptide, tetrapeptide, and so on. If many amino acids are linked together by peptide bonds to form a large linear polymer, the compound is a polypeptide. When the molecular weight of a polypeptide exceeds 5000 to 10,000, the polypeptide is called a protein. The end of the polypeptide chain bearing the free α-amino group is designated the amino-terminal or N-terminal end of the molecule; the end bearing the free α-carboxyl group is designated the carboxyl-terminal or C-terminal end. Peptides are named by sequentially designating each amino acid residue in the peptide, beginning with the amino-terminal amino acid. The ending *-yl* is employed on the names of all amino acid residues except the carboxyl-terminal one. This is illustrated here for a tetrapeptide.

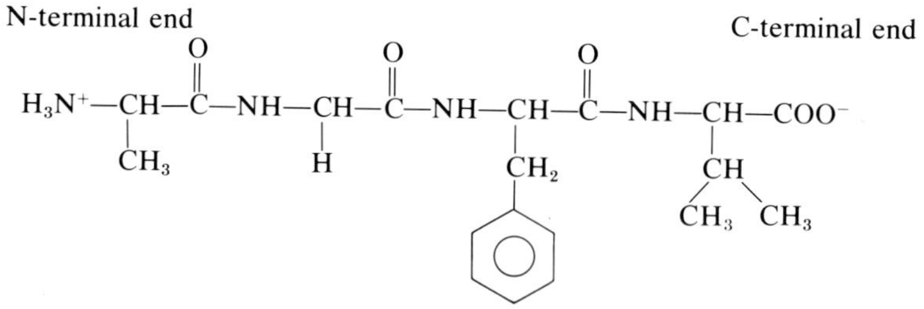

alanylglycylphenylalanylvaline
ala-gly-phe-val

One obvious consequence of peptide bond formation is that the positive charge on the α-amino group, and the negative charge on the α-carboxyl group, are no longer present when these groups are involved in a peptide bond. A second consequence is that the six atoms marked by asterisks in Figure 2–1 are coplanar, i.e., these six atoms at a peptide bond all lie in the same plane. The coplanarity of these atoms results because the peptide bond is a resonance-stabilized hybrid of the two contributing structures depicted in Figure 2–2. That is, the true structure of the peptide bond is not the same as either of the contributing structures but is really intermediate between the two. Therefore, the C—N peptide bond possesses some double bond character, and as a result all atoms directly attached to the carbon or nitrogen of this bond must lie in the same plane as the carbon and nitrogen. Furthermore, rotation is restricted about the C—N bond owing to its partial double bond character. Since every third bond in the backbone of a polypeptide chain is a peptide C—N bond, rotation will be restricted about every third bond in a polypeptide chain backbone. The consequence is that a polypeptide chain consists of a series of planar segments joined at their corners (Fig. 2–3). In this structure any planar segment can rotate with respect to its neighboring planar segments.

The Disulfide Bond. In addition to the peptide bond, a second type of covalent bond frequently found between amino acid residues in proteins and polypeptides is the disulfide bond, —S—S—, which is formed by the oxidation of the

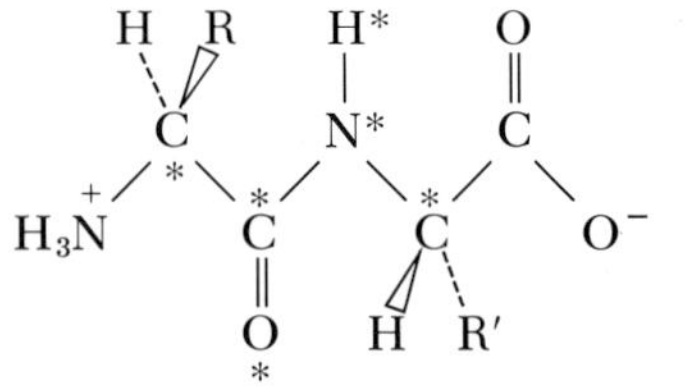

FIGURE 2–1 Projection formula for a dipeptide. Bonds in the plane of the page, projecting above the plane of the page and below the plane of the page, are represented by solid lines, wedges, and dashed lines, respectively. The asterisks designate the atoms, which must be coplanar.

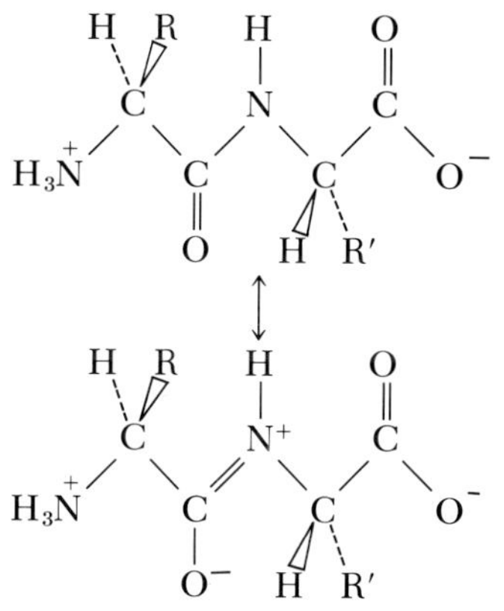

FIGURE 2–2 The peptide bond: a resonance-stabilized hybrid of two contributing structures.

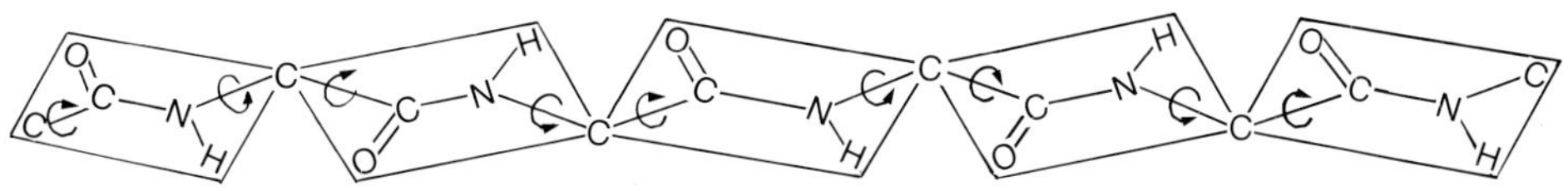

FIGURE 2–3 Planar segments in a polypeptide chain.

sulfhydryl groups of two cysteinyl residues. The polypeptide hormone insulin, for example, contains three disulfide bonds, as illustrated in Figure 2–4. Insulin consists of two polypeptide chains held together by two interchain disulfide bonds. In addition, there is one intrachain disulfide bond formed between two cysteinyl residues in the same polypeptide chain. Thus insulin provides an example of interchain and intrachain disulfide bonds, both of which are important in protein structures.

Polysaccharides—Polymers of Sugars

Polysaccharides are polymers of monosaccharides, or sugars. They can be divided into two classes, *homopolysaccharides* and *heteropolysaccharides.* Homopolysaccharides are polymers containing only one kind of monosaccharide; heteropolysaccharides are polymers containing more than one kind of monosaccharide. For the present, only the simpler class, the homopolysaccharides, will be considered.

Many plant and animal cells store the monosaccharide D-glucose as a polymer or polysaccharide. D-glucose exists in two isomeric cyclic forms, α-D-glucose and β-D-glucose. (See Appendix B for further discussion of the structure and properties of D-glucose.)

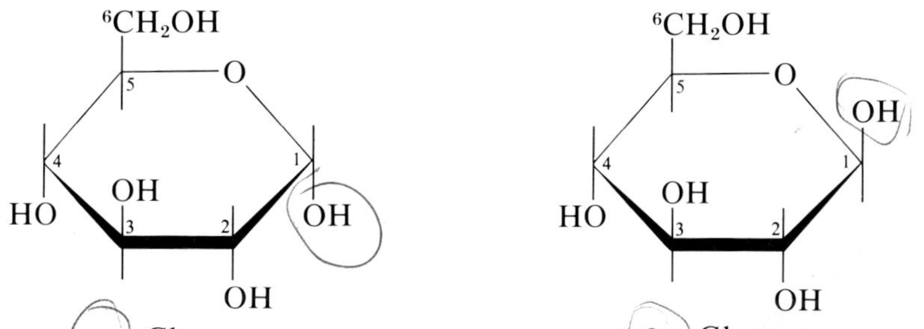

α-D-Glucose β-D-Glucose

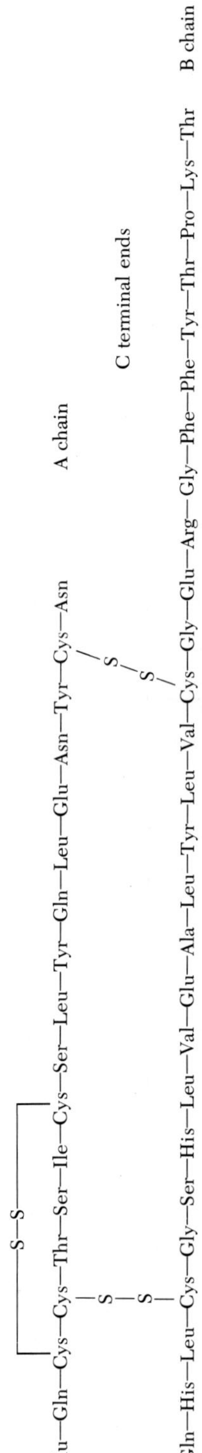

FIGURE 2–4 The amino acid sequence of human insulin.

These two forms differ only in the configuration of the asymmetric center at C-1; hence they are *anomers*.

The anomers α-D-glucose and β-D-glucose are hemiacetals and thus can react with other hydroxyl-containing compounds to form acetals. If the hydroxyl-containing compound is another monosaccharide molecule, the product is known as a *disaccharide*.

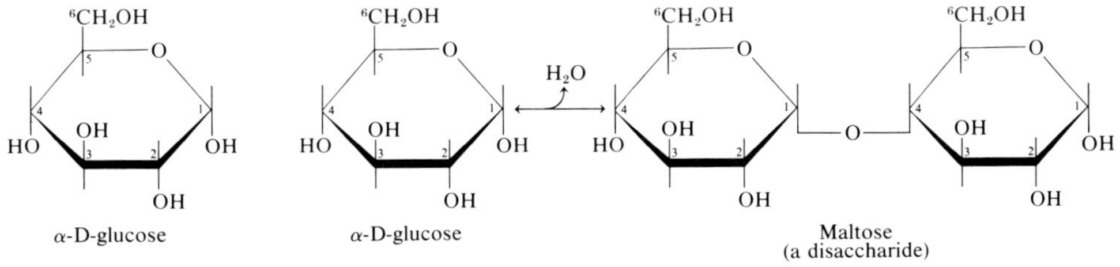

In the disaccharide maltose represented here, the C-1 of the left-hand residue is in an acetal linkage, while the C-1 of the right-hand residue remains in a hemiacetal linkage. The bond between the two glucose residues is called a *glycosidic bond*. More specifically, it is designated an α(1→4) glycosidic bond to indicate that the glycosidic bond is formed between C-1 of one monosaccharide and C-4 of the other, and that the configuration of the anomeric C-1 carbon participating in the glycosidic bond is α. When the anomeric carbon of a monosaccharide participates in a glycosidic bond, that monosaccharide residue is no longer capable of reducing Cu^{++}. Thus, maltose is a reducing disaccharide, but only the right-hand monosaccharide residue with the hemiacetal C-1, is readily oxidized.

Many monosaccharide molecules can be joined together by glycosidic bonds to form large polymers known as polysaccharides. The amylose fraction of starch, a segment of which is shown in Figure 2–5*A*, consists of linear polymers of D-glucose residues linked by α(1→4) glycosidic bonds to form molecules with molecular weights up to 500,000. Amylopectin, another fraction of starch, is a branched polymer of glucose with a molecular weight of approximately 1,000,000. Amylopectin consists of chains of α(1→4)-linked glucose residues with branches formed by α(1→6) linkages, as illustrated in Figure 2–5*B*. Glycogen, the polysaccharide of D-glucose units found in animal cells, has a molecular weight of over 1,000,000 and is much more highly branched than amylopectin. Like amylopectin, glycogen consists primarily of chains of D-glucose linked by α(1→4) glycosidic linkages; however, the branch points formed by α(1→6) glycosidic bonds are much more frequent in glycogen (Fig. 2–5*C*).

Storage of a large number of small molecules, such as glucose, in the form of large polymeric molecules may serve several functions. One obvious advantage is that conversion of many small molecules to a few large molecules reduces the osmotic effect of the stored substance because it reduces the total number of particles that are stored.

Glycoproteins

Glycoproteins are macromolecules that have carbohydrate covalently attached to the protein. Most secretory proteins are glycoproteins (Chapter 9), and glycoproteins are important constituents of the plasma membrane (Chapter 14). The carbohydrate moieties of glycoproteins consist of monosaccharides linked

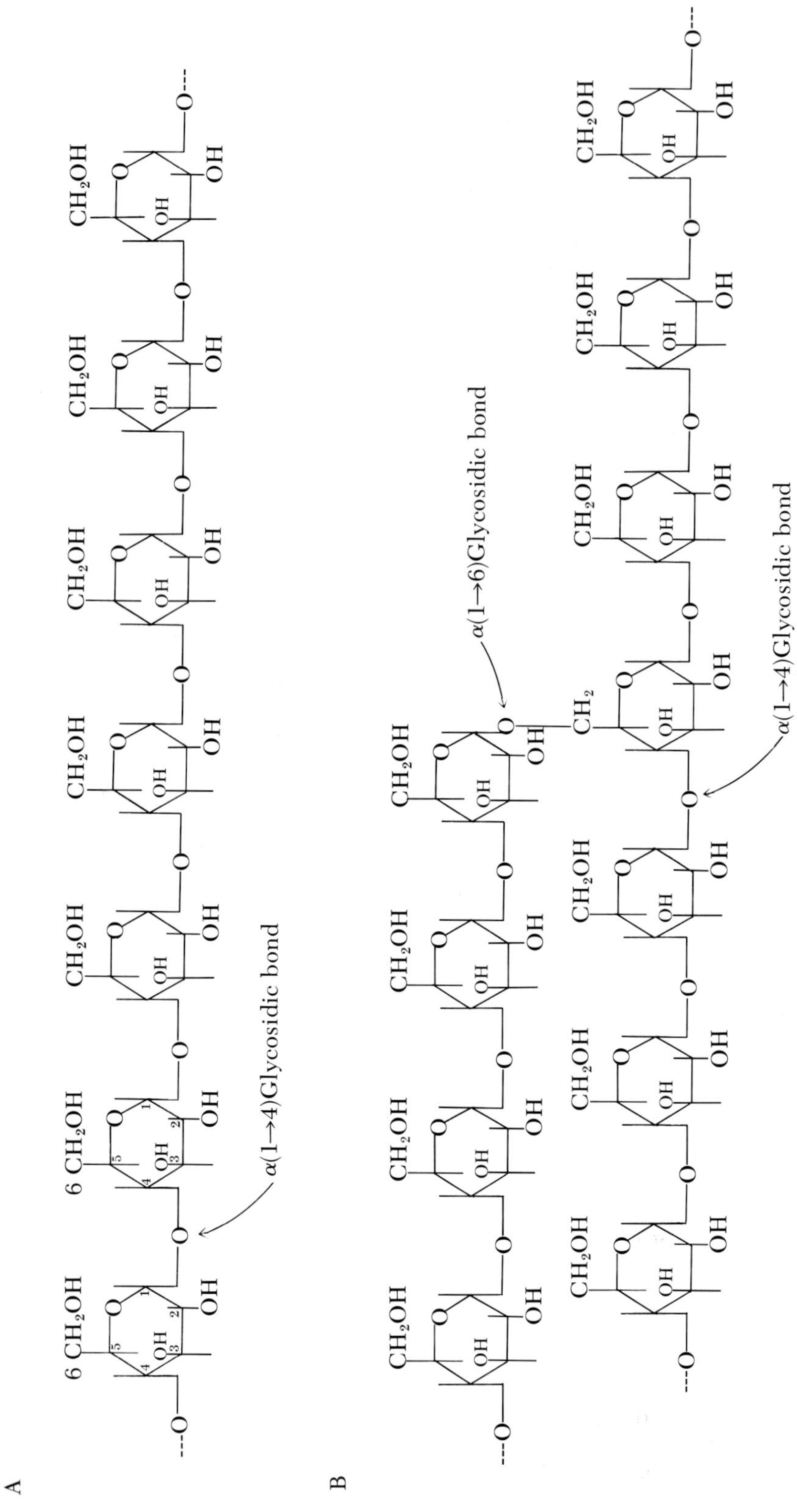

FIGURE 2–5 *See legend on the opposite page.*

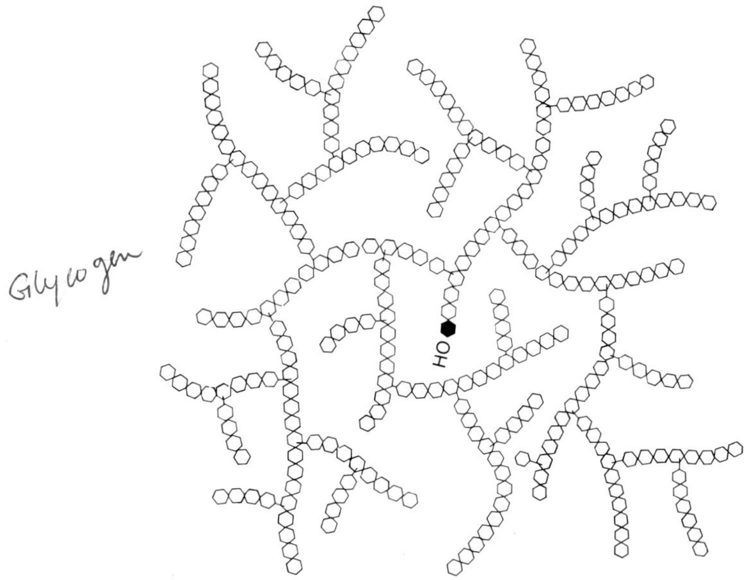

C

FIGURE 2–5 The structure of some polysaccharides. *A*. A segment of amylose, a linear polysaccharide consisting of *D*-glucose residues linked by $\alpha(1 \rightarrow 4)$ glycosidic bonds. *B*; A segment of amylopectin, a branched polysaccharide consisting of chains of *D*-glucose residues linked by $\alpha(1 \rightarrow 4)$ glycosidic bonds, with branch points formed by $\alpha(1 \rightarrow 6)$ glycosidic bonds. *C*. A schematic representation of glycogen with the *D*-glucose residues linked by $\alpha(1 \rightarrow 4)$ glycosidic bonds represented by the symbol ⬡⬡, and the branch points formed by $\alpha(1 \rightarrow 6)$ glycosidic bonds represented by the symbol ⬡. The single reducing glucosyl residue in glycogen is represented by ⬢OH.

by glycosidic bonds to form oligosaccharides. Typically, the oligosaccharide is linked to the polypeptide by a glycosidic bond to a seryl or threonyl residue, or by an N-glycosidic bond formed with an asparaginyl residue.

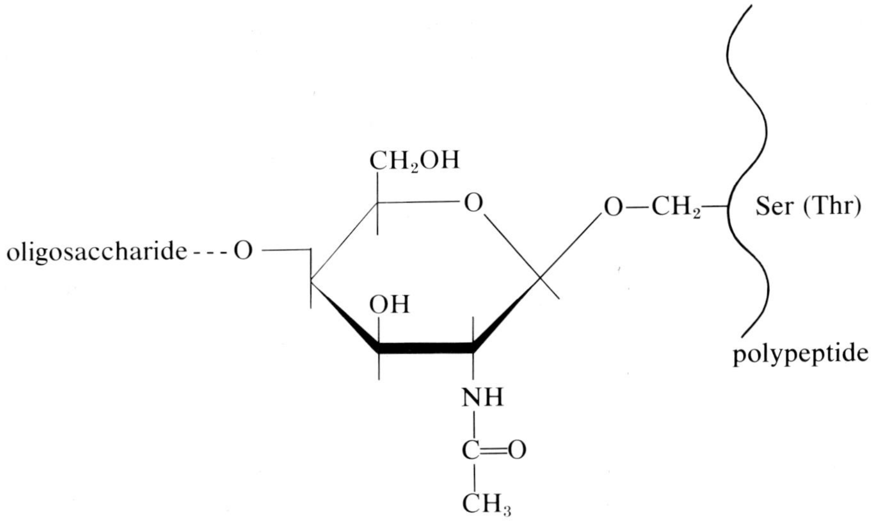

glycosidic link to Ser or Thr

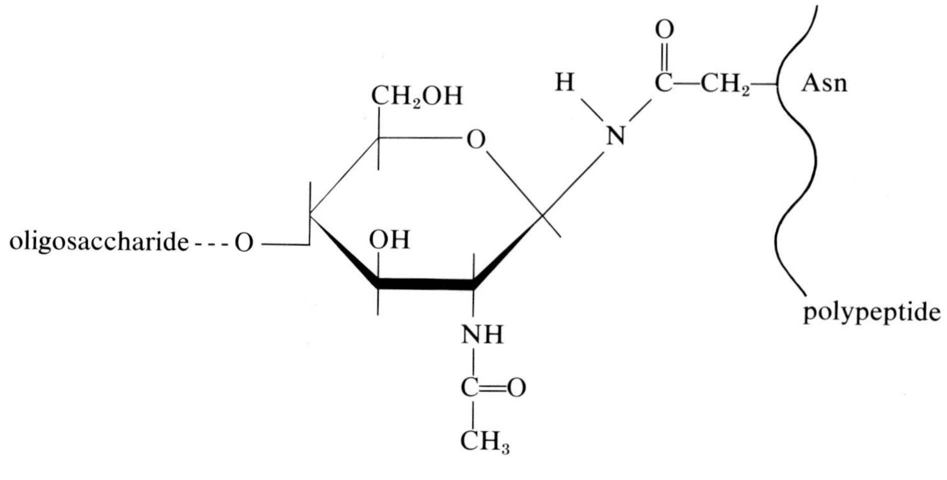

N-glycosidic link to Asn

In collagen, the carbohydrate is linked glycosidically to hydroxylysyl residues in the protein. The monosaccharides that make up the oligosaccharide moieties include glucose, galactose, mannose, N-acetylglucosamine, N-acetylgalactosamine, fucose, and N-acetylneuraminic acid.

Nucleic Acids—Linear Polymers of Nucleotides

The nucleic acids are divided into two classes on the basis of the chemical structure of their constituent monomeric units: deoxyribonucleic acid (DNA) and ribonucleic acid (RNA).

DNA—DEOXYRIBONUCLEIC ACID

DNA is found in nuclei, but small amounts are also present in mitochondria. Complete hydrolysis of any sample of DNA yields inorganic phosphate, 2-deoxy-D-ribose, and a mixture of nitrogenous bases. In DNA, the bases are the purines, adenine and guanine, and the pyrimidines, cytosine and thymine. The structures and names of the products resulting from the hydrolysis of DNA are shown in Figure 2–6.

In contrast to complete hydrolysis, partial hydrolysis of DNA yields a mixture of deoxynucleotides: deoxyadenylic acid, deoxyguanylic acid, deoxycytidylic acid, and deoxythymidylic acid. Each contains one phosphate group, one deoxy sugar molecule, and one nitrogenous base linked together, as shown in Figure 2–7. The linkage between the nitrogen of the purine or pyrimidine base and the C-1′ of the deoxyribose is a β-N-glycosidic bond. The phosphate is attached to the C-5′ position of the deoxyribose by a phosphate ester bond.

All DNA molecules are linear polymers of the four deoxyribonucleotides shown in Figure 2–7. The DNA molecules are very large, with molecular weights ranging from a few million to many million. This long, linear, unbranched polymer is formed by linking the deoxyribonucleotide subunits together via 3′, 5′-phosphodiester bonds, as illustrated in Figure 2–8. Thus, the covalent backbone of a DNA molecule consists of alternating deoxyribose and phosphate groups linked

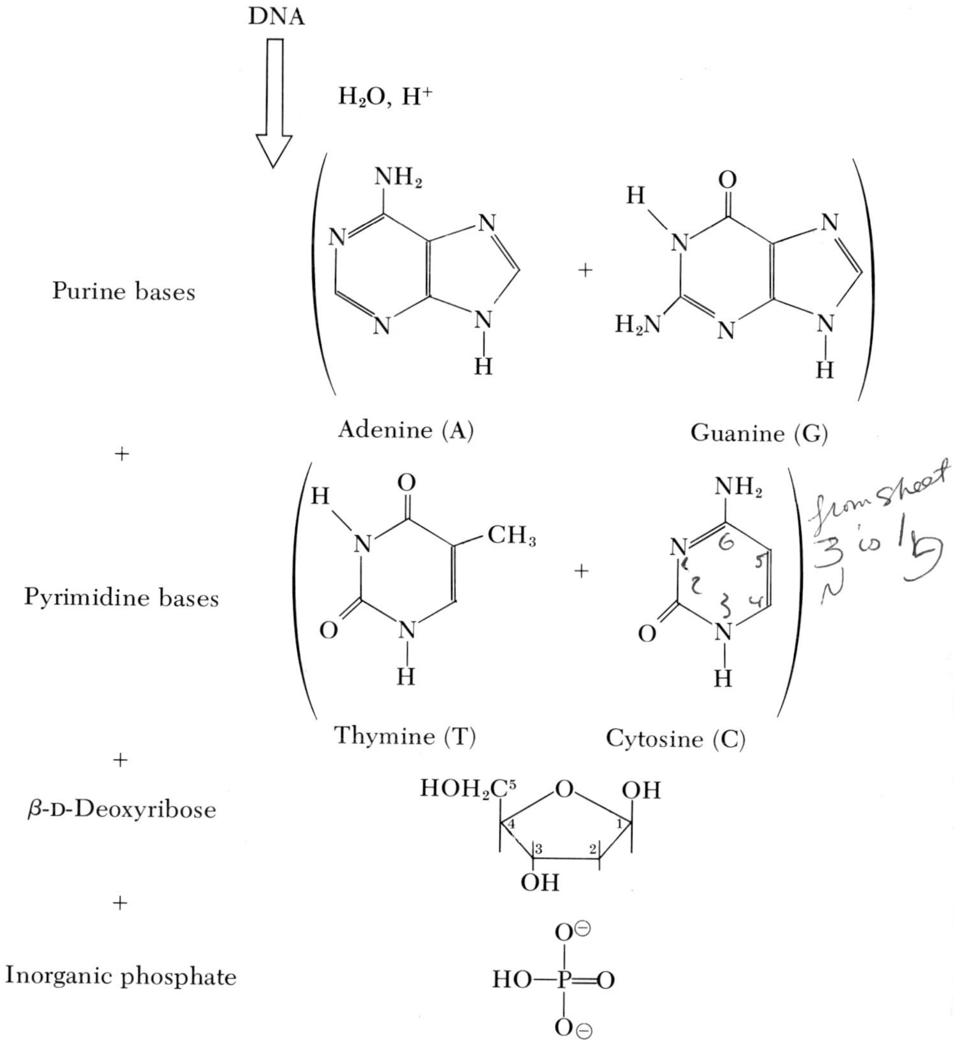

FIGURE 2–6 Products of the hydrolysis of DNA.

together by phosphodiester bonds, while the purine and pyrimidine bases form short side chains projecting from this backbone. The end of the DNA molecule that contains a free 3′-hydroxyl group is designated the 3′ end of the molecule; the other end is designated the 5′ end.

RNA—RIBONUCLEIC ACID

Approximately 90 per cent of the RNA is present in the cytoplasm of the cell, while the remaining 10 per cent is found in the nucleolus. The four major nucleotides, isolated after partial hydrolysis of RNA, are adenylic acid, guanylic acid, cytidylic acid, and uridylic acid (Fig. 2–9). As in DNA, each nucleotide consists of phosphate, sugar, and a base. In RNA these are all ribonucleotides, since the pentose is β-D-ribose, in contrast to the β-D-deoxyribose found in DNA. An additional difference between the major ribonucleotides isolated from RNA and the deoxyribonucleotides isolated from DNA is the presence of uracil as one of the major bases in RNA, as opposed to the thymine found in DNA.

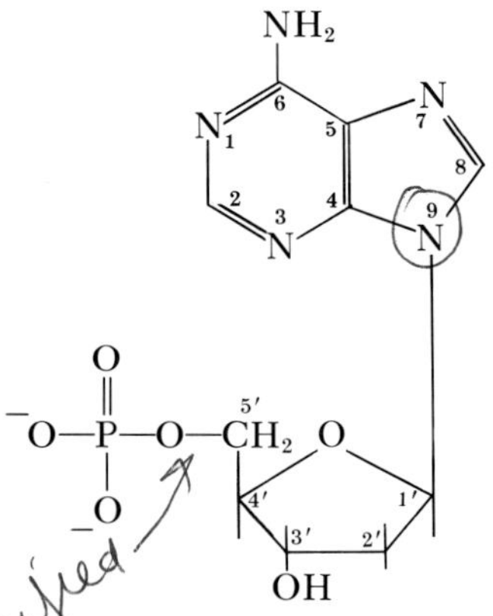

Deoxyadenosine-5′-phosphate
(Deoxyadenylate)
dAMP

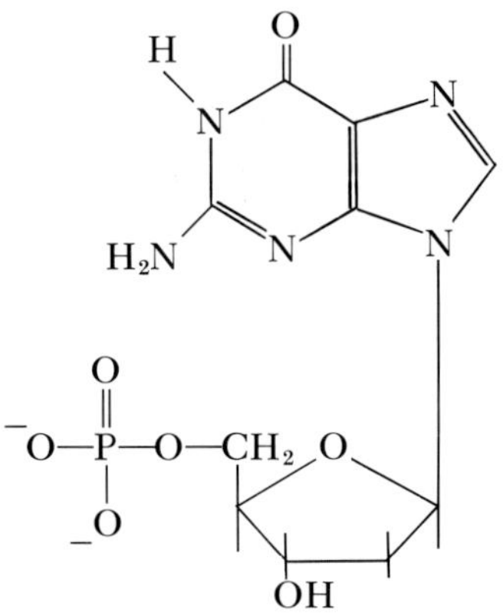

Deoxyguanosine-5′-phosphate
(Deoxyguanylate)
dGMP

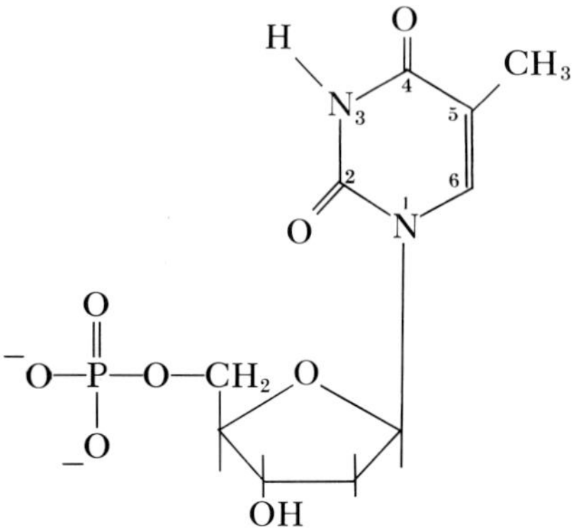

Deoxythymidine-5′-phosphate
(Deoxythymidylate)
dTMP

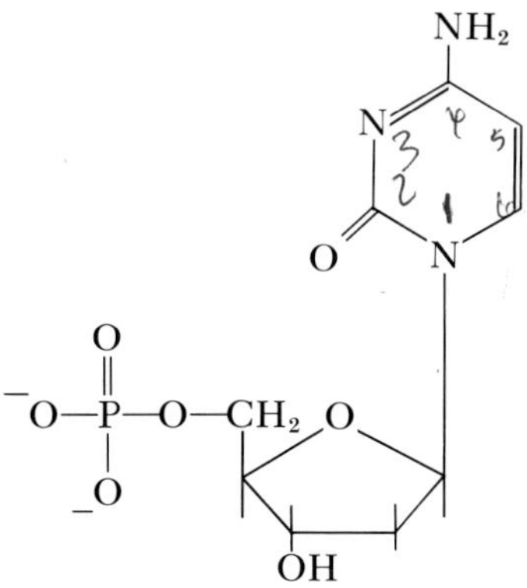

Deoxycytidine-5′-phosphate
(Deoxycytidylate)
dCMP

FIGURE 2–7 Deoxyribonucleotides in DNA.

Most RNA molecules consist of the four major nucleotides depicted in Figure 2–9, linked together by 3′, 5′-phosphodiester bonds to form long linear polymers. Thus, RNA molecules contain a covalently linked backbone of alternating ribose and phosphate groups joined by phosphodiester bonds, with the purine and pyrimidine bases appearing as side chains attached to this backbone. Although adenine, guanine, cytidine, and uracil are the major bases of RNA, some RNA molecules also contain small amounts of other bases.

STRUCTURE OF MACROMOLECULES

Proteins and nucleic acids are linear unbranched polymers of covalently linked monomers. There are hundreds of different kinds of protein molecules in a cell, each possessing its own specific functional properties; yet they are all made from the same 20 amino acids. Likewise, there are many different kinds of

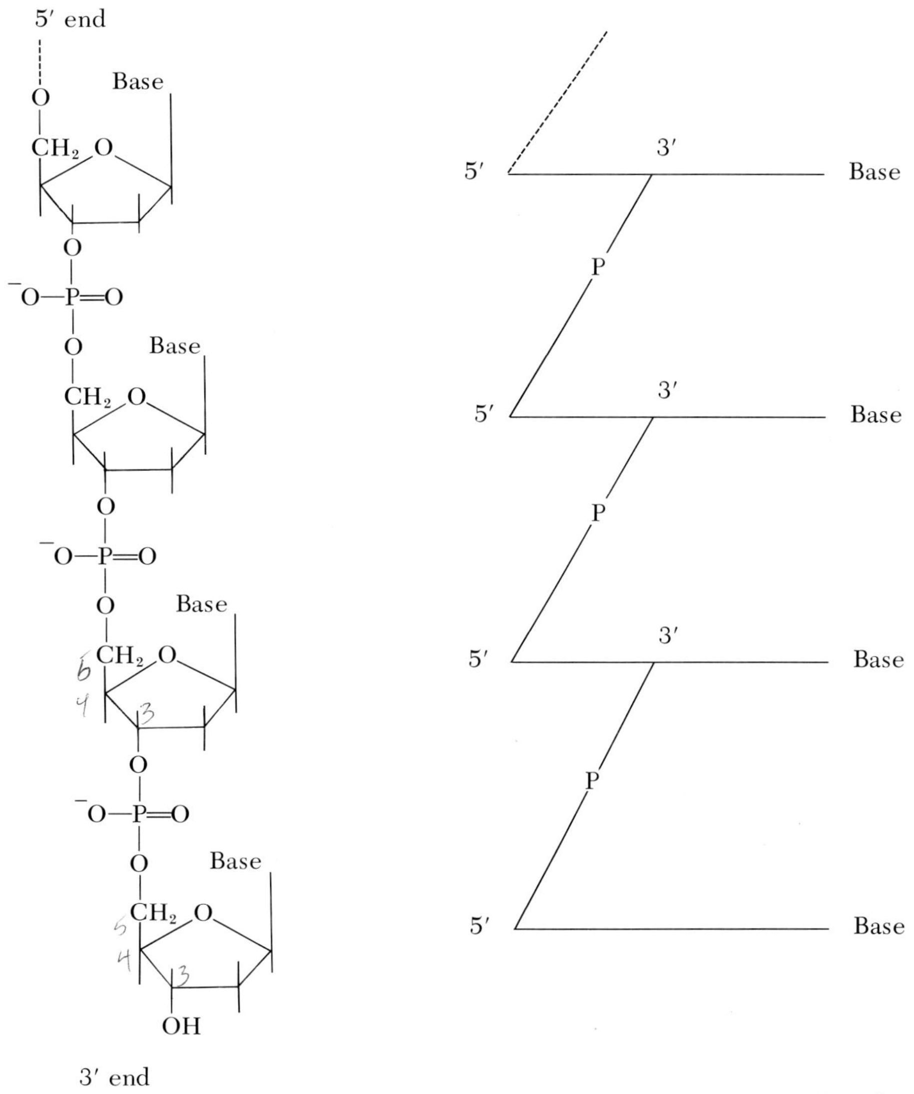

FIGURE 2–8 The linkage of nucleotides in DNA. A schematic representation is depicted at the right.

DNA molecules, and each kind has a different informational content, even though they are all composed of the same four nucleotides. All proteins are made from similar monomeric units linked together by peptide bonds, and all DNA molecules contain only four types of deoxyribonucleotides joined by phosphodiester bonds. Therefore, the specific properties of a given protein and the informational content of a particular nucleic acid must be functions of the amount and distribution of monomeric units in the molecule. The composition and spatial arrangement of monomeric units in specific proteins will now be considered. The composition and spatial arrangement of monomeric units in nucleic acids will be examined in Chapters 5 and 7.

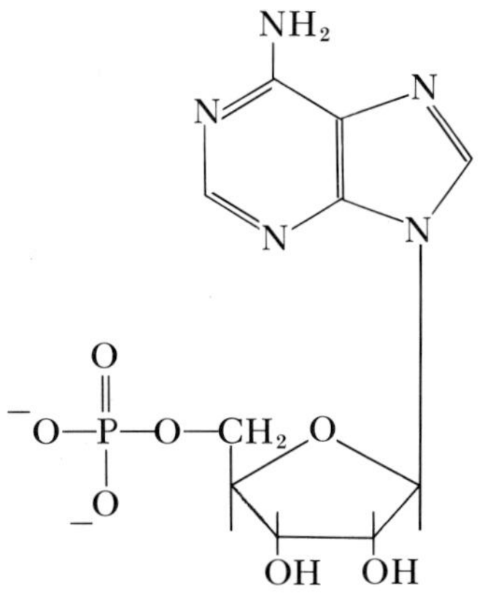

Adenosine-5′-phosphate
(Adenylate)
AMP

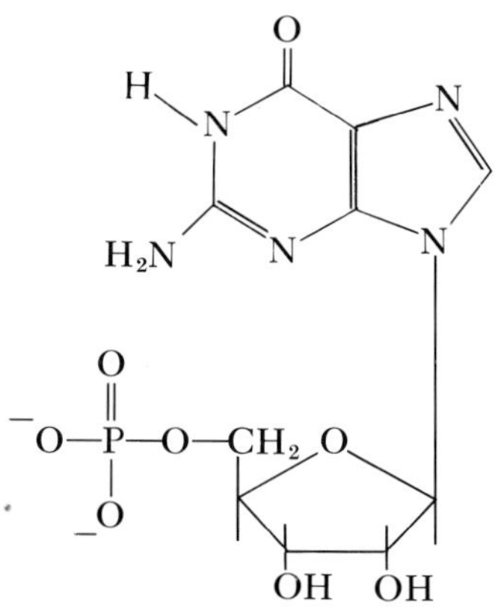

Guanosine-5′-phosphate
(Guanylate)
GMP

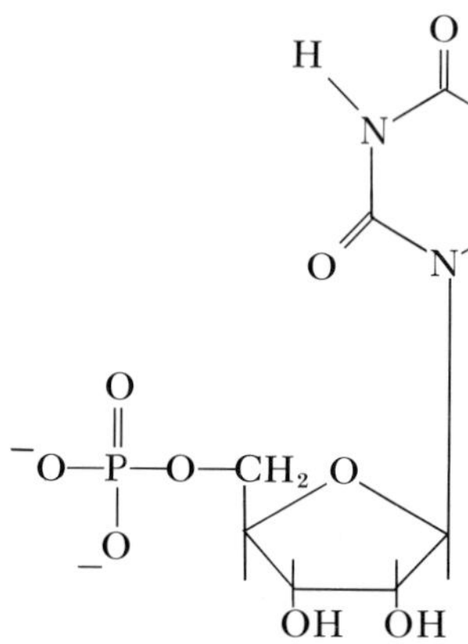

Uridine-5′-phosphate
(Uridylate)
UMP

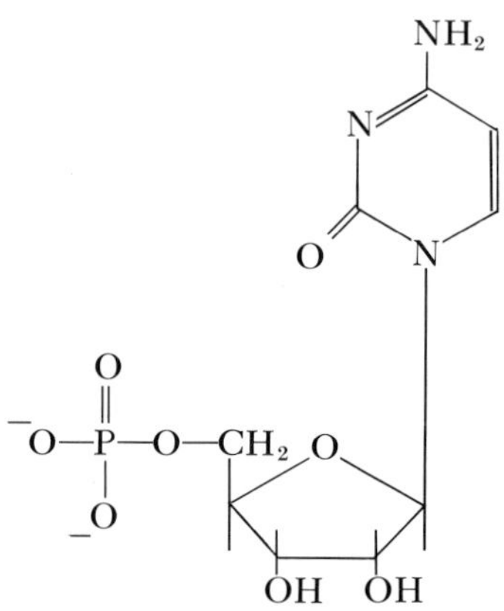

Cytidine-5′-phosphate
(Cytidylate)
CMP

FIGURE 2–9 Ribonucleotides in RNA.

Protein Structure

COMPOSITION OF SPECIFIC PROTEINS

Determination of the amino acid composition of a specific protein is a relatively simple task today, although this has not always been the case. Basically, the procedure involves isolation of a specific protein, hydrolysis of this protein to yield a mixture of amino acids, and determination of the quantity of each amino acid present in the hydrolysate. The determination of the quantity of each amino acid in the hydrolysate is now a fully automated procedure involving the use of an instrument known as an amino acid analyzer. The amino acid analyzer separates the amino acid mixture in the protein hydrolysate into its component amino acids by ion exchange chromatography and then determines the amount of each amino acid present colorimetrically. From these data, the relative amounts of all amino acids present in a specific protein can be obtained. If the amount of protein hydrolyzed and the molecular weight of the protein are also known, the number of molecules of each amino acid present in the native protein molecule can be calculated. The amino acid composition of several proteins appears in Table 2–3.

These proteins have very different functions. For example, cytochrome c is a mitochondrial protein involved in electron transport, while hemoglobin is found in erythrocytes and is essential for O_2 transport in the blood. Growth hormone affects the rate of skeletal growth, and lysozyme is an enzyme that catalyzes the hydrolysis of certain complex heteropolysaccharides. While there may be obvious differences in the sizes and amino acid composition of some proteins, other proteins which differ markedly in their functional properties may have very similar amino acid compositions and molecular sizes. Thus, factors in addition to amino acid composition must be important in determining the functional properties of specific proteins.

PRIMARY STRUCTURE OF PROTEINS

While the amino acid composition tells us how many residues of each amino acid are present in a specific protein, it affords no clue to the sequence of amino acid residues in the protein. Just as a sequence of letters determines a word, the sequence of amino acid residues in a protein is important in determining the structure and functional properties of the protein. This sequence of amino acid residues in a given protein molecule is known as the *primary structure* of the protein.

The first determination of the primary structure of a large polypeptide was carried out by Sanger and reported in 1953. The polypeptide was insulin, and its amino acid sequence, or primary structure, appears in Figure 2–4. The ability to determine the primary structure of a protein represents such an important advance in biochemistry that it warrants a brief review of the principles and methodology involved.

AMINO ACID COMPOSITION OF FOUR PROTEINS — **TABLE 2–3**

	Hemoglobin A_1 (human)	Cytochrome c (human)	Lysozyme (human)	Growth Hormone (human)
Gly	40	13	11	8
Ala	72	6	14	7
Val	62	3	9	7
Leu	72	6	8	26
Ile	0	8	5	8
Phe	30	3	2	13
Tyr	12	5	6	8
Trp	6	1	5	1
Ser	32	2	6	18
Thr	32	7	5	10
Cys	6	2	8	4
Met	6	3	2	3
Pro	28	4	2	8
Lys	44	18	5	9
Arg	12	2	14	11
His	38	3	1	3
*Asx	50	8	18	20
†Glx	32	10	9	26

*Composition expressed as number of amino acid residues per molecule of protein.
†Asx and Glx represent Asp + Asn and Glu + Gln, respectively.

Identification of the N-Terminal Amino Acid Residue of a Protein or Peptide

Two of the standard methods for determining the N-terminal amino acid residue of a protein or peptide are outlined in Figure 2–10. Method *A* was developed by Sanger and consists of reacting the protein or peptide with fluorodinitrobenzene (FDNB) to form a 2,4-dinitrophenyl (DNP) derivative of the N-terminal amino acid residue. The N-terminal amino acid is the only amino acid residue in the protein that is capable of reacting with FDNB to form a DNP derivative in which the DNP group is covalently attached to the nitrogen of the α-amino group. Hydrolysis of the dinitrophenylated protein or peptide yields a mixture of amino acids plus the DNP derivative of the N-terminal amino acid. The DNP derivative of the N-terminal amino acid is identified by chromatographic methods, thereby establishing the identity of the N-terminal amino acid residue in the original protein or peptide.

Method *B* makes use of the Edman reaction, in which phenylisothiocyanate reacts with the α-amino group of the N-terminal amino acid residue of the protein or peptide to form a phenylthiocarbamyl derivative of the protein or peptide. Treatment of this phenylthiocarbamyl derivative with acid in an organic solvent results in the formation of a phenylthiohydantoin from the N-terminal amino acid residue, leaving the remainder of the peptide as an intact peptide that is one amino acid residue shorter than the original protein or peptide. Chromatographic identification of the phenylthiohydantoin formed establishes the identity of the N-terminal amino acid residue of the original protein or peptide. Furthermore, the entire procedure can be repeated on the residual peptide, which is one amino acid shorter than the original peptide. Thus the N-terminal amino acid of the residual peptide can be identified, thereby establishing the identity of the amino acid adjacent to the N-terminal amino acid in the original protein or peptide. By repeated application of the Edman reaction, the amino acid sequence of a protein, beginning at the N-terminal end, can be elucidated. This procedure for determining the amino acid sequence of a protein has been successfully automated in recent years and should greatly facilitate the determination of the primary structures of proteins. In one instance it was possible to obtain the amino acid sequence of the first 50 amino acids of a protein, beginning at the N-terminal end, by this method.

Identification of the C-Terminal Amino Acid Residue of a Protein or Peptide

Carboxypeptidase A and carboxypeptidase B are exopeptidases that catalyze the hydrolysis of the C-terminal peptide bond only. Thus, these enzymes bring about the sequential release of the amino acid residues at the C-terminal end of the molecule. Since the first amino acid released is the C-terminal amino acid of the peptide or protein, identification of the amino acid that appears most rapidly after treatment of the protein with the carboxypeptidases establishes the C-terminal amino acid residue of the protein. Carboxypeptidase A catalyzes the hydrolysis of the C-terminal aromatic and bulky aliphatic amino acids most rapidly, while carboxypeptidase B is specific for the hydrolysis of C-terminal lysine and arginine residues.

Cleavage of Proteins to Form Smaller Peptides

At the present time the determination of the primary structure of a large protein requires that the protein be cleaved into smaller polypeptides whose primary structures can be elucidated. Generally, an enzyme or reagent that hydrolyzes only peptide bonds involving a few specific amino acids is employed. These include:

A. Sanger Method

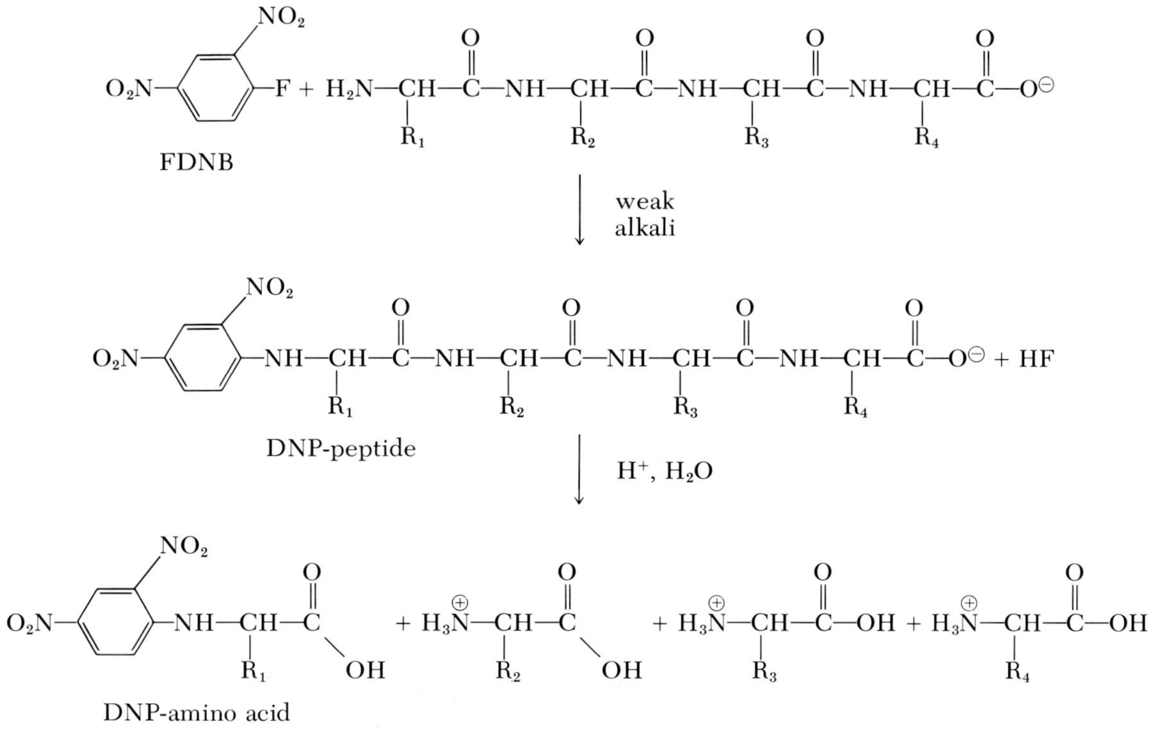

B. Edman Reaction

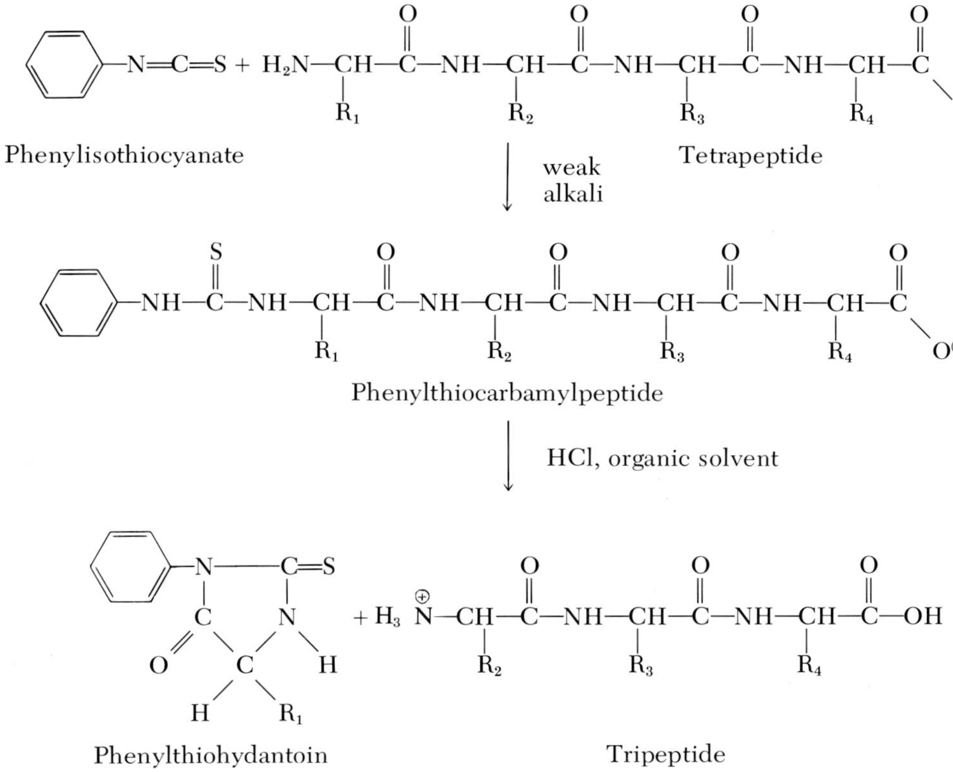

FIGURE 2–10 N-terminal amino acid analysis. *A*. Sanger method; *B*. Edman reaction.

TRYPSIN. Trypsin is an endopeptidase that catalyzes only the hydrolysis of peptide bonds in which the carboxyl group of lysine or arginine participates.

CHYMOTRYPSIN. This is an endopeptidase specific for the hydrolysis of bonds in which the carboxyl group of tyrosine, phenylalanine, or tryptophan participates.

PEPSIN. Pepsin is an endopeptidase that catalyzes the hydrolysis of peptide bonds in which one of the participating amino acids has an aromatic or bulky nonpolar R group.

CYANOGEN BROMIDE. This is a chemical reagent that specifically cleaves the polypeptide chain at peptide bonds in which the carboxyl group of methionine participates.

After the protein has been cleaved into a set of smaller polypeptides by one of the procedures described, the members of this set of smaller polypeptides are separated by chromatographic and electrophoretic techniques, and the amino acid sequence in each of the smaller polypeptides is determined by the Edman or some other method. This furnishes the amino acid sequence of all the constituent segments of the original protein but does not yield the information necessary to determine the order in which these segments occur in the original protein. This information is obtained by cleaving a second sample of the original protein by one of the other procedures listed, thereby generating a second set of polypeptides which overlap the first set whose amino acid sequences have been determined. Since the polypeptides in the second set overlap those in the first set, elucidation of the amino acid sequence of all the polypeptides in this second set permits the investigator to establish the order in which the polypeptides of the first set occur in the original protein. This approach is illustrated by an example in Appendix C.

INTERACTIONS THAT DETERMINE THE SECONDARY AND TERTIARY STRUCTURE OF PROTEINS

Many of the properties of native proteins, such as their behavior in the ultracentrifuge, indicate that proteins exist as well-defined structures in an aqueous solution and not as long, randomly coiled peptide chains. For proteins to exist as specific and often compact structures, there must be some interactions between the various segments of the polypeptide chain that stabilize a particular spatial arrangement, or conformation, of the polypeptide chain. The bonds and interactions that are involved in the stabilization of particular conformations of polypeptide chains include disulfide bonds, electrostatic interactions, hydrogen bonds, and hydrophobic interactions.

Disulfide Bonds, R—S—S—R. Examples of disulfide bond formation between two cysteine residues in a polypeptide are illustrated in Figure 2–4. A disulfide bond is characterized by a bond strength of approximately 50 kilocalories per mole and a bond length of about 2 Å between the two sulfur atoms. Hence, disulfide bond formation between two cysteine residues located some distance apart in the polypeptide chain requires that the polypeptide chain be folded back on itself to bring the sulfhydryl groups close together. Disulfide bonds make very important contributions to the stabilization of a particular conformation in some proteins.

Electrostatic Interactions. In considering the structures of the amino acids, it was pointed out that the R groups of glutamate and aspartate contain negatively charged carboxylate groups, and that the basic amino acids, lysine, arginine, and histidine, contain positively charged groups in the physiological pH range. Thus, these amino acids contribute negatively charged and positively charged side chains to the polypeptide chain backbone. When two oppositely charged groups are brought close together, electrostatic interactions lead to a strong attraction, resulting in the formation of an electrostatic, or ionic, bond. In a long polypeptide chain containing a large number of charged side chains, there are many opportunities for electrostatic interaction, and it is easy to visualize how these interactions can play a role in determining how the polypeptide folds.

Hydrogen Bonds. When a group containing a hydrogen atom that is covalently bonded to an electronegative atom, such as oxygen or nitrogen, is in the vicinity of a second group containing an electronegative atom, an energetically favorable interaction occurs which is referred to as a hydrogen bond. The hydrogen bond is usually represented by a dotted line, as in Figure 2–11. The strength of the hydrogen bond is only 5 to 8 kilocalories per mole and is maximal when the bond is linear, as depicted in Figure 2–11. Hydrogen bonding between amides or peptides of the type depicted in Figure 2–11*C* plays an important role in stabilizing some conformations of the polypeptide chain.

Hydrophobic Interactions. In considering the structures of amino acids, we stressed the importance of hydrophobic nonpolar R groups, such as those present in leucine, isoleucine, valine, and phenylalanine, in determining protein structure. Nonpolar groups such as these interact with each other only through very weak van der Waals forces. However, to put a nonpolar group of this type in water is energetically unfavorable by 3 to 5 kilocalories per mole because it brings about an increase in the structure of water surrounding the nonpolar group and a decrease in the entropy of the system. Therefore, a system consisting of water and hydrophobic nonpolar groups will be much more stable when the nonpolar groups are out of the aqueous environment. Polypeptide chains in aqueous solution achieve a stable conformation by folding in such a way that the nonpolar hydrophobic R groups of their constituent amino acids are placed together in the interior of the molecule and out of the aqueous environment. Therefore, the nonpolar hydrophobic R groups play an extremely important part in the determination of protein structure, even though they interact only weakly with one another.

These then are the forces that determine how a polypeptide will fold, and consequently what the structure of a given protein will be. Some of the specific types of protein structures brought about by the interplay of these forces will now be considered.

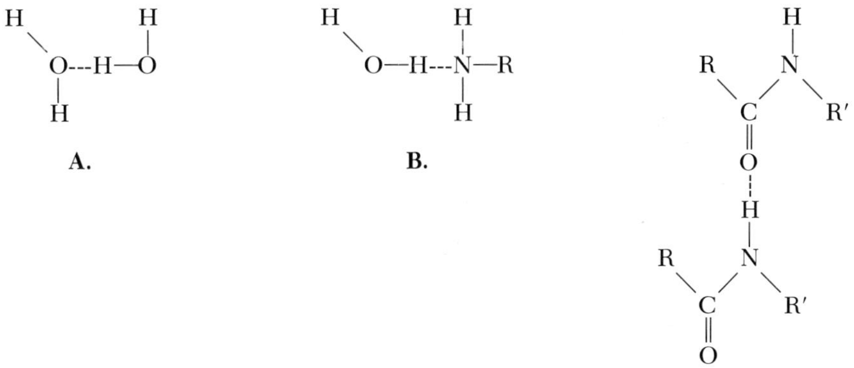

FIGURE 2–11 Examples of hydrogen bonding. Hydrogen bonds (*dotted lines*) between: A. water molecules, B. water and an amine, and C. two amide groups.

SECONDARY STRUCTURE OF PROTEINS

Secondary structure of proteins refers to the steric or spatial relationship of amino acids that are near to each other in the amino acid sequence. Two regular secondary structures are frequently encountered in proteins. These are the right-handed α helix and the β pleated sheet structure.

α Helix. Early x-ray diffraction studies of fibrous proteins, such as hair and wool, showed a major periodicity or repeat unit of 5.0 to 5.5 Å, indicating some regularity in the structure of these proteins. A minor repeat unit of 1.5 Å was also observed. Working with very accurate models based on x-ray diffraction studies of amino acids and simple dipeptides and tripeptides, Pauling and Corey found that a polypeptide chain with planar peptide bonds would form a right-handed helical structure by simple twists about the α-carbon-to-nitrogen and the α-carbon-to-carboxyl carbon bonds. This helical structure, now known as the α helix, had a pitch of 5.4 Å and contained 3.6 amino acids per turn of the helix, thereby giving a rise per residue of 1.5 Å.

In addition to accounting for the major and minor periodicities indicated by the x-ray diffraction studies, this structure possesses a very interesting feature that may account for its existence. As can be seen in Figure 2–12, the carbonyl group ($>C{=}O$) of every peptide bond is in a position to form a hydrogen bond with the $>N{-}H$ group of the peptide bond in the next turn of the helix, thereby contributing to the stability of this particular structure. The frequent occurrence

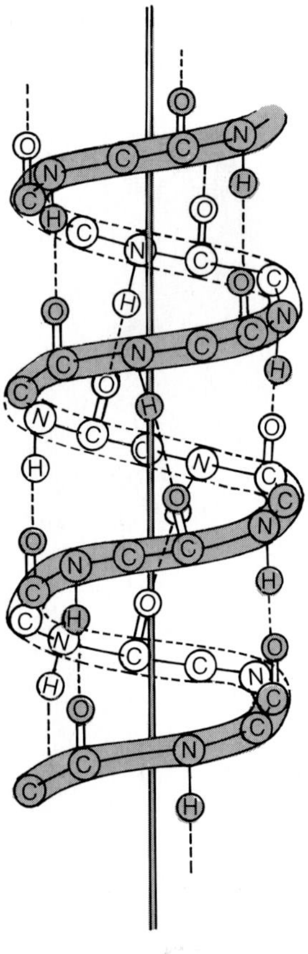

FIGURE 2–12 Diagram of a polypeptide chain backbone in a right-handed α helix.

of the α helix in globular as well as in fibrous proteins has been verified many times since its existence was first proposed in 1951.

Long polypeptide chains in the α-helical conformation are able to coil about one another, forming structures reminiscent of a multistranded rope. The protofibril of hair, for example, is 20 Å in diameter and appears to be made up of three right-handed α helices wound about each other to form a left-handed supercoil, as schematically illustrated in Figure 2–13.

The β Pleated Sheet. Another minimum-energy or stable conformation for the polypeptide chain is the more extended β conformation illustrated in Figure 2–14*A*. This conformation of the peptide chain leads to the β pleated sheet structure depicted in Figure 2–14*B*. The pleated sheet structure is formed by the parallel alignment of a number of polypeptide chains in a plane, with hydrogen bonds between the $>C{=}O$ and $>N{-}H$ groups of adjacent chains. The R groups of the constituent amino acids in one polypeptide chain alternately project above and below the plane of the sheet, leading to a two-residue repeat unit of 6.95 Å observable in x-ray diffraction studies.

If the N-terminal ends of all the participating polypeptide chains lie on the same edge of the sheet, with all C-terminal ends on the opposite edge, the structure is known as a parallel pleated sheet. In contrast, if the direction of the chains

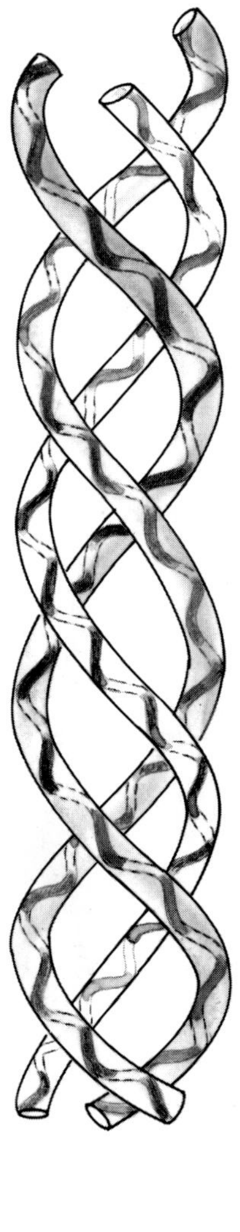

FIGURE 2–13 Diagram of a segment of a protofibril of hair.

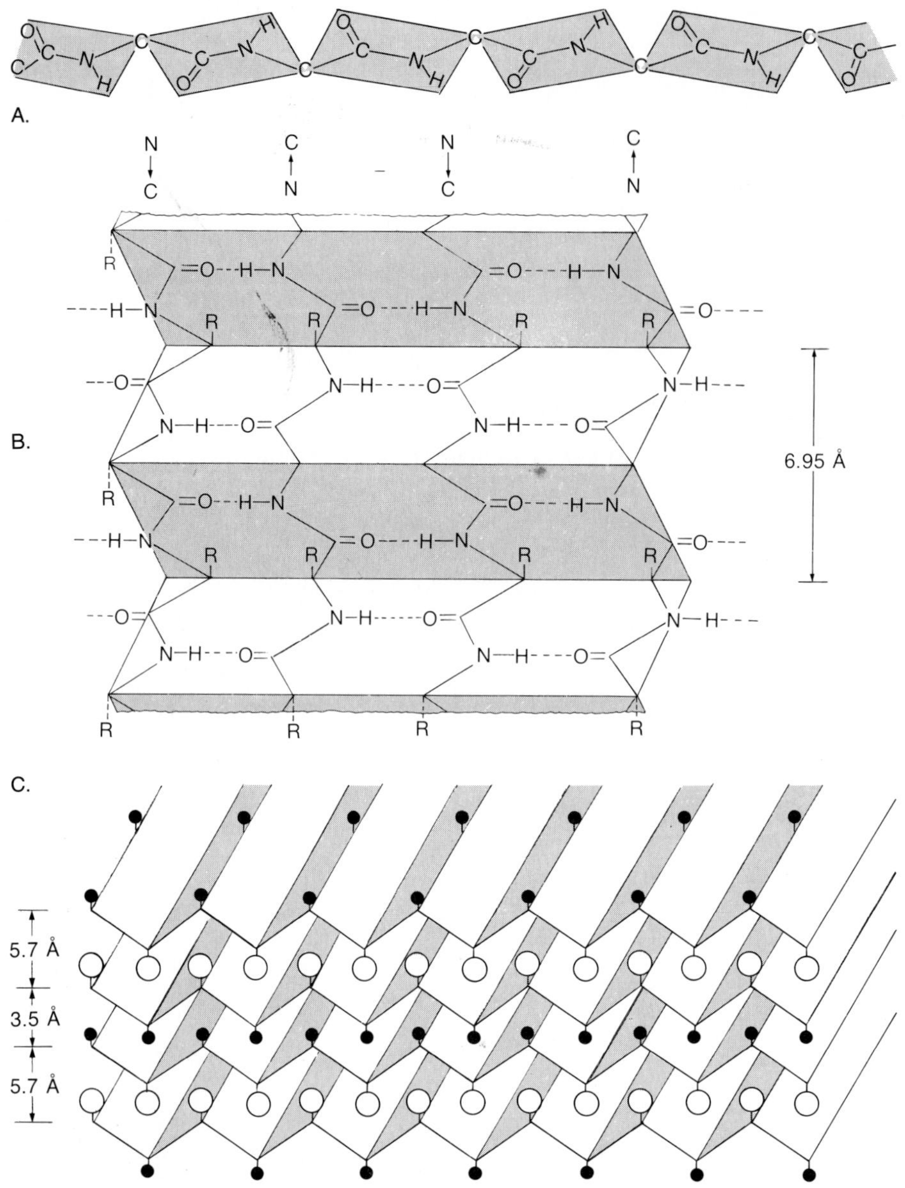

FIGURE 2–14 Diagrams of the β-pleated sheet. *A*. β conformation of a peptide chain. *B*. The β-pleated sheet structure. *C*. Stacking of β-pleated sheets in silk fibroin. The solid circles represent the R groups of glycine; the open circles are the R groups of alanine and serine.

alternates so that every other chain has its N-terminal end on the same side of the sheet, the structure is known as the antiparallel pleated sheet. Silk fibroin is one protein that has the antiparallel pleated sheet structure. Glycine constitutes approximately 45 per cent of the amino acid residues of silk fibroin, and alanine plus serine compose another 42 per cent. Hence, the R groups extending above and below the plane of the pleated sheet are small and allow the pleated sheets to stack, as indicated in Figure 2–14*C*. Pleated sheet structures can also be formed from a single polypeptide chain if the chain repeatedly folds back on itself.

The Collagen Triple Helix. Collagen fibers are the principal structural elements in connective tissue and make up one fourth to one third of all the protein in the human body. Glycine accounts for approximately one third of the amino acid residues present in collagen. Proline plus a modified proline known as hydroxyproline accounts for more than one fifth of the total amino acids of collagen. In the previous discussion of the α helix, it was noted that rotation about the C_α—N bond is required for the formation of an α helix. Proline is unique among the amino acids because its cyclic structure prevents free rotation about the α-carbon-to-nitrogen bond.

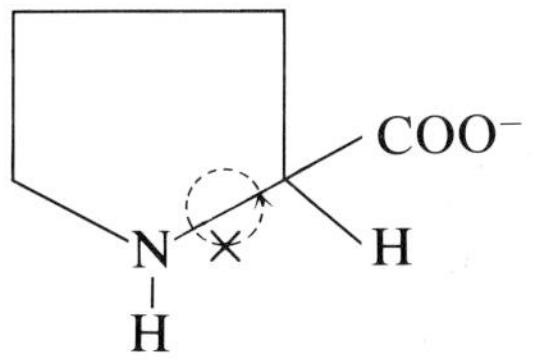

Therefore, proline can participate in an α helix only at the ends of the helix. Proline is called an α helix breaker because the α helix is disrupted when a proline or hydroxyproline residue is encountered in the sequence. Thus collagen, with its unusually high content of proline, cannot exist as an α helix. However, a polypeptide with a high proline content can form a left-handed helix with three amino acid residues per turn, and this type of helix is found in collagen.

In addition to being a left-handed helix containing only three amino acids per turn, this helix also differs from the α helix by lacking intrachain hydrogen bonds. Instead, three of these left-handed helices are twisted together to form a right-handed superhelix, which is stabilized by the *interchain* formation of hydrogen bonds between the >C=O group of one chain and the >N—H group of another chain. Proline does not have a hydrogen atom attached to its nitrogen when participating in a peptide bond and therefore cannot participate in this interchain hydrogen bonding. This three-stranded superhelix is known as the collagen triple helix and is depicted in Figure 2–15*A*.

It is interesting to note that the triple-stranded helical structure of collagen is made possible by its high content of glycine. Every third amino acid in each of the polypeptide chains is glycine. Therefore, when the left-handed helix with

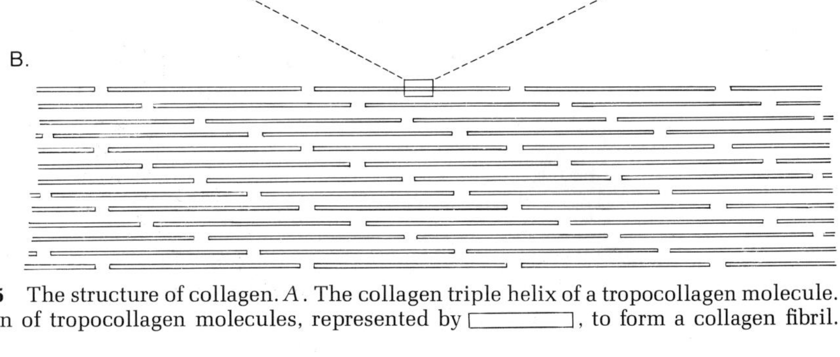

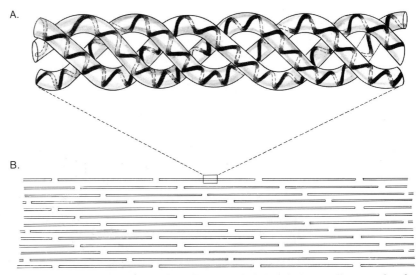

FIGURE 2–15 The structure of collagen. *A*. The collagen triple helix of a tropocollagen molecule. *B*. Association of tropocollagen molecules, represented by ▭, to form a collagen fibril.

three residues per turn is formed, only glycine residues are on one side of the helix. Since the R group of glycine is small, being only a hydrogen atom, the presence of a glycine face on each of the three strands permits the close apposition that is required for the formation of interchain hydrogen bonds.

The three polypeptide chains in the collagen triple helix are similar but not always identical in their primary structure. For example, in type I collagen from bone, dermis, and tendon, two of the chains are identical and are called α 1(I) type, while the third chain is different and is labeled α 2 type.

Each of the three polypeptide chains in the collagen triple helix contains approximately 1000 amino acids; therefore, the triple-stranded molecule is very long and thin, being about 3000 Å in length but only 15 Å in diameter. This triple-stranded helical molecule is known as tropocollagen and is a very strong, stiff molecule. The tropocollagen molecules spontaneously associate in a characteristic quarter-staggered array (Fig. 2–15*B*) to form collagen fibrils which are approximately 200 to 1000 Å in diameter. A series of complex covalent cross-links are formed within and between the tropocollagen molecules in the fibril, leading to the formation of strong mature collagen.

THE TERTIARY STRUCTURE OF PROTEINS

Tertiary structure refers to how a protein is folded in space. Thus consideration of tertiary structure takes into account the steric or spatial relationships of amino acids that are some distance apart in the protein's amino acid sequence. The specific proteins discussed earlier are all fibrous proteins. As a result of their structure, the fibrous proteins are tough, water-insoluble materials. Many other proteins, known as globular proteins, are much more folded and compact than fibrous proteins, and many are water-soluble. The globular proteins also may contain α-helical segments and some β pleated sheet structure, but these secondary structures are not continuous throughout the molecule.

There are several reasons why a molecule may contain some α helix, although the entire molecule does not exist in an α-helical form. One reason is that proline residues disrupt the α helix. In addition, there are a number of other amino acids that tend to destabilize an α helix. The stability of the α helix is a function of the amino acid sequence and is not everywhere the same. Therefore, there is a tendency for the α helix to bend at certain points if other forces are applied. These other forces, which have been discussed previously, include hydrophobic and electrostatic interactions. In the case of hydrophobic interactions, folding of the polypeptide chain may achieve a more energetically favorable conformation by shielding the hydrophobic nonpolar R groups from the aqueous environment. Folding of the chain can also maximize favorable electrostatic interactions and minimize unfavorable ones. In the simplest terms, the protein will spontaneously assume the most stable conformation, or that of minimum energy. Examples of the secondary and tertiary structure of several specific globular proteins will now be considered.

Myoglobin. *Myoglobin* is a relatively small protein of molecular weight 16,900. It contains 153 amino acid residues and a prosthetic group, the heme group, whose structure is depicted in Figure 2–16*A*. Myoglobin binds molecular oxygen reversibly, enabling it to function in oxygen storage in skeletal muscle cells of vertebrates. It has the distinction of being the first globular protein to have its three-dimensional structure elucidated by x-ray diffraction studies. This was accomplished by John Kendrew at a resolution of 6 Å in 1957, 2 Å in 1959, and 1.4 Å in 1962.

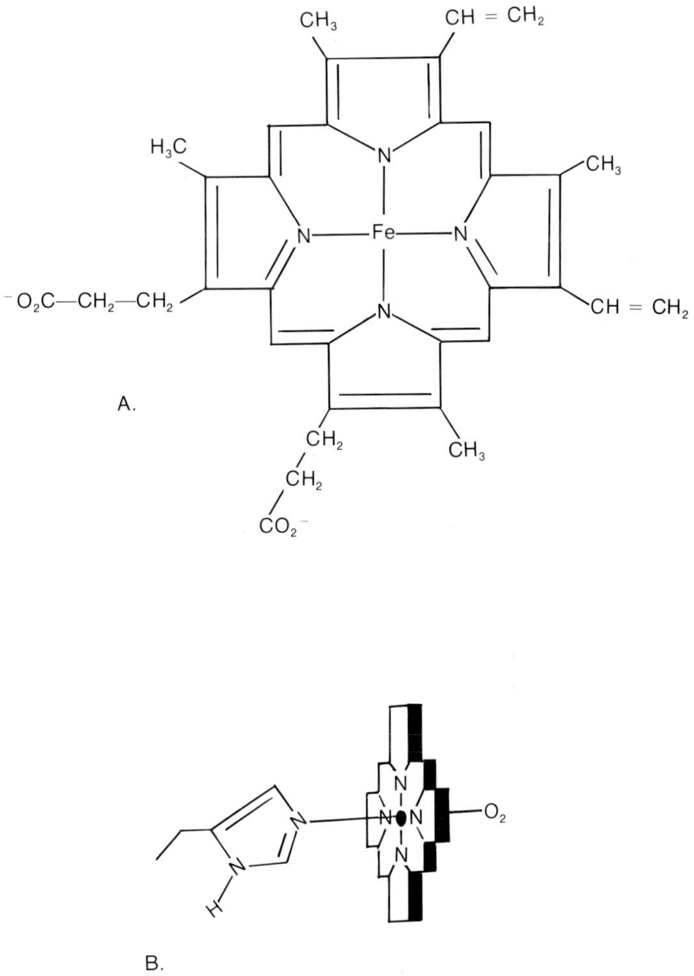

FIGURE 2–16 Parts of the myoglobin molecule. *A*. Heme, the prosthetic group of myoglobin, consists of a ferrous ion coordinated to the four nitrogens of a heterocyclic aromatic organic compound known as a porphyrin. *B*. The binding of the proximal histidine and molecular oxygen in the fifth and sixth coordination positions of the ferrous ion of heme.

A schematic representation of a three-dimensional model of the polypeptide chain backbone of myoglobin is shown in Figure 2–17. Of the 153 amino acid residues in the myoglobin molecule, 121, or 79 per cent, of the residues are involved in forming the eight helical regions labeled A to H. The remaining 32 amino acid residues are in the nonhelical segments that connect the helical regions or form the ends of the molecule. As a result of this folding, the myoglobin molecule assumes a very compact oblate spheroid shape, with the hydrophilic R groups of the constituent amino acids on the surface of the molecule and the hydrophobic R groups predominately in the interior of the molecule. Hydrophobic R groups of helices E and F form the sides of a pocket into which the hydrophobic heme group fits. The porphyrin ring of the heme is largely hydrophobic, except for the two propionic acid side chains which stick out of the pocket and into the aqueous environment.

The heme group is held in the hydrophobic pocket of myoglobin partially by van der Waals interactions with the hydrophobic amino acid R groups lining the sides of the pocket. The ferrous ion in heme has the capacity to form octahedral complexes by binding ligands in a fifth and sixth coordination position perpendicular to the plane of the porphyrin ring. In myoglobin, the fifth coordination site of the ferrous ion is occupied by a nitrogen of the imidazole group of a

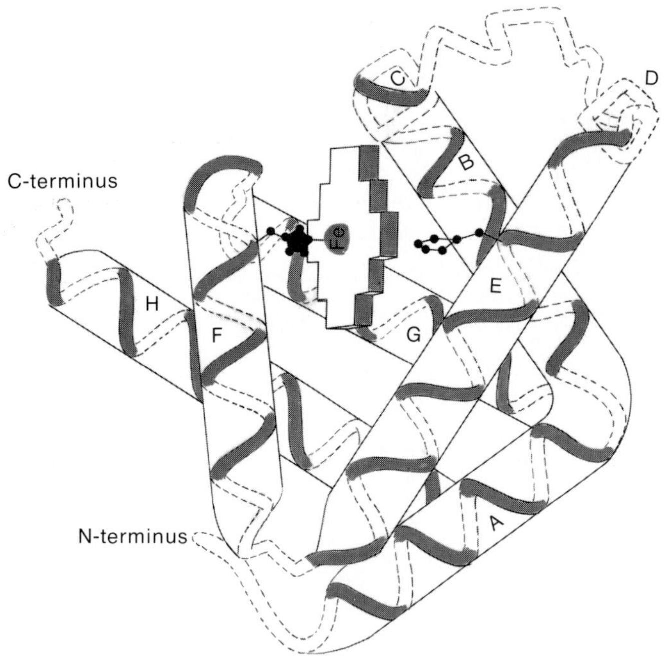

FIGURE 2–17 A model of the myoglobin molecule. (Adapted from Dickerson, R. E., and Geis, I.: *The Structure and Action of Proteins*. Harper & Row, 1969, p. 47.)

histidine in the F helix (Fig. 2–16*B*). The sixth coordination position on the ferrous ion is occupied by molecular oxygen. The histidine in the F helix, which supplies the imidazole group for the fifth coordination position of the ferrous ion, is known as the proximal histidine. In addition to this proximal histidine, another imidazole group from a histidine in the E helix is present on the other side of the heme group, but this imidazole group is too far removed to coordinate with the iron, even when the myoglobin is deoxygenated. This histidine in the E helix is known as the distal histidine. Its function is not well understood at present.

Myoglobin without its heme prosthetic group is designated apomyoglobin. The principal function of the apomyoglobin molecule is to provide a hydrophobic environment for the heme group and a properly oriented imidazole group to occupy the fifth coordination position of the iron. The presence of the hydrophobic environment and the proximal histidine enable the heme to combine reversibly with O_2 and prevent the oxidation of Fe^{++} to Fe^{+++}.

Carboxypeptidase A. The functional properties of *carboxypeptidase A* were considered earlier in this chapter. This enzyme is a globular protein, which contains 307 amino acid residues and has a molecular weight of 34,600. Its three-dimensional structure has been elucidated by Lipscomb and coworkers. The dimensions of the molecule are 52 Å × 44 Å × 40 Å, and it is roughly spherical in shape. This molecule is considerably larger than myoglobin and shows more variety in its secondary structures. Of the 307 amino acids present in the molecule, 35 per cent are involved in the formation of eight helical segments (designated A through H) and 20 per cent are involved in a β pleated sheet structure that extends through the molecule, forming a core. Helix D is connected to helix E by a segment containing 52 amino acids. The folding of this segment is very complex and contains no recognizable regular secondary structure. The remaining 25 percent of the amino acids connect the helices and the pleated sheet structure.

QUATERNARY STRUCTURE OF PROTEINS

Many important proteins are complexes formed by the association of smaller proteins, which may or may not be identical. These smaller proteins, which associate to form the larger protein complex, are termed subunits. The number of subunits in the complex and their steric relationship to each other is designated the quaternary structure of a protein. An excellent example of quaternary structure and its importance is found in the hemoglobin molecule, which will be considered in some detail in Chapter 3.

SUMMARY

Even though they are linear polymers, *proteins structure or organize very precisely a relatively large volume of space* as a result of their secondary, tertiary and quaternary structure. The way in which a protein folds is dictated by its primary structure, which in turn is determined by the genetic information stored in DNA, another linear polymer. The secondary, tertiary and quaternary structure of proteins is very important in determining their functional properties. For example, we have seen that the ability of myoglobin to generate a hydrophobic pocket of the correct dimensions, containing a properly oriented proximal histidine, enables heme to undergo a reversible oxygenation, rather than an oxidation, in the presence of molecular oxygen. The importance of a protein's ability to precisely structure a volume of space will become even more apparent when hemoglobin and some mechanisms of enzyme action are considered in Chapters 3 and 4.

ACID-BASE PROPERTIES OF MACROMOLECULES

Cells are very sensitive to small changes in the concentration of hydrogen ions in their immediate environment. The hydrogen ion concentration in the interstitial fluid and blood plasma is normally held remarkably constant at approximately 4×10^{-8}M (pH 7.4). A decrease in the plasma hydrogen ion concentration to values lower than 1.6×10^{-8} (pH 7.8), or an increase to values greater than 1×10^{-7}M (pH 7), can be fatal unless corrected physiologically or therapeutically. The sensitivity of biological systems to small changes in the hydrogen ion concentration is in large part due to the acid-base properties of the macromolecules that make up the cells.

Before considering the acid-base properties of these complex molecules, some relevant terms will be defined. A more detailed discussion of these terms and a review of the acid-base chemistry of simple monofunctional acids, bases, and buffers is presented in Appendix D.

Acid. An acid is any proton donor.

Base. A base is any proton acceptor.

K'_a. The ionization constant, K'_a, is the equilibrium constant for the dissociation of an acid, HA, in aqueous solution.

$$\underset{\text{acid}}{HA} \rightleftharpoons H^+ + \underset{\text{conjugate base}}{A^-}$$

$$K'_a = \frac{[H^+][A^-]}{[HA]}$$

A *strong acid* dissociates completely in aqueous solution; therefore K'_a for a strong acid is infinite.
A *weak acid* does not dissociate completely in aqueous solution; hence K'_a for a weak acid is finite.

pH. $pH = \log_{10} \frac{1}{[H^+]} = -\log_{10} [H^+]$

pK′a. $pK'_a = \log_{10} \frac{1}{K'_a} = -\log_{10} K'_a$

Henderson-Hasselbalch equation. $pH = pK'_a + \log_{10} \frac{[A^-]}{[HA]}$

From the Henderson-Hasselbalch equation, it is apparent that the pH of a solution containing equimolar concentrations of a weak acid and its conjugate base $\left(\frac{[A^-]}{[HA]} = 1\right)$ will be equal to the pK'_a of the weak acid.

Buffer. A solution containing a weak acid and its conjugate base is referred to as a buffer or buffer system because the solution resists large pH changes on addition of a strong acid or strong base. The maximum buffering power of a buffer occurs at a pH equal to the pK'_a of the weak acid.

TITRATION OF AMINO ACIDS

Proteins are one class of macromolecules whose properties may be markedly altered by changes in the pH of the system. Since proteins are polymers of amino acids, the acid-base chemistry of amino acids will be considered first.

A simple amino acid, such as glycine, exists in solution as the zwitterion, $H_3N^+—CH_2—COO^-$, with zero net charge. Addition of a strong acid to the solution containing the zwitterionic form of glycine converts the carboxylate anion to a carboxyl group, while addition of a strong base converts the $—NH_3^+$ group to $—NH_2$. Thus, as depicted in the following ionization scheme, a glycine molecule may bear a net charge of +1, 0, or −1, depending on the pH of the solution.

$$\underset{(+1 \text{ net charge})}{H_3N^+—CH_2COOH} \underset{+H^+}{\overset{-H^+}{\rightleftharpoons}} \underset{(0 \text{ net charge})}{H_3N^+—CH_2—COO^-} \underset{+H^+}{\overset{-H^+}{\rightleftharpoons}} \underset{(-1 \text{ net charge})}{H_2N—CH_2—COO^-}$$

Since there are two titratable groups on glycine, its titration curve, shown in Figure 2–18, is more complex than that of a simple monofunctional acid. Beginning the titration with the positively charged form of glycine that exists at low pH, the titration curve indicates the presence of two titratable groups, one with a pK'_a of 2.34 and the other with a pK'_a of 9.69. These are designated pK'_{a_1} and pK'_{a_2}, respectively. The first group titrated is the carboxyl group ($pK'_{a_1} = 2.34$), and the second group titrated is the $—NH_3^+$ group ($pK'_{a_2} = 9.69$). The carboxyl group of glycine is a stronger acid than the carboxyl

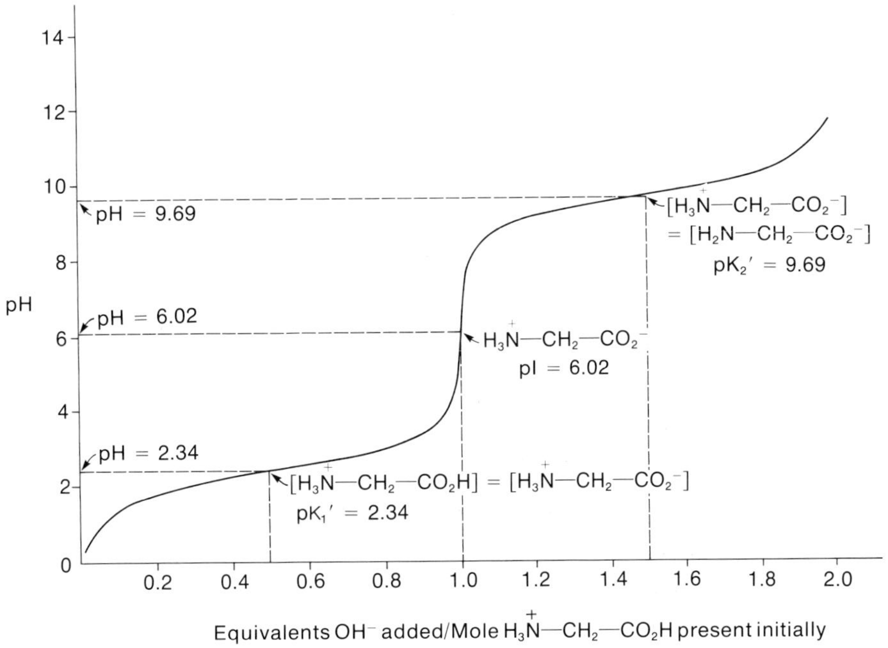

FIGURE 2–18 Titration of an amino acid, glycine.

group of acetic acid (pK'_a = 4.76) because the positive charge on the $—NH_3^+$ group facilitates the removal of the hydrogen from the carboxyl group as an H^+. At pH 6.02, glycine will not migrate in an electric field; therefore, the time average net charge on the glycine molecule must be zero. The pH at which a molecule will not migrate in an electric field is known as the isoelectric point, pI. For a simple amino acid such as glycine, it can be shown that $pI = \frac{pK'_{a_1} + pK'_{a2}}{2}$.

The titration curve for an amino acid with three titratable groups, such as cysteine, is even more complex. The three pK_a values indicated by the titration curve of cysteine are $pK'_{a_1} = 1.71$, $pK'_{a_2} = 8.36$, and $pK'_{a_3} = 10.53$. The pK'_{a_1} of 1.71 clearly represents the ionization of the carboxyl group. However, assignment of the pK'_a values of 8.36 and 10.53 to particular groups is not straightforward. In fact, it has been shown that cysteine follows the following ionization scheme:

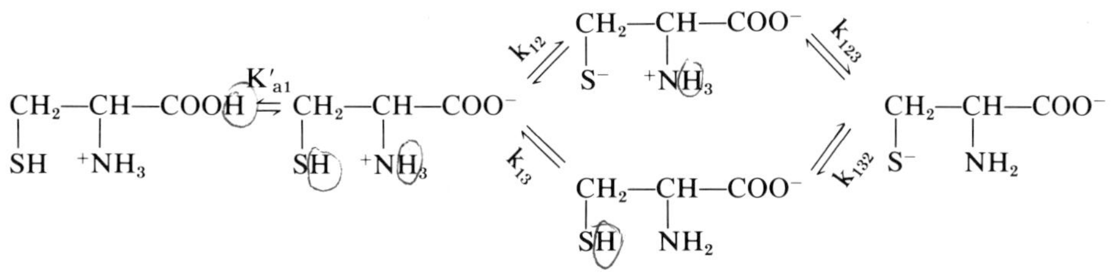

where k_{12}, k_{123}, k_{13}, and k_{132} represent the ionization constants of the species indicated. These have been resolved and found to be as follows: pk_{12} = 8.53; pk_{13} = 8.86; pk_{123} = 10.36; and pk_{132} = 10.03. Thus, for cysteine there are two ionization pathways of importance in the conversion of $\underset{SH}{CH_2}—\underset{^+NH_3}{CH}—COO^-$

to CH_2—CH—COO^-. Approximately two thirds of the molecules follow the upper
S^- NH_2
pathway, and one third follow the lower pathway. Furthermore, the pK'_{a_2} and the pK'_{a_3} values of 8.36 and 10.53, indicated by the titration curve, are hybrid ionization constants and bear the following relationships to the true intrinsic ionization constants:

$$K'_{a_2} = k_{12} + k_{13} \text{ and } K'_{a_3} = \frac{k_{12}\, k_{123}}{k_{12} + k_{13}}$$

TITRATION OF PROTEINS

As one might expect, titration curves for proteins are even more complex and difficult to resolve than those for amino acids, owing to the large number of titratable groups present in the molecule. The titratable groups found on proteins, arranged in the order of decreasing acid strength, are as follows: the C-terminal carboxyl group; carboxyl groups of aspartate and glutamate residues; imidazole groups of histidine residues; the N-terminal amino group; sulfhydryl groups of cysteine residues; phenolic groups of tyrosine residues; ε-amino groups of lysine residues; and the guanidino groups of arginine residues. To further complicate the interpretation of protein titration curves, the pK's of all groups of the same type are not identical in a given protein or from one protein to another, because the pK'_a of each group is affected by hydrogen bonding, participation in electrostatic bonds, and the nature of the environment in which the group resides. The ranges of pK'_a values that have been observed for these ionizable groups in several proteins are tabulated in Table 2–4.

Effect of pH on the Functional Properties of Proteins

The activity of enzymes is markedly affected by pH, as illustrated in the pH-rate profiles of Figure 2–19. For example, pepsin, which catalyzes the hydrolysis of dietary proteins in the stomach where the pH is 1.2 to 3.0, has a very acid pH optimum and is no longer functionally active at a neutral pH. In contrast, chymotrypsin, which catalyzes the hydrolysis of dietary proteins in the small intestine where the pH is higher, has an alkaline pH optimum.

TABLE 2–4 VALUES OF pK'_a FOR FUNCTIONAL GROUPS IN PROTEINS*

Functional Group	Range of pK'_a Values Observed
C-terminal carboxyl	3.6–3.75
Carboxyl (Asp and Glu)	4.0–4.7
Imidazole (His)	6.4–7.0
N-terminal amino	7.4–7.9
ε-amino (Lys)	10.1–10.6
Phenolic (Tyr)	8.5–10.9
Guanidino (Arg)	11.9–13.3

*Modified from Edsall, J. T., and Wyman, J.: Biophysical Chemistry. Chapter 9. New York, Academic Press, 1958.

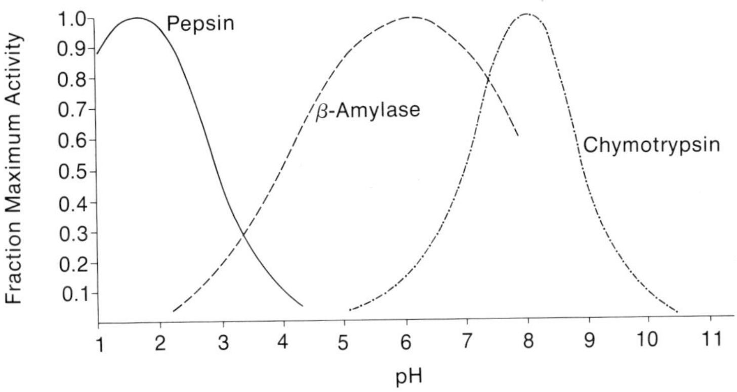

FIGURE 2–19 Activity of some enzymes as a function of pH.

There are several reasons for the dramatic effect of pH on the enzymatic properties of protein. Proteins are macroions as well as macromolecules, and the charge configuration on the protein is a function of the pH of the solution. Since electrostatic interactions and hydrogen bonding play an important role in determining the conformation of proteins, a change in the charge configuration of the protein, as a result of a change in the pH of the solution, can induce a change in the conformation of the protein, leading to an alteration in enzymatic activity. Furthermore, as will be discussed in more detail in Chapter 4, the mechanism by which an enzyme brings about its catalysis may require the participation of a specific acidic or basic functional group on the protein. If the state of ionization of this group is altered by a change in pH, the ability of the group to participate in the catalytic process may be destroyed.

FURTHER READING

Anfinsen, C. B., and Scheraga, H. A.: Experimental and theoretical aspects of protein folding. *In*: Anfinsen, C. B., Edsall, J. T., and Richards, F. M. (eds.) Advances in Protein Chemistry. Vol. 29. New York, Academic Press, 1975, pp. 205–300.

Davidson, J. N.: The Biochemistry of the Nucleic Acids. 8th Ed. New York, Academic Press, 1975.

Dayhoff, M. O. (ed.): Atlas of Protein Sequence and Structure. Vol. 5, 1972. Suppl. *1* (1973) and *2* (1976). National Biomedical Research Foundation, Silver Spring, Md.

Dickerson, R. E., and Geis, I.: The Structure and Action of Proteins. New York, Harper & Row, 1969.

Edman, P., and Begg, G.: A protein sequenator. Eur. J. Biochem., *1*:80, 1967.

Edsall, J. T., and Wyman, J.: Biophysical Chemistry. Vol. I. New York, Academic Press, 1958.

Guthrie, R. D., and Honeyman, J.: An Introduction to the Chemistry of Carbohydrates, 3rd ed. Oxford, Clarendon Press, 1968.

Jencks, W. P.: Forces in aqueous solution. *In*: Catalysis in Chemistry and Enzymology. New York, McGraw-Hill, 1969, pp.323–462.

Kendrew, J. C.: The three-dimensional structure of a protein molecule. Sci. Am., *205*:96, 1961.

Neurath, H., and Hill, R. L. (eds.): The Proteins. Vols. I–III. 3rd ed. New York, Academic Press, 1975–1977.

Pauling, L., and Corey, R. B.: Configurations of polypeptide chains with favored orientations around single bonds: two new pleated sheets. Proc. Natl. Acad. Sci. USA, *37*:729, 1951.

Pauling, L., Corey, R. B., and Branson, H. R.: The structure of proteins: two hydrogen-bonded helical configurations of the polypeptide chain. Proc. Natl. Acad. Sci. USA, *37*:205, 1951.

Quiocho, F. A., and Lipscomb, W. N.: Carboxypeptidase A: a protein and an enzyme. *In*: Afinsen, C. B., Edsall, J. T., and Richards, F. M. (eds.): Advances in Protein Chemistry. Vol. 25. New York, Academic Press, 1971, pp. 1–78.

Sanger, F.: The structure of insulin. *In*: Green, D. E. (ed.): Currents in Biochemical Research. New York, John Wiley and Sons, 1956, pp. 434–459.

3

THE RELATION BETWEEN MACROMOLECULAR STRUCTURE AND FUNCTION: PROPERTIES OF HEMOGLOBIN

In recent years, considerable progress has been made in understanding the way in which the structure of a protein generates the properties that enable it to carry out specific functions. In this chapter hemoglobin has been selected as an example to illustrate this relationship between structure and function. Attention will be focused in turn on the following: (1) the job to be done by hemoglobin, (2) the efficiency of hemoglobin in accomplishing the task, (3) the structure of hemoglobin, and (4) what is known about how the structure of hemoglobin generates the properties that enable the molecule to perform its function efficiently.

THE JOB TO BE DONE

The cell is a steady-state system, not a system at equilibrium. In order to maintain this steady-state position, the cell must expend energy, and aerobic organisms derive their energy from the oxidation of foodstuffs, such as glucose. The overall equation for the oxidation of glucose is

$$C_6H_{12}O_6 + 6\,O_2 \rightarrow 6\,CO_2 + 6\,H_2O + \text{energy}$$

The detailed reactions by which the cell traps this energy in utilizable form will be the subject of Chapters 10 and 11. From this equation, it is apparent that the cell must be supplied with molecular oxygen and that the end product, CO_2, must be removed. Single cells or small groups of cells bathed by an adequately aerated medium can obtain the necessary O_2 and eliminate the CO_2 by diffusion. However, the evolution of larger species with increased oxidative metabolism necessitated a circulatory system for transporting the O_2 from the atmosphere to the fluid bathing the cells as well as mechanisms for removing the CO_2 produced by the cells.

The partial pressure of O_2, Po_2, in the alveoli of the human lung is 100 to 112 mm Hg. The amount of oxygen that will dissolve in H_2O is proportional to the partial pressure of oxygen. At a Po_2 of 100 mm Hg, only 0.003 ml of O_2 will dissolve in 1 ml of H_2O at 37° C; hence the transport of O_2 by simple solution in an aqueous medium would be very inefficient. One way to increase the efficiency of O_2 transport is via a circulating carrier that combines reversibly with O_2.

$$O_{2(gas)} \rightleftharpoons O_{2(dissolved)} + \text{Carrier} \rightleftharpoons \text{Carrier} \cdot O_2$$

The amount of O_2 transported in a milliliter of solution then becomes equal to that which is dissolved in the solution plus that which is bound to the carrier present in the solution.

The basic requirements for an efficient carrier are as follows: (1) The combination of the carrier with O_2 must be reversible if the carrier is to take on O_2 in the lung and release O_2 to the fluid bathing the tissue; and (2) the carrier should take on a full load of O_2 in the lungs and should unload a large fraction of O_2 at the site of delivery.

THE EFFICIENCY OF HEMOGLOBIN IN ACCOMPLISHING THE TASK

Hemoglobin, which functions as a carrier in the circulatory system, is contained within the red blood cells, or erythrocytes. The mature erythrocyte is an enucleate cell shaped like a biconcave disc. The erythrocyte has a life span of approximately 125 days. It consists of 65 per cent H_2O and 35 per cent solids, of which 90 to 95 per cent is hemoglobin.

Oxygen Binding by Hemoglobin and Myoglobin

Plots known as oxygen saturation curves are constructed by plotting the per cent saturation of a carrier with oxygen *versus* the oxygen concentration expressed as the partial pressure of oxygen. A carrier that is fully loaded with O_2 is said to be 100 per cent saturated. The oxygen saturation curves for hemoglobin at pH 7.0, 7.2, 7.4, and 7.6, and the oxygen saturation curve for myoglobin, which is independent of pH, are shown in Figure 3–1. This pH range of 7.0 to 7.6 is of interest because the pH of plasma is normally about 7.4. The saturation curves for myoglobin and hemoglobin differ not only in their pH dependence but also in their shape, with myoglobin showing a hyperbolic saturation curve and hemoglobin showing S-shaped, or sigmoidal, saturation curves. The hyperbolic curve of myoglobin (Mb) is the type of curve that one would expect for a simple association reaction, as in

$$\text{Mb} + O_2 \rightleftharpoons \text{MbO}_2$$

The S-shaped curve of hemoglobin is fundamentally different, and its significance will be considered later.

Both myoglobin and hemoglobin meet the first basic requirement of a good carrier in that the combination of O_2 with either of these proteins is freely reversible. In other words, each of these proteins will bind oxygen when the oxygen concentration, Po_2, is high, and then release this bound oxygen when the oxygen concentration is low (Fig. 3–1). Furthermore, the process of oxygen uptake and release by these proteins follows the saturation curves depicted in Figures 3–1, regardless of the number of times the process is repeated with a given sample of protein.

With regard to the second basic requirement, that the carrier take on a full load of oxygen in the lungs and release a large fraction of this oxygen at the delivery site, only hemoglobin qualifies as a good carrier. Inspection of the myoglobin curve and the pH 7.4 hemoglobin curve of Figure 3–1 reveals that while both hemoglobin and myoglobin would be essentially saturated with oxygen in the lungs, only hemoglobin would release an appreciable fraction of its bound oxygen in the capillaries of a working muscle in which the Po_2 falls to 20 mm Hg. Since hemoglobin is capable of delivering a much larger fraction of its bound oxygen to tissues than myoglobin, hemoglobin is much more efficient as a transport system for oxygen. Now the significance of the S-shaped saturation curve for hemoglobin should be apparent, because it is the sigmoidal shape of this curve that enables hemoglobin to take on a full load of O_2 at a Po_2 of 100 mm Hg and then to release most of this bound O_2 at the still appreciable Po_2 of 20 mm Hg.

The Bohr Effect and CO_2 Transport

The pH of the solution has a marked effect on the position of the oxygen saturation curve of hemoglobin, as is evident in Figure 3–1. At a given Po_2, the amount of oxygen bound by hemoglobin decreases as the pH is lowered. This is known as the *Bohr effect*. Hemoglobin that contains bound O_2 is referred to as oxyhemoglobin, and hemoglobin that does not contain bound oxygen is termed deoxyhemoglobin. The basis for the Bohr effect is that oxyhemoglobin is a stronger acid than deoxyhemoglobin. Consider the following equilibria:

a. deoxyhemoglobin + $O_2 \rightleftharpoons$ oxyhemoglobin

b. oxyhemoglobin $\rightleftharpoons$ oxyhemoglobin$^-$ + H^+

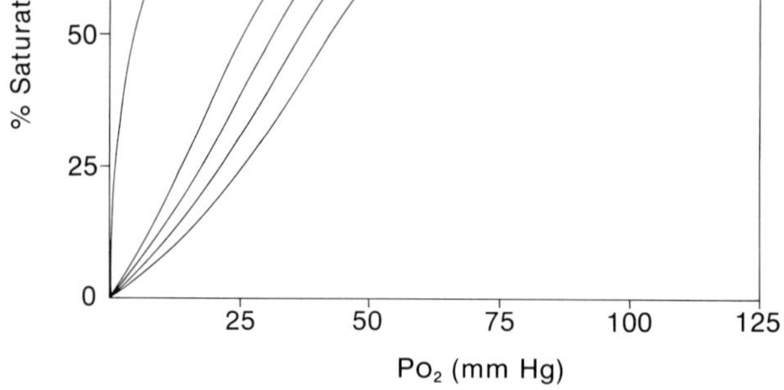

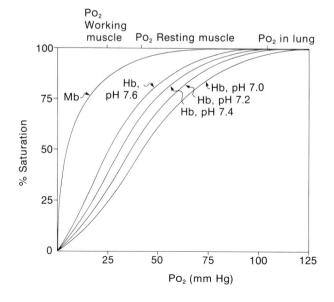

FIGURE 3–1 Oxygen saturation curves for myoglobin and hemoglobin.

If reactions *a* and *b* are at equilibrium, and the concentration of H^+ is then increased, reaction *b* will be driven to the left according to the law of mass action. This shift to the left decreases the concentration of oxyhemoglobin$^-$ and increases the concentration of oxyhemoglobin which in turn drives reaction *a* to the left, resulting in the release of O_2. The difference in the acid strengths of deoxyhemoglobin and oxyhemoglobin is such that at pH 7.4, 0.7 moles of H^+ are released for each mole of O_2 bound.

The Bohr effect is not just a scientific curiosity; it is of great physiological significance because it enables hemoglobin to play an essential role in CO_2 transport. As shown previously, for every mole of O_2 consumed in the oxidation of glucose, a mole of CO_2 is produced. Therefore, an actively metabolizing tissue produces large quantities of CO_2 which must be transported to the lungs in the venous circulation. The CO_2 produced in the tissue diffuses into the erythrocyte where its hydration to H_2CO_3 is catalyzed by the enzyme carbonic anhydrase. The H_2CO_3 produced is a weak acid which dissociates to yield H^+ and HCO_3^-.

$$\text{c.}\quad CO_2 + H_2O \xrightleftharpoons{\text{carbonic anhydrase}} H_2CO_3$$

$$\text{d.}\quad H_2CO_3 \rightleftharpoons H^+ + HCO_3^-$$

As a result of the high HCO_3^- concentration generated inside the erythrocyte, some of the HCO_3^- diffuses out of the erythrocyte and into the blood plasma. The fate of the H^+ generated by reaction *d* is different. The pH of blood is approximately 7.4 ($[H^+] = 3.98 \times 10^{-8}M$) and must be maintained within narrow limits, since relatively small increases or decreases in plasma hydrogen ion concentration, known respectively as acidosis and alkalosis, can be life-threatening. Considering the large amounts of CO_2 that must be carried to the lungs in the venous circulation, one might expect equations *c* and *d* to lead to the accumulation of toxic concentrations of hydrogen ion. That this does *not* occur is due primarily to the Bohr effect of hemoglobin. For every mole of CO_2 produced by a tissue metabolizing glucose, a mole of O_2 must be delivered to the tissue by the conversion of a mole of oxyhemoglobin to a mole of deoxyhemoglobin, which requires the uptake of H^+ (see reactions *a* and *b*). Hence, the H^+ generated by the ionization of H_2CO_3 (reaction *d*) is largely taken up by the conversion of oxyhemoglobin$^-$ to deoxyhemoglobin and O_2 via reactions *b* and *a*. An overall equation for the process occurring in the capillaries of the tissue can be obtained by summing up *a, b, c,* and *d* to yield *e*.

$$\text{e.}\quad CO_2 + H_2O + \text{oxyhemoglobin}^- \rightleftharpoons HCO_3^- + \text{deoxyhemoglobin} + O_2$$

The reaction is driven to the right by the high P_{CO_2} and low P_{O_2} existing in the actively metabolizing tissue. Thus, the Bohr effect of hemoglobin enables the blood to take up a large fraction of the CO_2 produced by tissues and transport it to the lungs as HCO_3^- without a major change in the pH of the blood.

When the blood reaches the lungs, the high P_{O_2} and low P_{CO_2} in the lungs drives reaction *e* in the reverse direction.

$$O_2 + \text{deoxyhemoglobin} + HCO_3^- \rightleftharpoons CO_2 \uparrow + \text{oxyhemoglobin}^- + H_2O$$

The CO_2 is blown off via the lungs as indicated by the upward pointing arrow. The H^+ that is released in the conversion of deoxyhemoglobin to oxyhemoglobin$^-$ (reactions *a* and *b*) is taken up in the conversion of HCO_3^- to CO_2 (reactions *c* and *d*), enabling the entire process to occur with no significant change in the blood pH.

Although most of the CO_2 is transported from the tissues to the lungs as HCO_3^-, approximately 20 per cent is transported as a carbamino derivative of hemoglobin. The α-amino groups of the N-terminal valine residues of hemoglobin combine reversibly with the CO_2 to form carbaminohemoglobin.

$$\underset{\text{hemoglobin}}{CO_2 + H_2N \cdot \text{Val—P}} \rightleftharpoons \underset{\text{carbaminohemoglobin}}{H^+ + {}^-O—\overset{\overset{\displaystyle O}{\|}}{C}—NH \cdot \text{Val—P}}$$

where *P* represents the rest of the polypeptide chain. In a rapidly metabolizing tissue where the P_{CO_2} is high, the reaction is driven to the right; in the lungs where the P_{CO_2} is low, the reaction is driven to the left, and the CO_2 is removed.

Thus, the hemoglobin molecule is a very efficient and finely tuned transport machine which not only carries molecular O_2 to the interstitial fluid bathing the cells but also participates in the removal of CO_2, the end product of cellular metabolism. The primary, secondary, tertiary, and quaternary structure of hemoglobin will now be examined in an attempt to learn how the structural properties of the molecule give rise to the functional properties just considered.

THE STRUCTURE OF HEMOGLOBIN

Hemoglobin consists of four subunits, each of which is very similar in many respects to the myoglobin molecule. For example, each subunit contains a heme group identical to that found in myoglobin (See Chapter 2, Figure 2–16). In addition, each subunit contains a hydrophobic pocket for the heme group which is very similar to the hydrophobic pocket of myoglobin. The sides of the pocket are formed by two helices. One helix supplies the proximal histidine which occupies the fifth coordination position of the iron, while the other helix supplies the distal histidine group. In fact, the entire tertiary structure of each hemoglobin subunit is reminiscent of the tertiary structure of myoglobin. The most obvious difference between myoglobin and hemoglobin is that hemoglobin consists of subunits; therefore, the quaternary structure of hemoglobin will be considered first.

Quaternary Structure of Hemoglobin

Hemoglobin has a molecular weight of 64,500 and consists of four polypeptide chains. Two of these polypeptide chains are of one kind and are designated α chains; the other two polypeptide chains are of a different kind and are designated β chains. An α chain contains 141 amino acid residues, and a β chain contains 146 amino acid residues; therefore, they are each slightly smaller than the polypeptide chain of myoglobin, which contains 153 amino acid residues.

Through x-ray crystallographic studies, Perutz and coworkers elucidated the secondary, tertiary, and quaternary structures of oxyhemoglobin and deoxyhemoglobin. From the schematic representation of the structure of oxyhemoglobin in Figure 3–2, it can be seen that the two identical α-chains, designated α_1 and α_2, and the two identical β chains, designated β_1 and β_2, are closely packed in a roughly tetrahedral manner to form a molecule that is approximately spherical. The subunits are held together by hydrophobic and electrostatic interactions between the subunits. The predominant force holding α_1 to β_1 and α_2 to

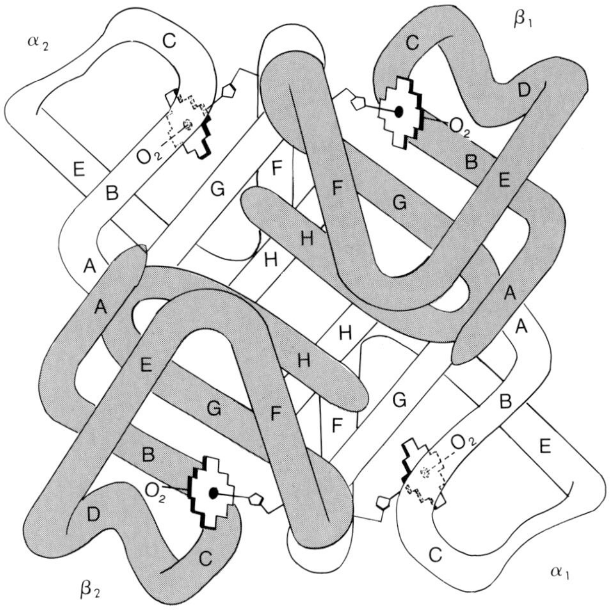

FIGURE 3–2 A diagram of the structure of oxyhemoglobin. (Based on Dickerson, R. E., Ann. Rev. Biochem., *41*:820, 1972.)

β_2 appears to be the result of the hydrophobic interactions between the subunits at points of contact.

The quaternary structures of oxyhemoglobin and deoxyhemoglobin are very similar but are not identical. The largest change in the quaternary structure of the molecule in going from deoxyhemoglobin to oxyhemoglobin is a movement that brings the β_1 and β_2 chains closer together, thereby decreasing the distance between the heme group in the β_1 chain and the heme group in the β_2 chain from 39.9 Å in deoxyhemoglobin to 33.4 Å in oxyhemoglobin. The possible significance of the change in quaternary structure upon binding oxygen will be discussed later, when models proposed to explain the S-shaped saturation curve and the Bohr effect are considered.

Primary, Secondary, and Tertiary Structures of Hemoglobin

The amino acid sequences of sperm whale myoglobin and the α and β chains of human hemoglobin are shown in Figure 3–3. A comparison reveals that the secondary structures, i.e., α-helical regions, are very similar in all three proteins, with the exception that the short D helix is missing in the α chain. A comparison of the sequence of myoglobin with that of the α chain indicates that 38 residues are identical, and comparison of myoglobin with the β chain shows that 37 residues are identical. Comparison of the α and β chains reveals a greater correspondence, with 64 residues being identical. Moreover, at many positions where the amino acids differ in one or more of the chains, the difference results from a conservative substitution, e.g., a nonpolar leucine instead of a nonpolar valine, a basic arginine in place of a basic lysine, or an acidic glutamate for an acidic aspartate. At a few positions the amino acids are very different, such as an aspartate instead of a histidine. In spite of these differences, however, both the α and the β chains are folded in much the same way as the myoglobin chain.

												A														B					
Mb	VAL	—	LEU	SER	GLU	GLY	GLU	TRP	GLN	LEU	VAL	LEU	HIS	VAL	TRP	ALA	LYS	VAL	GLU	ALA	ASP	VAL	ALA	GLY	HIS	GLY	GLN	ASP	ILE	LEU	ILE
α	VAL	—	LEU	SER	PRO	ALA	ASP	LYS	THR	ASN	VAL	LYS	ALA	ALA	TRP	GLY	LYS	VAL	GLY	ALA	HIS	ALA	GLY	GLU	TYR	GLY	ALA	GLU	ALA	LEU	GLU
β	VAL	HIS	LEU	THR	PRO	GLU	GLU	LYS	SER	ALA	VAL	THR	ALA	LEU	TRP	GLY	LYS	VAL	ASN	—	—	VAL	ASP	GLU	VAL	GLY	GLY	GLU	ALA	LEU	GLY

		B								C														D						E
Mb	ARG	LEU	PHE	LYS	SER	HIS	PRO	GLU	THR	LEU	GLU	LYS	PHE	ASP	ARG	PHE	LYS	HIS	LEU	LYS	THR	GLU	ALA	GLU	MET	LYS	ALA	SER	GLU	ASP
α	ARG	MET	PHE	LEU	SER	PHE	PRO	THR	THR	LYS	THR	TYR	PHE	PRO	HIS	PHE	—	ASP	LEU	SER	HIS	—	—	—	—	—	GLY	SER	ALA	GLN
β	ARG	LEU	LEU	VAL	VAL	TYR	PRO	TRP	THR	GLN	ARG	PHE	PHE	GLU	SER	PHE	GLY	ASP	LEU	SER	THR	PRO	ASP	ALA	VAL	MET	GLY	ASN	PRO	LYS

									E																				F	
Mb	LEU	LYS	LYS	HIS	GLY	VAL	THR	VAL	LEU	THR	ALA	LEU	GLY	ALA	ILE	LEU	LYS	LYS	LYS	GLY	HIS	HIS	GLU	ALA	GLU	LEU	LYS	PRO	LEU	ALA
α	VAL	LYS	GLY	HIS	GLY	LYS	LYS	VAL	ALA	ASP	ALA	LEU	THR	ASN	ALA	VAL	ALA	HIS	VAL	ASP	ASP	MET	PRO	ASN	ALA	LEU	SER	ALA	LEU	SER
β	VAL	LYS	ALA	HIS	GLY	LYS	LYS	VAL	LEU	GLY	ALA	PHE	SER	ASP	GLY	LEU	ALA	HIS	LEU	ASP	ASN	LEU	LYS	GLY	THR	PHE	ALA	THR	LEU	SER

		F																		G										
Mb	GLN	SER	HIS	ALA	THR	LYS	HIS	LYS	ILE	PRO	ILE	LYS	TYR	LEU	GLU	PHE	ILE	SER	GLU	ALA	ILE	ILE	HIS	VAL	LEU	HIS	SER	ARG	HIS	PRO
α	ASP	LEU	HIS	ALA	HIS	LYS	LEU	ARG	VAL	ASP	PRO	VAL	ASN	PHE	LYS	LEU	LEU	SER	HIS	CYS	LEU	LEU	VAL	THR	LEU	ALA	ALA	HIS	LEU	PRO
β	GLU	LEU	HIS	CYS	ASP	LYS	LEU	HIS	VAL	ASP	PRO	GLU	ASN	PHE	ARG	LEU	LEU	GLY	ASN	VAL	LEU	VAL	CYS	VAL	LEU	ALA	HIS	HIS	PHE	GLY

																H																	
Mb	GLY	ASN	PHE	GLY	ALA	ASP	ALA	GLN	GLY	ALA	MET	ASN	LYS	ALA	LEU	GLU	LEU	PHE	ARG	LYS	ASP	ILE	ALA	ALA	LYS	TYR	LYS	GLU	LEU	GLY	TYR	GLN	GLY
α	ALA	GLU	PHE	THR	PRO	ALA	VAL	HIS	ALA	SER	LEU	ASP	LYS	PHE	LEU	ALA	SER	VAL	SER	THR	VAL	LEU	THR	SER	LYS	TYR	ARG						
β	LYS	GLU	PHE	THR	PRO	PRO	VAL	GLN	ALA	ALA	TYR	GLN	LYS	VAL	VAL	ALA	GLY	VAL	ALA	ASN	ALA	LEU	ALA	HIS	LYS	TYR	HIS						

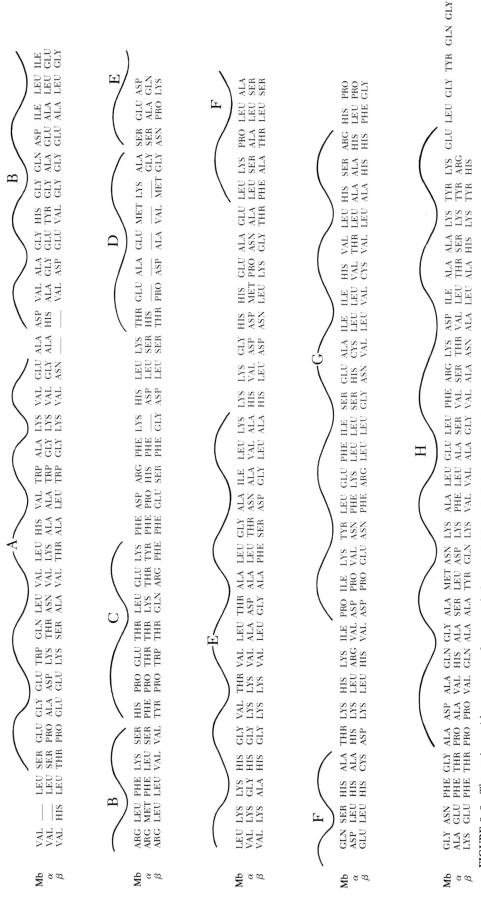

FIGURE 3–3 The amino acid sequences of sperm whale myoglobin and of the α and β chains of human hemoglobin. The letters A-H and the helix symbol ∿∿ identify those amino acids involved in α-helices. Since the α and β polypeptide chains are shorter than the myoglobin chain, blank spaces are left in the shorter chains, when comparing their sequences to the sequence of myoglobin, in order to achieve the best match of amino acids.

RELATIONSHIP BETWEEN THE STRUCTURES OF MYOGLOBIN AND HEMOGLOBIN AND THEIR FUNCTIONAL PROPERTIES

The relationship between the structure of hemoglobin and its properties is complicated and to a large extent hypothetical. The following discussion of this relationship is not presented so that it may be learned in detail; rather it is intended to provide an illustration of how detailed analysis of the structure of a macromolecule can help explain some of its seemingly mysterious functional features. In addition, this discussion serves as background for consideration of the abnormal hemoglobins in the following section, which will illustrate how detailed knowledge of a macromolecule can not only explain the basis of a disease but can also help to formulate treatment for a pathological condition.

The prosthetic heme group can be removed from hemoglobin and myoglobin, and the properties of the isolated heme group can be investigated. Such studies have shown that when the isolated heme group is placed in an aqueous environment in the presence or absence of ligands and then exposed to O_2, the ferrous ion of heme is rapidly oxidized to the ferric ion. Thus, an oxidation of the heme group occurs, rather than the reversible binding of O_2 to the heme group that occurs when the heme group is associated with the proteins of myoglobin or hemoglobin. However, when the isolated heme group is placed in a *nonpolar* environment containing ligands, such as derivatives of imidazole, and then is exposed to O_2, the heme group binds the O_2 reversibly much as myoglobin does and does not undergo oxidation to the ferric form. Therefore, an important role of the protein in both myoglobin and hemoglobin is to provide a hydrophobic environment for the heme group and a proximal histidine to coordinate with the ferrous ion of the heme, thereby enabling the heme to undergo oxygenation rather than oxidation.

Both myoglobin and hemoglobin provide a very similar hydrophobic pocket and proximal histidine for the heme. However, myoglobin displays a hyperbolic oxygen saturation curve and no Bohr effect, while hemoglobin displays an S-shaped saturation curve and a Bohr effect. These two important differences in the functional properties of hemoglobin and myoglobin must be the result of differences in structure. The most obvious difference is that myoglobin consists of one polypeptide chain and one heme group, while hemoglobin consists of four polypeptide chains and four heme groups. If the α subunits and the β subunits of hemoglobin are separated, the isolated α subunits and β subunits show a hyperbolic oxygen saturation curve similar to that of myoglobin, even though the isolated β subunits aggregate to form a tetramer containing four β subunits. Hence, subunit interaction between the α and β subunits appears to be involved in the generation of the S-shaped oxygen saturation curve. Before pursuing this further, let us consider the effect of diphosphoglycerate, a small molecule present in the erythrocyte, on the oxygen saturation curve of hemoglobin.

The S-shaped, or sigmoidal, oxygen saturation curve of hemoglobin implies that the binding of O_2 to hemoglobin is *cooperative*. That is, the binding of oxygen to some fraction of the heme groups in a hemoglobin molecule facilitates the binding of oxygen to other heme groups in that molecule. When the binding of a ligand at one site on a protein molecule affects some property of another site on that same protein molecule, the effect is known as an *allosteric effect*. Another molecule that has an allosteric effect on hemoglobin is 2,3-diphosphoglycerate.

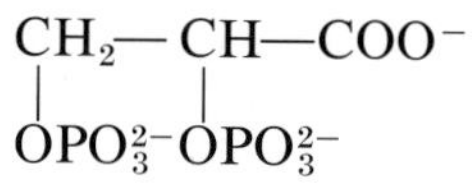

2,3-diphosphoglycerate

This compound is a normal constituent of the erythrocyte, and each mole of deoxyhemoglobin in the erythrocyte contains one mole of 2,3-diphosphoglycerate bound by electrostatic bonds in a cavity between the two β subunits. This bound 2,3-diphosphoglycerate has a marked allosteric effect on the oxygen affinity of deoxyhemoglobin. As can be seen in Figure 3–4, removal of the 2,3-diphosphoglycerate from deoxyhemoglobin shifts the oxygen saturation curve for hemoglobin to the left, toward the myoglobin curve. Thus the oxygen affinity of hemoglobin is markedly increased in the absence of 2,3-diphosphoglycerate. On the other hand, the removal of 2,3-diphosphoglycerate does not significantly affect the cooperativity of oxygen binding or the Bohr effect. It is apparent from Figure 3–4 that hemoglobin would be little better than myoglobin as an oxygen transport system were it not for the allosteric effect of 2,3-diphosphoglycerate which decreases the oxygen affinity of hemoglobin, thereby enabling hemoglobin to efficiently maintain a relatively high Po_2 in the medium bathing the tissues.

Next, the way in which the structure of hemoglobin generates the S-shaped saturation curve and the Bohr effect will be considered, as well as the mechanism by which 2,3-diphosphoglycerate alters the oxygen affinity of hemoglobin. It should be stressed at the outset that our understanding in some of these areas is not complete. As a result, several hypotheses have been proposed to explain the cooperative binding of O_2 by hemoglobin. Attention will be focused on an intriguing hypothesis advanced by Perutz in 1970.

Possible Mechanisms for the Cooperativity of Oxygen Binding and the Allosteric Effect of 2,3-diphosphoglycerate

The x-ray crystallographic studies of Perutz and coworkers established that the tertiary structure of each subunit and the quaternary structure of the molecule

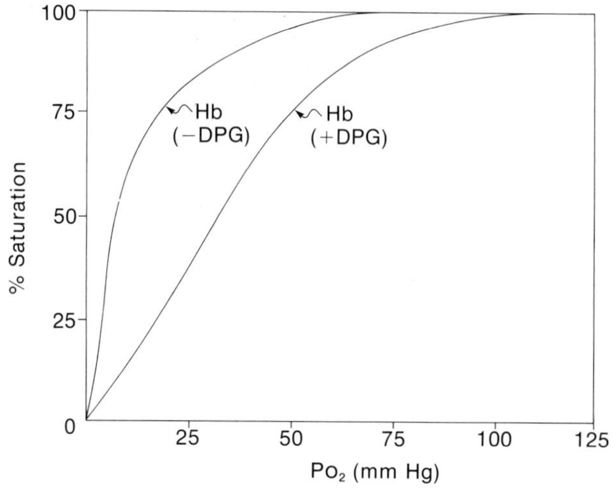

FIGURE 3–4 Oxygen saturation curves for human hemoglobin in the presence and absence of 2,3-diphosphoglycerate (DPG).

undergo a change when deoxyhemoglobin binds four oxygen molecules to form oxyhemoglobin. In an hypothesis developed to account for the cooperative binding of O_2 by hemoglobin, Perutz proposed that the observed conformational changes occur via a stepwise pathway. One possible pathway is represented diagrammatically in Figure 3–5. In accordance with this proposed pathway, the coordination of an oxygen molecule to the heme group of a hemoglobin subunit (Fig. 3–5, 1→2) pulls the iron atom into the plane of the porphyrin ring of the heme group, resulting in an 0.75Å shift in the position of the iron atom. This small shift in position triggers a change in the tertiary structure of the subunit, leading to the conversion of the subunit from a *deoxy* conformation to an *oxy* conformation. The conversion of a subunit from a *deoxy* to an *oxy* conformation requires that two electrostatic bonds be broken. After two of the four subunits of the hemoglobin molecule have bound oxygen and have undergone a conversion to the *oxy* conformation (Fig. 3–5, 1→3), the *deoxy* quaternary structure of the hemoglobin molecule is no longer the most stable, and the molecule spontaneously rearranges to the *oxy* quaternary structure (Fig. 3–5, 3→4). This spontaneous rearrangement to the *oxy* quaternary structure requires the breaking of additional electrostatic bonds and the expulsion of the 2,3-diphosphoglycerate from the cavity between the two β subunits. The driving force for this spontaneous rearrangement to the *oxy* quaternary structure is the greater stability of this structure relative to the deoxy structure after two of the subunits have been oxygenated. The binding of the two final oxygen molecules by the two remaining *deoxy* subunits converts these two subunits to the *oxy* conformation (Fig. 3–5, 4→6), a process that requires the breaking of only one electrostatic bond per subunit when the hemoglobin molecule is in the *oxy* quaternary structure.

How does a model such as the one outlined in Figure 3–5 account for the cooperative binding of O_2? To break an electrostatic bond requires energy, and to break two similar electrostatic bonds requires more energy than is needed to break one. The binding of the first two O_2 molecules to deoxyhemoglobin necessitates the breaking of two electrostatic bonds for each O_2 bound, whereas the binding of the last two O_2 molecules involves the disruption of only one electrostatic bond for each O_2 bound. The energy required to break these electrostatic bonds is obtained at the expense of the binding energy of the O_2 combining with the heme. Therefore, the binding of the last two oxygens will be energetically more favorable than the binding of the first two oxygens. Hence the binding of the last two O_2 molecules is facilitated by the binding of the first two O_2 molecules, and as a result, the binding of O_2 is cooperative. This model also accounts for the allosteric effect of 2,3-diphosphoglycerate, since this compound stabilizes the *deoxy* quaternary structure. In the absence of 2,3-diphosphoglycerate, the conversion of the *deoxy* quaternary structure to the *oxy* quaternary structure should occur at a lower Po_2. However, the model depicted in Figure 3–5 is only one of several models that have been advanced to account for the cooperative binding of oxygen by hemoglobin. On the basis of the experimental data currently available, it is not possible to state which of the models for the cooperative binding of O_2 is correct.

A number of important enzymes also display cooperative binding of substrates and allosteric modifiers. Thus far, all enzymes displaying cooperative binding have been found to consist of two or more polypeptide chains or subunits. The common feature of most models proposed to explain cooperative and allosteric effects is that they employ conformational changes in the tertiary or quaternary structure of the protein as the communication system between distance sites on the protein molecule.

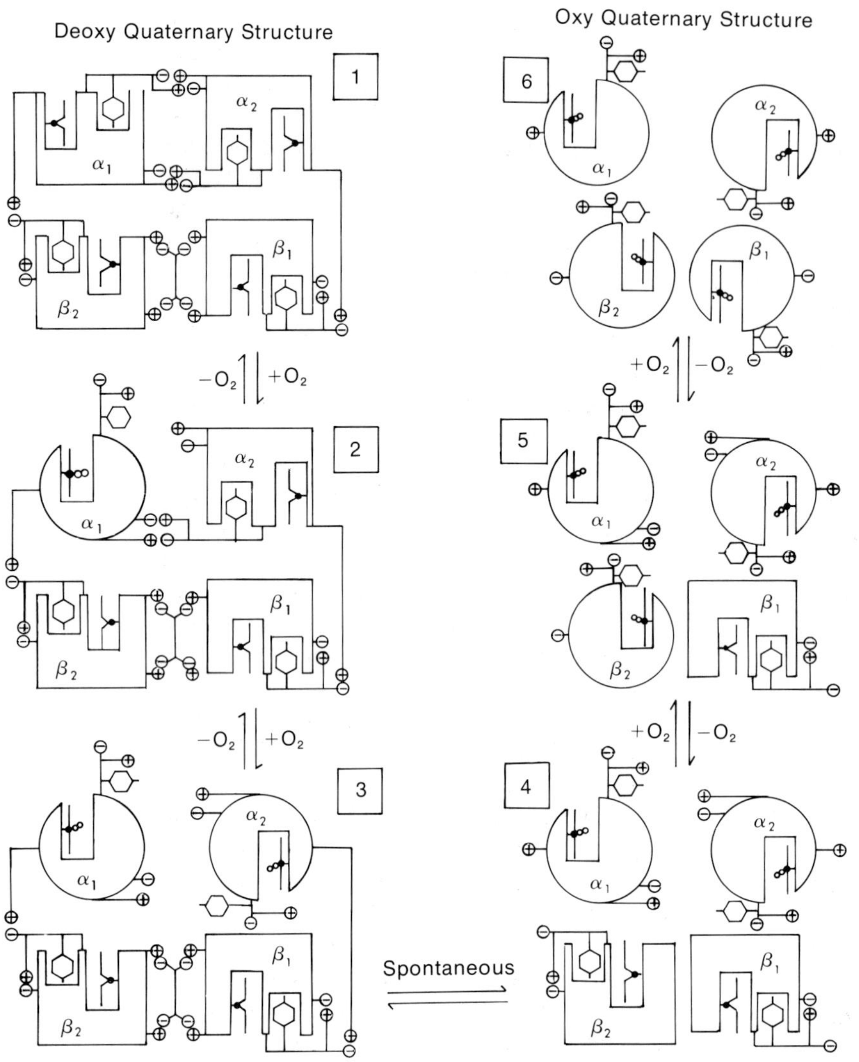

FIGURE 3–5 Diagram of a possible pathway proposed by Perutz to account for the cooperative binding of oxygen by hemoglobin and the allosteric effect of 2,3-diphosphoglycerate on the binding of oxygen by hemoglobin. *Structure 1*. Deoxyhemoglobin with each subunit in the *deoxy* conformation in which the iron atom is displaced approximately 0.75 Å from the plane of the porphyrin ring of the heme group and the phenolic group of the penultimate tyrosine residue is in a pocket formed by the F and H helices . The deoxyhemoglobin molecule also possesses the *deoxy* quaternary structure which is characterized by a large gap between the two β-subunits. This *deoxy* quaternary structure is stabilized by hydrophobic and electrostatic interactions between the subunits, which include the 12 electrostatic bonds indicated in the diagram. Four of these electrostatic bonds are formed between the two subunits and the 2, 3-diphosphoglycerate occupying the gap between the two β subunits. The remaining eight electrostatic bonds involve the interaction of the C-terminal residue of each subunit (arginine in the α chains and histidine in the β chains) with charged groups on the same or other subunits of the molecule, as indicated. *Structure 2*. Coordination of O_2 to the heme group in the α_1 subunit causes the iron to move into the plane of the porphyrin ring . This displacement of the iron produces a similar displacement in the position of the proximal histidine to which it is coordinated. Since the proximal histidine is attached to the F-helix, the F-helix is also pulled toward the heme group. This movement of the F-helix forces the R group of the penultimate tyrosine residue out of a pocket between the F-helix and the H-helix. The penultimate tyrosine is attached to the C-terminal arginine. When the penultimate tyrosine is forced out of the pocket between the F-helix and the H-helix, it drags the C-terminal arginine with it, breaking the two electrostatic bonds between the C-terminal arginine of the α_1 subunit and the two charged groups on the α_2 subunit, one of which is the $-NH_3^+$ group of the N-terminal valine of the α_2 subunit. This leads to Structure 2 in which the α_1 subunit is in the *oxy* conformation . *Structure 3*. The

Legend continued on the following page

The Molecular Basis of the Bohr Effect

The Bohr effect occurs because oxyhemoglobin is a stronger acid than deoxyhemoglobin. This increase in acid strength upon oxygenation must be a reflection of the increase in the acid strength of some acidic group or groups in the molecule. Two of the groups responsible for the Bohr effect have now been identified by chemical and x-ray crystallographic studies. They are the N-terminal valine residues in the α chains and the C-terminal histidine residues of the β chains. These residues participate in electrostatic bonds when the subunits are in the *deoxy* conformation. Conversion of the subunits to the *oxy* conformation by the binding of oxygen breaks these electrostatic bonds by increasing the distance between these positively charged amino or imidazole groups and the negatively charged groups involved in the electrostatic bonds. When the positively charged amino or imidazole groups are in the immediate vicinity of a negatively charged group, as they are in the deoxy conformation of hemoglobin, the extent of dissociation, or acid strength, of these groups is decreased. This is because more work must be done to remove the H^+ owing to the electrostatic attraction between the H^+ and the negatively charged group. Hence, the protonated amino group of valine and the protonated imidazole group of histidine are stronger acids when hemoglobin is oxygenated. This mechanism for the Bohr effect is illustrated for the C-terminal histidine residues of the β chains in Figure 3–6.

In conclusion, the ability of hemoglobin to organize or structure very precisely a large volume of space, and to change the organization of this space upon combining with O_2, bestows upon the hemoglobin molecule the ability to combine with oxygen in a reversible and cooperative manner and the ability to function in CO_2 transport via the Bohr effect.

Figure 3–5 *Continued*
next proposed step involves the addition of the O_2 to the heme of the α_2 subunit, whereupon the α_2 subunit undergoes the same conformational changes outlined above for the α_1 subunit. This results in the formation of structure 3. *Structure* 4. In going from structure 1 to structure 3, two oxygens have been bound by the heme groups of the α subunits, and four electrostatic bonds, which were stabilizing the deoxy quaternary structure, have been broken. As a result of the tertiary structure changes in the two α subunits and the disruption of the four stabilizing electrostatic bonds, the deoxy quaternary structure is no longer the most stable quaternary structure for the molecule. Hence, structure 3 undergoes a spontaneous rearrangement to the oxy quaternary structure represented by structure 4. This rearrangement from the deoxy to the oxy quaternary structure involves slight shifts of the subunits relative to one another. These shifts in the positions of the subunits result in a narrowing of the cavity between the two β subunits, which forces out the bound 2,3-diphosphoglycerate. In addition, an electrostatic bond between the α_2 and β_2 subunits and another between the α_1 and β_2 subunits are disrupted by the shift in the positions of the subunits. *Structure* 5. The conversion of structure 4 to structure 5 involves the binding of an O_2 to the heme of the β_2 subunit. This produces a change in the conformation of the subunit which results in the disruption of one electrostatic bond within the subunit. *Structure* 6. The generation of the fully oxygenated molecule, structure 6, requires the addition of O_2 to heme of the β_1 subunit which induces a conformational change in the β_1 subunit, resulting in the breaking of one electrostatic bond in the β_1 subunit. (Based on Perutz, M. F.: Nature, *228*:726, 1970. For a discussion of refinements of these proposed mechanisms, see Perutz, M. F.: Sci. Am., *239*:92, 1978.)

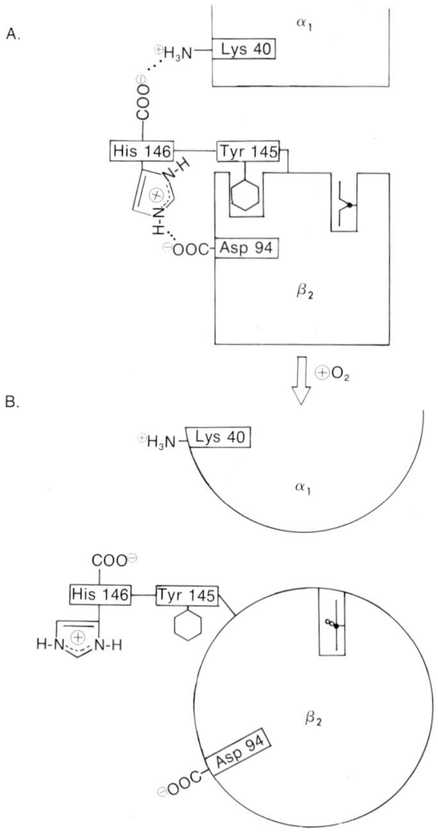

FIGURE 3–6 Diagram of the molecular basis for the contribution of the C-terminal histidine residues of the β subunits to the Bohr effect. *A.* A schematic diagram of the electrostatic bonds or salt bridges (····) formed by the C-terminal histidine residue of a β subunit in the deoxy conformation. One bond is formed by the interaction of the positively charged ϵ-amino group of lysine 40 of the α_1 subunit and the negatively charged carboxylate group of the C-terminal histidine residue of the β_2 subunit. Another electrostatic bond is formed between the positively charged imidazole group of the C-terminal histidine residue of the β subunit and the negatively charged carboxylate group of aspartate 94 in the same subunit. Similar electrostatic bonding would exist between the C-terminal histidine residue of the β_1 subunit and the α_2 subunit. *B.* A schematic diagram of the position of the C-terminal histidine residue of the β subunit in the oxy conformation, showing the disruption of the electrostatic bonds present in the deoxy conformation (Fig. 3–6*A*) as a result of the changes in the relative positions of the charged groups. This shift of the protonated imidazole group of histidine 146 from the immediate vicinity of the negatively charged carboxylate group of aspartate 94 upon oxygenation of the heme groups leads to an increase in the extent of dissociation of the proton from the protonated imidazole group. Approximately half of the protons released from hemoglobin upon oxygenation appear to result from this increased dissociation of protons from the protonated imidazole groups of histidine 146 of the β subunits. (Modified from Perutz, M. F.: Nature, *228*:734, 1970.)

HEMOGLOBINOPATHIES—EXAMPLES OF MOLECULAR DISEASES

Normal Hemoglobins

Before the abnormal or mutant hemoglobins are considered, the types of hemoglobins normally found in the human will be reviewed.

The hemoglobin discussed earlier, containing two α and two β chains ($\alpha_2\,\beta_2$), is designated hemoglobin A_1 and is abbreviated HbA_1. Hemoglobin A_1 constitutes over 90 per cent of the hemoglobin in the normal adult. Hemoglobin A_2 consists of two α and two δ chains ($\alpha_2\ \delta_2$) and accounts for 2 to 3 per cent of the hemoglobin in the normal adult. The sequence of amino acids in the δ polypeptide chain differs from that found in the β chain at 10 positions. Fetal hemoglobin (HbF) consists of two α and two γ chains ($\alpha_2\ \gamma_2$). The amino acid sequence in the γ chain differs slightly from that of the β or δ chains. HbF constitutes only a small percentage of the hemoglobin found in the normal adult, but it makes up approximately 85 per cent of the hemoglobin of the newborn. HbF has a lower affinity than HbA_1 for 2,3-diphosphoglycerate. Therefore, the O_2 affinity of HbF is greater than the O_2 affinity of HbA_1. This greater oxygen affinity of HbF facilitates the transfer of O_2 via the placenta from the mother's circulatory system to the circulatory system of the fetus.

The three hemoglobins normally found in the human, HbA_1, HbA_2, and HbF, are all capable of functioning as O_2 transport systems even though their primary structures differ at a number of positions. When the primary structures of the polypeptide chains of normal hemoglobins from eleven species are compared, only twenty-one residues are found to be identical in all of the structures. These 21 common residues include the proximal and distal histidines and the penultimate tyrosine residues, which have been discussed in some detail. Even though many of the substitutions observed in these structures are conservative in nature, the fact that all of these different hemoglobins are functional molecules raises a legitimate question as to the importance of the primary structure. Perhaps the best way to approach this question is to look at the properties of some of the more than 150 different mutant hemoglobins that have been detected and studied.

A point mutation in a gene coding for one of the polypeptide chains of hemoglobin can lead to the synthesis of an abnormal polypeptide chain that differs from a normal chain by the substitution of one amino acid for another at only one position. If an individual is homozygous for an abnormal gene coding for the β chain, all of the β chains synthesized by this individual will be abnormal. If the individual is heterozygous, both normal and abnormal β chains will be synthesized. The most thoroughly studied molecular disease involving abnormal hemoglobin is sickle cell anemia.

Sickle Cell Anemia—A Disease due to Replacement of an Amino Acid on the Surface of Hemoglobin

Many people living in a zone extending across central Africa, and descendants of people from this area, carry an abnormal gene coding for the β chain of hemoglobin. The product of this abnormal gene differs from the normal β chain by having a valine substituted for glutamate in the sixth position. This abnormal β chain is represented by $\beta^{6\mathrm{Glu} \rightarrow \mathrm{Val}}$. When the abnormal β chains combine with the normal α chains present in these individuals, hemoglobin S ($\alpha_2\ \beta_2{}^{6\mathrm{Glu} \rightarrow \mathrm{Val}}$) results. Hemoglobin S has one unfortunate property. The substitution of valine for glutamate in position 6 of the β chain changes the conformation of the N-terminal portion of the β chain. As a result of this conformational change, deoxyhemoglobin S tends to aggregate or stack up in ordered arrays, which precipitate as long insoluble strands. The presence of these strands of deoxyhemoglobin S in the erythrocyte forces the cell into a characteristic crescent, or sickle, shape. Oxyhemoglobin S does not aggregate and form the long insoluble strands.

Individuals who are homozygous for this abnormal gene have *sickle cell anemia* and usually die at an early age. Often this is the result of damage done to various organs because the sickled cells are less distortable than normal red cells and become trapped in small capillaries, impairing the circulation. The anemia occurs because the sickled cells are more fragile than normal red blood cells and tend to break. Individuals who are heterozygous for this abnormal gene have the *sickle cell trait*, and their erythrocytes contain a mixture of normal HbA_1 and the abnormal HbS. The decreased concentration of HbS and its admixture with normal HbA_1 in individuals with sickle cell trait markedly lowers the frequency of cell sickling. As a result, individuals with sickle cell trait may be unaware of their condition for as long as they remain at low altitudes where the Po_2 is high. However, the low Po_2 encountered at high altitudes increases the fraction of HbS in the *deoxy* form and may precipitate a dangerous crisis in the heterozygous individual.

Sickle cell anemia and sickle cell trait are not rare conditions, since the frequency of the abnormal gene is over 20 per cent in some populations in Africa. Approximately 9 per cent of blacks in America have the sickle cell trait. Why should a mutation that is usually lethal to the homozygote survive in a population? The answer to this question appears to be that the mutation protects the heterozygote against a malarial parasite that spends a part of its life cycle in the red blood cell.

The molecular basis of sickle cell anemia is well defined. One medical goal is to take advantage of this knowledge to formulate treatment and thus, in this case, to determine if the deleterious effect of this single amino acid substitution can be overcome. One promising approach to this problem, which is now being investigated, involves chemical modification of the N-terminal portion of the polypeptide chains of hemoglobin S by sodium cyanate. In aqueous solution the cyanate ion, NCO^-, is in equilibrium with isocyanic acid, $HN{=}C{=}O$. Isocyanic acid carbamylates the N-terminal valines of the α and β chains of hemoglobin.

$$\underset{\text{N-terminal amino group}}{R{-}NH_2} + H{-}N{=}C{=}O \rightleftharpoons \underset{\text{carbamyl derivative}}{R{-}\overset{\overset{H}{|}}{N}{-}\overset{\overset{O}{\|}}{C}{-}NH_2}$$

Carbamylation of the N-terminal valines of HbS inhibits the deleterious aggregation and precipitation of deoxyhemoglobin S. The sickling of red blood cells is reduced even when an average of only one of the four subunits of HbS is carbamylated. This level of carbamylation increases the oxygen affinity of HbS, which also contributes to the prevention of sickling. Although isocyanic acid will carbamylate the N-terminal amino acids of other proteins, it appears to react more rapidly with the N-terminal valines of hemoglobin. This specificity for hemoglobin may be the result of its similarity to CO_2, which also rapidly reacts with the N-terminal valines of hemoglobin to form carbamino compounds (see page 70). Although the efficacy of cyanate treatment has not yet been established by large-scale clinical trials, it should be apparent that a knowledge of the structure and chemistry of HbS has made possible a rational approach to the problem of therapy.

Hemoglobin S is just one of the many mutant hemoglobins that involve an amino acid substitution on the external surface of the molecule. Hemoglobin S is unique in this group, however, since most known mutations involving external amino acids do not seriously impair the functional properties of the hemoglobin molecule. However, as might be expected, mutations involving replacements of amino acid residues in the hydrophobic pocket and in the contact regions between subunits often have very adverse effects on the functional properties of the hemoglobin molecule. Several of the known mutations that result in amino acid replacement in these very critical regions of the molecule will now be considered.

Replacements Involving Amino Acids in the Hydrophobic Pocket

Hemoglobin M Iwate ($\alpha_2^{87\ \mathrm{His} \rightarrow \mathrm{Tyr}}\ \beta_2$) is a mutant hemoglobin in which the proximal histidine of the α chain is replaced by a tyrosine. The phenolic group of tyrosine probably occupies the fifth coordination position of the iron, and as a result the iron undergoes an oxidation to the ferric state in the presence of O_2 instead of a reversible oxygenation. Hemoglobin in which the iron has been oxidized to the ferric state is known as methemoglobin, abbreviated HbM. Since methemoglobin does not bind O_2, this mutation leads to cyanosis, a bluish discoloration of the skin due to lack of oxygen.

Other mutant hemoglobin M variants are known in which the distal histidine in the α chain is replaced by tyrosine (HbM Boston), the proximal histidine in the β chain is replaced by tyrosine (HbM Hyde Park), and the distal histidine in the β chain is replaced by tyrosine (HbM Saskatoon). These three hemoglobin M variants also result in cyanosis and methemoglobinemia. Individuals in whom these mutant hemoglobins have been found have all been heterozygotes. A homozygote for any of these mutant hemoglobins would probably not survive.

Hemoglobin Sydney ($\alpha_2^{67\ \mathrm{Val} \rightarrow \mathrm{Ala}}\ \beta_2$) results from a substitution of an alanine for valine in the hydrophobic pocket. As a result of this substitution, the heme group is not as tightly held in the pocket. Inclusion bodies are formed in the red cell, and hemolytic anemia results.

Replacements Involving Amino Acid Residues in the Contact Regions between Subunits

Although some mutations leading to amino acid replacements in the contact regions between subunits do not alter the functional properties of the hemoglobin molecule, many others do. As might be expected, one of the properties of hemoglobin that is frequently altered by amino acid replacements in these contact regions is the cooperativity in the binding of O_2. For example, Hb Chesapeake ($\alpha_2^{92\ \mathrm{Arg} \rightarrow \mathrm{Leu}}\ \beta_2$) and Hb Yakima ($\alpha_2\ \beta_2^{99\ \mathrm{Asp} \rightarrow \mathrm{His}}$) both show an increased oxygen affinity and a decreased cooperativity in oxygen binding. The clinical symptoms brought about by each of these mutations is polycythemia, an increase in the number of circulating red blood cells. Polycythemia may be an adaptive adjustment to compensate for the fact that the mutant hemoglobin is a much less efficient transporter of O_2 as a result of its increased oxygen affinity and decreased cooperativity of oxygen binding.

SUMMARY

In conclusion, it is apparent that replacing one amino acid with another at a single position in either polypeptide chain of hemoglobin can markedly alter the functional properties of hemoglobin, rendering the molecule less effective or even totally ineffective as a transport system of O_2. Therefore the primary structure is critical to the functioning of the molecule, even though there are only a few invariant residues found in the many functioning hemoglobin molecules. The fact that only a few invariant residues are present merely indicates that similar functional properties can be generated by protein molecules with different primary structures.

FURTHER READING

Antonini, E., and Brunori, M.: Hemoglobin and myoglobin in their reactions with ligands. Frontiers of Biology. Vol. 21. Amsterdam, North-Holland, 1971.

Benesch, R. E., and Benesch, R.: The mechanism of interaction of red cell organic phosphates with hemoglobin. *In*: Anfinsen, C. B., Edsall, J. T., and Richards, F. M. (eds.): Advances in Protein Chemistry. Vol. 28. New York, Academic Press, 1974, pp. 211–237.

Dickerson, R. E.: X-ray studies of protein mechanisms. Annu. Rev. Biochem., *41*:815, 1972.

Dickerson, R. E., and Geis, I.: The Structure and Action of Proteins. New York, Harper & Row, 1969.

Edelstein, S. J.: Cooperative interactions of hemoglobin. Annu. Rev. Biochem., *44*:209, 1975.

Kilmartin, J. V., and Rossi-Bernardi, L.: Interaction of hemoglobin with hydrogen ions, carbon dioxide and organic phosphates. Physiol. Rev., *53*:836, 1973.

Pauling, L., Itano, H. A., Singer, S. J., et al.: Sickle cell anemia, a molecular disease. Science, *110*:543, 1949.

Perutz, M. F.: Stereochemistry of cooperative effects in haemoglobin. Nature, *228*:726, 1970.

Perutz, M. F.: The hemoglobin molecule. Sci. Am., *211*:67, 1964.

Perutz, M. F., and Lehmann, H.: Molecular pathology of human haemoglobin. Nature, *219*:902, 1968.

4

ENZYMES

Enzymes are proteins that function as catalysts in the cell. Essentially every chemical reaction occurring in the cell is catalyzed by a specific enzyme. In fact, many of the chemical reactions that are absolutely essential to the cell's existence would not proceed at a significant rate under the cellular conditions of temperature and pH were it not for the presence of these enzymes.

Two striking features about enzymes make them the best catalysts known. These are their *effectiveness* and their *specificity*. Effectiveness refers to the number of moles of substrate a mole of enzyme converts to product per minute. This is known as the turnover number for the enzyme. The turnover numbers for different enzymes range from 10^3 to 10^7 per minute. Thus some enzymes are more effective than others, and in some cases one molecule of enzyme can catalyze the conversion of millions of molecules of substrate to product every minute.

Equally striking is the great specificity of enzymes. Not only are they specific as to the type of reaction they will catalyze, but also they are often very specific as to the substrates they will utilize. Some enzymes show a greater substrate specificity than others, and the most specific utilize only one particular compound as a substrate. Others, such as the proteolytic enzymes discussed in Chapter 2, show a much broader specificity for substrate. For example, trypsin will catalyze the hydrolysis of any peptide bond in which the carboxyl group of lysine or arginine participates. An important and characteristic feature of all enzymatic reactions is the virtual absence of any side products. Therefore, just as hemoglobin is precisely tailored to transport oxygen, an enzyme is precisely adapted to catalyze a particular reaction.

THE ROLE OF CATALYSTS

Before focusing on the molecular basis for the remarkable catalytic properties of enzymes, let us consider the general role played by all catalysts. A catalyst is a substance that increases the rate of a chemical reaction without being consumed in the reaction. A catalyst does not alter the equilibrium constant or equilibrium position of a reversible reaction, but it does alter the rate at which equilibrium is achieved. Catalysts bring about this acceleration in rate by lowering the activation energy, E_a, for the reaction.

Consider the simple bimolecular reaction $A+B \rightarrow C+D$. According to the transition state theory, the reactants A and B must cross an energy barrier before becoming products. The height of the energy barrier is known as the energy of activation, E_a. This is depicted in the reaction coordinate diagram represented by the solid line in Figure 4–1. In the course of the reaction, the electrons and nuclei of A and B pass through an arrangement known as the transition state, which is represented by AB‡ in Figure 4–1.

The transition state is the least stable arrangement of nuclei and electrons occurring during the progress of the reaction; therefore it is the state whose potential energy is represented by the peak of the energy barrier. The difference between the potential energy of the reactants A and B and the potential energy of the transition state AB‡ is the activation energy of the reaction. The potential energy of the transition state AB‡ is achieved by conversion of the kinetic energy of A and B into potential energy upon collision. Only those collisions involving A and B molecules with sufficient kinetic energy to cross the energy barrier will be fruitful.

Catalysts increase the rate of a chemical reaction by lowering the energy barrier between reactants and products. This is depicted in the reaction coordinate diagram represented by the dashed line in Figure 4–1. Hence, at any given temperature, a much larger fraction of the molecules will possess a total kinetic energy sufficient to surmount the energy barrier. How enzymes bring about this decrease in the activation energy of a chemical reaction will be considered after some of the important aspects of enzyme kinetics have been presented.

STEADY-STATE ENZYME KINETICS

Reaction Velocities

Much has been learned about enzyme catalysis by studying the rates of enzyme-catalyzed reactions. In the enzyme-catalyzed conversion of substrate S to product P, S → P, the velocity of the reaction, v, is the rate of appearance of product or the rate of disappearance of substrate

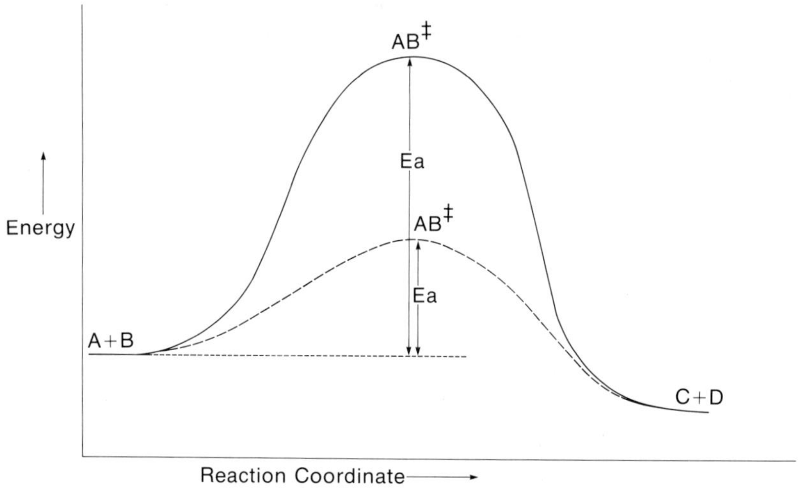

FIGURE 4–1 A reaction coordinate diagram for the reaction A + B → C + D in the absence of a catalyst (*solid line*) and in the presence of a catalyst (*dashed line*).

$$v = \frac{d[P]}{dt} = -\frac{d[S]}{dt}$$

where t = time. Therefore, the velocity of a reaction can be determined from the slope of a plot of [S] or [P] versus t. As shown in Figure 4–2, these plots are often curved because the velocity decreases as the reaction proceeds. For reasons that will become apparent, the velocity employed in most enzyme kinetic studies is the initial velocity of the reaction, v_0, which is the velocity of the reaction when the extent of the reaction is small. Since only a small amount of S has been converted to P when v_0 is measured, the approximation can be made that the concentration of S is equal to the initial concentration of substrate, $[S_0]$.

If the initial velocity of an enzyme-catalyzed reaction is plotted versus $[S_0]$, it is seen to have a somewhat unusual dependence on substrate concentration. With some enzymes, the plot of v_0 versus $[S_0]$ is hyperbolic, as depicted in Figure 4–3*A*. With other enzymes, the plot of v_0 versus $[S_0]$ is sigmoidal, or S-shaped, as shown in Figure 4–3*B*. The significance of a hyperbolic curve will be considered first.

The curve depicted in Figure 4–3*A* is a rectangular hyperbola. At very low initial substrate concentrations, v_0 shows an approximate first-order dependence on $[S_0]$, that is,

$$v_0 \approx k_1 [S_0]$$

where k_1 is a constant. At very high initial substrate concentrations, v_0 appears to be nearly constant and independent of $[S_0]$, that is,

$$v_0 = k_0$$

where k_0 is a constant. In 1913, Michaelis and Menten proposed a simple model to explain the hyperbolic dependence of v_0 on $[S_0]$.

Michaelis-Menten Theory

According to the Michaelis-Menten theory, the substrate S is bound at the "active site" of the enzyme, forming an enzyme-substrate complex designated

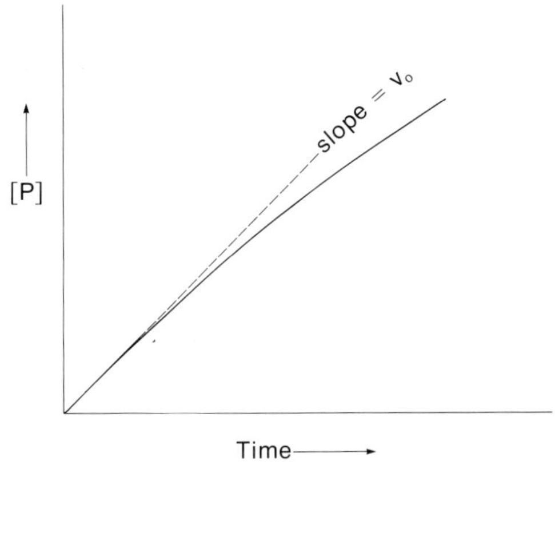

FIGURE 4–2 A plot of the concentration of product [P] versus time for an enzyme-catalyzed reaction.

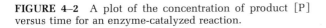

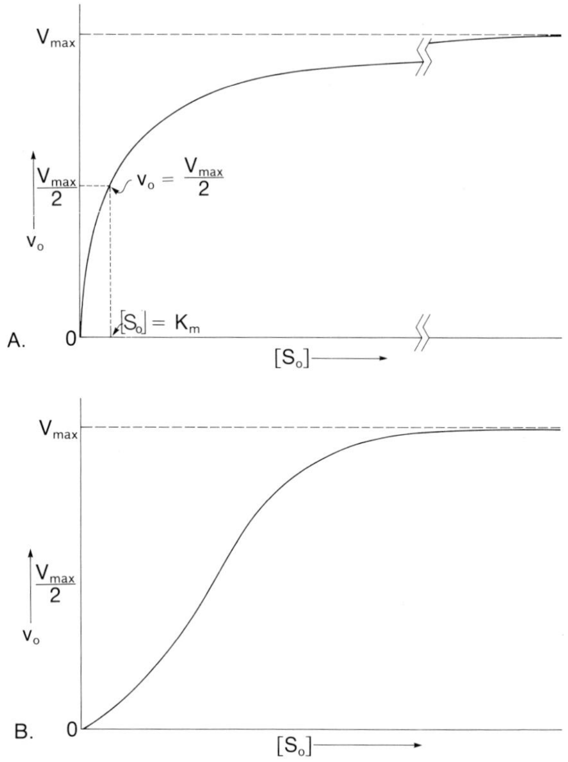

FIGURE 4–3 Effect of substrate concentration $[S_0]$ on the initial velocity, v_0, of enzyme-catalyzed reactions. *A*. Typical plot for an enzyme-catalyzed reaction that follows Michaelis-Menten kinetics. *B*. Typical plot for a reaction catalyzed by an enzyme displaying positive cooperativity.

ES, before being converted to product P. Therefore, we can represent an enzyme-catalyzed reversible reaction by

(1) $$E + S \underset{k_2}{\overset{k_1}{\rightleftharpoons}} ES \underset{k_4}{\overset{k_3}{\rightleftharpoons}} P + E$$

when k_1, k_2, k_3, and k_4 are specific rate constants, and E and ES represent free enzyme and enzyme-substrate complex, respectively. The velocity, or net rate of appearance of P, may be expressed as the rate of appearance of P minus the rate of disappearance of P.

$$v = \frac{d[P]}{dt} = k_3[ES] - k_4[P][E]$$

However, when the initial velocity, v_0, is determined, the concentration of P is very small. If the concentration of P is sufficiently small so that $k_4[P][E] \ll k_3[ES]$, the initial velocity may be approximated by equation 2.

(2) $$v_0 = k_3[ES]$$

Employing the approximation represented by equation 2 and the steady-state assumption, the Michaelis-Menten equation (equation 3) can be derived (see Appendix E for the derivation of the Michaelis-Menten equation).

(3) $$v_0 = \frac{V_{max}[S_0]}{K_m + [S_0]}$$ Michaelis-Menten equation

where K_m, the Michaelis constant, is given by equation 4

$$K_m = \frac{k_2 + k_3}{k_1} \tag{4}$$

and V_{max}, the maximum initial velocity, is the product of k_3 and $[E_T]$, the total enzyme concentration, as expressed in equation 5

$$V_{max} = k_3[E_T] \tag{5}$$

Some of the important properties of the Michaelis-Menten equation are the following: When $[S_0]$ is much smaller than K_m, $[S_0] \ll K_m$, then $v_0 \approx \frac{V_{max}}{K_m}[S_0]$; thus, the velocity shows an approximate first-order dependence on $[S_0]$. When $[S_0]$ is much larger than K_m, $[S_0] \gg K_m$, then $v_0 \approx V_{max}$, and the initial velocity is independent of $[S_0]$. As shown in Figure 4–3*A*, V_{max} is the asymptote to the curve. When $[S_0]$ is equal to K_m, $v_0 = \frac{1}{2} V_{max}$. Hence, K_m is equal to the substrate concentration that produces an initial velocity equal to $\frac{1}{2} V_{max}$. One way of estimating the value of K_m is to determine the substrate concentration that produces an initial velocity equal to $\frac{1}{2} V_{max}$. This can be done graphically, as illustrated in Figure 4–3*A*.

A Linear Transform of the Michaelis-Menten Equation

The Michaelis-Menten equation (equation 3) can be transformed into a linear equation by taking the reciprocal of both sides of the equation. This gives equation 6,

$$\frac{1}{v_0} = \frac{K_m}{V_{max}}\left(\frac{1}{[S_0]}\right) + \frac{1}{V_{max}} \tag{6}$$

which is often referred to as the Lineweaver-Burk equation. Inspection of equation 6 reveals that a plot of $\frac{1}{v_0}$ versus $\frac{1}{[S_0]}$ will be linear and that the slope of the line will be $\frac{K_m}{V_{max}}$, with an intercept on the $\frac{1}{v_0}$ axis equal to $\frac{1}{V_{max}}$ (Fig. 4–4). If the line is extrapolated to the abscissa, the intercept is equal to $\frac{-1}{K_m}$. Thus, K_m and V_{max} can be readily determined by measuring v_0 at different initial substrate concentrations and then plotting the data, as shown in Figure 4–4. All other conditions, such as pH, temperature, ionic strength, and total enzyme concentration, must be kept constant when determining initial velocities as a function of substrate concentration, because these variables can also affect the initial velocity. The main advantage of a linear plot, such as the Lineweaver-Burk plot, over the hyperbolic plot of Figure 4–3*A* is that the determination of K_m and V_{max} does not require determination of an asymptote.

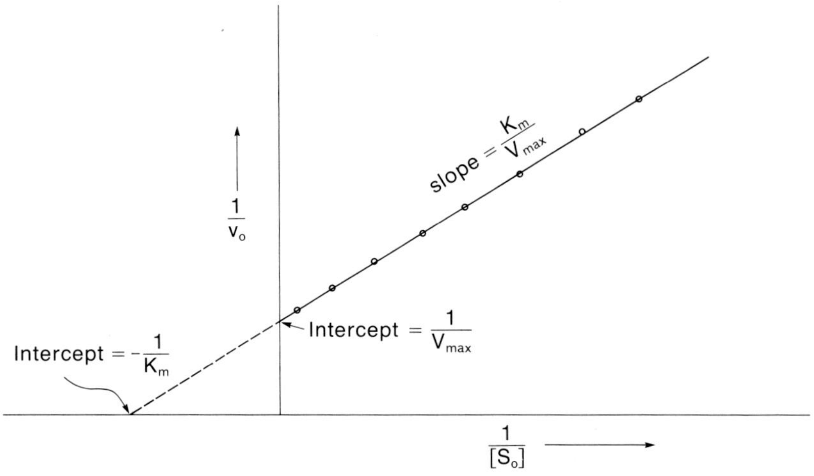

FIGURE 4–4 A Lineweaver-Burk plot of kinetic data for an enzyme-catalyzed reaction.

Regulatory or Allosteric Enzymes

The substrate concentration-initial velocity curves for some enzymes are sigmoidal, as shown in Figure 4–3*B*. It is apparent that the simple Michaelis-Menten theory cannot account for this nonhyperbolic type of curve. Many of the enzymes that display this kind of substrate-velocity relationship are regulatory enzymes and consist of subunits. Regulatory enzymes usually catalyze an early reaction in a sequential, multireaction biosynthetic or degradative pathway, and they are classified as regulatory enzymes because they play a key role in controlling or regulating the overall rate of the pathway.

The S-shaped substrate concentation-velocity curve of Figure 4–3*B* is reminiscent of the oxygen saturation curve of hemoglobin (see Figure 3–1, Chapter 3). As with hemoglobin, the sigmoidal shape of the curve results from the cooperative binding of substrate. That is, the binding of substrate to one site on the enzyme facilitates the binding of substrate at another site on the enzyme. This phenomenon is known as *positive cooperativity*. Another phenomenon, known as *negative cooperativity*, is displayed by certain enzymes. In negative cooperativity the initial binding of a substrate or ligand at one site on the enzyme decreases the affinity of a second site on the enzyme for the substrate or ligand.

Effects of Some Other Factors on Enzymatic Catalysis

Since $v_0 = k_3[ES]$ and $V_{max} = k_3[E_t]$, the velocity of an enzyme-catalyzed reaction and V_{max} will be proportional to the concentration of enzyme. K_m, on the other hand, should be independent on enzyme concentration unless the enzyme undergoes a concentration-dependent change in its state of aggregation, in which case K_m as well as V_{max} may show a dependence on enzyme concentration.

As discussed in Chapter 2, pH can have a major effect on enzymatic catalysis. A typical pH-rate profile is depicted in Figure 4–5. Most enzymatic reactions show a pH optimum. A number of explanations can be advanced for the sensitivity of enzyme-catalyzed reactions to pH. A change in pH can change the charge configuration on the protein and substrate, thereby altering the K_m. A change in the charge configuration on the protein can also lead to a conformational change

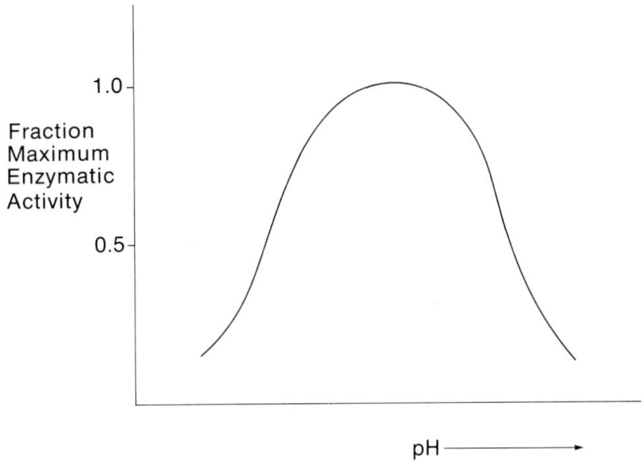

FIGURE 4–5 A typical pH-rate profile for an enzyme-catalyzed reaction.

that alters K_m, V_{max}, or both. A change of pH can alter the ionization state of a functional group on the protein that is essential in the catalytic mechanism, thereby changing k_3 or V_{max}. Studies of the effect of pH on the kinetics of enzymatic reactions have contributed to our understanding of the mechanisms of enzyme action.

As with the other chemical reactions, the rate of the enzyme-catalyzed reaction increases with increasing temperature. However, as shown in Figure 4–6, enzyme-catalyzed reactions reach a temperature optimum above which the rate of the reaction rapidly falls. The decrease in rate at high temperatures is usually a reflection of the destruction of the enzyme by heat, a process known as heat denaturation.

Some enzymes are also inhibited by certain other molecules or ions. The inhibitory substance may be a small molecule or ion, or it may be a macromolecule. Inhibition and activation of enzymes by other cellular components are physiologically very important processes, since they constitute the most important control mechanisms for the rapid regulation of metabolic processes within the cell. Studies of enzyme inhibitors and enzyme inhibition have contributed to our understanding of the mechanisms of enzyme action and the nature of the active site. Thus, the importance of enzyme inhibition warrants a more detailed consideration of the process.

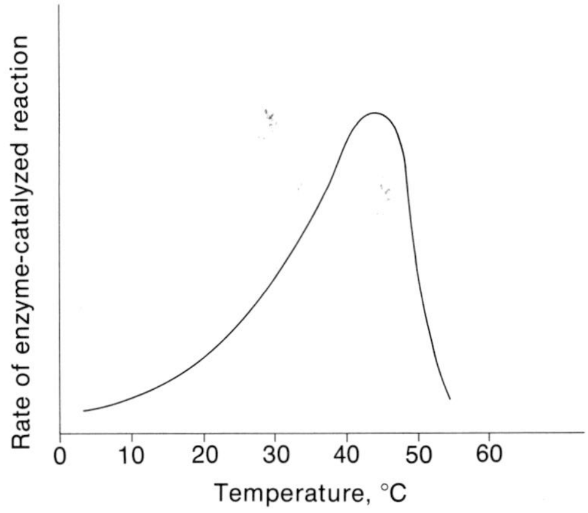

FIGURE 4–6 Effect of temperature on an enzyme-catalyzed reaction.

THE INHIBITION OF ENZYMES

Inhibitors of enzymes can be divided into two classes: reversible inhibitors and irreversible inhibitors. A *reversible* inhibitor produces an inhibition that can be removed by dilution, dialysis, or chromatography. Reversible inhibitors bind to the enzyme in much the same manner as substrates; that is, their binding involves hydrogen bonds, electrostatic interactions, or hydrophobic interactions. If a covalent bond is involved in the binding of a reversible inhibitor, the bond is of the readily reversible type, such as the C=N linkage found in imines and the C—O linkage found in hemiacetals.

In contrast, *irreversible* inhibitors produce an inhibition that cannot be relieved by dilution, dialysis, or chromatography. Irreversible inhibitors usually react chemically with some functional group or groups on the enzyme, forming a covalent linkage. In so doing, they decrease or abolish the enzymatic activity of the enzyme. An example of an irreversible inhibitor is iodoacetate, which reacts with the sulfhydryl groups of cysteinyl residues in the enzyme.

$$\mathcal{E}\text{—SH} + \text{I—CH}_2\text{—COO}^- \longrightarrow \mathcal{E}\text{—S—CH}_2\text{—COO}^- + \text{HI}$$

Although an irreversible inhibitor may produce its inhibition by reacting with a functional group at the active site of the enzyme, inhibition may also result from a conformational change produced by the reaction of the inhibitor with a functional group far removed from the active site of the enzyme.

Two major classes of *reversible* inhibitors are competitive inhibitors and noncompetitive inhibitors. Both types combine reversibly with the enzyme to form an enzyme-inhibitor complex, EI.

$$\text{E} + \text{I} \rightleftharpoons \text{EI}$$

The equilibrium constant for this reaction may be written as

$$K_I = \frac{[E][I]}{[EI]}$$

where K_I is the dissociation constant of the enzyme-inhibitor complex.

Competitive inhibitors often bear a structural resemblance to the enzyme's normal substrate. For example, the enzyme succinate dehydrogenase catalyzes the following reaction, in which the electrons resulting from the oxidation of succinate are transferred to an electron acceptor.

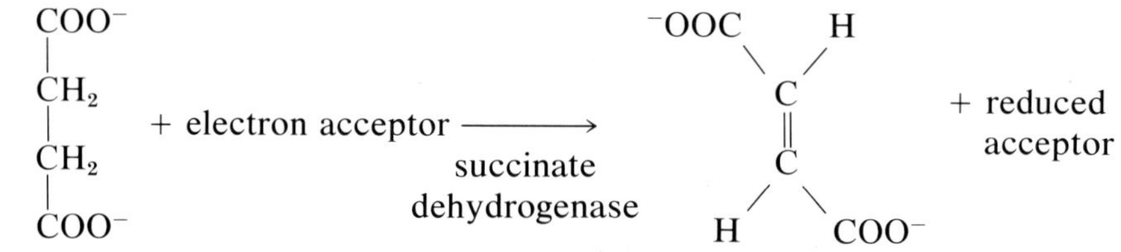

Succinate Fumarate

Three of the competitive inhibitors of succinate dehydrogenase are malonate, oxaloacetate, and pyrophosphate.

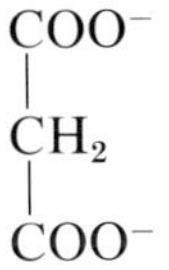
Malonate

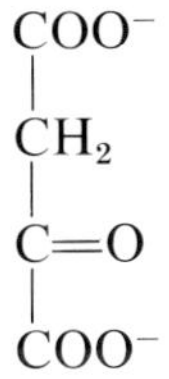
Oxaloacetate

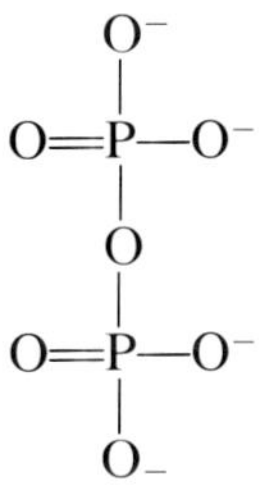
Pyrophosphate

These three competitive inhibitors of succinate dehydrogenase resemble succinate, the enzyme's substrate, both in size and in their content of negatively charged groups which are approximately the same distance apart. Each of these three inhibitors appears to compete with the substrate succinate for a binding site on the enzyme, suggesting that the binding of substrate or inhibitor to the enzyme may involve electrostatic bonds between the negatively charged groups of these molecules and two positively charged groups at the active site of the enzyme.

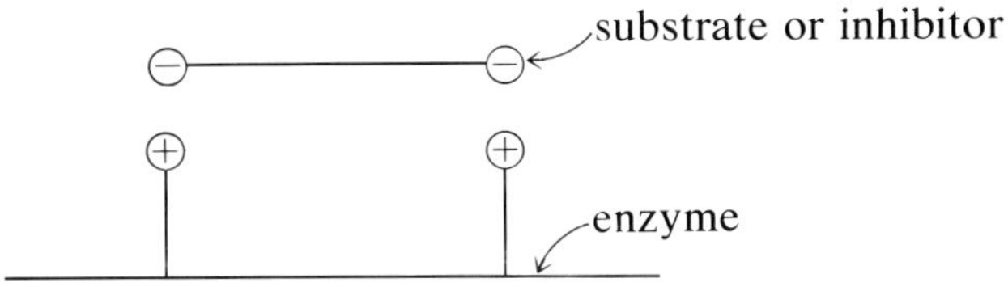

Thus, an insight into the possible nature of the active site and the type of bonds involved in the formation of the ES complex can sometimes be achieved by studying competitive inhibitors of the enzyme. However, considerable caution must be exercised in interpreting the results of such studies, since not all competitive inhibitors bear a close structural resemblance to the substrate, and not all competitive inhibitors necessarily bind at the active site of the enzyme.

The one characteristic feature of competitive inhibition is the ability of large concentrations of substrate to overcome the inhibition. The linear double-reciprocal form of the steady-state rate equation in the presence of a competitive inhibitor is given in equation 7,

$$\frac{1}{v_0} = \frac{K_m}{V_{max}}\left(1 + \frac{[I]}{K_I}\right)\left(\frac{1}{[S_0]}\right) + \frac{1}{V_{max}} \tag{7}$$

where [I] is the inhibitor concentration. As shown in Figure 4–7*A*, an extrapolation of the plot to infinite substrate concentration $\left(\frac{1}{[S_0]} = 0\right)$ yields an intercept on the $\frac{1}{v_0}$ axis equal to $\frac{1}{V_{max}}$ at all finite concentrations of inhibitor. Furthermore, an extrapolation of the plot to the $\frac{1}{[S_0]}$ axis intercepts the abscissa at $\frac{1}{[S_0]} = -\frac{1}{K_m\left(1+\frac{[I]}{K_I}\right)}$. Therefore, by determining the intercepts on the abscissa in the absence of inhibitor and at finite concentrations of inhibitor, one can calculate both K_m and K_I.

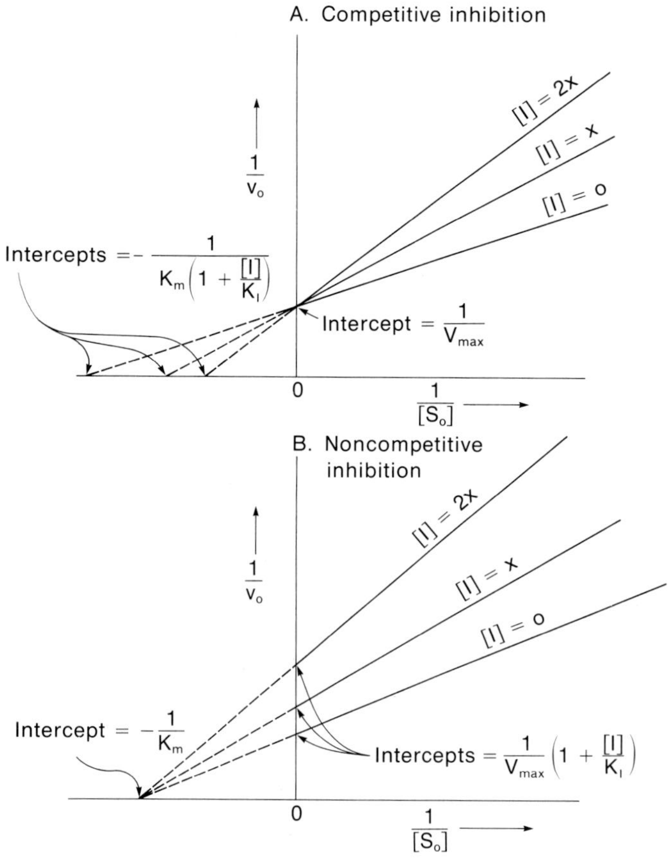

FIGURE 4–7 Lineweaver-Burk plots of kinetic data for an enzyme-catalyzed reaction in the absence and presence of a competitive inhibitor (*A*) and in the absence and presence of a noncompetitive inhibitor (*B*).

In contrast to a competitive inhibitor, which combines only with free enzyme, a noncompetitive inhibitor combines equally well with both the free enzyme and the enzyme-substrate complex ES.

$$E + I \rightleftharpoons EI$$

and

$$ES + I \rightleftharpoons ESI$$

Since the ESI complex is either totally inactive or less active than the ES complex, no amount of substrate will overcome the inhibition produced by a noncompetitive inhibitor—hence the name, noncompetitive. The linear double-reciprocal form of the steady-state rate equation in the presence of a noncompetitive inhibitor is given by equation 8.

$$(8) \qquad \frac{1}{v_0} = \frac{K_m}{V_{max}}\left(1 + \frac{[I]}{K_I}\right)\left(\frac{1}{[S_0]}\right) + \left(1 + \frac{[I]}{K_I}\right)\frac{1}{V_{max}}$$

Thus, as shown in Figure 4–7*B*, a double-reciprocal plot, $\frac{1}{v_0}$ versus $\frac{1}{[S_0]}$, has intercepts on the $\frac{1}{v_0}$ axis equal to $\frac{1}{V_{max}}\left(1 + \frac{[I]}{K_I}\right)$ in the presence of a noncompetitive inhibitor. A determination of the intercepts on the $\frac{1}{v_0}$ axis in the absence and presence of known concentrations of a noncompetitive inhibitor allows one to calculate the K_I for the inhibitor.

MECHANISMS OF ENZYME ACTION

The effectiveness of enzymes as catalysts is truly remarkable. This can be seen, for example, if the rate of hydrolysis of urea when bound to the enzyme urease (reaction *a*) is compared with the rate for the noncatalyzed hydrolysis of urea (reaction *b*).

$$\text{a. Urease} \cdot H_2N\text{—}\overset{\overset{\displaystyle O}{\|}}{C}\text{—}NH_2 + H_2O \longrightarrow \text{Urease} + CO_2 + 2NH_3$$

$$\text{b.} \quad \underset{\text{urea}}{H_2N\text{—}\overset{\overset{\displaystyle O}{\|}}{C}\text{—}NH_2} + H_2O \longrightarrow CO_2 + 2NH_3$$

The half-life of each reaction is the time required for one half of the urea present to be converted to CO_2 and NH_3. Since these are both pseudo–first-order reactions, their half-lives can be compared. At 20.8° C, the urea bound to urease has a half-life of 23 microseconds, whereas the half-life of urea in the absence of any catalyst other than water is approximately 73 years. Therefore, the enzyme urease accelerates the hydrolysis of urea by a factor of 10^{14}. Furthermore, urease is not an exceptional enzyme in this respect; many enzymes produce rate enhancements of this magnitude. This raises the important and intriguing question of how to account for the tremendous effectiveness of enzymes as catalysts. In other words, how does an enzyme work? Like many interesting questions in science, the answer is not definitely known. However, our understanding of the mechanisms of enzyme action has advanced during the past 25 years to the point at which we can at least begin to account for their remarkable catalytic properties.

One approach that has been very fruitful in gaining an understanding of enzyme catalysis has been the study of simple organic reactions and their catalysis by small molecules. The conclusions reached from these studies are that at least four different factors might contribute to the effectiveness of enzymatic catalysis. These factors are (1) general acid-base catalysis, (2) covalent catalysis, (3) approximation and orientation of reactants, and (4) induction of strain or distortion in the substrate or enzyme on forming the enzyme-substrate complex. (A brief discussion of these catalytic mechanisms is presented in Appendix F for those readers not familiar with the terms.)

The magnitudes of the rate enhancements produced by each of the four factors discussed are such that no single factor can account for the remarkable catalytic properties of enzymes. In addition to the approximation and orientation factor, which is a part of all enzyme mechanisms, it appears likely that most enzymes employ at least two other factors as well. The way in which an enzyme can combine these factors will be illustrated by considering two of the better-understood enzyme reaction mechanisms, the actions of chymotrypsin and lysozyme.

EXAMPLES OF MECHANISMS OF ENZYME ACTION

Chymotrypsin

Chymotrypsin is one of a number of hydrolytic enzymes that have a serine residue at the active site. Included in this group of so-called serine hydrolases are trypsin, thrombin, elastase, acetylcholinesterase, subtilisin, liver aliesterase, α-lytic protease, and plasmin. Much of what is known about the mechanism of chymotrypsin action was elucidated by kinetic and chemical studies on the enzyme before the three-dimensional structure was determined by x-ray studies. A proposed mechanism of chymotrypsin action and some of the experimental observations upon which it is based will be considered briefly.

A schematic representation of this proposed mechanism of chymotrypsin action is presented in Figure 4–8. The important functional groups at the active site of the enzyme are the hydroxyl group, contributed by the serine residue occupying position 195 in the amino acid sequence, and the imidazole group of the histidine residue in position 57 in the sequence. In addition, the carboxyl group of the aspartate residue in position 102, which is buried within the protein, is believed to play an important part. As explained in the figure, catalysis by chymotrypsin involves general acid-base catalysis, covalent (nucleophilic) catalysis by Ser 195, and approximation and orientation through formation of an ES complex. It is also possible that strain or distortion may be involved, but the evidence in support of this is not so clear.

The mechanism of chymotrypsin action is one of the most thoroughly studied of all enzyme mechanisms; consequently, there are a number of experimental findings to support a mechanism such as that presented in Figure 4–8. One of the key points of the proposed mechanism is the involvement of Ser 195 and the formation of an acyl-enzyme intermediate. The experimental data in support of this point are very strong. The first observation implicating Ser 195 in the catalytic action of chymotrypsin was the finding that diisopropylfluorophosphate (DFP) reacts stoichiometrically with chymotrypsin to yield a modified enzyme that is totally and irreversibly inhibited. This modified inactive enzyme differs structurally from the native enzyme at only one amino acid residue, Ser 195. The inactive DFP-treated enzyme contains a diisopropylphosphoryl group covalently linked to the oxygen of Ser 195.

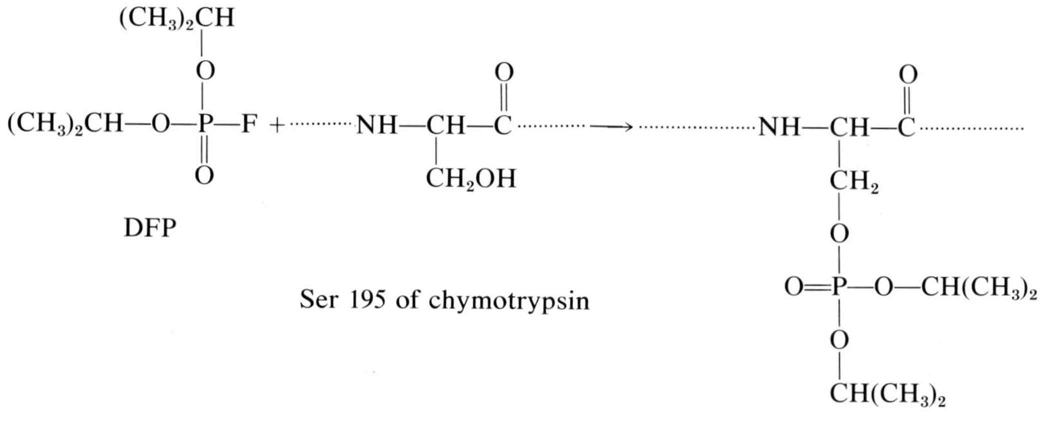

Diisopropylfluorophosphate derivative of chymotrypsin

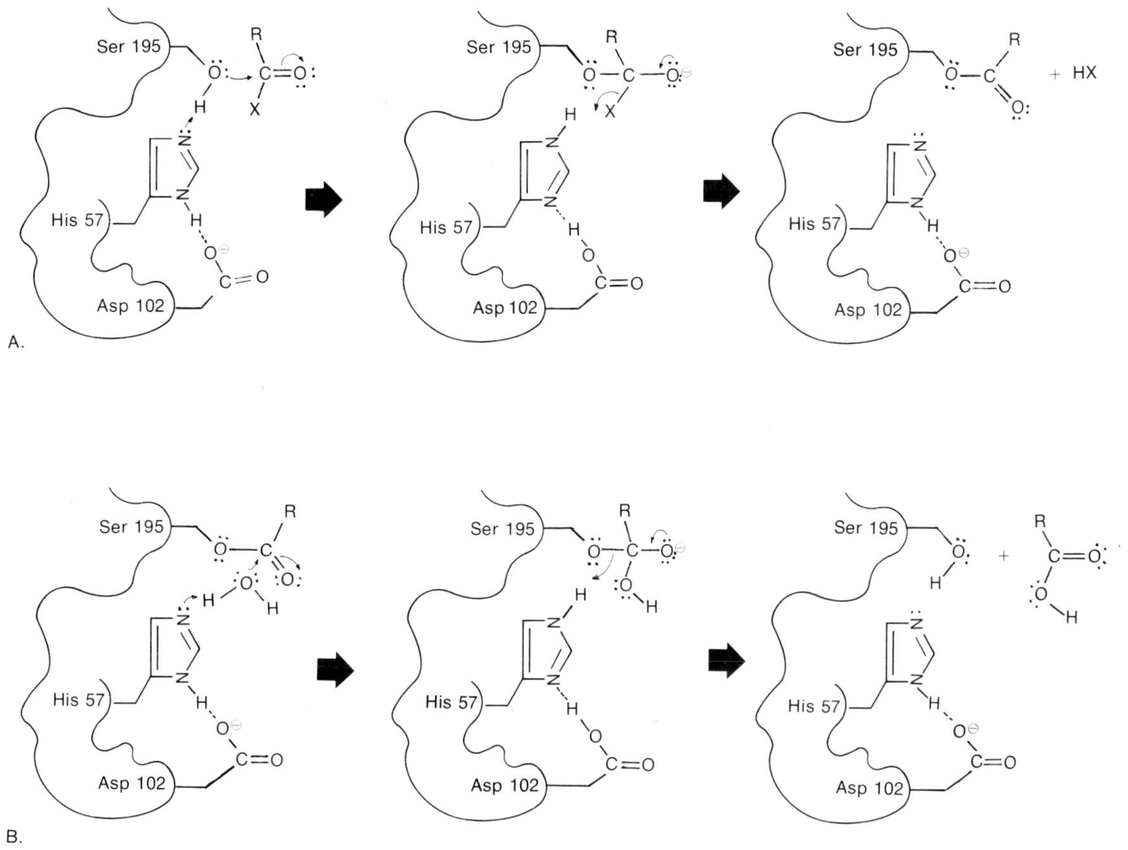

FIGURE 4–8 A proposed mechanism for catalysis by chymotrypsin. The substrate is represented by $R\text{-}\overset{\overset{\displaystyle O}{\|}}{C}\text{-}X$, where C-X is the peptide bond to be cleaved by chymotrypsin. The carboxyl group of the aspartate residue in position 102 is buried in the protein. Its position in a nonaqueous environment is believed to enable it to accept a proton from His 57, when His 57 functions as a general base to abstract a proton from the hydroxyl group of Ser 195. *A*. The structure to the left shows the substrate $R\text{-}\overset{\overset{\displaystyle O}{\|}}{C}\text{-}X$ bound to the "active site" to form the ES complex. The first reaction is the formation of a tetrahedral intermediate, $195\ ser)\text{-}O\text{-}\overset{\overset{\displaystyle R}{|}}{\underset{\underset{\displaystyle X}{|}}{C}}\text{-}O^{-}$, resulting from the nucleophilic attack of the serine hydroxyl group on the carbonyl carbon of the substrate. The nucleophilic attack of the serine hydroxyl group is facilitated by the imidazole group of His 57 and the carboxylate group of Asp 102, which function together to abstract the proton from the serine hydroxyl group as it attacks the carbonyl carbon of the substrate. The second reaction is the collapse of the tetrahedral intermediate, eliminating the C-terminal portion of the peptide substrate as HX and generating the acyl–enzyme intermediate, shown in the right-hand structure. In this structure, the N-terminal portion of substrate molecule is covalently bonded to the serine hydroxyl via an ester linkage. The collapse of the tetrahedral intermediate to form the acyl–enzyme intermediate is catalyzed by the imidazole group of His 57 and the carboxyl groups of Asp 102. They function together as a general acid to donate a proton to the peptide nitrogen of the bond undergoing cleavage, thereby converting X into a good leaving group, HX. *B*. The acyl–enzyme intermediate is cleaved by H_2O, and the N-terminal portion of the substrate is released as $R\text{-}\overset{\overset{\displaystyle O}{\|}}{C}\text{-}OH$, thereby regenerating the "active site" of the enzyme. Again, the imidazole group of His 57, in conjunction with Asp 102, facilitates the reaction, serving first as a general base catalyst, then as a general acid catalyst.

All of the other serine hydrolases are inhibited in like manner by DFP. In fact, DFP, which was developed as a chemical warfare agent, owes its great toxicity to the fact that it inhibits acetylcholinesterase. Kinetic data, and the fact that the acyl-enzyme intermediate can be isolated under certain experimental conditions, also strongly support the conclusion that acylation of Ser 195 is an essential part of the reaction mechanism of chymotrypsin.

The possible involvement of a histidine residue in the mechanism of action of chymotrypsin was first suggested by the pH-rate profile for the enzyme. As can be seen in Figure 4–9, the ascending limb of this pH-rate profile resembles a titration curve for an acid with a pK_a of approximately 7. One possible explanation for the ascending limb is that the enzyme is active only when some acidic group on the enzyme with a $pK_a \approx 7$ has lost its proton. The acidic group on a protein with a pK_a normally in this range is the imidazole group of a histidine residue. More conclusive evidence for involvement of a histidine residue at the active site was obtained by reacting chymotrypsin with the following chloroketone:

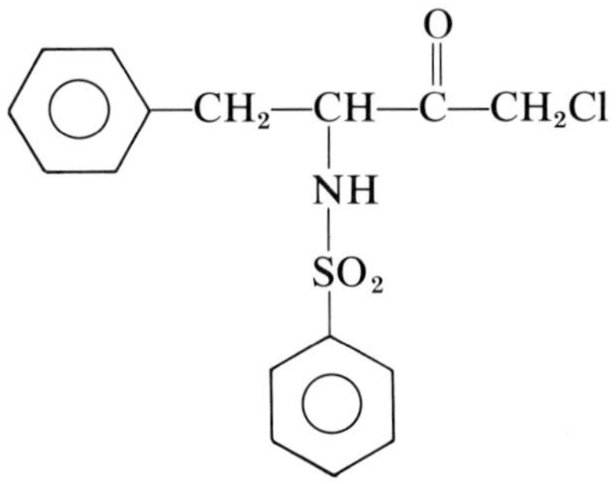

Chloroketones are strong alkylating agents capable of alkylating amino groups, sulfhydryl groups, imidazole groups, and others. Chymotrypsin is specific for the hydrolysis of peptide bonds in which the carboxyl group of an aromatic amino acid participates. It was reasoned, therefore, that this particular chloroketone, which bears a structural resemblance to phenylalanine, should rapidly bind at the active site of chymotrypsin and specifically alkylate only those groups present at the active site. Indeed, this proved to be the case. Treatment of chy-

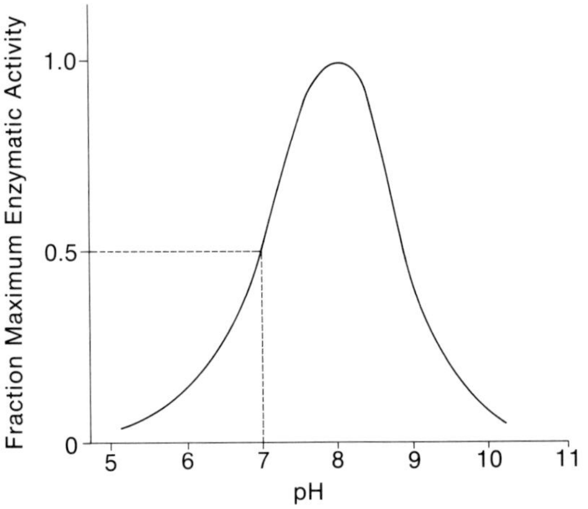

FIGURE 4–9 The pH-rate profile for the chymotrypsin-catalyzed hydrolysis of substrate.

motrypsin with low concentrations of the chloroketone resulted in the alkylation of a single amino acid residue of the enzyme and an irreversible loss of enzymatic activity. Through amino acid analysis and peptide mapping studies, it was found that the single amino acid residue covalently modified by the chloroketone was His 57. These results not only place His 57 at the active site of the enzyme but also implicate His 57 in the reaction mechanism of the enzyme, since all enzymatic activity was lost when His 57 was alkylated. Reagents such as this chloroketone that make use of the specificity of an enzyme are known as *active-site-directed reagents*, and modification of the active site by such a reagent is called *affinity labeling*.

The three-dimensional structure of chymotrypsin, as determined by x-ray crystallography, verified the presence of Ser 195 and His 57 at the active site of the enzyme. This three-dimensional structure also revealed the presence of Asp 102 in a position to hydrogen-bond to His 57. Interestingly, this same arrangement of a serine, a histidine, and a buried aspartate has now been found in a number of the other serine hydrolases, including elastase, subtilisin, and trypsin.

In summary, approximation and orientation, nucleophilic catalysis, and general acid-base catalysis appear to be essential factors in the catalytic mechanism of chymotrypsin. All of these essential factors in the catalytic mechanism result from the ability of the protein to organize or structure precisely a volume of space.

Lysozyme

The mechanism of lysozyme action furnishes an example of the role of strain in enzymatic catalysis. Lysozyme is an enzyme of 14,600 molecular weight, containing 129 amino acid residues. Lysozyme catalyzes the hydrolysis of $\beta(1 \rightarrow 4)$ glycosidic bonds of a complex heteropolysaccharide occurring in bacterial cell walls. From studies of the hydrolysis of simple glycosides, it has been deduced that acid-catalyzed hydrolysis of the glycosidic bond proceeds via the formation of an unstable carbonium ion, $\rangle C^{+}—$, as depicted in Figure 4–10.

As shown in Figure 4–11, lysozyme binds six of the monomeric units of its polysaccharide substrate, and strain induced by the binding facilitates the formation of the carbonium ion intermediate. Thus, as described in Figure 4–11, the proposed mechanism of lysozyme catalysis employs (1) orientation and approximation through formation of the ES complex, (2) strain, (3) general acid-base catalysis, and (4) electrostatic stabilization of a carbonium ion intermediate. The importance of the abilty of proteins to structure precisely a volume of space is again evident.

SPECIFICITY OF ENZYMES

The striking specificity displayed by enzymes also depends upon the ability of a protein to structure precisely a volume of space. As might be expected, a large part of the specificity of enzymes for a particular substrate, or class of substrates, resides in the specificity of the binding process whereby the enzyme-substrate complex is formed, which is an essential first step in all enzyme-catalyzed reactions. The arrangement of charged groups, groups capable of hydrogen bonding, hydrophobic groups, large bulky groups, and small groups all contribute

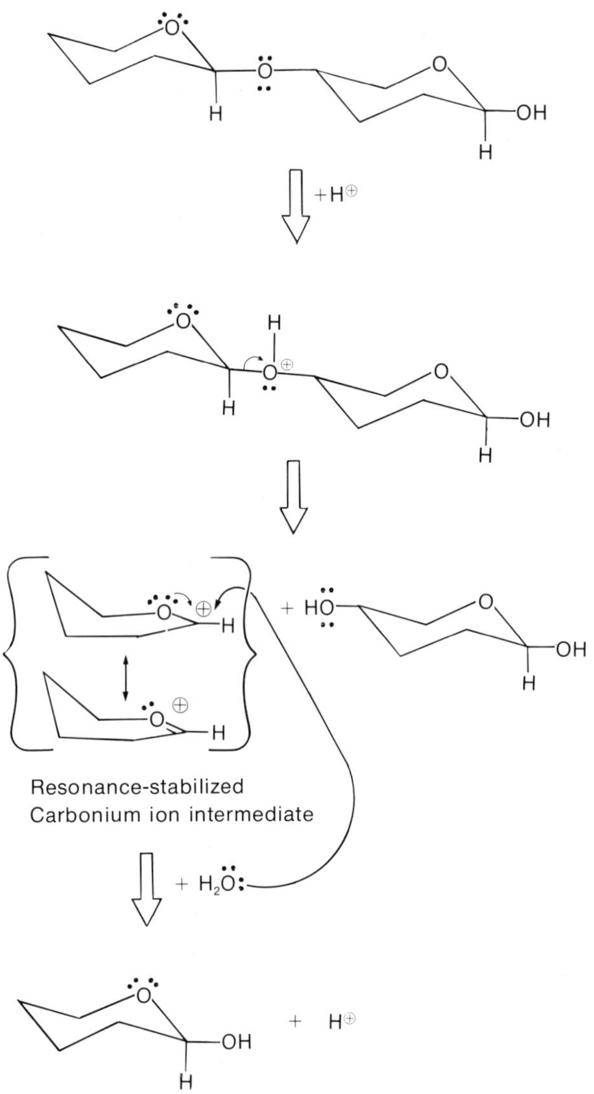

FIGURE 4–10 Mechanism for the acid-catalyzed hydrolysis of a glycosidic bond. The carbonium ion is a very unstable, highly reactive intermediate which cannot be isolated. In order for a carbonium ion to exist even fleetingly, the three bonds to the positively charged, electron-deficient carbon must be coplanar. The coplanarity of these three bonds permits some resonance stabiliziation of the very unstable carbonium ion. Hence, a monosaccharide, which normally exists in a chair conformation, must assume a less stable half-chair conformation if C-1 of the monosaccharide is to be converted to a carbonium ion.

to the binding site, and the topography of the binding site on the enzyme determines which molecules the enzyme will bind to form the enzyme-substrate complex.

Although the specificity in binding undoubtedly plays an important role, enzyme specificity often appears to involve additional complex factors. Binding of substrate to form the ES complex is only the first step in the process; the second step involves the actual bond-breaking and bond-making processes. For the second step to occur rapidly and efficiently, that part of the substrate in which the bond-making and bond-breaking reactions occur must bear the proper spatial relationship to the groups on the protein that participate in the catalytic process. There is now evidence that the binding of substrate to the active site of some enzymes actually induces conformational changes in the enzyme. This is known as the *induced-fit* hypothesis. Thus a ''poor'' substrate might bind very well to the enzyme to form an enzyme-substrate complex and yet not induce the subsequent precise conformational changes required for maximum catalytic effectiveness that a ''good'' substrate induces.

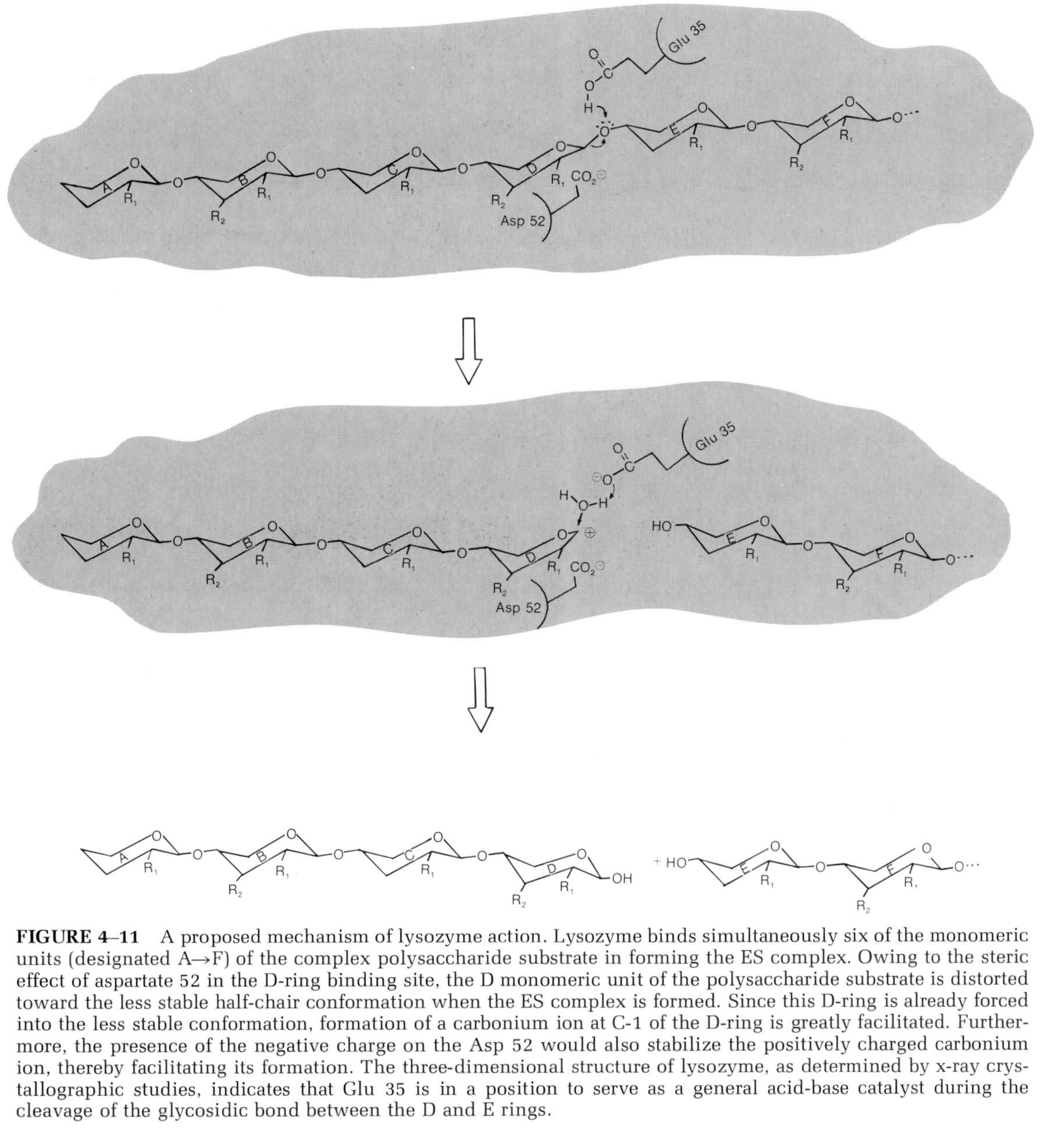

FIGURE 4–11 A proposed mechanism of lysozyme action. Lysozyme binds simultaneously six of the monomeric units (designated A→F) of the complex polysaccharide substrate in forming the ES complex. Owing to the steric effect of aspartate 52 in the D-ring binding site, the D monomeric unit of the polysaccharide substrate is distorted toward the less stable half-chair conformation when the ES complex is formed. Since this D-ring is already forced into the less stable conformation, formation of a carbonium ion at C-1 of the D-ring is greatly facilitated. Furthermore, the presence of the negative charge on the Asp 52 would also stabilize the positively charged carbonium ion, thereby facilitating its formation. The three-dimensional structure of lysozyme, as determined by x-ray crystallographic studies, indicates that Glu 35 is in a position to serve as a general acid-base catalyst during the cleavage of the glycosidic bond between the D and E rings.

PHENYLKETONURIA—ONE OF THE MANY GENETIC DISEASES RESULTING FROM AN ENZYME DEFICIENCY

A number of diseases are due to inborn errors of metabolism. In these diseases the genetic information in the DNA of the affected individual is such that a particular enzyme is not synthesized, or if synthesized is not normal. One of the most common diseases of this type is phenylketonuria. The classic form of this disease results from a deficiency in phenylalanine hydroxylase, an enzyme that is normally present in the endoplasmic reticulum of the liver cell. Phenylalanine hydroxylase catalyzes the following reaction.

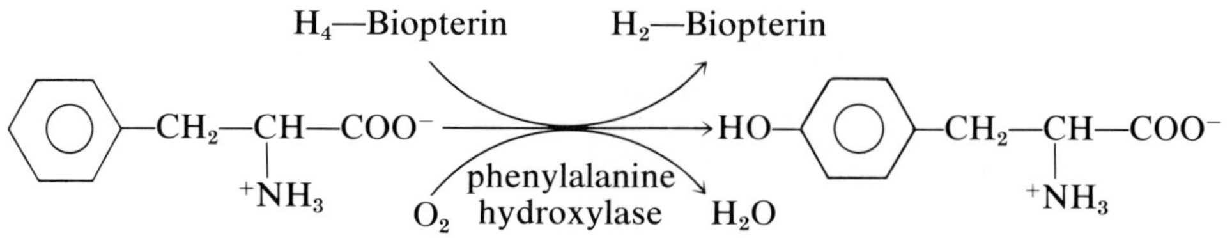

Phenylalanine Tyrosine

In addition to phenylalanine and the enzyme, the reaction also requires molecular oxygen and a reducing agent, tetrahydrobiopterin.

Phenylalanine is an essential amino acid and must be present in the human diet for protein synthesis to occur. Most diets contain more phenylalanine than is required to meet the demands of protein synthesis, and the excess phenylalanine is metabolized to CO_2, H_2O, and urea via a pathway that leads to the generation of energy. The first step in this metabolic pathway is the conversion of phenylalanine to tyrosine, which requires phenylalanine hydroxylase. If phenylalanine hydroxylase is absent or markedly deficient, phenylalanine cannot be metabolized by this pathway and accumulates in the body. This accumulation leads to an increased rate of conversion of phenylalanine to other metabolites, such as phenylpyruvate,

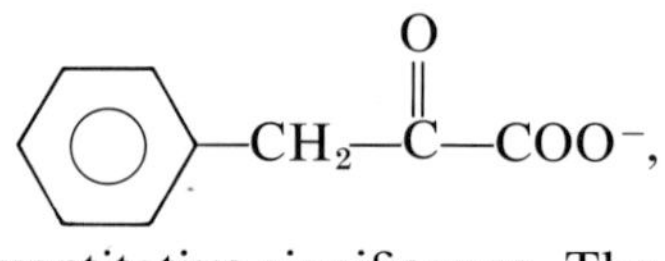

via metabolic pathways that are normally of minor quantitative significance. The increased concentrations of phenylalanine, phenylpyruvate, and other metabolites in the blood lead to an increased concentration of these substances in the urine. Therefore, the disorder is named phenylketonuria.

The onset of the disease is in the neonatal period, and if treatment is not begun in the first few weeks after birth, severe mental retardation can result. Mental retardation probably results from deficient myelination in the central nervous system, although the way in which elevated levels of phenylalanine and its metabolites interfere with myelination is not completely understood at present.

Treatment of the disease consists of limiting the dietary intake of phenylalanine to an amount just sufficient to meet the demands of protein synthesis. At the age of five years, when myelination is complete, a return to a normal diet does not appear to seriously interfere with subsequent normal mental development. However, it is important that a phenylketonuric mother return to the restricted diet during pregnancy to prevent possible damage to the developing fetus.

THE USE OF ENZYMES AS DIAGNOSTIC AIDS

Damage to tissues and cell death both result in a release of intracellular enzymes from damaged cells into the blood. Therefore, the concentrations of these enzymes increase in the blood serum and can serve as a valuable diagnostic aid in a number of diseases, including myocardial infarction, pancreatitis, liver disease, and prostatic cancer.

The serum concentrations of three enzymes are often used in the diagnosis of myocardial infarction. These are creatinine phosphokinase (CPK), glutamate-oxaloacetate transaminase (GOT), and lactate dehydrogenase (LDH). In more than 95 per cent of patients with myocardial infarction, serum levels of GOT rise rapidly and then return to normal in four to five days. The serum levels of CPK characteristically rise and fall even more rapidly than GOT levels. However, the serum levels of LDH typically rise and fall more slowly, returning to normal levels in approximately 10 days.

Lactate dehydrogenase is a particularly interesting and useful enzyme for diagnostic purposes because it can be separated by electrophoresis into five different LDH enzymes. These are referred to as *isozymes*, and they are designated LDH-1, LDH-2, LDH-3, LDH-4, and LDH-5 on the basis of their electrophoretic mobility. The LDH isozymes each contain four subunits. There are two types of polypeptide chains, designated H and M, that can form an LDH subunit. The five isozymes of LDH arise because there are five ways in which two different subunits can be combined to form a tetramer. The structures of the five isozymes are

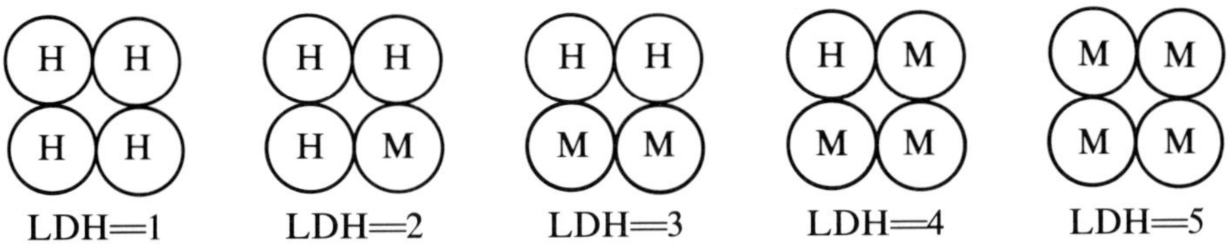

The relative concentrations of the LDH isozymes differ from one type of tissue to another, and the relative composition is characteristic of the tissue. For example, LDH-1 plus LDH-2 makes up approximately 60 per cent of the lactate dehydrogenase of cardiac muscle, while LDH-4 plus LDH-5 constitutes approximately 80 per cent of the lactate dehydrogenase of liver. Thus a diagnosis based on elevated serum LDH can be made more specific with respect to the organ involved by an electrophoretic analysis of the LDH isozymes. For example, the elevated serum level of LDH in a myocardial infarct is primarily due to an increase in LDH-1 and LDH-2, while in liver disease the elevated serum LDH is primarily due to increases in LDH-4 and LDH-5.

FURTHER READING

Bernhard, S. A.: The Structure and Function of Enzymes. Menlo Park, Calif., W. A. Benjamin, 1968.

Boyer, P. D. (ed.): The Enzymes. Vols. I and II. New York, Academic Press, 1970.

Bruice, T. C.: Some pertinent aspects of mechanism as determined with small molecules. Annu. Rev. Biochem., *45*:331, 1976.

Bruice, T. C., and Benkovic, S. J.: Bioorganic Mechanisms. Vols. I and II. Menlo Park, Calif., W. A. Benjamin, 1966.

Dickerson, R. E., and Geis, I.: The Structure and Action of Proteins. New York, Harper & Row, 1969.

Fersht, A.: Enzyme Structure and Mechanism. San Francisco, W. H. Freeman, 1977.

Gutfreund, H.: Enzymes: Physical Principles. New York, John Wiley and Sons, 1972.

Jencks, W. P.: Catalysis in Chemistry and Enzymology. New York, McGraw-Hill, 1969.

Koshland, Jr., D. E., Nemethy, G., and Filmer, D.: Comparison of experimental binding data and theoretical models in proteins containing subunits. Biochemistry, *5*:365, 1966.

Monod, J., Wyman, J., and Changeux, J.-P.: On the nature of allosteric transitions: a plausible model. J. Mol. Biol., *12*:88, 1965.

Phillips, D. C.: The three-dimensional structure of an enzyme molecule. Sci. Am., *215*:78, 1966.

Piszkiewicz, D.: Kinetics of Chemical and Enzyme-Catalyzed Reactions. New York, Oxford University Press, 1977.

Segel, I. H.: Enzyme Kinetics. Behavior and Analysis of Rapid Equilibrium and Steady-State Enzyme Systems. New York, John Wiiley and Sons, 1975.

Structure and Function of Proteins at the Three-Dimensional Level. Cold Spring Harbor Symp. Quant. Biol., *36*, 1972.

5

DNA SYNTHESIS, THE CELL CYCLE, AND CELL DIVISION

THE FUNCTIONS OF THE NUCLEUS: DNA AS THE GENETIC MATERIAL.

The nucleus is the cellular organelle primarily responsible for the storage and transmission of genetic information. Inside each nucleus is a complete set of all the instructions required for the construction of an individual organism and for its proper functioning. Furthermore, these instructions are contained in a form that can be copied exactly and transmitted to succeeding generations. At the level of the whole organism or of the cell this is a familiar observation. Like begets like; children are found to resemble their parents, and the daughters produced by division of a single cell bear a close resemblance to the original cell. At the molecular level, however, complex processes underlie the storage, expression and transmission of genetic information. These processes will be described in some detail in this chapter and the next. Study of the biochemical basis of heredity has come to be known as molecular biology or molecular genetics, and it is currently a very active field of research.

Our understanding of the molecular basis of inheritance is usually thought of as beginning with the identification of the genetic material itself, the chemical form in which hereditary information is stored and replicated. By the early part of this century it was known, as a result of the work of Mendel, that expression of individual phenotypic traits is under the control of paired factors or genes which are transmitted in unchanged form from one generation to the next. Early cytologists, including Sutton and Weismann, had observed that the intranuclear elements called chromosomes behave during cell division and sexual recombination in just the way that Mendel's genes were postulated to behave. Interphase cells were found to contain two copies of each chromosome, and these were replicated exactly and distributed equally to the two daughter cells during mitosis. It was common knowledge to early molecular geneticists that this chromosomal behavior was not accidental, but reflected the fact that genes were associated in some way with chromosomes. The basic problem they confronted, therefore, was to understand in what chemical form the genetic information is carried on chromosomes and how it is replicated. At the same time, it was of great interest to know how genetic information is expressed. That is, what do genes do, and how do chromosomal genes influence phenotypic traits, such as eye color and body structure?

In principle, the chemical nature of the genetic material ought to have been the easier of these problems to solve. Since chromosomes were known to contain only deoxyribonucleic acid (DNA) and protein, it was clear that the genetic material had to be contained in one or the other of these substances. In practice, however, it proved to be very difficult to determine which of the two it was. The question was not really answered until 1944, and the results were not universally accepted until the early 1950's.

The crucial experiments proving that DNA is the hereditary material involved genetic transformation of the bacterium *Streptococcus pneumoniae*. These organisms can form colonies of two morphological types, smooth and rough, depending upon whether individual cells have a polysaccharide capsule. Smooth cells have a capsule and form smooth colonies, while rough cells lack a capsule and form rough colonies. This property is related to their behavior *in vivo*; smooth cells are virulent and cause pneumonia, while rough cells do not cause disease. The presence of a capsule is a genetically determined trait in smooth cells. Loss or alteration of the genes governing capsule synthesis can result in the formation of rough cells that breed true.

In 1928 Griffith showed that if rough strains were mixed with heat-killed (boiled) smooth cells, smooth cells could be recovered from the otherwise rough population. Smooth cells produced in this way were found to give rise only to smooth progeny. Clearly, something from the heat-killed smooth cells had genetically transformed rough cells to the smooth phenotype. In the early 1940's, Avery, MacLeod, and McCarty chemically fractionated the heat-inactivated smooth cell preparation to determine which component was responsible for the ability to transform rough cells. They found that transformation could be accomplished with purified DNA and that cellular fractions lacking DNA would not transform. Transformation activity could be abolished by treatment of heat-killed cells with the enzyme deoxyribonuclease, which specifically degrades DNA. These experiments proved conclusively that DNA itself is the chemical form in which genetic information is carried and that information can be passed from one cell to another in this form.

A very general answer to the question, How are genes expressed? also became available in the mid-1940's. Although many lines of investigation pointed toward the same result, the most conclusive experiments were carried out by Beadle and Tatum using the bread mold *Neurospora crassa*. These investigators examined the effect of genetic mutations on the functioning of biosynthetic pathways leading to the production of low molecular weight metabolic products, such as vitamins, amino acids, purines, and pyrimidines; an example of a hypothetical pathway of this type is shown here. Beadle and Tatum observed that mutations in single genes

$$\mathrm{A} \xrightarrow{\text{Gene a}} \mathrm{B} \xrightarrow{\text{gene b}} \mathrm{C} \xrightarrow{\text{gene c}} \mathrm{D}\ \text{(product)}$$

could result in the accumulation of one, and only one, of the metabolic intermediates in this type of pathway. For example, a mutation in gene a would cause accumulation of A, and strains carrying that mutation would be unable to make

B, C, or D. Such strains were therefore metabolically dependent for growth (auxotrophic) on exogeneously added B, C, or D. Similarly, mutations in gene b would result in accumulation of B, and these strains would require either C or D.

At the time it must have been very exciting to appreciate this very simple relationship between genes and metabolism. One gene controlled one biosynthetic step. Such a simple relationship must have seemed then to be much less likely than a more complicated situation in which, for example, many genes might interact to produce the observed biochemical (phenotypic) effect. However, since biochemical conversions of the type studied by Beadle and Tatum were known to be mediated by enzymes, their results suggested a simple relationship between genes and enzymes. One gene specified one enzyme. The clear implication was that the business of genes is to specify the structure of enzymes. Since enzymes were known to be proteins, it followed that genes function by determining the structure of proteins, that is, genetic information is expressed in the form of proteins.

This chapter will consider in more detail the structure of DNA, its replication, and the transmission of genetic information. The way in which the information in genes is expressed, and particularly the role of RNA, will be discussed further in Chapter 6.

THE CHEMICAL STRUCTURE OF DNA

Once it was appreciated that DNA is the genetic material and that genes function by specifying the structure of proteins, it became possible to ask much more detailed questions about the biochemical basis of heredity. For example, it was important to know the biochemical details of how genetic information is stored in DNA and how the DNA is replicated. It was also urgent to understand how information stored in DNA could be expressed as protein. Clearly, a change in the biochemical form of the genetic program was required; the language of the nucleic acids had to be "translated" into the language of proteins. This problem was complicated in the case of eukaryotic organisms by the fact that while DNA is ordinarily sequestered in the nucleus, most protein synthesis was known to take place in the cytoplasm.

In attempting to answer these questions it is reasonable to begin with the chemical structure of the genetic material itself, the DNA, since this is the ultimate reservoir of genetic information. The correct structure of DNA was first proposed in 1953 by Watson and Crick, who had analyzed the results of x-ray diffraction experiments carried out by Wilkins on purified fibers of DNA. Determination of the DNA structure in this way proved to be one of the most important contributions to the field of molecular biology. Watson and Crick found that individual DNA molecules consist of two polynucleotide chains. Each chain consists of a backbone containing alternating sugar and phosphate groups, and attached to each sugar is either a purine or a pyrimidine base (see Chapter 2). The sugars are all 2-deoxyribose, and they are connected to each other by phosphoric acid groups which extend from the 5 carbon of one sugar to the 3 carbon of the next, thereby forming a phosphodiester bond. The individual polynucleotide chains therefore have direction; one end contains deoxyribose with a free 5 hydroxyl group, while the other end has deoxyribose with a free 3 hydroxyl. Individual chains also have a net negative charge owing to the phosphate groups.

Bases are attached to the 1 carbon of deoxyribose, and they are of the four types described in Chapter 2; adenine, guanine, thymine, and cytosine. The two purine bases, adenine (A) and guanine (G), are attached to deoxyribose at N7, while the pyrimidines, thymine (T) and cytosine (C), are attached to N3 (Fig. 5–1). The carbon and nitrogen atoms in DNA are numbered just as they are in the separate sugar and purine bases, except that the atoms in deoxyribose are indicated with a primed number. Therefore, one speaks of 2′-deoxyribose or adenosine-5′-monophosphate.

In the overall structure of DNA the two individual polynucleotide chains are associated by hydrogen bonds which link, in a very precise way, purine bases on one strand with pyrimidine bases on the other. A binds only to T, and G binds only to C. Since the bases are conjugated aromatic molecules, all their atoms are coplanar and the individual bases are flat. In DNA the bases are associated with each other in such a way that the overall base pairs are also flat. The base pairs are stacked on top of one another at the center of the DNA molecule, with the planes of the base pairs perpendicular to the long axis of the DNA (Fig. 5–2). The two sugar-phosphate backbones lie outside the stacked base pairs and run in opposite directions. The two chains are twisted into a helix, so that the overall DNA molecule is a very long cylinder approximately 20 Å in diameter. Each turn of the helix includes 10 base pairs, and it involves a translation of 34 Å along the cylinder; base pairs are therefore separated from each other by a translation of 3.4 Å. Whole DNA molecules are very long and contain many base pairs. For example, in bacteria all the genes are found on one DNA molecule, which may be 1 mm in length and contain 3×10^6 base pairs. Similarly, the DNA in a single human chromosome may be 4 to 5 cm long and contain 15×10^7 base pairs.

The most important feature of DNA structure is the exact way in which the purine and pyrimidine bases are hydrogen-bonded to each other. Since A always

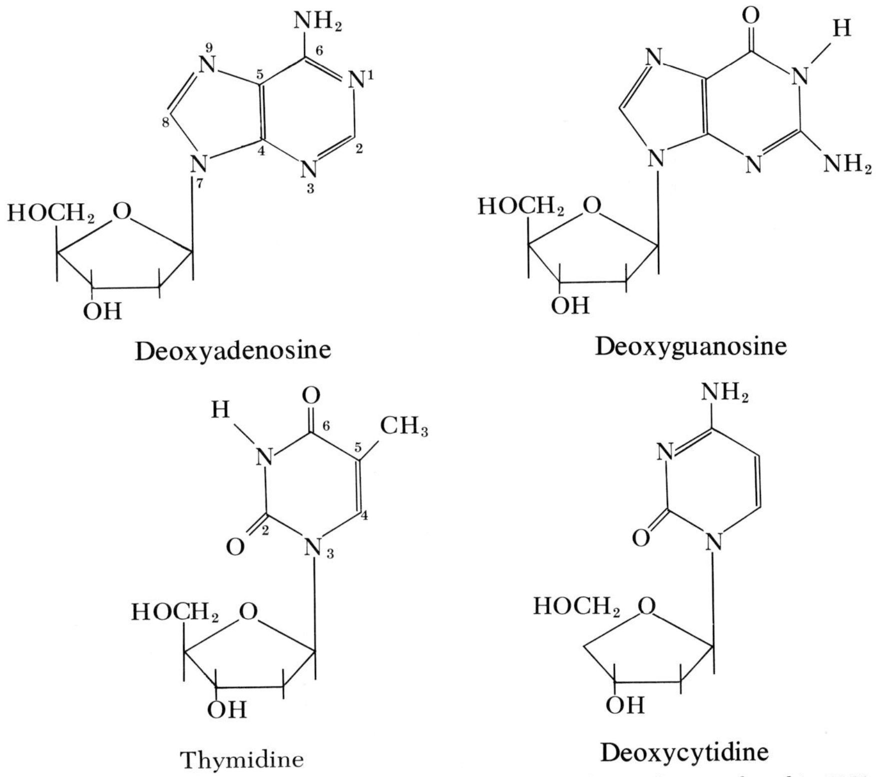

FIGURE 5–1 Attachment of the purine and pyrimidine bases to 2-deoxyribose, as found in DNA.

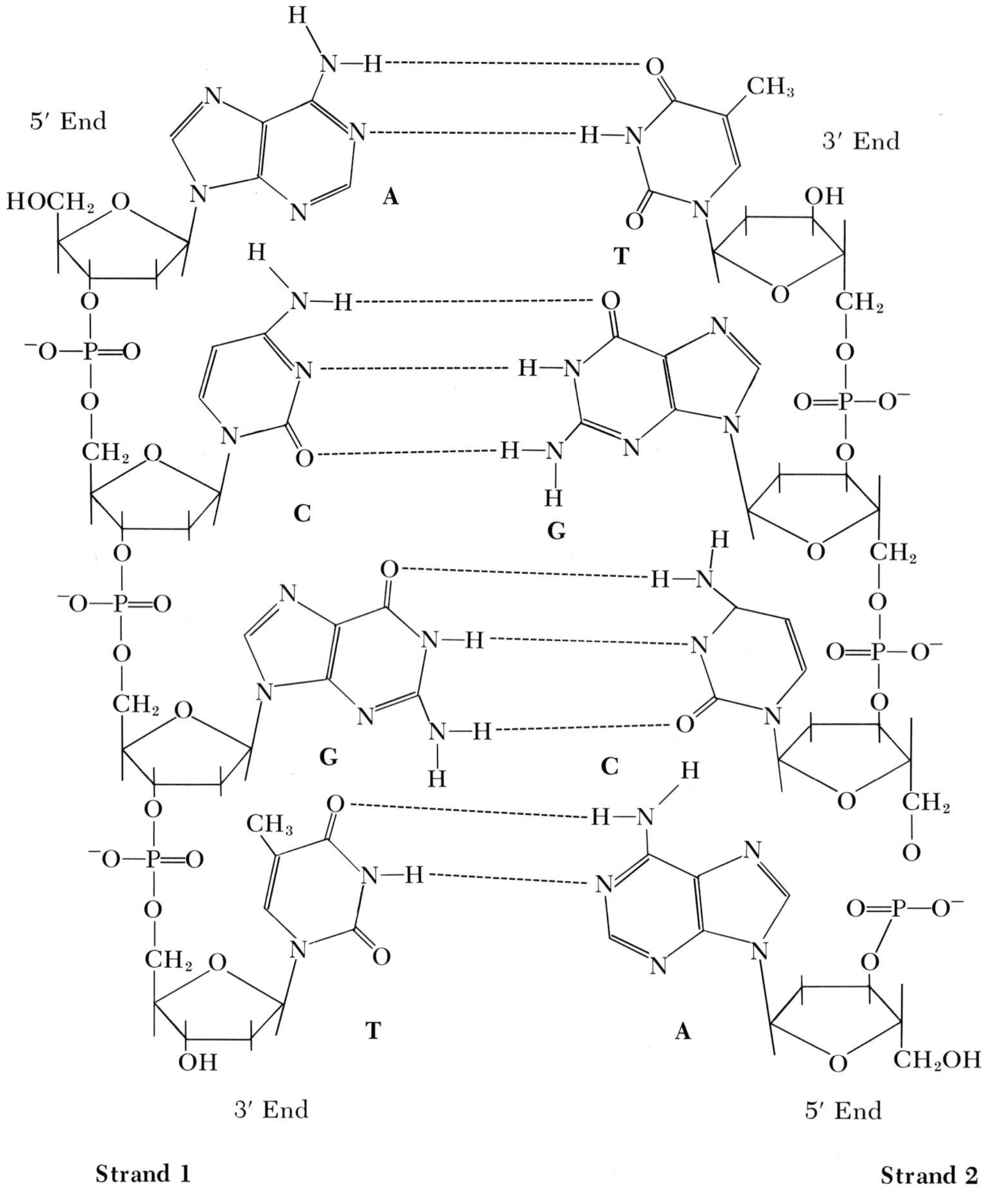

FIGURE 5–2 The chemical structure of DNA. Note that the two polynucleotide strands run in opposite directions and that base pairs are formed only between A and T or between G and C.

pairs with T, and G always pairs with C, the overall content of A in DNA must be equal to T, and the content of G must equal C. With that restriction, however, the base compositions of different DNA's are found to vary widely; different DNA's have different proportions of GC compared with AT base pairs (Table 5–1). More important, however, the exact sequence of bases in DNA also varies enormously from one organism to another. Different DNA's with the same overall base composition have bases arranged in quite different sequences.

It is now common knowledge that these sequence differences are not accidental, but reflect the fact that genetic information is contained in the exact sequence of nucleotide bases in DNA. The order of the four possible DNA bases

TABLE 5–1

BASE COMPOSITION OF DNA

DNA Source	% G + C
Staphylococcus aureus (bacterium)	33%
Cow	39%
Human	41%
Mouse	45%
Escherichia coli (bacterium)	50%
Herpes simplex virus	65%

is related by the genetic code to the sequence of the 20 possible amino acids found in proteins. A series of three bases is required to specify each amino acid, and these triplet code words or "codons" are arranged in a linear, nonoverlapping fashion along the DNA in the same order as the amino acids in the protein they specify. It is clear from the structure of DNA, however, that this information must be contained on both polynucleotide strands and that these two forms of the genetic information are related to each other by the rules of base complementarity. Although only one of these two forms of the genetic information, the "sense" strand, can be properly interpreted by the cell's protein synthetic machinery, the total information content of both strands is obviously the same (Fig. 5–3).

REPLICATION OF DNA

The biochemistry of DNA replication has been most extensively studied in the bacterium *Escherichia coli*. The *E. coli* genome is a covalently closed circle of DNA approximately 1 mm in length (1000 times the length of the organism itself), containing 3×10^6 base pairs. Its molecular weight is 2×10^9, and it can code for approximately 2000 to 5000 proteins. DNA replication in *E. coli* begins from a fixed point on the circular genome and proceeds in two directions at once.

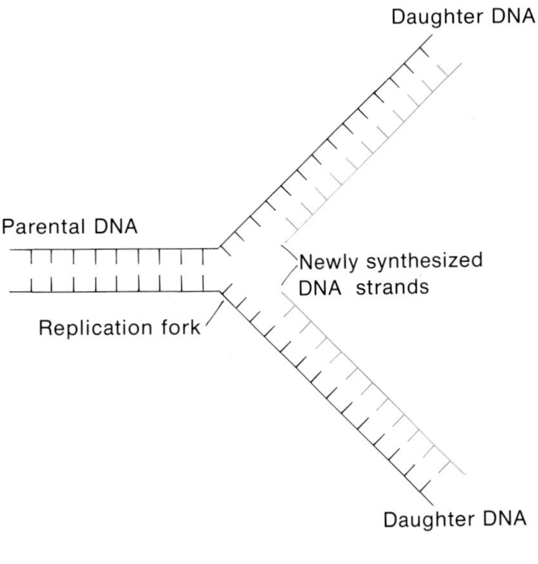

FIGURE 5–3 Separation of parental DNA strands and their use as templates during DNA replication.

There are, therefore, two replication forks of the type described earlier, and these proceed around the circular DNA in opposite directions, as shown in Figure 5–4. Since both parental DNA strands are copied at each replication fork, the entire genome has been replicated when the two forks meet halfway around the circle. The two daughter DNA molecules then separate and are distributed to different cells. At each replication fork the parental DNA is copied by the addition of new nucleoside-5′-triphosphates to the growing progeny polynucleotide chain, as shown in Figure 5–5. One molecule of inorganic pyrophosphate is lost at each step, leaving a 3′-5′ phosphodiester bond in the extended DNA chain. DNA chain growth is always in the 5′ to 3′ direction, and new bases are added very rapidly; approximately 800 nucleotides per second are added to each growing polynucleotide strand.

As a result of recent experimental studies a great deal is now known about the complicated sequence of molecular events that occurs at the two DNA replication forks in *E. coli*. For example, it is now well established that DNA replication is discontinuous, in the sense that new DNA strands are synthesized in short segments of approximately 1000 nucleotides each. After their synthesis, these "Okazaki fragments" (named in recognition of their discoverer) are joined enzymatically onto nascent DNA strands, as shown in Figure 5–6.

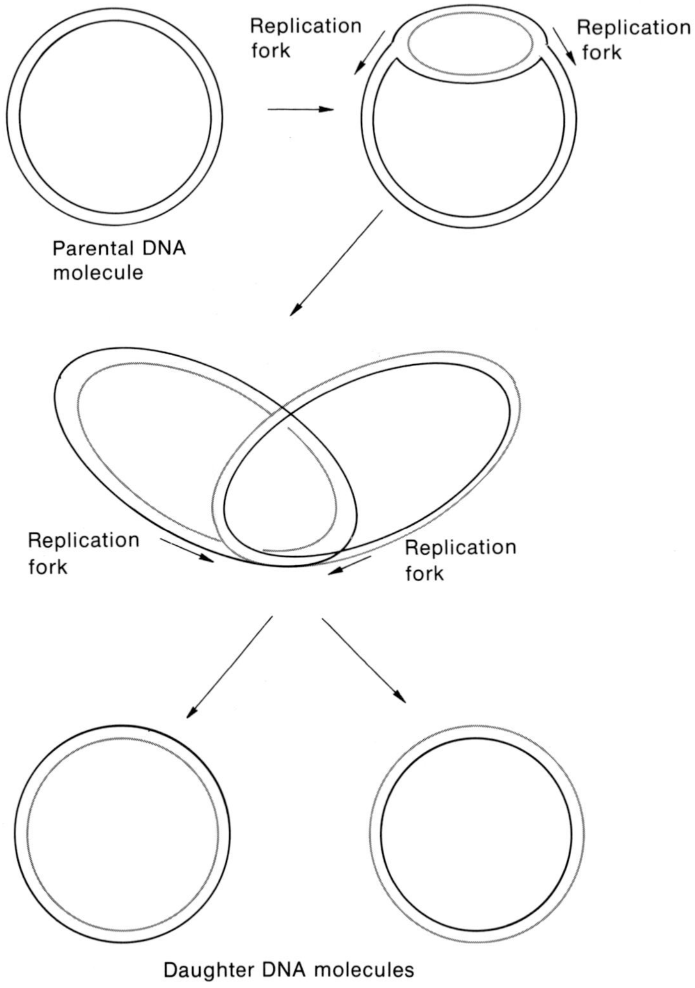

FIGURE 5–4 Bidirectional replication of DNA in *E. coli*.

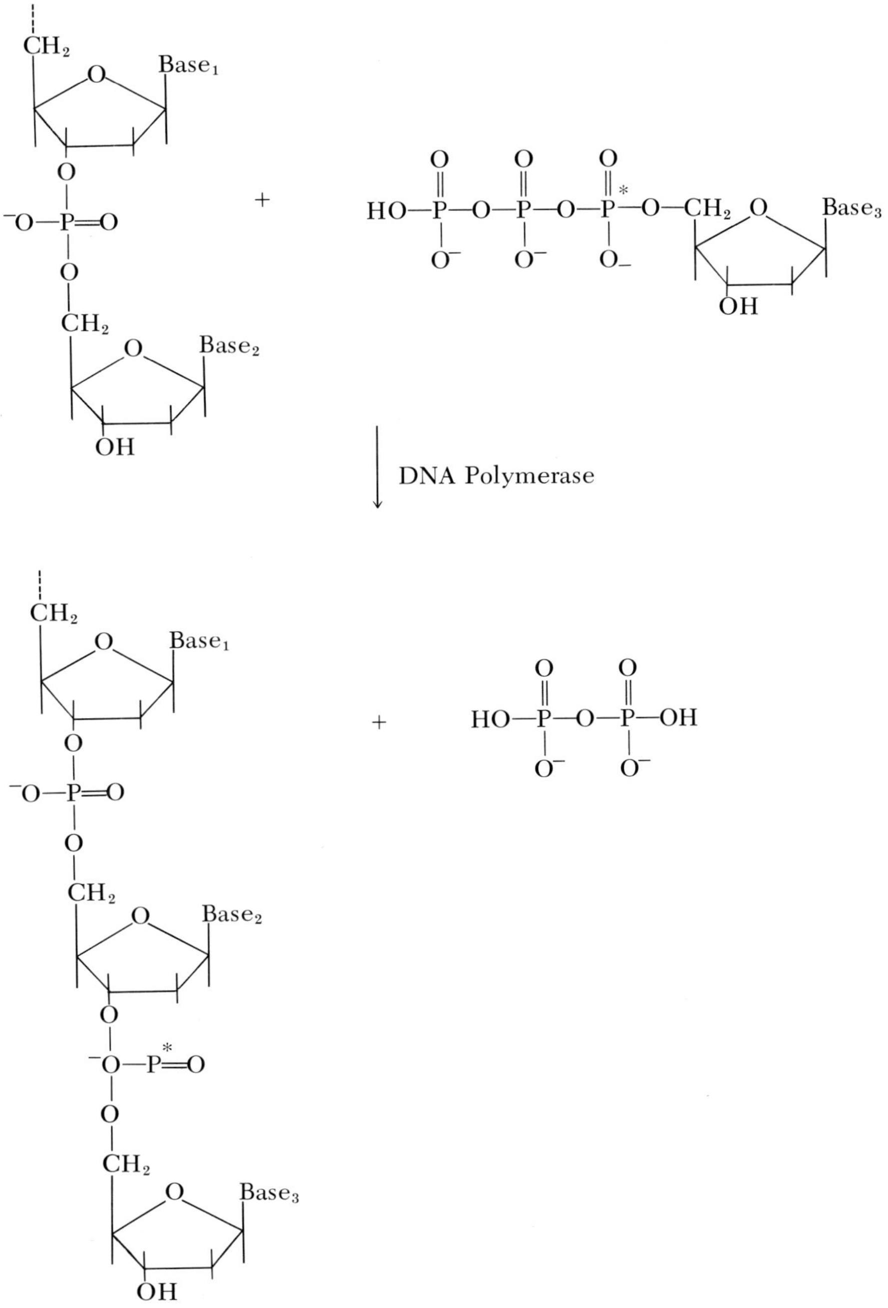

FIGURE 5–5 Extension of a DNA chain by one nucleotide as it occurs in DNA replication. The asterisk indicates the phosphate in the precursor nucleoside-5′-triphosphate that forms the new 3′-5′ phosphodiester bond.

Elongation of a DNA chain is accomplished by cyclic repetition of this process of template-dependent fragment synthesis, followed by ligation of the fragment onto a growing polynucleotide strand. Each cycle is initiated when the two parental DNA strands are separated for a distance (approximately 1000 nucleotide pairs or 0.34 μm) corresponding to the length of the fragments to be synthesized.

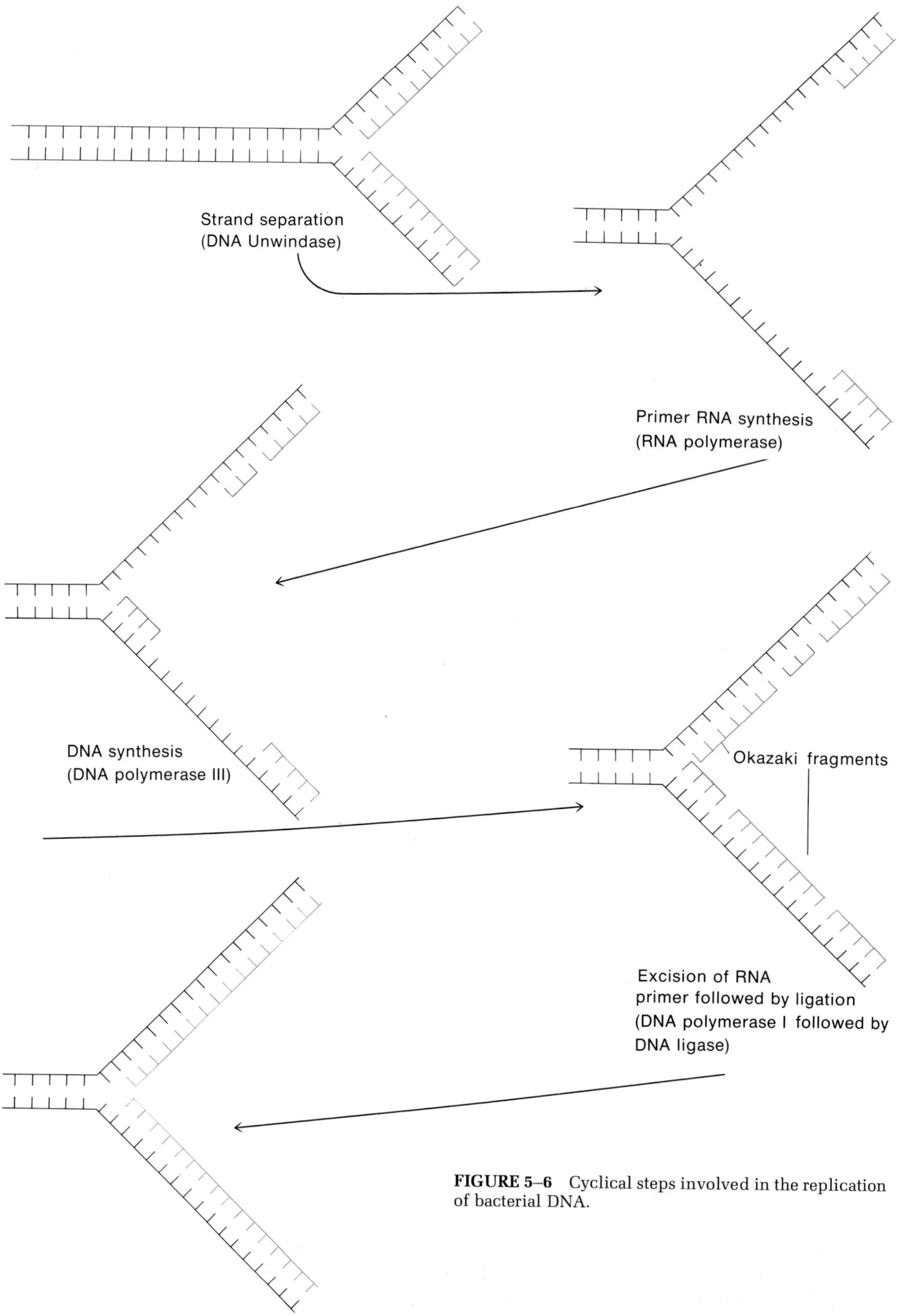

FIGURE 5–6 Cyclical steps involved in the replication of bacterial DNA.

A special protein called DNA unwindase is involved in physically separating the two parental DNA strands. After strand separation, the two parental DNA chains are copied in a template-dependent fashion by an enzyme, DNA polymerase III, which synthesizes DNA strands beginning at the 5′ end and proceeding toward the 3′ end. Since the two parental DNA strands run in opposite directions, this implies that one progeny DNA fragment must be synthesized in the same direction as overall chain growth, while the other is synthesized in the opposite direction. The newly synthesized DNA chains remain base-paired to the parental template DNA strand at all times after their synthesis.

To insure that a proper AT or GC base pair has been made at each step of polymerization, each newly formed base pair is tested by an "editing" function also carried out by DNA polymerase III. If a mistake has been made and an incorrect nucleotide has been added onto the growing polynucleotide chain, the incorrect nucleotide is removed by a 3′-5′ exonuclease activity associated with the DNA polymerase III molecule. The proper nucleotide is then inserted, and template-dependent DNA synthesis continues. When the complete ~1000 nucleotide fragment has been polymerized, it is joined onto the longer progeny DNA chain by an enzyme called DNA ligase, whose mechanism of action is shown here.

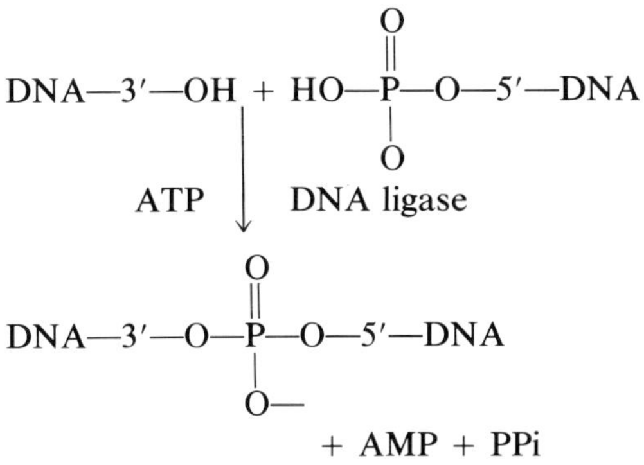

After ligation, the two parental DNA strands can be separated over a further 1000 nucleotide segment in preparation for another round of fragment synthesis and ligation. Successive cycles of this process, therefore, result in the production of two progeny DNA double helices, each of which contains one intact parental and one intact progeny DNA polynucleotide chain.

The mechanism of DNA replication described here is complicated at the stage of the initiation of DNA fragment synthesis by the fact that DNA polymerase III cannot by itself begin polymerization of new DNA chains using single-stranded DNA as template. It can only extend previously initiated polynucleotide chains. This problem is solved in *E. coli* by the use of RNA polymerase; this enzyme can initiate new *RNA* chain synthesis by using a region of single-stranded *DNA* as template. At each step of DNA replication, synthesis of each 1000-nucleotide Okazaki fragment is initiated by the polymerization of a short region (approximately 100 nucleotides in length) of RNA in the 5′ to 3′ direction. DNA polymerase III then uses this RNA sequence as a "primer" and completes synthesis of the fragment. Before ligation can occur, however, the primer RNA must be removed and replaced by DNA nucleotides. Both of these functions are now thought to be carried out by a single remarkable enzyme called DNA polymerase I. This single polypeptide chain has a 5′-3′ exonuclease activity thought to be responsible for the degradation of the primer RNA and also for the two activities (deoxynucleotide polymerization and 3′-5′ exonuclease activity) required for accurate template-dependent DNA synthesis.

DNA polymerase I, which apparently plays such a versatile role in DNA replication, is also thought to be involved in repair of damaged DNA. An *E. coli* mutant called Pol Al is severely depressed in DNA polymerase I activity. It is found to be very sensitive to mutagenesis by ultraviolet light, and this is believed to be caused by its reduced ability to repair radiation-induced damage.

There is a third enzyme, DNA polymerase II, which is capable of polymerizing deoxynucleotides in a template-dependent fashion in *E. coli*, but its role in DNA replication is unclear at present. A summary of the properties of the three *E. coli* DNA polymerases is shown in Table 5–2.

The overall mechanism for DNA replication in mammalian cells is similar to that in *E. coli* except that more initiation sites are involved. The chromosomes are the basic units of DNA replication in mammalian cells, and these consist of single DNA molecules which may be several centimeters long. DNA replication begins simultaneously at many initiation sites spaced at intervals of approximately 100 μm along the DNA. Therefore, there are several hundred initiation sites in a single chromosome. As in *E. coli,* DNA replication occurs in both directions at once from each initiation site, and it proceeds until replication forks from adjacent initiation sites meet, as shown in Figure 5–7. Replication forks move at a rate of approximately 2.5μm per minute, so that replication of each segment requires approximately 20 minutes. The result of this process, when all segments have been replicated and joined, is that the original DNA molecule has been copied in a semiconservative fashion.

In spite of the overall similarity between bacterial and mammalian DNA replication, the enzymatic machinery is sufficiently different that it has been possible to identify antibacterial drugs that selectively inhibit bacterial DNA synthesis. Nalidixic acid, for example, inhibits prokaryotic but not eukaryotic DNA replication.

DNA synthesis occurs during part of interphase. The regular repetition of DNA synthesis, cell division, and other metabolic events in proliferating cells constitutes the cell cycle.

TABLE 5–2 PROPERTIES OF THE THREE *E. COLI* DNA POLYMERASES

	DNA Polymerase		
Property	*I*	*II*	*III*
Molecular weight	109,000	120,000	180,000
Molecules per cell	400	100	10
Nucleotides polymerized per minute (37°C)	~1,000	~50	~15,000
3′-5′ Exonuclease function	+	+	+
5′-3′ Exonuclease function	+	–	–
Intracellular role	DNA replication DNA repair	?	DNA replication

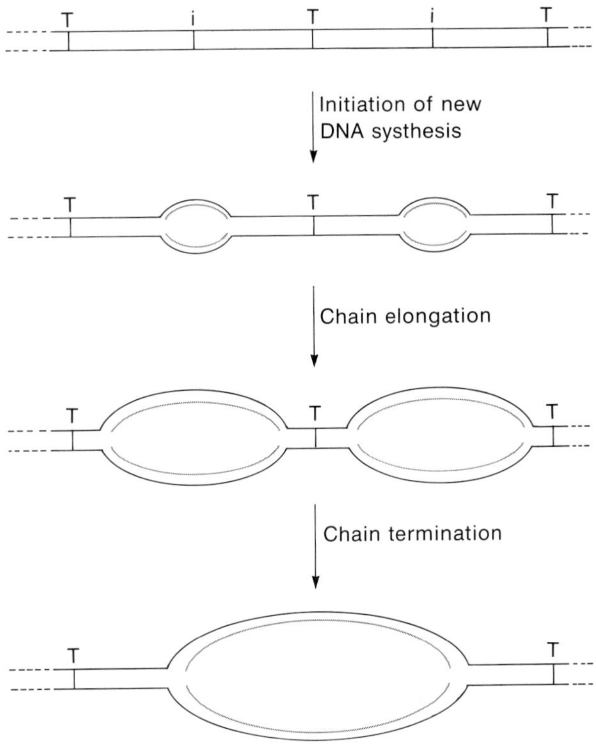

FIGURE 5–7 Multiple sites for the initiation (I) and termination (T) of DNA synthesis, as found in the chromosomes of eukaryotic cells.

THE CELL CYCLE

The life history of actively multiplying cells consists of repetitive divisions with interphases between them. Since this is a repeating cyclic process, it has become customary to represent it diagrammatically as a circle or the face of a clock (Fig. 5–8). The length of time between two successive divisions, represented by a complete revolution around the circle, is designated the generation time, T. For mammalian cells growing in a culture, the generation time is often about 12 to 24 hours.

The changes in the nucleus at the time of cell division are the most impressive morphological changes of the cell life cycle. Therefore, for many years the study of cell reproduction was focused on the events of mitosis. In the time between division, little seemed to be happening, at least so far as could be ascertained with the microscope, so interphase was sometimes referred to as resting phase. This was a particularly unfortunate choice of terminology, because it is now known that many important activities occur during interphase and that cells are more active metabolically during interphase than during mitosis.

The most important event of interphase is DNA synthesis. In mammalian cells, DNA synthesis requires several hours and occupies an interval near the middle of interphase, designated the S phase. During the S phase, DNA duplication in different chromosomes, and even in different parts of a given chromosome, proceeds asynchronously. That is, duplication of DNA in a particular region of a given chromosome takes up only part of the total S phase, and the S phase may be regarded as being composed of a number of shorter, overlapping synthetic periods. The S phase is separated from division by two intervals during which there is no DNA synthesis. These "gaps" are designated G_1, the pre-DNA

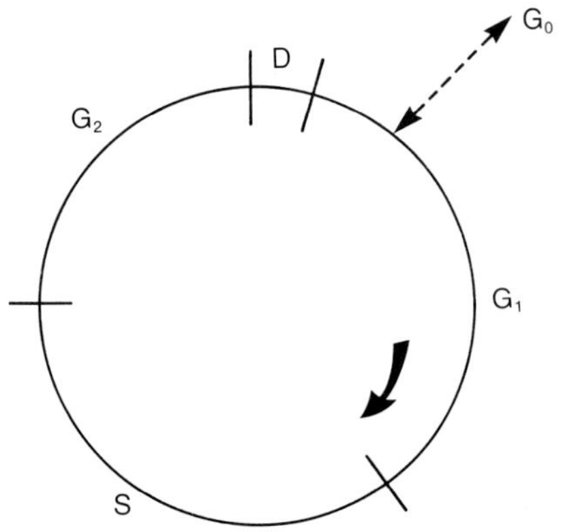

FIGURE 5–8 Diagram of the cell cycle. D, division; G_1, the pre-DNA synthesis period; S, DNA synthesis; G_2, the post-DNA synthesis period. The generation time, T, is represented by a complete revolution around the circumference of the circle. G_0 is the hypothetical state occupied by cells that are no longer dividing and thus are "withdrawn" from the proliferating population.

synthesis period, and G_2, the post-DNA synthesis period. In most mammalian cells, G_1 is longer than G_2. Examples of the duration of the main parts of the cell cycle for a mammalian cell type growing in culture are as follows: G_1, 6 to 9 hours; S, 6 to 8 hours; G_2, 2 to 5 hours; D, 0.5 to 2 hours; and T, the generation time, 12 to 24 hours. It should be noted, however, that there are many variations on this basic plan. Cleaving eggs, for example, have a very short interphase occupied largely by S, and they undergo successive rapid divisions. In lower forms there are many differences in the pattern of parts of the cell cycle. Some lack a G_1 or a G_2 phase, and others lack both G_1 and G_2. A generalization that applies to many cell types is that G_1 is the most variable in duration, while S, G_2, and D remain relatively constant.

This description is applicable to the behavior of populations of cells that are actively growing and dividing with a relatively constant generation time. These ideal conditions may obtain temporarily in a culture of cells, but in the tissues of an organism different kinds of cells divide with varying frequencies, and some cell types, such as mature neurons and skeletal muscle cells, normally do not divide at all. Thus cells often stop in their progress through the cell cycle. When this occurs, most cells stop in the G_1 phase of the cell cycle, although it has been reported that the progress of some cells is arrested in G_2. When a cell stops in G_1, this state is sometimes designated G_0 (Fig. 5–8). It implies that a cell has withdrawn from the proliferative cycle for some time. If the cell is capable of being stimulated to divide once again, it is thought to enter the proliferative cycle at the same point, after which it proceeds through the remainder of G_1, S, G_2, and division at the usual rate.

The cell cycle can be studied using radioautography after the administration of thymidine-^{3}H, which is a precursor of DNA. Cells are exposed to thymidine-^{3}H by injecting it into an animal or by adding it to the medium in which cells are growing *in vitro*. If a given cell is synthesizing DNA during the time that this precursor is available, it will take up the thymidine-^{3}H and incorporate it into its DNA. The presence of DNA containing radioactive thymidine can then be detected by radioautography (Fig. 5–9). If a sample of the cells is fixed and prepared within a few minutes of the administration of the precursor, it will be observed that some nuclei are labeled, representing cells that were synthesizing DNA, while other nuclei, those of cells that were in G_1, G_2, or D, will not be labeled. The percentage of cell nuclei labeled corresponds approximately to the precentage of the cell cycle that is occupied by the S phase (provided that all the cells in the population under consideration are proliferating).

It is possible to estimate the duration of the parts of the cell cycle in the

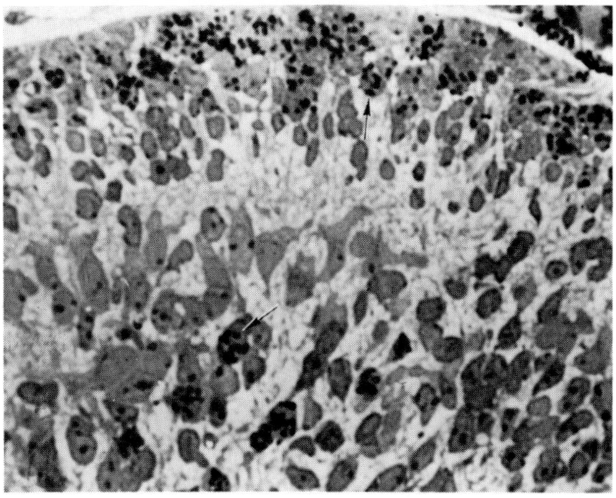

FIGURE 5–9 Light microscope radioautograph of part of the cerebellum of a young mouse following the injection of ^{3}H-thymidine. Silver grains overlie the nuclei of cells in some locations, indicating that these cells were synthesizing DNA. Other cells that were not synthesizing DNA are unlabeled. × 510. (Micrograph courtesy of Langman, J., Shimada, M., and Rodier, P.: Pediatr. Res., 6:758, 1972. Reproduced by permission of Williams & Wilkins.

following way. A population of cells growing and dividing asynchronously is administered a short pulse (e.g., 10 minutes) of thymidine-^{3}H by exposing the cells to a medium containing thymidine-^{3}H for 10 minutes and then replacing the radioactive medium with nonradioactive medium for the remainder of the experiment. At intervals after the end of the pulse, samples of cells are withdrawn, fixed, and prepared for light microscope radioautography. In each of these samples, the percentage of *mitotic figures* (nuclei in the division phase of the cell cycle) that are labeled is determined. If the percentage of labeled mitotic figures is plotted against time, a curve similar to that shown in Figure 5–10 is obtained, and the durations of parts of the cell cycle correspond to the intervals shown in the figure.

Differences in various tissues in the proportion of cells that are synthesizing DNA in preparation for division have led to the recognition of three main types of cell populations with respect to the properties of growth and renewal. *Static* cell populations do not undergo DNA synthesis and division; thus there is no increase in the total amount of DNA in the population. Neurons of the adult nervous system are an example of a static population. In *expanding* cell populations, a small proportion of cells undergo DNA synthesis and division. The increase in number of cells and in their total amount of DNA is just sufficient to account for growth of the organ. Examples of expanding cell populations are numerous and include the liver, kidney, and many endocrine and exocrine glands. In some organs with expanding cell populations, such as the liver, the fraction of cells synthesizing DNA may increase greatly following removal of part of the organ until the approximate original size of the organ is restored. In *renewing* cell populations, the individual cells have a finite lifetime, and continuous DNA synthesis and cell divisions are necessary to replace dying cells. Thus the fraction of cells synthesizing DNA and dividing is much greater than that required for growth of the tissues alone. Examples of renewing populations include cells in the bone marrow that are precursors of red blood cells, since the life of red blood cells in the circulation normally averages only about 120 days. The epithelial lining of the intestine and the epidermis of the skin offer additional examples of renewing populations, with cells continuously being produced by division to replace those constantly being shed from the epithelial surface.

Knowledge of the events of the cell cycle during interphase has been important both in study of the regulation of normal cell proliferation and in efforts to understand the alterations that lead to the abnormal cell proliferation in cancer.

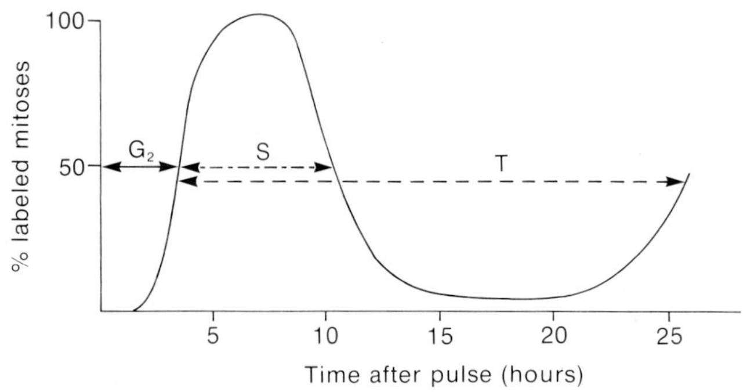

FIGURE 5–10 One method for determining the length of parts of the cell cycle. Cells growing in culture were exposed to a brief pulse of thymidine-^{3}H. At intervals after the end of the pulse, samples were fixed and prepared for radioautography. The percentage of the mitotic metaphases labeled in each sample was determined and plotted against time. To understand the graph, it is important to remember that it is the percentage of labeled *mitotic figures*, not the percentage of labeled nuclei, that is being studied here. The time designated G_2 in the figure represents the time required for those cells that were toward the end of S when the pulse was given to progress through G_2 and to enter division. The interval occupied by S reflects the time required for those cells that were in all parts of S when the pulse was given to pass through mitosis. The period designated T reflects the time required for those cells that were at the end of S when the pulse was given to enter division once, to go through an entire cell cycle, and to enter division once again. G_1 is obtained by subtraction of the length of D, S, and G_2 from the length of T. There is variation among cells in the population in the length of time it takes to complete a cell cycle; the 50 percent level represents the average length of the parts of the cell cycle.

Awareness of the synthetic events of interphase has directed attention from the microscopically visible events at the initiation of mitosis to the earlier chain of events that leads to mitosis. As noted earlier, many cells stop in G_1 when they cease to proliferate. Thus, when a cell enters the S phase it appears to be committed to proceed through S and G_2 and then to divide. Therefore, much effort is being directed toward studying the G_1 to S transition in an attempt to understand how cell proliferation is controlled.

Study of the cell cycle is also of medical importance in understanding the effects of various anticancer agents on cells. Nonsurgical methods of treating cancer include the use of chemotherapeutic drugs and physical agents, such as radiation, which act on rapidly multiplying cells often by interfering with some aspect of DNA synthesis. It is important to realize that these agents generally act nonspecifically on rapidly multiplying cells. Thus they are able to kill cancer cells. However, important side effects of these agents result from their destruction of normally rapidly proliferating cells of renewing cell populations. Such cells include those in the bone marrow that are precursors of blood cells and those in the crypts of the intestine that are responsible for renewal of the intestinal lining.

MITOSIS

Having considered the interphase parts of the cell cycle, including replication of the DNA of the chromosomes, let us now turn to the events of cell division. The term *mitosis* refers to a process of *nuclear* division observed in many eukaryotes. *Cytokinesis* refers to division of the cell *cytoplasm*. Mitosis and cytokinesis are both parts of the overall process of *cell division*. The following discussion considers in particular nuclear division in mammalian cells, which occurs by mitosis. It should be kept in mind, however, that other modes of nuclear division are known and are important in the cell biology of other kinds of organisms. For example, variations of mitosis are found in certain fungi and protozoa in which a spindle forms within an intact nuclear envelope. The polyploid macronuclei of some ciliated protozoans divide by amitosis, an apparent pinching in-two of the nucleus in the absence of a typical mitotic apparatus. Certain protozoa are thought to represent an evolutionary transition in the form of nuclear division between the nuclear division in prokaryotes in the absence of a mitotic apparatus and the typical mitotic divisions of higher eukaryotes. The genetic importance of mitosis in eukaryotes, of course, is that it permits the distribution of identical sets of genes contained in the chromosomes to the two daughter cells produced by cell division.

There appear in mammalian cells at the time of cell division several structures that carry out the separation of the sets of genes. Collectively these structures are designated the mitotic apparatus. They include the mitotic chromosomes, the spindle, and the mitotic centers.

Mitotic Chromosomes

During interphase the DNA of the chromosomes is extended and is visible microscopically as chromatin (see Chapter 6). Parts of it are metabolically active in RNA synthesis. During the S phase the DNA is duplicated, and at mitosis it becomes condensed into the cylindrical mitotic chromosomes. The formation of the mitotic chromosomes can be thought of as a means of packaging long threads of DNA in order to separate and transport the two sets of genes into the daughter cells. This is necessary because of the great length of DNA in a chromosome; if extended, the DNA of a human chromosome may be several centimeters long, but it is condensed at mitosis into a chromosome only 5 to 10 μm long.

In the light microscope, each chromosome at metaphase of mitosis is seen to be composed of two daughter chromatids, one destined for each daughter cell. The two chromatids are joined together at the primary constriction or kinetochore, at which the spindle fibers attach to the chromosome. Different types of chromosomes are identified according to the location of the kinetochore, as shown in the following table.

Some chromosomes also have a secondary constriction. The nucleolus forms during early interphase at some secondary constrictions, which are therefore known as nucleolar organizers. Different members of a set of chromosomes can be distinguished on the basis of size, the location of the kinetochore, and the presence or absence of a secondary contriction (Figs. 5–11 to 5–13). In recent years, application of staining methods, such as the Giemsa stain, to the study of mitotic chromosomes has permitted recognition of a complex pattern of banding in human chromosomes (Fig. 5–14). This has made it possible to distinguish more

IDENTIFICATION OF CHROMOSOMES BY LOCATION OF THE KINETOCHORE **TABLE 5–3**

Name	Location of Kinetochore
Metacentric	Middle
Acrocentric	End
Telocentric	Near the end
Submetacentric	Near the middle

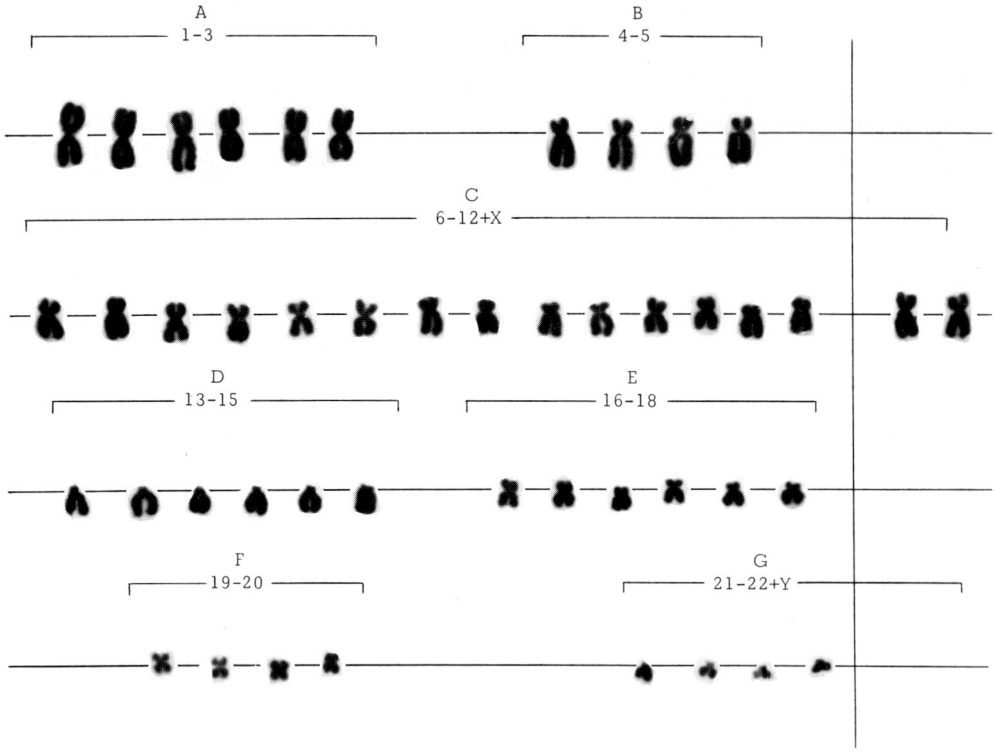

FIGURE 5–11 Normal human female karyotype, containing 22 pairs of autosomes. The sex chromosomes at the right consist of two X chromosomes; no Y chromosome is present. (Courtesy of J. Q. Miller and the University of Virginia Chromosome Laboratory.)

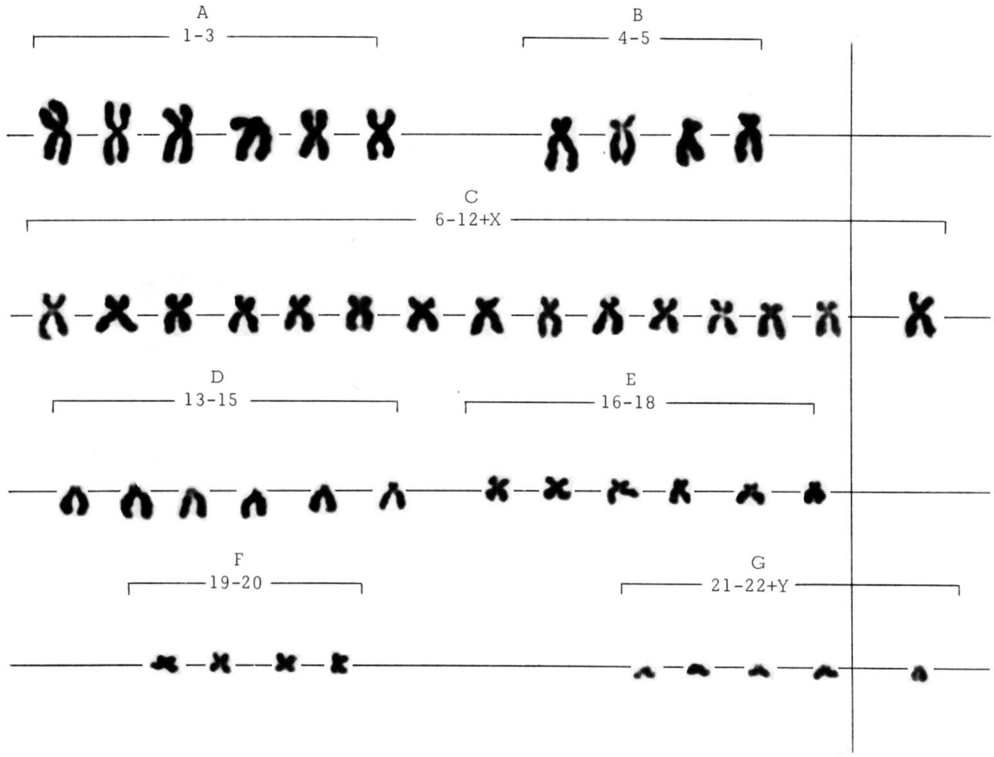

FIGURE 5–12 Normal human male karyotype. The sex chromosomes, consisting of one X and one Y chromosome, are shown at the right of the figure. (Courtesy of J. Q. Miller and the University of Virginia Chromosome Laboratory.)

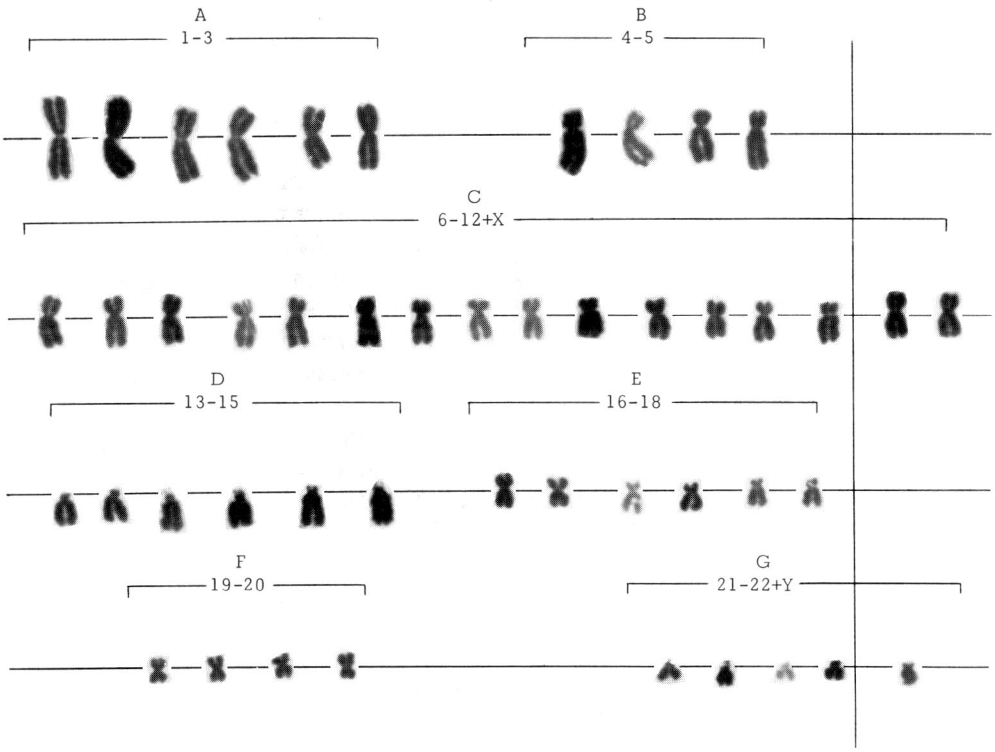

FIGURE 5–13 Human karyotype in Klinefelter's syndrome. The sex chromosome complement consists of two X chromosomes and one Y chromosome. Thus, there is an extra sex chromosome, and since the normal 22 pairs of autosomes are present the total number of chromosomes is 47. (Courtesy of J. Q. Miller and the University of Virginia Chromosome Laboratory.)

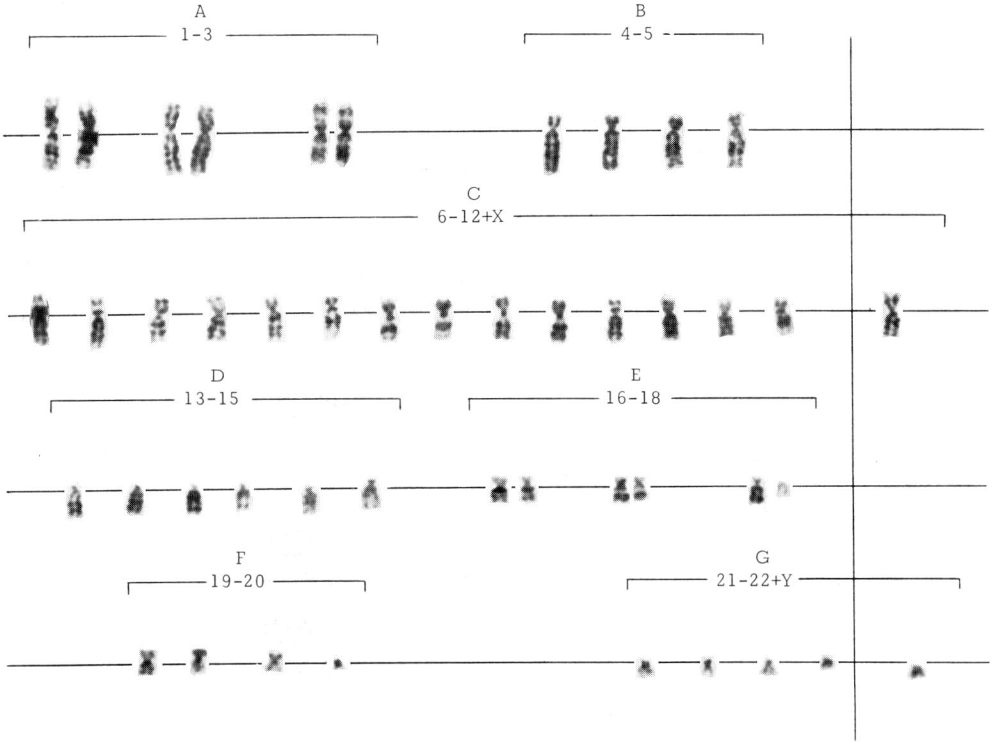

FIGURE 5–14 Normal human male karyotype stained with the Giemsa stain. A pattern of bands on the chromosomes is demonstrated by this and similar methods. (Courtesy of J. Q. Miller and the University of Virginia Chromosome Laboratory.)

accurately the members of a set of chromosomes from one another as well as to detect abnormalities of the chromosomes that occur in disease.

The chromosomal complement of a given species is known as the karyotype. For convenience in study, photographs of a set of condensed metaphase chromosomes from a single cell are commonly arranged in order of decreasing size to form a picture known as an idiogram (Figs. 5–11 and 5–12). Pairs of homologous chromosomes are present, one member of each pair being derived from the individual's male parent and one from the female parent. The human karyotype consists of 46 chromosomes; there are 22 pairs of autosomes and two sex chromosomes. The sex chromosomes of females consist of a pair of X chromosomes of equal size (Fig. 5–11). Males possess one X chromosome and one smaller Y chromosome (Fig. 5–12).

The human karyotype can be studied in white blood cells obtained from a blood sample and stimulated to divide *in vitro*. Increased numbers of mitotic nuclei can be obtained by arresting cells in metaphase of mitosis, using the drug colchicine. Metaphase chromosomes are then visible in samples of cells squashed on the surface of a slide. Study of the karyotype is important in medicine because certain diseases are associated with abnormalities of the karyotype (Fig. 5–13), some examples of which will be noted later in this chapter. Study of the karyotype of cells obtained from amniotic fluid has proved useful in the diagnosis of genetic diseases of fetuses *in utero*.

In thin sections viewed with the electron microscope, the chromosomes of dividing cells appear to be composed of a complex mass of fibrils approximately 100 to 250 Å in diameter (Fig. 5–15). However, it has been very difficult to deduce

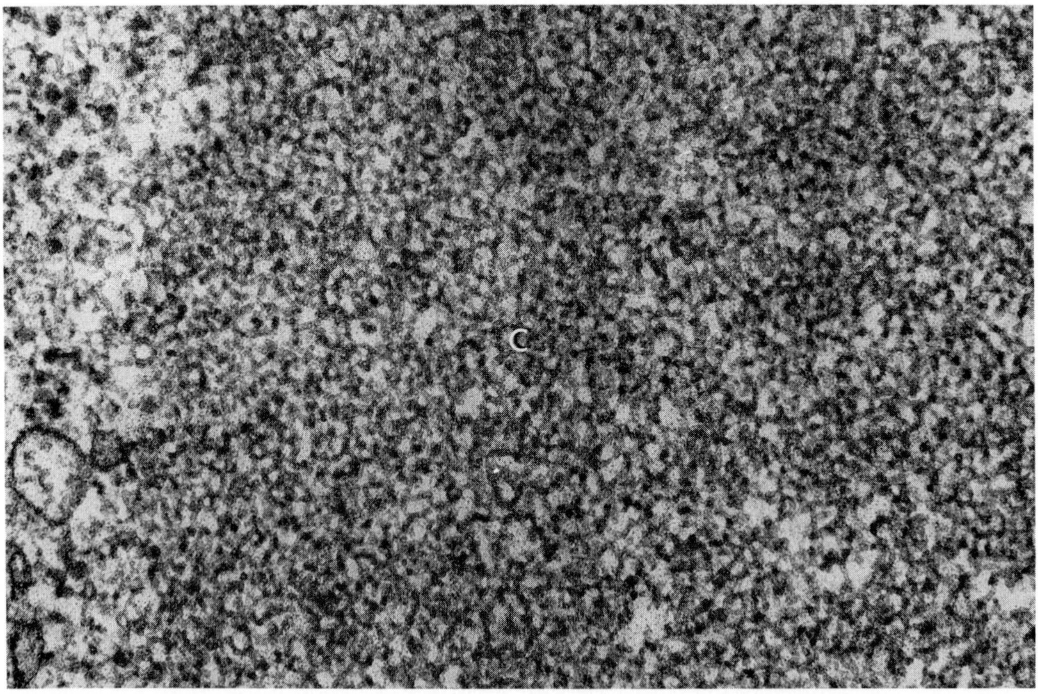

FIGURE 5–15 High magnification electron micrograph of a portion of a chromosome during meiosis in a male germ cell of the cricket. The chromosome (C) contains many fibrils slightly less than 100 Å in diameter. × 98,000. (Micrograph courtesy of Kaye, J. S., and McMaster-Kaye, R.: J. Cell Biol., *31*:159, 1966. Reproduced by permission of the Rockefeller University Press.)

features of the three-dimensional organization of these fibrils within the chromosomes from their representation in thin sections. Study of whole chromosomes with the electron microscope (Fig. 5–16) has permitted visualization of a complex pattern of loops with no free ends at the periphery of mitotic chromosomes. It is thought that the DNA-containing chromatin fibrils are packed into mitotic chro-

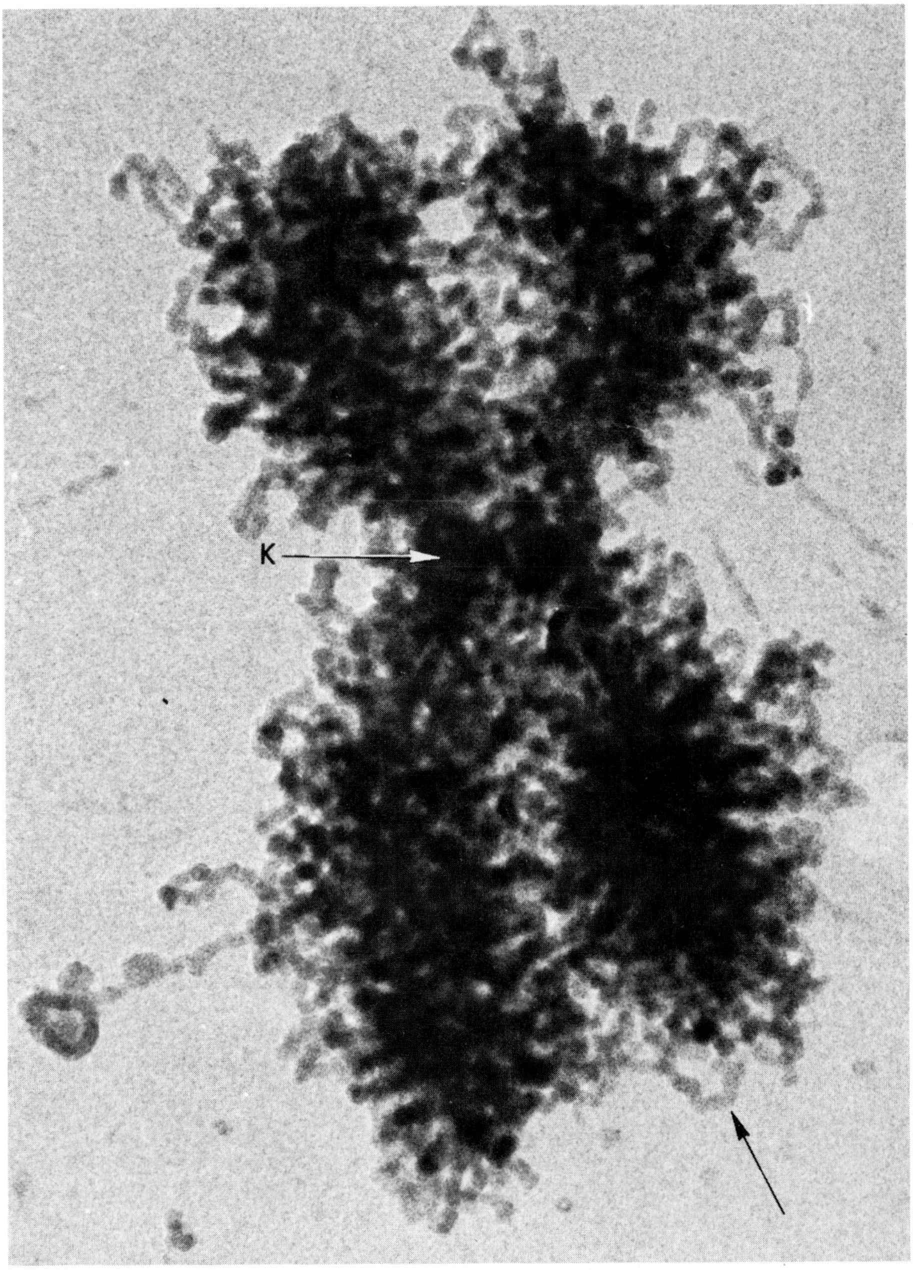

FIGURE 5–16 Electron micrograph of a whole mount of human chromosome 16 isolated at metaphase. The two chromatids consist of chromatin fibers 200 to 500 Å thick. Few free ends but many loops (*arrow*) are seen, in accordance with the notion that each chromatid is composed of a single folded fiber. Two dense kinetochores (K) are visible in this specimen. × 57,000. (Micrograph courtesy of E. J. DuPraw. From DuPraw, E. J.: DNA and Chromosomes. New York, Holt, Rinehart and Winston, 1970. Reproduced by permission of the publisher.)

mosomes by several orders of coiling, folding, and formation of loops, but the details of DNA organization in mitotic chromosomes remains to be elucidated. Use of the high-voltage electron microscope, which permits viewing of thicker sections than conventional instruments, may aid in deciphering the structure of mitotic chromosomes.

The fine structure of the kinetochore region differs from the remainder of the mitotic chromosome. Often it appears in sections as a dense plaquelike structure, composed of a convex outer dense layer, a middle layer of low density, and an inner dense layer continuous with the bulk of the chromosome (Figs. 5–18, and 5–19). Some investigators have interpreted the configuration of the kinetochore as an oval plate applied to the surface of the chromosome. Others have reported that it has a more complicated lamellar appearance and contend that the dense portions visualized in sections are actually parts of two dense filaments, each of which is composed of two intertwined fibrils. The chemical nature of the microscopically visible portions of the kinetochore is still in doubt. It has been suggested that the kinetochore may represent a product of the genes in the region of the primary constriction. The microtubules of the spindle attach to the chromosome by inserting on the outer dense layer of the kinetochore.

The Spindle

The mitotic spindle (Fig. 5–17) is a device that functions in separation and transport of the chromosomes. In the living cell it has a gel-like consistency, being more viscous than the surrounding cytoplasm. With the light microscope, the spindle has a threadlike appearance owing to the presence of numerous spindle fibers (Fig. 5–27*C* and *D*). Two types are usually distinguished. "Pole-to-pole" or interpolar fibers are arranged in parallel and course in the direction of the axis between the two mitotic centers, or poles, toward which the chromosomes move. Others, the "chromosome-to-pole," or chromosomal spindle fibers, attach to the chromosomes at the kinetochores and appear to extend from there to one of the poles of the spindle.

With the electron microscope, the spindle is seen to contain a large number of straight microtubules, each about 200 Å in diameter (Figs. 5–18 and 5–19). Since individual microtubules are too small to be resolved with the light microscope, it is evident that bundles of parallel microtubules constitute the spindle "fibers" seen in light microscope preparations. Microtubules of a similar appearance radiate outward in all directions from the mitotic centers in animal cells and comprise the asters. In light microscope preparations, a spherical clear region immediately surrounding the centers is known as the centrosome.

The presence of large numbers of microtubules oriented parallel to one another accounts for the birefringence of the spindle in living cells viewed with the polarizing microscope (Fig. 5–20). This property is important in permitting the visualization of the spindle in living cells, and changes in birefringence during mitosis have been intensively studied in relation to the movements of chromosomes.

High-resolution studies of spindle microtubules *in situ* and in isolated preparation have shown that the wall of each tubule is usually composed of approximately 13 longitudinally disposed protofilaments, each of which has a beaded appearance and consists of a chain of globular subunits about 50Å in diameter (see Chapter 13). A series of fine projections from the surface of spindle microtubules has also been described. In sectioned material these sometimes appear

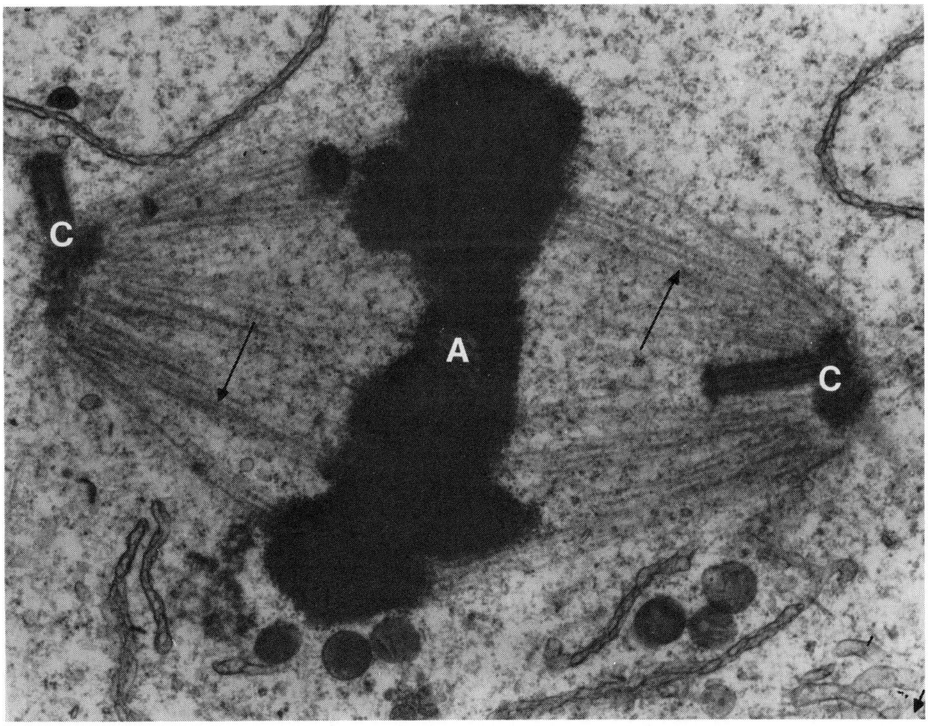

FIGURE 5–17 Electron micrograph of the spindle during meiosis in the chicken testis. Centrioles (C) are found at the poles of the spindle. The dark mass of material (A) represents the chromosomes. The spindle fibers (*arrows*) are bundles of microtubules. × 19,000. (Micrograph courtesy of McIntosh, J. R.: J. Cell Biol., *61*:166, 1974. Reproduced by permission of the Rockefeller University Press.)

to extend from one microtubule to another in the vicinity, forming a bridge between the two. It has been suggested that these bridges, by permitting microtubules to slide in relation to one another, enable them to interact and to play a role in chromosome movement.

The spindle has been isolated and analyzed chemically. It is composed mainly of proteins with small amounts of other substances. An important component is a 6S protein, tubulin. As discussed further in Chapter 13, similar proteins are components of microtubules isolated from a variety of other locations and from many different kinds of cells. The 6S tubulin molecule is a dimer of about 120,000 molecular weight, and it is thought to be the basic subunit of spindle microtubules. It binds one molecule of the antimitotic drug colchicine, which is commonly used to disrupt the mitotic apparatus and block mitosis.

Mitotic Centers and Centrioles

The mitotic centers are the two points toward which the chromosomes move during mitosis. In animal cells each center contains a pair of centrioles (Fig. 5–21). Each centriole is a hollow cylindrical structure, and when cut in a cross section its wall is seen to be composed of nine triplet tubules (Fig. 5–22). One end of each centriole often displays a set of interior densities that imparts a cartwheel appearance to its cross section. The members of a pair of centrioles are usually oriented at right angles to one another (Fig. 5–21). Spindle micro-

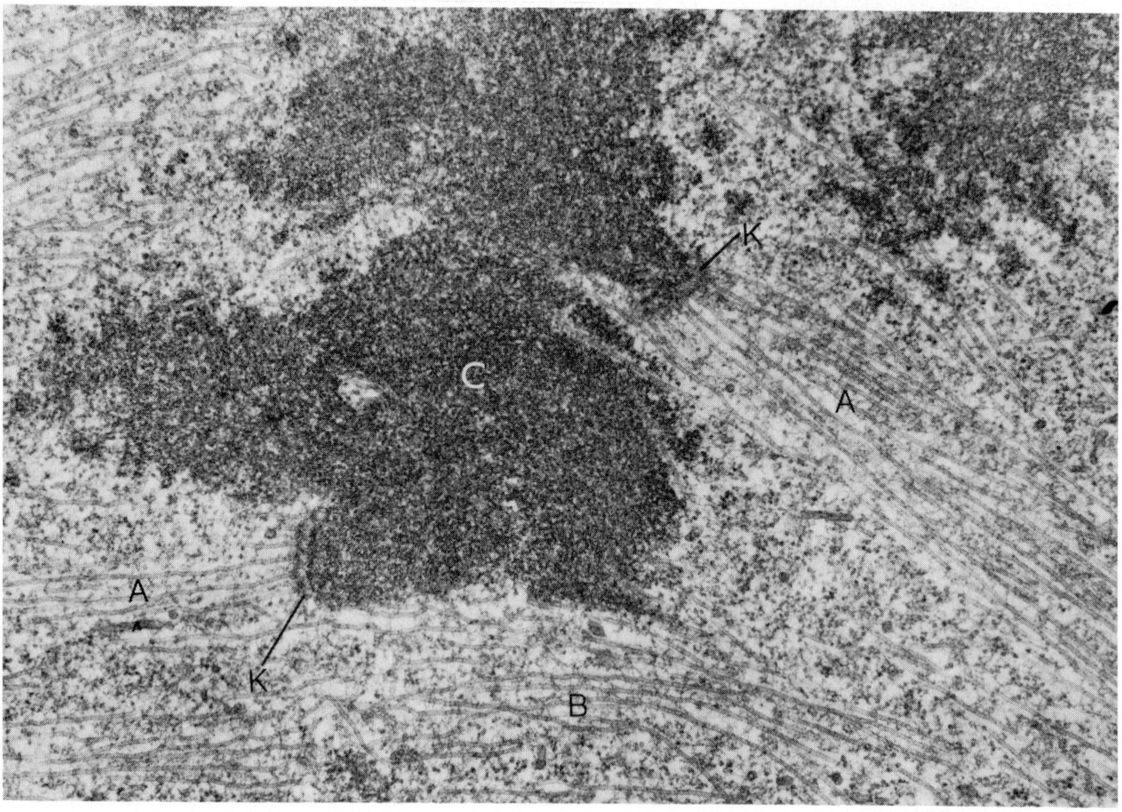

FIGURE 5–18 Metaphase of rat kangaroo cell grown in culture. Spindle microtubules occupy much of the field. The chromosomal microtubules (A) attach to the chromosomes (C) at the kinetochores (K). Other microtubules, which appear to bypass the chromosomes, may be the interpolar variety (B). × 27,000. (Micrograph courtesy of Brinkley, B. R., and Cartwright, J.: J. Cell Biol., *50*:416, 1971. Reproduced by permission of the Rockefeller University Press.)

tubules run toward the poles and become closely associated with the centrioles but do not necessarily make direct contact with them. Often the microtubules appear to end near the centrioles in small masses of dense material termed pericentriolar satellites (Fig. 5–22).

The role of centrioles in the mitotic centers of animal cells is not clear. They were long thought to have a role in organization of the spindle. This may yet prove to be the case, but it should be recalled that the cells of higher plants undergo mitotic nuclear division even though they lack centrioles. Thus centrioles do not appear to be essential for the functioning of mitotic centers and mitotic division of the nucleus. Recent studies have implicated small masses of dense material in various locations in the organization of microtubules, suggesting that the pericentriolar satellites (Fig. 5–22) may be important structures in organization of the spindle.

The centrioles seen at the poles of the spindle of animal cells are structurally indistinguishable from the basal bodies of cilia and flagella, and instances are known in which a given organelle functions alternately as a centriole in a mitotic center and as the basal body of a cilium. In contrast to the relation between centrioles and spindle microtubules, however, the doublet microtubules of the shaft of the cilium are continuous with two of the three tubules in the wall of the

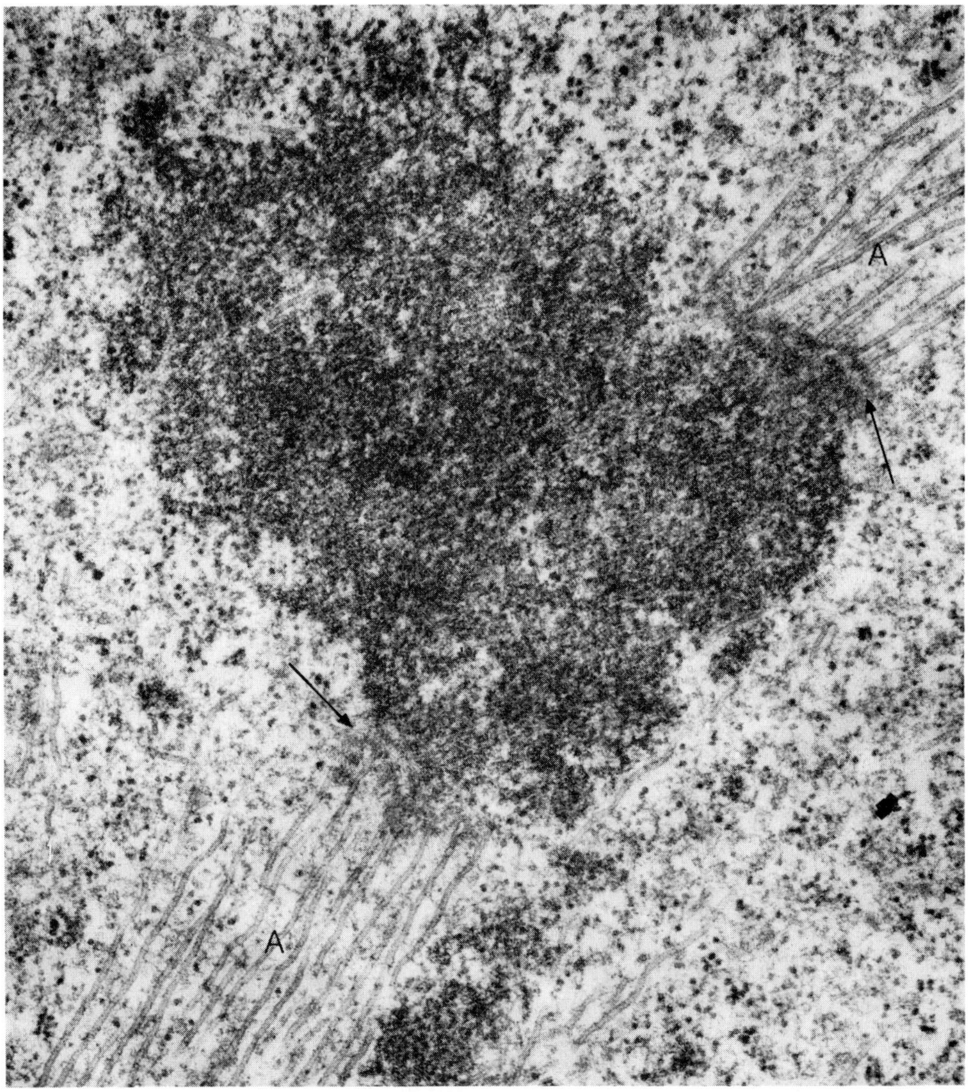

FIGURE 5–19 A small chromosome of a rat kangaroo fibroblast at metaphase. The microtubules of the spindle (A) attach to the sister chromatids at the dense kinetochores (*arrows*). × 48,000. (Micrograph courtesy of Brinkley, B. R., and Stubblefield, E.: Adv. Cell Biol., *1*:119, 1970.)

basal body (see Chapter 13), and the ciliary tubules appear to extend from the basal body during genesis of a cilium.

In cells that are proliferating, duplication of the centrioles occurs during the preceding cell cycle, the specific time varying with the cell type. In eggs that are dividing rapidly with a short interphase, duplication of the centrioles occurs as early as telophase of the previous division. In HeLa cells, mammalian cells grown in culture, centriole duplication occurs during the S phase (Fig. 5–23). Duplication of centrioles in proliferating cells occurs by generation of a smaller procentriole. This structure forms at right angles to the parent centriole, close to the parent but not in direct contact with it. The procentriole then increases in length until it reaches the dimensions of a mature centriole. Thus the ends of a centriole differ

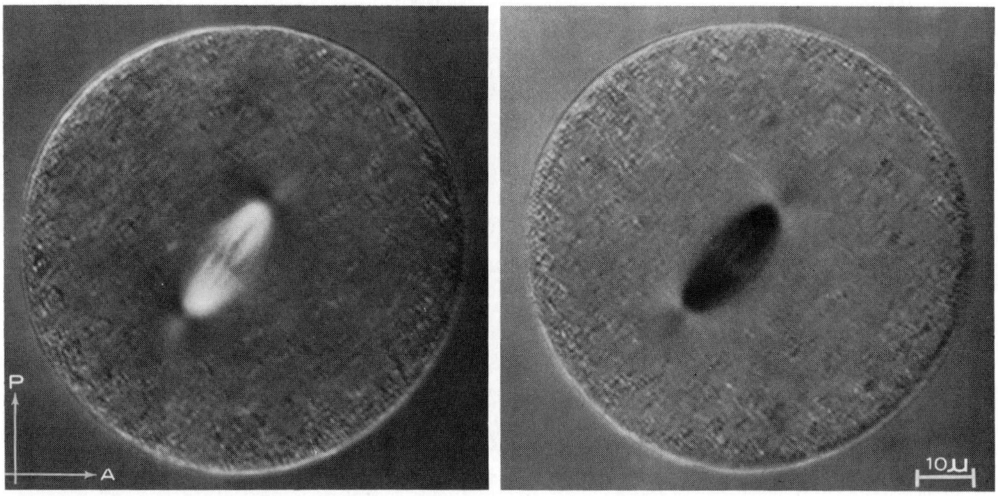

FIGURE 5–20 The mitotic spindle in a dividing egg viewed with the polarizing microscope. For technical reasons, the spindle appears bright in one micrograph and dark in the other. (Micrographs courtesy of Sato, H. and Inoué, S.: J. Gen. Physiol., *50*:259, 1967. Reproduced by permission of the Rockefeller University Press.)

from each other. A proximal end is formed initially and a distal end formed subsequently; the internal cartwheel is found in the proximal end. When functioning as a basal body, the ciliary microtubules grow from the distal end. Little is known about the mechanism of procentriole formation, except that the mature centriole somehow seems to direct the formation of the new procentriole in geometrical relation to the parent.

Since centrioles and basal bodies in many cells appear to be self-duplicating organelles, much effort has been directed toward determining whether they contain nucleic acid, but the results have been conflicting. Centrioles are difficult to isolate and purify satisfactorily, and the small amount of material they contain

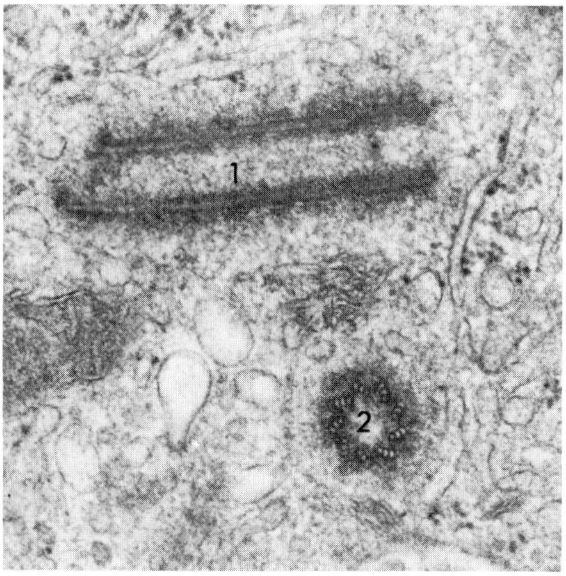

FIGURE 5–21 A pair of centrioles in the hamster epididymal epithelium. The cylindrical centrioles lie at right angles to one another. One of them (1) is sectioned parallel to its long axis, while the other (2) is shown in cross section. × 50,000.

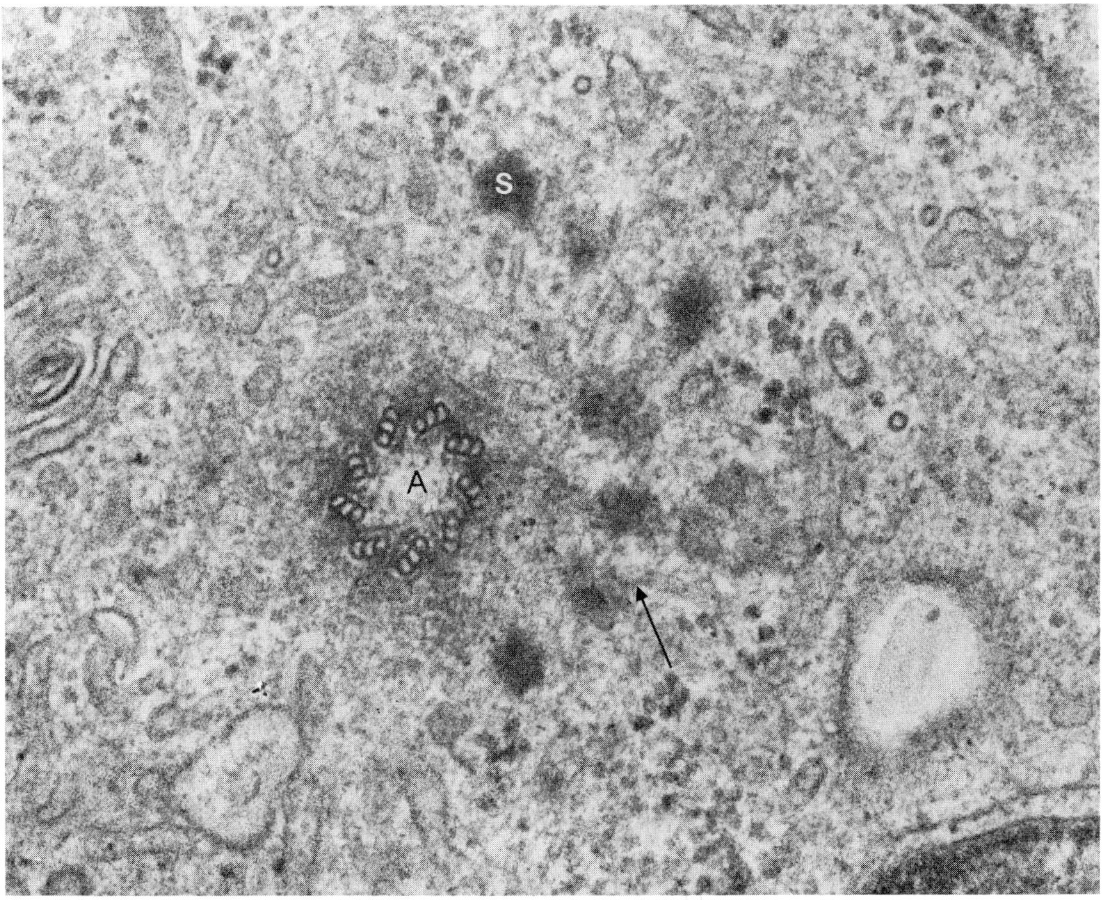

FIGURE 5–22 Cross section of a centriole (A), accompanied by several dense pericentriolar satellites (S). The end of a microtubule is just visible near a satellite (*arrow*). × 92,000. (Micrograph courtesy of Brinkley, B. R., and Stubblefield, E.: Adv. Cell Biol, 1:119, 1970.)

makes radioautographic studies difficult. At present the question of nucleic acid in centrioles is unsettled, but it appears that if DNA or RNA is present it comprises only a small fraction of the organelle.

Formation of centrioles by one-to-one generation of procentrioles is observed in duplication of mitotic centers and in replication of basal bodies in some cells. Under certain circumstances, however, procentrioles are known to be formed in different ways, usually when larger numbers of centrioles are required to serve as basal bodies. This situation is seen in mammals during development of a ciliated epithelium, such as that lining the trachea or oviduct (Figs. 5–24 and 5–25). The developing procentrioles often occupy the shell of a hollow sphere, the center of which is occupied by a smaller sphere of fibrillar and granular material. This material is termed a deuterosome or condensation form and is thought to contribute to the growth of the procentrioles. Sometimes a pair of mature centrioles also occupy the center of the sphere, but other times they are absent. Thus intimate contact with mature centrioles does not appear to be essential for formation of large numbers of new centrioles in this way.

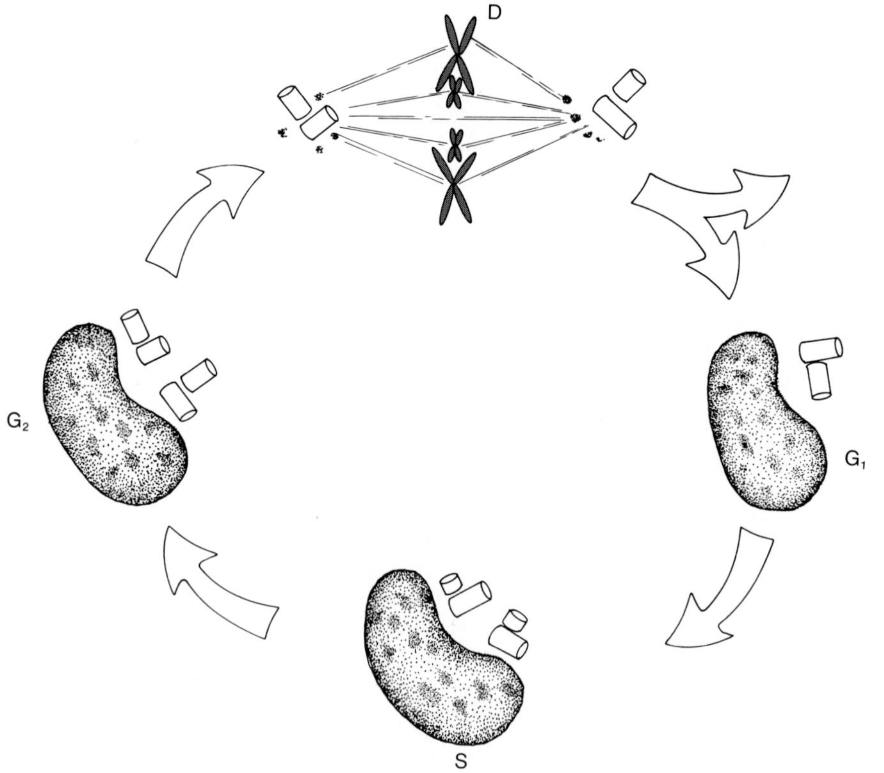

FIGURE 5–23 The behavior and replication of centrioles during the cell cycle of a mammalian cell. A pair of centrioles is present at both poles of the spindle during division. Therefore, after cytokinesis each daughter cell in G_1 possesses one pair of centrioles, with the individual centrioles oriented at right angles to one another. During S, the two centrioles separate slightly from one another, and a small procentriole begins to form at right angles to each parent. The procentrioles increase in size until they reach the size of a mature centriole. Thus, in G_2 the cell contains two pairs of centrioles. In prophase of mitosis, the pairs of centrioles separate from one another and the spindle forms between them, so that a pair of centrioles comes to lie at each pole of the mitotic apparatus. The size of the centrioles is exaggerated with respect to that of the nucleus.

Stages of Mitosis

Since the stages of mitosis are well known to most students of cell biology, they will be outlined only briefly. The stages are prophase, metaphase, anaphase, and telophase (Figs. 5–26 and 5–27).

Prophase begins with the appearance of condensed chromosomes as thread-like structures visible with the light microscope. As prophase proceeds, the chromosomes become thicker and shorter. The two mitotic centers, formed by duplication during the preceding interphase, migrate to take up positions on the opposite sides of the nucleus. The primary spindle forms between the centers, and the nuclear envelope breaks down (Fig. 5–28). The nucleolus diminishes in size and usually disappears by the end of prophase, although in some cell types it may remain throughout mitosis. During the short portion of mitosis sometimes designated as prometaphase, the progressively condensing chromosomes attach to the spindle and migrate to the equatorial plane of the spindle, midway between the two poles and perpendicular to the spindle axis.

In metaphase the chromosomes are aligned in the equatorial plane to form the metaphase plate. Each chromosome can now be seen to be composed of two

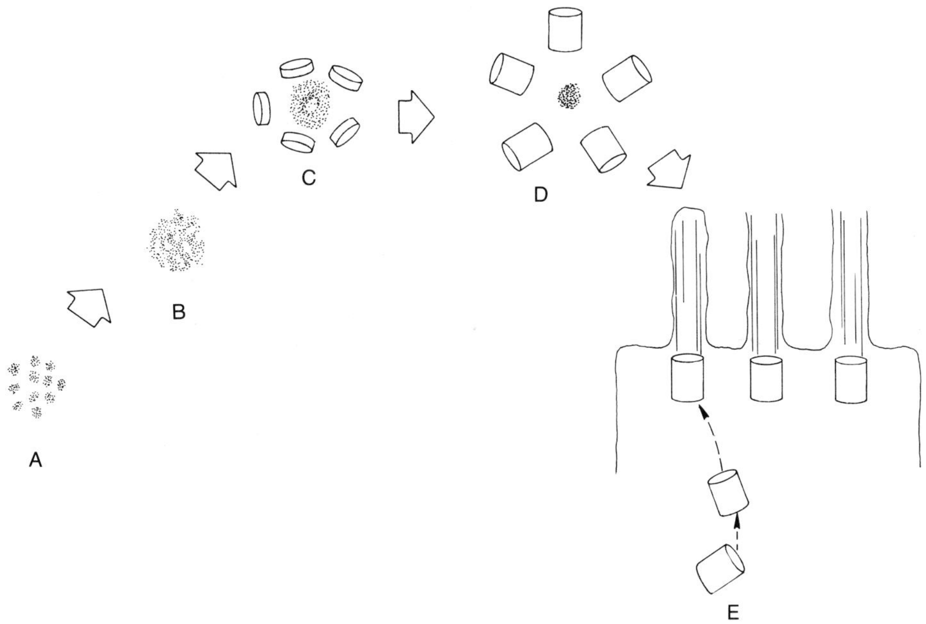

FIGURE 5–24 Centriole formation during ciliogenesis in ciliated epithelia. Masses of a fibrillar–granular material (*A*), the proliferative elements, appear to condense to form a spherical structure (*B*), sometimes designated a deuterosome or condensation form. Procentrioles appear around the condensation form in the structure termed a generative complex (*C*). When the procentrioles reach a mature size (*D*), they move toward the cell surface and serve as the basal bodies for cilia (*E*). A decrease in the size of the condensation form concomitant with the growth of the procentrioles has led to the suggestion that the condensation form contributes material to the developing procentrioles. In this instance, procentrioles are apparently formed without the proximity of a mature centriole. In other cases, a mature centriole occupies the center of the sphere of forming procentrioles in place of the deuterosome, as shown here. During the normal cell cycle, centrioles are duplicated by one-to-one formation of procentrioles in relation to mature centrioles, as illustrated in Figure 5–23.

daughter chromatids, the result of duplication of the chromosomes during the preceding S phase. The chromosomes appear to be attached to the spindle by their kinetochores and to be subjected to pulls from chromosome-to-pole spindle fibers that are directed toward opposite poles. At the beginning of anaphase, the two daughter chromatids of each chromosome separate, and as new daughter chromosomes they move apart toward the poles, with the kinetochores leading the way and the arms trailing. Telophase begins when the chromosomes reach the poles and progressively decondense and resume their interphase appearance. In telophase the nuclear envelope is reconstructed, and the nucleolus is reformed.

Reconstruction of the nuclear membranes is thought to occur by fusion of membranous elements resembling pieces of endoplasmic reticulum around the chromosomes (Fig. 5–29), but recent evidence indicates that some lipid components of the old nuclear envelope, dispersed at prophase to the cytoplasm, are conserved and returned quantitatively to the nuclear envelopes of daughter cells in telophase. Many details of the formation of the nucleolus in telophase remain uncertain. The possible contribution of conserved material from the old nucleolus of the preceding interphase, as distinct from nucleolar material newly synthesized during telophase or G_1, is disputed. The nucleolus is formed at the site of secondary constrictions or nucleolar organizers, which are known to contain genes

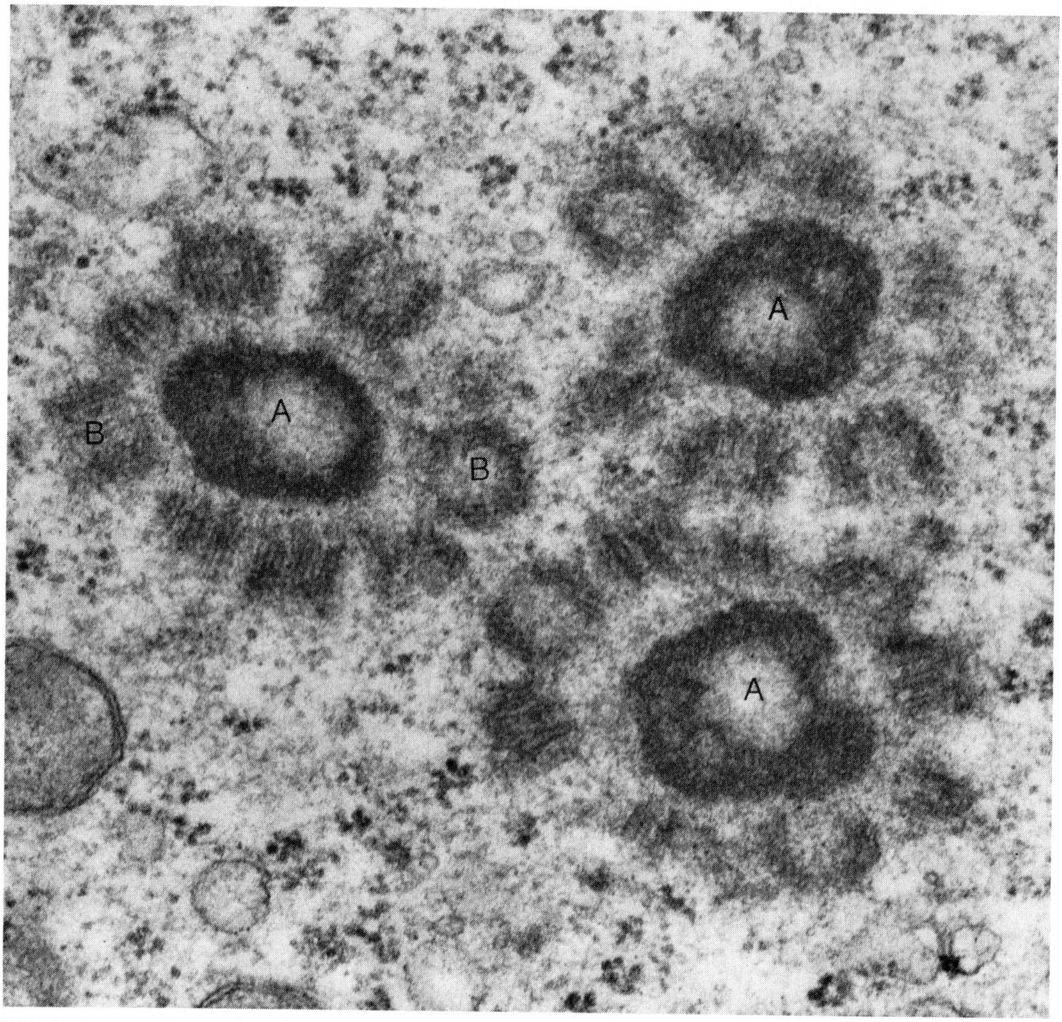

FIGURE 5–25 Formation of centrioles in the epithelium of the mouse oviduct. Three generative complexes are shown. In each, a hollow condensation form (*A*) is surrounded by several developing procentrioles (*B*). × 63,000. (Micrograph courtesy of Dirksen, E. R.: J. Cell Biol., *51*:286, 1971. Reproduced by permission of the Rockefeller University Press.)

that code for the major kinds of ribosomal RNA. Some investigators believe, however, that synthesis of nucleolar material also takes place at other sites dispersed among the chromosomes and that the material is then collected in the nucleolar organizer regions.

Chromosome Movement

The mechanism by which chromosomes move is one of the most important problems of mitosis. It is not yet completely understood, but there is currently a great deal of activity in the study of cell motility in general, and recent developments promise new insights. The history of theories of chromosome movement is particularly interesting because the problem has stimulated investigators to exercise a great deal of imagination. Attempts to explain chromosome movement have included such ideas as microscopic jet engines located in the kinetochores,

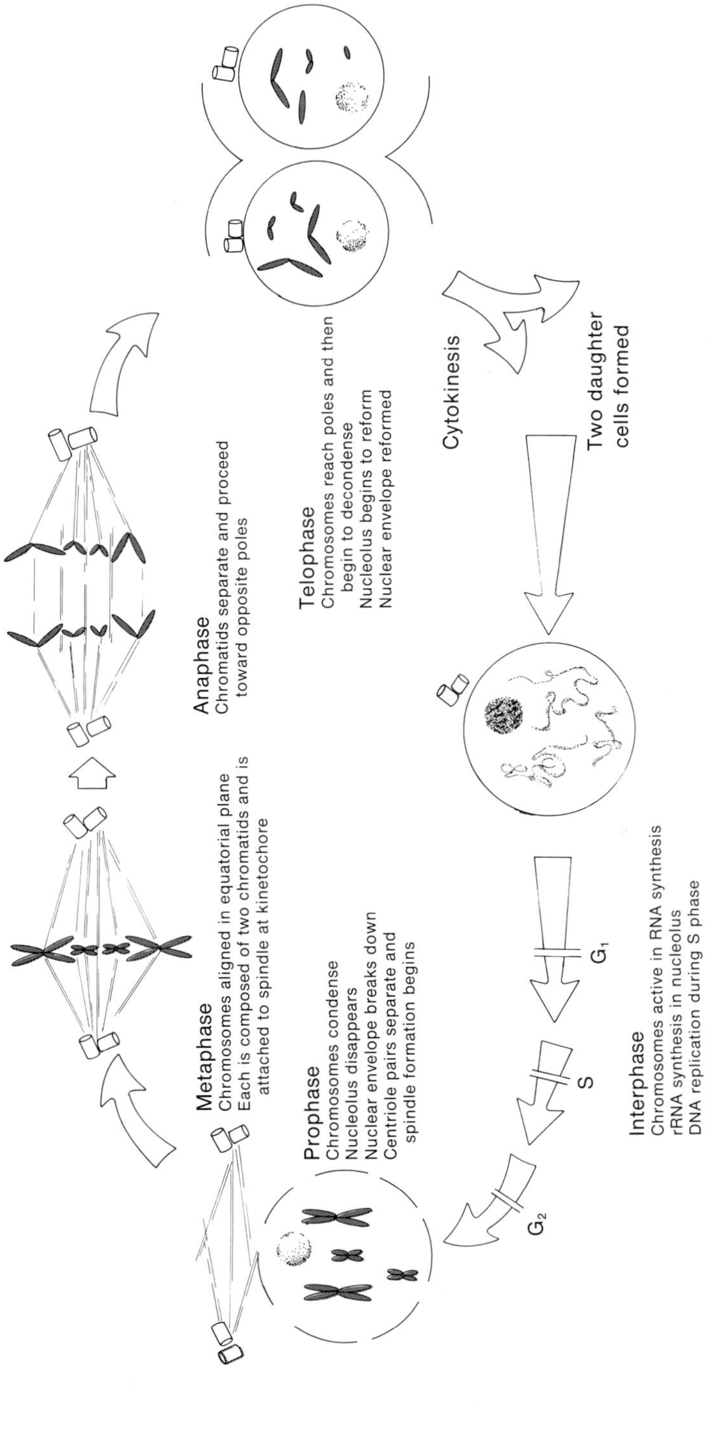

Prophase
Chromosomes condense
Nucleolus disappears
Nuclear envelope breaks down
Centriole pairs separate and spindle formation begins
Metaphase
Chromosomes aligned in equatorial plane
Each is composed of two chromatids and is attached to spindle at kinetochore
Anaphase
Chromatids separate and proceed toward opposite poles
Telophase
Chromosomes reach poles and then begin to decondense
Nucleolus begins to reform
Nuclear envelope reformed
Cytokinesis
Two daughter cells formed
Interphase
Chromosomes active in RNA synthesis
rRNA synthesis in nucleolus
DNA replication during S phase
G1
S
G2

FIGURE 5–26 A diagram of the stages of mitosis in a mammalian cell.

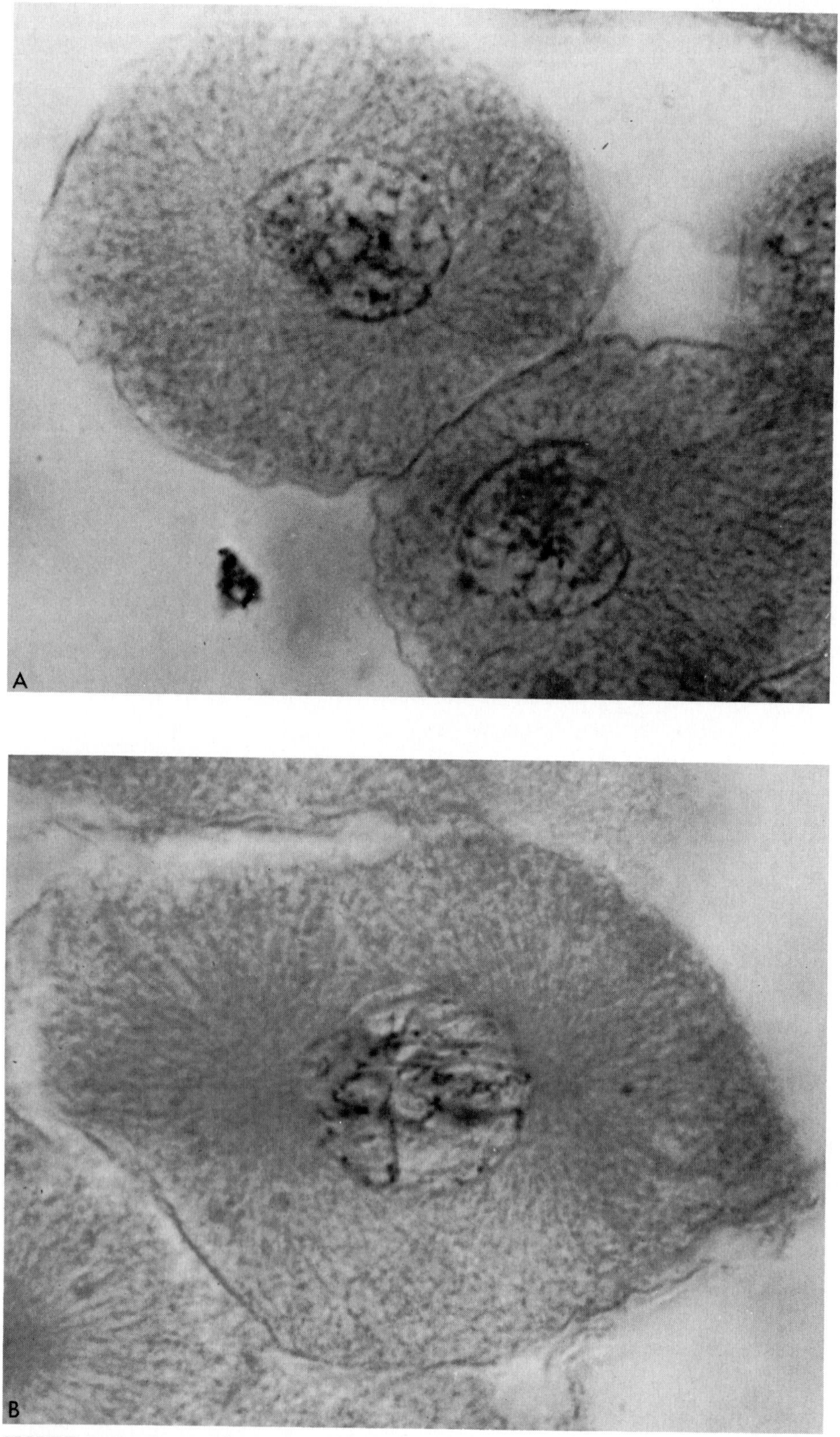

FIGURE 5–27 Stages of mitosis in dividing cells of the whitefish blastula. *A*, interphase; *B*, prophase. Threadlike chromosomes are visible within the nucleus.

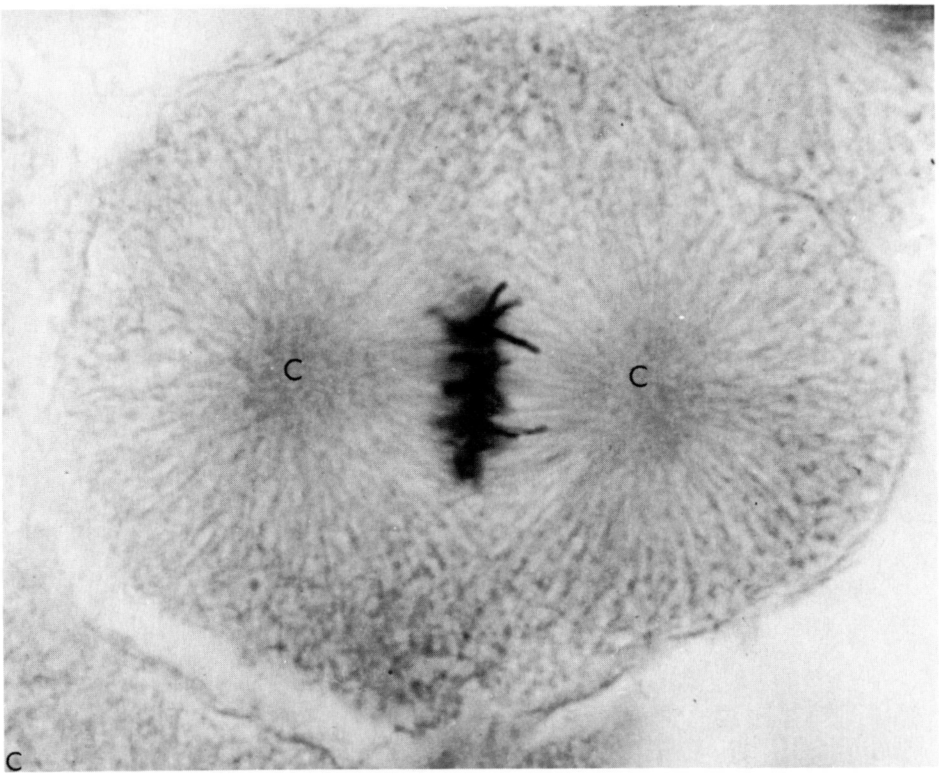

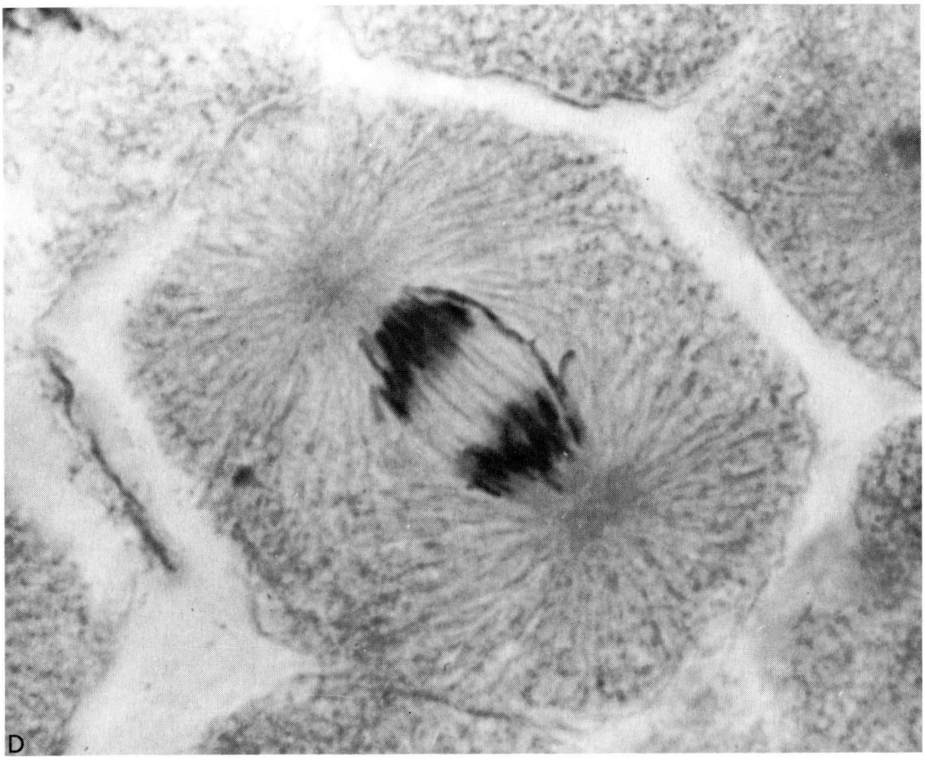

FIGURE 5–27 *Continued* *C*, metaphase. The chromosomes are aligned in the equatorial plane of the spindle. The locations of the mitotic center (C) are made evident by the radiating fibers of the aster. *D*, early anaphase. Separation of the chromosomes has begun.

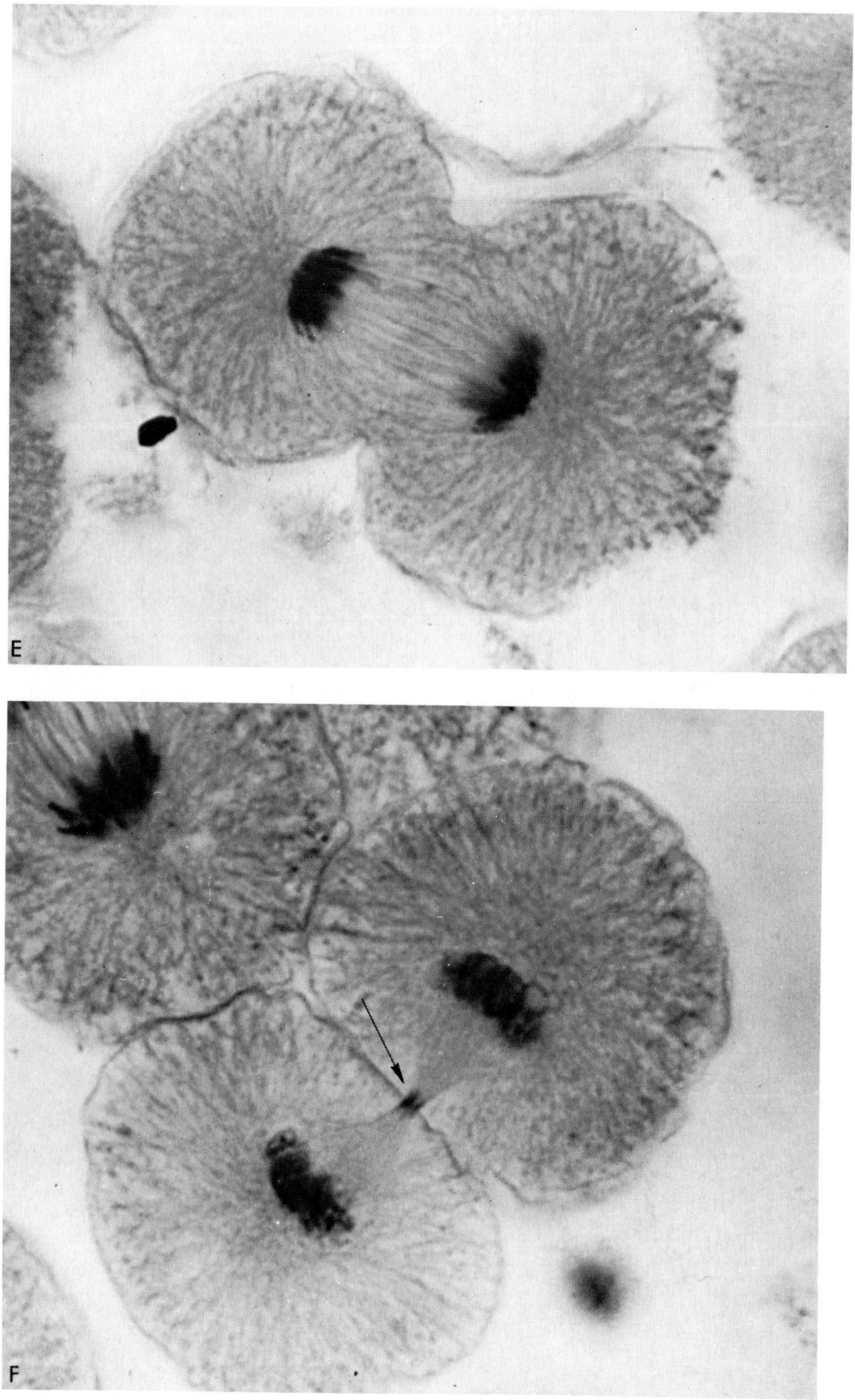

FIGURE 5–27 *Continued* *E*, late anaphase. Note the cleavage furrow which forms as cytokinesis begins. *F*, telophase. Division of the cell is almost complete. Spindle fibers and associated material form the dense midbody (*arrow*). (Reproduced by permission of the Upjohn Company, Kalamazoo, Michigan.)

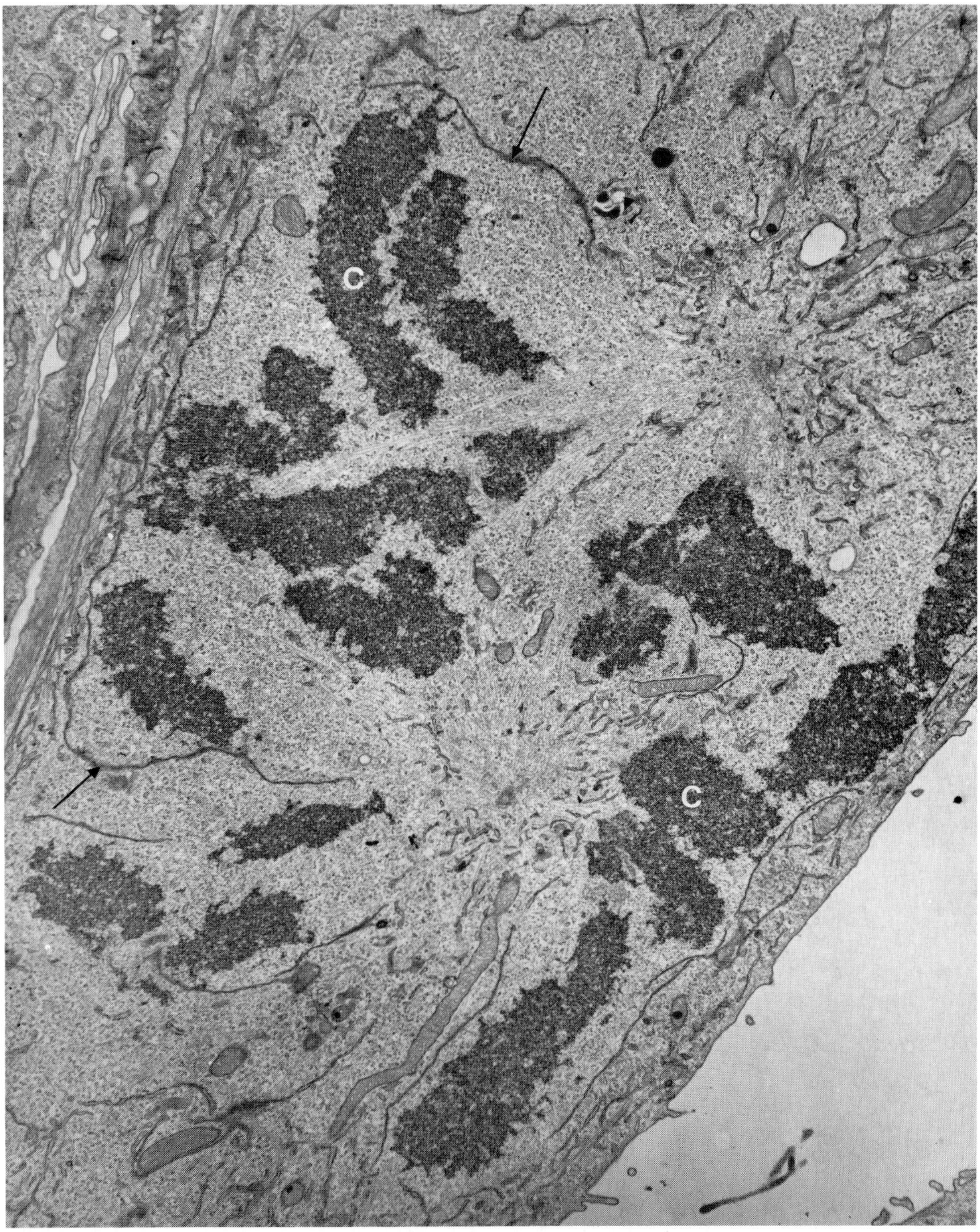

FIGURE 5–28 Low-power electron micrograph of prophase in a rat kangaroo cell. The nuclear envelope was breaking down and remnants of it (*arrows*) surround the condensing chromosomes (C). × 6700. (Micrograph courtesy of Brinkley, B. R., Fuller, G. M., and Highfield, D. P.: In: Goldman, R., Pollard, T., and Rosenbaum, J. (eds.): Cell Motility, Book A. Cold Spring Harbor Laboratory, 1976, p. 435. Reproduced by permission of the publisher.)

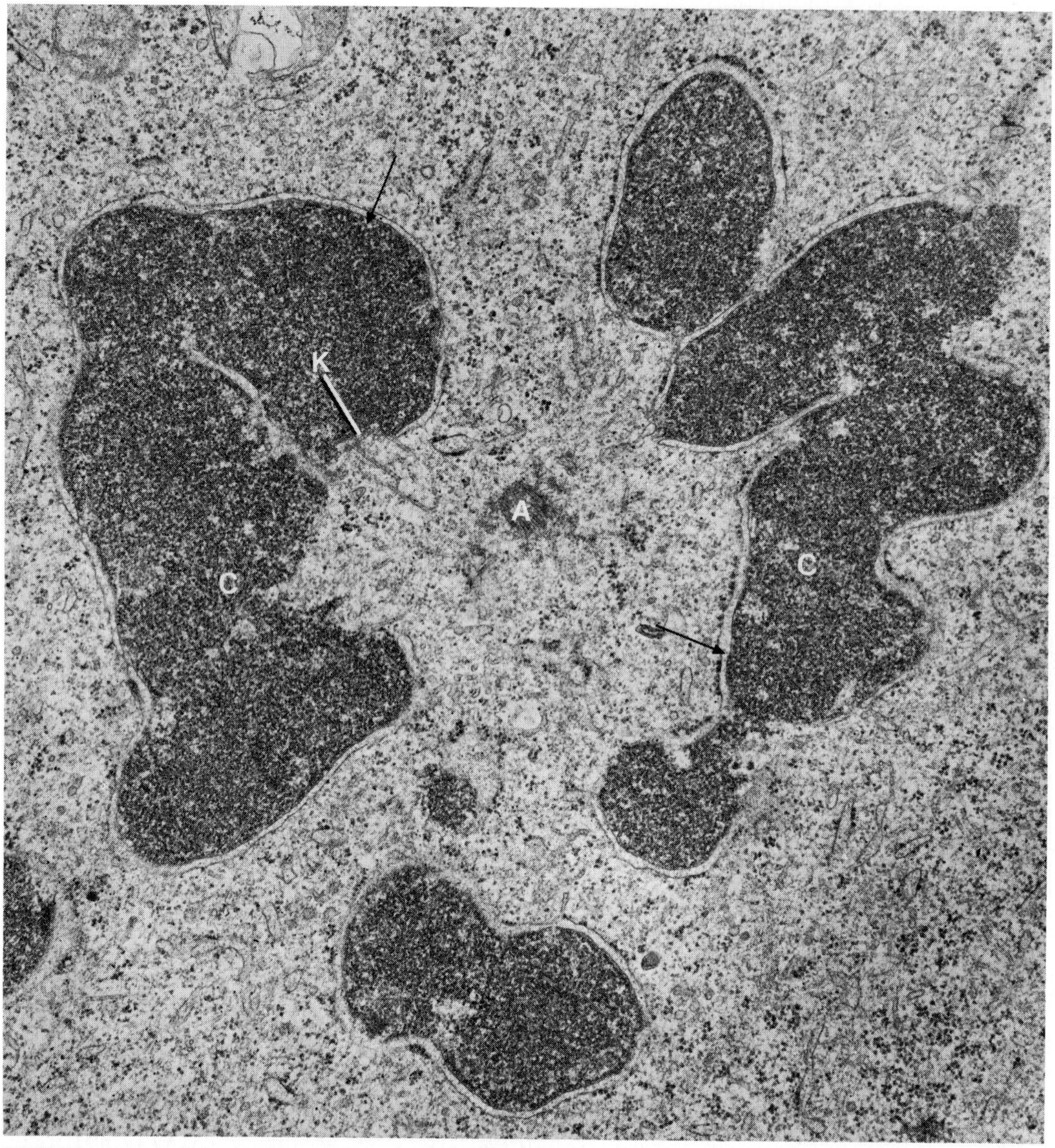

FIGURE 5–29 Polar view of telophase showing chromosomes (C) close to one centriole (A). The nuclear envelope was in the process of reformation by alignment of membranous elements on the surface of the chromosomes (*arrows*). The nuclear envelope almost completely surrounds the chromosomes except at the kinetochores (K). Approximately × 20,000. (Micrograph courtesy of Brinkley, B. R., and Stubblefield, E.: Adv. Cell Biol., *1*: 119, 1970.)

magnetic attraction between chromosomes and centers, and rolling of chromosomes along tracks furnished by spindle fibers.

Observations made on numerous cell types support the idea that the spindle has an important role in chromosome movement. Chromosome-to-pole fibers connect kinetochores to the centers, and they appear to shorten and drag the chromosomes toward the poles during anaphase. The chromosomes behave as if they were being pulled by their kinetochores, since if chromosomes are broken, acentric fragments (those lacking a kinetochore) do not move, while those with kinetochores do. Concomitantly, pole-to-pole fibers seem to lengthen, and by doing so move the poles further apart, thus helping to separate the chromosomes. However, if the microtubules that comprise the spindle fibers are studied with the electron microscope, it is seen that the microtubules undergo no change in caliber. That is, shortening of chromosome-to-pole fibers is accomplished without an increase in thickness of the component microtubules, and lengthening of pole-to-pole fibers occurs without a decrease in diameter of the microtubules.

According to one model proposed to account for the generation of forces in chromosome movement, microtubules slide in relation to one another. It is suggested that microtubules have a polarity and that two tubules of opposite polarity, interacting by means of bridges between them, exert a mutual force that results in their sliding relative to each other. The model predicts that the number of tubules in different parts of the spindle should vary at different stages of mitosis, but efforts to confirm the prediction have had only limited success.

Attempts to understand the behavior of the spindle during mitosis are currently focused on its dynamic nature. The tubulin molecules that are subunits of spindle microtubules are believed to be in equilibrium with a pool of free tubulin molecules.

$$\underset{\text{free}}{\text{T}} \rightleftharpoons \underset{\text{microtubule}}{\text{T—T—T—T—T—}}$$

Thus spindle microtubules are constantly being formed and dismantled. For example, if the spindle in living cells is subjected to cold or to high pressure, it dissolves and then reappears when conditions return to normal. This is thought to be the result of shifting the equilibrium first toward free subunits and then back again toward the polymerization of subunits into tubules. The action of colchicine can be readily understood in this context since colchicine, by combining with free tubulin molecules, shifts the equilibrium away from the polymerized form and dissolves the spindle. A dramatic illustration of the dynamic nature of the spindle is furnished by microsurgical studies. Chromosomes can be detached from the spindle at metaphase by prodding the kinetochore, moved about the cell, and then returned to the spindle region. Chromosomes manipulated in this way reattach to the spindle, and subsequent stages of division proceed normally. According to this view of the spindle as a plastic structure, the shortening of chromosome-to-pole fibers can be accounted for by disassembly of microtubules, while lengthening of pole-to-pole fibers may occur by growth of their constituent tubules.

The mitotic centers and the kinetochores are believed to be sites of polymerization of microtubules and are thought to play a role in organization of the spindle. The ability of kinetochores to function in this way is illustrated in dramatic fashion by *in vitro* experiments in which polymerization of microtubules from a solution of tubulin occurs at the kinetochores of isolated chromosomes.

Questions remain of how forces can be generated by polymerization and depolymerization of microtubules, particularly in shortening of chromosome-to-pole fibers. Interactions between microtubules and other elements also continue to figure in discussions of chromosome movements. Some workers have pointed out that substances besides microtubules are present in the spindle, and that in isolated spindles microtubles are exceeded in amount by other materials. It has been reported that the protein actin is present in the spindle between microtubules and could participate in chromosome movements by interacting with microtubules either directly or indirectly by way of other as yet uncharacterized components.

Abnormalities of Mitosis

One defect that can occur during mitosis is nondisjunction of chromosomes. In this condition, two daughter chromatids fail to separate from one another at the beginning of anaphase, with the result that one daughter cell gets two copies of that chromosome while the other daughter cell gets none. Depending on the tissues involved, mitotic nondisjunction in adult life may be of little consequence because few cells are involved. However, if it occurs during early development, the individual can become a mosaic, a mixture of patches of normal and abnormal cells. He or she may exhibit one of the syndromes more commonly associated with meiotic nondisjunction, which are noted later in this chapter.

Unusual mitotic figures are frequently observed in the rapidly proliferating cells of malignant tumors. The interphase nuclei of cancer cells are often larger than normal in proportion to cell size. They may be aneuploid, which means that the nuclei do not contain the normal number of chromosomes. Usually they have more chromosomes than normal, although not a complete additional set. Perhaps as a result of the additional chromosomes, bizarre mitotic figures with multipolar divisions are common. These abnormal mitotic figures can sometimes be identified in histological sections. Recognition of normal mitotic figures in sections is also important, because it aids in estimating the proliferative activity of a tissue and contributes to the diagnosis of certain pathological processes.

CYTOKINESIS

In animal cells, division of the cytoplasm, cytokinesis, normally occurs by formation of a cleavage furrow which constricts the cell in a plane perpendicular to the long axis of the mitotic spindle (Fig. 5–27*E* and *F*). Division is usually into two approximately equal daughter cells, but in some instances division is asymmetrical, as in the divisions of the oocyte. Inside the advancing cleavage furrow at later stages of cytokinesis is the midbody (Fig. 5–27*F*), a bundle of persisting spindle microtubules surrounded by a dense material, which connects the two daughter cells until cytokinesis is complete.

As is the case with chromosome movement, various theories to explain cytokinesis have been popular over the years. Currently, the "contractile ring" hypothesis is most widely discussed. This postulates that there is a donut-shaped ring of a contractile material located at the margin of the cell which progressively constricts and pinches the cell in two. This theory is supported by electron microscope observations of a ring of fine filaments in the cytoplasm underlying the cleavage furrow in the position expected of the contractile ring (Fig. 5–30). Recently these filaments have been shown to bind heavy meromyosin, indicating that actin is one of their components, and lending credence to their proposed contractile function.

The location of the cleavage furrow seems to be determined in some way by the spindle. Normally the furrow forms at right angles to the spindle axis. If the spindle is moved by microsurgery at a stage of mitosis prior to metaphase, the location of the furrow is altered correspondingly. However, if the spindle is moved after metaphase, the furrow retains its original location, suggesting that at metaphase the spindle somehow "signals" the location of the cleavage furrow.

It should be recalled, as for mitosis, that the description of cytokinesis given here applies primarily to animal cells. Cytokinesis in higher plant cells, for example, occurs by a different mechanism, perhaps as a consequence of their rigid

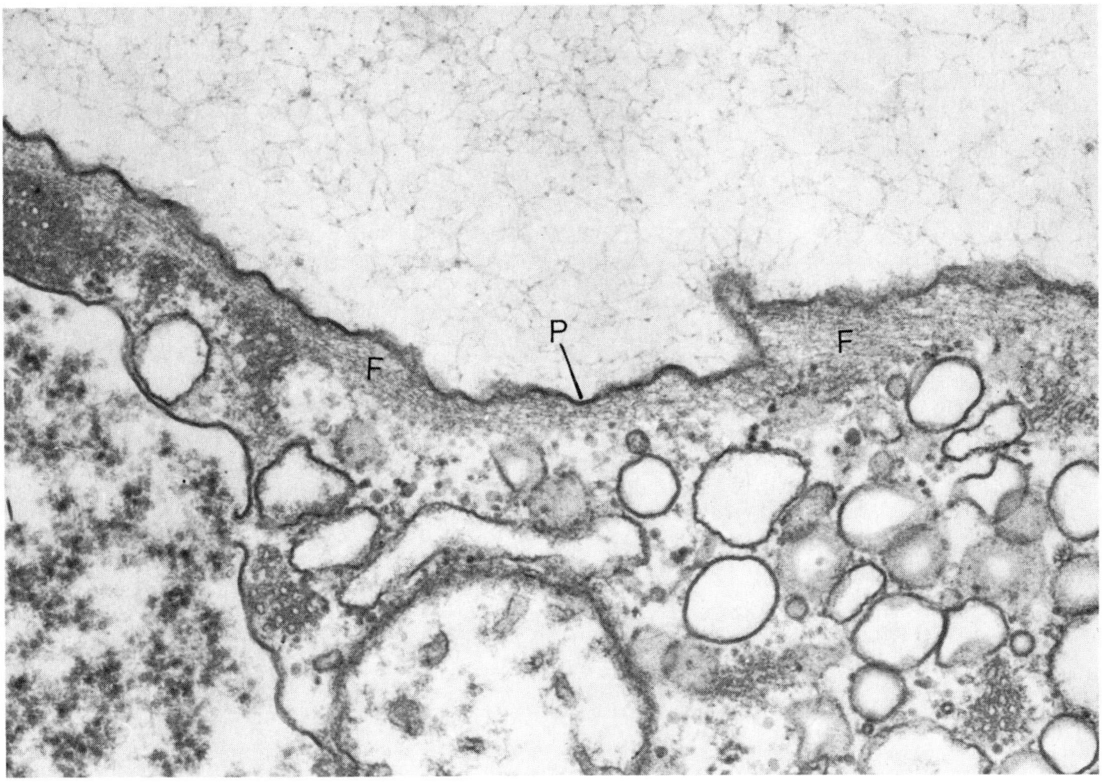

FIGURE 5–30 Filaments in a cleaving marine egg. The filaments (F), which are 50 to 70 Å in diameter, are oriented parallel to the plasma membrane (P) of the cleavage furrow. Their location corresponds to the "contractile ring" which has been proposed to accomplish cytokinesis. × 56,000. (Micrograph courtesy of Szollosi, D.: J. Cell Biol., *44*:192, 1970. Reproduced by permission of the Rockefeller University Press.)

cell wall. In cells of higher plants, cytoplasmic vesicles derived from the Golgi apparatus coalesce to form two membranes that divide the daughter cells from each other. The plane of separation between the daughters is visible in the light microscope as the progressively enlarging cell plate.

MEIOSIS

The genetic significance of meiosis is that by this process the number of chromosomes is reduced from the diploid of somatic cells to the haploid of the gametes. In humans the number is reduced from 46 to 23. This is accomplished by one replication of DNA, followed by two cell divisions without an S phase between them. In humans and other mammals, meiosis takes place in the gonads, the ovaries and the testes, and results in the production of haploid eggs and sperm. The diploid number of chromosomes is restored at fertilization when an egg and a sperm fuse to form a zygote.

Stages of Meiosis (Fig. 5–31)

DNA synthesis occurs during interphase preceding meiosis. If the normal diploid amount of DNA is designated 2N, then primary oocytes and primary

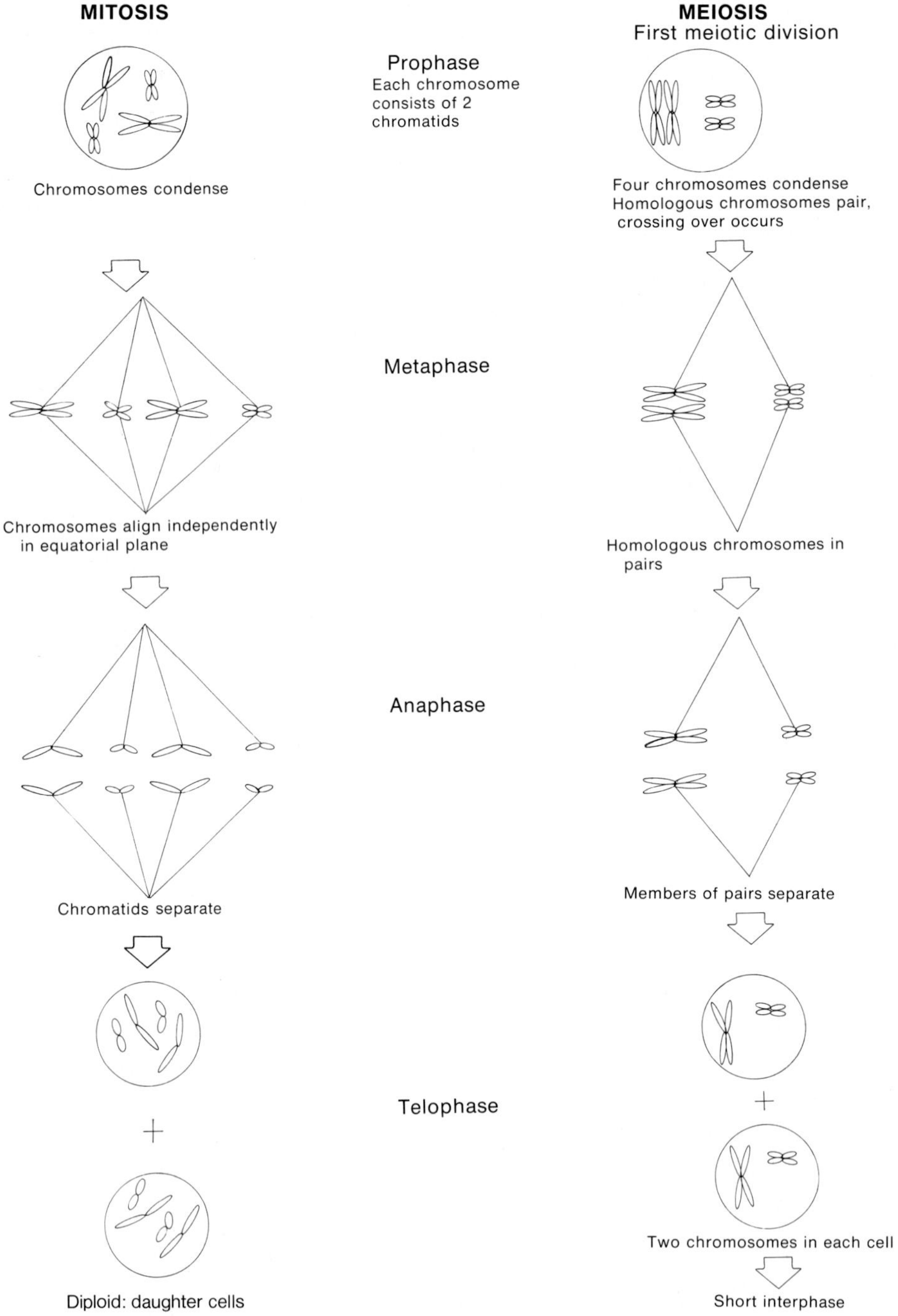

FIGURE 5–31 A diagram of the behavior of chromosomes during meiosis in an organism with a diploid chromosome number of four. The disposition of the chromosomes during meiosis is compared with that during mitosis in a somatic cell of the same organism.

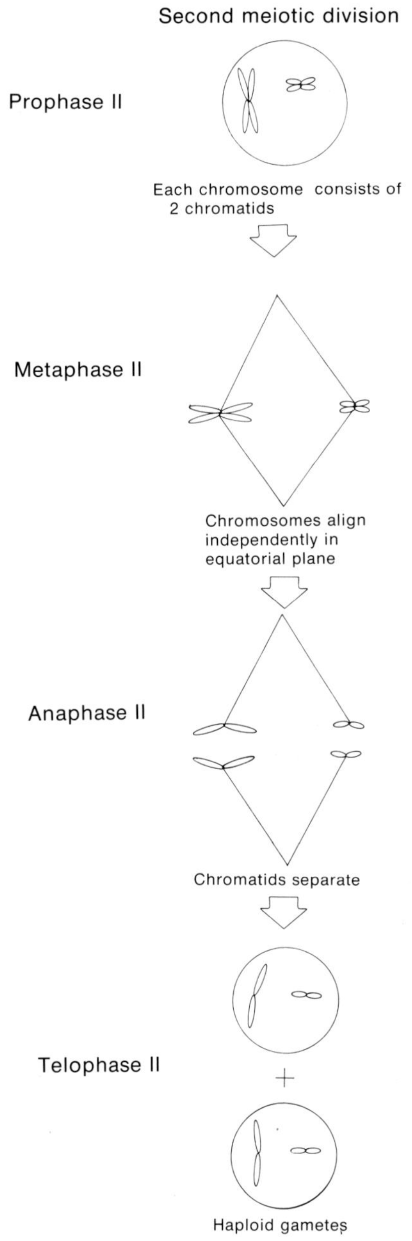
Second meiotic division
Prophase II
Each chromosome consists of 2 chromatids
Metaphase II
Chromosomes align independently in equatorial plane
Anaphase II
Chromatids separate
Telophase II
+
Haploid gametes

Figure 5–31 *Continued*

spermatocytes enter prophase of the first meiotic division with a 4N amount of DNA. Prophase of the first meiotic division is prolonged, lasting approximately three weeks in primary spermatocytes and as long as forty years in primary oocytes. Several stages are distinguished by cytologists on the basis of degree of condensation of chromosomes and the relation between the members of the pairs of homologous chromosomes. In leptotene, the chromosomes appear as the diploid number of thin threads. During zygotene, the homologous chromosomes form the haploid number of pairs. In the next stage, pachytene, the chromosomes become thicker and are readily visible as dense staining bodies in microscopic preparations. During diplotene, crossing-over occurs, resulting in the exchange of segments between chromatids; in diakinesis, shortening and thickening of the chromosomes continues. Detailed descriptions of the morphologies of the meiotic prophase are left for specialized textbooks of cytology. It is important to appreciate, however, the two essential events that occur during prophase of the first meiotic division (Fig. 5–31). First, the homologous chromosomes form pairs, one member of each pair being derived from the individual's father and one member from the mother. Each chromosome contains two chromatids as the result of the preceding DNA synthesis. Thus each pair of chromosomes, or bivalent, contains four chromatids and is sometimes also referred to as a tetrad. Second, the crossing-over that occurs between two chromatids in a bivalent results in exchange of genetic material between the original paternal and maternal chromosomes, providing for reshuffling of genes and their arrangement in new combinations.

At metaphase of the first meiotic division, the pairs of homologous chromosomes (23 pairs in humans) are disposed in the equatorial plane of the spindle. Thus an important difference between meiosis and mitosis occurs in anaphase (Fig. 5–31). At anaphase of the first meiotic division, the members of each pair of homologous chromosomes separate, so that the paternal chromosome goes to one pole while the maternal member goes to the other pole. Thus by telophase of the first meiotic division the number of chromosomes in each daughter nucleus has been reduced to the haploid 23. Since each chromosome is composed of two chromatids, however, the amount of DNA in each nucleus is 2N. The different pairs of homologous chromosomes segregate independently in the first meiotic division. That is, the paternal chromosome of a particular pair goes to a given pole, but the paternal member of another pair may go to the same or to the opposite pole. Thus a large number of different combinations of chromosomes can result in the gametes. Although for simplicity the behavior of paternal and maternal "chromosomes" during anaphase has been described, it may be more precise to speak of the paternal kinetochore going to one pole and the maternal kinetochore to the other at prophase of the first meiotic division. This is because the exchange of material in the crossing-over during prophase results in a mixture of segments of paternal and maternal DNA in any given chromosome by the time anaphase is reached.

Interphase between the two meiotic divisions is short. There is no DNA synthesis. The second meiotic division is rapid and resembles a mitotic division. At metaphase the 23 chromosomes, each composed of two chromatids, are aligned singly in the equatorial plane. At anaphase the two daughter chromatids of each chromosome separate from each other and proceed to opposite poles as daughter chromosomes. The two daughter nuclei at telophase of the second meiotic division thus contain the haploid number of chromosomes, 23, and the amount of DNA is 1N.

Meiosis and Sex Determination

During spermatogenesis in the seminiferous tubules of the male, spermatogonia develop into primary spermatocytes, which divide at the first meiotic division into two secondary spermatocytes of approximately equal size. As a result of the second meiotic division, each secondary spermatocyte produces two spermatids, and each of these develops into a spermatozoon. Thus in the male, four sperm are obtained from one primary spermatocyte. Spermatogenesis begins at puberty and normally continues into old age.

In contrast, in the female, division of the primary oocyte is unequal, forming one secondary oocyte and the smaller first polar body which degenerates. Similarly, the second meiotic division is also asymmetrical, resulting in the production of one ovum and the small second polar body which also degenerates. The final DNA synthesis and prophase of the first meiotic division begin in embryos; however, the primary oocytes are then arrested in prophase of the first meiotic division. Each month from puberty until the menopause, one oocyte is stimulated to resume progress through meiosis. The first meiotic division is usually completed prior to ovulation. The second meiotic division proceeds up to metaphase and then, if the egg is fertilized, the second meiotic division continues.

As described previously, human females possess a pair of X chromosomes, while the cells of males contain one X and one Y chromosome. During prophase of the first meiotic division in women, the two X chromosomes form a pair and separate from each another at anaphase of the first meiotic division. The X chromosome of the secondary oocyte divides at the second meiotic division into two daughter X chromosomes. Thus all normal ova contain one X chromosome. In men, however, the X and Y chromosome separate from each another at the first meiotic division. Each divides at the second meiotic division. Thus spermatozoa contain either an X or a Y chromosome. Fertilization of an egg that contains an X chromosome by a sperm containing an X chromosome results in a female zygote with an XX chromosome complement. If an egg is fertilized by a sperm containing a Y chromosome, an XY zygote with the male chromosome complement is formed.

Synaptonemal Complexes

The behavior of chromosomes during meiosis is more easily followed with the light microscope than with the electron microscope. However, distinctive structures known as synaptonemal complexes are visible with the electron microscope in the nuclei of cells during prophase of the first meiotic division (Fig. 5–32). Each synaptonemal complex consists of three linear densities, a central element flanked by two parallel lateral elements. Filamentous structures bridging the densities have been described.

Synaptonemal complexes have excited the interest of many investigators because their presence coincides with the pairing of homologous chromosomes. Although the exact nature of the components of the synaptonemal complexes has not been definitely established, it is thought that they represent the region of contact between two homologous chromosomes, and that crossing-over occurs within them. One view of their structure is that the two homologous chromosomes contribute to the lateral elements, but that the bulk of the complex is a protein framework that facilitates interaction between the homologues.

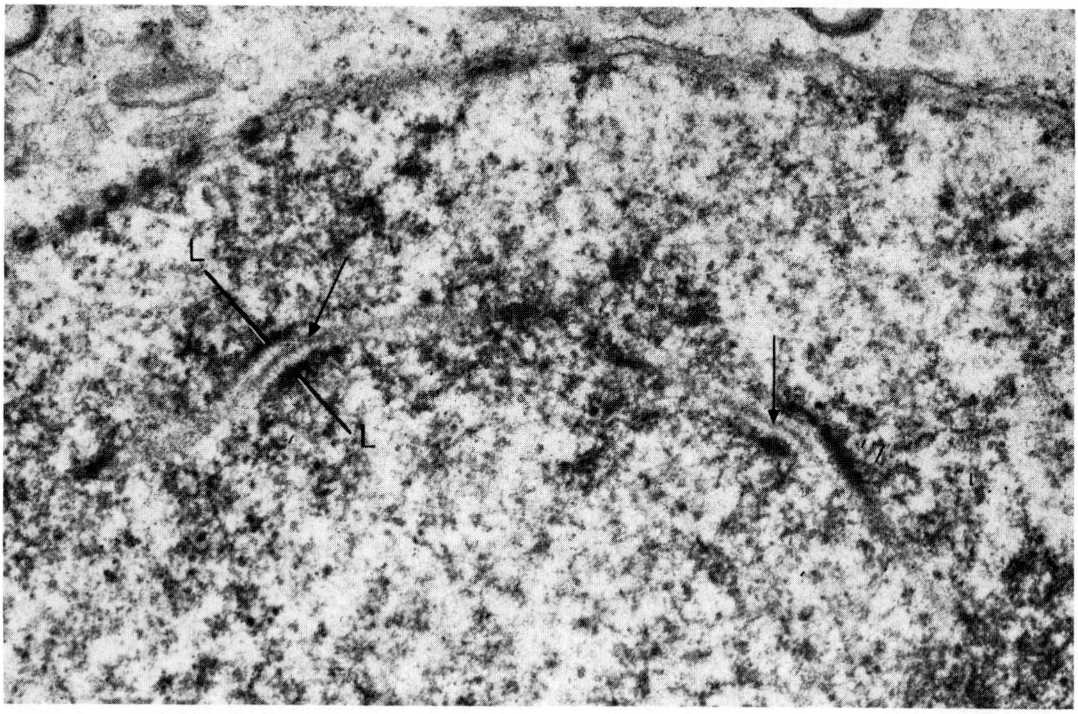

FIGURE 5–32 Synaptonemal complex in the nucleus of a rat spermatocyte. The complex consists of two lateral elements (L) flanking a central structure (*arrow*). Only a part of the complex is visible in this plane of section, and additional complexes are found in other parts of the nucleus. The synaptonemal complexes are believed to represent sites in pairing between two homologous chromosomes during meiosis. × 34,000.

A small amount of DNA synthesis, corresponding to about 0.3 per cent of the genome, occurs during prophase of meiosis. This DNA synthesis is scattered throughout the chromosomes, and although small in amount, it is apparently essential for normal behavior of the chromosomes during later meiotic stages. It has been suggested that this DNA synthesis represents repair of small segments of DNA as part of the mechanism of crossing-over and recombination. Of interest in this regard is the observation that inhibition of DNA synthesis during prophase blocks the formation of synaptonemal complexes.

Abnormalities of Meiosis

Nondisjunction or failure of chromosomes to separate in the first meiotic division results in one daughter cell receiving both members of a pair of homologous chromosomes, while the other daughter cell gets none. Thus gametes with a missing chromosome or an extra chromosome can be produced. At fertilization, combination with a normal gamete results in a zygote with an extra chromosome or a missing one.

If the autosomes are affected, one of the following situations results. Absence of an autosome is usually lethal early in development. Presence of an extra autosome is designated a trisomy because the individual possesses three examples of a particular chromosome—the normal pair of homologous chromosomes and the extra one. Trisomies are usually associated with mental retardation. An example is Down's syndrome, which is associated with trisomy of chromosome 21. The affected individuals have mental retardation, increased susceptibility to certain diseases, and a characteristic facial structure. The condition occurs in 1 in

1000 births, and it is more common in the offspring of mothers over the age of 35.

Since the absence of a sex chromosome as well as the presence of an extra example is compatible with life, numerous syndromes due to abnormalities of the sex chromosomes are known. An example of this is ovarian dysgenesis, also known as Turner's syndrome. This occurs in females with a single X chromosome and in individuals with a variety of other abnormalities of one X chromosome. These patients have hypoplastic ovaries that lack germinal follicles and thus are sterile. Other abnormalities described include primary amenorrhea, short stature, and a webbed neck. Klinefelter's syndrome is a sex chromosome abnormality seen in males with 47 chromosomes and an XXY sex chromosome composition (Fig. 5–13). It also is associated with other karyotypes having an excess of X chromosomes over Y chromosomes. It is said to occur as frequently as once in every 500 births, with the affected individual being sterile and having small testes, gynecomastia, and a high-pitched voice.

FURTHER READING

Avery, O. T., MacLeod, C. M., and McCarthy, M.: Studies on the chemical nature of the substance inducing transformation of pneumococcal types. Induction of transformation by a deoxyribonucleic acid from pneumococcus type III. J. Exp. Med., *79*:137, 1944.

Beadle, G. W., The Genes of Men and Molds. Sci. Am. Sept. 1948. Offprint No. 1, W. H. Freeman, San Francisco.

Beams, H. W., and Kessel, R. G.: Cytokinesis: a comparative study of cytoplasmic division in animal cells. Am. Sci., *64*:279, 1976.

Brinkley, B. R., and Stubblefield, E. T.: Ultrastructure and interaction of the kinetochore and centriole in mitosis and meiosis. *In*: Prescott, D. M., Goldstein, L., and McConkey, E. (eds.): Advances in Cell Biology, Vol. 1. New York, Appleton-Century-Crofts, 1970, p. 119.

Chromosome structure and function. Cold Spring Harbor Symp. Quant. Biol., *38*, 1973.

DeHarven, E.: The centriole and mitotic spindle. *In*: Dalton, A. J., and Haguenau, F. (eds.): Ultrastructure in Biological Systems. Vol. 3. New York, Academic Press, 1968.

Dirksen, E. R.: Centriole morphogenesis in developing ciliated epithelium of the mouse oviduct. J. Cell Biol., *51*:286, 1971.

Fulton, C.: Centrioles. *In*: Reinert, J., and Ursprung, H. (eds.): Origin and Continuity of Cell Organelles vol. 2. Heidelberg, Springer-Verlag, 1971, p. 170.

Gerald, P. S.: Sex chromosome disorders. N. Engl. J. Med., *294*:706, 1976.

Goss, R. T.: Turnover in cells and tissues. *In*: Prescott, D. M., Goldstein, L., and McConkey, E. (eds.): Advances in Cell Biology. Vol. 1. New York, Appleton-Century-Crofts, 1970, p. 233.

Huberman, J. A., and Riggs, A. D.: On the mechanism of DNA replication in mammalian chromosomes. J. Mol. Biol., *32*:327, 1968.

Inoué, S.: Organization and function of the mitotic spindle. *In*: Allen, R., and Kamiya, N. (eds.): Primitive Motile Systems in Cell Biology. New York, Academic Press, 1964, pp. 549–598.

Kornberg, A.: DNA Synthesis. San Francisco, W. H. Freeman, 1974.

Luykx, P.: Cellular mechanisms of chromosome distribution. Int. Rev. Cytol. (Suppl.), *2*:1, 1970.

Mazia, D.: The cell cycle. Sci. Am. *230:1*:55, 1974.

McIntosh, J. R., Hepler, P. K., and VanWie, D. G.: Model for mitosis. Nature, *224*:659, 1969.

Meselson, M., and Stahl, F. W.: The replication of DNA in *Escherichia coli*. Proc. Natl. Acad. Sci. USA, *44*:671, 1958.

Nicklas, R. B.: Mitosis. *In*: Prescott, D. M., Goldstein, L., and McConkey, E. H. (eds.): Advances in Cell Biology. Vol. 2. New York, Appleton-Century-Crofts, 1970, p. 225.

Pickett-Heaps, J. D.: The autonomy of the centriole: fact or fallacy? Cytobios, *3*:205, 1971.

Prescott, D. M. Structure and replication of eukaryotic chromosomes. *In*: Prescott, D. M., Goldstein, L., and McConkey, E. (eds.): Advances in Cell Biology. Vol. 1. New York, Appleton-Century-Crofts, 1970, p. 57.

Prescott, D. M. (ed.): Reproduction of Eukaryotic Cells. New York, Academic Press, 1976.

Prescott, D. M., and Keumpel, P. L.: Bidirectional replication of the chromosome in *E. coli*. Proc. Natl. Acad. Sci., USA, *69*:2842, 1972.

Stubblefield, E.: The structure of mammalian chromosomes. Int. Rev. Cytol., *35*:1, 1973.

Taylor, J. H.: The duplication of chromosomes. *In*: von Sette, P. (ed.): Probleme der biologischen Reduplikation. Berlin, Springer-Verlag, 1966.

Watson, J. D., and Crick, F. H. C.: Molecular structure of nucleic acid. A structure for deoxyribose nucleic acid. Nature, *171*:737, 1953.

6

THE INTERPHASE NUCLEUS

The main functions of the nucleus are storage and replication of DNA; synthesis of RNA, or transcription; and transport of RNA to the cytoplasm. The preceding chapter was concerned with how genetic information is stored, replicated, and distributed to daughter cells during cell division. It was seen that information is stored in the sequences of nucleotide bases in DNA, that the DNA molecule is replicated by a semiconservative mechanism, and that DNA is distributed in the condensed mitotic chromosomes during cell division. In this chapter the study of nuclear functions will be continued by considering the way the genetic information in DNA is expressed and the relation of the main structural components of the nucleus—chromatin, the nucleolus, and the nuclear envelope—to this activity.

It is known from the work of Beadle and Tatum in the 1940's that many genes are ultimately expressed in the functions of protein enzymes. However, many proteins are found in the cytoplasm, and protein synthesis is known to take place there. The problem then is to explain how DNA, which does not leave the nucleus, can specify the exact structure of proteins. Clearly, DNA is in the wrong place to be a direct template for protein synthesis. Therefore, there must be an intermediate form of the genetic information that can bridge the gap between the nuclear DNA and the cytoplasm, where protein synthesis takes place.

By 1961 it had been demonstrated that the intermediate stage of the genetic information is the metabolically unstable form of RNA called messenger RNA. Upon reflection, it can be seen that RNA is well suited to perform this messenger function. RNA is structurally similar to DNA in that it consists of an unbranched backbone of alternating sugar and phosphate groups, with purine and pyrimidine bases attached to the sugars. The sugar-phosphate backbone is the same as that in DNA except that the sugar is ribose instead of 2-deoxyribose, and RNA consists of a single polynucleotide chain instead of the two chains in DNA. The bases in RNA are the same as in DNA except that uracil, which forms base pairs with adenine, is present instead of thymine. Genetic information can therefore be carried in the sequence of RNA bases in the same way that it is carried in DNA. RNA also has the proper metabolic properties to function as a carrier of genetic information. Virtually all cellular RNA is synthesized in the nucleus, and much of it is then transported to the cytoplasm where it functions in protein synthesis.

The overall flow of genetic information in living cells, therefore, takes place from DNA (the reservoir and replicative form of genetic information) to RNA (the messenger) to protein (the functional form of the genetic program). This generalization is often identified as the central dogma of molecular biology, and it may be summarized by this drawing, in which the arrows indicate the flow of genetic information.

$$\text{DNA} \rightarrow \text{RNA} \rightarrow \text{Protein}$$

The crucial events are the transfer of genetic information from DNA into RNA and then from RNA into the amino acid sequences of proteins. This chapter deals with the former process, which is called transcription, or the transcription of DNA into RNA, since information contained in the nucleotide sequences of DNA is transferred to similar nucleotide sequences in RNA. The latter process is identified as translation, because the language of nucleotide sequences in RNA must be translated into the language of amino acid sequences in protein. Translation will be discussed in Chapter 7.

TRANSCRIPTION

General Features of RNA Synthesis

Transcription, or DNA-dependent RNA synthesis, is the process by which nucleotide sequences in DNA are faithfully copied into corresponding nucleotide sequences in RNA. The end products of transcription are the three types of cellular RNA, all of which are involved in protein synthesis. These are messenger RNA (mRNA), ribosomal RNA (rRNA), and transfer RNA (tRNA).

MESSENGER RNA (mRNA)

Messenger RNA's contain the sequences of nucleotides that directly specify the sequences of amino acids found in proteins. They vary between approximately five hundred and several thousand nucleotides in length. Generally, mRNA's contain only the usual four RNA bases, A, G, U, and C, with the exception of the 5′ ends of some eukaryotic mRNA's, which contain a 7-methyl G group. Intracellularly, mRNA's are associated with ribosomes, but mRNA is not an integral part of the ribosomal structure. Messenger RNA's are metabolically unstable in the sense that they do not have an indefinite life span inside the cell. Most are degraded after a definite interval following their synthesis, and they must be replaced if synthesis of the proteins for which they code is to continue.

RIBOSOMAL RNA (rRNA)

The ribosomal RNA's are an integral and permanent part of the ribosome structure, yet they do not code for protein amino acid sequences. All cytoplasmic ribosomes in a given species of organism have identical sets of rRNA's, and these rRNA's are identified by their sedimentation coefficients. Ribosomes from prokaryotic organisms contain one copy each of three kinds of rRNA, designated 23S RNA (MW 1.2×10^6), 16S RNA (MW 6×10^5), and 5S RNA (MW 4×10^4). Ribosomes from eukaryotic organisms contain 28S RNA (MW 1.7×10^6), 18S RNA (MW 7×10^5), 5.8S RNA (MW 5×10^4), and 5S RNA (MW 4×10^4). In both prokaryotic and eukaryotic organisms the largest of the rRNA species (23S or 28S) is associated with the larger of the two ribosomal subunits, while the next largest rRNA (16S or 18S) is associated with the small ribosomal subunit. The 5S and 5.8S rRNA's are found in the large subunit. The two largest species of rRNA (23 and 16S, or 28 and 18S) contain 1 to 2 per cent of nucleotides that are methylated either on the 2′ hydroxyl of ribose or on the base itself to yield derivatives such as N^2-methylguanine, 2-methyladenine, or 5-methylcytosine.

Ribosomal RNA is by far the most abundant class of intracellular RNA; 60 to 80 per cent of a cell's total RNA is ordinarily found to be rRNA. In spite of this abundance, however, the biological function of rRNA is not yet known. Possibly rRNA's simply have a structural function; they may hold the ribosomal proteins together in the proper orientation for protein synthesis.

TRANSFER RNA (tRNA)

Transfer RNA's are the smallest of the intracellular RNA's, being approximately 80 nucleotides in length. They function in protein synthesis as the direct link between the triplet nucleotide code in mRNA and the amino acid sequence in proteins. There is at least one and usually three or more tRNA species for each of the twenty amino acids found in proteins. Each tRNA molecule has a region, called the anticodon, which is complementary to 1 of the 64 possible triplet code words or codons in the genetic code. At the same time, each tRNA molecule also contains a site on its 3′ end for the covalent attachment of 1 of the 20 amino acids. Only the amino acid for which its anticodon is specific can be attached to a particular tRNA species, and this amino acid is eventually donated in a specific fashion onto the end of the growing polypeptide chain during protein synthesis. Transfer RNA chains always contain approximately 10 per cent of unusual nucleotides, including ribothymidine, pseudouridine, dihydrouridine, inosine, and 2′-O-methylguanine.

All three types of celluar RNA are synthesized by the same mechanism, which resembles in many ways the molecular mechanism for DNA replication. Growing RNA strands are built up one step at a time from the four ribonucleoside-5′-triphosphates, beginning at the 5′ and finishing at the 3′ end of the growing RNA chain. At each step the correct nucleotide is identified by its ability to base pair with a corresponding nucleotide in one of the two DNA strands. In order to accommodate base pair formation with the ribonucleoside-5′-triphosphates, the two DNA strands must separate from each other in the local region where transcription is taking place.

Once the correct ribonucleoside-5′-triphosphate is identified in this way, it is covalently joined to the growing polynucleotide chain enzymatically, with the loss of inorganic pyrophosphate, as shown in Figure 6–1. The enzyme involved is called DNA-dependent RNA polymerase or simply RNA polymerase. Like DNA polymerase, it can connect any of the four ribonucleoside-5′-triphosphates onto the end of the growing RNA chain; it has no nucleotide specificity. Unlike the situation in DNA replication, however, only one of the DNA strands, the sense strand, is copied during transcription; the other strand, the nonsense strand, is not copied. Which of the two DNA polynucleotide chains serves as the sense strand and which as the nonsense strand is found to vary from one gene or unit of transcription to another. The sense strand for one gene may be the nonsense strand for another. The final product of transcription is therefore complementary in base sequence to one of the two DNA strands in its gene and identical to the other DNA strand (except for the fact that U in RNA replaces T in DNA).

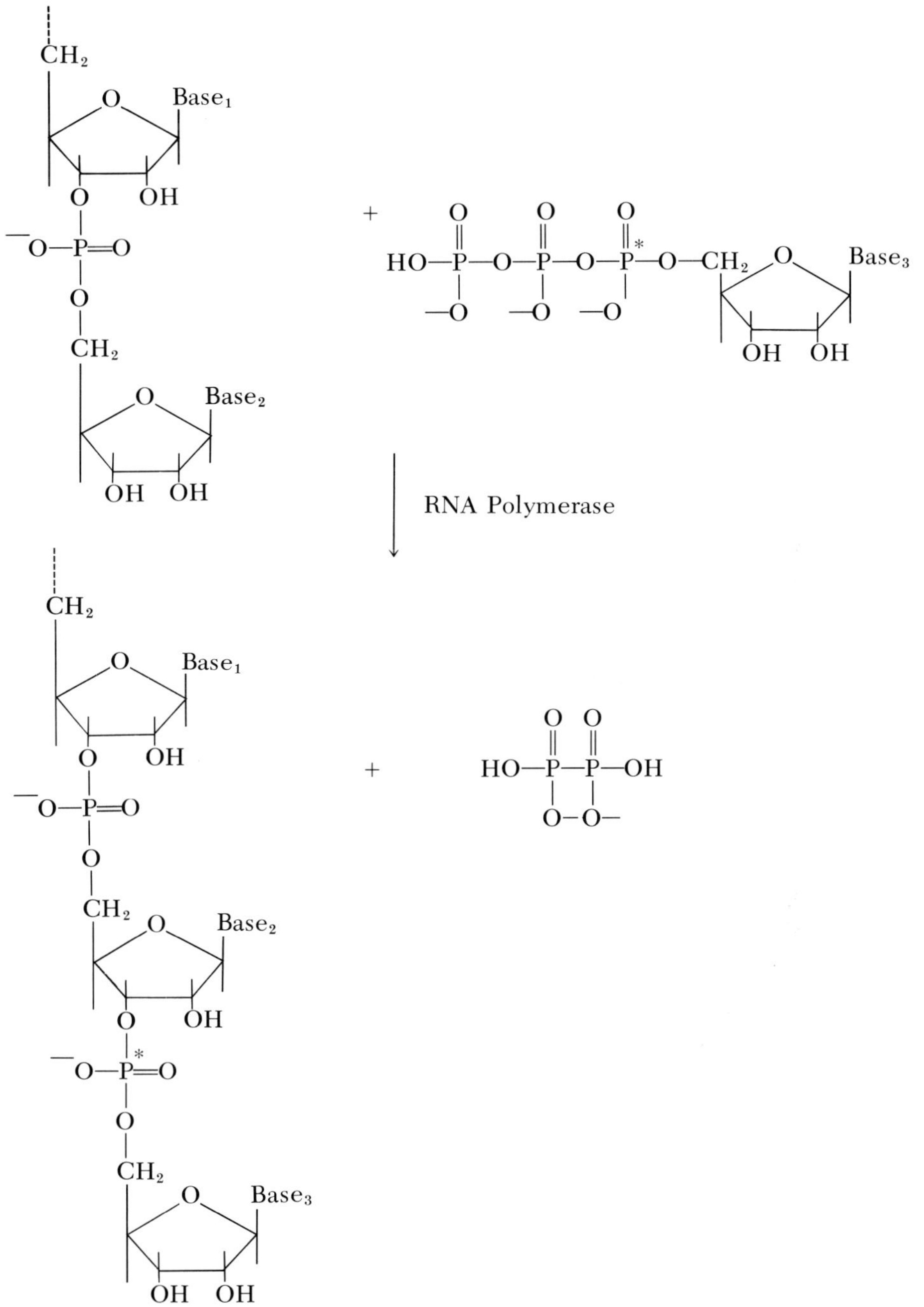

FIGURE 6–1 Extension of an RNA chain by one nucleotide as it occurs in DNA-dependent RNA synthesis (transcription). The identity of the added nucleotide is determined by its complementarity to the template DNA. An asterisk (*) indicates the phosphate in the precursor nucleoside-5′-triphosphate that forms the new 3′-5′ phosphodiester bond.

Precursor RNA's

Transcription of all three kinds of cellular RNA is characterized by the fact that the product of the transcription process is always larger than the final functional form of the RNA. The initial product of transcription must therefore be partially degraded by ribonucleases to produce the mature RNA species. Furthermore, since only the four ribonucleoside-5′-triphosphates can be incorporated into RNA during transcription, the primary product of transcription must also be modified to produce the methylated and other chemically altered nucleotides found in RNA.

Together these two functions, nuclease digestion and chemical modification of the primary transcription product, are referred to as processing or tailoring of RNA. Some combination of these functions is involved in the maturation of all functional RNA species. For example, three of the four ribosomal RNA species found in mammalian cells are derived from the same primary transcription product. This precursor rRNA is produced in the nucleolus, and it has a sedimentation coefficient of 45S. After its synthesis, the 45S RNA (MW 4.1×10^6) is processed in at least three steps in the nucleus to yield the 28S, 18S, and 5.8S rRNA's, as shown in Figure 6–2. In the first step an endonuclease cleaves the 45S precursor rRNA into a 41S precursor rRNA (MW 3.1×10^6) and a fragment of molecular weight 1×10^6, which is degraded. Two other degraded fragments are produced in the final stages of rRNA maturation, and in the end approximately 43 per cent of the 45S precursor rRNA is lost before the 28S, 18S, and 5.8S rRNA's are produced. Once the 28S and 18S rRNA species have been cleaved to the proper size, they serve as substrates for methylase enzymes which introduce methyl groups at the proper places along the RNA chains. All rRNA species then as-

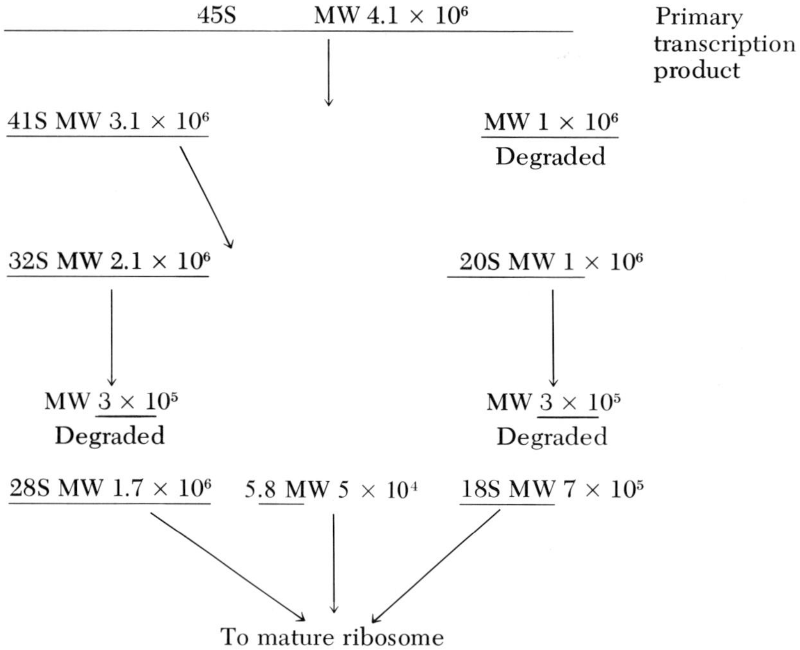

FIGURE 6–2 Processing of the 45S "pre-ribosomal" RNA to give three (28S, 18S, and 5.8S) of the four RNA species found in mature ribosomes.

sociate with proteins to form the intact ribosomal particles that function in protein synthesis.

Similar processing steps are involved in the maturation of transfer RNA. The best-studied case of tRNA biosynthesis is that of the *E. coli* tyrosine tRNA, which in its mature form consists of a single polynucleotide chain 85 nucleotides in length. Although only a single polynucleotide chain is present, the sequence loops back on itself in five places where base pairs form. This results in local double helical regions and provides the molecule with a cloverleaf configuration (Fig. 6–3). Similar base-pairing schemes and cloverleaf configurations can be drawn for all tRNA molecules. In addition to the usual four nucleotides, the *E. coli* tyrosine tRNA contains two pseudouridine (ψ) residues and one each of ribothymidine (T), 4-thiouridine (TU), 2′-O-methylguanosine (2′ OMG), and 2-methyl-6-isopentenyl-adenosine (A2M6I). This mature tRNA is derived from a primary transcription product 128 nucleotides in length, which contains the tyrosine-tRNA sequence plus an additional 41 nucleotides at the 5′ end and 2 nucleotides at the 3′ end. After the complete precursor has been synthesized, these extra nucleotides are removed by endonucleolytic cleavage, as shown in Figure 6–4. Chemical modifications are then introduced into the remaining structure to yield the mature tyrosine-tRNA.

In general, much less is known about the exact steps involved in the processing of mRNA. In eukaryotic cells at least, a great deal of processing is believed to take place. The precursors of eukaryotic mRNA's are thought to be a class of very large RNA's (up to 200S) called heterogeneous nuclear RNA or hnRNA. The hnRNA's are synthesized in the nucleus and cleaved there to produce much smaller structures having sedimentation constants of approximately 6S to 30S. These structures contain the nucleotide sequences that code for proteins. After they are produced in the nucleus, poly A is added to the 3′ end and a 7-methyl G group may be added to the 5′ end. The mature mRNA's are then transported to the cytoplasm where they become associated with ribosomes and direct the synthesis of protein.

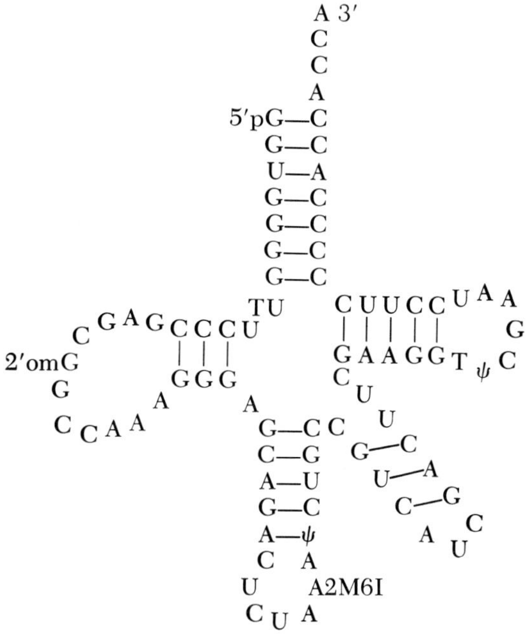

FIGURE 6–3 Nucleotide sequence and cloverleaf base pairing scheme in *E. coli* tRNA^{tyr}.

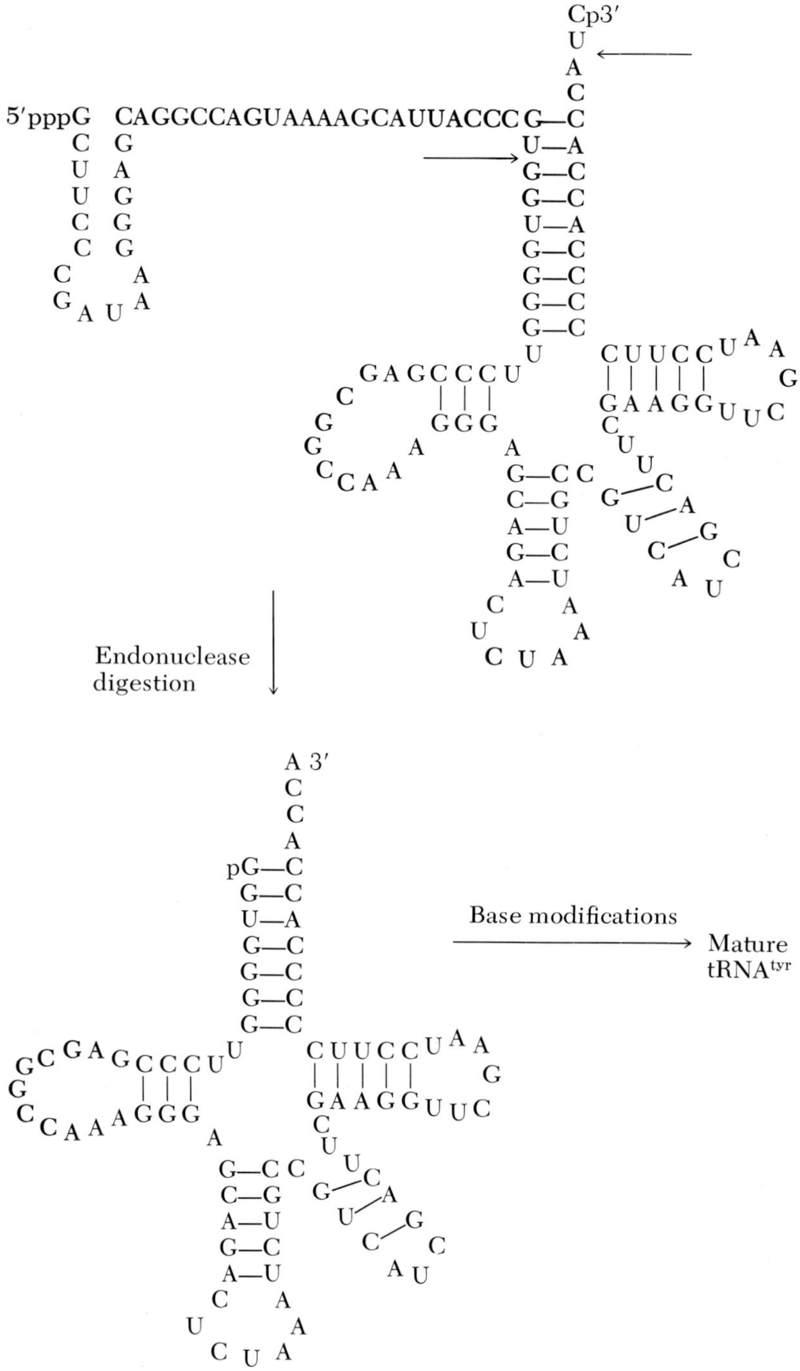

FIGURE 6–4 Processing of the precursor tRNA^{tyr} to yield the mature form of tRNA^{tyr}. The sites of endonucleolytic cleavages are indicated by arrows.

Comparison of Eukaryotic and Prokaryotic Transcription

As in the case of DNA replication, the mechanism of transcription differs somewhat in eukaryotic cells compared with prokaryotic organisms. For example, the DNA-dependent RNA polymerases of eukaryotic cells are all found in the nucleus, whereas in prokaryotic cells, which have no nuclear envelope, the RNA polymerase molecules are free in the cytoplasm. Furthermore, whereas prokaryotes have only one type of RNA polymerase, eukaryotic organisms have three. These are called RNA polymerases I, II, and III, respectively, and each has a different function. RNA polymerase I is found only in the nucleolus, and its function is to synthesize ribosomal RNA. Its main product, therefore, is the 45S precursor rRNA. RNA polymerase II is specialized to synthesize mRNA, while RNA polymerase III synthesizes all the tRNA's. The three eukaryotic RNA polymerases can be distinguished operationally on the basis of their sensitivity to the drug α-amanitin, the toxic factor present in the mushroom *Amanita phalloides*. RNA polymerase II is completely inhibited by low concentrations of α-amanitin, RNA polymerase III is inhibited only by high concentrations, and RNA polymerase I is insensitive to the drug. The properties of the three eukaryotic RNA polymerases are summarized in Table 6–1.

Although the vast majority of RNA synthesis in eukaryotic cells takes place in the nucleus, a small amount of transcription also occurs in the mitochondria, which contain small amounts of DNA. Mitochondrial RNA synthesis is mediated by a fourth type of RNA polymerase, sometimes called RNA polymerase IV, which is identical to the prokaryotic enzyme. This enzyme may be distinguished from the three nuclear eukaryotic RNA polymerases by its sensitivity to inhibition by the drugs rifampicin and streptolydigin; the prokaryotic enzyme is sensitive, while the eukaryotic nuclear enzymes are not.

Eukaryotic and prokaryotic transcription also differ in the way mRNA is synthesized and processed. Whereas in eukaryotic cells the mRNA's are thought to be produced by the cleavage of much larger precursors, the hnRNA's, this is often not the case in prokaryotic cells. In fact, translation of an mRNA in prokaryotic cells may begin even before its synthesis is complete. Ribosomes can begin translating the 5′ end of an mRNA before RNA polymerase has completed synthesis of the entire mRNA molecule. This is possible because in prokaryotic cells there is no nuclear membrane separating the cytoplasmic ribosomes from the nuclear sites of transcription. The result of this process is that transcription and translation are coupled events in prokaryotic cells, and the mRNA never exists in an intact form, free of ribosomes.

The Mechanism of Transcription in E. coli

By far the clearest view of the molecular events involved in transcription comes from studies of RNA synthesis in *E. coli*. Like other prokaryotes, *E. coli* has only one species of RNA polymerase. This enzyme, however, exists in two different forms, holoenzyme and core enzyme. The holoenzyme is made up of

PROPERTIES OF THE THREE NUCLEAR EUKARYOTIC RNA POLYMERASES **TABLE 6–1**

Enzyme	Location	Sensitive to α-Amanitin	Product
RNA polymerase I	Nucleolus	No	Ribosomal RNA
RNA polymerase II	Nucleus	Yes	Messenger RNA
RNA polymerase III	Nucleus	High concentration only	Transfer RNA

five different types of polypeptide chains called β' (MW 160,000), β (MW 150,000), σ (MW 90,000), α (MW 40,000), and ω (MW 10,000). All are associated with each other by weak noncovalent interactions. In the intact holoenzyme, there is one copy each of the five different polypeptides, except for α which is represented twice. The subunit composition of holo- RNA polymerase is therefore $\alpha_2\beta'\beta\ \omega\ \sigma$, and its molecular weight is approximately 500,000. Core RNA polymerase has the same subunit composition as the holoenzyme except that the core enzyme lacks σ. Its composition is $\alpha_2\beta'\beta\ \omega$, and its molecular weight is about 400,000, as shown in Table 6–2.

RNA polymerases isolated from cells other than *E. coli* also have this multisubunit composition. For example, all three eukaryotic RNA polymerases contain two large subunits with molecular weights comparable to those of β' and β in the *E. coli* enzyme plus several smaller subunits.

In vitro studies with purified RNA polymerase have shown that both core enzyme and holoenzyme will synthesize RNA in a DNA-dependent, template-specific fashion. It is clear, therefore, that the σ subunit is not involved either in identifying the correct ribonucleoside-5′-triphosphate at each step or in adding it onto the nascent RNA chain; the β', β, α, and ω subunits must perform these functions. The core enzyme and holoenzyme do, however, differ in *where* they will initiate DNA-dependent RNA synthesis. *In vitro,* the core enzyme can initiate RNA synthesis at any site along its template DNA molecule, but the holoenzyme does so only at selected regions of the genome. These regions are called promoters; they are found at the beginning of each gene, and they serve as specific sites for the initiation of transcription *in vivo.* The σ subunit is therefore thought to be involved in recognizing the promoter DNA base sequences and in insuring that transcription starts only at these very specific places. Since it would be very wasteful for a cell to initiate RNA synthesis at random places throughout the genome, the core enzyme is thought not to initiate RNA synthesis *in vivo* independently of σ.

The σ-dependent initiation of RNA synthesis *in vivo* is thought to take place in three steps. In the *first or "entry" step*, RNA polymerase becomes weakly associated with DNA. There is little or no specificity for a particular base sequence in the template DNA at this stage and σ factor is not required, although it may be present. The DNA strands probably do not come apart, and RNA polymerase molecules are relatively free to slide along the DNA. The association constant of RNA polymerase and DNA is relatively low during this period (Ka $\simeq 10^{10}M^{-1}$).

In the *second step*, RNA polymerase becomes tightly bound to specific promoter regions in DNA. The σ factor is absolutely required for recognizing the proper DNA base sequence; there is a high association constant between RNA polymerase and DNA (Ka $\simeq 10^{13}M^{-1}$); and the two DNA strands separate during tight binding. In prokaryotic organisms this stage of the overall transcription process is inhibited by rifampicin; no tight binding can take place in its presence.

TABLE 6–2 PROPERTIES OF *E. coli* RNA POLYMERASE

RNA Polymerase	Subunit Composition	Molecular Weight
Holoenzyme	$\alpha_2\beta'\beta\ \omega\ \sigma$	490,000
Core enzyme	$\alpha_2\beta'\beta$	400,000

The exact sequence of DNA nucleotides recognized by RNA polymerase during specific binding has now been determined for several genes in *E. coli*, including those involved in lactose utilization (lac operon) and the gene coding for the precursor of tyrosine-tRNA (Fig. 6–5). In every case studied, the RNA polymerase binding site has been shown to be a set of seven contiguous base pairs having a sequence similar, but not necessarily identical, to the sequences shown in Figure 6–5. These are the sequences that must be recognized by σ factor. They are found either five or six base pairs removed from the first transcribed nucleotide, the site of the actual initiation of RNA synthesis.

The *third and final step* in the initiation of transcription occurs when the first few nucleotides of the new RNA chain are polymerized together. In *E. coli* all transcription begins with either A or G, and since the second nucleotide is added onto the 3′ hydroxyl group of the first, all nascent RNA chains have a free triphosphate group at the 5′ end, as shown here.

$$5'\text{ppp}\!-\!\underset{\substack{\text{or}\\ \text{G}}}{\text{A}}\!-\!\text{p}\!-\!\text{N}_2\!-\!\text{p}\!-\!\text{N}_3 \ldots 3'$$

After polymerization of nucleotides begins, there is no further requirement for RNA polymerase to recognize particular regions of the genome as initiation sites, so the σ factor is lost. Thus, only the core RNA polymerase is involved in RNA chain extension. The σ factor is then free to associate with another core RNA polymerase molecule and to initiate synthesis of another RNA chain.

The specificity involved in the initiation of RNA synthesis is crucial to living organisms for at least two reasons. First, it is important to begin transcription and translation only at the beginning of genes and not in the middle. To initiate RNA synthesis at random in the genome would be wasteful and might also result in the synthesis of many fragments of proteins that would not be functional. Second, it is important to transcribe only the correct set of genes at any particular time. Clearly, it would not be practical for an organism to transcribe and translate all its genes all the time. It is essential for organisms to be able to modulate and control expression of the genetic program in order to generate developmental programs and responses to environmental changes.

The initiation of transcription is an obvious step for this type of regulation to take place. Perhaps there are different σ-like factors available in the cell to

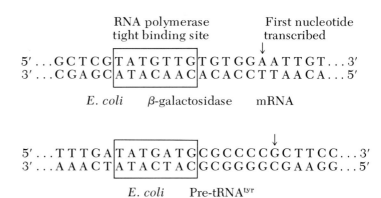

RNA polymerase tight binding site — First nucleotide transcribed ↓

5′ . . . G C T C G [T A T G T T G] T G T G G A A T T G T . . . 3′
3′ . . . C G A G C [A T A C A A C] A C A C C T T A A C A . . . 5′

E. coli β-galactosidase mRNA

↓

5′ . . . T T T G A [T A T G A T G] C G C C C C G C T T C C . . . 3′
3′ . . . A A A C T [A T A C T A C] G C G G G G C G A A G G . . . 5′

E. coli Pre-tRNA^{tyr}

FIGURE 6–5 RNA polymerase specific or tight binding sites (promoters) in the *E. coli* genes coding for the β-galactosidase mRNA and the precursor of tRNA^{tyr}.

recognize different DNA base sequences as binding sites for RNA polymerase. Changes in expression of the genetic program could then be effected by simply changing the σ-like factor associated with the core RNA polymerase. This would change the initiation sites recognized by RNA polymerase without affecting its ability to synthesize RNA. Although this is an appealing hypothesis, there is not yet sufficient evidence to support it as a general mechanism used in the regulation of gene expression.

Once synthesis of a new RNA strand has begun, it is elongated by the cyclic addition of nucleotides as previously described. The two DNA strands separate in the local region where transcription is taking place, and the next nucleotide to be added onto the nascent RNA chain is identified by its ability to form a base pair with the next nucleotide in the template DNA strand. When the correct ribonucleoside-5′-triphosphate is identified by this process, it is enzymatically joined to the nascent RNA strand by RNA polymerase. After a particular region of DNA has been transcribed, the two DNA strands reanneal with each other; the nascent RNA chain does not remain base paired to DNA. RNA synthesis in *E. coli* proceeds at a rate of approximately 40 nucleotides per second or about 20-fold more slowly than the corresponding rate of DNA synthesis. This rather slow rate of RNA synthesis may be the rate-limiting process for the growth of *E. coli* in some situations. The antibiotic streptolydigin specifically inhibits chain elongation by the prokaryotic type of RNA polymerase.

Termination of RNA synthesis is the least understood aspect of transcription. It is clear that transcription stops at very precisely defined points along the DNA, and both the completed RNA chain and RNA polymerase dissociate from DNA. The information for where this occurs must be contained in DNA nucleotide sequences, that is, there must be "stop" signals for transcription. However, it now seems likely that the core RNA polymerase cannot recognize these sites by itself and that other protein factors are required either to recognize the "stop" sequences or to dissociate the core enzyme and nascent RNA from DNA. Some possible factors of this type have been described in the literature. For example, one called ρ (rho) and another called SF (sizing factor) have both been implicated in the termination of RNA synthesis. Just how they function, however, is not yet known. Processing of newly synthesized RNA chains as described earlier can take place immediately after the termination step. A summary of the events involved in transcription of the *E. coli* genome is given in Figure 6–6.

In eukaryotic cells, transcription takes place in chromatin and in the nucleolus. Their structures and the sites of synthesis of different types of RNA are considered in the following sections.

CHROMATIN

Chromatin is the name applied to microscopically visible nuclear material that contains DNA and proteins. It is the structural representation of the chromosomes during interphase. Chromatin is basophilic in light microscope preparations owing to the binding of positively charged molecules of basic dyes by the negative charges on phosphate groups of DNA. Chromatin also stains with the Feulgen reaction, which is specific for DNA. In this procedure, mild hydrolysis with HCl removes purines from DNA, exposing aldehyde groups on deoxyribose. The Schiff reagent combines with the aldehydes to produce a characteristic pink color.

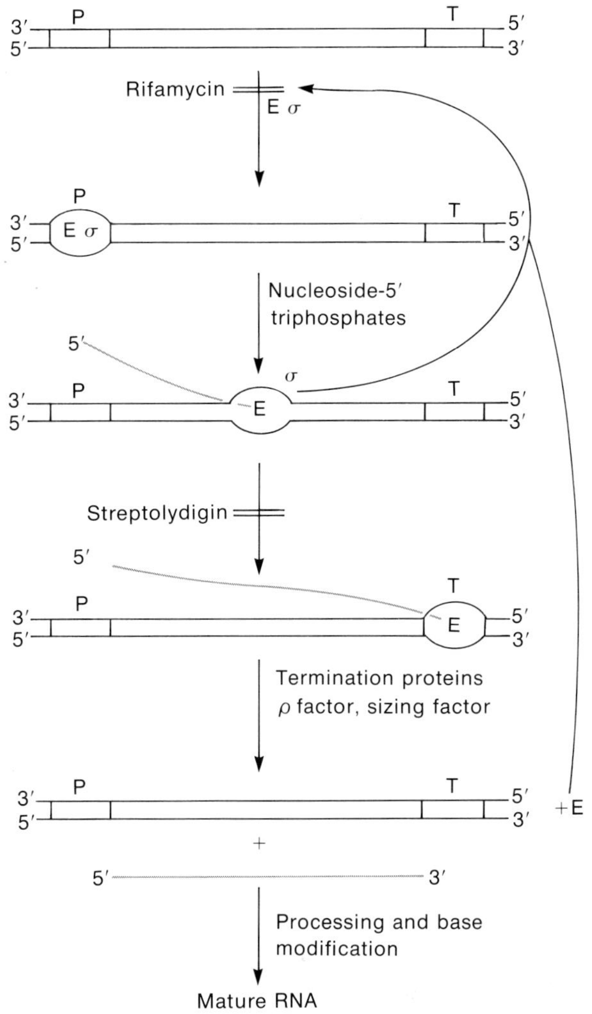

FIGURE 6–6 Summary of the steps involved in RNA synthesis by *E. coli* RNA polymerase. The promoter (P) and terminator (T) sites in the DNA are indicated. E represents the core form of RNA polymerase.

Euchromatin and Heterochromatin

The distribution of chromatin through the nucleus is uneven (Fig. 6–7). Thus two types of chromatin can be distinguished: dispersed chromatin, or euchromatin, and condensed chromatin, or heterochromatin. Euchromatin is loosely packed and thus is only lightly basophilic. Similarly, its staining with the Feulgen reaction is light or unapparent. However, euchromatin is metabolically active in the synthesis of RNA. If cells are exposed to uridine-^{3}H, which is a precursor of RNA, the label is incorporated mainly into the euchromatic regions of the nucleus. In contrast, heterochromatin is condensed, tightly packed, and intensely basophilic. As a result, it stains heavily with the Feulgen reaction. Heterochromatin, however, is relatively inactive metabolically and shows little or no incorporation of uridine-^{3}H. Heterochromatin usually replicates later in the S phase than euchromatin. The basis for the condensation and inactivity of heterochromatin in comparison with euchromatin is not well understood, but it may be related to differences in complexing of DNA with proteins or other molecules, such as RNA. The problem of compositional differences between heterochro-

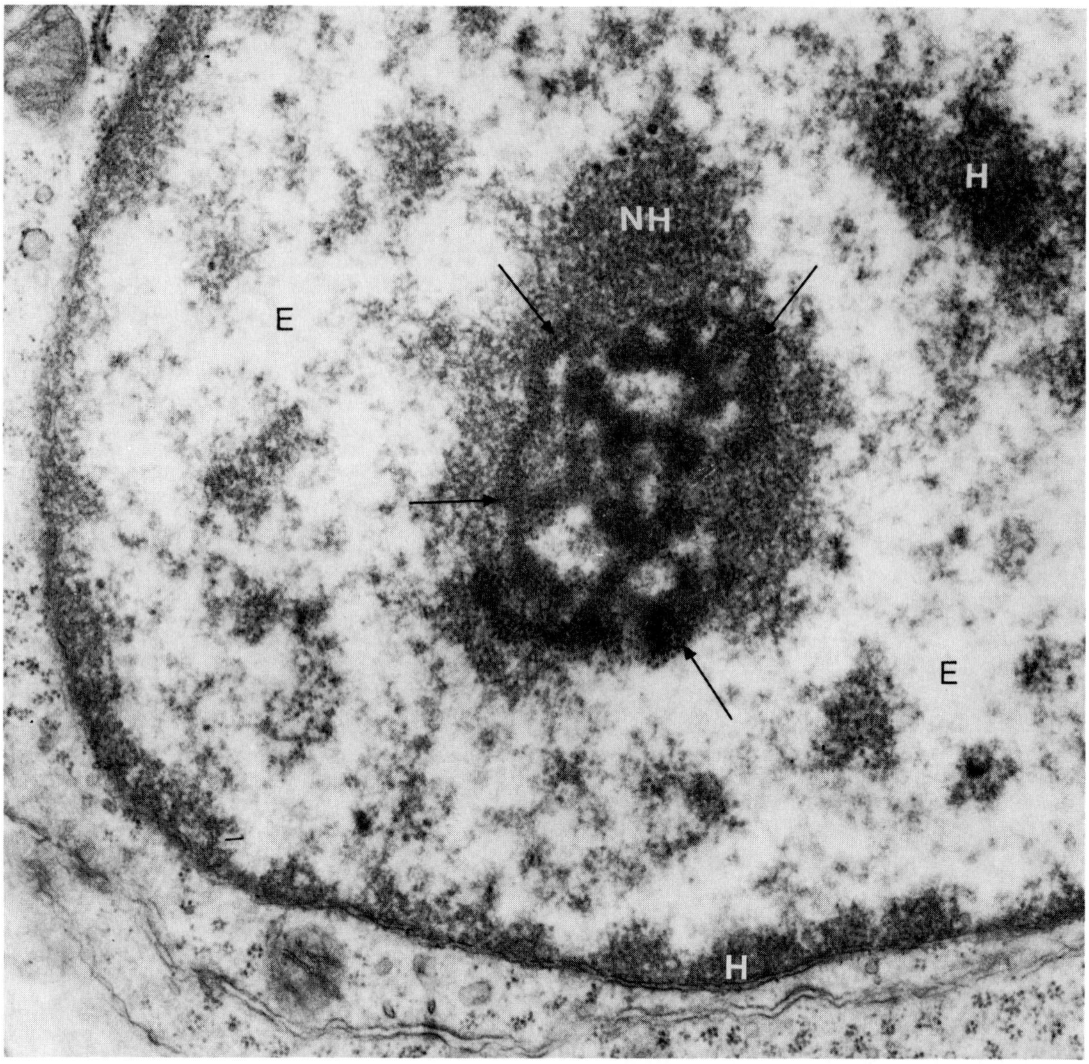

FIGURE 6–7 Nucleus of an epithelial cell from the fetal rat wolffian duct. Condensed heterochromatin (H) is found immediately inside the nuclear envelope and in clumps in the interior of the nucleus. The nucleolus is outlined by arrows. Some heterochromatin (NH) is associated with the margin of the nucleolus. The filamentous euchromatin (E) is dispersed throughout much of the remainder of the nucleus. × 35,000.

matin and euchromatin is a fundamental one, because it is related to the question of how gene expression is regulated in eukaryotic cells.

Since euchromatin is active in RNA synthesis, and RNA is used in protein synthesis in the cytoplasm, the proportion of euchromatin in the nucleus can often be taken as a rough index of the metabolic activity of a given cell. That is, much euchromatin and a large nucleolus are observed in synthetically active cells (Fig. 6–8). A large proportion of heterochromatin is associated with low metabolic activity (Fig. 6–9). An extreme degree of chromatin condensation is seen in the nuclei of mature spermatozoa (Fig. 6–10), in which compaction of the chromatin aids in its transportation in these motile cells.

Apparent exceptions to the generalization that euchromatin is active and heterochromatin inactive can be rationalized from knowledge of the life history

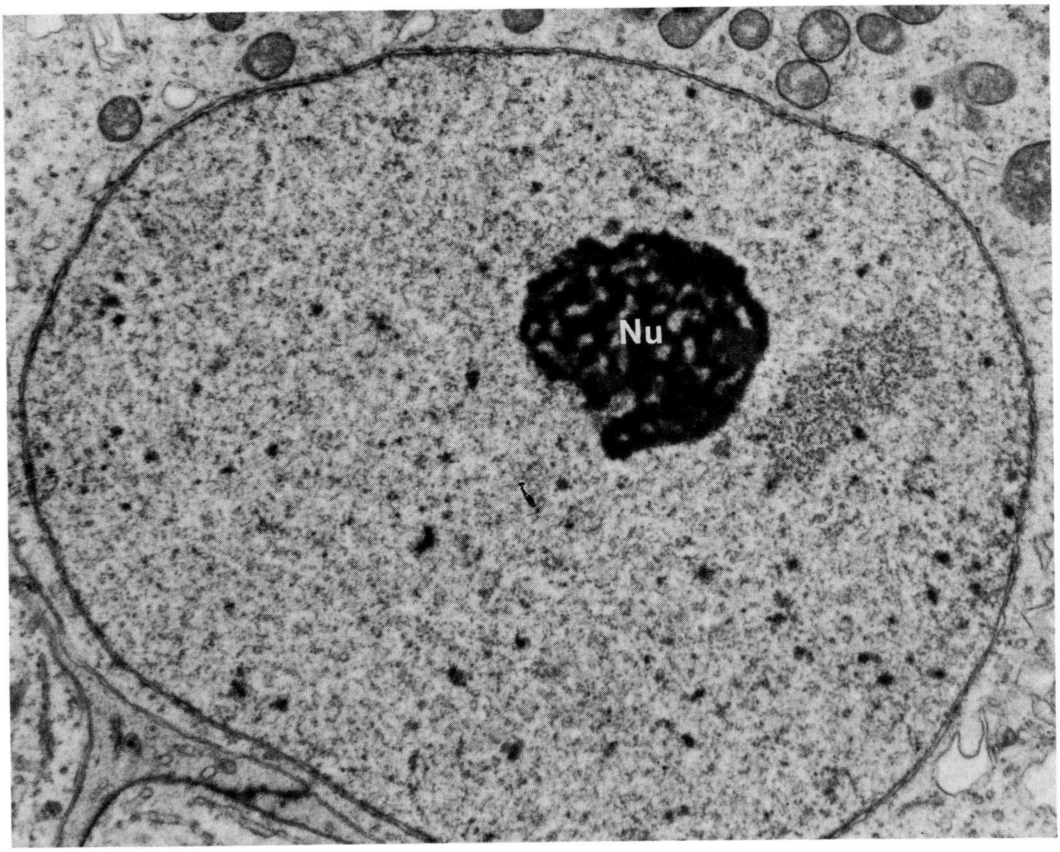

FIGURE 6–8 Nucleus of a germinal cell in the rabbit testis. Most of the interior of the nucleus is occupied by euchromatin and a large nucleolus (Nu). × 13,000.

and specialized activity of the cell in question. For example, in the later stages of development of erythroblasts in the bone marrow, the nucleus contains heterochromatin almost entirely, yet the cytoplasm of these cells actively synthesizes hemoglobin. In this case, the mRNA and ribosomes needed for hemoglobin production are synthesized at an earlier stage of development of the cell when the nucleus is more euchromatic. Being relatively long-lived molecules, they persist and continue to function. Plasma cell nuclei also contain a large amount of heterochromatin, but the cells synthesize much protein antibody. The explanation for this apparent paradox may be that these cells synthesize only a small number of different proteins in large quantities. Thus it is necessary for only a small proportion of the genome to be in the euchromatic state and to be transcribed.

Chromatin can transform from heterochromatin to euchromatin and vice versa. A striking example of such a change is seen when human lymphocytes are activated *in vitro*. Normally, the nucleus of a peripheral blood lymphocyte contains a high proportion of heterochromatin. When stimulated with phytohemagglutinin, however, the nucleus enlarges, most of the heterochromatin is converted to euchromatin, and the nucleolus increases in size. Concomitantly, the nucleus begins to synthesize much RNA.

Two main types of heterochromatin are sometimes distinguished, constitutive heterochromatin and facultative heterochromatin. Constitutive heterochromatin is that which is condensed at all times in any given cell, whereas facultative heterochromatin may or may not be in a condensed state.

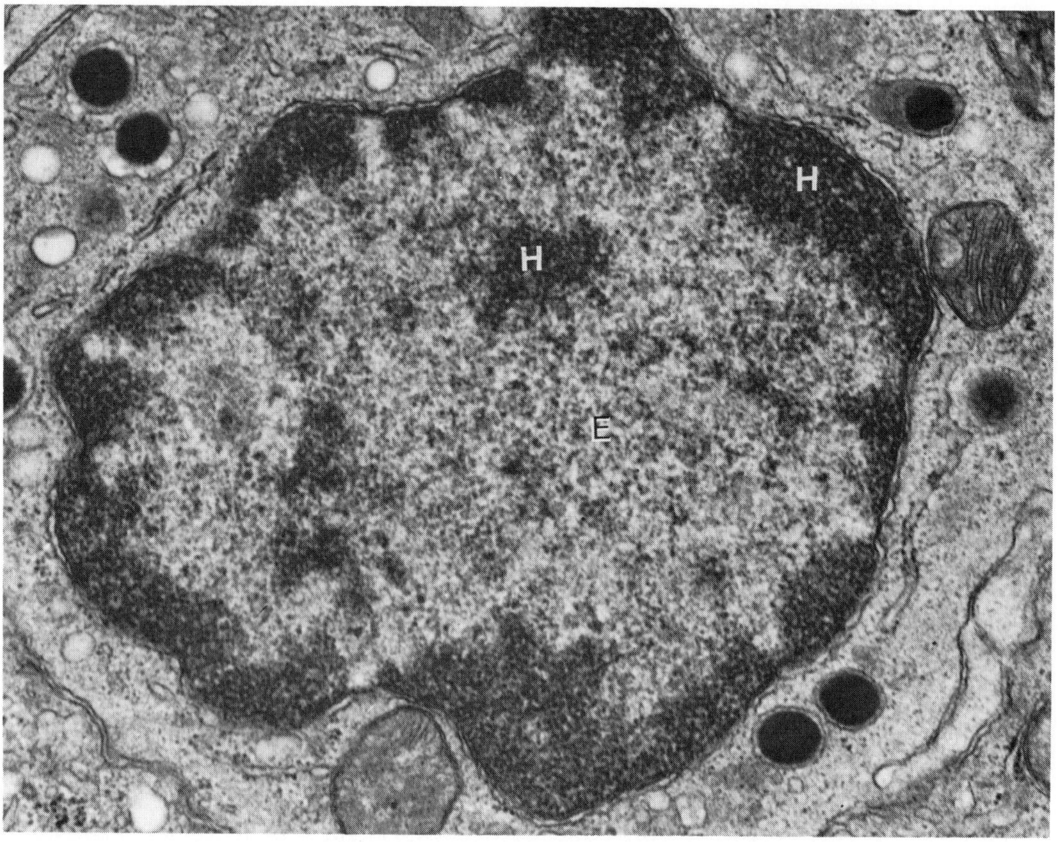

FIGURE 6–9 Nucleus of an intraepithelial leukocyte from the rat epididymis. Heterochromatin (H) is conspicuous immediately inside the nuclear envelope and in clumps in the interior of the nucleus. Part of the nucleus contains euchromatin (E). × 30,000.

One kind of constitutive heterochromatin is centromeric heterochromatin, which occupies the region of the kinetochore on the chromosomes. Centromeric heterochromatin remains condensed in interphase as well as in mitosis, replicates late in the S phase, and contains highly repetitive sequences of DNA. This DNA

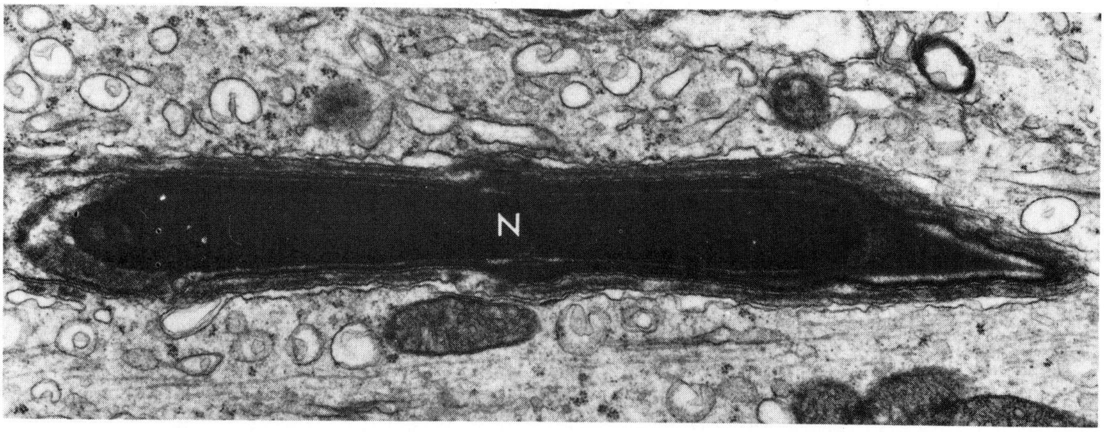

FIGURE 6–10 Head region of a nearly mature rat spermatid. In this precursor of a spermatozoon, the chromatin in the nucleus (N) is extremely condensed. × 19,000.

is also known as satellite DNA because it sediments in a centrifugal field apart from the bulk of the DNA, apparently as a result of its greater G—C content.

The best-known example of facultative heterochromatin is the inactive X chromosome of females. In female mammals, most of one X chromosome is inactive and is replicated late in the S phase. The particular one of the normal pair that will be inactivated is determined early in development and is random in each cell at that time. The inactive X chromosome is visible microscopically and is referred to as the sex chromatin or Barr body (Fig. 6–11). Its location within the nucleus varies. In neutrophils of the blood it often appears as a projection, or "drumstick," attached to one of the lobes of the nucleus. A Barr body is also visible as a dense clump of heterochromatin in a portion of cells in smears of the buccal mucosa from females, and study of the nuclei of these cells can be used as a test of the genetic sex of an individual. Since all but one of the X chromosomes in a given nucleus are inactivated, in cases of abnormalities of the sex chromosomes the number of Barr bodies per nucleus is equal to one less than the number of X chromosomes. For example, the nuclei of cells from an individual with the sex chromosome complement XXX display two Barr bodies.

Polytene and Lampbrush Chromosomes

The giant polytene chromosomes of insect salivary glands and the lampbrush chromosomes of amphibian oocytes have been widely studied and are familiar to many students of cell biology because their unusual features illustrate fundamental aspects of chromosome structure and function particularly well.

The giant polytene chromosomes of insect salivary glands (Fig. 6–12) contains many parallel strands of DNA formed by reduplication of the DNA of a chromosome. Since they are so large, different regions of condensed and less condensed DNA are readily visible with the light microscope, and consequently each chromosome displays a characteristic pattern of bands. At stages in the development of the organism, certain bands undergo reversible "puffing," with the formation of microscopically visible enlargements, the larger of which are known as Balbiani rings. RNA synthesis occurs in the puffs, and like the loops of lampbrush chromosomes, they can be regarded as regions of the DNA that are euchromatic, extended, and active in the synthesis of RNA. The insect hormone ecdysone induced a characteristic pattern of puffing, and thus this system furnishes an example of activation of specific genes in response to a hormone.

Lampbrush chromosomes (Fig. 6–13) occur in the nuclei of amphibian oocytes during prophase of meiosis. Each chromosome, which consists of two chromatids as the result of preceding DNA duplication, has a central axis, with loops extending laterally in pairs from the axis. RNA synthesis is known to occur on the loops. Thus the loops can be considered euchromatic regions of the chromosomes which are extended and active in transcription, while the DNA of the axis has been likened to condensed, inactive, heterochromatic DNA.

Fine Structure of Chromatin

Chromatin often appears in thin sections viewed in the electron microscope as a mass of fine fibrils (Fig. 6–14). The diameter of the fibrils varies with the cell and with the method of preparation in the range of approximately 80 to 200 Å, frequently being about 100 Å. The chromatin fibrils appear similar in euchromatin

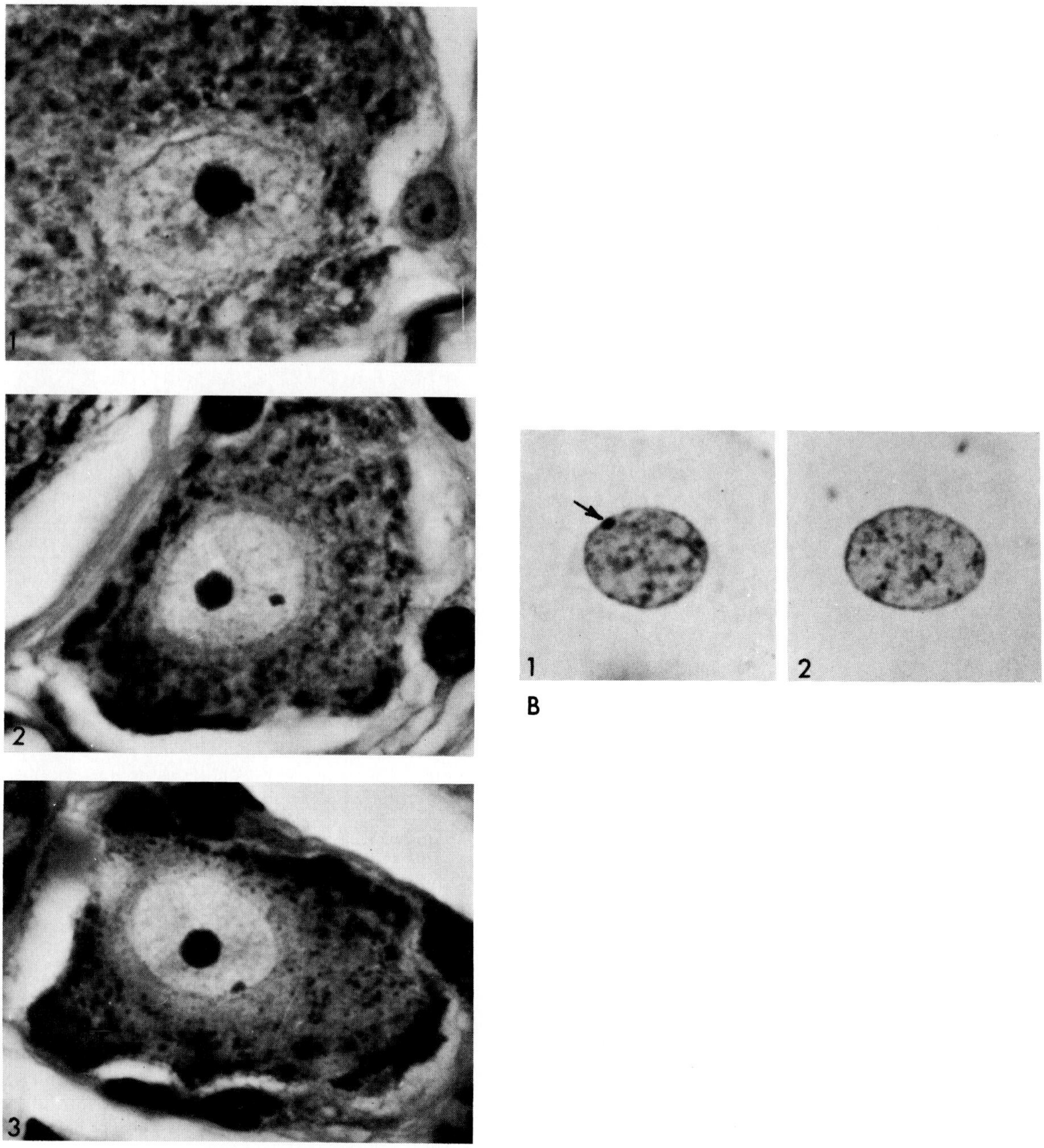

FIGURE 6–11 *A*. Sex chromatin in varying locations in the nuclei of neurons of a female cat: (1) adjacent to the nucleus; (2) in the nucleoplasm; (3) in a peripheral position just inside the nuclear envelope, which is typical of neurons of women. 1, dorsal root ganglion; 2 and 3, sympathetic ganglia. × 1600. *B*. Nuclei of cells from a buccal smear: (1) The sex chromatin is visible as a densely staining mass immediately inside the nuclear envelope (*arrow*) in a cell from a female; (2) The sex chromatin is absent in a cell from a male. Cresyl violet stain. × 2000. (*A*. Micrographs courtesy of Barr, M. L.: Int. Rev. Cytol., *19*:35, 1966. Reproduced by permission of Academic Press. *B*. Micrographs courtesy of M. L. Barr.)

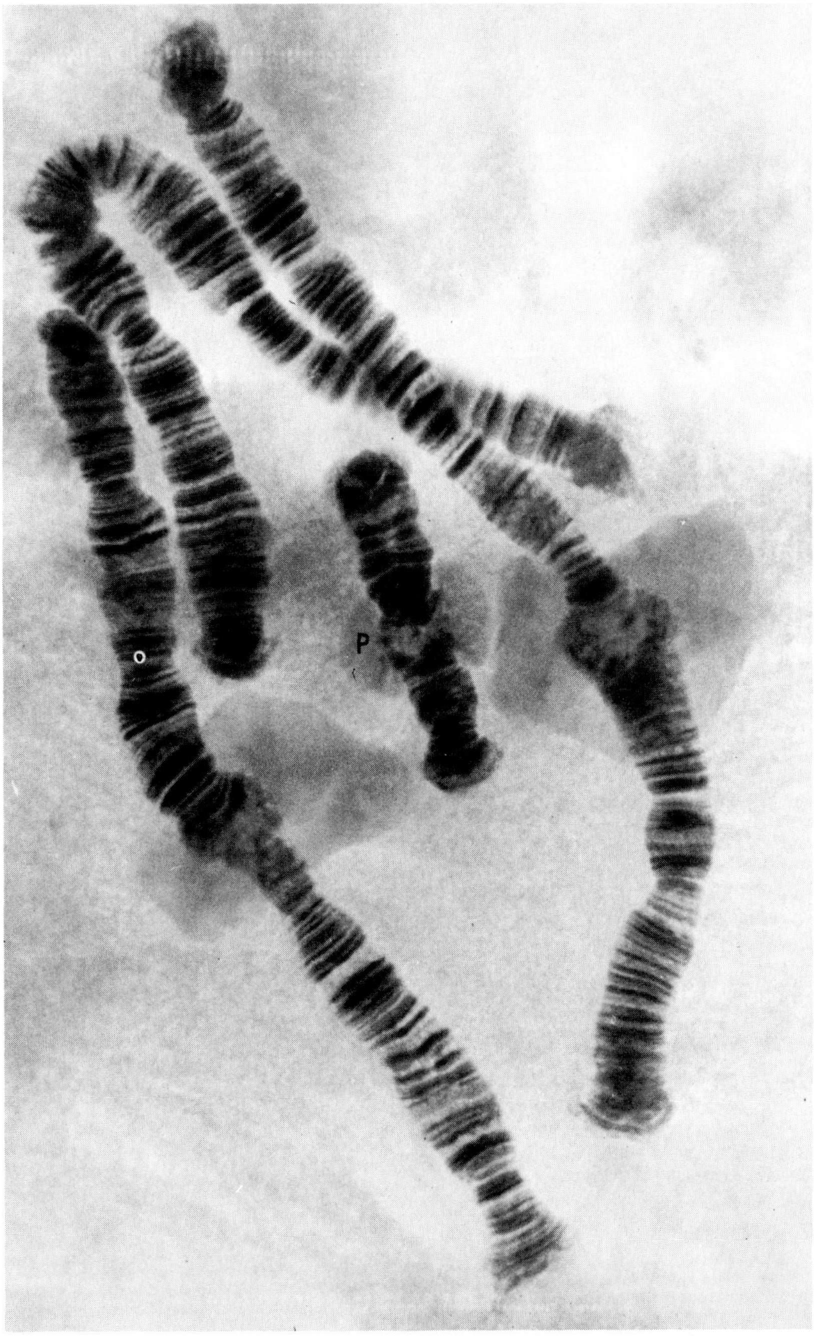

FIGURE 6–12 Giant polytene chromosomes from the salivary gland of *Chironomus tentans*. Note the conspicuous pattern of bands on the chromosomes. P indicates a large puff known as a Balbiani ring. Each puff represents a gene that is active in the synthesis of RNA. × 750. (Micrograph courtesy of Edström, J. E., and Beermann, W.: J. Cell Biol., *14*: 371, 1962. Reproduced by permission of the Rockefeller University Press.)

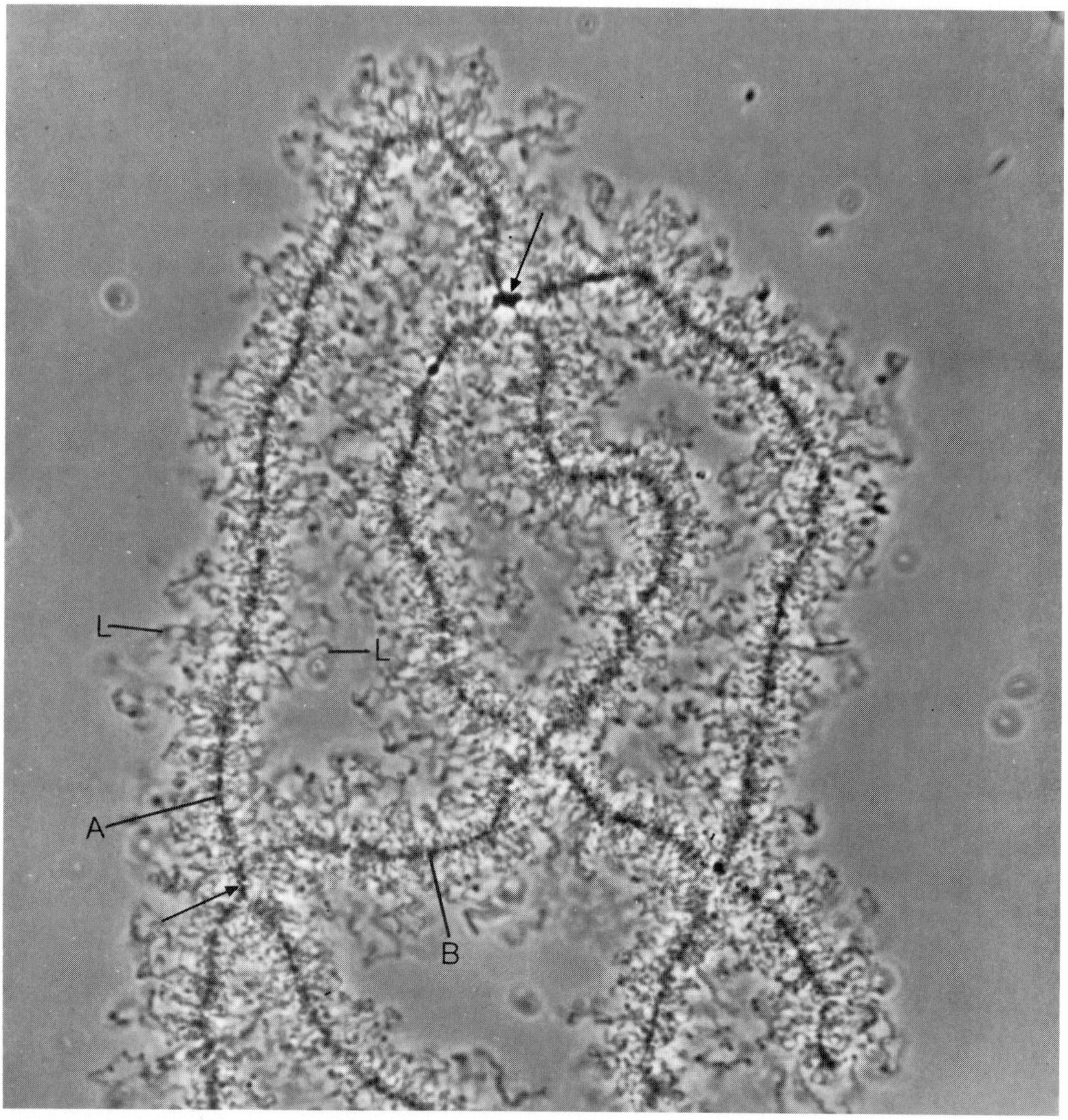

FIGURE 6–13 Lampbrush chromosomes of a newt oocyte viewed with the phase contrast light microscope. The two members of a pair of homologous chromosomes (A and B) are held together at diplotene of prophase in the first meiotic division by chiasmata (*arrows*). Fine paired loops (L) extend outward from the axis of each chromosome. The loops represent extended portions of the DNA axis that are active in RNA synthesis. × 680. (Micrograph courtesy of J. G. Gall.)

and heterochromatin, but they are more closely packed in heterochromatin. In some preparations the chromatin is reported to have a granular appearance. The chromatin is known to contain DNA and protein, including histones, but the precise organization of these molecules in the microscopically visible chromatin fibrils is uncertain. The fundamental problem is that chromatin fibrils are much thicker than DNA molecules. Although chromatin fibrils in sections are in the order of 100 Å or more in diameter, the DNA double helix is only 20 Å in diameter. Thus, the chromatin fibril cannot represent simply an extended DNA helix, rather, the DNA must be packed in the chromatin fibril in some more complicated way. This problem has been approached through the study of thin

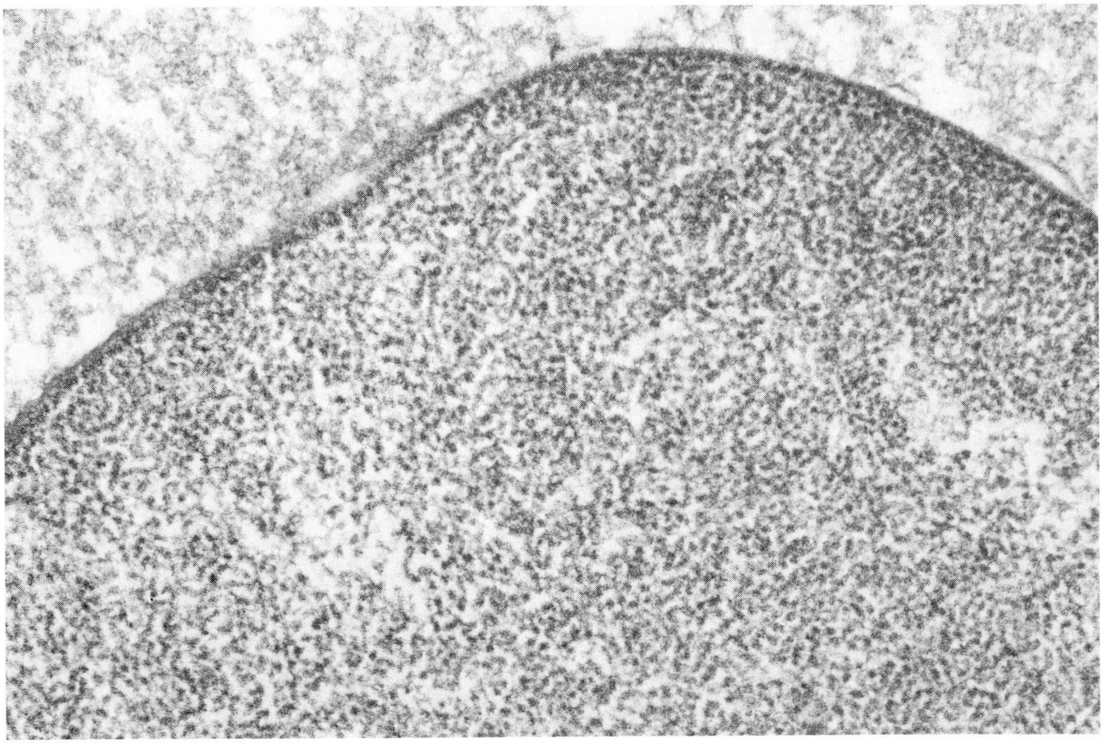

FIGURE 6–14 Thin section of the nucleus of a frog erythrocyte. Chromatin fibrils fill most of the field. × 43,000. (Micrograph courtesy of H. Ris. *From*: The Structure and Function of Chromatin. Amsterdam, Ciba Foundation Symposium, 1975, pp. 7–23. Reproduced by permission of the publisher.)

sections, isolated chromatin, and isolated mitotic chromosomes. An additional difficulty is that isolated chromatin fibrils often appear to be thicker (200 to 500 Å) than those observed in sections.

Despite the expenditure of a great deal of effort, the details of the structure of chromatin fibrils remain unresolved. Results and their interpretation vary from one investigator to another, and a review of this complicated subject is beyond the scope of this survey. However, the main ideas regarding the substructure of the chromatin fibril may be summarized as follows (Fig. 6–15). The view that currently enjoys the greatest popularity is that the DNA molecule with its associated proteins may form a coil. Several orders of coiling may occur, accounting for the differences in diameter of chromatin fibrils observed under different conditions. Another proposal is that two or more DNA helices lie parallel to one another, that is, the chromatin fibrils are multistranded. This would account for the thickness of the chromatin fibrils, but objections to this idea have been raised because of the difficulty in explaining well-known genetic phenomena, such as mutations and the semiconservative segregation of DNA, with a multistranded model. Yet another possibility is that the DNA molecule may be folded back and forth upon itself.

Some recent studies of isolated chromatin prepared in a particular way have shown it to contain particles that are connected by a fine thread (Fig. 6–16). The particles are approximately 100 Å in diameter and are known as *nu* bodies or nucleosomes. DNA has been shown to be present both in the threads and in the particles. Each particle is associated with about 200 base pairs of DNA and an octamer of histone, which is composed of two molecules each of four of the major

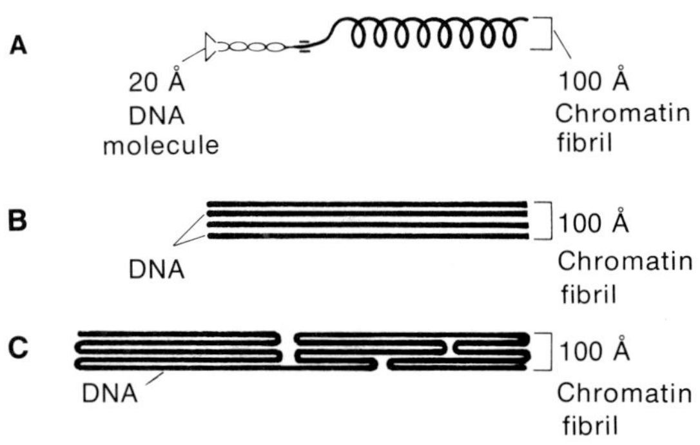

FIGURE 6–15 Diagram of models for the structure of the chromatin fibrils. *A*. DNA molecules undergo supercoiling. Additional orders of coiling may occur beyond that depicted. Recent studies indicate that nucleosomes (Fig. 6–16) play an important role in the coiling. *B*. A multistranded model in which identical DNA molecules lie parallel to one another. *C*. The DNA molecule is folded back upon itself.

classes of histones. It is proposed that the histone within each nucleosome serves as a structural support for the DNA, which lies on the surface of the histones. This is an especially active area of research, in which current efforts focus on the relation between nucleosomes in the chromatin fibril and the state of the nucleosomes during transcription and DNA replication.

Early studies suggested that chromosomes maintained positions during interphase, and in recent years much attention has been directed toward the attachment of chromatin to the inside of the nuclear envelope. This relationship between DNA and a membrane in eukaryotes may be analogous to the attachment of DNA to the plasma membrane in prokaryotes. Several chromatin fibrils appear to converge and attach to a single site on the nuclear membrane, and there are indications that particular regions of the chromosomes may be preferentially attached to the membrane. The significance of these attachments is not fully understood, but it is speculated that they could aid in orientation and maintenance of

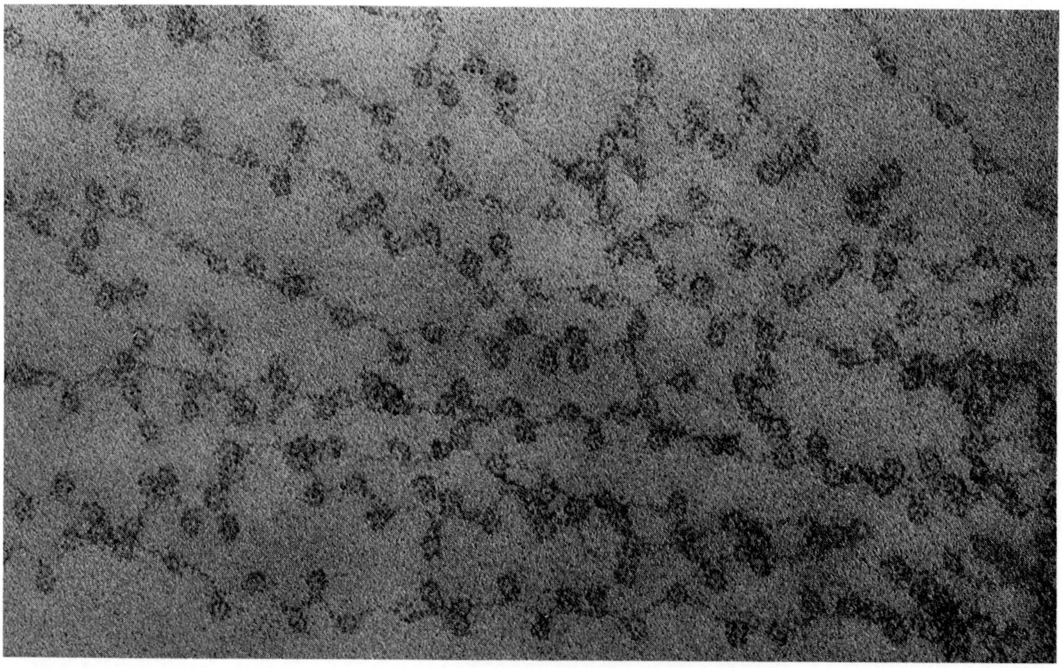

FIGURE 6–16 Electron micrograph of isolated negatively stained chromatin from a chicken erythrocyte. Globular "nu bodies" 70 Å in diameter are connected by a slender strand. × 285,000. (Micrograph courtesy of A. L. Olins and D. E. Olins.)

structural continuity of the chromosomes in transitions from interphase to mitosis and vice versa, or that they might have a role in chromosome replication.

SITES OF SYNTHESIS OF DIFFERENT TYPES OF RNA WITHIN THE NUCLEUS

The different types of RNA described earlier in this chapter are synthesized in particular locations within the nucleus (Table 6–3). The euchromatic extranucleolar parts of the nucleus are the sites of synthesis of the following: mRNA, which is synthesized as heterogeneous nuclear RNA, tRNA, and the 5S RNA component of the large subunit of ribosomes. Morphologically, the extranucleolar regions of the nucleus display a variety of RNA-containing particles (Fig. 6–17). For descriptive purposes these are designated perichromatin granules and interchromatin granules, according to their respective locations either near to or between aggregations of chromatin. In certain cells, fibrils or helical arrays of ribonucleoprotein are also found in the interior of the nucleus. These different kinds of RNA-containing granules and fibrils are presumed to represent the varieties of RNA synthesized in the extranucleolar regions, but the relation of specific kinds of RNA to particles of particular morphology is poorly understood.

It has been suggested that the perichromatin granules may represent messenger RNA because they are found in a region of high activity in RNA synthesis and because they have a morphological similarity to the Balbiani granules of the large puffs of insect salivary gland chromosomes which are known to be active in synthesis of mRNA. Helical or filamentous forms of ribonucleoprotein of cells, such as amoebae and insect salivary glands, are sometimes seen within nuclear pores (Fig. 6–18), suggesting that these forms may represent mRNA in transit from the chromatin to the cytoplasm.

The 28S, 18S, and 5.8S RNA molecules found in ribosomes are synthesized in the nucleolus. They will be considered in the following section.

NUCLEOLUS

The nucleolus is composed of 5 to 10 per cent RNA. Most of the remainder is protein, with a small amount of DNA being present as well. With the light microscope and in low-power electron micrographs (Fig. 6–7), the nucleolus usually appears as a spherical structure, frequently surrounded by a rim of condensed chromatin referred to as nucleolus-associated chromatin. The nucleolus is baso-

TABLE 6–3 SITES OF SYNTHESIS OF RNA IN THE NUCLEUS OF A EUKARYOTIC CELL

Site of Synthesis in Nucleus	Type of RNA	Location in Cytoplasm
Extranucleolar euchromatin	hnRNA ⟶ mRNA tRNA	⟶ Associated with ribosomes for protein synthesis
	5S rRNA	
Nucleolus	28S rRNA 5.8S rRNA	⟶ Large ribosomal subunit (with 5S rRNA)
	18S rRNA	⟶ Small ribosomal subunit

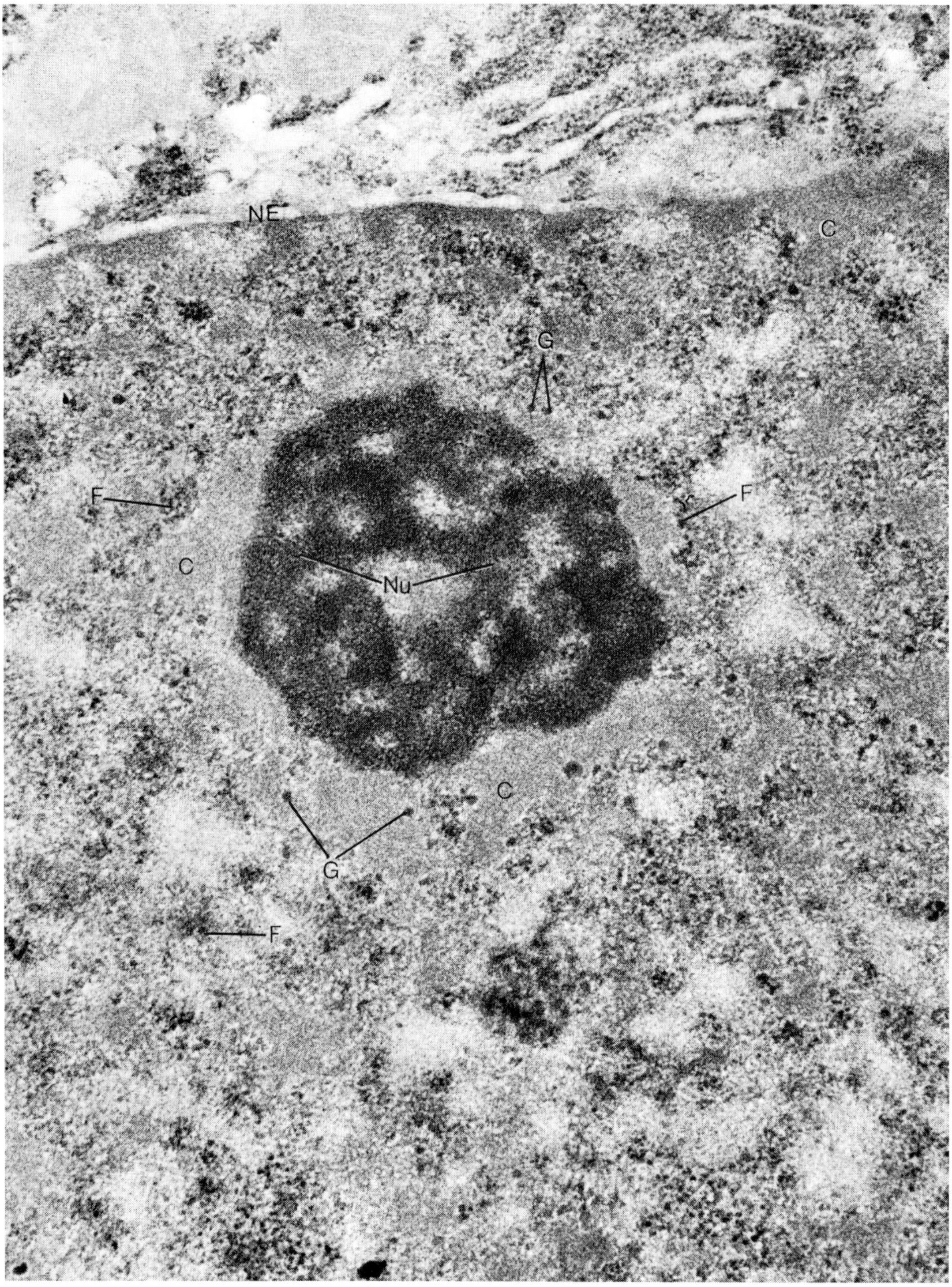

FIGURE 6–17 A portion of the nucleus of a rat liver cell, which has been stained so that ribonucleoprotein appears dense while the chromatin (C) is light. Various forms of ribonucleoprotein are present in the nucleoplasm. Perichromatin fibrils (F) probably represent newly synthesized heterogeneous nuclear RNA, while perichromatin granules (G) are believed to be transport forms of messenger RNA. The ribonucleoprotein of the nucleolus (Nu) is darkly stained. NE, nuclear envelope. × 50,000. (Micrograph courtesy of W. Bernhard.)

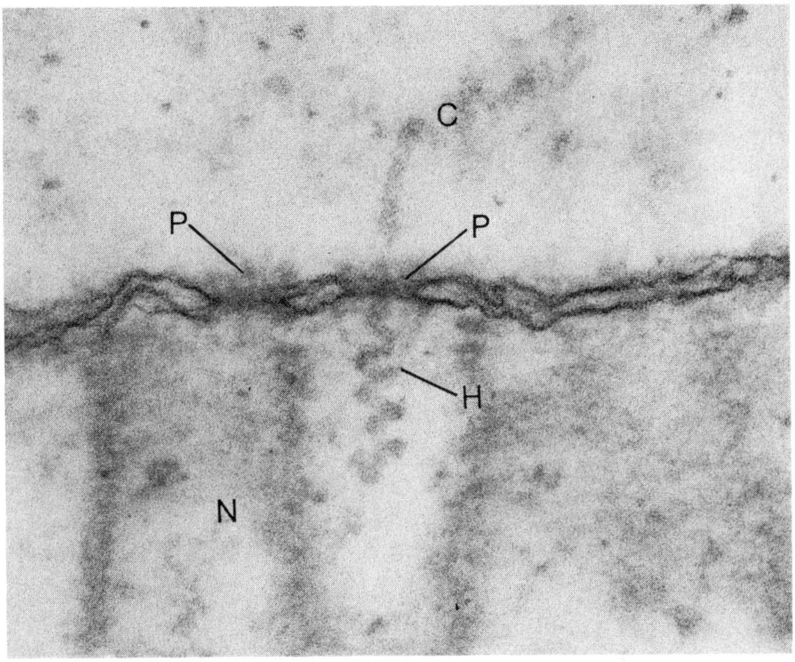

FIGURE 6–18 Nuclear envelope of an amoeba. Two pores are visible, and a helical structure, which is known to contain RNA, appears to have been passing through one of the pores. N, nucleus; C, cytoplasm. × 131,000.

philic owing to the presence of RNA, and it may stain metachromatically with dyes such as toluidine blue. It is usually Feulgen-negative because the DNA in the interior of the nucleolus proper is present in low concentration. The rim of nucleolus-associated chromatin, however, is intensely Feulgen-positive.

The fine structure of the nucleolus is complex. To understand its organization, it is important to realize that the same basic components are present in the nucleolus of most cells, but that the topographical distribution of these components varies, resulting in differences in nucleolar morphology. There are four basic structural components of the nucleolus: granules, fibrils, chromatin, and a proteinaceous substance sometimes referred to as a matrix.

The granules (Figs. 6–19 and 6–20) are 120 to 150 Å in diameter and contain RNA and protein. The fibrils also contain ribonucleoprotein and are about 50 Å in diameter. Since the fibrils are often very closely packed, the texture of the fibrillar portions of the nucleolus can be appreciated only at high magnification in the electron microscope. Chromatin is represented by the nucleolus-associated chromatin at the periphery of the nucleolus and by the intranucleolar chromatin present in the interior of the nucleolus. The intranucleolar chromatin consists of fine loops of chromatin, which extend from the nucleolus-associated chromatin into the interior of the nucleolus. The intranucleolar chromatin is important because it contains the genes from which ribosomal RNA is transcribed. Special methods may be necessary for its demonstration, however, since the presence of DNA in the interior of the nucleolus is often obscured in routine electron micrographs by the more numerous RNA-containing granules and fibrils. The proteinaceous component is variable in its extent; in some cells it forms distinct aggregations with a filamentous or amorphous texture.

These basic components are disposed in various ways in the nucleoli of different cells. In many mammalian cells, the nucleolar substance forms a pattern

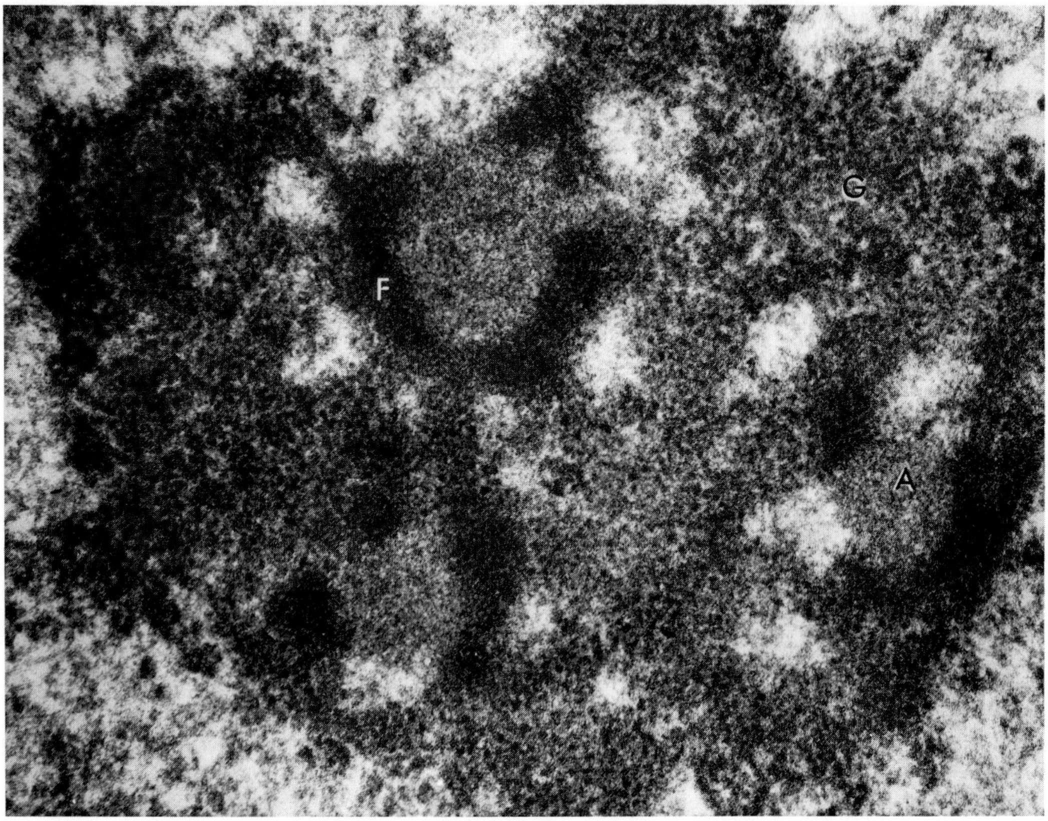

FIGURE 6–19 Nucleolus of a cell from human endometrium. Different regions are visible: those that contain predominantly granules (*G*), a dense fibrillar material (*F*), or an amorphous substance (*A*). × 72,000. (Micrograph courtesy of Terzakis, J. A.: J. Cell Biol., *27*:293, 1965. Reproduced by permission of the Rockefeller University Press.)

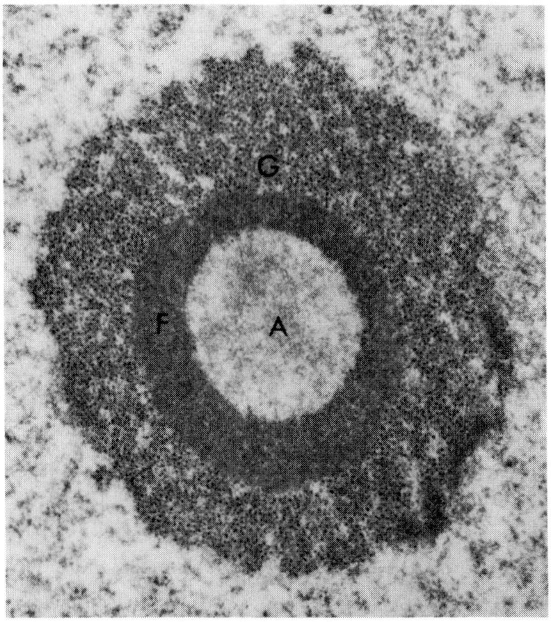

FIGURE 6–20 Nucleolus in a germinal cell of the rat testis. The fibrillar portion (F) in this spherical nucleolus is surrounded by a granular region (G). In the center is an amorphous region (A). × 28,500.

of thick anastomosing cords referred to as a nucleolonema (Figs. 6–7 and 6–19). Separate segments or patches along the cords contain either granules or fibrils. In other cells the nucleolus is a solid structure designated a compact nucleolus. The distribution of the basic components in compact nucleoli varies. There may be a core of fibrils and a shell or a cap of granules. In the compact nucleoli of some protozoa the granules and fibrils are intermixed. Ring-shaped nucleoli have also been observed. In these the ribonucleoprotein granules and fibrils occupy the periphery around a light center of protein and chromatin. Whatever their topographical disposition, the presence of the fundamental components in nucleoli of widely varying shapes is illustrated by the segregation of nucleolar components into distinct regions of granules, fibrils, proteinaceous material, and chromatin under the influence of certain drugs, such as the antibiotic actinomycin D.

The function of the nucleolus is the synthesis of ribosomal RNA, including the 28S, 18S, and 5.8S ribosomal RNA molecules. These are all synthesized in the nucleolus as parts of the 45S ribosomal RNA precursor molecule. Many lines of evidence for participation of the nucleolus in the formation of ribosomes have accumulated, and it is possible to mention only a few of these. It was known for many years, for example, that the nucleolus is large in cells that are actively synthesizing protein, and some of the 19th-century cytologists thought that the similarity in basophilic staining between the nucleolus and the basophilic bodies of the cytoplasm suggested a relationship between the two. Later, irradiation of the nucleolus with a microbeam of ultraviolet light was shown to abolish ribosomal RNA synthesis, and similarly the anucleolate mutant of *Xenopus laevis* was found to synthesize no rRNA. More recently, use of DNA-RNA hybridization methods have shown that DNA that codes for rRNA is present in isolated nucleoli.

The genome of most eukaryotes contains not a single rRNA gene, but multiple copies of the genes for rRNA. For example, several hundred copies per haploid genome are estimated to be present in HeLa cells, and *Xenopus* may contain as many as 1000 copies. In addition to these multiple copies which are part of the genome of all cells of an organism, certain cells, such as oocytes, display an *amplification* of genes for rRNA. In these cells, there is a selective extra synthesis of DNA of the genes for rRNA without replication of the remainder of the genome. These extra copies detach from the chromosomes and form rings of DNA. They synthesize rRNA and are visible in the cell as many small nucleoli. Production of these extra nucleoli apparently occurs in response to demands for many ribosomes for the cytoplasm of the egg cells, which are very large, and the numbers of nucleoli in a single nucleus may reach the thousands.

These extra nucleoli have been isolated and studied with the electron microscope, and remarkable micrographs of them have been obtained (Fig. 6–21). Each one consists of an axial fiber coated with a filamentous material. Enzymatic digestion has shown that the axial fiber is DNA and that the filaments are RNA attached by one end to the DNA. It is believed that the DNA axis represents the genes for rRNA and that the fibrils are rRNA molecules in the process of transcription. A small granule at the base of each fibril is thought to represent an RNA polymerase molecule. Additional features of these preparations are noted in the legend of Figure 6–21.

As described earlier, the major RNA components of the ribosomes, the 28S and 18S rRNA, are synthesized as parts of the larger 45S rRNA precursor molecule, which is methylated and cleaved in a series of steps to produce 28S and 18S rRNA plus fragments that are discarded. Processing of the 18S rRNA and its

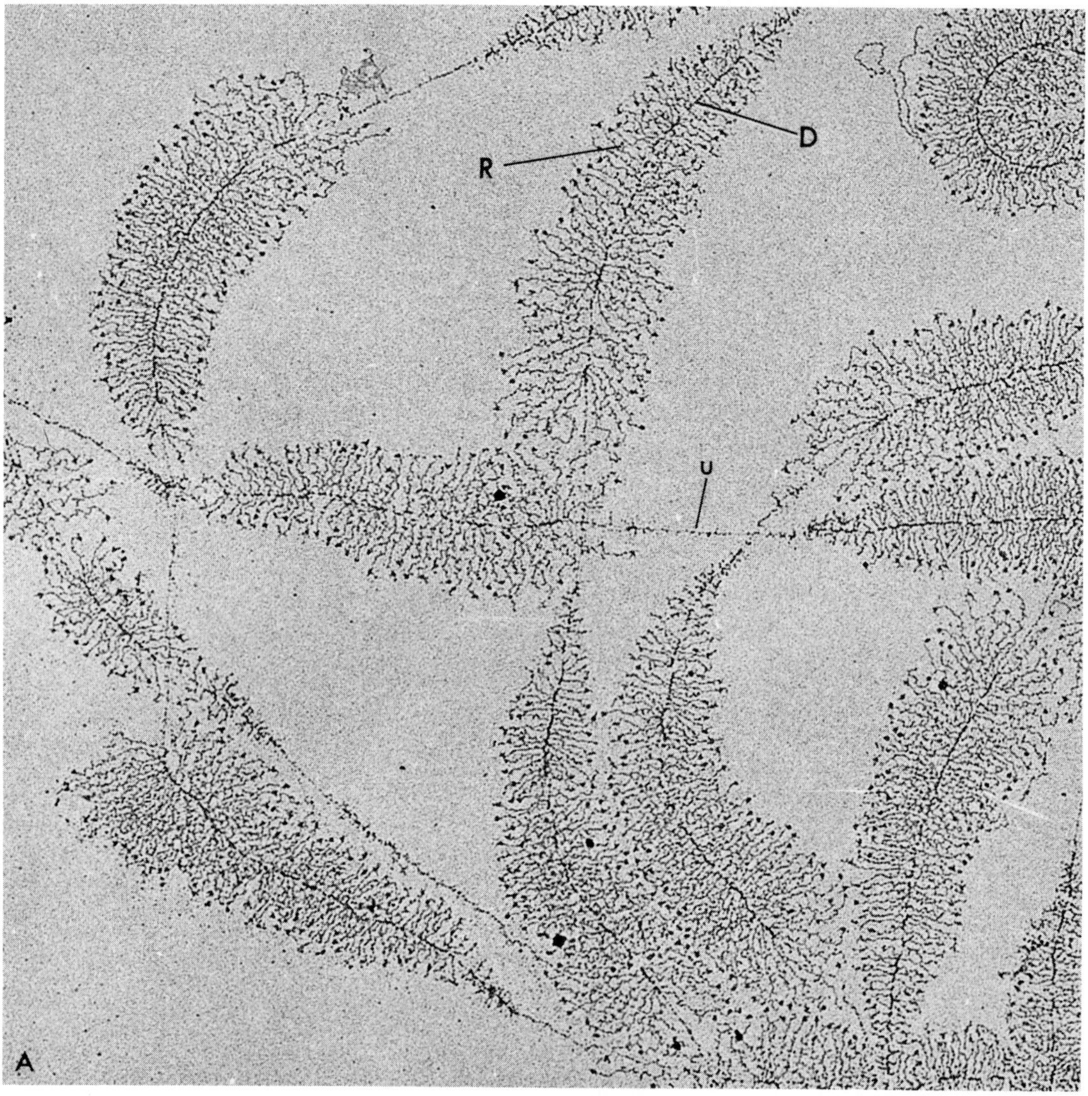

FIGURE 6–21 *A*. Isolated nucleolar core from a salamander oocyte. Enzymatic digestions and radioautographic studies have shown the long fibers (D) to be DNA, while the attached filamentous matrix material is RNA (R). The RNA fibrils form a gradient in length within each matrix unit, and the orientation of the units is the same along the length of a DNA fiber. Each unit is thought to represent one gene, and within a unit the length of the RNA fibrils increases as more of the molecule is transcribed. Note the presence of apparently untranscribed regions (u) between the genes.

transport to the cytoplasm as part of the small ribosomal subunit occur very rapidly, within a few minutes. Processing of the 28S rRNA that is part of the large ribosomal subunit is much slower. Thus 28S rRNA molecules and their precursors are normally found within the nucleolus in much greater amounts than the 18S molecules.

Just as the 28 and 18S rRNA's are associated with proteins in the large and small subunits of the ribosome, respectively, the 45S rRNA precursor molecule becomes associated with protein shortly after its synthesis, and the intermediates in the formation of 28 and 18S RNA are also associated with protein. Since the proteins are synthesized in the cytoplasm, their synthesis and transport to the nucleolus must be coordinated in some manner with rRNA synthesis. Each large

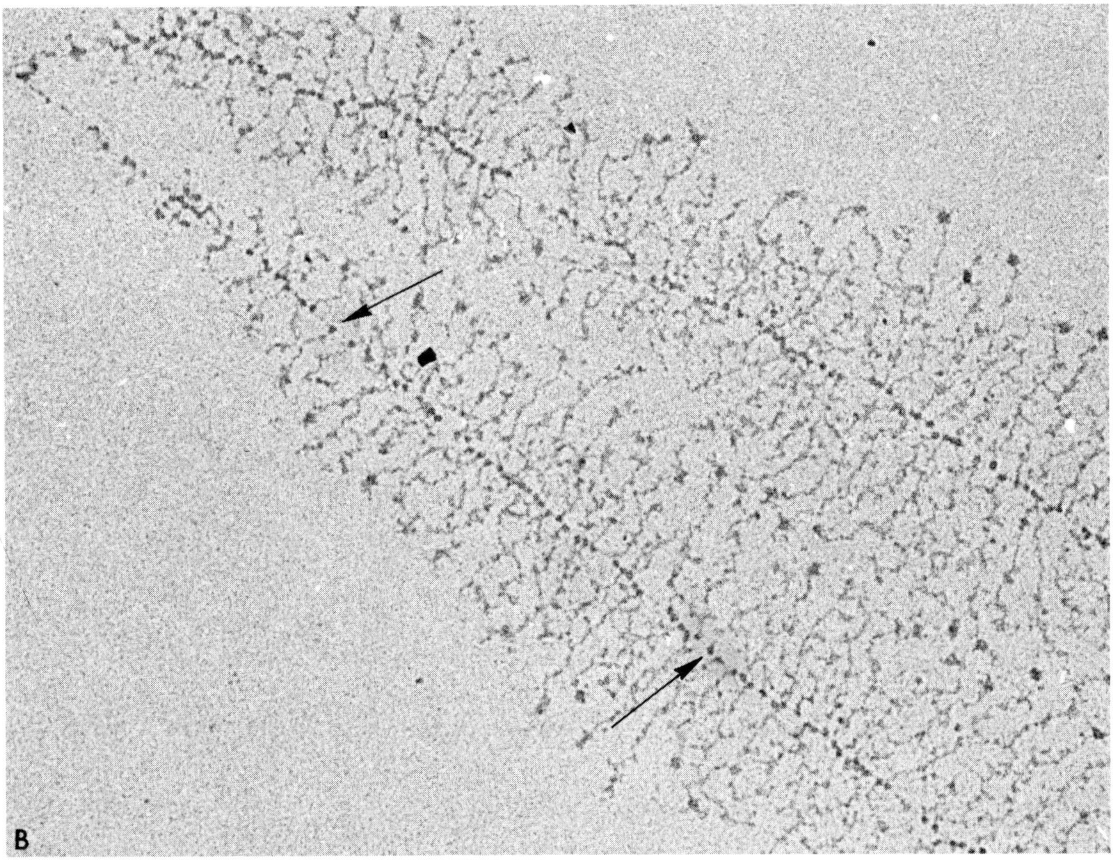

FIGURE 6–21 *Continued* *B*. Higher magnification of a similar preparation of two genes. A small granule (*arrow*) where each RNA fibril contacts the DNA is thought to represent an RNA polymerase molecule. *A*. × 27,000. *B*. × 72,500. (*A*. Courtesy of Miller, O. L., and Beatty, B. R.: J. Cell Physiol., *74* (1): 225, 1969. Reproduced by permission of the Wistar Institute Press. *B*. Courtesy of Miller, O. L., and Hamkalo, B. A.: Int. Rev. Cytol., *33*:1, 1972. Reproduced by permission of Academic Press.)

ribosomal subunit contains, in addition to a 28S rRNA molecule, a 5.8S and a 5S RNA molecule. The 5.8S molecule is synthesized as part of the 45S rRNA precursor, and it accompanies the 28S molecule in its processing. In contrast, the 5S RNA is synthesized outside the nucleolus, and it apparently enters the nucleolus where it joins other RNA's in assembly of the large ribosomal subunit (Table 6–3).

The fundamental features of nucleolar ultrastructure and the molecular biology of ribosomal RNA synthesis are both reasonably well understood. The correspondence between the steps in rRNA processing and the structures visible within the nucleolus is less certain, but the two seem to be related in the following way. If cells are exposed to a very brief pulse of uridine-^{3}H and electron microscope radioautographs are prepared at intervals thereafter, it is found that initially the fibrillar regions of the nucleolus are labeled and that subsequently the label is found in the granular regions. This suggests that the synthesis of 45S rRNA precursor takes place on fine strands of DNA that extend into the fibrillar regions of the nucleolus. As further processing of the RNA takes place, it is found in the granular regions. Since the 18S RNA of the small ribosomal subunit leaves the nucleolus rapidly and does not accumulate, it seems likely that the granules in the nucleolus represent large ribosomal subunits or their precursors in the processing of 28S rRNA.

NUCLEAR ENVELOPE

Nuclear Membranes and the Fibrous Lamina

The nuclear envelope surrounds the nucleus and forms the boundary between the nucleus and the cytoplasm in eukaryotic cells. In prokaryotes, it is absent, and the DNA-containing nucleoid is in direct contact with the cytoplasm. The nuclear envelope itself is too thin to be visible with the light microscope. However, the interface between nucleus and cytoplasm, coupled with the presence of condensed chromatin on the inner aspect of the nuclear envelope, makes the margin of the nucleus visible in most light microscope preparations. With the electron microscope (Figs. 6–22 and 6–23), the nuclear envelope is seen to be composed of two membranes separated by a space, the perinuclear cisterna. The outer surface of the nuclear envelope faces the cytoplasm and is studded with ribosomes, while the inner surface, facing the interior of the nucleus, lacks ribosomes.

The nuclear envelopes of many cells display an additional layer, the fibrous lamina, which is apposed to the inner aspect of the inner nuclear membrane, facing the interior of the nucleus (Fig. 6–23). It is composed of fine protein filaments, and it is thought to have a structural function of providing mechanical reinforcement to the nuclear membranes. Some investigators have also proposed that the fibrous lamina may influence the exchange of materials between nucleus and cytoplasm. The degree of development of the fibrous lamina varies greatly in different kinds of cells. In mammalian cells it is a thin mat that is only a few

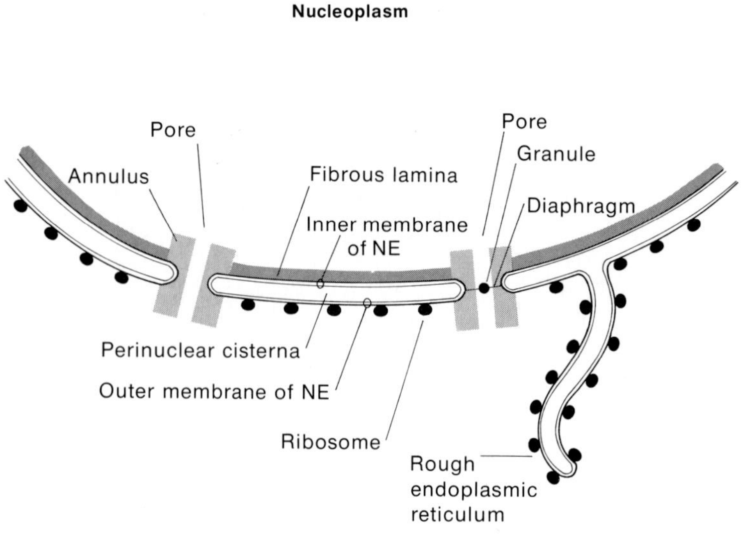

FIGURE 6–22 Diagram of the nuclear envelope. Note that both the inner and outer membranes of the nuclear envelope are "unit membranes," i.e., each has a three-layered appearance (dark–light–dark) when viewed at high magnification in the electron microscope. Notice also that the perinuclear cisterna is continuous with the lumen of the rough endoplasmic reticulum.

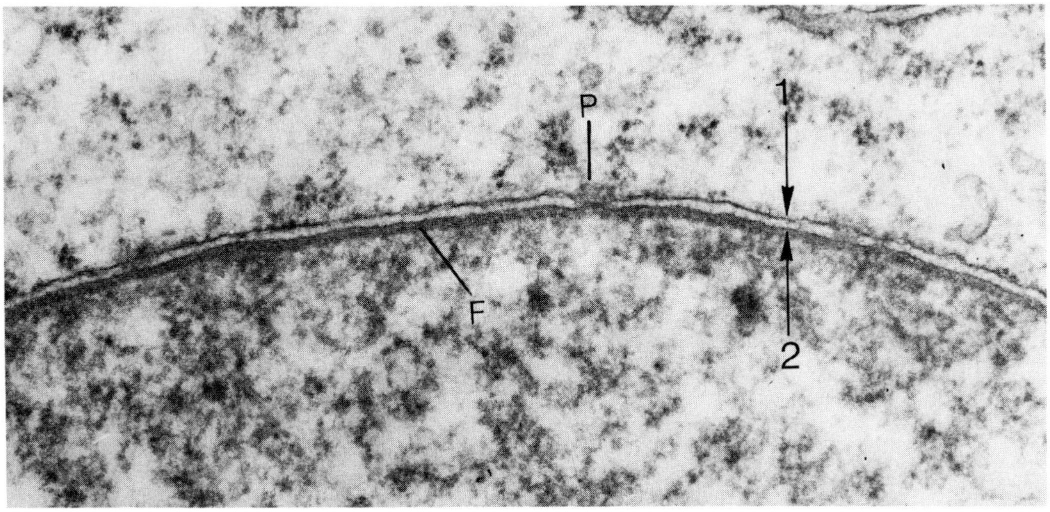

FIGURE 6–23 Nuclear envelope of an epithelial cell of the fetal rat urogenital sinus. The outer (1) and inner (2) membranes of the nuclear envelope bound the perinuclear cisterna. The fibrous lamina (F) of the nuclear envelope is a thin dense band applied to the inner nuclear membrane. N, interior of the nucleus; C, cytoplasm; P, nuclear pore. × 54,000.

hundred Å thick. However, the fibrous lamina attains a very high degree of development in amoebae and certain other invertebrates. In these instances, it has a honeycomb-like configuration and is 1000 to 1500 Å in thickness (Fig. 6–27).

Nuclear Pores

The two nuclear membranes are fused together at intervals to form pores approximately 600 Å in diameter (Figs. 6–23 and 6–24). They sometimes appear to have an octagonal rather than a circular outline in face view. It is estimated that pores occupy about 10 per cent of the surface area of the nuclear envelope in mammalian cells, although this figure may vary greatly in different cell types and under different conditions.

The importance of nuclear pores is that they provide channels for the possible exchange of material between nucleus and cytoplasm. Thus much effort has been directed toward determining the detailed ultrastructure of the pores and its relation to function. Because pore structure is difficult to perceive in routine sectioned preparations for electron microscopy, a variety of other techniques must be applied. It is generally agreed that in most cells the pores are not simple holes, but show structural complexity. The most frequently described structure associated with nuclear pores is the annulus (Figs. 6–25 and 6–26). This is a cylinder of dense material that lies partially within the pore and also overlaps the nuclear membranes at the margins of the pore. Different models for the structure of the annulus and its relation to the rest of the pore complex have been proposed, and it has been debated whether the wall of the annulus is composed of filaments, rodlets, or small tubules. In addition, in some cells a thin diaphragm spanning the pore has been described (Fig. 6–22). The diaphragm is thinner than a unit membrane and appears to be comparable in thickness to a single dense lamina of a unit membrane. In other instances a granule, centrally located within the pore, has been observed (Figs. 6–22 and 6–26). The significance of the diaphragms and

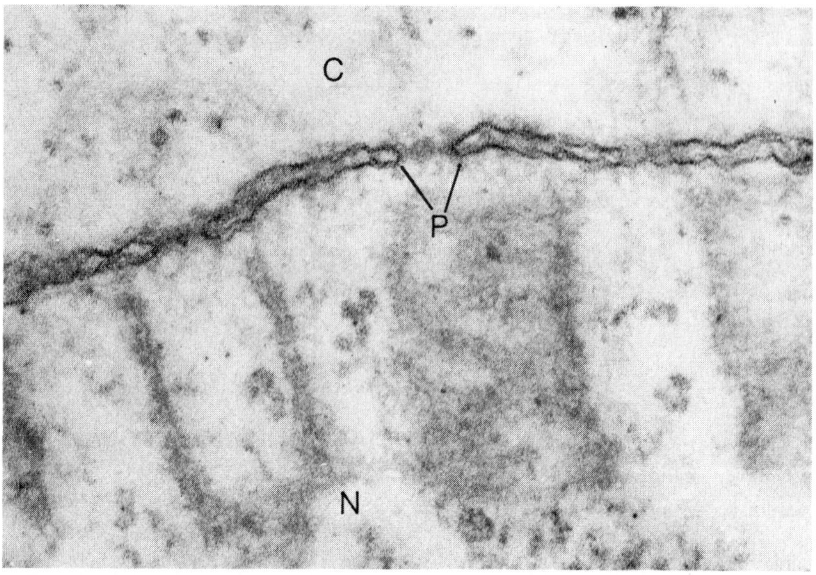

FIGURE 6–24 Nuclear envelope of *Amoeba proteus*. Nuclear pores (P), formed by fusion of the two membranes of the nuclear envelope, provide a channel between the interior of the nucleus (N) and the cytoplasm (C). × 108,000. (From Flickinger, C. J., Exp. Cell Res., *60*:225, 1970. Reproduced by permission of Academic Press.)

granules is uncertain, since they do not appear to be universal constituents of nuclear pores. Perhaps they are transient structures which are present in certain physiological states and not in others. It has been suggested, for example, that the central granules represent material in transit through the pores.

The fine structure of the pores is of particular functional significance because the components of the pore complex are in a position to influence the exchange of materials between nucleus and cytoplasm. Investigation of transport of materials through nuclear pores has indicated that the pores are not always open holes, but rather that they can impose constraints on the movement of materials

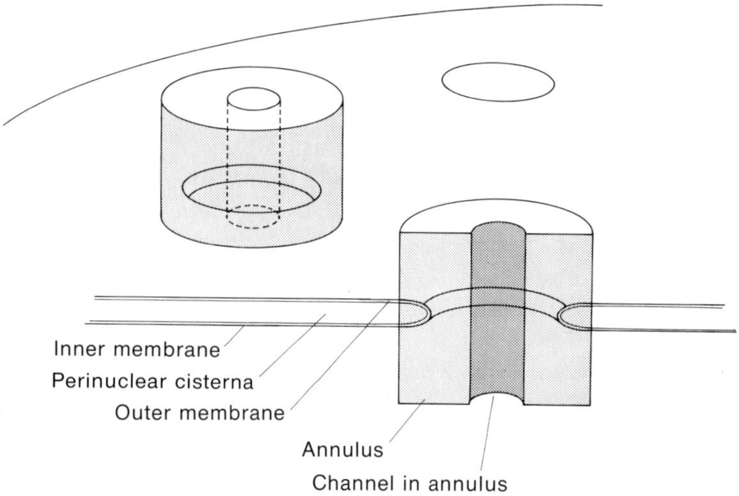

FIGURE 6–25 A diagram of an oblique view of the surface of the nuclear envelope. The annuli of the nuclear pores are depicted as cylinders that lie within pores and overlap the nuclear membranes at the margins of the pores.

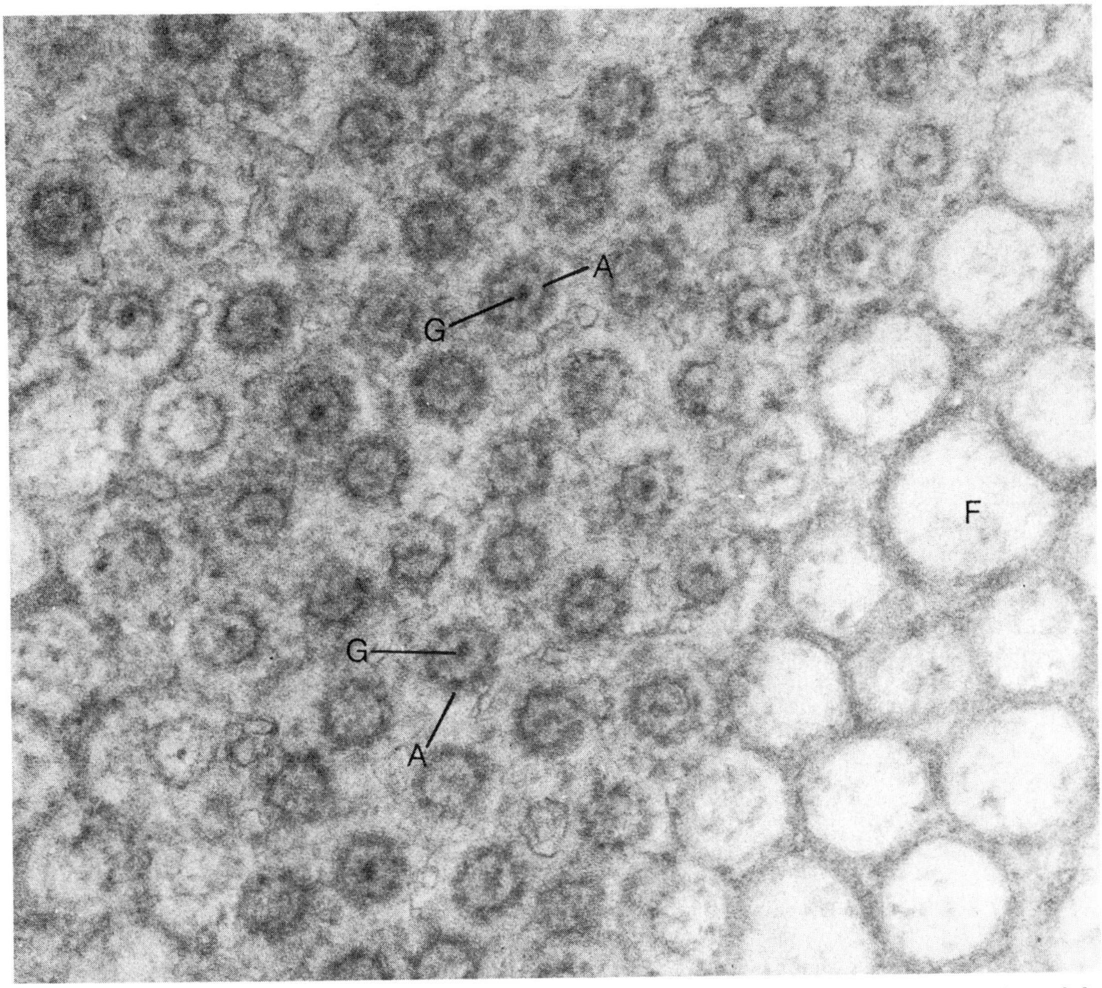

FIGURE 6–26 Nuclear envelope of *Amoeba proteus*. The plane of section is tangential to the surface of the nucleus, so nuclear pores are shown in face view. The wall of the annulus of the pore complex appears as a ring (A), and there is a granule (G) in the center of the pore. The unusual honeycomb-like fibrous lamina (F) of these cells is visible where the plane of section has passed to the nuclear side of the nuclear envelope. × 82,000. (From Flickinger, C. J.: Exp. Cell Res., *60*:225, 1970. Reproduced by permission of Academic Press.)

between nucleus and cytoplasm. A potential difference across the nuclear envelope of about 15 mV has been found in some cells, which implies that the pores are not available for free diffusion of ions. The absence of a detectable potential difference in other cells, however, underscores the diversity that exists with respect to the functional characteristics of nuclear pores.

Other experiments have shown that not all of the cross-sectional area of the pore is available for transport between nucleus and cytoplasm. By virtue of their large size, amoebae can be injected into the cytoplasm with colloidal gold particles, which are visible in the electron microscope. The particles vary in size. Those less than 85 Å enter the nucleus rapidly, while particles of approximately 89 to 106 Å enter only slowly. However, gold particles larger than about 106 Å do not enter the nucleus at all, despite the fact that the diameter of the pore as reckoned from the membrane margins is about 600 Å. Furthermore, particles lying within the pores in apparent transit between cytoplasm and nucleus are confined to a central channel through the pore. Thus the annular material of the pore complex appears to restrict the area available for transport. It has been suggested that the annulus of the pore could play a role in regulating nucleocytoplasmic exchanges.

Although these observations provide some examples of possible roles of the pore complex, it should be remembered that the permeability of the nuclear envelope not only varies in different cells, but also may change in the same cell under different physiological states and at different times in the cell cycle.

Although many nuclei have a circular profile, in some cells the contour of the nuclear envelope is highly irregular (Fig. 6–27). This has the effect of greatly increasing the surface area of the nuclear envelope, but its physiological significance is not clear.

Relation to the Endoplasmic Reticulum

The nuclear envelope has been regarded by many as a regional specialization of the endoplasmic reticulum. Some reasons for this line of thought follow. Ribosomes are present on the surface of the rough endoplasmic reticulum and the outer surface of the nuclear envelope, although they are absent from the inner membrane of the nuclear envelope. In certain cells a distinctive granular or crystalline content accumulates in the lumen of the rough endoplasmic reticulum, and a similar content is often found in the perinuclear cisterna. Connections between the rough endoplasmic reticulum and the outer membrane of the nuclear envelope are frequently observed (Fig. 6–28) and provide a continuous channel from the perinuclear cisterna to the interior of the rough endoplasmic reticulum. The idea that the nuclear envelope is re-formed after mitosis by fusion of membranous elements resembling rough endoplasmic reticulum has been widely accepted. Finally, the chemical compositions of isolated rough endoplasmic reticulum and nuclear envelopes are similar, and differences between the two might reflect the

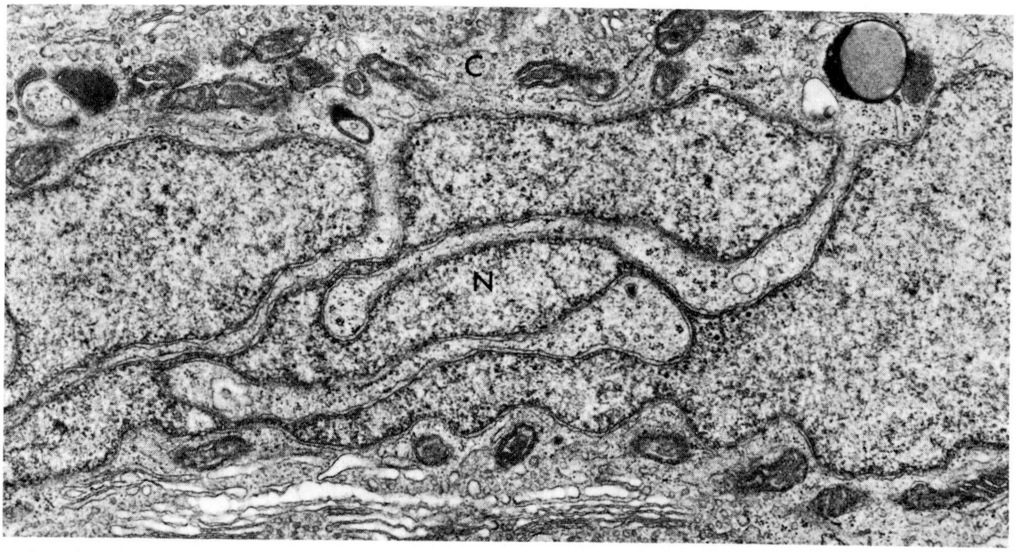

FIGURE 6–27 Nucleus of an epithelial cell in the human epididymis. The extremely irregular contour of the nuclear envelope in this cell type is evident. N, nucleus; C, cytoplasm. × 17,000.

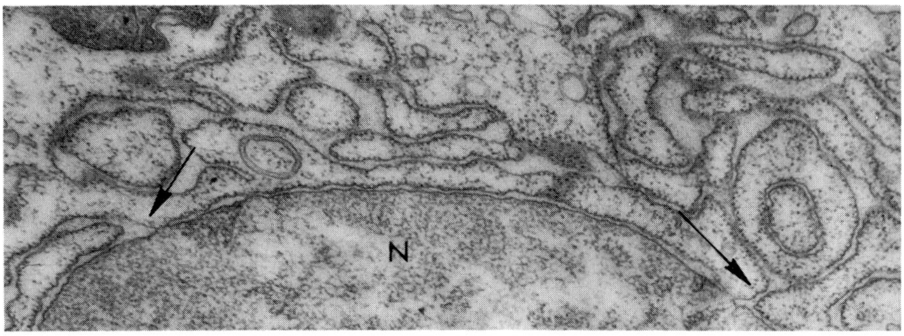

FIGURE 6–28 Mouse salivary gland epithelium. Connections between the nuclear envelope and the rough endoplasmic reticulum are shown at the arrows. Note that the perinuclear cisterna is continuous with the lumen of the endoplasmic reticulum at these points. N, interior of the nucleus. × 23,000. (Micrograph courtesy of H. F. Parks.)

presence of the additional structures associated with the nuclear envelope, such as pores and the fibrous lamina.

PATHOLOGICAL CHANGES IN THE NUCLEUS

Alterations in nuclear structure occur in many normal and pathological processes. The large nuclei and unusual mitotic figures of malignant tumors were noted previously, as were changes in chromatin and the nucleolus with the functional state of the cell. Study of the nucleus in the presence of intranuclear viruses offers other examples of striking structural alterations that are related to disease. The term intranuclear virus refers to viruses that are formed or assembled in large part in the nucleus, and thus whole viruses or their parts often accumulate in the nucleus. Most of the intranuclear viruses are DNA viruses, but some RNA types are also represented. Intranuclear viruses are known to cause a variety of diseases ranging from fever blisters and measles to neurological disorders and tumors.

Following attachment of the virus to the cell membrane, uptake into the cytoplasm, and transfer to the nucleus, a variety of nuclear changes can be observed (Fig. 6–29). Alterations may occur in the main parts of the nucleus, including redistribution or disruption of the chromatin and changes in the size and structure of the nucleolus. In addition, many different morphological changes can result from the accumulation of viral precursors, metabolic products, or mature virions. In the light microscope these may appear as either acidophilic or basophilic bodies of variable size and number, depending upon their content. The location and type, however, are often distinctive of a particular virus or group of viruses and thus may be useful diagnostically. In the electron microscope a great variety of structures can be seen, such as large numbers of mature virions in crystal-like arrays, or granular, fibrillar, and crystalline inclusions composed of proteins or nucleic acids (Fig. 6–30). A classic example of an intranuclear inclusion is provided by the Negri bodies which occur in neurons in rabies.

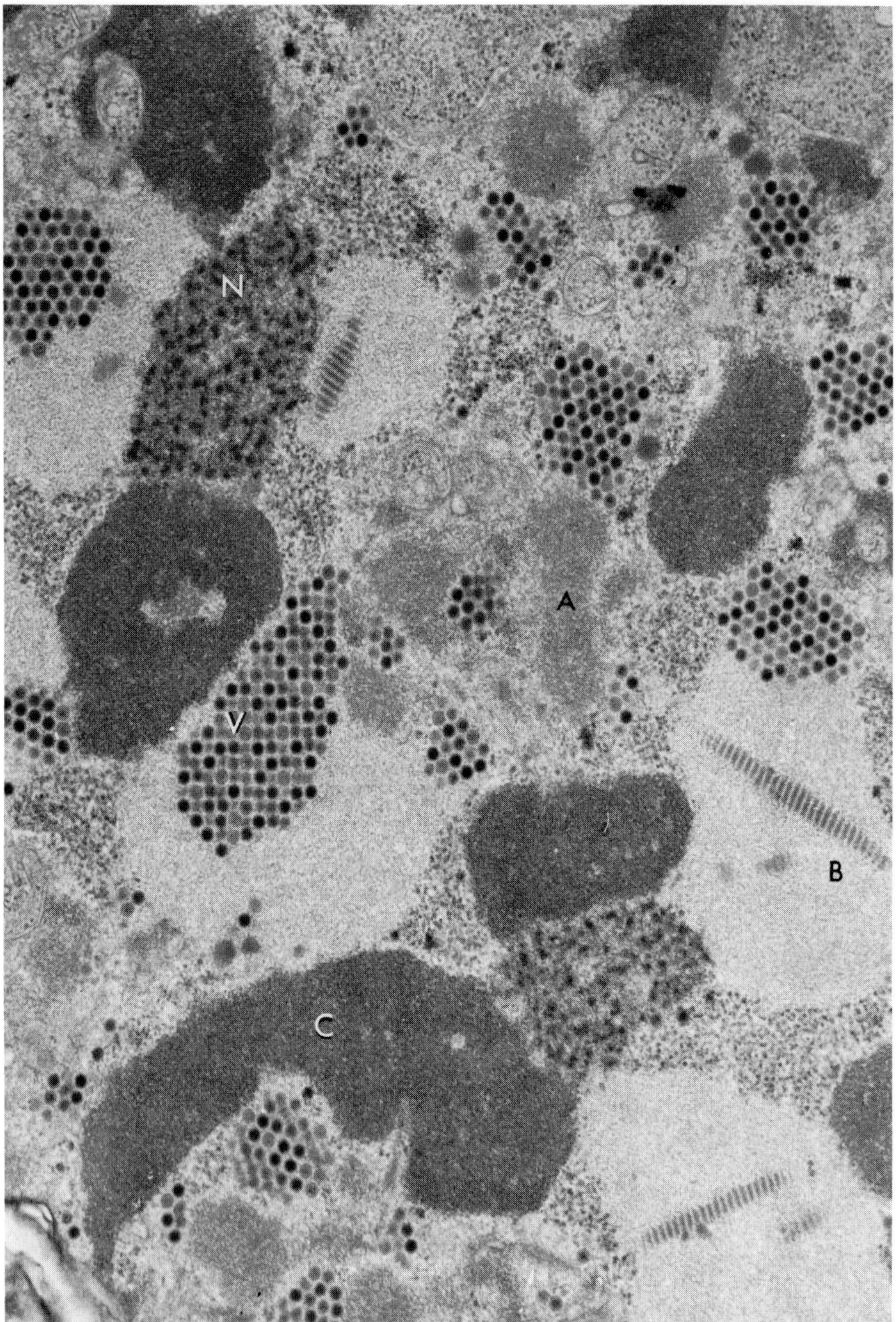

FIGURE 6–29 Portion of the nucleus of a human cell in culture that has been infected with an adenovirus. Many virus crystals (V) and several types of inclusions (A, B, and C) are visible. The nucleolus (N) displays a structure that has come to be known as spotted nucleolus. × 13,000. (Micrograph courtesy of Martinez-Palomo, A.: Path. Microbiol., *31*: 147, 1968. Reproduced by permission of S. Karger, AG, Basil.)

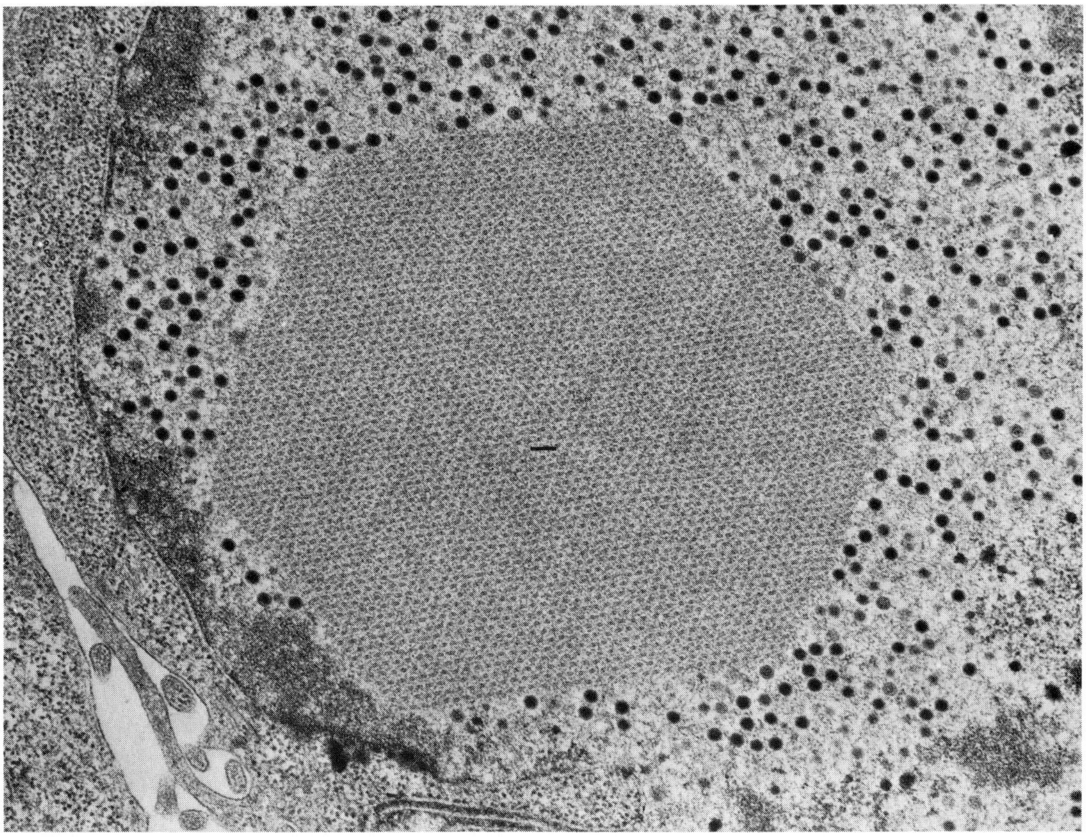

FIGURE 6–30 A paracrystalline inclusion (I) in the nucleus of a cell infected with a human adenovirus. The paracrystals are related to adenovirus core proteins. Dense virus particles are also visible in the nucleus. × 39,000. (Micrograph courtesy of Marusyk, R., Norrby, E., and Marusyk, H.: J. Gen. Virol., *14*:261, 1972. Reproduced by permission of the Cambridge University Press.)

FURTHER READING

Barr, M. L.: Sex chromatin and phenotype in man. Science, *130*:679, 1959.

Beerman, W., and Clever, U.: Chromosome puffs. Sci. Am. *210*:150, 1964.

Bernhard, W., and Granboulan, N.: Electron microscopy of the nucleolus in vertebrate cells. *In*: Dalton, A. J., and Haguenau, F. (eds.): The Nucleus. New York, Academic Press, 1968, p. 81.

Brenner, S., Jacob, F., and Meselson, M.: An unstable intermediate carrying information from genes to ribosomes for protein synthesis. Nature, *190*:576, 1961.

Burgess, R., Travers, A. A., Dunn, J. J., et al.: Factor stimulating transcription by RNA polymerase. Nature, *221*:43, 1969.

Busch, H. (ed.): The Nucleus. Vols. 1–3. New York, Academic Press, 1974..

Busch, H., and Smetana, K. (eds.): The Nucleolus. New York, Academic Press, 1970.

Doi, R. H.: Role of RNA polymerases in gene selection in prokaryotes. Bacteriol Rev., *41*:568, 1977.

DuPraw, E. J.: DNA and Chromosomes. New York, Holt, Rinehart & Winston, 1970.

Fawcett, D. W.: On the occurrence of a fibrous lamina on the inner aspect of the nuclear envelope in certain cells of vertebrates. Am. J. Anat., *119*:129, 1966.

Feldherr, C. M.: The nuclear annulli as pathways for nucleocytoplasmic exchanges. J. Cell Biol., *14*:65, 1962.

Feldherr, C. M.: Structure and function of the nuclear envelope. Adv. Cell Mol. Biol., *2*:273, 1972.

Franke, W. W.: Structure, biochemistry, and functions of the nuclear envelope. Int. Rev. Cytol. (Suppl.), *4*:72, 1974.

Hamkalo, B. A., and Miller, O. L.: Electronmicroscopy of genetic activity. Annu. Rev. Biochem., *42*:379, 1973.

Hay, E. D.: Structure and function of the nucleolus in developing cells. *In*: Dalton, A. J., and Haguenau, F. (eds.): The Nucleus. New York, Academic Press, 1968, p. 2.

Kornberg, R. D.: Structure of chromatin. Annu. Rev. Biochem., *46*:931, 1977.

Losick, R., and Chamberlin, M.: RNA Polymerase. New York, Cold Spring Harbor Laboratory, 1975.

Maul, G. G.: The nuclear and cytoplasmic pore complex: structure, dynamics, distribution, and evolution. Int. Rev. Cytol. (Suppl.), *6*:76, 1977.

Miller, O. L.: The visualization of genes in action. Sci. Am. *228*:34, 1973.

Monneron, A., and Bernhard, W.: The structural organization of the interphase nucleus in some mammalian cells. J. Ultrastruct. Res., *27*:266, 1969.

Perry, R. P.: Processing of RNA. Annu. Rev. Biochem., *45*:605, 1976.

Perry, R. P.: The nucleolus and the synthesis of ribosomes. Prog. Nucleic Acid Res. Mol. Biol., *6*:219, 1967.

Ris, H. (ed.): Chromosomal structure as seen by electron microscopy. *In*: The Structure and Function of Chromatin. Ciba Foundation Symposium. Amsterdam, North-Holland, 1975, pp. 7–23.

Smith, J. D.: Transcription and processing of transfer RNA precursors. Prog. Nucleic Acid Res. Mol. Biol., *16*:25, 1976.

Wischnitzer, S.: The lampbrush chromosomes: their morphology and physiological importance. Endeavour, *35*:27, 1976.

7

THE MECHANISM OF PROTEIN SYNTHESIS

The final step in expression of the hereditary program occurs when genetic information contained in mRNA is decoded or "translated" into protein. It is the sequence of nucleotides in mRNA that determines the sequence of amino acids in proteins. The process of protein synthesis takes place on the ribosomes in the cytoplasm, and its fundamental features are characteristic of all living cells. Once the correct amino acids have been polymerized in the proper sequence, however, no further input from the genetic program is required to produce a fully functional protein molecule. The newly synthesized polypeptide spontaneously arranges itself into the proper shape (conformation) for performing its function. No gene activity is required, for example, to influence the folding of individual polypeptide chains or to cause them to assemble into the proper multi-subunit structures.

Protein synthesis requires a change in the chemical nature of the genetic information, and a very sophisticated set of specific biochemical interactions is involved in this change. Therefore, the following discussion of protein synthesis is divided into three parts. The first is devoted to the genetic code and the way in which genetic information for the construction of proteins is encoded in mRNA. In the second section the structure of the cellular components involved in protein synthesis is considered and in the third the interactions of the cellular components that produce the finished polypeptide chain are discussed.

THE GENETIC CODE

The problem of protein synthesis is to translate the information contained in a sequence of nucleotides in mRNA into a sequence of amino acids in a protein. Perhaps the most straightforward solution would be for RNA somehow to form twenty different pockets, each of which would exactly fit one and only one of the twenty amino acids found in proteins. If such pockets were arranged in the proper order, each one could be occupied by just the right amino acid, and the amino acids could be enzymatically polymerized to generate a specific protein molecule. In fact, however, this does not occur. Although RNA molecules are well suited to form specific base pairs with each other, their chemical construction does not include the ability to form specific pockets that could be recognized by amino acids. Therefore, amino acids do not directly recognize particular regions of mRNA structure. Instead, an adaptor molecule is involved. The adaptor is the type of RNA called adaptor, soluble, or transfer RNA. Each transfer RNA molecule has the property of recognizing, by base pairing, a specific sequence of bases in mRNA. In addition, each transfer RNA molecule binds covalently to only one of the twenty amino acids. The position of a particular amino acid in a protein sequence is determined by the ability of its adaptor to base pair specifically with a particular sequence of bases in mRNA; once the correct amino acids have been identified in this way, they are polymerized to form a protein chain.

The number of nucleotides in mRNA required to specify one of the twenty amino acids can be calculated directly. Since there are only four possible bases in mRNA, a sequence of two bases is insufficient; it would give only 4^2 or 16 possible combinations or code words. Thus a sequence of no fewer than three nucleotides is required, and in fact sequences of three bases in mRNA are known to be the individual code words, or codons. The codons are arranged linearly in mRNA, they do not overlap, and there are no spacer regions between them. Messenger RNA is decoded from the 5′ toward the 3′ end, and codons read in this order correspond to amino acids beginning at the N-terminal and proceeding toward the C-terminal end of the finished protein. The sequence of mRNA nucleotides is therefore colinear with the sequence of amino acids in the protein for which the mRNA codes.

The particular sequences of mRNA bases that specify each of the 20 amino acids are shown in Table 7–1. Since there are 4^3 or 64 possible mRNA code words, each amino acid can in principle be encoded by more than one triplet. In fact, there is more than one code word for all the amino acids except methionine (AUG) and tryptophan (UGG). Frequently codons that specify the same amino acid (synonym codons) differ only in the third position. Examples are the codons for phenylalanine (UUU and UUC), valine (GUU, GUC, GUA, and GUG), and several other amino acids. This is not always the case, however. Serine, arginine, and leucine, for example, all have sets of codons that differ in the first or second

TABLE 7–1 THE GENETIC CODE

First Base \ Second Base	*U*	*C*	*A*	*G*	Third Base
U	Phe Phe Leu Leu	Ser Ser Ser Ser	Tyr Tyr Termination Termination	Cys Cys Termination Tryp	*U* *C* *A* *G*
C	Leu Leu Leu Leu	Pro Pro Pro Pro	His His GluN GluN	Arg Arg Arg Arg	*U* *C* *A* *G*
A	Ileu Ileu Ileu Met	Thr Thr Thr Thr	AspN AspN Lys Lys	Ser Ser Arg Arg	*U* *C* *A* *G*
G	Val Val Val Val	Ala Ala Ala Ala	Asp Asp Glu Glu	Gly Gly Gly Gly	*U* *C* *A* *G*

position. Protein synthesis begins at some AUG methionine codons. Methionine is therefore the first amino acid incorporated into all nascent polypeptide chains. In addition to codons for the twenty amino acids, there are three codons (UAA, UAG, and UGA) that specify the termination of translation. One or more of these codons must always be present at the end of an mRNA sequence coding for a protein.

There are several curious features of the overall structure of the genetic code that relate the chemistry of individual amino acids to the codons used to specify them. For example, all codons with U in the second position specify hydrophobic amino acids (phenylalanine, leucine, isoleucine, methionine, and valine), while all the charged amino acids except arginine have codons with A in the second position. All the aromatic amino acids have codons beginning with U. Although the exact significance of these observations is not clear at the present time, it is probably related to the chemical details of how the present genetic code was derived evolutionarily from a simpler coding scheme.

THE CELLULAR COMPONENTS INVOLVED IN PROTEIN SYNTHESIS

Messenger RNA

Although all three types of cellular RNA are involved in protein synthesis, mRNA is the most significant from the point of view of information flow. Messenger RNA's contain a complementary copy of the DNA nucleotide sequences that code for protein structures, and they are translated on ribosomes to yield the specified protein molecule. Since each mRNA molecule can be translated many times, each may be used to direct the synthesis of many identical protein molecules.

In addition to the codons that specify the sequence of amino acids in a protein, it is clear that an mRNA must also contain signals indicating where to begin and where to end translation. Understanding the code for initiating protein synthesis is complicated by the fact that the same codon, AUG, is used both for the initiation of protein synthesis with methionine and for the methionine residues at internal positions in the polypeptide chain. Some mechanism must therefore exist to distinguish initiator AUG codons from chain extension AUG's. Furthermore, since it is possible to decode a sequence of RNA nucleotides in three possible reading frames or "phrases," depending upon the exact nucleotide at which translation is begun, it is clear that the initiation mechanism must be able to distinguish initiator AUG codons from AUG's that may exist in the two incorrect reading frames (Fig. 7–1).

Since messenger RNA is decoded from the 5′ toward the 3′ end, it is reasonable to expect that the site of initiation of protein synthesis will be located at or near the 5′ end and that the termination site will be at or near the 3′ end. This has been found to be the case in all the mRNA molecules for which direct sequence information is available. The best-studied mRNA sequence is the RNA from the small *E. coli* bacteriophage MS2. As in the related phages R17 and f2, the MS2 viral RNA itself serves as a messenger RNA during viral infection. The RNA codes for three proteins, a structural A protein, the virus coat protein, and a viral RNA replicase. The genes for these proteins are arranged on the viral mRNA, as shown in Figure 7–2. MS2 RNA is known as a polycistronic mRNA since the same RNA molecule codes for more than one protein. At the beginning

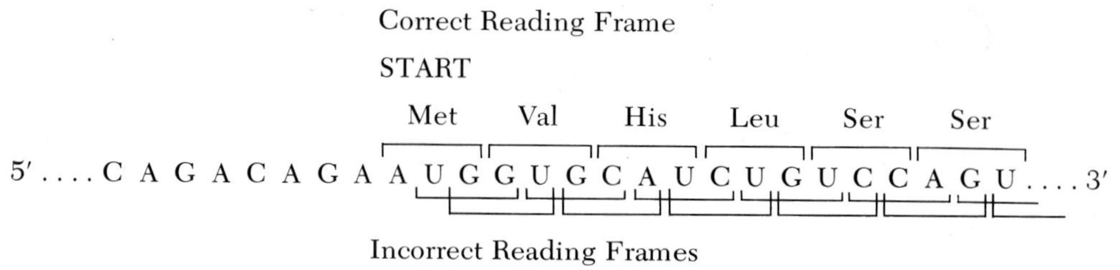

FIGURE 7–1 Selection of the correct reading frame in protein synthesis, as illustrated by the nucleotide sequence at the initiation site for translation in the mRNA coding for rabbit β-globin.

of each gene, a "start" signal is located just ahead of and in phase with the codons that specify the amino acids of the three viral proteins. After the codon for the C-terminal amino acid, one of the chain termination codons is present (Fig. 7–2). The first "start" signal does not occur directly at the 5′ end of the MS2 RNA molecule, but is approximately 130 nucleotides in from the end. Similarly, the last "stop" signal is approximately 170 nucleotides removed from the 3′ end, and there are 20 to 30 nucleotides between the internal stop and start signals (Fig. 7–2).

Although the function of all these "extra" nucleotides is not known, it is clear that they are not translated into protein. At least some of them must be involved in identifying the correct AUG to be used for the initiation of protein synthesis, and some evidence is available to show how this may occur. It seems likely that a set of base pairs transiently forms on the ribosome surface between a pyrimidine-rich sequence of bases at the 3′ end of the 16S ribosomal RNA and a complementary purine-rich sequence that is invariably found approximately 10 nucleotides removed (toward the 5′ end) from AUG codons that signal the initiation of protein synthesis (Fig. 7–2). The formation of these base pairs is thought in some way to enable the initiator AUG codon to be properly positioned on the ribosome surface so that it may interact specifically with the initiator tRNA.

Although MS2 RNA is the most thoroughly studied mRNA, several other mRNA species are now available in purified form. These include the mRNA's coding for hemoglobin, immunoglobulin light chain, ovalbumin, myosin, lens crystallin, histones, and several viral proteins. In general, the mRNA's isolated from eukaryotic sources differ from prokaryotic mRNA's in that they are all monocistronic, that is, all have only one start and one stop signal for protein synthesis. Prokaryotic mRNA's, on the other hand, may have internal stop and restart signals, as is the case with MS2 RNA. Eukaryotic mRNA's have two other structural features not found in prokaryotic mRNA. First, the 5′ end of many eukaryotic mRNA's is blocked or "capped" with a 7-methylguanosine

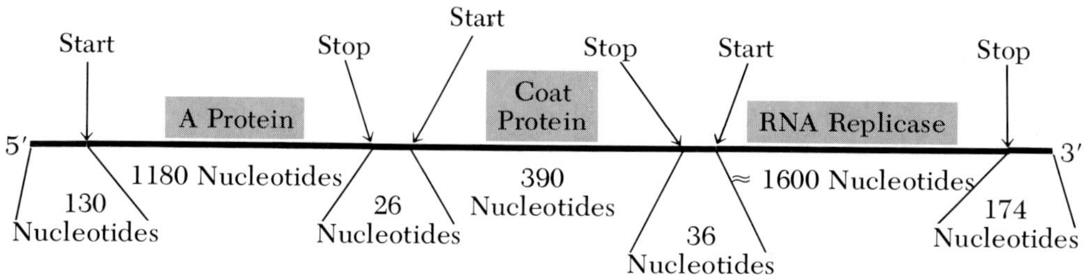

FIGURE 7–2 Schematic diagram of phage MS-2 RNA. Translated regions are shown in gray, and untranslated regions are shown in black.

group. This modified G residue is attached after transcription in an unusual 5′–5′ triphosphate linkage (Fig. 7–3). The biological function of the 7-methylguanosine group is not yet fully understood, although it is thought to be required for translation of mRNA in some situations. Second, most eukaryotic mRNA's contain at their 3′ ends a region of 50 to 200 A nucleotides, called the poly A "tail." The function of the poly A tail is also uncertain, although it may be involved in determining how many times an mRNA molecule may be translated.

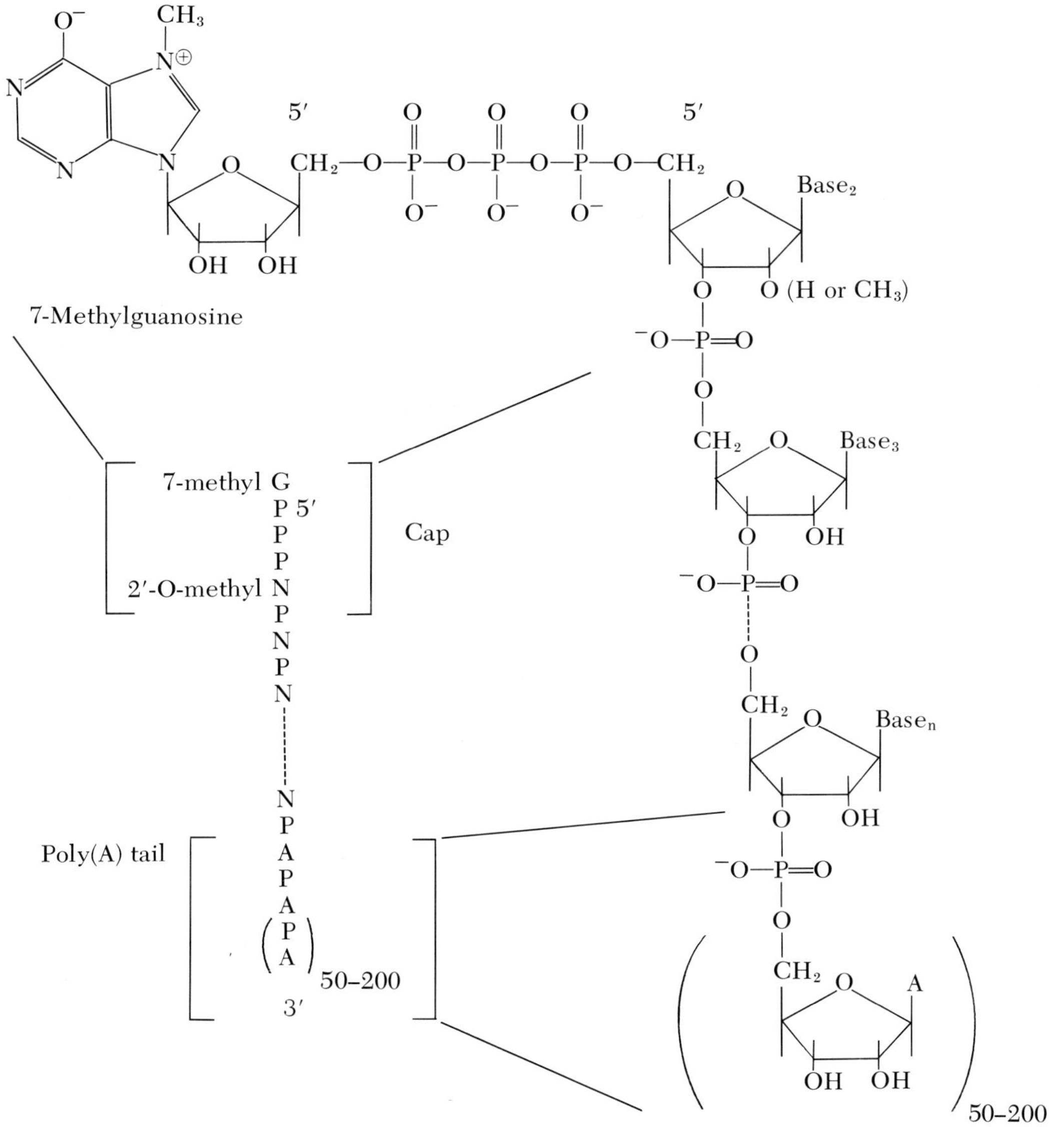

FIGURE 7–3 General structure for eukaryotic mRNA showing the "cap" at the 5′ end and the poly (A) "tail" at the 3′ end.

Transfer RNA

STRUCTURE

Transfer RNA molecules play a unique and crucial role in the translation of genetic information. They provide the direct molecular link between the language of nucleic acid base sequences and the language of amino acid sequences. Each transfer RNA molecule has a sequence of three nucleotides, the anticodon, which is complementary to and interacts specifically with one of the triplet codons in mRNA. At the same time, the amino acid specified by that codon is covalently bound to the tRNA in a state that permits its addition onto the growing polypeptide chain. Individual tRNA molecules can bind only one of the twenty amino acids, and amino acids cannot be incorporated into protein unless they are first bound to the correct tRNA molecules. There is at least one and often as many as four or five different species of tRNA molecules for each amino acid.

Some details of tRNA structure have already been mentioned (Chapter 6). Transfer RNA's are short RNA chains, ordinarily between 73 and 93 nucleotides in length. They may contain 10 per cent or more of bases other than the usual four found in RNA. The sequence of most tRNA molecules begins at the 5′ end with pG, and all have the sequence pCpCpA—OH at the 3′ end. The amino acid is attached enzymatically to the 3′ hydroxyl of the terminal A nucleotide, as shown in Figure 7–4. From there it is donated directly onto the C-terminal end of the nascent polypeptide during protein synthesis.

In between their common 5′ and 3′ end groups, different tRNA's vary greatly in their nucleotide sequences. In every case, however, the sequence folds back on itself in four places at which base pairing can occur. This folding results in the formation of four local regions of antiparallel double helix, each of which consists of four to eight base pairs. More than half of the bases in tRNA may therefore be present in these double helical regions, as shown in Figure 7–5. Here two tRNA sequences are drawn in the familiar cloverleaf structure, which emphasizes

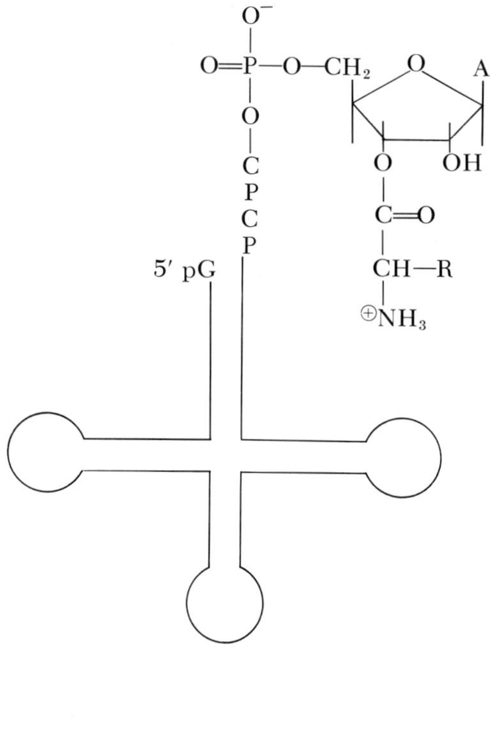

FIGURE 7–4 General structure for the attachment of amino acids to tRNA molecules by ester linkages.

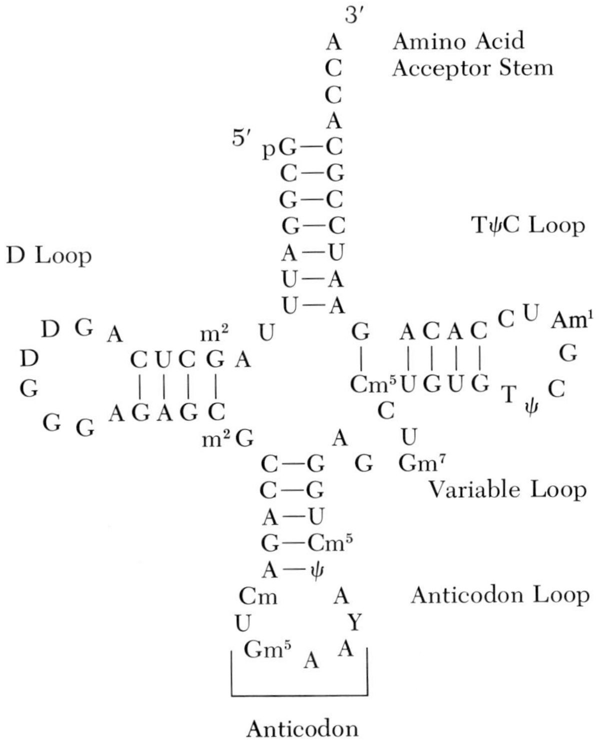

Yeast phenylalanine tRNA

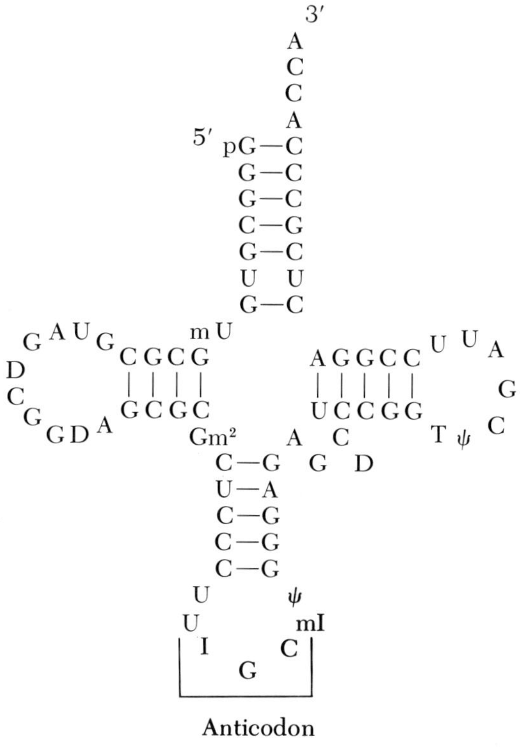

Yeast alanine tRNA

FIGURE 7–5 Cloverleaf structure for transfer RNA as illustrated by yeast phenylalanine and alanine tRNA's. The positions of the D, anticodon, and TψC loops are indicated. The nucleotide labeled Y in phenylalanine tRNA is a modified purine whose structure has not yet been established.

the similarities in base pair formation found in all tRNA molecules. The nucleotides not involved in double helical regions are found in four loops of unpaired nucleotides. The loop of eight unpaired bases to the left in both tRNA's shown in Figure 7–5 often contains one or more dihydrouridine residues, and it is therefore called the dihydrouridine or D loop. The set of four base pairs leading to the D loop is identified as the D stem. The anticodon sequence is the middle three bases in the bottom or anticodon loop. For example, in yeast phenylalanine tRNA the anticodon sequence is 5′ methyl GAA 3′, and it can form complementary base pairs with both phenylalanine codons 5′ UUU 3′ and 5′ UUC 3′, as described further on. Of the two remaining unpaired regions in tRNA, the variable loop takes its name from the fact that it contains quite different numbers of nucleotides in different tRNA molecules. The TΨC loop to the right nearly always contains the sequence 5′ TΨC 3′ at the position indicated in Figure 7–5.

Although the cloverleaf model for tRNA structure provides an accurate picture of the bases found in double helical regions, this model is derived from nucleotide sequence information only and therefore does not illustrate how the double helical regions are oriented with respect to each other. This information has to be obtained experimentally from x-ray crystallographic studies of purified tRNA. Studies of this type show that, as expected from the similarities in base pairing, most tRNA's have structures similar to each other and to the structure of yeast phenylalanine tRNA, shown in Figure 7–6. In this tRNA, the set of four base pairs leading to the D loop (the D Stem) are stacked on top of the five base pairs leading to the anticodon loop (the anticodon stem) to form an extended double helical region containing nine base pairs. Similarly, the amino acid acceptor stem is stacked on top of the TΨC stem to form a second extended region of double helix that contains 12 base pairs. These two extended double helical structures are roughly cylindrical, and each has a diameter of approximatley 20Å. They are oriented at right angles to one another and are joined by the bases comprising the D and TΨC loops to form an overall structure that has the shape of an L. The anticodon sequence is found at the end of the arm made up of the D and anticodon stems, while the amino acid itself is located at the end of the other arm, some 80Å removed from the anticodon. Careful inspection of this tRNA structure suggests that it may be possible for tRNA to change shape during protein synthesis by bending in the region where the two extended double helices are joined (the vertex of the L).

CODON-ANTICODON RECOGNITION

The rules for base pairing in the recognition of the mRNA codon by the tRNA anticodon differ slightly from the standard Watson-Crick rules which specify that A pairs only with U, and G pairs only with C. The Watson-Crick rules apply strictly to recognition of the first two bases in the codon by the last two bases in the anticodon (when read in the 5′ to 3′ direction). For example, the codon 5′ GAU 3′ (aspartic acid) must be recognized by an anticodon sequence that ends in 5′ UC 3′. However, more flexibility exists for base pairing between the third base of the codon and the first base of the anticodon; these bases may be held less rigidly in place and may be free to undergo a limited flex or "wobble" on the ribosome surface. The rules for base pair formation at this position therefore allow for a greater variability in the kinds of base pairs that may be formed, as shown in Table 7–2. These rules were first described by Crick in 1966 and are identified as the "wobble hypothesis."

According to this scheme, five types of the tRNA's may exist, depending

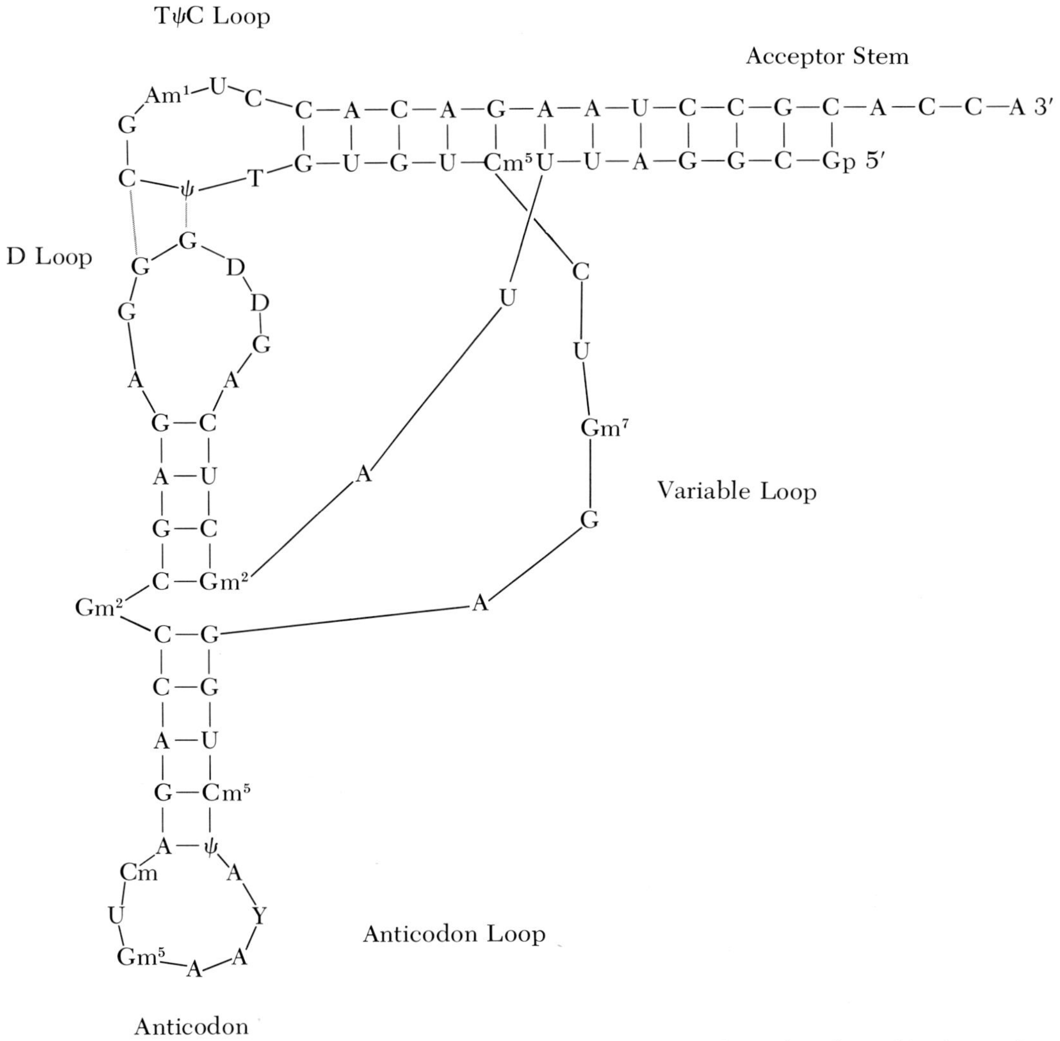

FIGURE 7–6 Yeast phenylalanine tRNA drawn to emphasize the nucleotide stacking interactions that exist in the native molecule. Progress of the polynucleotide chain is indicated in black, and some of the bases linked by hydrogen bonds are shown in gray. Note that the D stem and the anticodon stem interact to form an extended double helical region and that the acceptor stem and the TψC stem form a similar extended segment of double helix. The two extended double helical regions are linked by H bond interactions between the TψC loop and the D loop in such a way that the overall molecule has the shape of an L. The anticodon loop and the amino acid are found at the two ends of the L.

THE WOBBLE HYPOTHESIS* **TABLE 7–2**

Anticodon Base	Codon Base
A	U only
C	G only
G	C or U
U	A or G
I	A, C, or U

*Base pairs as described in the Wobble hypothesis can be formed between the third (3′) base in the codon and the first (5′) base in the anticodon on the ribosome surface.

upon the first base in the anticodon. If the first base is G, the tRNA will recognize codons ending in C or U, while tRNA's whose anticodon begins with the unusual base I will recognize codons ending in U, C, or A, and so on (Table 7–2). For instance, there is a species of yeast alanine tRNA that has the anticodon sequence 5′ IGC 3′, and it recognizes the alanine codons 5′ GCU 3′, 5′ GCC 3′, and 5′ GCA 3′, but not 5′ GCG 3′. Thus, although there may be more than one tRNA for a given amino acid, the multiple tRNA's must each obey the wobble codon recognition rules.

There may be a valine tRNA, for example, that recognizes *either* the codons 5′ GUG 3′ and 5′ GUA 3′ *or* the codons 5′ GUU 3′ and 5′ GUC 3′, but there is not a tRNA that recognizes 5′ GUC 3′ and 5′ GUG 3′. Furthermore, since the Watson-Crick base pairing rules apply strictly to the first and second positions of the codon, separate tRNA's must exist to read codons for the same amino acid that differ in the first or second position. Codons and tRNA's for leucine, serine, and arginine are affected by this rule. For example, separate leucine tRNA's must recognize the two leucine codons 5′ UUA 3′ and 5′ CUA 3′. All known tRNA species have codon recognition schemes that correspond to the rules of the wobble hypothesis.

TRANSFER RNA AMINOACYLATION

The amino acid is attached covalently to the free 3′ hydroxyl group of tRNA by an ester linkage, as shown in Figure 7–4. Transfer RNA's carrying an amino acid in this way are said to be aminoacylated or to exist in the aminoacylated form. The amino acid is joined to its tRNA by an enzyme called an aminoacyl-tRNA synthetase or simply a tRNA synthetase. Each aminoacyl-tRNA synthetase is specific for only one of the twenty amino acids, so one may speak of leucyl-tRNA synthetase or valyl-tRNA synthetase and so on. These enzymes are specific in that they will add only their own amino acid onto its proper tRNA(s). In prokaryotic organisms there is only one species of tRNA synthetase for each amino acid, even if an amino acid has many codons and many different species of tRNA. The same enzyme must therefore be able to distinguish different tRNA's for its own amino acid from tRNA's for different amino acids. In eukaryotic organisms there may be more than one species of tRNA synthetase for the same amino acid.

All aminoacyl-tRNA synthetases operate by the same general mechanism. In the first step the amino acid and adenosine triphosphate (ATP), react to form an adenylated amino acid, or aminoacyladenylate, and inorganic pyrophosphate, as shown in Figure 7–7. Only the correct amino acid can be adenylated in this way by each aminoacyl-tRNA synthetase, and once it is formed the aminoacyladenylate remains bound to the enzyme surface.

The second step in tRNA aminoacylation or "charging" occurs when the amino acid is transferred from its aminoacyladenylate to the 3′ end of tRNA; adenosine monophosphate (AMP) is released at this stage. The enzyme-bound aminoacyladenylate will interact in this step only with tRNA's meant to carry the activated amino acid. Other tRNA's are not bound to the enzyme surface. Thus some structural feature of the correct tRNA molecules must specifically be recognized by the synthetase enzyme. Although the exact location in tRNA of the region recognized by the synthetase is not yet known, the anticodon sequence is thought to be somehow involved in the recognition process.

The amino acid is said to be activated when it is bound to tRNA, because approximately 6 kcal/mole of energy is released when the ester bond between the

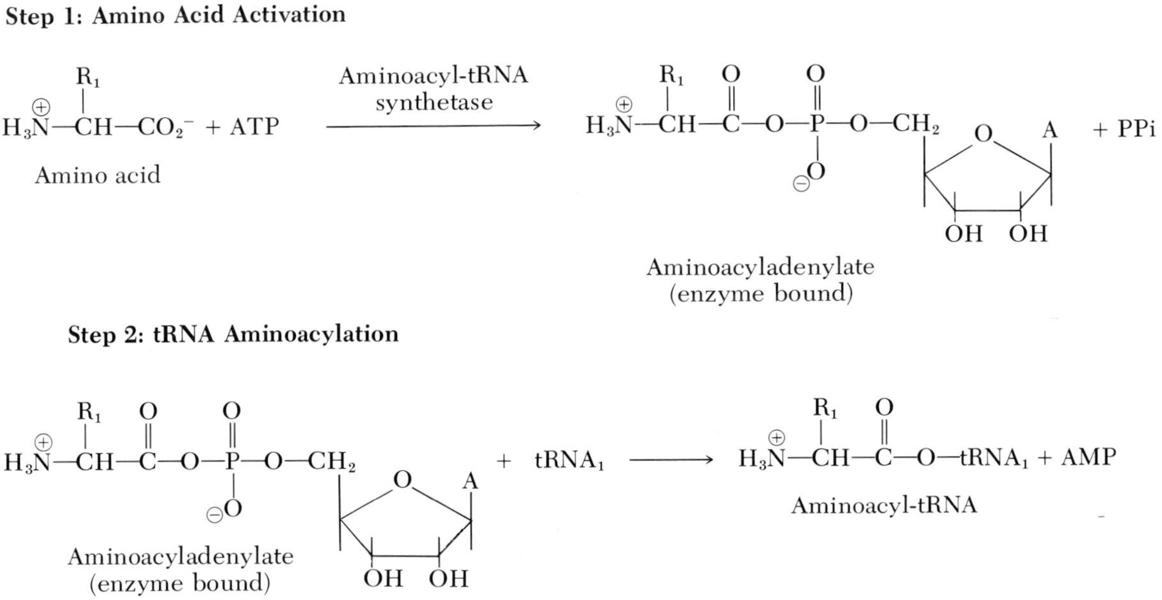

FIGURE 7–7 Two-step reaction sequence for the aminoacylation of tRNA by aminoacyl–tRNA synthetases.

amino acid and tRNA is broken. The energy for activation comes ultimately from the splitting of ATP during aminoacyladenylate formation, and it is used when the amino acid is added onto the end of the growing polypeptide chain during protein synthesis.

Ribosomes

Protein synthesis takes place in the cytoplasm on ribosomes. These structures consist only of RNA and protein, and they have a roughly spherical shape with a diameter of 150Å to 250Å. Their overall composition by weight is approximately 60 per cent RNA and 40 per cent protein. Ribosomes have two sites for binding tRNA and one for mRNA. One ribosome can translate only one mRNA molecule at a time. The amino acid sequence of a protein being synthesized on a ribosome is determined by the mRNA and not by the ribosome itself, so ribosomes may be regarded as nonspecific "workbenches" on which any protein may be produced, depending on the mRNA. In any species of organism all ribosomes are identical or very nearly so. In a bacterial cell there may be as many as 20,000 ribosomes which constitute as much as 25 per cent of the total cell mass.

Complete ribosomes can be dissociated into two smaller subribosomal particles which are identified by their sedimentation coefficients. Complete prokaryotic ribosomes have a sedimentation coefficient of 70S, and they dissociate into small and large subunits having sedimentation coefficients of 30S and 50S. Like the intact ribosome, the ribosomal subunits consist only of RNA and protein in roughly the same proportions as the 70S structures.

The smaller 30S ribosomal subunit contains the binding site for mRNA. Codon-anticodon recognition therefore takes place on the 30S subunit. This particle is probably also involved in the mechanism used for moving the mRNA along, three nucleotides at a time, after each step of polypeptide chain elongation. The 30 S subunit is composed of one 16S ribosomal RNA molecule (MW 6×10^5) plus 21 proteins. The 16S RNA is a single polynucleotide chain approximately 1800 nucleotides in length. It contains the usual four RNA bases plus approximately 1 per cent of methylated residues. A great deal of the 16S rRNA nucleotide sequence has now been determined, including a pyrimidine-rich sequence at the 3′ end which is thought to be involved in the identification of initiator AUG codons in mRNA, as previously described. The 21 proteins found in the 30S ribosomal subunit are all different, and only one copy of each is present per 30S subunit. They tend to be small positively charged proteins with molecular weights between 10,000 and 30,000. Just how these proteins are associated with RNA or how they function in protein synthesis is not known.

The larger 50S ribosomal subunit forms portions of the binding sites for the tRNA molecules, and it also contains the peptidyl transferase, the enzyme that actually joins amino acids to the growing polypeptide chain during protein synthesis. The 50S subunit contains two species of ribosomal RNA, the 23S rRNA (MW 1.2×10^6) and the 5S rRNA (MW 4×10^4), plus 34 different proteins of which there is one copy in each 50S subunit. The 5S rRNA has now been completely sequenced and found to contain only the usual four RNA bases. The 23S rRNA, however, contains approximately 2 per cent of methylated nucleotides. Very little sequence information is available for the 23S rRNA. The 34 different 50S subunit proteins resemble the 30S proteins in that they are small positively charged molecules, but none of the 50S subunit proteins is identical to any of the 30S proteins.

Both the 30S and the 50S *E. coli* ribosomal subunits can now be reassembled

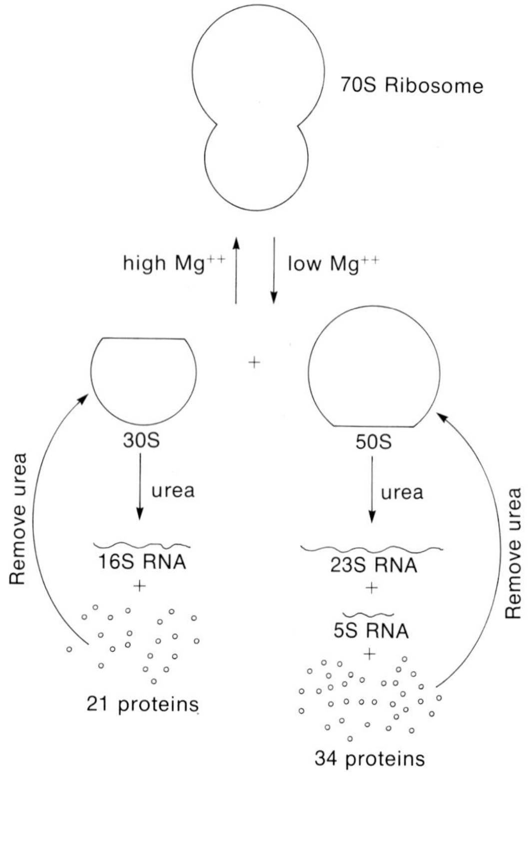

FIGURE 7–8 Dissociation and reassociation of prokaryotic (70S) ribosomes.

from their component rRNA and proteins. For example, the 30S subunit can be assembled from 16S rRNA and its 21 protein species to yield subunits that are fully functional in protein synthesis (Fig. 7–8). This may prove to be an important experimental method for future structural and functional studies of ribosomes.

Very little is now known about the detailed structure of the ribosome. This is, in fact, one of the most important unsolved problems in the field of molecular biology. An important question is how the proteins and RNA interact with each other to form a functioning ribosome. Of particular interest are how the mRNA binding site is formed, how the two tRNA molecules are bound to the ribosome surface, and where the peptidyl transferase is located with resepct to these other sites. The great diversity of protein and RNA components present in ribosomes leads one to expect that the overall structure will prove to be quite complex. However, it is difficult to imagine why such a complex structure is required for the relatively simple sequence of events involved in protein synthesis. Unfortunately, ribosomes are not well suited for structural studies by the traditional methods of molecular biology. Electron microscope pictures do not reveal much about the details of ribosome structure, and crystals of ribosomes suitable for x-ray analysis are not yet available. Some information about the relative location of different ribosomal proteins has been obtained from studies employing chemical cross-linking agents, but so far our knowledge about ribosome structure is very rudimentary.

Significant differences are found to exist between prokaryotic ribosomes and ribosomes found in the cytoplasm of eukaryotic cells. Eukaryotic ribosomes are larger than prokaryotic ribosomes and have a sedimentation coefficient of 80S, rather than 70S; the 80S eukaryotic ribosomal subunits sediment at 60S and 40S, rather than at 50S and 30S. Similarly, the major eukaryotic rRNA's are 28S and 18S, rather than 23S and 16S. Other differences are shown in Table 7–3.

In addition, whereas all prokaryotic ribosomes are found free in the cytoplasm, eukaryotic ribosomes are often attached to cytoplasmic membranes to form the rough endoplasmic reticulum (Chapter 8), and a great deal of protein synthesis in eukaryotic cells takes place on these membrane-bound ribosomes.

As in the case of the transcription machinery, it is possible to exploit the difference between prokaryotic and eukaryotic ribosomes with drugs that selectively inhibit prokaryotic protein synthesis and thereby limit bacterial cell growth. Many of the drugs currently in use for combating bacterial infections function by inhibiting bacterial protein synthesis. These include the tetracyclines, erythromycin, chloramphenicol, and streptomycin. Their exact mechanism of action will be described in a later section.

COMPONENTS OF EUKARYOTIC AND PROKARYOTIC RIBOSOMES **TABLE 7–3**

	Ribosome Type	
	Eukaryotic	*Prokaryotic*
Sedimentation coefficient	80S	70S
Subunits	60S, 40S	50S, 30S
RNA species	28S, 18S, 5.8S, 5S	23S, 16S, 5S

THE MECHANISM OF PROTEIN SYNTHESIS

An Overview

Now that the most important components involved in protein synthesis have been discussed, the interactions of these components to produce protein will be considered. It is important to note that the overall process of protein synthesis must be extremely accurate. All molecules of a single type of protein, such as hemoglobin, must have exactly the same amino acid sequence. In fact, there is less than 1 error for every 10,000 amino acids polymerized into protein. Another characteristic worth noting is that the mechanism of protein synthesis is cyclical. Proteins are linear polymers composed of monomer units which are each attached to their neighbors by identical peptide linkages. Therefore, this type of structure can be synthesized by a mechanism in which monomer units are added sequentially to the growing polymer, with essentially the same events at each step. It would be less efficient, probably less accurate, and certainly wasteful of genetic material if an entirely separate mechanism were used to add each amino acid onto the nascent polypeptide.

The cycling mechanism in protein biosynthesis operates on the surface of the ribosome. Proteins are built up one amino acid at a time while the nascent polypeptide is still attached to a tRNA in a structure called peptidyl-tRNA. The peptidyl-tRNA is itself bound noncovalently to the ribosome at a special site, the peptidyl-tRNA or P site. Immediately adjacent to the P site on the ribosome surface is another site, the A site, for the codon-specific binding of one aminoacyl-tRNA molecule; only the aminoacyl-tRNA corresponding to the mRNA codon exposed at the A site can be bound. Once the correct aminoacyl-tRNA has been identified, however, a peptide bond is formed between the free amino group of aminoacyl-tRNA and the carboxy terminus of peptidyl-tRNA, as shown in Figure 7–9. This chemical reaction is catalyzed by the peptidyl tranferase, and it results in extension of the nascent polypeptide by one amino acid unit. As a result of

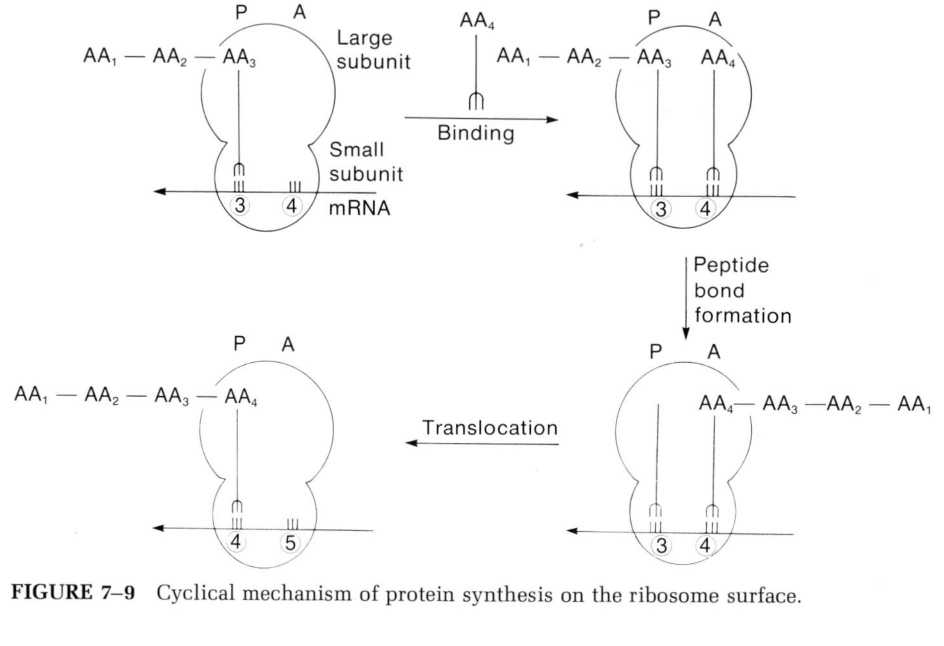

FIGURE 7–9 Cyclical mechanism of protein synthesis on the ribosome surface.

this process, the nascent polypeptide is transferred from one tRNA molecule to another.

After peptide bond formation has taken place, the newly extended peptidyl-tRNA is moved or "translocated" from the A to the P site, the mRNA moves along the ribosome surface by one triplet, and the molecule of free tRNA is lost from the P site. The ribosome, therefore, ends this cyclical process in the same state that it began, except that the nascent polypeptide is one amino acid unit longer and it is attached to a different tRNA. The A site is again free for binding the next aminoacyl-tRNA specified by mRNA, and the whole cycle can be repeated. When the last amino acid specified in mRNA has been incorporated, the ester bond linking the polypeptide to tRNA is hydrolyzed and the completed protein leaves the ribosome surface.

Energy is required at each step in protein synthesis, and it is supplied by guanosine triphosphate (GTP) which is hydrolyzed to GDP, and Pi. One molecule of GTP is hydrolyzed when aminoacyl-tRNA is bound to the ribosome surface, and a second is consumed during the translocation process. Two molecules of GTP are therefore hydrolyzed for each peptide bond formed. The exact steps that require energy in binding and translocation have not yet been identified.

A closer look at the chemical events taking place during peptide bond formation, the crucial step in protein synthesis, reveals that the peptidyl transferase must catalyze a nucleophilic attack of the free amino group in aminoacyl-tRNA onto the ester bond linking the nascent polypeptide and its tRNA, as shown in Figure 7–10. The ester bond is broken and a new amide bond is formed, linking the newly bound aminoacyl-tRNA to the pre-existing nascent polypeptide. Since energy is released in this transfer reaction, no new energy needs to be supplied to drive it toward formation of the peptide bond.

A similarly detailed examination of other aspects of protein synthesis requires a dissection of the overall process into its component parts, initiation, chain propagation, and termination. Although these functions are similar in all cells, they have been most thoroughly studied in *E. coli.* Therefore, the following discussion will be focused on that system, and important differences between prokaryotic and eukaryotic cell protein synthesis will be noted.

Initiation

As pointed out previously, all protein synthesis begins with the incorporation of methionine into the N-terminal position of the polypeptide chain. The initiating methionine residue is derived from a special species of methionyl-tRNA, called met-$tRNA_F$, which is involved only in the initiation of protein synthesis. Chain extension methionine residues are all donated by an entirely separate met-tRNA called met-$tRNA_M$. Both met-$tRNA_F$ and met-$tRNA_M$ read the same codon, AUG. In prokaryotic, but not in eukaryotic, organisms the initiating methionine residue contains an N-formyl group, as shown in Figure 7–11. The formyl group is derived from N^{10}-formyltetrahydrofolic acid, and it is added to methionine after methionine has already been incorporated into met-$tRNA_F$. Formylation is catalyzed by an enzyme called transformylase which can distinguish between met-$tRNA_F$ and met-$tRNA_M$; only met-$tRNA_F$ is formylated. The role of the formyl group in initiation of prokaryotic protein synthesis is not now known. No formyl group is present on the initiating met-$tRNA_F$ in eukaryotic cells.

Although all protein synthesis is initiated with methionine, it is clear that not all mature proteins have methionine at the amino terminus. In these cases, the

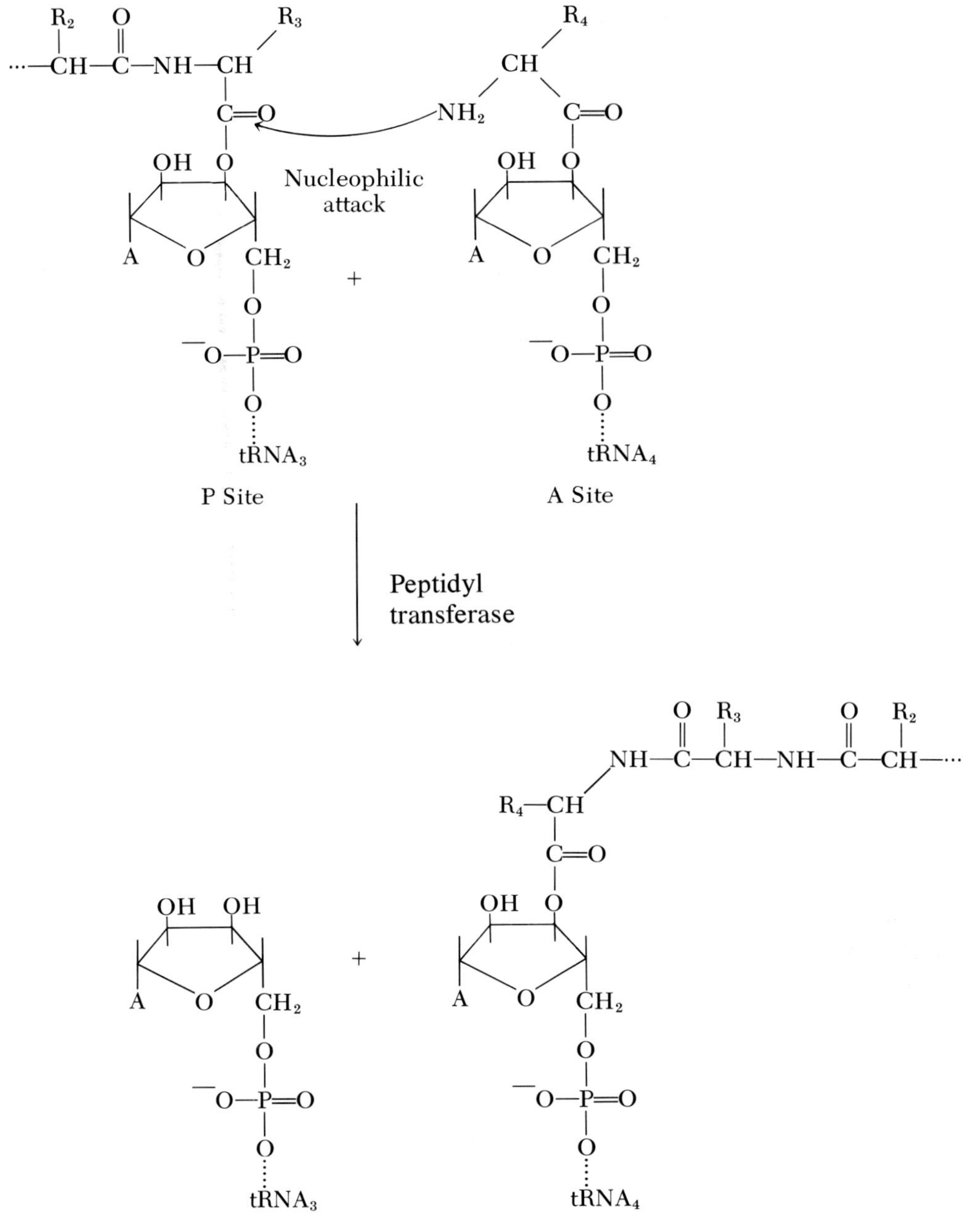

FIGURE 7–10 Peptide bond formation catalyzed by the peptidyl transferase on the ribosome surface.

N-terminal methionine and often a few additional amino acids are removed from the nascent polypeptide even before its synthesis is complete. This serves to expose a new N-terminal amino acid which is found in the mature protein molecule.

In addition to the ribosome, mRNA, and N-formylmethionyl-$tRNA_F$, initiation of protein synthesis in *E. coli* requires the participation of three protein "initiation factors," called IF_1 (MW 9000), IF_2 (MW 80,000), and IF_3 (MW 22,000). Similar but less well characterized factors are found in eukaryotic cells. IF_3 is required for the binding and proper positioning of mRNA on the ribosome,

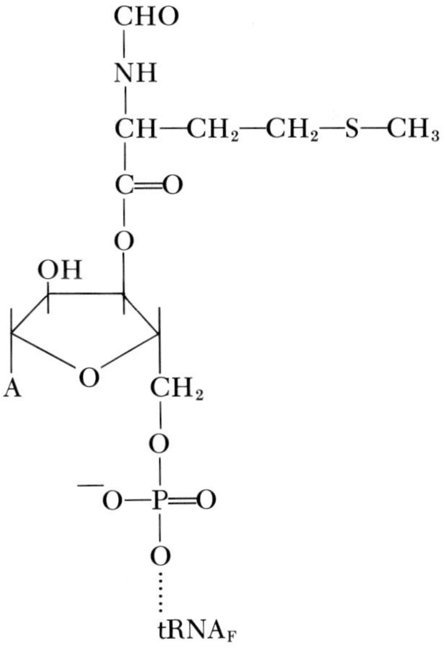

FIGURE 7–11 N-formylmethionyl-tRNA$_F$, the chain initiating tRNA in prokaryotic protein synthesis.

while IF_1 and IF_2 are involved in attachment of N-formylmet-tRNA$_F$ (f-met-tRNA$_F$) to the mRNA-ribosome complex. IF_2 forms a weak association with f-met-tRNA$_F$ plus GTP, and this ternary complex migrates to the ribosome where f-met-tRNA$_F$ is specifically bound. GTP is hydrolyzed during this "enzymatic" binding process, but it is not known how the energy released from GTP hydrolysis is used. After f-met-tRNA$_F$ is securely bound to the mRNA-ribosome complex, IF_1 is required to remove IF_2 from the ribosome surface.

Unlike polypeptide chain elongation steps, the steps in the initiation of protein synthesis do not take place on the intact 70S ribosome, but rather on the 30S ribosomal subunit. Messenger RNA is first bound in the presence of IF_3 to the 30S subunit and then f-met-tRNA$_F$ is added to form an mRNA-30S subunit–f-met-tRNA$_F$ complex, as described earlier. Only then does the 50S subunit join this complex to form an "initiation complex" as shown in Figure 7–12.

Chain Propagation

The ribosome is now in the position to begin the cyclical process of polypeptide chain elongation. It is not clear at present whether a translocation step is required to move f-met-tRNA$_F$ from the A to the P site or whether f-met-tRNA$_F$ binds directly to the P site. It is clear, however, that soon after formation of the initiation complex the aminoacyl-tRNA corresponding to the second codon in mRNA is bound to the A site on the ribosome surface. A peptide bond is then formed between f-met and the second amino acid. The free initiator tRNA produced by this process is dissociated from the ribosome surface. The dipeptide, now attached to the tRNA for the second amino acid, is translocated from the A to the P site, and a second cycle of chain elongation can be begun.

Two features of this overall chain propagation mechanism deserve emphasis. First, the growing polypeptide is attached to tRNA at every stage. The specific tRNA changes at each step—the polypeptide is attached to the tRNA that donated the most recently added amino acid—but it is clear that the growing polypeptide remains attached to tRNA until its synthesis is complete. Second, although not

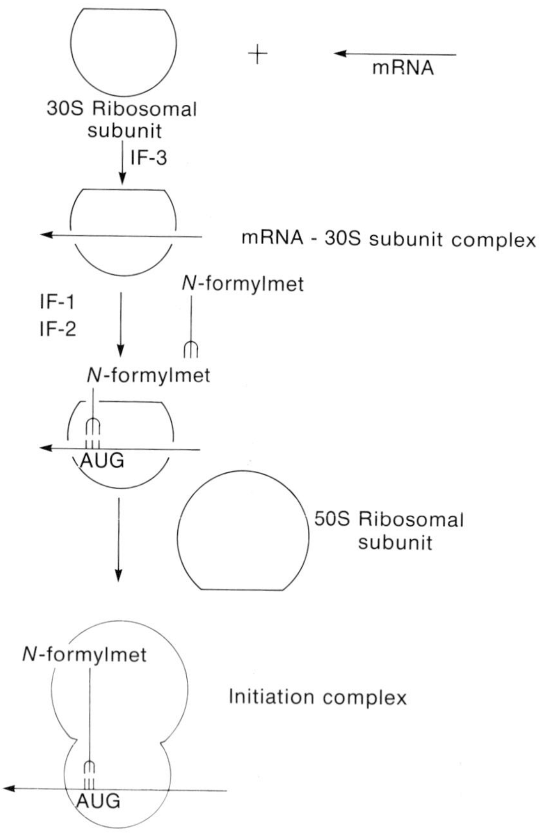

FIGURE 7–12 Early steps in the initiation of protein synthesis on 30S ribosomal subunits.

enough is known about ribosome structure to define the locations of the A and P sites, these sites can be clearly distinguished according to their functions. All decoding of mRNA takes place at the A site. The P site is not involved in the initial interaction between codon and anticodon, and it is not even clear that base pairs between codon and anticodon need to exist at the P site. The P site, however, is *never* occupied by ordinary aminoacyl-tRNA. Before peptide bond formation takes place the P site is occupied by initiator or by peptidyl-tRNA, and it contains free tRNA briefly just after peptide bond formation. In none of these steps, however, does the P site contain ordinary aminoacyl-tRNA.

As in the case of initiation, protein factors called elongation factors are involved in polypeptide chain propagation. In *E. coli* there are three elongation factors: EF-Tu (MW 47,000), EF-Ts (MW 34,000) and EF-G (MW 83,000). EF-Tu and EF-Ts can form a complex that is often identified as EF-T. The three elongation factors have been isolated in purified form, and together they account for approximately 2 per cent of the total *E. coli* soluble protein mass. Similar elongation factors called EF-1 and EF-2 have been identified in eukaryotic cells, and these correspond to EF-T and EF-G, respectively, in the *E. coli* system.

EF-Tu is involved in attaching aminoacyl-tRNA's to the ribosome surface. Thus it performs the same function for aminoacyl-tRNA that IF_2 is thought to perform for f-met-$tRNA_F$. A ternary complex is first formed, involving EF-Tu, aminoacyl-tRNA, and GTP. This complex then migrates to the ribosome surface, where aminoacyl-tRNA becomes specifically bound (noncovalently) and GTP is hydrolyzed. Although GTP hydrolysis is required for functional binding of aminoacyl-tRNA, it is not known at what stage energy is actually consumed. Unlike

IF_2, which is specific for f-met-$tRNA_F$, EF-Tu is not specific for any one particular aminoacyl-tRNA; all EF-Tu molecules are identical, and all can donate any aminoacyl-tRNA (except initiator tRNA) to the ribosome surface. Once the incoming aminoacyl-tRNA is bound and GTP is hydrolyzed, a complex consisting of EF-Tu and GDP leaves the ribosome surface and is dissociated by EF-Ts.

EF-G participates in chain elongation after the peptide bond has been formed. It is a GTPase and it functions in the translocation step; no translocation takes place in the absence of EF-G and GTP. The GTP hydrolyzed by EF-G in conjunction with translocation is clearly different from the GTP hydrolyzed at the binding step. Two molecules of GTP are therefore consumed for each peptide bond formed.

Polypeptide chain propagation is a very rapid and efficient process. Polypeptides are extended at a rate of 10 to 20 amino acid units per second per ribosome. To increase the overall efficiency, more than one ribosome can translate a single mRNA molecule at a time. While a polypeptide is being synthesized by the interaction of one ribosome and the other necessary molecules with mRNA, another ribosome may bind and translation may begin again at the initiator codon on the messenger. Repetition of this process results in a row of ribosomes along the mRNA molecule, all the ribosomes being engaged in the synthesis of polypeptide chains in varying stages of completion (Fig. 7–13). These complexes of mRNA and ribosomes are known as polyribosomes, or polysomes, and they are characteristic of cells active in protein synthesis.

The number of ribosomes in a polysome is proportional to the length of the mRNA being translated. Since prokaryotes produce polycistronic messages, i.e., long messenger RNA molecules consisting of the transcripts of multiple genes, large polysomes containing up to 20 or more ribosomes can result. Polysomes in eukaryotes that have monocistronic messages (transcripts of single genes) generally contain a smaller number of ribosomes; most of the polysomes engaged in the synthesis of hemoglobin, for example, consist of five ribosomes. In this instance, therefore, five hemoglobin polypeptides are produced simultaneously from a single mRNA molecule.

Prokaryotes lack a nuclear envelope. Since mRNA is synthesized in a 5′ to 3′ direction, and translation proceeds in the same direction, translation of the mRNA in prokaryotes can begin even before its synthesis is complete. In eukaryotes, the mRNA must cross the nuclear envelope and reach the cytoplasm before translation begins. The structure of polysomes as seen in preparations for electron microscopy is considered further in Chapter 9.

Chain Termination

Termination of protein synthesis takes place when the ribosome encounters one of the three terminator codons, UAA, UAG, or UGA. These occur in mRNA just after the codon specifying the C-terminal amino acid in the finished polypeptide, as shown for MS2 RNA in Figure 7–2. Terminator codons are not read by

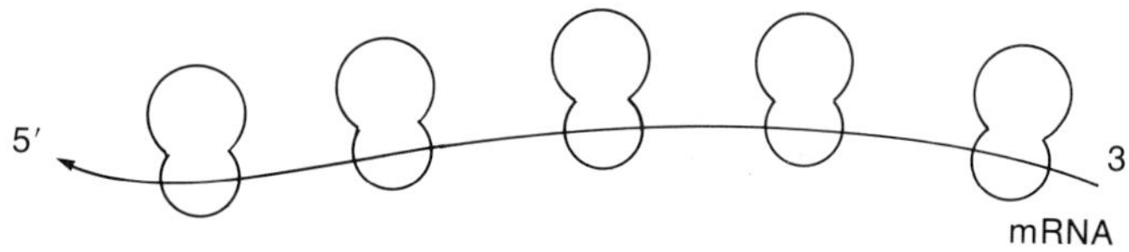

FIGURE 7–13 Schematic diagram showing how several ribosomes may be involved in translating a single mRNA molecule at once.

tRNA molecules, but rather by one of two protein "release factors" called RF-1 (MW 44,000) and RF-2 (MW 47,000). RF-1 can read the codons UAA and UAG, while RF-2 reads UAA and UGA. This is the only situation in which mRNA codons are read by proteins rather than by tRNA molecules. Recognition of a terminator codon by one of the release factors is followed by hydrolysis of the bond linking the completed polypeptide chain to tRNA. The finished protein is then free to leave the ribosome surface. Soon after loss of the completed protein, the tRNA and mRNA also dissociate from the ribosome surface. Finally, the ribosome itself dissociates into its 30S and 50S subunits in preparation for initiating another cycle of protein synthesis. A pool of free 30S and 50S subunits exists inside the cell, so it is likely that the same 30S and 50S subunits will not reassociate when a new round of protein synthesis is begun. A schematic diagram summarizing all the events taking place during protein synthesis is shown in Figure 7–14.

Posttranslational Processing

The immediate product of the translation process is often chemically modified to produce the mature functional protein molecule. This tailoring is similar in principle to the processing steps that occur during RNA synthesis. For example, the initiating methionine residue and a few other amino acids are removed from the N-terminal end of most polypeptide chains, occasionally even before synthesis of the whole protein is complete. Also, proteolytic enzymes are often synthesized as inactive precursors called zymogens, which must be partially degraded to produce their active form. The cleavage of inactive trypsinogen to trypsin or of pepsinogen to pepsin are examples of this process.

Similarly, polypeptide hormones are often synthesized as large prohormones which must be cleaved to produce the active hormone molecules as shown for insulin in Figure 7–15. Chemical modifications may also be introduced into amino acids after synthesis of the primary polypeptides is complete. For example, the carbohydrate groups in glycoproteins are added onto asparagine, serine, or threonine residues during or shortly after synthesis of the polypeptide backbone. Many proline and lysine residues in collagen are modified to produce hydroxyproline and hydroxylysine after synthesis of the collagen polypeptide has been completed.

Antibiotic Drugs and Toxins

The relatively minor differences between eukarytotic and prokaryotic systems of protein synthesis have been exploited in the use of antibiotic drugs that inhibit prokaryotic 70S ribosome protein synthesis without affecting eukaryotic 80S ribosome function. Since protein synthesis is required for cell growth, such drugs ought to retard bacterial growth without affecting the human host.

Many drugs of this type are now in clinical use, including the following: tetracyclines, which inhibit binding of aminoacyl-tRNA to the ribosome surface; erythromycin, which inhibits translocation; chloramphenicol, which blocks peptidyl transferase function; and streptomycin, which blocks initiation of bacterial protein synthesis and also causes mistakes in codon-anticodon recognition. Many other drugs that specifically inhibit particular steps in protein synthesis are available, but for a variety of reasons are not clinically useful. These compounds have,

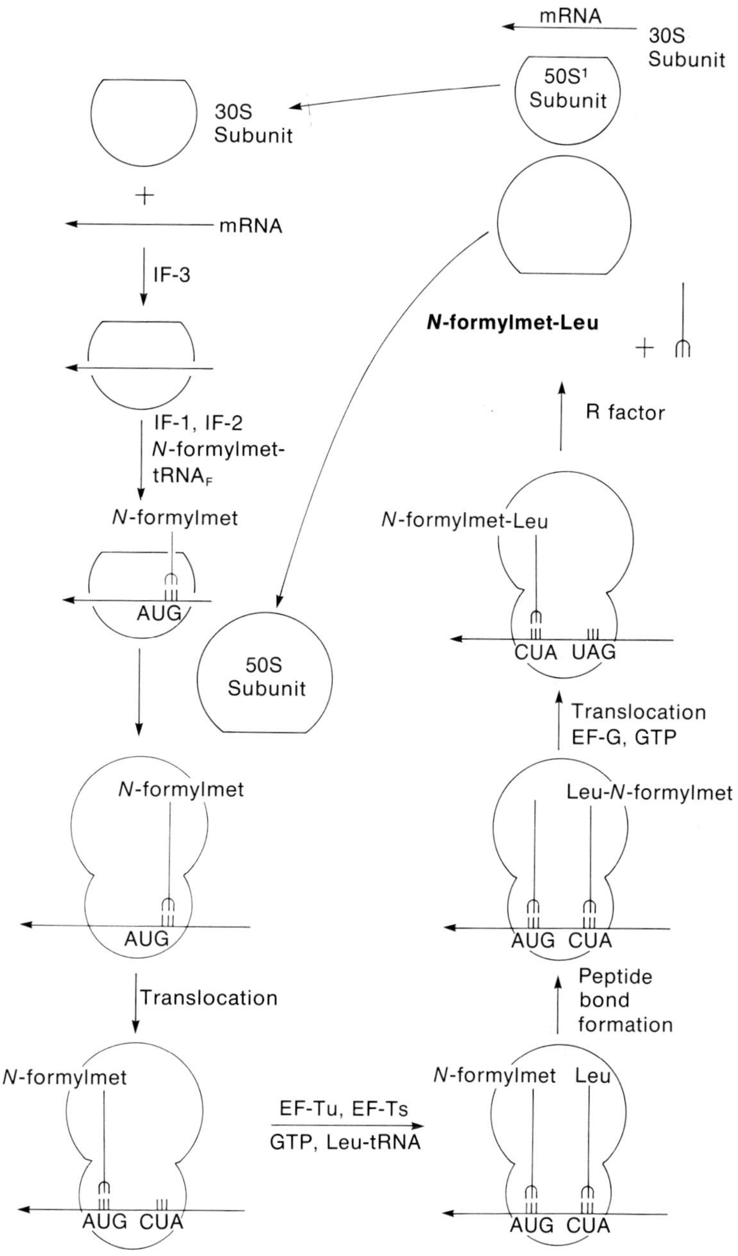

FIGURE 7–14 Summary of the events involved in ribosomal protein synthesis, as illustrated by the synthesis of the peptide N-formylmethionylleucine. The peptide product is shown in bold type.

however, found wide use in basic studies on the mechanism of protein synthesis. For example, fusidic acid blocks EF-G GTPase activity; puromycin and cycloheximide cause premature termination of protein synthesis; thiostrepton inhibits translocation; pactamycin stops initiation of protein synthesis without affecting chain propagation; and sparsomycin inhibits the peptidyl transferase activity.

A unique case of a bacterial toxin that affects eukaryotic cell protein synthesis is found in *Corynebacterium diphtheriae,* the causative agent of diphtheria. Strains of this bacterium, lysogenic for a phage called β, secrete a protein exotoxin that affects eukaryotic cell protein synthesis in a very specific way. Secreted diphtheria toxin (MW 62,000) binds to the surface of a target cell where it is split by proteolytic cleavage into two fragments, A (MW 24,000) and B (MW 38,000).

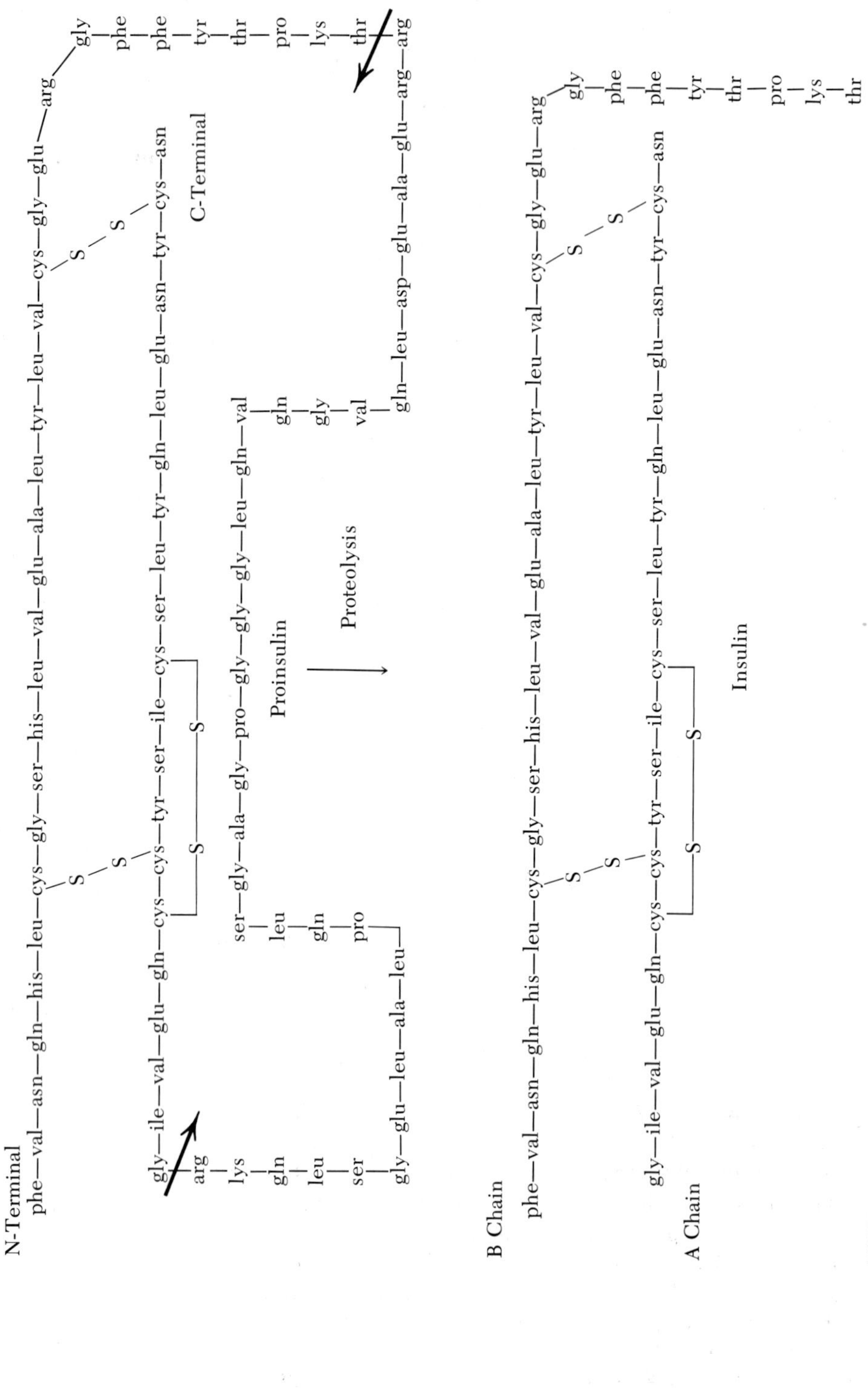

FIGURE 7–15 Proteolytic cleavage of human proinsulin to yield biologically active insulin. Proteolytic hydrolysis takes place at the two sites indicated by the arrows.

The B fragment remains outside the cell, but A enters and enzymatically catalyzes the hydrolysis of NAD^+ into ADP-ribose and nicotinamide (Fig. 7–16). ADP-ribose is then bound by a covalent linkage to a molecule of EF-2, the eukaryotic "translocase," and inactivates it. The A fragment is not consumed in this process, but is free to act catalytically to inactivate more EF-2 molecules until the cell can no longer synthesize protein. Full EF-2 activity can be recovered if ADP-ribose is removed from the inactivated EF-2 factor.

In addition to diphtheria toxin, two protein toxins derived from plants have been found to act by inhibiting eukaryotic cell protein synthesis. Ricin, the toxic factor from castor beans, and abrin, a toxic protein found in seeds of *Abrus precatorius,* both exert their toxic effects by inhibiting the translocation step in protein synthesis. Their exact mechanism of action, however, is not as well understood as that of diphtheria toxin.

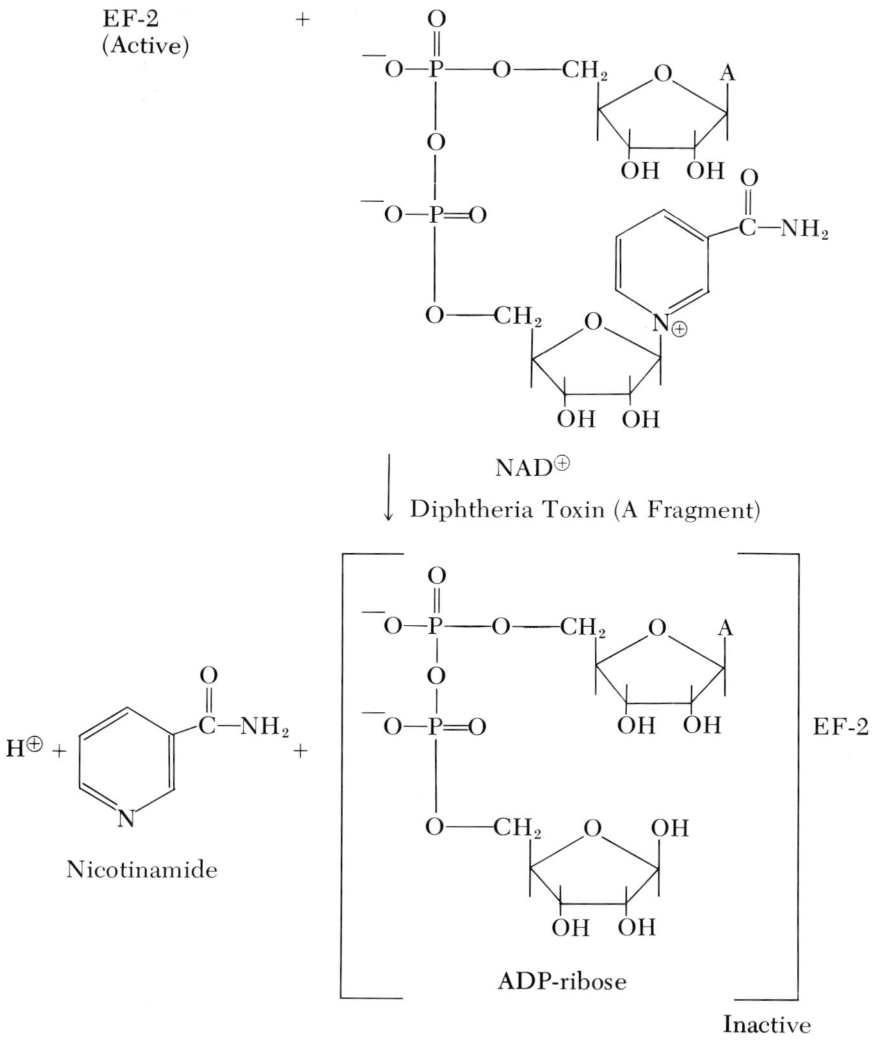

FIGURE 7–16 Inactivation of EF-2, the eukaryotic translocase, by ADP-ribosylation catalyzed by diphtheria toxin.

FURTHER READING

Brown, J. C., and Smith, A. E.: Initiator codons in eukaryotes. Nature, *226*:610, 1970.

Capecchi, M. R., and Klein, H. A.: Characterization of three proteins involved in polypeptide chain termination. Cold Spring Harbor Symp. Quant. Biol., *34*:469, 1969.

Crick, F. H. C.: Codon-anticoding pairing: the wobble hypothesis. J. Mol. Biol., *19*:548, 1966.

Crick, F. H. C.: The genetic code III. Sci. Am., *215*(4):55, 1966.

Crick, F. H. C.: The genetic code: Yesterday, today and tomorrow. Cold Spring Harbor Symp. Quant. Biol., *31*:3, 1966.

Fiers, W., Contreras, R., Duerinck, F., et al.: Complete nucleotide sequence of phage MS2 RNA. Nature, *260*:500, 1976.

Haselkorn, R., and Rothman-Denes, L. B.: Protein synthesis. Annu. Rev. Biochem., *42*:397, 1973.

Kaempfer, R.: Dissociation of ribosomes on polypeptide chain termination. Nature, *228*:534, 1970.

Kim, S. H., Suddath, F. L., Quigley, F. L., et al. Three dimensional tertiary structure of yeast phenylalanine-tRNA. Science, *185*:435, 1974.

Lewin, B.: Gene Expression. Vol. 1: Bacterial Genomes. New York, John Wiley and Sons, 1974.

Nirenberg, M. W., and Matthaei, J. H.: The dependence of cell free protein synthesis in *E. coli* upon naturally occurring or synthetic polyribonucleotides. Proc. Natl. Acad. Sci. USA, *47*:1588, 1961.

Nomura, M., Tissieres, A., and Lengyel, P.: Ribosomes. New York, Cold Spring Harbor Laboratory, 1974.

Pappenheimer, A., and Gill, D. M.: Diphtheria. Science, *182*:353, 1973.

Robertus, J. D., Ladner, J. E., Finch, J. T., et al.: Structure of yeast phenylalanine-tRNA at 3Å resolution. Nature, *250*:546, 1974.

Shatkin, A., Banerjee, A., Both, G., et al.: Dependence of translation on 5′-terminal methylation of mRNA. Fed. Proc., *35*:2214, 1976.

Shine, J., and Dalgarno, L.: The 3′-terminal sequence of *E. coli* 16S ribosomal RNA: complementarity to nonsense triplets and ribosome binding sites. Proc. Natl. Acad. Sci. USA, *71*:1342, 1974.

Watson, J. D.: Molecular Biology of the Gene. 3rd ed. Menlo Park, Calif. W. A. Benjamin, 1976.

Weissbach, H., and Ochoa, S.: Soluble factors required for eukaryotic protein synthesis. Annu. Rev. Biochem., *45*:191, 1976.

8

REGULATION OF GENE EXPRESSION

The mechanism of gene expression has been discussed in previous chapters. It has been shown how genetic information for the construction of proteins is stored and replicated in the form of DNA, how it is transcribed into RNA, and how it is translated into protein. Recent experimental studies, mostly with microorganisms, have made it possible to discuss these processes in great chemical detail. It is clear, however, that a higher level of organization must exist, since chaos would result if every cell were to express all its genes all the time. Obviously, some mechanism must exist for controlling which genes are expressed and the extent to which they are expressed at any particular time. Muscle cells must produce large amounts of the muscle proteins myosin and actin, but not hemoglobin or antibody molecules, despite the fact that hemoglobin and antibody genes are present in the muscle cell genome. Similarly, reticulocytes (precursors of erythrocytes) must produce much hemoglobin and little or none of the products of other genes. Thus there is a large class of genes, called regulated genes, whose expression is subject to regulatory control, and there must be biochemical mechanisms for controlling the function of these genes. In higher organisms with complicated developmental sequences, the mechanisms involved in the regulation of gene expression may prove to be extremely complex.

In contrast to genes whose expression is under regulatory control, there is a class of genes whose expression is not affected by regulatory mechanisms. These genes are called constitutive, and they are expressed continuously regardless of environmental or developmental factors. The protein products of regulatory genes in bacteria, for example, are not themselves subject to regulatory control; they are expressed constitutively regardless of other variables. There are probably examples of constitutive genes in eukaryotic cells as well. For instance, the genes for histones or for ribosomal components are probably expressed constitutively in all continuously dividing cells.

Considering only the regulated genes, there are two situations that depend on selective gene expression. The first has to do with compensating for changes in a cell's environment. For example, it would be inefficient for a bacterial cell to make its own histidine if the medium in which it was growing already contained plenty of histidine. To synthesize histidine would necessitate using protein synthetic machinery and expending energy to make something it could get from the environment. It would be far more efficient for the cell to make histidine only if it were not available from the surroundings. Similarly, it would be wasteful for a cell to make the enzymes involved in the utilization of lactose if it never encountered any lactose. On the other hand, it would be useful for the cell to be able to use lactose if it were to become available. This would necessitate expressing the genes for lactose utilization only if lactose were actually present.

The second situation requiring selective gene expression is generation of the complicated developmental programs found in many higher organisms. Although all cells in an adult animal arise from the same single cell (the zygote) and all have exactly the same complement of genes, different cells express quite different functions. Cells become differentiated to express their specialized functions in a stepwise fashion which requires careful control over the genes that are expressed at each step. The question of how individual cells become specialized or "differentiated" to express one set of their genes and not another lies at the center of the overall problem of development. It now seems likely that different molecular mechanisms apply to these situations than apply to the regulation of gene expression in response to environmental changes.

Both prokaryotic and eukaryotic organisms modify their gene expression in response to environmental changes. A great deal is now known about the molecular mechanisms employed for the regulation of certain *E. coli* genes, including those of the lactose, histidine, and arabinose operons. Therefore, this discussion of gene regulation will begin with a description of some prokaryotic systems. Considerably less is known about the mechanisms used for selective gene expression during development in eukaryotic organisms, but some factors that may be relevant will be considered in the second part of this chapter.

PROKARYOTIC SYSTEMS

Repression and Induction

There are two types of responses a cell can make to a change in the nutrients available from its environment. One is illustrated by the example of the genes for lactose utilization. If a cell is growing in the absence of a *nonessential* nutrient, such as lactose, it would be superfluous for it to express the genes for utilization of lactose. Yet if lactose were to become available from the environment, it might be advantageous for the cell to use it as a carbon or energy source. This involves expressing the genes for lactose utilization, which are then said to be induced. Genes whose expression is subject to induction are called *inducible*.

The converse of this situation is illustrated by the example of the genes for histidine. If a cell is to grow in the absence of a *required* compound (histidine), then it must make its own. Therefore, it must produce the enzymes necessary for biosynthesis of the required product. In the bacterium *Salmonella typhimurium,* histidine synthesis involves nine different enzymes to catalyze the steps in making histidine from readily available precursors. However, if the required end product becomes available from the environment, the cell ought to be able to stop synthesizing its own. This involves turning off the genes coding for the enzymes involved in end-product biosynthesis. Genes in this state are said to be repressed, and the class of genes subject to repression is called *repressible*.

An important generalization that applies to both repression and induction in prokaryotic systems is that genes with related functions are often located together on the chromosome. In this way their expression can be controlled by the same regulatory elements. Such clusters of genes with related functions are called operons, and they are identified by their function. For example, the genes involved in lactose utilization are called the lactose operon, and the genes involved in tryptophan biosynthesis are called the tryptophan operon. Genes in the same operon are often transcribed together into a single polycistronic mRNA. For example, the three genes in the lactose operon are transcribed into one polycistronic mRNA, as are the nine genes in the histidine operon. Furthermore, the detailed molecular mechanisms involved in induction and repression of different operons are similar. These mechanisms will be illustrated by discussing the induction of the genes for lactose utilization in *E. coli,* which is by far the best-studied example of regulated gene expression.

THE LACTOSE OPERON, AN INDUCIBLE OPERON

When wild-type *E. coli* cells are grown in the presence of lactose, three genes are expressed that are not expressed in the absence of lactose. They are required for getting lactose into the cell and for hydrolyzing it into its component monosaccharides, glucose and galactose. The three genes are: (1) β-galactosidase, which catalyzes the hydrolysis of lactose into glucose and galactose; (2) galactoside permease, a membrane protein involved in transporting lactose into the cell; and (3) transacetylase, whose function is unknown. These genes are called *z, y,* and *a,* respectively, and they are located together on the *E. coli* chromosome, as shown in Figure 8–1. This cluster of three genes is called the lactose operon, or lac operon, and expression of all three genes is controlled as a unit; one of them cannot ordinarily be expressed without the others. All three lac genes are transcribed at once onto a single polycistronic messenger RNA.

Expression of the lac operon genes is controlled by the product of a fourth closely linked gene, the regulatory gene, *i*. The protein product of the *i* gene is called the repressor. In the absence of lactose, it binds to a fifth genetic locus called the *o* gene, or "operator," and blocks transcription of the lac genes. The operator differs from the *z, y, a,* and *i* genes in that it does not code for a protein product. Its sole function is to serve as a binding site for the repressor protein. Since binding of repressor to the operator locus blocks transcription of the lac genes, it stops their expression. It now seems likely that the repressor blocks transcription by physically preventing RNA polymerase from gaining access to the transcription initiation site. The operator is located between the promoter,

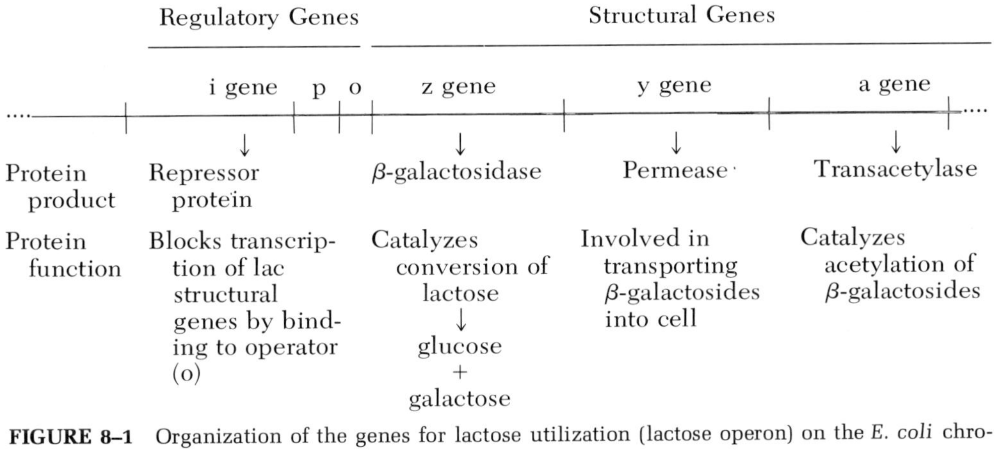

FIGURE 8–1 Organization of the genes for lactose utilization (lactose operon) on the *E. coli* chromosome.

the tight binding site for RNA polymerase, and the beginning of the *z* gene. The operator locus actually contains the site for the initiation of transcription, so it is reasonable to expect that blockage of this site by the repressor would stop lac transcription.

Transcription of the lac genes can take place when the repressor is removed from the operator locus. This occurs when an inducer is bound to the repressor molecule. Inducers of the lac operon are β-galactosides, which can be specifically bound to a site on the repressor surface. The physiological inducer is allolactose, a metabolic product of lactose, but other β-galactosides will also serve as inducers. When an inducer is bound to its site on the repressor molecule, a conformational change takes place in the repressor which results in its losing its affinity for the operator. Repressor is then lost from the operator region, and RNA polymerase can transcribe the lac genes. This accounts for the induction of their synthesis in the presence of inducer, as shown in Figure 8–2.

Genetic studies of *E. coli* strains carrying mutations in the *i* and *o* genes have clarified many of the basic features of gene regulation in the lac operon. Mutations in the regulatory elements have been shown to yield three possible states for expression of the lac genes. They are the following:

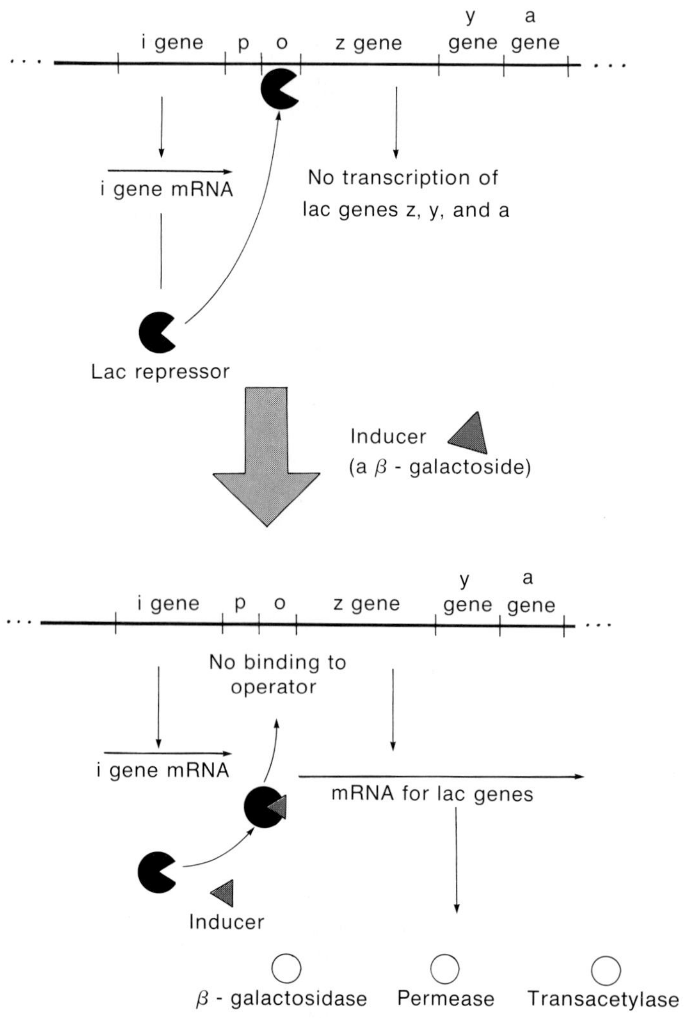

FIGURE 8–2 Effect of inducer to derepress synthesis of the lac genes in *E. coli*.

1. *Inducible*. This is the state found in wild-type *E. coli*. The lac genes are expressed only in the presence of inducer (lactose in the growth medium).

2. *Constitutive*. In this case, the lac genes are always expressed regardless of the presence or absence of inducer.

3. *Noninducible*. Normal lac genes are present in the genome, but they are not expressed even in the presence of inducer.

The *constitutive* phenotype can be produced by mutations in either the *i* or the *o* genes. If the repressor is not produced (i^-), or if a mutant repressor that will not bind to the *o* locus (i^c) is produced, then expression of *z, y,* and *a* will be constitutive. Also, if the *o* locus is altered in base sequence so that the repressor does not recognize it (o^c), expression of the lac genes will be constitutive. In contrast, *z, y,* and *a* will be *noninducible* if repressor is made that recognizes and binds to *o* normally, but does not bind inducer (i^s). In this case, the repressor will bind to *o* and will not be removed by inducer, so no transcription of *z, y,* and *a* will take place.

The interactions among these mutant genes in the lac operon were first determined from studies of *E. coli* strains that are partially diploid for the genes in the lac region. These "merodiploid" strains are haploid, like wild-type *E. coli*, for most of their genome, but they contain two copies of the lac genes. Studies with such strains indicated for the repressor that the i^+ allele is dominant to i^- and that i^s is dominant to both i^+ and i^-. Therefore, diploids containing both i^- and i^+ alleles (i^+/i^-) are found to be inducible, whereas any strain carrying i^s (i^s/i^+ or i^s/i^-) is noninducible. In a similar way it was determined that the o^c operator allele is dominant to o^+. The o^c/o^+ strains produce *z, y,* and *a* constitutively. Expression of the *i* and *o* genes differs in that the *i* gene product, the repressor, can affect the expression of *z, y,* and *a* not only on its own DNA molecule but also on any other genetic elements that may be present. It is a diffusible substance, and it can bind to all lac operators in the cell. The repressor is therefore said to have a "trans-dominant" effect on lac gene expression. For example, a partially diploid strain containing $i^-o^+z^+y^+a^+/i^+o^+z^-y^-a^-$ will be inducible and *not* constitutive. In contrast, the *o* gene can affect only those lac genes on the same genetic element and it is said to have a "cis-dominant" effect. Thus, the strain $i^+o^+z^+y^+a^+/i^+o^cz^-y^-a^-$ will be inducible and *not* constitutive (Fig. 8–3). These relationships among mutations in the lac regulatory genes are consistent with the molecular model described earlier.

The lac repressor has been isolated in purified form. It consists of four identical polypeptide chains. Each has a molecular weight of 37,000, so the overall repressor has a molecular weight of 148,000. The individual polypeptide chains consist of 347 amino acids, and their sequence has been determined. The repressor gene (*i* gene) is expressed constitutively in *E. coli*; that is, the *i* gene itself is not subject to control by regulatory elements other than its own promoter. Under these circumstances, *E. coli* cells are found to contain 10 to 20 molecules of lac repressor per chromosome.

The region of *E. coli* DNA bound by the lac repressor was recently identified by binding the repressor specifically to DNA and then digesting the unbound DNA regions with deoxyribonuclease. The protected DNA was then isolated free from the repressor, and its nucleotide sequence was determined. The results of this analysis showed that the lac repressor binds to a segment of DNA 24 base pairs in length (Fig. 8–4). Twenty-one of these 24 base pairs are found together in a region whose sequence is almost exactly symmetrical about the GC base pair indicated at position 11 in Figure 8–4. Of the possible 20 base pairs, 16 are

Genotype	*Phenotype*
$\frac{i^+o^+z^+y^-a^-}{i^-o^+z^-y^+a^+}$	z, y, and a inducible
$\frac{i^+o^+z^+y^-a^-}{i^+o^cz^-y^+a^+}$	z inducible; y and a constitutive
$\frac{i^-o^+z^+y^-a^-}{i^+o^cz^-y^+a^+}$	z inducible; y and a constitutive
$\frac{i^so^+z^+y^+a^+}{i^+o^+z^+y^+a^+}$	z, y, and a noninducible
$\frac{i^so^+z^+y^+a^+}{i^+o^cz^+y^+a^+}$	z, y, and a constitutive
$\frac{i^+o^+z^-y^+a^+}{i^+o^cz^+y^-a^-}$	z constitutive; y and a inducible

Alleles at *i* and *o* loci

- i^+—Wild type (normal) repressor
- i^-—No repressor produced
- i^c—Mutant repressor that will bind inducer but will not bind to the operator *(o)*
- i^s—Mutant repressor that will bind normally to the operator *(o)* but will not bind inducer
- o^+—Wild type operator locus
- o^c—Mutant operator locus that will not bind repressor

FIGURE 8–3 Interactions among normal and mutant lac regulatory elements in partially diploid strains of *E. coli*.

symmetrical in this region (boxes), and these base pairs are thought to be involved in the specific recognition of repressor molecules. Transcription of the lac operon is begun at the first AT base pair in this symmetrical region, as shown at position 1 of Figure 8–4.

More extensive nucleotide sequence analysis in the lac region has shown that the repressor binding site, the operator, is located between the promoter (the RNA polymerase tight binding site) and the beginning of the sequence that codes for the N-terminal amino acid of β-galactosidase. The operator is found only two base pairs from the RNA polymerase tight binding site and 17 base pairs from the *z* gene initiator AUG codon (Fig. 8–5). The terminator sequence for the *i* gene (repressor) product is 81 base pairs removed from the operator site in the direction of the promoter. There are, therefore, approximately 72 base pairs between the end of the *i* gene and the beginning of the RNA polymerase tight binding site. The function of these 72 base pairs is not completely understood, but it now seems likely that a portion of this region may be involved in catabolite repression.

THE HISTIDINE OPERON, A REPRESSIBLE OPERON

Repression of gene expression is found to operate by molecular mechanisms very similar to those used in induction. Genes with related functions are located together on the chromosome, and their expression is controlled by a separate regulatory gene whose expression is itself not subject to regulatory control (constitutive). However, in contrast to induction, if the regulatory gene product is a repressor it cannot repress operon function by itself; the end product of the pathway to be repressed must also be present. The end product is called the corepressor, and it must be noncovalently bound to the repressor before the

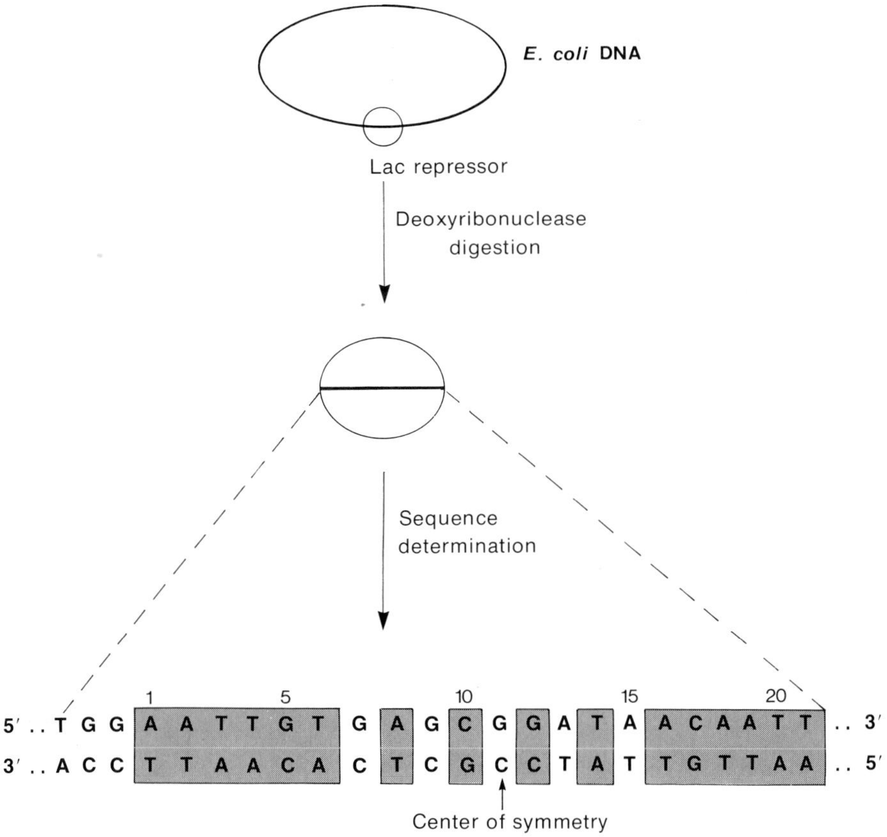

FIGURE 8–4 Nucleotide sequence of the region of *E. coli* DNA bound by the lac repressor. The boxes indicate sequences symmetrical with respect to the GC base pair shown at position 11.

repressor can interact with the operator to block operon transcription. For example, the nine genes of the histidine operon in *Salmonella typhimurium* (Fig. 8–6) are subject to repression by the end product of their pathway, histidine, and repression is mediated by the product of a separate regulatory gene. The repressor alone, however, cannot block transcription of the histidine operon. It requires the simultaneous presence of histidyl-tRNA, which presumably is bound to the repressor surface. In this case, histidyl-tRNA, and not histidine itself, serves as the corepressor.

NEGATIVE AND POSITIVE CONTROL

Both of the regulatory systems we have discussed so far are examples of negative control mechanisms. This means that in both cases something must be done to *stop* functioning of the genes in question. Otherwise they will be expressed. If no lac repressor is bound to the operator, the lac mRNA will be synthesized, and if the complex of repressor plus corepressor is not bound to the histidine operator, the histidine operon will be expressed. It is clear, however, that another possibility exists. Genes could have the property of *not* being expressed unless something positive is done. In this case, a gene "activator" would have to be present to induce operon expression. This is called positive control

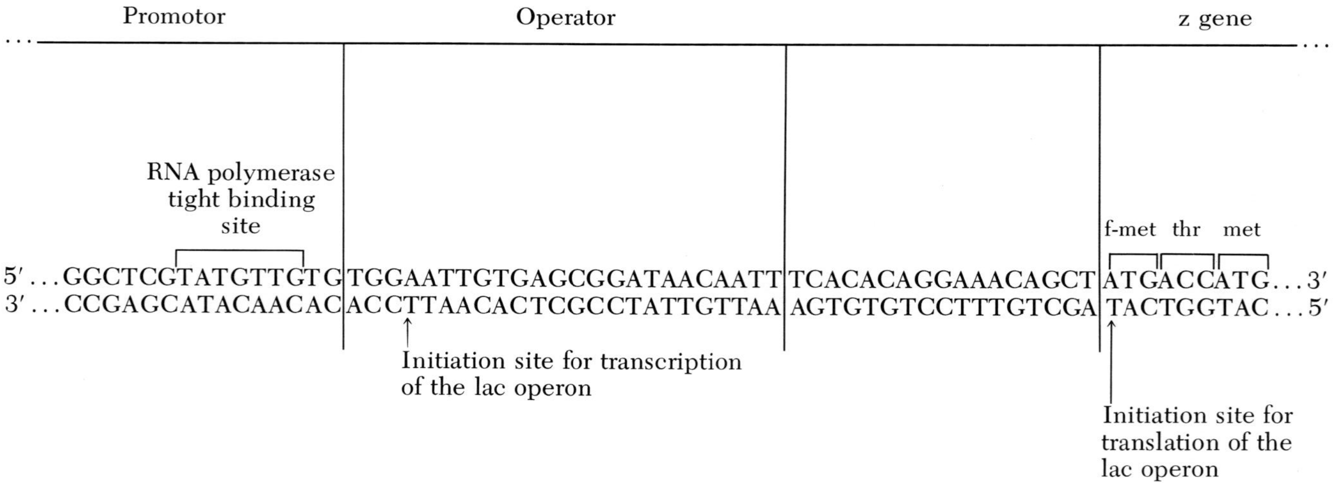

FIGURE 8–5 Nucleotide sequence at the promotor and operator regions of the lac operon in *E. coli*. Sites for the initiation of transcription and translation are indicated.

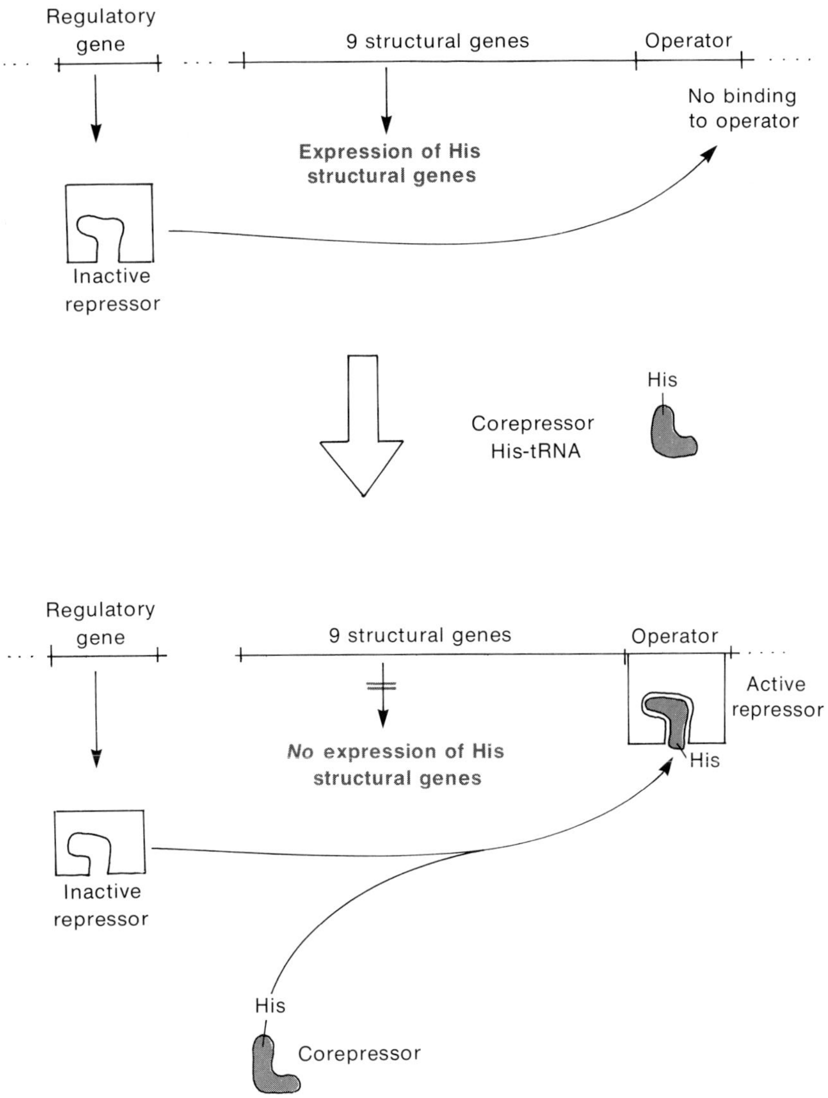

FIGURE 8–6 Repression of gene expression in the histidine operon of *Salmonella typhimurium*.

of gene expression, and both inducible and repressible operons could, in principle, be regulated in this way. Systems of positive control are known in prokaryotic organisms, and in eukaryotes most regulation of gene expression is thought to involve positive control mechanisms.

The best-studied example of positive control is the arabinose operon in *E. coli*. Like the lac operon, the arabinose operon consists of three genes involved in arabinose utilization: A, arabinose isomerase; B, L-ribulokinase; and D, L-ribulose-5-phosphate epimerase. They are located together on the *E. coli* genome, as shown in Figure 8–7. Their expression is under the control of a regulatory gene C, called the activator, which must be present for transcription of the arabinose operon to take place. The C gene product binds to the operator in the presence of arabinose, and in some way this results in transcription of the arabinose genes. If activator is not bound, the operon is not expressed. Therefore, mutations (c^-) in the arabinose regulatory gene are found to be noninducible, not constitutive, as in the lac operon.

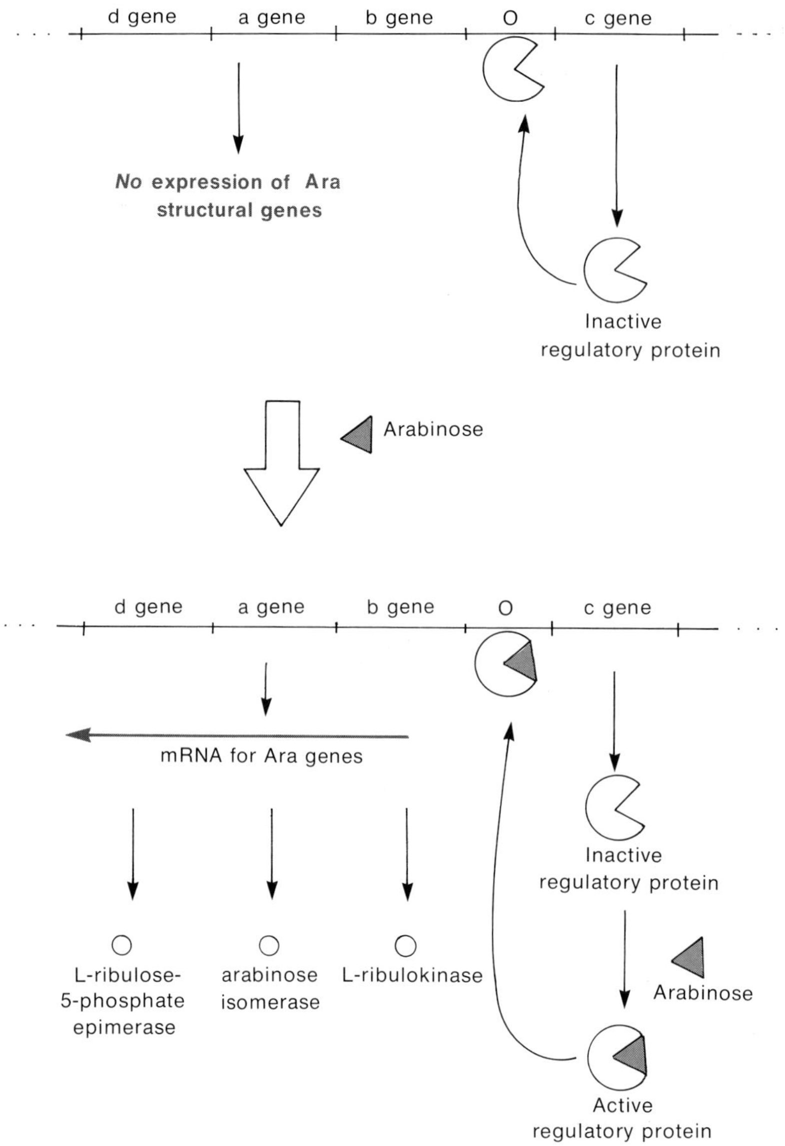

FIGURE 8–7 Positive control of gene expression in the arabinose operon of *E. coli*. The regulatory protein product of the c gene is required in the presence of arabinose to activate transcription of the three structural genes d, a, and b.

Catabolite Repression

A second type of positive control system has been found to operate in *E. coli* to regulate carbohydrate utilization. This system insures that *E. coli* will use all the glucose present in its environment before other possible carbohydrates are metabolized. The system operates at the level of initiation of transcription, and cyclic 3′-5′ AMP (cAMP) is involved. There is a set of operons in *E. coli* whose expression in every case is dependent on the presence of a high intracellular concentration of cAMP. These include the operons containing genes for the utilization of sugars, such as lactose, maltose, arabinose, and galactose, but other genes, such as those involved in flagella formation, are also affected.

When *E. coli* is growing in the presence of glucose, an unknown catabolic product unique to glucose degradation decreases the intracellular level of cAMP.

The lower level of intracellular cAMP makes transcription of cAMP-dependent operons impossible, and these genes are not expressed. Such genes are said to be catabolite-repressed or subject to catabolite repression. The result is that catabolite-repressed genes, such as those of the lactose operon, are not expressed even in the presence of their inducer. Catabolite repression therefore operates as an override system on other types of control that may be present in cAMP-dependent operons. It has the effect of suppressing the genes for utilization of other possible carbohydrates while glucose is present in the medium. When glucose is depleted, the intracellular level of cAMP rises and catabolite repressible genes can be expressed. In the case of the lactose and arabinose operons, this shows clearly that more than one regulatory mechanism may affect expression of a particular operon.

The molecular mechanisms involved in catabolite repression are not yet fully understood. The exact product of glucose degradation required for catabolite repression has not been identified, and it is not known how this catabolite affects the intracellular concentration of cAMP. It is reasonable to speculate, however, that the glucose catabolite either depresses the activity of adenyl cyclase, the enzyme that forms cAMP from ATP, or that it stimulates the activity of cAMP phosphodiesterase, the enzyme that degrades cAMP to 5′ AMP, as shown in Figure 8–8.

Intracellular cAMP exerts its effects on transcription through a positive control element, a protein called catabolite activator protein (CAP), which will bind to the promoter region of catabolite-repressible operons in the presence of cAMP. Binding of CAP plus cAMP to the promoter is necessary for expression of catabolite-repressible operons, in addition to any other regulatory control elements that may be required. For example, transcription of the lac operon requires the presence of CAP plus cAMP, and it also requires the presence of inducer to remove the repressor from its binding site on the lac operator. The CAP protein is a dimer of two identical polypeptides, each of which has a molecular weight of 22,000. The 44,000 MW dimer binds to cAMP and also to a specific region of the lac promoter to the left of the RNA polymerase tight binding site. The CAP binding site consists of 16 base pairs, which are almost exactly symmetrical about the center, as shown in Figure 8–4. Just how binding of CAP at this site can affect initiation of transcription by RNA polymerase is not yet clear, although it is speculated that CAP may facilitate the local separation of the two DNA strands required for entry of RNA polymerase onto its template.

EUKARYOTIC SYSTEMS

Much less is known about the regulation of gene expression in eukaryotic compared to prokaryotic organisms. This is primarily because genetic experiments, which are the source of much of our basic information about gene control, are significantly more difficult to perform in eukaryotic organisms. No system of regulated gene expression in a eukaryotic organism has been studied as thoroughly as the lac operon in *E. coli*. Nevertheless, these mechanisms probably do function in eukaryotes, even though none has been examined in such great detail. For example, there are probably cases of negative control, positive control, induction, repression, and cAMP effects in eukaryotic as well as in prokaryotic systems. It now seems likely that novel mechanisms not found in prokaryotic organisms exist for the regulation of eukaryotic gene expression, and these will be emphasized in the following discussion.

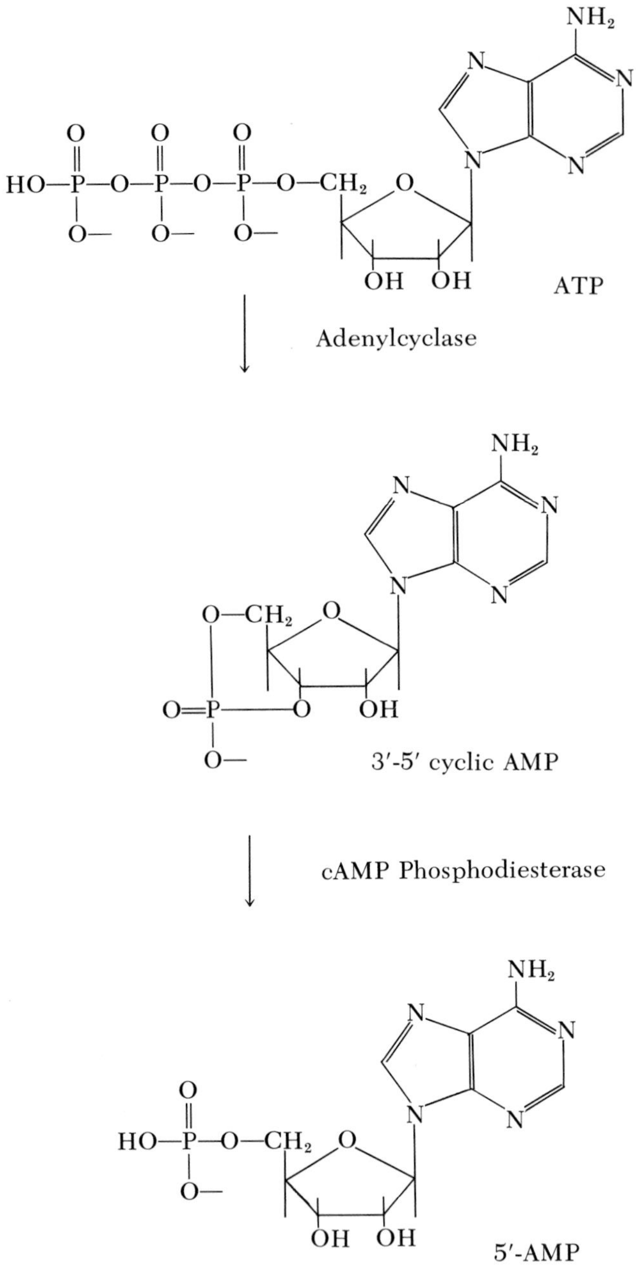

FIGURE 8–8 Formation and degradation of 3′-5′ cyclic AMP as found in *E. coli*.

Organization of the Eukaryotic Genome

One of the most striking differences between the prokaryotic and eukaryotic genomes is that eukaryotic cells have much more DNA per cell than prokaryotes. For example, the prokaryotic *E. coli* genome contains 4×10^6 nucleotide pairs, or enough DNA to code for approximately 4000 proteins. The eukaryotic yeast genome contains five times that amount of DNA (2×10^7 base pairs), and the fruit fly *Drosophila melanogaster* contains almost forty times as much (1.5×10^8 base pairs). The haploid human genome contains almost 800 times as much DNA as *E. coli* (3×10^9 base pairs), enough DNA to code for approximately 3×10^6 proteins. Furthermore, the amount of DNA per cell in prokaryotic organisms varies only over approximately a fourfold range, with the *E. coli* value being typical. In contrast, eukaryotic organisms are found to vary over at least a 200-fold range in the amount of DNA per cell (Fig. 8–9).

In general, the variable pattern in the amount of DNA per cell for different eukaryotic organisms might be expected, since more highly organized or sophisticated species should have more DNA per cell than less sophisticated ones. Fungi are not as a rule as highly organized as sponges, which are not as complex as reptiles, and this fact is reflected in the amount of DNA per cell in these species. There are exceptions to the rule, however. Certain plant and amphibian species contain at least 20 times as much DNA per cell as highly developed mammalian species. Also some very closely related plant species differ by 50-fold or more in their haploid DNA content. Clearly, the haploid amount of DNA reflects more than the overall complexity of the species, indicating that other, as yet unknown, factors must be involved.

An additional problem in interpreting the significance of the amount of DNA per haploid genome (C value) in different eukaryotic species is that the amount of DNA often seems to be in excess of the number of proteins required by the organism. Geneticists estimate that even the most complex species could not have more than a few tens of thousands (e.g., 30,000) of genes coding for proteins; otherwise the target for deleterious mutations would be too great, and individual organisms could not survive. Species above approximately yeast on the evolutionary scale, however, have more DNA than is required to code for this number of proteins. There is enough human DNA, for example, to code for approximately 3×10^6 proteins, and some species have even more DNA than that. The problem is, therefore, to account for the function of this "extra" DNA. So far no entirely satisfactory explanation is available.

Another major difference between eukaryotic and prokaryotic genomes is in the location of genes on the chromosome. The prokaryotic genome is often divided neatly into operons, in which genes with related functions are clustered together. This does not appear to be the case with eukaryotes, in which genes with related functions may be found at different locations. The *Neurospora crassa* (bread mold) genome illustrates this point. Genes for various enzymes are local-

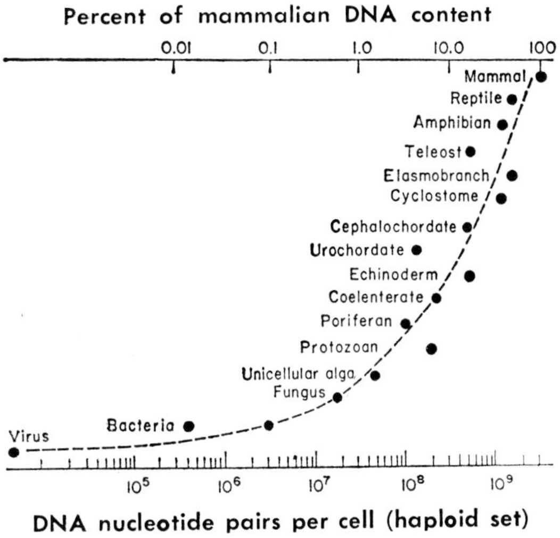

FIGURE 8–9 The minimum haploid content of DNA per cell is indicated for organisms at various levels of evolutionary development. (From Britten, R., and Davidson, E.: Science, *165*:349, 1969. Reproduced by permission of the American Association for the Advancement of Science.)

ized on the seven *Neurospora* chromosomes (Fig. 8–10). In the case of the genes for histidine biosynthesis (his-1, his-2, and so on), it can be seen that whereas they are all clustered into one operon in *Salmonella typhimurium,* they occur at separate locations in the *Neurospora* genome. Other examples of this type can be identified. This apparent disorder of the eukaryotic genome presents formidable difficulties if genes with related functions are to be regulated coordinately, as they are in bacteria. It implies that there are similar regulatory elements near each related gene, instead of only one operator and one promoter for an entire set of related genes (operon).

Finally, to make the situation even more complicated, a substantial fraction of the eukaryotic cell DNA occurs in segments that are present in many copies in the overall genome. This is called repeated, repetitious, or repetitive (but *not* redundant) DNA, and it is characteristic of eukaryotic organisms. Virtually all eukaryotic species have some repeated DNA in their genome, while prokaryotic cells never do. The proportion of the total genome allotted to repeated DNA varies in different eukaryotic species (Table 8–1), but it may account for as much as 50 per cent of more. Except for the fact that very simple eukaryotes, such as yeast, have a comparatively small amount of repeated DNA, there is no satisfactory evolutionary or other explanation to account for the variable proportions of repeated, compared to unique, DNA sequences in different eukaryotes.

Repeated DNA's may be of two different types, depending on the number of copies per genome. These two types will be referred to as highly repeated and middle repeated DNA, respectively. These together with the unique or single-

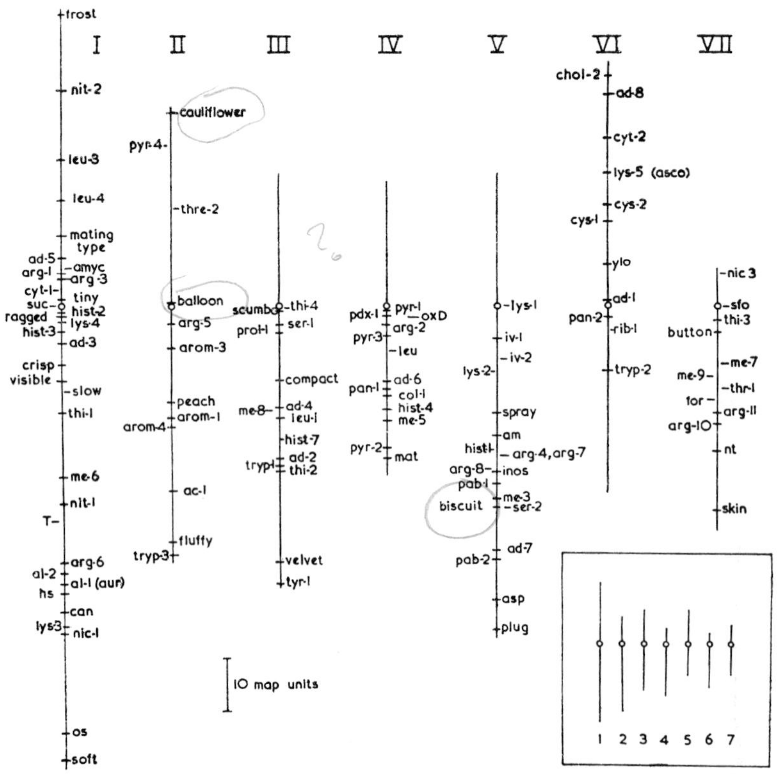

FIGURE 8–10 Genetic map of *Neurospora crassa* showing the positions of genes on the seven chromosomes.The position of the centromere is indicated on each chromosome by a small open circle. (From Fincham, J., and Day, P.: Fungal Genetics. New York, Blackwell Scientific Publications, Ltd., 1963. Reproduced by permission of the publisher.)

PROPORTIONS OF SINGLE-COPY AND REPEATED DNA IN SELECTED EUKARYOTIC SPECIES **TABLE 8–1**

Organism	% of Single-Copy DNA	% of Repeated DNA
Sea urchin	62	38
Snail	40	60
Toad	54	46
Mouse	60	40
Calf	55	45
Human	70	30

copy sequences account for all the DNA in the genome. The *highly repeated DNA* sequences generally account for 3 to 10 per cent of the total cell DNA. They consist of rather short DNA segments, having lengths of less than 100 nucleotides, but they are repeated many times in the genome. Depending on the species, there may be 10^6 or more copies of these sequences. As with most repeated DNA, no biological function is yet known for the highly repeated DNA. It is known to be located only at the centromere region of chromosomes, and it is therefore occasionally identified as centromeric DNA. These sequences are *not* transcribed into RNA, so they cannot code for proteins in the way previously described. Many biologists now feel that the highly repeated DNA sequences are involved either in chromosome pairing during mitosis and meiosis or in the attachment of chromosomes to spindle fibers during cell division. Although these theories are attractive, they still require proof.

The *middle repeated DNA* sequences may themselves be divided into different classes, depending upon the number of times they are repeated. The repeat frequency in this type of DNA ranges from 10 to approximately 10^5 times. Different species of eukaryotes have classes of DNA with very different repeat frequencies. In general, there is no way to predict which classes of middle repeated DNA sequences will be present in a given species or how much of the total DNA will be present in any of them. The length of the repeating unit among the middle repeated DNA sequences is much greater than in the highly repeated DNA. The lengths vary among different middle repeated DNA classes, but they are seldom less than several thousand nucleotide pairs in length, and they may be as many as 10^7 nucleotide pairs long.

The cellular functions of some of the middle repeated DNA sequences are known. For example, in *Xenopus laevis,* the African clawed toad, the gene coding for the 45S precursor of ribosomal RNA exists in 450 copies per genome. This gene is 13,000 base pairs long, and in all its copies it accounts for approximately 0.2 per cent of the total *Xenopus* DNA. Similarly, the gene for 5S RNA together with an associated "spacer" region is repeated approximately 24,000 times in *Xenopus*. The length of the repeating unit in this case is around 800 nucleotide pairs, and together the 5S RNA genes account for 0.7 per cent of the total *Xenopus* DNA. The function of most middle repeated DNA, however, is unknown. Many and perhaps all of the middle repeated DNA sequences are transcribed into RNA, so it is possible that some of them may code for proteins. Most biologists today, however, believe that this is not likely, because the mRNA's presently available have almost all been found to correspond to unique sequence DNA. Instead, it is speculated that most middle repeated DNA is somehow involved in the regulation of gene expression. This view is consistent with the fact that the middle repeated DNA sequences, like the genes coding for proteins, are located in all parts of the genetic map and in all regions of the chromosomes. They are not restricted to any one region of the genome as are the highly repeated DNA sequences.

The unique sequence or *single-copy DNA* corresponds to those DNA sequences represented only once in the entire genome. Single-copy DNA exists in all eukaryotic species, and it accounts for 35 to 90 per cent of the total genome. Prokaryotic cell genomes consist entirely of unique sequence DNA, with the single exception of the gene specifying ribosomal RNA, which may be present in six to ten copies. In eukaryotic organisms, the unique sequence DNA is transcribed into RNA, and firm evidence shows that at least some of the unique sequence DNA codes for proteins. Because of the large amount of DNA in eukaryotes, however, it is possible that not all the single-copy DNA codes for protein sequences—especially in higher eukaryotic species which contain a great deal of unique sequence DNA. In these cases, some or even a great deal of the unique sequence DNA may have some other function, such as the regulation of gene expression.

From this discussion of the organization of the eukaryotic genome it can be seen that our knowledge of this crucial subject is still in its infancy. Many more basic genetic and biochemical studies on the structure of the eukaryotic genome are required before firm clues about the regulation of eukaryotic gene expression can be obtained. It is especially important to know more about the function of repeated DNA and why most eukaryotic organisms have so much DNA per cell.

Proteins Associated with DNA

Rather more substantial clues about how eukaryotic gene expression may be regulated come from studies of the proteins, the histones and the nonhistone proteins, which are associated with DNA in the nucleus of all eukaryotic cells. The chromatin of typical eukaryotic cells in interphase is found to consist by weight of approximately one-third DNA, one-third histone, and one-third nonhistone proteins. Clearly, if there are regulatory protein components, such as repressors and gene activators, in eukaryotic cells, then they are most likely to be among the histone and nonhistone proteins that are closely associated with DNA.

HISTONES

Much more is known about histones than about nonhistone proteins. Histones are found in all eukaryotic chromatin, from which they can be extracted with dilute mineral acid or with 1 M NaCl. The histones are a set of five small, highly basic, closely related proteins. They serve as the counter ions for DNA in the cell, the negative charges on DNA phosphates being neutralized by the positively charged lysine and arginine groups of histones. The histones are therefore very tightly bound to DNA by salt linkages. Although prokaryotic cell DNA is also associated with positively charged small molecules, these compounds are not the same as histones, which are found only in eukaryotic cells.

The five histone components are called H1, H2A, H2B, H3, and H4. All five histones have been identified in virtually every eukaryotic cell type and in every eukaryotic species yet examined. Each histone constitutes approximately 20 per cent by weight of the total histone present, and this proportion does not vary significantly from one tissue or from one species to another. Thus the results shown in Table 8–2 for calf thymus histones are typical of all eukaryotic cells. Furthermore, detailed amino acid sequence analysis of purified histone molecules has revealed striking similarities in the structure of corresponding histone components from evolutionarily divergent species. Histone H4 molecules from pea and from cow, for instance, differ in only 2 of 102 positions in their amino acid sequences. Similar instances of evolutionary conservation have emerged from sequence studies of the other histones.

The limited number of distinct histone types and their relative lack of tissue- and species specificity argue strongly that histones are not responsible for the highly specific interactions required for the regulation of gene expression in eukaryotic cells. Instead, there is now good evidence that the histones are actually structural components of the chromosomes and possibly are involved in the cyclic steps of chromosome condensation and unraveling that take place during mitosis. For example, as noted in Chapter 5, the basic element of chromosome structure seen in certain preparations of isolated chromatin is the *nu* body or nucleosome (Fig. 8–11). Recent studies have shown that the nu bodies, which are spheres about 100 Å in diameter, consist of a segment of DNA approximately 200 base pairs long associated with a protein core composed of two copies each of histones H2A, H2B, H3, and H4. The spheres are joined to each other like "beads on a string" by regions of DNA approximately 34 base pairs in length. These interbead regions are probably associated with histone H1 (Fig. 8–12). In the overall chromosome structure a higher level of nucleosome organization is required, but at present there is no general agreement about how this occurs.

NONHISTONE PROTEINS

The nonhistone proteins are likely candidates for involvement in the specific aspects of gene regulation. Nonhistone proteins are defined as all the proteins other than histones that are associated with DNA in chromatin. This class of proteins includes RNA polymerase and its associated factors, the enzymatic machinery required for DNA replication, and the proteins involved in repair of damaged DNA. It may also, however, contain proteins, such as repressors and gene activators, which are involved in the control of gene expression. There is now some experimental evidence that this is the case. For example, in contrast to the situation with histones, there is a wide variety of different nonhistone proteins. At least 30 different nonhistone proteins have been detected in virtually every homogeneous cell population yet examined. Because of the limitations inherent in the experimental methods employed, the true number may be even

TABLE 8–2

PROPERTIES AND RELATIVE ABUNDANCE OF CALF THYMUS HISTONES

Histone	Molecular Weight	Number of Amino Acid Residues	Lysine/Arginine Ratio	% of Total Histone Mass	Molar Ratio
H1	21,000	220	20	16.4	0.52
H2A	14,500	129	1.25	21.7	1.00
H2B	13,800	125	2.50	28.4	1.37
H3	15,300	135	0.72	17.7	0.77
H4	11,300	102	0.79	15.5	0.92

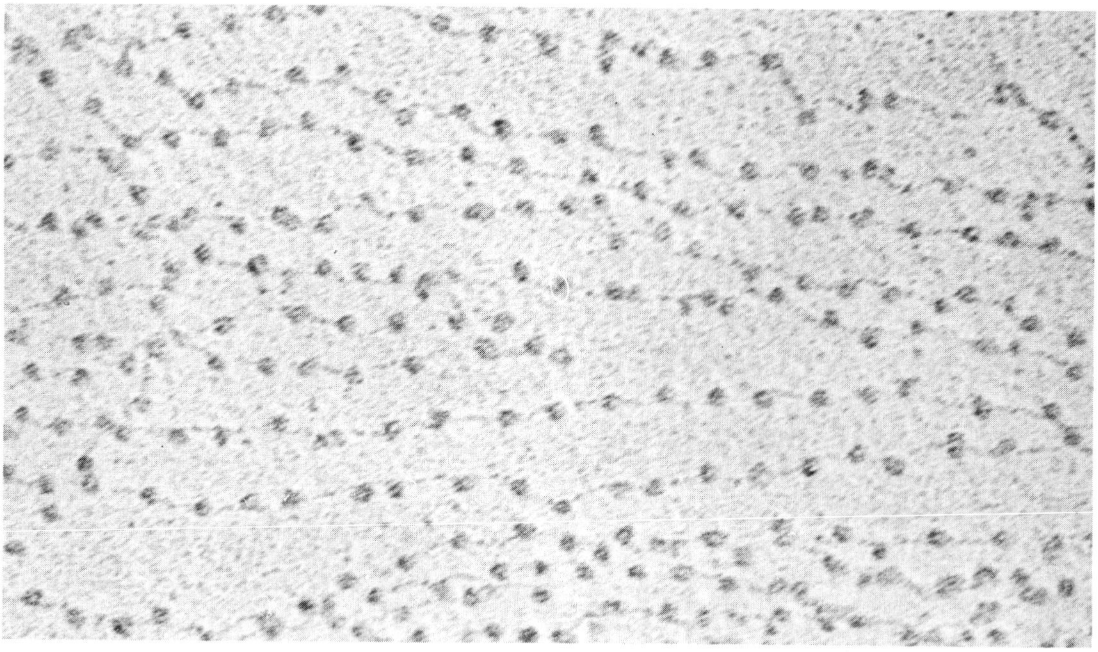

FIGURE 8–11 Electron micrograph of chromatin fibers showing characteristic nucleosome structures. × 200,000. (Micrograph courtesy of S. Pruitt and R. Grainger.)

greater. Furthermore, there appears to be a considerable degree of species and tissue specificity among the nonhistone proteins.

The spectrum of nonhistone proteins varies greatly when different species are compared, and there are usually significant differences even in separate tissues within the same species. Therefore, it is attractive to speculate that the nonhistone proteins may play a role in the regulation of gene expression. Definite evidence is now needed to support this speculation, and experimental studies with this goal are in progress. The magnitude of the problem and the difficulty in focusing on a single system of regulated gene expression suggest, however, that it may be a long time before there is a satisfactory understanding of the precise functions of nonhistone proteins.

Hormone Effects and Protein Stability

The most straightforward way to study the regulation of gene expression in eukaryotic cells is to focus on a gene or system of genes whose expression is

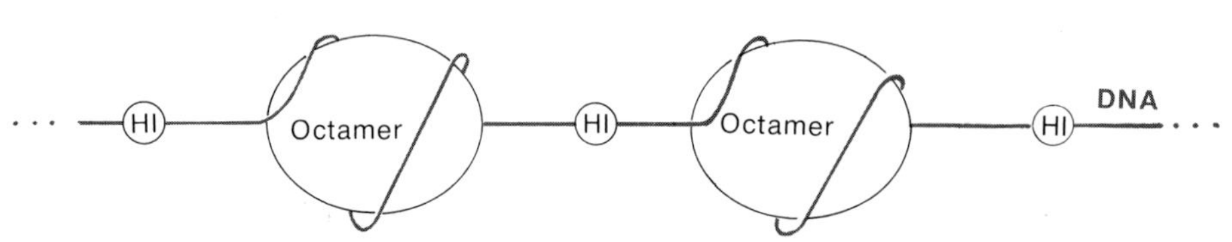

FIGURE 8–12 Schematic diagram illustrating how histones and DNA are associated to form nucleosomes, as found in eukaryotic chromosomes. The spherical octamers consist of two copies each of histones H2A, H2B, H3, and H4.

under regulatory control and to examine it in detail. This is the strategy that has been so conspicuously successful with prokaryotic systems. Several studies of this type have clarified some of the characteristic mechanisms used for the regulation of gene expression in eukaryotic cells. For example, expression of the gene coding for the enzyme tryptophan pyrrolase in rat liver has been studied by Schimke and his colleagues.

Tryptophan pyrrolase is involved in the degradation of tryptophan when this amino acid is present in excessive amounts in the liver. The enzyme catalyzes the following reaction.

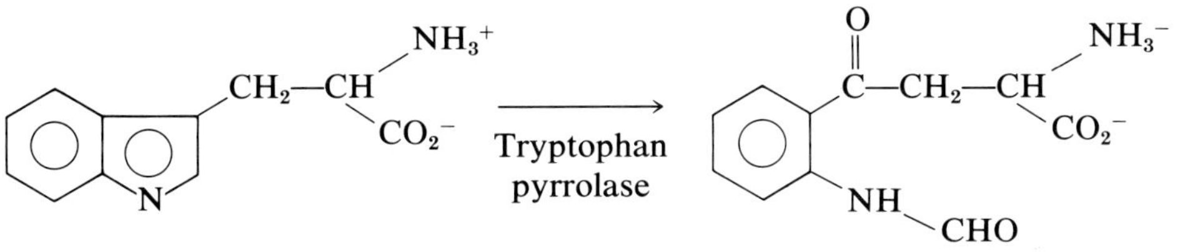

Tryptophan pyrrolase gene function illustrates two features of regulatory processes found in eukaryotic, but not in prokaryotic, systems. First, hormones can affect gene expression in their target tissues in eukaryotes. Second, proteins in eukaryotic cells do not have an indefinite lifetime in the cell after they are synthesized. Rather, since they are regularly degraded, they have a finite life span. If no new synthesis of a protein occurs, its function will eventually be lost from the cell.

The exact half-life varies among different proteins and for the same protein when it is present in different cell types. The half-life for tryptophan pyrrolase in rat liver was determined experimentally by feeding rats a diet containing radioactive amino acids for a considerable time and then switching to a nonradioactive diet. The amount of radioactively labeled tryptophan pyrrolase present in the liver was then determined as a function of time after change in diet. The results of this experiment are shown in Figure 8–13 (solid curve). They indicate that the half-life for tryptophan pyrrolase in normal rat liver is only about five hours.

The metabolic instability of this enzyme suggests that the overall level of tryptophan pyrrolase activity in the liver can be regulated either by increasing the rate of enzyme synthesis or by decreasing the rate of its degradation. In fact,

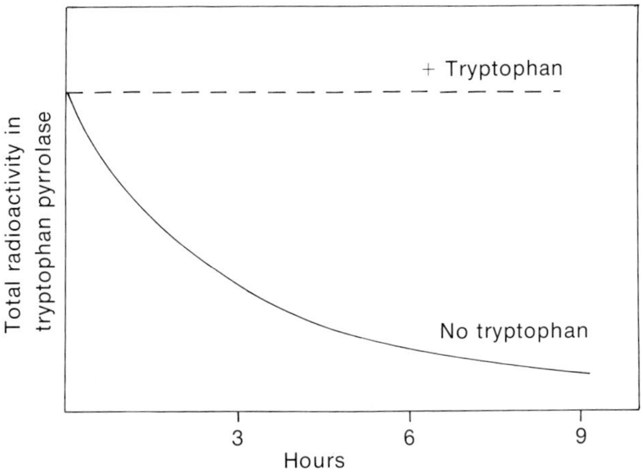

FIGURE 8–13 Effect of tryptophan on the metabolic stability of tryptophan pyrrolase. In the presence of its substrate the enzyme is not digested by intracellular proteases responsible for protein turnover.

both processes have been observed experimentally. For instance, the rate of tryptophan pyrrolase biosynthesis can be stimulated in the liver by hydrocortisone treatment.

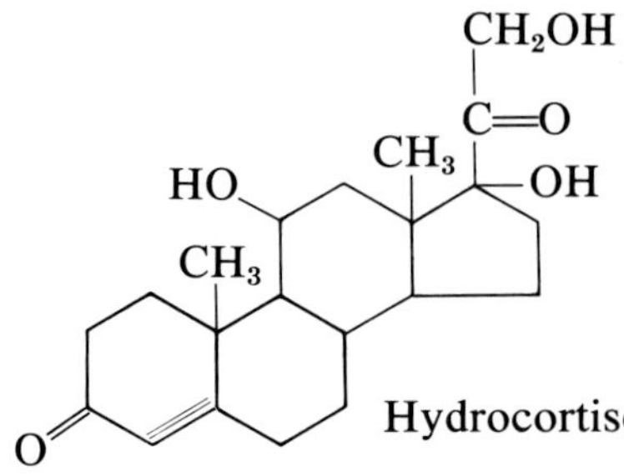

Hydrocortisone

This hormone is produced in the adrenal cortex and secreted into the circulation where it affects metabolic activity in various organs, including the liver. When the liver is exposed to hydrocortisone there is an increase in tryptophan pyrrolase activity. This has been shown to be the result of new enzyme synthesis, because drugs such as cycloheximide, which specifically inhibit protein synthesis, block the effect of hydrocortisone to increase liver tryptophan pyrrolase activity. Enzyme newly synthesized in response to the hormone continues to be degraded at the usual rate, however, so that after sufficient time a new equilibrium between synthesis and degradation is reached (Fig. 8–14).

It is clear that the total level of tryptophan pyrrolase activity in liver cells can also be elevated by decreasing the rate of enzyme degradation. This can be

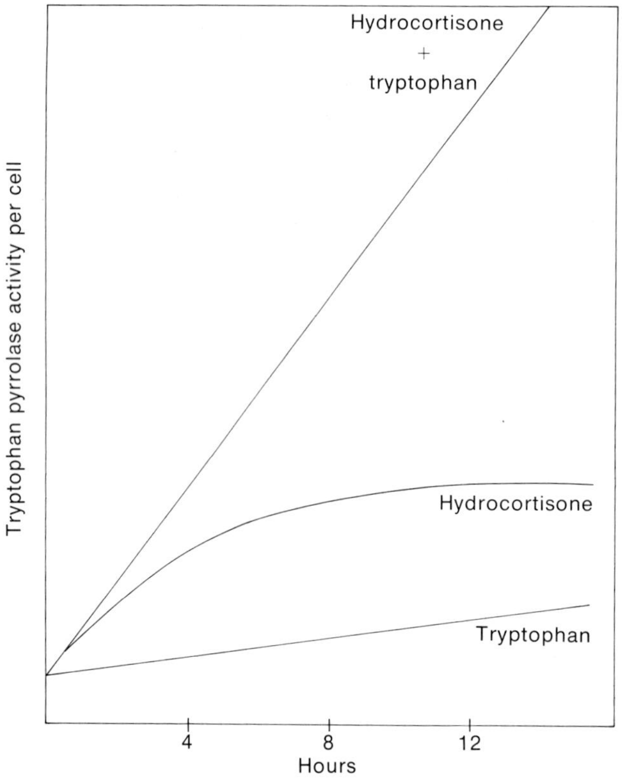

FIGURE 8–14 Synergistic effects of hydrocortisone and tryptophan on the intracellular level of tryptophan pyrrolase activity.

accomplished by providing the liver with high levels of tryptophan, the substrate for tryptophan pyrrolase. Evidently, if the enzyme is actively engaged in digesting tryptophan, it is not itself degraded (Fig. 8–12, dashed curve). Thus the level of tryptophan pyrrolase activity in the liver increases after tryptophan administration. The enzyme continues to be produced at its usual rate, but it is not degraded. Finally, since hydrocortisone and tryptophan increase the level of tryptophan pyrrolase by completely different mechanisms, the effect of administering both should be additive; that is, tryptophan pyrrolase activity should increase more than when either compound is provided separately. In practice, this is the case, as shown in Figure 8–14.

From this brief consideration of the regulation of gene expression, it can be appreciated that our knowledge of this subject is still in a primitive state. It contrasts markedly, for instance, with the advanced stage of our knowledge about the mechanics of gene expression. One can be confident, however, that future studies in the field of molecular biology will be increasingly devoted to understanding how gene expression is regulated, so that we will eventually be rewarded with a clearer understanding of how these processes operate and how they are integrated during embryological development.

FURTHER READING

Beckwith, J. R., and Zipser, D. (eds.): The Lactose Operon. New York, Cold Spring Harbor Laboratory, 1970.

Britten, R. J., and Kohne, D. E.: Repeated sequences in DNA. Science, *161*:529, 1968.

Dickson, R. E., Abelson, J., Barnes, W. M., et al.: Genetic regulation: the lac control region. Science, *187*:27, 1975.

Gilbert, W., and Müller-Hill, B.: Isolation of the lac repressor. Proc. Nat. Acad. Sci. USA, *56*:1891, 1966.

Greenblatt, J., and Schleif, R.: Arabinose C protein: regulation of the arabinose operon *in vitro*. Nature New Biol., *233*:166, 1971.

Hood, L. E., Wilson, J. H., and Wood, W. B.: Molecular Biology of Eukaryotic Cells. Menlo Park, Calif., W. A. Benjamin, 1975.

Lewin, B.: Gene expression. Vol. 1: Bacterial Genomes. New York, John Wiley and Sons, 1974.

Roth, J. R., Anton, D. R., and Hartman, P. E.: Histidine regulatory mutants in *S. typhimurium*. J. Mol. Biol., *22*:305, 1966.

Schimke, R. T., Sweeney, E. W., and Berlin, C. M.: The roles of synthesis and degradation in the control of rat liver tryptophan pyrrolase. J. Biol. Chem., *240*:322, 1965.

Watson, J. D.: Molecular Biology of the Gene. 3rd ed. Menlo Park, Calif., W. A. Benjamin, 1976.

9

ENDOPLASMIC RETICULUM AND THE GOLGI APPARATUS

RIBOSOMES AND CYTOPLASMIC BASOPHILIA

As long ago as the 1880's cytologists noted the presence of basophilic regions of the cytoplasm, which were referred to by various names, such as ergastoplasm, Nissl bodies, and basophilic bodies. When it was observed that their staining was similar to that of the nucleolus, it was hypothesized that the ergastoplasm obtained some substance from the nucleolus. The prominence of the basophilic regions in gland cells, as well as changes in their extent with variations in activity of the gland, suggested that they might have a synthetic function. Later it became known that the basophilia of the ergastoplasm is caused by the presence of RNA, mostly that in ribosomes.

The ribosomes that are responsible for cytoplasmic basophilia are found either bound to membranes of the endoplasmic reticulum or unattached and free in the cytoplasmic matrix. The ribosomes themselves in both locations have a similar morphology. In routine sections for electron microscopy at low to medium magnification they appear as roughly oval dense granules 150 to 200 Å in diameter (Fig. 9–1), and they stain intensely with uranyl salts. In negatively stained or shadowed preparations of isolated ribosomes, the two ribosomal subunits, the large and the small, can be visualized (Fig. 9–2).

In electron micrographs, free ribosomes usually form clumps or rosettes (Fig. 9–1). Similarly, bound ribosomes are found in groups on the surface of endoplasmic reticulum membranes. They form spiral patterns which are visible in sections that pass tangentially to the surface of the membrane of the endoplasmic reticulum (Fig. 9–3). In both cases, the aggregates of ribosomes visible in sections are polysomes (or polyribosomes), and each represents a group of ribosomes that is translating an mRNA molecule (see Chapter 7). Further detail of the structure of polysomes can be seen in isolated preparations viewed in the electron microscope. If stained with an electron-opaque substance (Fig. 9–4) the ribosomes appear to form a chain linked by a fine thread which represents messenger RNA.

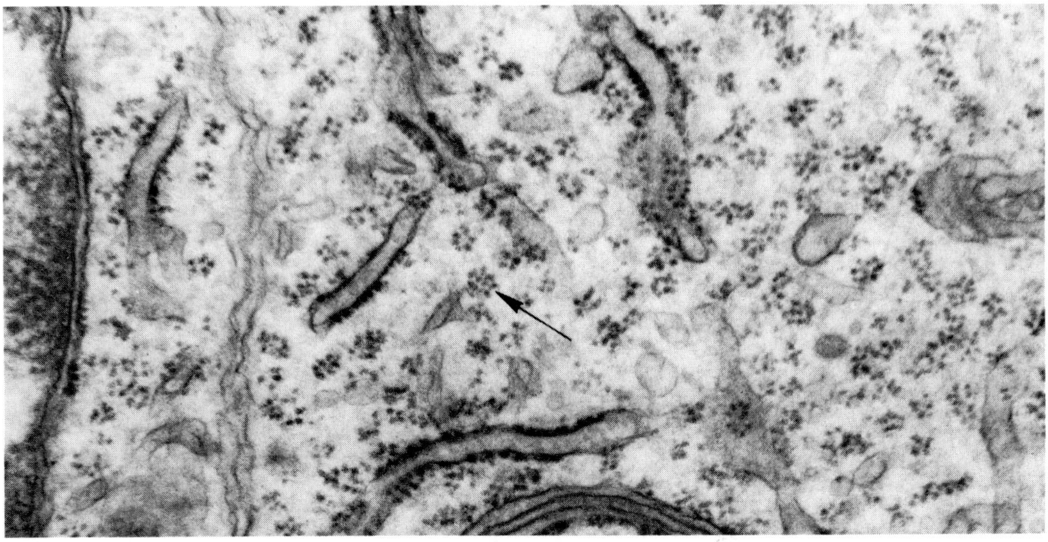

FIGURE 9–1 Portion of the cytoplasm of an epithelial cell of the fetal rat wolffian duct. Free (unattached) ribosomes are arranged in clumps or rosettes (*arrow*), which represent polysomes. × 53,000.

The ribosomes have a center-to-center spacing of about 300 Å. The longer the mRNA, the more ribosomes can fit on it at any given time. Thus the number of ribosomes in a polysome corresponds roughly to the size of the mRNA being translated.

Both free and bound ribosomes form polysomes and participate in protein

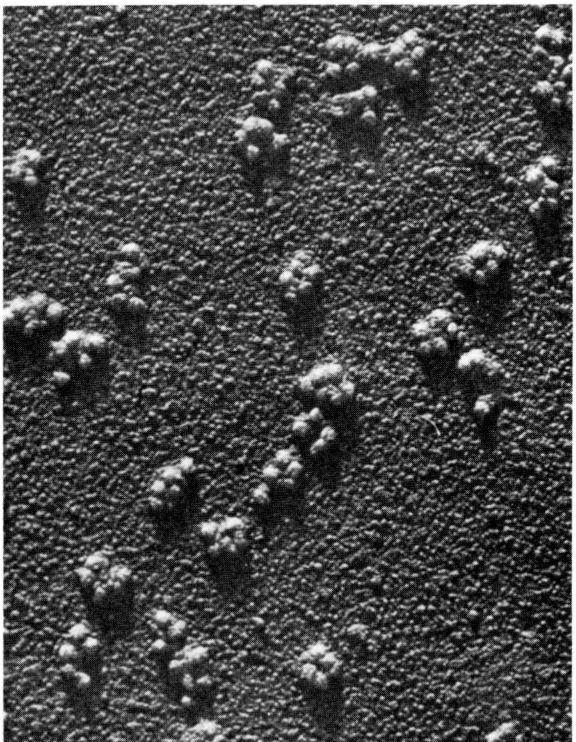

FIGURE 9–2 Preparation of isolated polysomes shadowed with platinum. The ribosomes occur in clusters of four to six units, which represent polysomes. These polysomes are active *in vivo* in the synthesis of hemoglobin. × 79,000. (Micrograph courtesy of Warner, J. R., Rich, A., and Hall, C. E.: Science, *138*:1399, 1962. Reproduced by permission of the American Association for the Advancement of Science.)

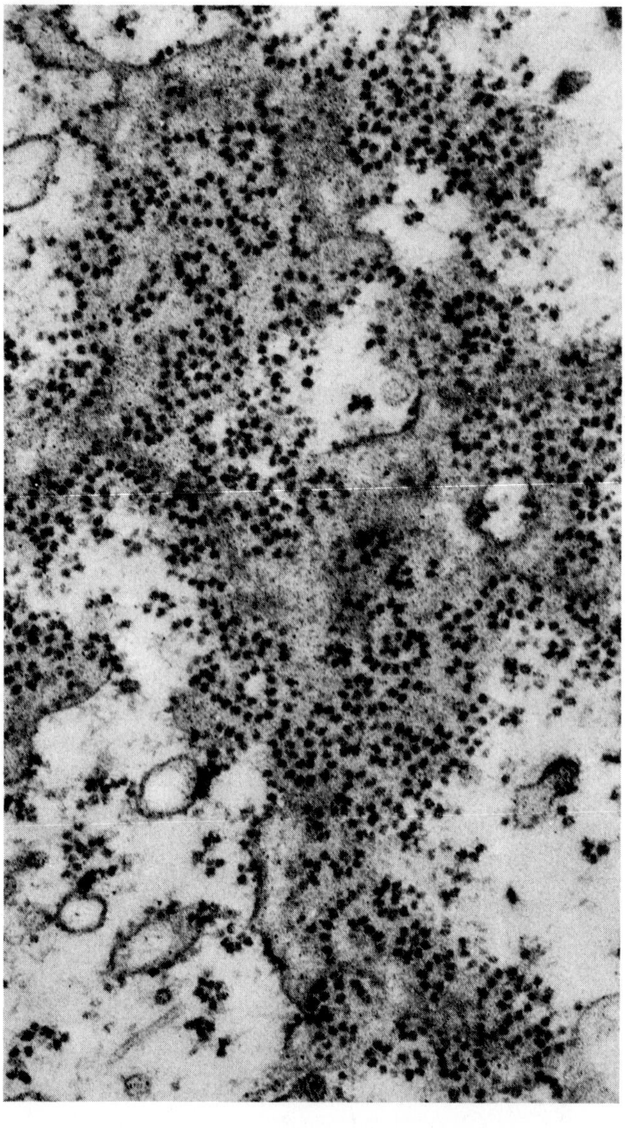

FIGURE 9–3 Rough endoplasmic reticulum in epidermal cells of a plant root tip. The arrangment of bound polyribosomes in spiral patterns on the surface of the endoplasmic reticulum is evident in this section, which is tangential to the membranes of the reticulum. × 57,000. (Micrograph courtesy of Bonnett, H. T., and Newcomb, E. H.: J. Cell Biol., *27*: 423, 1965. Reproduced by permission of the Rockefeller University Press.)

synthesis. In general, however, the fates of the protein products of free and bound ribosomes are different. Free polysomes usually are active in the synthesis of proteins for use within the cell. Thus free polysomes are abundant in rapidly growing cells which are mainly synthesizing proteins for their own growth. Examples include embryonic cells and certain cancer cells. In contrast, bound polysomes synthesize proteins for secretion or for use within membrane-bound structures, such as lysosomes. By virtue of their position on the membranes of the endoplasmic reticulum, they are favorably located to transfer their product to the interior of the endoplasmic reticulum. As will be seen later, a possible exception to this generalization about the fate of proteins synthesized by free and attached polysomes may be the synthesis of proteins that form part of cytoplasmic membranes by polysomes attached to the rough endoplasmic reticulum.

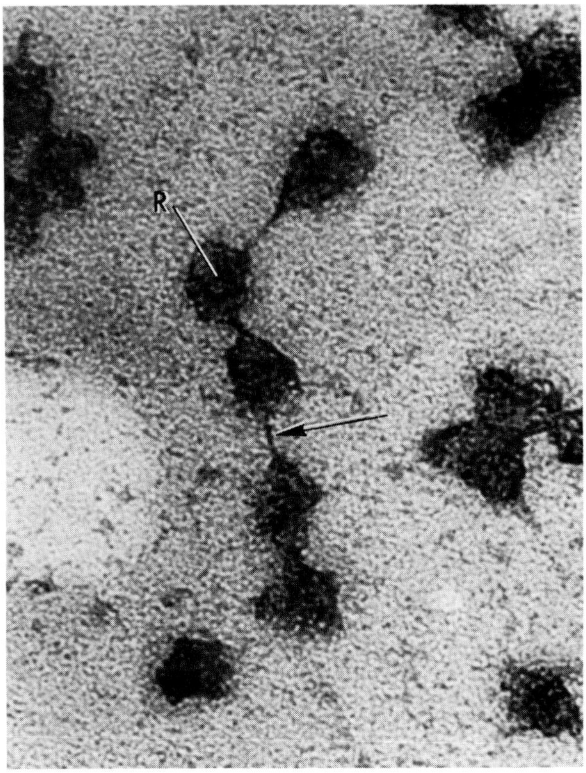

FIGURE 9–4 Polysomes from rabbit reticulocytes stained in positive contrast with uranyl acetate. The polysome in the center contains five individual ribosomes (R) linked to one another by a fine thread (*arrow*) believed to represent messenger RNA. × 420,000. (Micrograph courtesy of Slayter, H. S., Warner, J. R., Rich, A., et al.: J. Mol. Biol., 7: 652, 1963. Reproduced by permission of Academic Press.)

THE SECRETORY PROCESS

The process of secretion of proteins appears to be very similar in many cell types. The secretory process will be described in a general way in this section as an introduction to the study of the endoplasmic reticulum and Golgi apparatus. The steps in secretion will then be examined in more detail as the membranous organelles of the cytoplasm are studied.

The secretory process in an exocrine gland cell is summarized in a diagram (Fig. 9–5), and the cell organelles involved are illustrated in an electron micrograph (Fig. 9–6). Proteins for secretion are synthesized on ribosomes attached to the rough endoplasmic reticulum. They are then transferred through the membrane of the rough endoplasmic reticulum into the interior. In this way the secretory proteins are segregated from the rest of the cytoplasm. The secretory proteins are then transported from the endoplasmic reticulum to the Golgi apparatus within small smooth-surfaced vesicles, which bud from the surface of the endoplasmic reticulum, and are moved to the Golgi region of the cell. The smooth vesicles fuse with a component of the Golgi apparatus depositing their content of secretory protein in the Golgi apparatus. Many secretory proteins are actually glycoproteins and contain some carbohydrate linked to the protein part of the molecule. Addition of the carbohydrate to the protein part of the molecule may begin in the endoplasmic reticulum, but much of the addition takes place in the Golgi apparatus. Within the Golgi apparatus, the secretory product is concentrated and packaged into membrane-bound secretory vacuoles. After their release from the Golgi apparatus, the secretory vacuoles migrate toward the surface of the cell. Their content of secretory protein is discharged to the outside of the cell

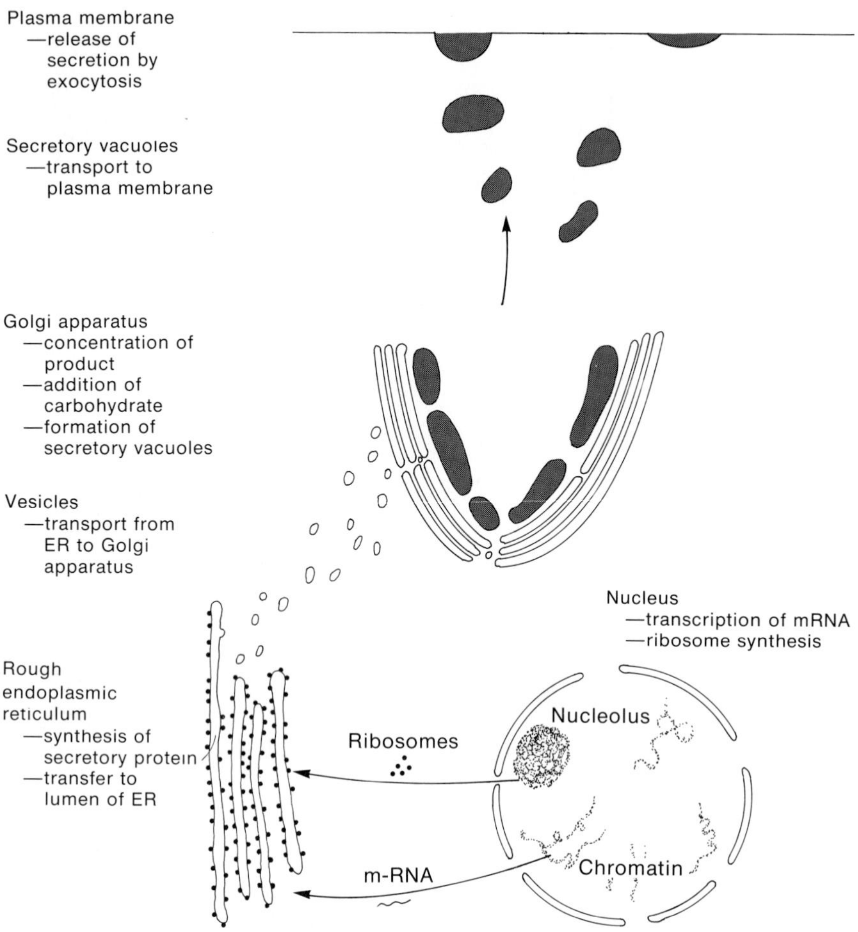

FIGURE 9–5 A diagram summarizing the steps in secretion of a protein or a glycoprotein in an exocrine gland cell.

when the membrane surrounding the secretory vacuole fuses with the plasma membrane, a process referred to as exocytosis.

ROUGH ENDOPLASMIC RETICULUM

Structure

The rough endoplasmic reticulum was detected with the light microscope by virtue of its staining with basic dyes, as described previously. It has been possible in more recent times to observe the endoplasmic reticulum with the light microscope in living cells in culture and in fixed preparations of certain cells by using phase contrast microscopy. The first observation of endoplasmic reticulum with the electron microscope was made in the late 1940's by Porter and his associates in the attenuated peripheral portions of unsectioned cells growing on a surface in culture (Fig. 9–7). This preparation was used because at that time techniques were not sufficiently advanced to allow the preparation of sections thin enough to be penetrated by the electron beam. Numerous beaded strands of material formed a network, or reticulum, in the "endoplasm" of the interior of the cell

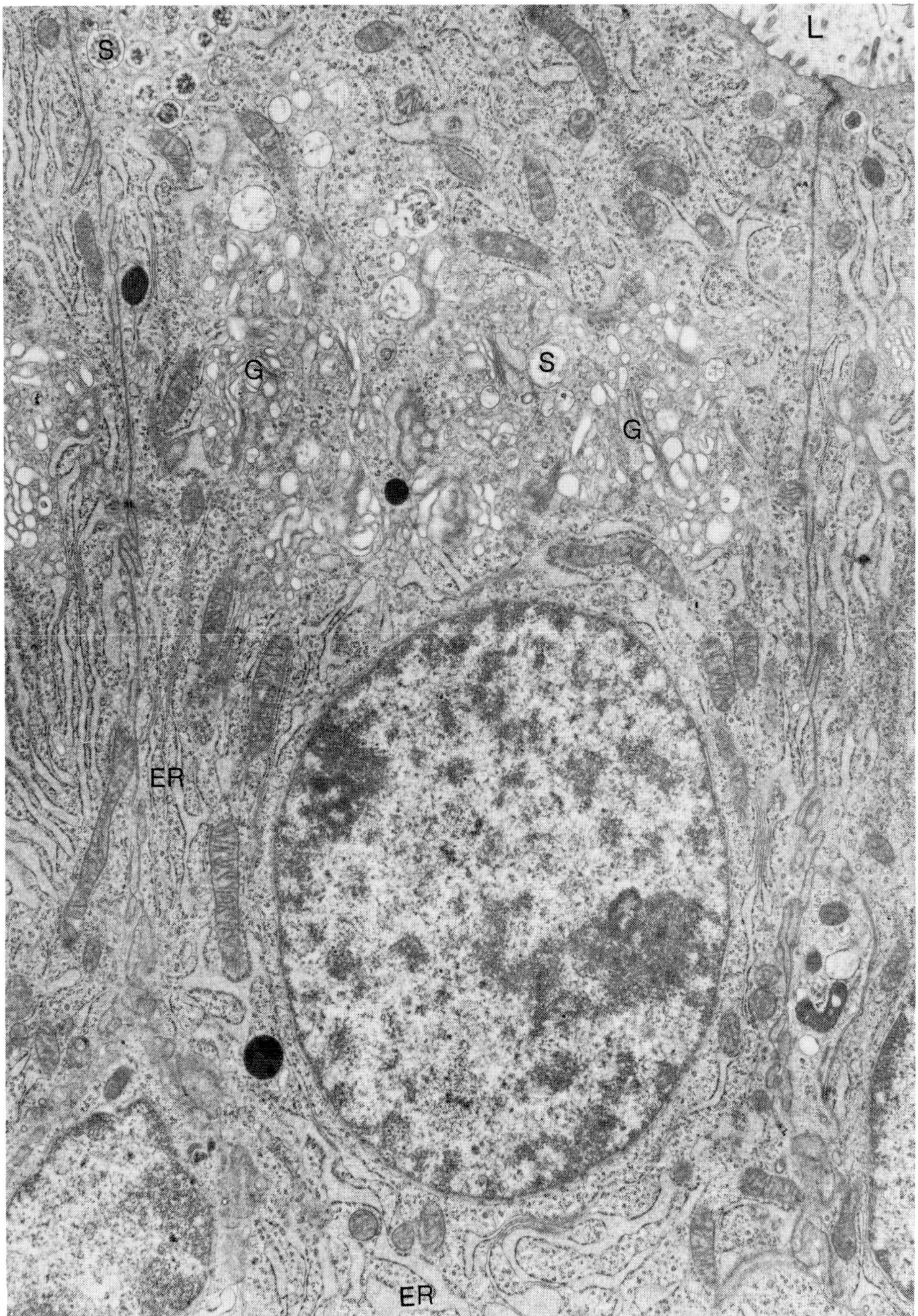

FIGURE 9–6 Epithelial cell from the rat ventral prostate gland. The basic organelles prominent in protein-secreting cells are rough endoplasmic reticulum (ER), Golgi apparatus (G), and secretory vacuoles (S). The protein secretions of these cells are released from the apical ends of the cells into the lumen (L). × 12,000. (From Greep, R. O., and Weiss, L.: Histology. 3rd ed. Reproduced by permission of the McGraw-Hill Book Company.)

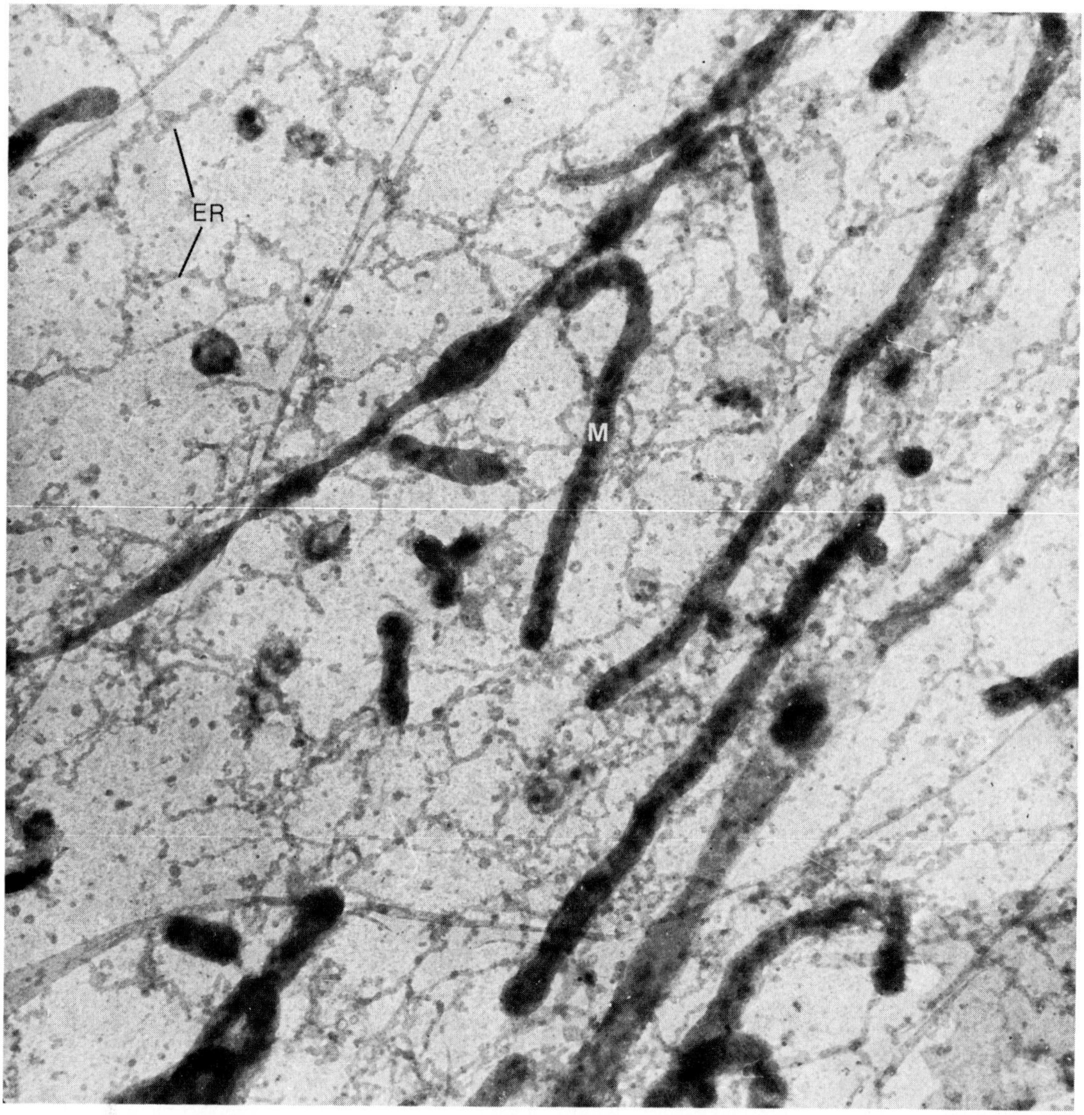

FIGURE 9–7 One of the early micrographs of the endoplasmic reticulum in an unsectioned mouse fibrocyte grown in culture. The endoplasmic reticulum (ER) appears as an irregular grayish network throughout the field. The dense elongated bodies are mitochondria (M). Magnification approximately × 7,000 (Micrograph courtesy of K. R. Porter.)

but were absent from the "ectoplasm" at the margin of the cell. Therefore, the name endoplasmic reticulum was applied to the structure.

Rough endoplasmic reticulum is visible in thin sections of virtually all eukaryotes. Its most familiar form is that of large flattened membranous sacs, known as cisternae (Fig. 9–3 and 9–8). The rough endoplasmic reticulum also occurs in the form of small cisternal elements, tubules, or rounded vesicles. Whatever the shape, the various forms have in common a continuous membrane that delimits a space and separates it from the cytoplasmic matrix. The space bounded by the membrane of the endoplasmic reticulum is called variously the lumen, the interior of the endoplasmic reticulum, or the intracisternal space. It often has a flocculent content of moderate density which represents newly synthesized secretory protein. Ribosomes are attached to the outer surface of the membranes of the endoplasmic reticulum, and thus they are in contact with the cytoplasmic matrix. The membrane of the rough endoplasmic reticulum appears trilaminar in high-magnification electron micrographs, and it is usually about 70 Å thick. In many exocrine gland cells that are synthesizing large quantities of secretory protein, numerous large cisternae of the rough endoplasmic reticulum lie parallel to one another and form large arrays in the cytoplasm (Fig. 9–8). In some instances the

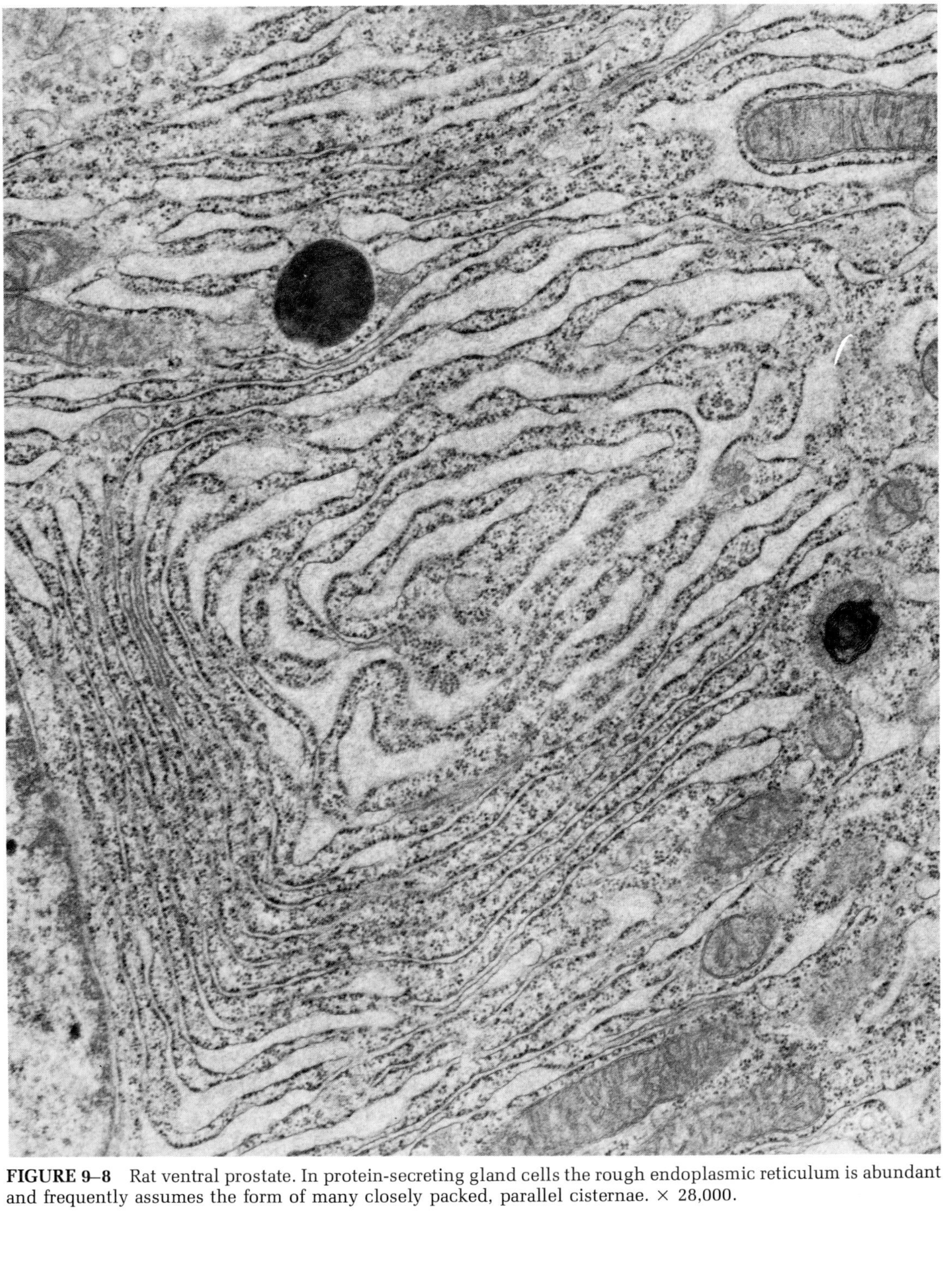

FIGURE 9–8 Rat ventral prostate. In protein-secreting gland cells the rough endoplasmic reticulum is abundant and frequently assumes the form of many closely packed, parallel cisternae. × 28,000.

elements of rough endoplasmic reticulum can be seen to anastomose, constituting a network that extends through much of the cytoplasm (Fig. 9–9).

Relation to Microsomes

Since the activities of the endoplasmic reticulum are studied biochemically in the microsomal fraction of cells, it is important to understand the relation between endoplasmic reticulum and microsomes. Although the microsomes represent the endoplasmic reticulum in a cell homogenate, the characteristic form of the endoplasmic reticulum is not preserved (Fig. 9–10). When the cell is homogenized, the endoplasmic reticulum is fragmented. However, small sheets and patches of membranes are not seen, since after fragmentation the membranes form small vesicles known as microsomes, which have a continuous bounding membrane. This process can be visualized as a pinching-off of small spheres of the endoplasmic reticulum. Thus the interior of microsomal vesicles corresponds to the lumen or intracisternal space of the endoplasmic reticulum. The content of the endoplasmic reticulum is not released but is contained within the microsomes. Ribosomes stud the outer surface of the microsomal vesicles. The ribosomes can be isolated from the other components of the microsomes by dissolving the membrane with a detergent. This procedure also frees the content of the vesicles. Isolation of microsome fractions has been widely used to determine the

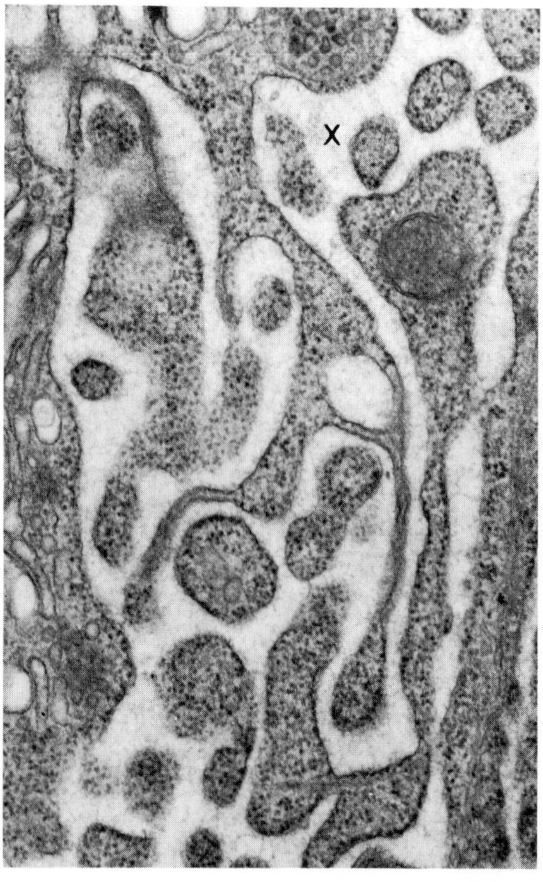

FIGURE 9–9 Dorsal lobe of rat prostate gland. In these secretory cells the rough endoplasmic reticulum is a system of irregularly arranged interconnecting channels. The interior of the endoplasmic reticulum (X) is lighter than the dense cytoplasmic matrix in this preparation. × 33,000. (From Flickinger, C. J., Z. Zellforsch, *113*:157, 1971. Reproduced by permission of Springer-Verlag.)

composition of the endoplasmic reticulum in various cells and under different conditions. The enzyme, lipid, and carbohydrate compositions of endoplasmic reticulum membranes from rat liver, for example, are shown in Table 9–1.

Since some endoplasmic reticulum is rough and some is smooth, corresponding rough and smooth microsomes are formed (Fig. 9–10). If microsomes are prepared from cells such as hepatocytes, which contain both kinds of endoplasmic reticulum, a mixture of rough and smooth microsomes is obtained, and the two types can be separated by centrifugation on a gradient. In certain types of cells, the endoplasmic reticulum is predominantly of one type, either rough or smooth, and microsomes from these cells are sometimes studied without further fractionation.

Functions

The main function of the rough endoplasmic reticulum is the synthesis of secretory protein and its segregation from the rest of the cytoplasm in the lumen of the rough endoplasmic reticulum. This was first established by Siekevitz and Palade in the mid-1950's in a series of cell fractionation studies on the pancreas. Guinea pigs were injected with leucine-^{14}C, and at intervals thereafter subcellular

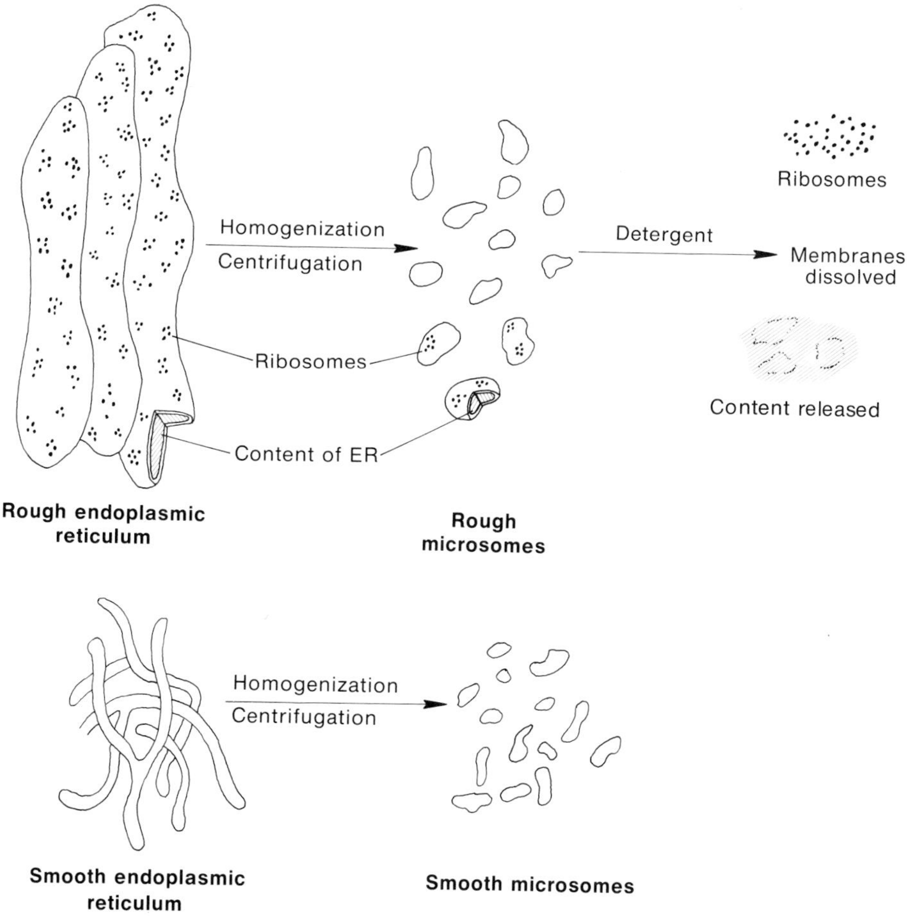

FIGURE 9–10 The relation between endoplasmic reticulum and microsomes.

TABLE 9–1 COMPOSITIONS OF MEMBRANE FRACTIONS OF RAT LIVER*

A. Relative Enzyme-Specific Activities

Enzyme or Constituent†	*Endoplasmic Reticulum*	*Golgi Apparatus*	*Plasma Membrane*
Glucose-6-phosphatase	10	1	0
UDP glucuronyl transferase	10	1	0
NADPH cytochrome-c reductase	10	1	0
Cytochrome P_{450}	4	1	0
Terminal phosphoglyceride biosynthetic enzymes for PC, PE, and PI	4	1	0
O-Demethylase	2	1	0
Glycolipid glycosyl transferases (monosialoganglioside pathway)	1	10	0
Glycoprotein glycosyl transferases (exogenous acceptors)	1	10	0
NADH cytochrome-c ferricyanide reductase	9	2	1
Cytochrome b_5	3	1	1
NADH juglone reductase	2	1	1
Nucleoside diphosphatase (IPD, GDP, and UPD as substrates)	2	1	1
Thiamine pyrophosphatase	3	10	1
UDP galactose hydrolase	1	1.5	6
Nucleoside diphosphatase (ACP, CDP, and TDP as substrates)	1	2	5
Mg^{2+}-ATPase	1	1	5
5′-Nucleotidase (5′-AMP as substrate	1	2	40
Xanthine oxidase	0	1	8
Adenylate cyclase	2	1	4
Na^+, K^+, Mg^{2+}-ATPase	1	0	7

†PC, Phosphatidylcholine; PE, phosphatidylethanolamine; PI, phosphatidylinositol.

B. Comparative Lipid Compositions

Constituent	*Grams per 100 Grams Total Membrane*			
	Nuclear Envelope	*Endoplasmic Reticulum*	*Golgi Apparatus*	*Plasma Membrane*
Total lipid	30	30	35	42
Neutral lipid	4	4	6	11
Sterols	2	2	3	7
Cholesterol	1.8	1.2	2.0	>6
Cholesterol esters	0.2	0.8	1.0	<1
Triglycerides	2	2	3	3–4
Glycolipids	0.025	0.05	0.2	0.6
Phospholipids	26	27	29	30
Phosphatidylcholine	16	16	15	14
Phosphatidylethanolamine	6	7	7	7
Phosphatidylserine	2	1	1	1
Phosphatidylinositol	2	2	2.5	2
Sphingomyelin	<1	1	3.5	6

COMPOSITIONS OF MEMBRANE FRACTIONS OF RAT LIVER* (Continued) — TABLE 9–1

C. Comparative Carbohydrate Compositions

Constituent (Units)	*Nuclear Envelope*	*Endoplasmic Reticulum*	*Golgi Apparatus*	*Plasma Membrane*
Total carbohydrate (% dry weight)	<1	1	2	4(2–7)
Constituent sugars (μg carbohydrate/mg protein)				
Sialic acid	Trace	2.0	13.6	18.0
Hexosamine	5.0	5.5	6.5	19.0
Fucose	Trace	0.5	3.0	3.0
Galactose	2.0	2.5	7.0	14.0
Glucose	2.0	0.8	0.3	Trace
Mannose	9.5	9.0	8.4	13.0
Total sialic acid (nmoles/mg protein)	2	2–5	20	30–50
Ganglioside sialic acid (nmoles/mg protein)	<0.3	<0.3	2	5
Protein sialic acid (nmoles/mg protein)	<2	2–5	18	30–45

*From: Morré, D. J., and Ovtract, L. P. Int. Rev. Cytol., Suppl. 5:61, 1977. Reproduced by permission of Academic Press.

fractions of the pancreas were isolated and assayed for radioactivity (Fig. 9–11). Within a few minutes, an initial peak of radioactivity was found associated with ribosomes bound to the rough endoplasmic reticulum. Shortly afterwards, while the radioactivity of the ribosomes was declining, the radioactivity of the content of the microsomes rose. At a later time, radioactivity appeared in the zymogen granule fraction, which is composed of secretory vacuoles and contains the digestive enzymes produced by the gland. These observations were interpreted as reflecting the synthesis of radioactive proteins by the ribosomes of the endoplasmic reticulum and the transfer of the newly synthesized protein to the interior of the endoplasmic reticulum where it formed the content of the microsomes. Later the radioactive protein appeared to be transported from the endoplasmic reticulum to the zymogen granules.

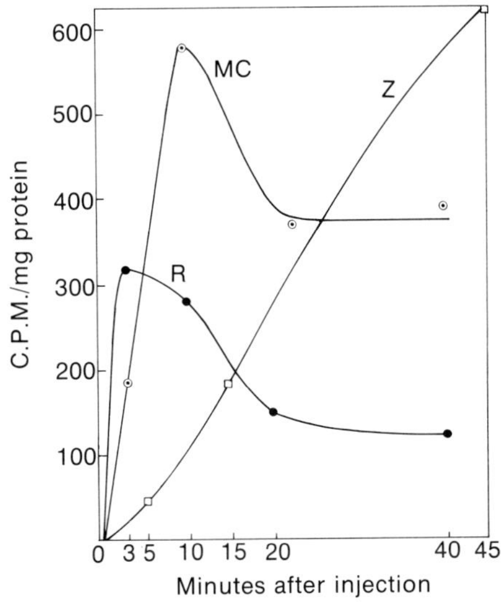

FIGURE 9–11 The radioactivity of proteins in the ribosomes (R), microsomal contents (MC), and zymogen granules (Z) of the pancreas after injection of ^{14}C-leucine into fasted guinea pigs. ((After Siekevitz, P. and Palade, G. E.: J. Biophys. Biochem. Cytol., 4:557, 1958. Reproduced by permission of the Rockefeller University Press.)

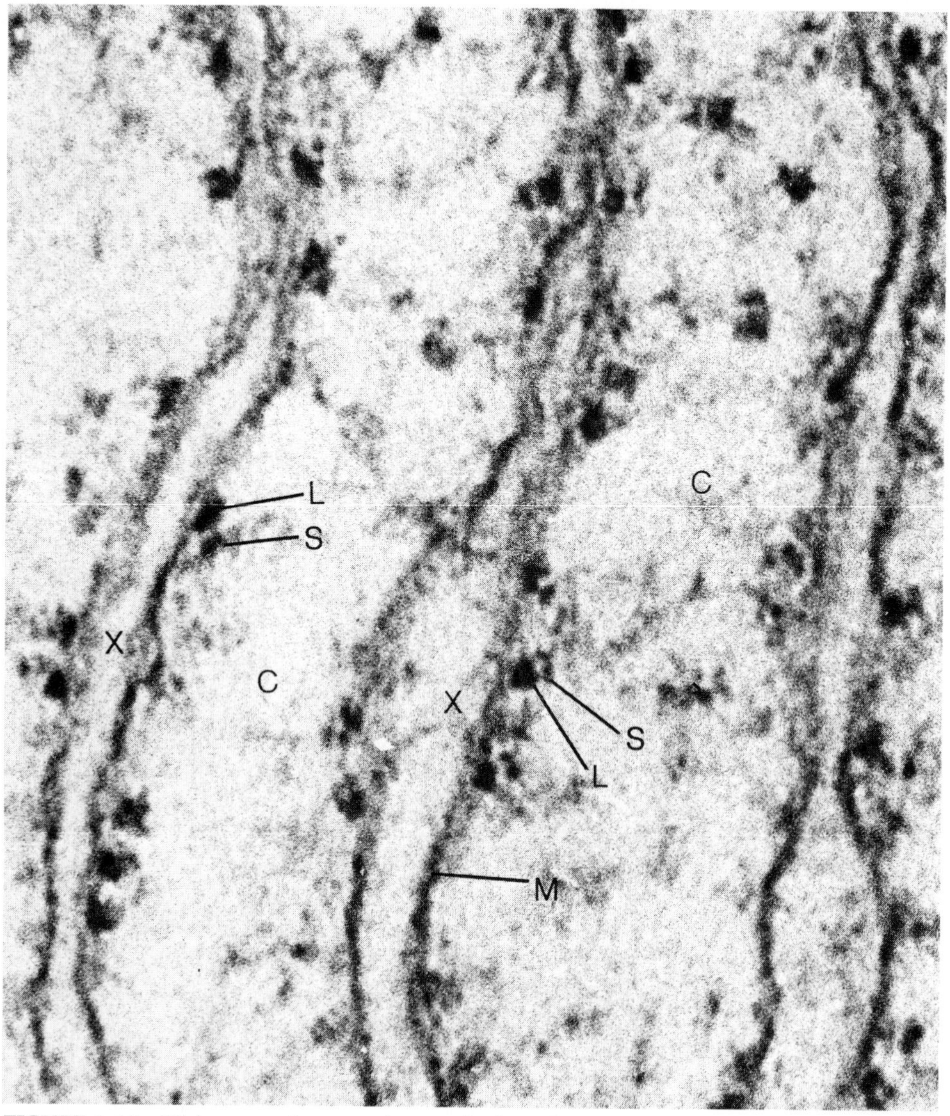

FIGURE 9–12 High-power electron micrograph of rough endoplasmic reticulum of a mouse liver cell. The large (l) and the small (s) ribosomal subunits can clearly be visualized. The ribosomes are attached to the membranes (M) of the endoplasmic reticulum by their large subunits. C, cytoplasmic matrix; X, lumen of the endoplasmic reticulum. × 200,000. (Micrograph courtesy of Florendo, N. T.: J. Cell Biol., *41*:335, 1969. Reproduced by permission of the Rockefeller University Press.)

This interpretation of the role of the rough endoplasmic reticulum received abundant confirmation through the use of more refined cell fractionation techniques and radioautography at the light and electron microscope levels, not only in the exocrine pancreas but also in a great variety of other protein-secreting cells. After the Golgi apparatus has been discussed, further examples of the roles of the rough endoplasmic reticulum and Golgi apparatus in the synthesis and processing of secretory products will be considered.

Although it is known that secretory proteins are synthesized on bound ribosomes and shortly thereafter are found in the interior of the endoplasmic reticulum, it is not yet clear how the newly synthesized protein is transferred across the membrane of the endoplasmic reticulum. The large ribosomal subunit is known to be in contact with the membrane of the endoplasmic reticulum, while the small subunit faces the cytoplasmic matrix (Figs. 9–12 and 9–13). The newly synthesized peptide chains do not appear to be free in the cytoplasmic matrix at any time. Some investigators have proposed that the newly synthesized peptides are formed in the groove between the two ribosomal subunits, enter a hole through the large subunit, and then pass through the membrane.

Another feature of the rough endoplasmic reticulum structure and function that is incompletely resolved is the means of attachment of polysomes to the membrane. Several factors may be involved (Fig. 9–13). It is reported that a sequence of the messenger RNA near the 3′ end, which contains polyadenylic acid, binds to the membrane. Some studies suggest also that secretory proteins have common sequences at their N-terminal ends, which could be recognized by a membrane component and aid in the attachment to membranes of polysomes synthesizing secretory proteins. The large subunits of the ribosomes may also bind directly to the membranes. Microsomal membrane proteins that bind ribosomes are currently being studied.

One additional function of the rough endoplasmic reticulum should be noted. Although concentration of secretory products is a function usually attributed to the Golgi apparatus, in some cells concentration of a product also occurs in the rough endoplasmic reticulum. In the exocrine pancreas of the guinea pig, which has been widely used in studies of the secretory process, concentration of secretory protein begins in the rough endoplasmic reticulum, as indicated by the appearance of electron-dense aggregates within the lumen of the reticulum (Fig. 9–14). An extreme example of the ability of the endoplasmic reticulum to concentrate cell products is seen in the liver of certain amphibians in which secretory

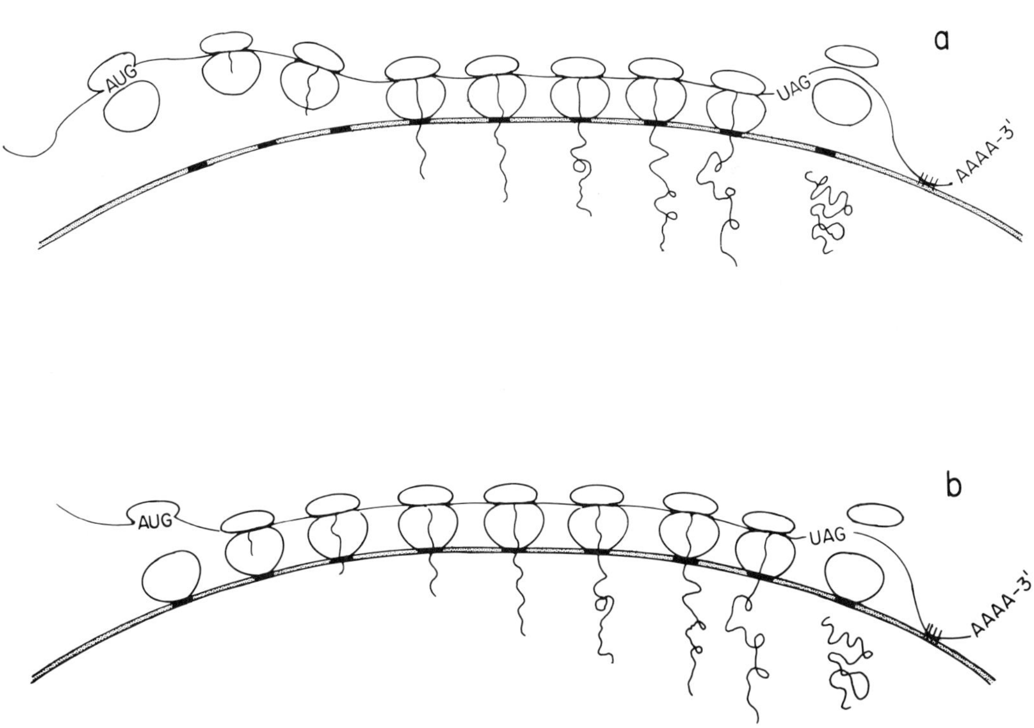

FIGURE 9–13 Diagram showing the attachment of ribosomes and messenger RNA to membranes of the endoplasmic reticulum. The ribosomes are attached by their large subunits. This binding is thought to be stabilized by nascent polypeptide chains that penetrate through the membrane. The messenger RNA is attached to the membrane by a segment near the 3′ end. (From Lande, M. A., Adesnik, M., Sumida, M., et al.: J. Cell Biol., 65:513, 1975. Reproduced by permission of the Rockefeller University Press.)

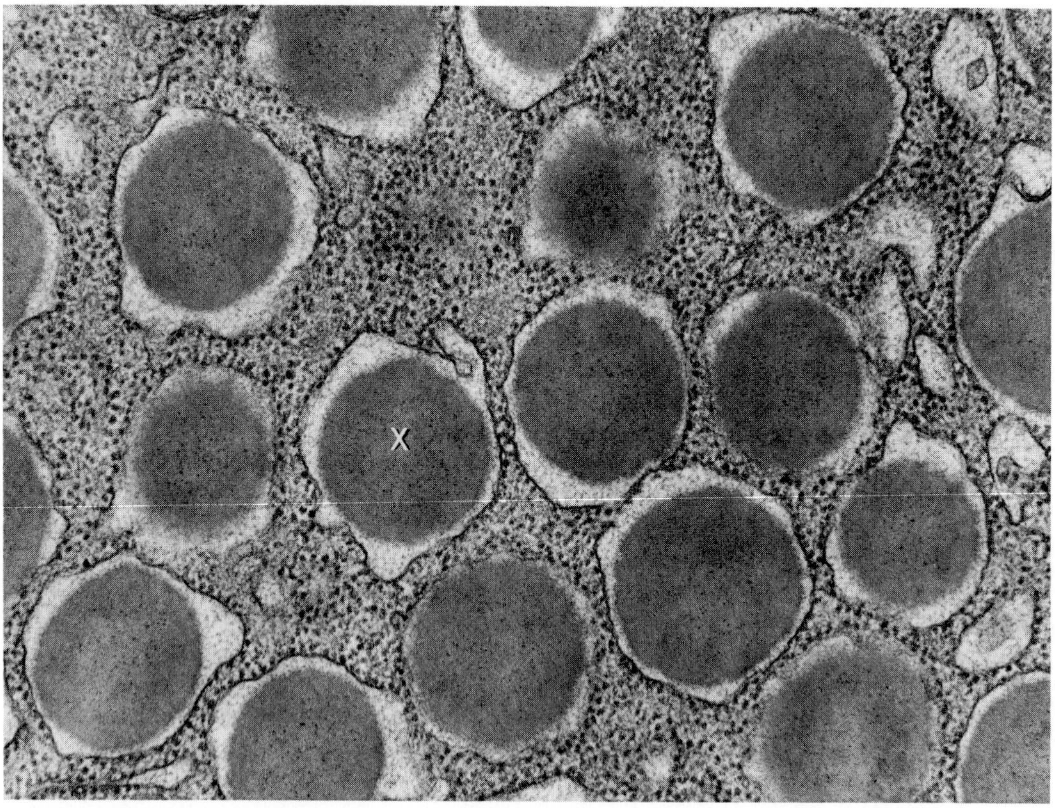

FIGURE 9–14 Rough endoplasmic reticulum in the guinea pig exocrine pancreas. Large dense granules (X) are present within the cisternae, apparently as the result of concentration of newly synthesized secretory proteins within the endoplasmic reticulum. × 47,000. (Micrograph courtesy of Bolender, R. P.: J. Cell Biol., *61*:269, 1974. Reproduced by permission of the Rockefeller University Press.)

products, presumably blood proteins, form crystalline masses within the lumen of the rough endoplasmic reticulum.

Relation to Other Organelles

The general relationships between the rough endoplasmic reticulum and other membranous organelles are summarized in Figure 9–15. It should be recalled that the rough endoplasmic reticulum is frequently connected to the outer membrane of the nuclear envelope; hence the interior of the rough endoplasmic reticulum is continuous with the perinuclear cisterna. There are also numerous connections between the rough and smooth endoplasmic reticulum. Direct structural connections between the rough endoplasmic reticulum and the Golgi apparatus have been reported, but these organelles appear to communicate functionally with one another primarily by means of vesicles that move between the two. Connections between the rough endoplasmic reticulum and the plasma membrane are not now regarded as general features of most cells, although such connections have been reported for some cell types.

Annulate Lamellae

Annulate lamellae are flat membranous cisternae which are apparently related to the rough endoplasmic reticulum and the nuclear envelope. Their name is

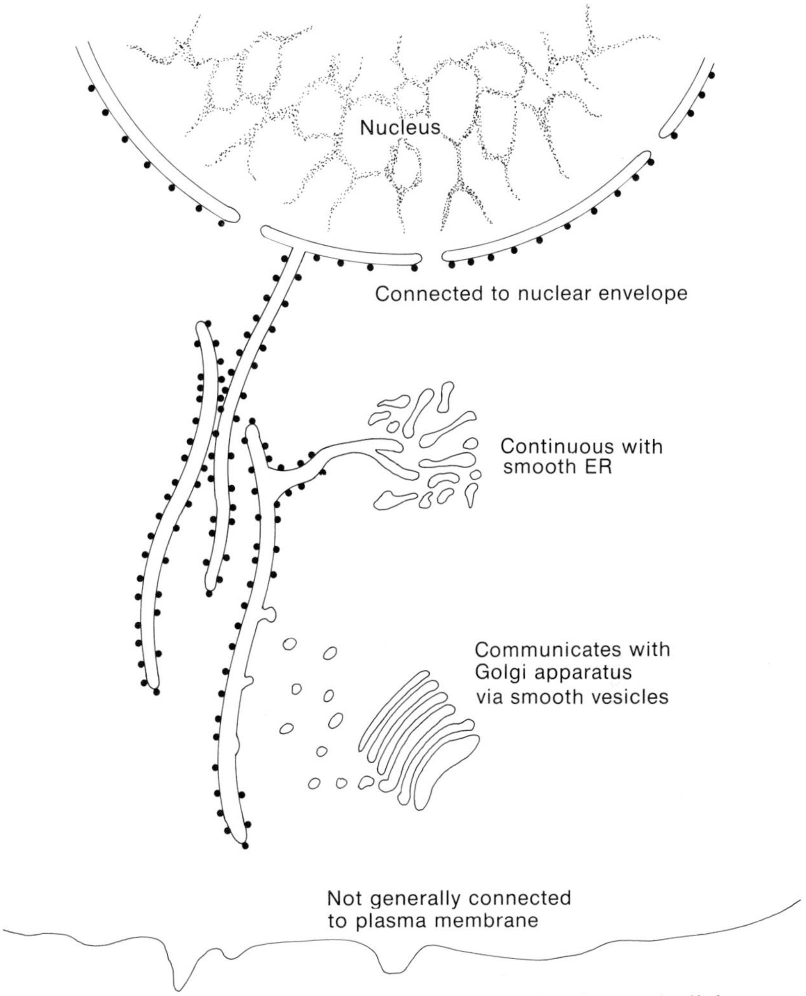

FIGURE 9–15 The relation of the rough endoplasmic reticulum to other classes of cellular membranes.

derived from the fact that the membranes are interrupted by numerous regularly spaced pores with associated material, which closely resemble the pore complexes of the nuclear envelope (Fig. 9–16). The lamellae occur singly or in stacks in the cytoplasm. In rare instances they are found inside the nucleus. The annulate lamellae are not usually studded with ribosomes but do have basophilic staining characteristics. They may be surrounded by a dense fibrillar or finely granular material which often is preferentially associated with the pores. Connections between the annulate lamellae and the rough endoplasmic reticulum have often been observed.

Annulate lamellae have been reported most frequently in rapidly growing cells, such as germ cells, early embryos, and tumor cells, but they occur sporadically in a variety of other cell types and may make a transient appearance during the life of a particular kind of cell. Mainly because of their remarkable resemblance to pieces of nuclear envelope, annular lamellae are thought to be derived from the nuclear envelope. It has been speculated that they may function in nucleocytoplasmic interactions by conveying material from the nucleus to the cytoplasm or that they might serve as localized sites of some special synthetic activity, but the significance of the annulate lamellae remains to be definitely established.

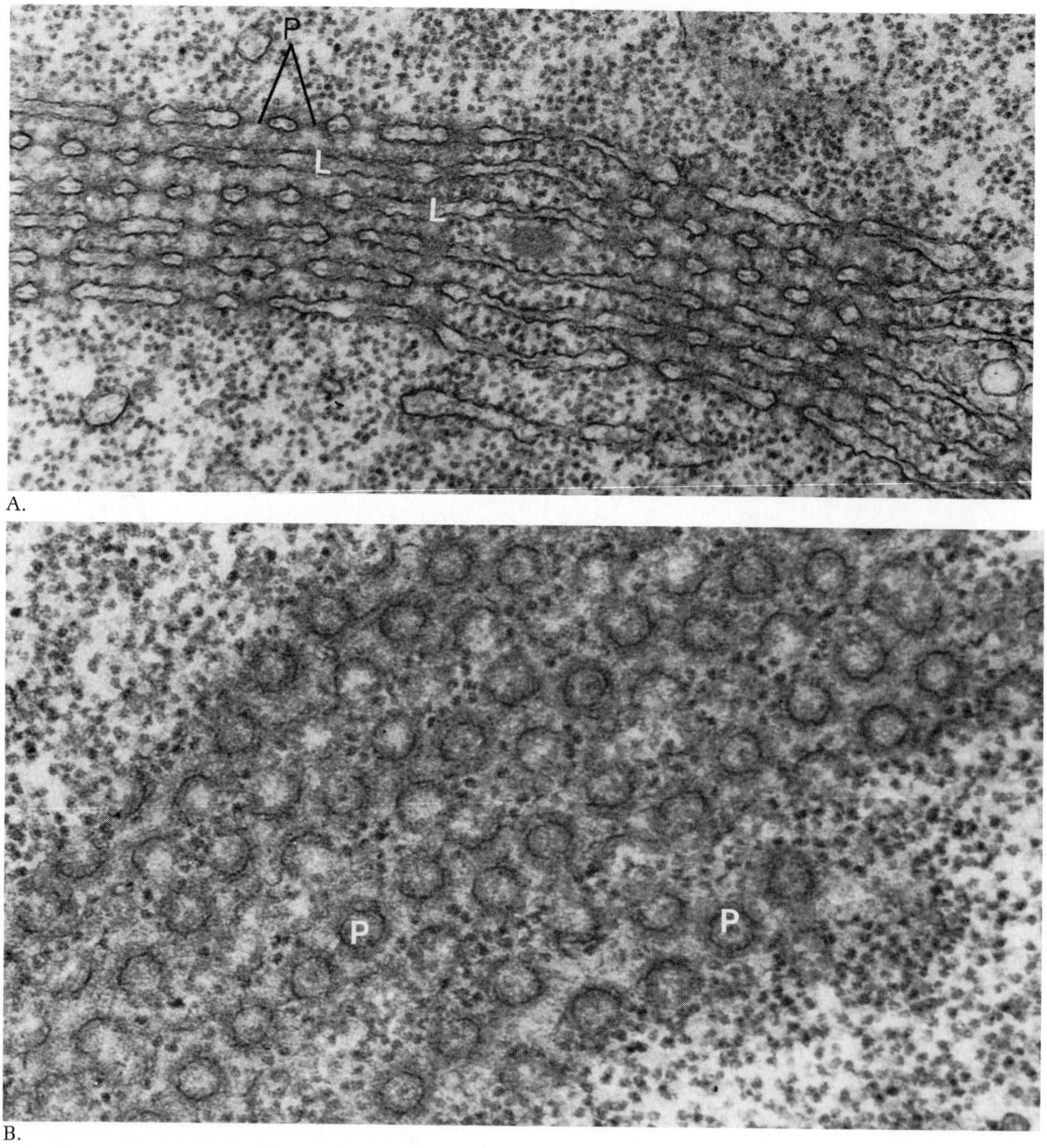

FIGURE 9–16 Annulate lamellae in a tunicate oocyte. *A*. The section is perpendicular to the surfaces of the flat lamellae (L), which contain pores (P) and form a stack. *B*. The section is tangential to the surface of the lamellae and thus displays the pores (P) in face view. *A*. × 50,000. *B*. × 73,000. (Micrographs courtesy of Kessel, R. G.: J. Ultrastruct. Res., Suppl. 10, 1968. Reproduced by permission of Academic Press, Inc.)

GOLGI APPARATUS

In 1898 Camillo Golgi described a netlike structure in the cytoplasm of neurons of the barn owl. He named it the internal reticular apparatus, but this structure was soon demonstrated in other forms in a variety of cells and became more widely known as the Golgi apparatus. Unfortunately, the metal impregnation methods that were used in those days to demonstrate the Golgi apparatus were very capricious, and the next 50 years of study on the Golgi apparatus were largely spent in a debate over whether it was a real structure or merely an artefact. Some cytologists noted that in gland cells it underwent changes with the activity

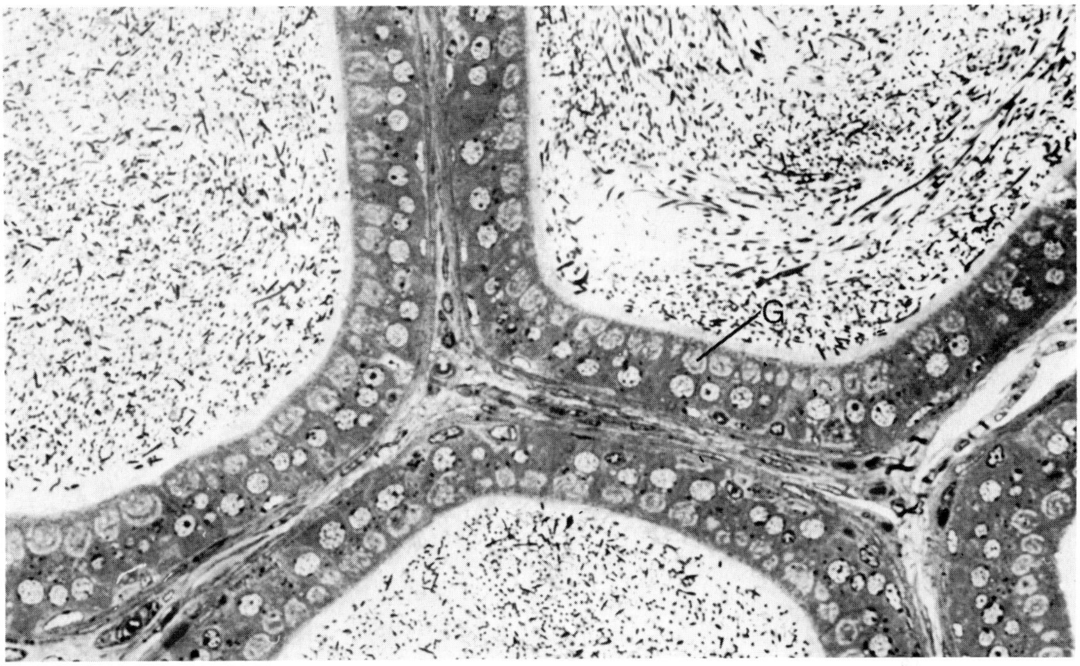

FIGURE 9–17 Hamster epididymis, toluidine blue stain. In light microscope preparations stained with basic dyes, the Golgi apparatus (G) is frequently visible as a light or clear area in the supranuclear cytoplasm of epithelial cells. × 344.

of the organ. However, as late as 1947 it was claimed that the Golgi apparatus was an artefact, because images similar to those seen in cells with the metal impregnation methods could be produced *in vitro* by staining a mixture of lipids and water. The reality of the Golgi apparatus was finally established through electron microscopy in the early 1950's. It is now known to be ubiquitous and to play an important role in secretion and other cellular activities.

Structure

As suggested by this brief historical review, the Golgi apparatus can be detected with the light microscope, but appreciation of the details of its structure

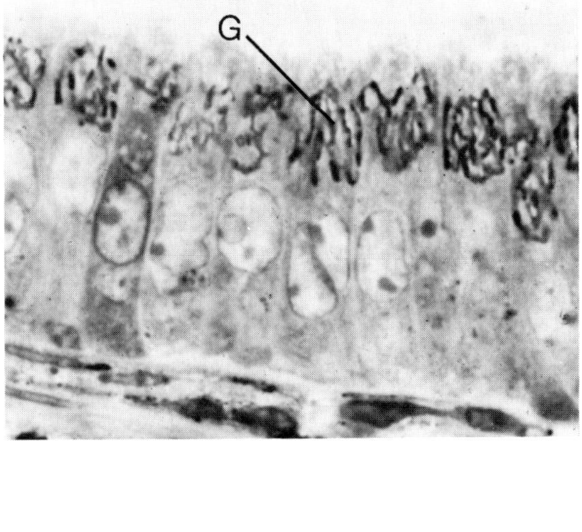

FIGURE 9–18 Epithelium of the mouse epididymis. The Golgi apparatus (G), which occupies a supranuclear position, has been blackened by deposits of reduced osmium following prolonged exposure of the tissue to aqueous osmium tetroxide. (Micrograph courtesy of Friend, D. S., and Murray, M. J.: Am. J. Anat., *117*:135, 1965. Reproduced by permission of the Wistar Institute Press.)

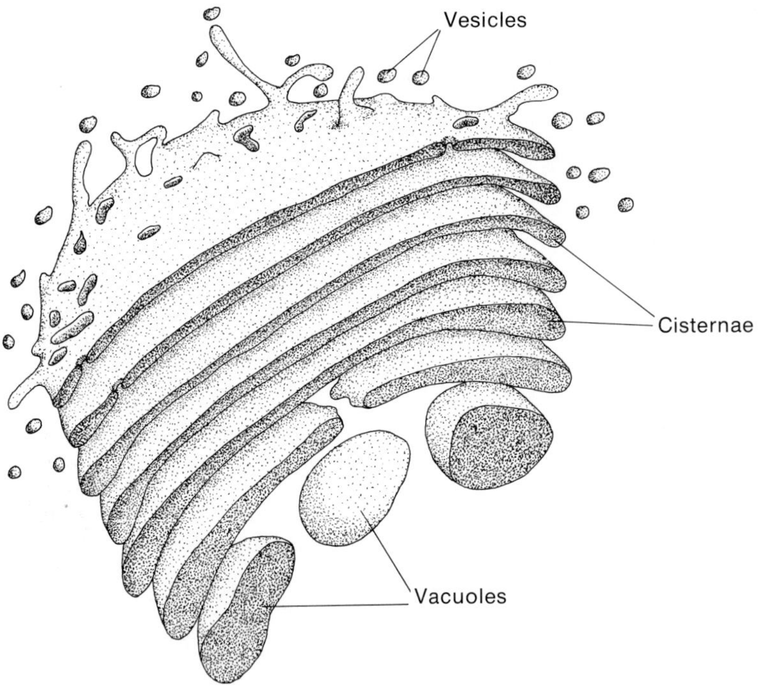

FIGURE 9–19 A diagram of the membranous components of the Golgi apparatus: vesicles, cisternae, and vacuoles.

requires electron microscopy. The Golgi apparatus can sometimes be visualized with the light microscope because, lacking ribosomes, it fails to stain with basic dyes, while the rough endoplasmic reticulum comprising the basophilic bodies stains intensely. Thus the Golgi apparatus appears as a clear area and is said to display a "negative image" (Fig. 9–17). In most columnar epithelia, the Golgi apparatus is located apical to the nucleus in each cell, and a row of light areas can often be seen parallel to the row of nuclei. Although metal impregnation techniques are not widely used to study the Golgi apparatus nowadays, they blacken the organelle so that it stands in great contrast to the rest of the cytoplasm (Fig. 9–18).

With the electron microscope it can be seen that the Golgi apparatus has three main membranous components: cisternae or saccules, small vesicles, and larger vacuoles (Figs. 9–19 and 9–20). All three forms of Golgi membranes lack attached ribosomes. The cisternae are the most distinctive components of the Golgi apparatus, and they usually form the basis for its morphological identification. The flat Golgi cisternae are aligned in parallel to form stacks. They are separated by a remarkably uniform space several hundred Å wide, which appears structureless in most mammalian cells but is occupied by a series of dense fibers in some plant cells. Golgi cisternae are often depicted as flattened platelike structures, but their organization is actually more complex. A cisterna usually does have a central disclike region, but the more peripheral parts are often expanded and contain numerous pores or fenestrations (Fig. 9–21). The margins of the cisternae may assume an almost tubular character. Numerous protuberances are visible on the margins of Golgi cisternae and are thought to represent vesicles in the process of fusion with the cisternae or fission from them. Golgi vesicles are spheres approximately 400 Å in diameter. Some are smooth-surfaced, while others have the bristlelike coat that makes them known as coated vesicles. Golgi vacuoles are commonly found at one side of the stacks of cisternae, and as will be seen later in more detail, they often contain a visible content of secretory material.

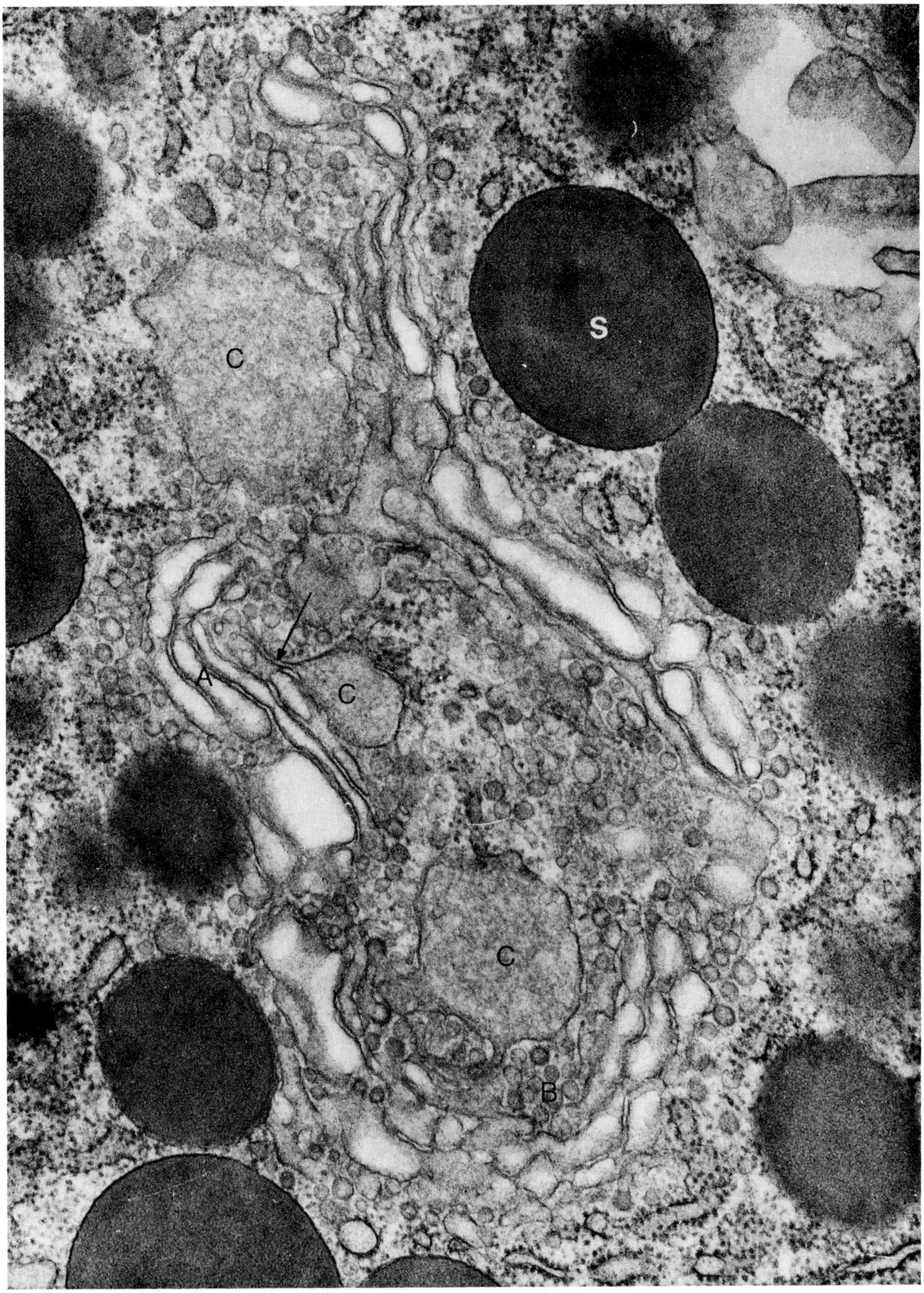

FIGURE 9–20 Golgi apparatus in a rat parotid gland acinar cell. The three main components of the Golgi apparatus are cisternae (A), vesicles (B), and condensing vacuoles (C). The condensing vacuoles have a moderately dense flocculent content, and one appears to be arising from a Golgi cisterna (*arrow*). The content of the condensing vacuoles increases in density as they mature into secretory granules (S). × 44,000. (Micrograph courtesy of Hand, A. H.: Am. J. Anat., *130*:141, 1971. Reproduced by permission of the Wistar Institute Press.)

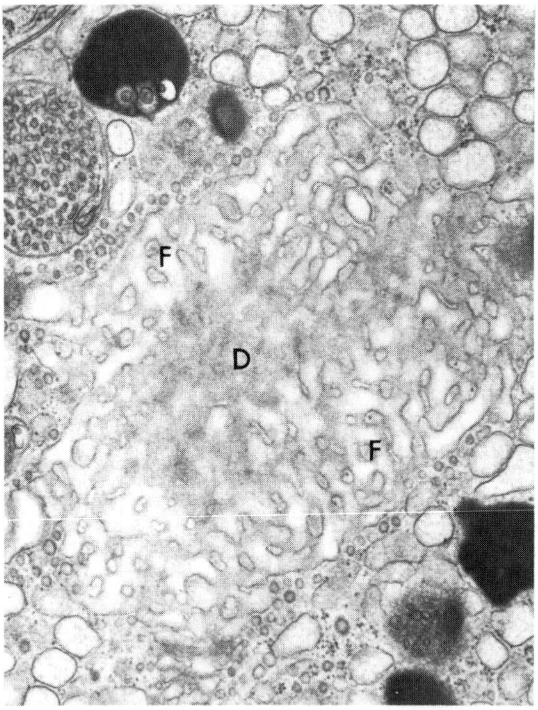

FIGURE 9–21 Mouse epididymis. A cisterna of the Golgi apparatus has been sectioned parallel to its surface so it is shown in face view. The cisterna has a central disclike portion (D) and a fenestrated peripheral region (F). × 27,000. (From Flickinger, C.: Anat. Rec., *163*:39, 1969. Reproduced by permission of the Wistar Institute Press.)

The Golgi apparatus varies greatly in its degree of development in different cell types. Although it attains its largest size and has been most widely studied in gland cells, it is a component of almost all eukaryotic cells. The disposition of the Golgi apparatus in three dimensions can be complex, and it is not readily appreciated by viewing thin sections. In columnar cells, the Golgi apparatus often has the shape of a flask, with its base near the nucleus and its open end directed toward the apex of the cell. In neurons it has the form of a reticulated hollow sphere surrounding the nucleus. In other cells in which the Golgi apparatus is not as well developed, only small stacks of cisternae are observed. It has been demonstrated in some instances that although the individual stacks of cisternae appear isolated from one another in thin sections, they are actually related by means of tubular and cisternal connections. Thus the entire Golgi apparatus of a neuron, for example, may be accurately visualized as a single continuous organelle.

Polarity of Organization

The Golgi apparatus has a polarity to its organization. That is, the cisternae within the stacks are not uniform but differ from one side, or pole, of the stack to the other. In addition, the distribution of vesicles and vacuoles with respect to the stacks is often asymmetrical. It is frequently observed that the cisternae at one pole are highly flattened and that they become more distended, particularly at their margins, toward the other pole (Fig. 9–22). Cytochemical differences among cisternae within the stacks have also been found; those at one pole may contain a particular enzyme, whereas those at the opposite pole may lack the enzyme (Fig. 9–23). In actively secreting cells, the content of the Golgi cisternae sometimes changes, becoming progressively denser from one side of the stack to the other. Variations in the appearance of the membranes of Golgi cisternae according to their location in the stacks has also been described. The membranes

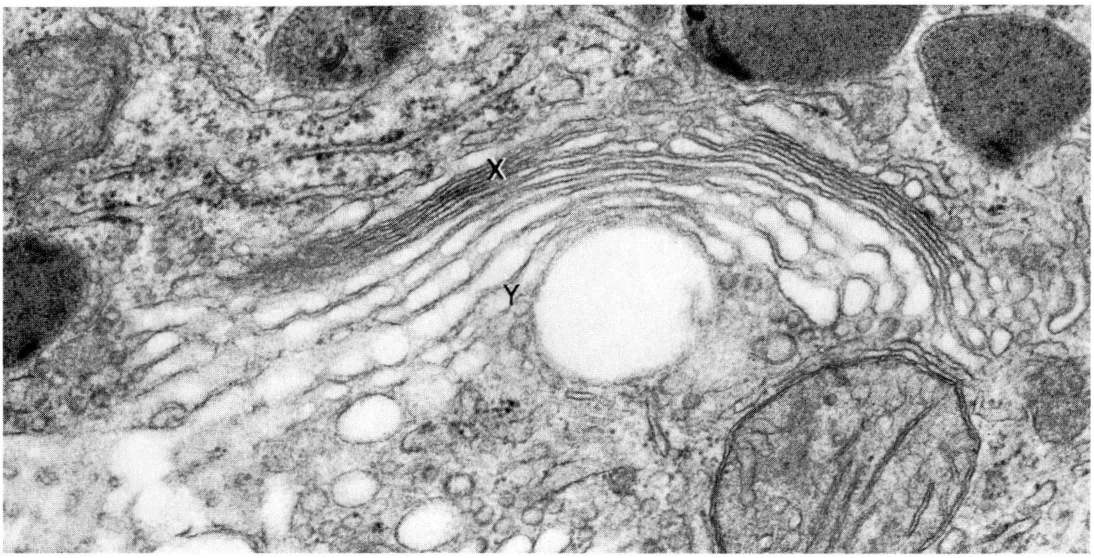

FIGURE 9–22 Rat epididymis. The cisternae of the Golgi apparatus differ in their width; at one pole of the stack (X) they are relatively narrow, but they become progressively more expanded toward the other pole (Y). × 48,000. (From Flickinger, C.: Anat. Rec., *163*:39, 1969. Reproduced by permission of the Wistar Institute Press.)

change progressively from a thin type, which is about 70 Å in thickness and is similar to the endoplasmic reticulum, to a thicker variety which measures about 100 Å and resembles the plasma membrane (Fig. 9–24).

These and similar observations on the polarity of the Golgi apparatus have led to a hypothesis about the transport of secretory products through the structure. One pole of the Golgi apparatus, which is more closely associated with the endoplasmic reticulum, has been called the forming face, input side, or cis-side. The opposite face, from which secretory material appears to leave the Golgi apparatus, has correspondingly been named the mature face, output side, or trans-side.

Transport To and From the Golgi Apparatus

In most protein-secreting cells, the secretory product is known to be transferred from the endoplasmic reticulum to the Golgi apparatus, where secretory vacuoles are formed. The transport of secretory proteins from the endoplasmic reticulum to the Golgi apparatus in small smooth vesicles is understood best in the exocrine cells of the pancreas. Since comparable structures are seen in other protein-secreting cells, a similar mechanism is presumed for them. The small smooth vesicles, about 400 Å in diameter, form from the surface of the endoplasmic reticulum through the production of small evaginations or "buds" that pinch off from the endoplasmic reticulum (Fig. 9–25). The elements of endoplasmic reticulum that participate in this activity have partly rough and partly smooth regions and are called transitional elements. Since the vesicles bud from the smooth areas, they lack ribosomes on their surfaces, but they enclose some of the content of the endoplasmic reticulum. Analysis of this process in slices of pancreas incubated *in vitro* has shown that transport from endoplasmic reticulum to Golgi apparatus requires energy derived from respiration, and it is thought that the membrane fission necessary to form vesicles is the energy-requiring step.

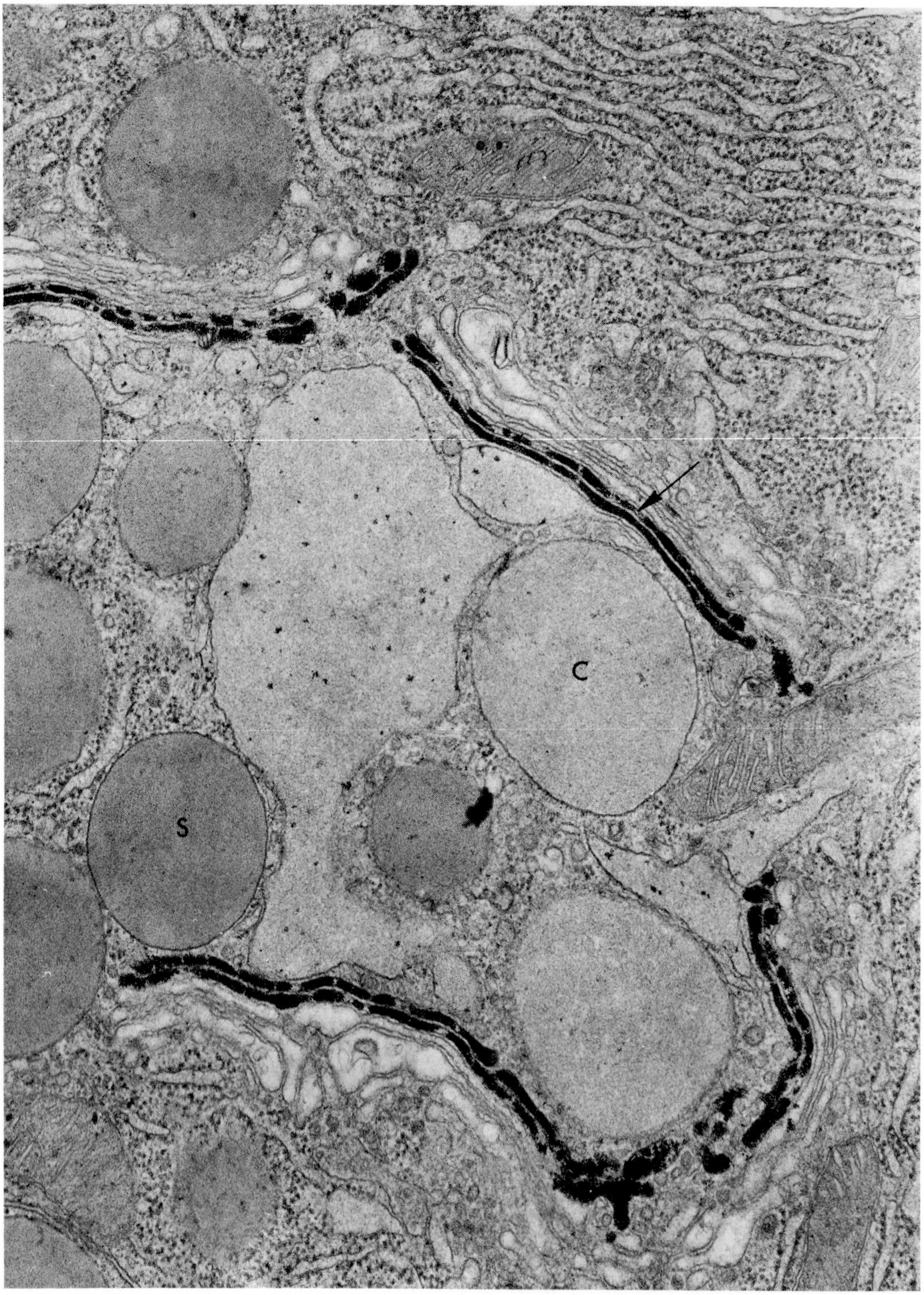

FIGURE 9–23 Golgi apparatus in a rat parotid acinar cell. Electron-dense reaction product for the enzyme thiamine pyrophosphatase is present in the inner Golgi cisternae (*arrow*) but is absent from the outer cisternae. Condensing vacuoles (C) and secretory granules (S) are also visible. × 41,000. (Micrograph courtesy of Hand, A. H.: Am. J. Anat., *130*:141, 1971. Reproduced by permission of the Wistar Institute Press.)

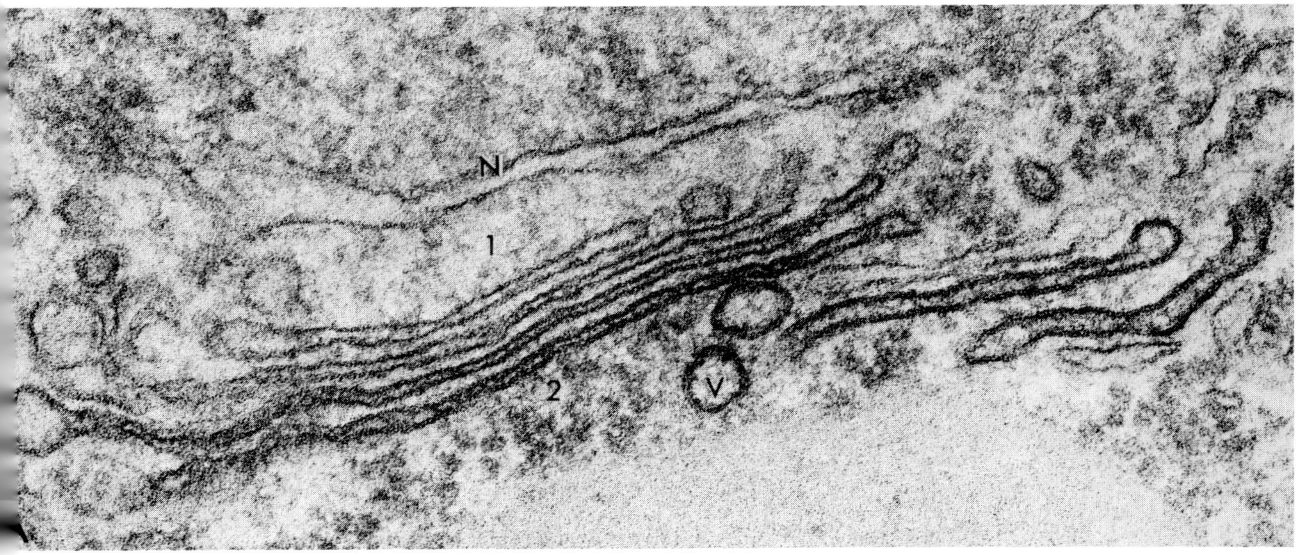

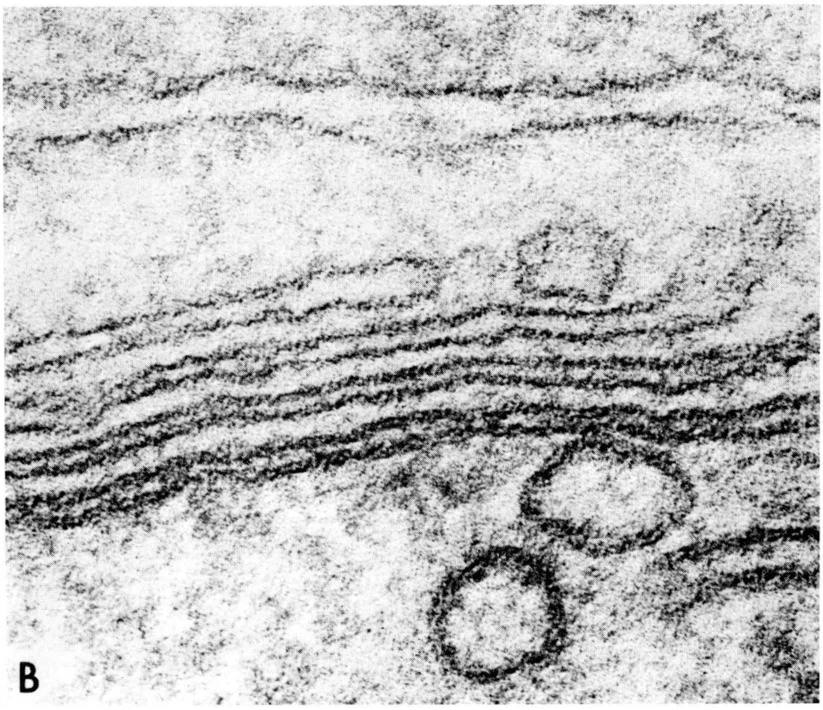

ɜURE 9–24 Golgi apparatus from hy-ae of a fungus. *A*. the cisternae at one ·e of the stack (1) resemble membranes :he nuclear envelope (N). the Golgi cis-nae increase in thickness and staining ensity toward the opposite pole (2), ere they resemble the membrane of se-tory vesicles (V). Part of the Golgi ap-atus is magnified in *B* to illustrate the mbrane changes in greater detail. *A*. 137,000. *B*. × 280,000. (Micrographs rtesy of Grove, S. N., Bracker, C. E., and ırré, D. J.: Science, *161*:171, 1968. ɔyright 1968 by The American Associ-ɔn for the Advancement of Science.

The small vesicles, each containing a small amount of secretory protein, migrate to the Golgi apparatus and empty their contents into it by fusing with a Golgi membrane. The exact Golgi element involved is not yet determined in all instances. In many actively secreting cells, such as goblet cells and the stimulated pancreas, Golgi cisternae appear to receive the product. In the unstimulated pancreas, the secretory product appears to be conveyed to Golgi vacuoles known as condensing vacuoles. In the hepatocyte, tubular connections between the endoplasmic reticulum and the Golgi apparatus have been implicated in the transport of very low density lipoproteins (VLDL's).

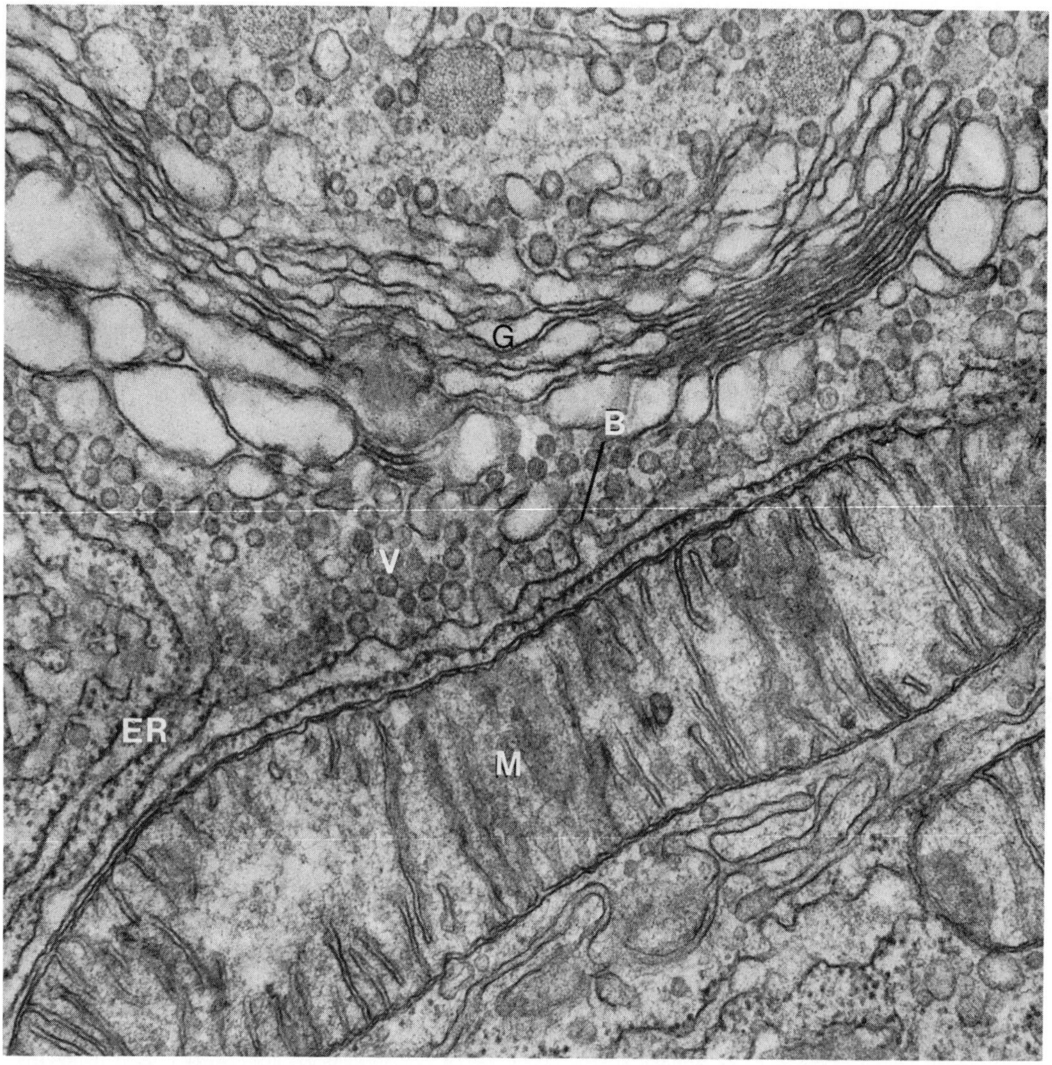

FIGURE 9–25 Portion of a Brunner's gland cell of the mouse duodenum. Cisternae of the rough endoplasmic reticulum (ER) generally are studded with ribosomes, but some portions facing the Golgi apparatus (G) have a smooth surface and "buds" (B) appear to form in these regions. Many smooth-surfaced vesicles (V), formed by budding from the endoplasmic reticulum, transport secretory material from the endoplasmic reticulum to the Golgi apparatus. M, mitochondrion. × 47,000. (Micrograph courtesy of Friend, D. S.: J. Cell Biol., *25*:563, 1965. Reproduced by permission of the Rockefeller University Press.)

Secretory vacuoles are formed in the Golgi apparatus (Figs. 9–20 and 9–23). This may occur by a "budding" or pinching-off of the expanded margin of a Golgi cisterna, or by a detachment of Golgi vacuoles from the apparatus. The secretory vacuoles generally move away from the Golgi apparatus toward the periphery of the cell, where they are retained until their content of secretory material is released by exocytosis. This consists of fusion of the membrane bounding the secretory vacuole with the plasma membrane (Fig. 9–26). As a result, the interior of the secretory vacuole becomes continuous with the extracellular space, and the content of the vacuole is released from the cell.

The content of secretory vacuoles may be released continuously, or it may be brought about or accelerated in response to a stimulus, such as a hormone. In some cells the stimulus may bring about an increase in the concentration of

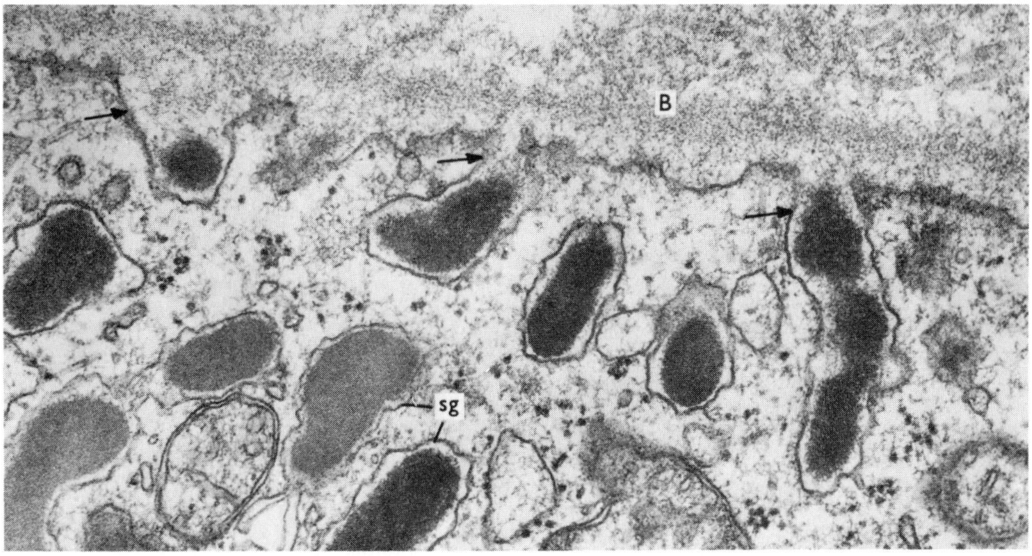

FIGURE 9–26 Part of a secretory cell (mammotroph) in the anterior pituitary gland of a lactating rat. Secretory vacuoles (sg) containing the dense secretory product are present in the peripheral cytoplasm. The membranes bounding several secretory vacuoles have fused with the plasma membrane (*arrows*), permitting discharge of the secretory product from the cell. B, basal lamina. × 45,000. (Micrograph courtesy of Farquhar, M. G.: Mem. Soc. Endocrinol., *19*:79, 1971. Reproduced by permission of the Cambridge University Press.)

cyclic AMP and in the concentration of calcium ions. It has been suggested that in the insulin-secreting cells of the pancreas, these events in turn activate a system of microtubules and filaments in the cytoplasm that convey secretory vacuoles to the cell surface and facilitate their release.

In the ciliated protozoan *Tetrahymena,* rings of particles termed fusion rosettes have been detected by freeze-cleave methods in the plasma membrane overlying secretory vacuoles (Fig. 9–27). It has been proposed that these specializations represent sites at which the secretory vacuole membrane can fuse with the plasma membrane when exocytosis occurs. In mammalian cells, however, the portions of the secretory vacuole and plasma membranes that fuse with one another appear to be devoid of intramembranous particles.

Functions

CONCENTRATION OF SECRETORY PRODUCTS

The Golgi apparatus has been widely implicated in the concentration of secretory proteins and in their packaging into secretory vacuoles, but a detailed understanding of the nature of the changes involved is incomplete. "Concentration" is usually understood to mean that the secretory material becomes visibly more dense and closely packed. For this reason, vacuoles at one pole of the Golgi apparatus, which contain secretory material of considerable density, have been referred to as condensing vacuoles (Figs. 9–20 and 9–23). The process of condensation of product in these vacuoles does not appear to involve an ion pump on the membrane of the vacuole. Instead, recent observations suggest that the condensation process involves a chemical change, with the synthesis of highly sulfated polyanions that bind ionically to secretory proteins and reduce their osmotic activity, resulting in the outflow of water from the condensing vacuoles.

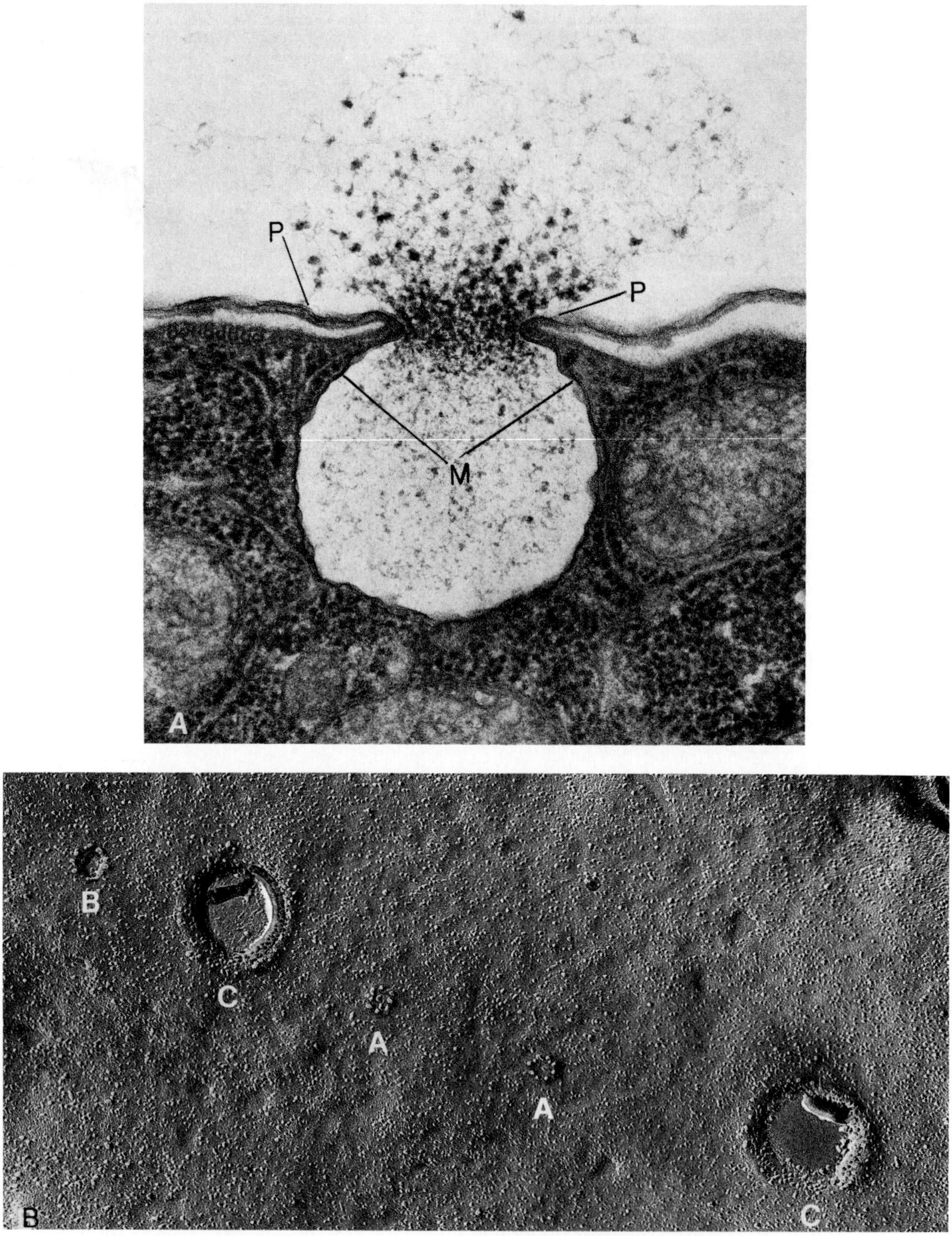

FIGURE 9–27 Steps in the discharge of mucocysts, secretory granules of the ciliate Tetrahymena. *A* is a thin section, showing fusion between the membrane of the mucocyst (M) and the plasma membrane (P), and release of the contents of the mucocyst from the cell. *B* is a freeze-cleave preparation of the plasma membrane (P face). Rosettes of particles (A) represent sites of plasma membrane specialization at which mucocysts will discharge. The membrane changes from A to B as depression develops within the rosette, and then to C as the mucocyst discharges. An array of particles is also present in the membrane of the mucocyst as it approaches the plasma membrane. The specialization of the mucocyst membrane and the rosettes of the plasma membrane are believed to represent matching sites of rearrangement of membrane components that facilitate fusion of the membranes and release of the secretion. *A*. × 68,000. *B*. × 72,000. (Micrographs courtesy of Satir, B., Schooley, C., and Satir, P.: Nature, *235*:53, 1972. Reproduced by permission of Macmillan Journals, Ltd.)

SYNTHESIS OF CARBOHYDRATES

An important advance in understanding the function of the Golgi apparatus was the realization that it contributes to the synthesis of the carbohydrate portions of glycoproteins. Some role for the Golgi apparatus in carbohydrate metabolism was suggested years ago by the observation that the Golgi apparatus in some cells stained with the PAS reaction, indicative of the presence of a carbohydrate-rich material. More recently it was shown by radioautography that sugar precursors of glycoproteins are incorporated in the Golgi apparatus. For example, glucose-^{3}H is incorporated into mucus synthesized by goblet cells of the intestine. If an animal is injected with glucose-^{3}H and radioautographs are prepared at intervals after the injection, the Golgi apparatus is labeled within a few minutes. Mucigen granules, which are the secretory vacuoles in these cells, are labeled 20 to 40 minutes after the injection, and after several hours radioactive mucus is released into the intestinal lumen.

In a similar way, using other radioactive sugars, the participation of the Golgi apparatus in the synthesis or assembly of the carbohydrate parts of glycoproteins produced by many other cells has been determined. Enzymes necessary for carbohydrate synthesis have now been detected in Golgi apparatus isolated from liver and other cells. These Golgi preparations are rich in glycosyl transferases, (Table 9–1*A*) enzymes that can add a sugar to a molecule and are known to be involved in the terminal synthesis of glycoproteins. Using radioactive sulfate, the Golgi apparatus has also been shown to have a role in the sulfation of certain cell products.

The synthesis of thyroglobulin by the follicular cells of the thyroid gland illustrates the roles of the endoplasmic reticulum and Golgi apparatus in secretion particularly well. Thyroglobulin is a glycoprotein composed of polypeptide subunits with carbohydrate side chains. The carbohydrate side chains contain several sugars. Mannose and N-acetylglucosamine are closer to the protein than is galactose, and the chain ends with a fucose or a sialic acid. If thyroid cells are labeled with different radioactive precursors of thyroglobulin, it can be shown that the protein part of the thyroglobulin molecule is synthesized in the rough endoplasmic reticulum. Before the thyroglobulin leaves the rough endoplasmic reticulum, mannose and glucosamine are added to the molecule. The addition of galactose, however, occurs only after the thyroglobulin has been transported to the Golgi apparatus, and carbohydrate chains are terminated in the Golgi apparatus by the addition of fucose. The thyroglobulin is later found in secretory vacuoles at the apical end of the cells and subsequently is discharged to the lumen of the follicle. The picture that emerges is that secretory products are transported through the cell, appearing sequentially in the endoplasmic reticulum, Golgi apparatus, and secretory vacuoles; during the transport, enzymes in the membranes of the different cell compartments act on the product to complete its synthesis.

A special case of the role of the Golgi apparatus in the synthesis of carbohydrate-rich materials is its participation in the production of glycoprotein of the cell coat. If it is well developed, the cell coat appears as a filamentous material on the external surface of the plasma membrane. A similar material can sometimes be visualized with the electron microscope in the Golgi apparatus, and studies with radioactive tracers have succeeded in labeling sequentially the Golgi apparatus and the cell surface coat.

PRODUCTION OF LYSOSOMES

Primary lysosomes are formed in the Golgi apparatus in at least some cells, as will be discussed further in Chapter 12. The acid hydrolases contained in

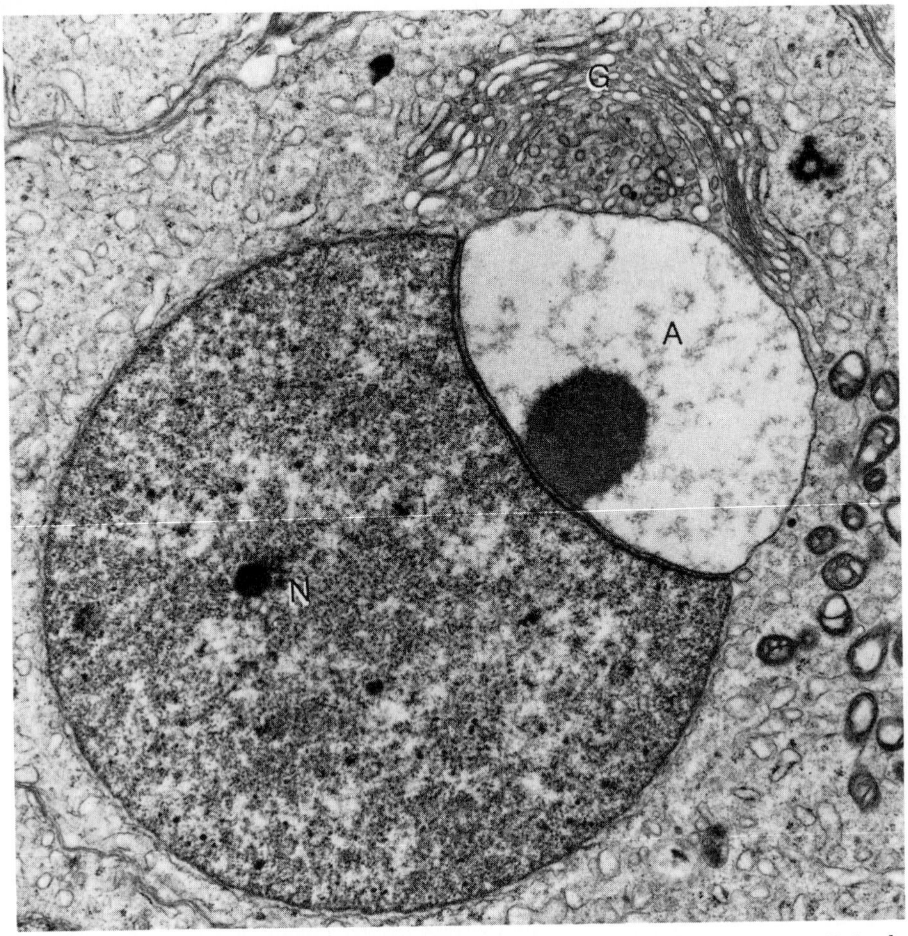

FIGURE 9–28 Spermatid of the rat testis. The product of the Golgi apparatus (G) in these developing germinal cells is the large acrosomal vacuole (A), which is applied to one pole of the nucleus (N). After further development, the acrosome of the mature sperm plays a role in penetration of the coats of the egg. × 14,000.

lysosomes are protein enzymes, and the formation of primary lysosomes may be considered analogous to the formation of secretory vacuoles. That is, the hydrolases are synthesized on ribosomes bound to the rough endoplasmic reticulum, transferred to the interior of the endoplasmic reticulum, and transported to the Golgi apparatus. In the Golgi apparatus, primary lysosomes in the form of hydrolase-containing vesicles or vacuoles are believed to be formed by budding from the margins of Golgi cisternae.

The Golgi apparatus also has a role in the formation of certain complex cell products, such as the acrosome of sperm (Fig 9–28), which contains enzymes that aid in penetration of the sperm through the coats of the egg. The acrosome can be considered a single large secretory vacuole, since it is formed in a way similar to the smaller secretory vacuoles of other cells and its contents are ultimately released from the cell.

GERL

In some cells a form of smooth endoplasmic reticulum located near the trans-side of the Golgi apparatus has been implicated in the formation of lysosomes and possibly secretory vacuoles as well. Because of its nature, location, and presumed function it has been termed GERL, which stands for Golgi, endoplasmic reticulum, and lysosome. The elements of smooth endoplasmic reticulum of the GERL complex have cisternal and tubular portions, and lysosomes or immature secretory vacuoles have been observed in continuity with GERL. The GERL stains for the presence of acid phosphatase, while nearby cisternae of the trans- side of the Golgi apparatus contain thiamine pyrophosphatase but not acid phosphatase. Some immature secretory vacuoles also contain acid phosphatase. If interpretations of the nature of the GERL are correct, lysosomes and possibly secretory vacuoles may be formed in at least some instances from the endoplasmic reticulum without the participation of the Golgi apparatus. In this case, the "packaging" function of the Golgi apparatus is assumed by the GERL complex.

THE GOLGI APPARATUS AND RELATIONSHIPS BETWEEN MEMBRANOUS ORGANELLES

The pathway of secretory proteins and glycoproteins through gland cells in which secretory vacuoles are formed is well established. Radioautographic and cell fractionation methods have convincingly demonstrated transport of *secretory material* from the endoplasmic reticulum to the Golgi apparatus to the secretory vacuole and ultimately to the outside of the cell. The behavior of the *membranes*

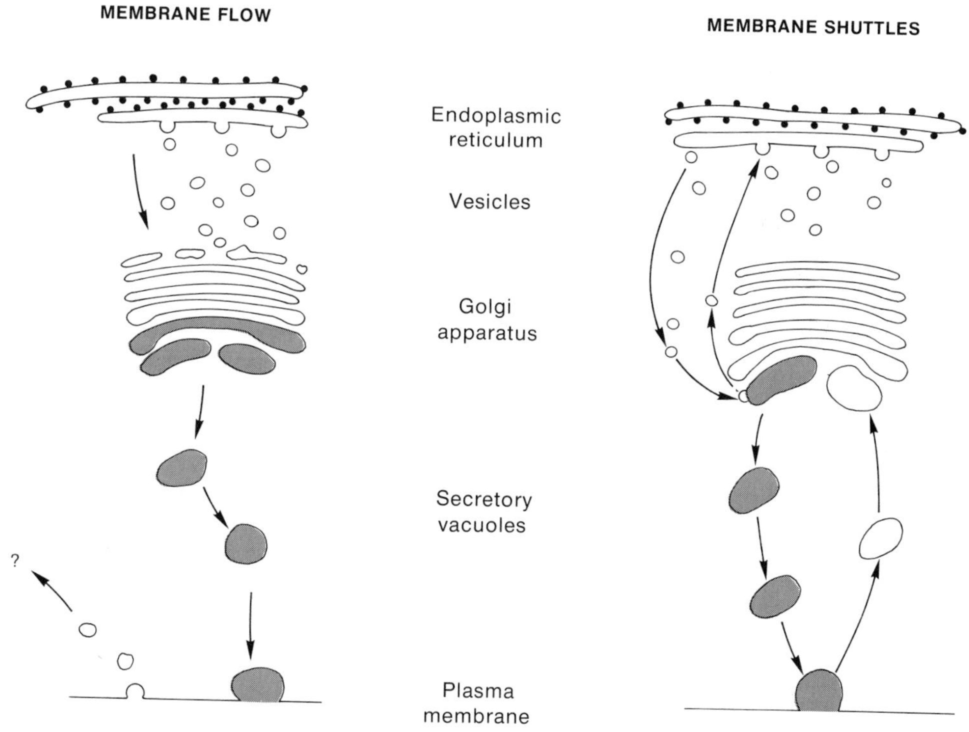

FIGURE 9–29 Two hypotheses for the behavior of membranes during the secretory process. According to the membrane flow hypothesis, membranes move through the cell from the endoplasmic reticulum to the Golgi apparatus, secretory vacuoles, and plasma membrane. In the membrane shuttle scheme, vesicles are thought to shuttle between the endoplasmic reticulum and the Golgi apparatus, while secretory vacuoles move back and forth between the Golgi apparatus and the plasma membrane.

of the organelles that enclose the secretory products, however, is uncertain and is the subject of current investigation. The two main hypotheses are those of "membrane flow" and "membrane shuttle" (Fig. 9–29).

According to the hypothesis of membrane flow, small vesicles bud from the surface of the endoplasmic reticulum and move to the Golgi region. At the forming face of the Golgi apparatus the vesicles fuse with one another to form a cisterna of the Golgi apparatus. Golgi cisternae are continually being formed in this way, and they move in a train from the forming face to the mature face of the Golgi apparatus. During this movement the membranes are thought to undergo maturation from an endoplasmic reticulum type to a plasma membrane type. On arriving at the mature face, each cisterna detaches from the Golgi apparatus as the membrane around one or more secretory vacuoles. The secretory vacuoles migrate to the margin of the cell, fuse with the plasma membrane, and release their content of secretory material. Since repetition of this process would in time lead to an increase in the area of the plasma membrane, some of the plasma membrane is thought to be internalized by endocytosis and degraded, possibly by digestion in lysosomes.

According to the membrane flow concept, the secretory product and the membranes could move simultaneously from the endoplasmic reticulum to the Golgi apparatus as vesicles, through the Golgi apparatus in a train of cisternae, and to the surface as secretory vacuoles. Alternatively, membranes might move through the stacks, while the secretory product could bypass the stacks and go directly from the endoplasmic reticulum to condensing vacuoles or to the expanded margins of cisternae at the mature face of the Golgi apparatus. In any event, the important points of the membrane flow hypothesis are that membranes move from endoplasmic reticulum to Golgi apparatus to plasma membrane, undergoing modification in the process, and that the Golgi apparatus is continually turning over, with membrane input from the endoplasmic reticulum balancing output as membrane surrounding secretory vacuoles. Proponents of this view emphasize the progressive morphological changes in membrane appearance from an endoplasmic reticulum type to a plasma membrane type from one pole of the Golgi stacks to the other (Fig. 9–24). They further observe that the composition of isolated Golgi membranes is intermediate between that of the endoplasmic reticulum and the plasma membrane (Table 9–1).

In contrast to this, supporters of the membrane shuttle hypothesis emphasize the fact that the composition of membranes of the endoplasmic reticulum, Golgi apparatus, secretory vacuoles, and plasma membrane differ. It is maintained, therefore, that these classes of membranes do not mix with one another. Transport of secretory material from the endoplasmic reticulum to the Golgi apparatus is visualized as occurring by means of small vesicles that bud from the endoplasmic reticulum, move to the Golgi region, and then fuse with a Golgi vacuole or cisterna. However, the visicle membrane is then thought to detach from the Golgi apparatus and move back to the endoplasmic reticulum where the process is begun anew. Thus the transport vesicles would shuttle back and forth between the endoplasmic reticulum and the Golgi apparatus, carrying secretory proteins but always returning to the endoplasmic reticulum and never mixing with Golgi membranes. A similar shuttle mechanism is thought to operate in the transport of secretory material in secretory vacuoles from the Golgi apparatus to the plasma membrane. Secretory vacuoles would move from the Golgi apparatus and fuse with the plasma membrane to release their contents to the outside. However, they are then thought to invaginate, detach from the plasma membrane, and return to the Golgi apparatus. Thus secretory vacuole membranes would not mix with plasma membranes. Proponents of this view emphasize the differences between the membranes of the various organelles, and point to reports that proteins of intracellular membranes turn over much more slowly than do secretory proteins, suggesting that membranes are reutilized, as demanded by the shuttle hypothesis.

SMOOTH ENDOPLASMIC RETICULUM

The smooth endoplasmic reticulum consists of smooth-surfaced membranes which are distinguished from those of the rough endoplasmic reticulum by the

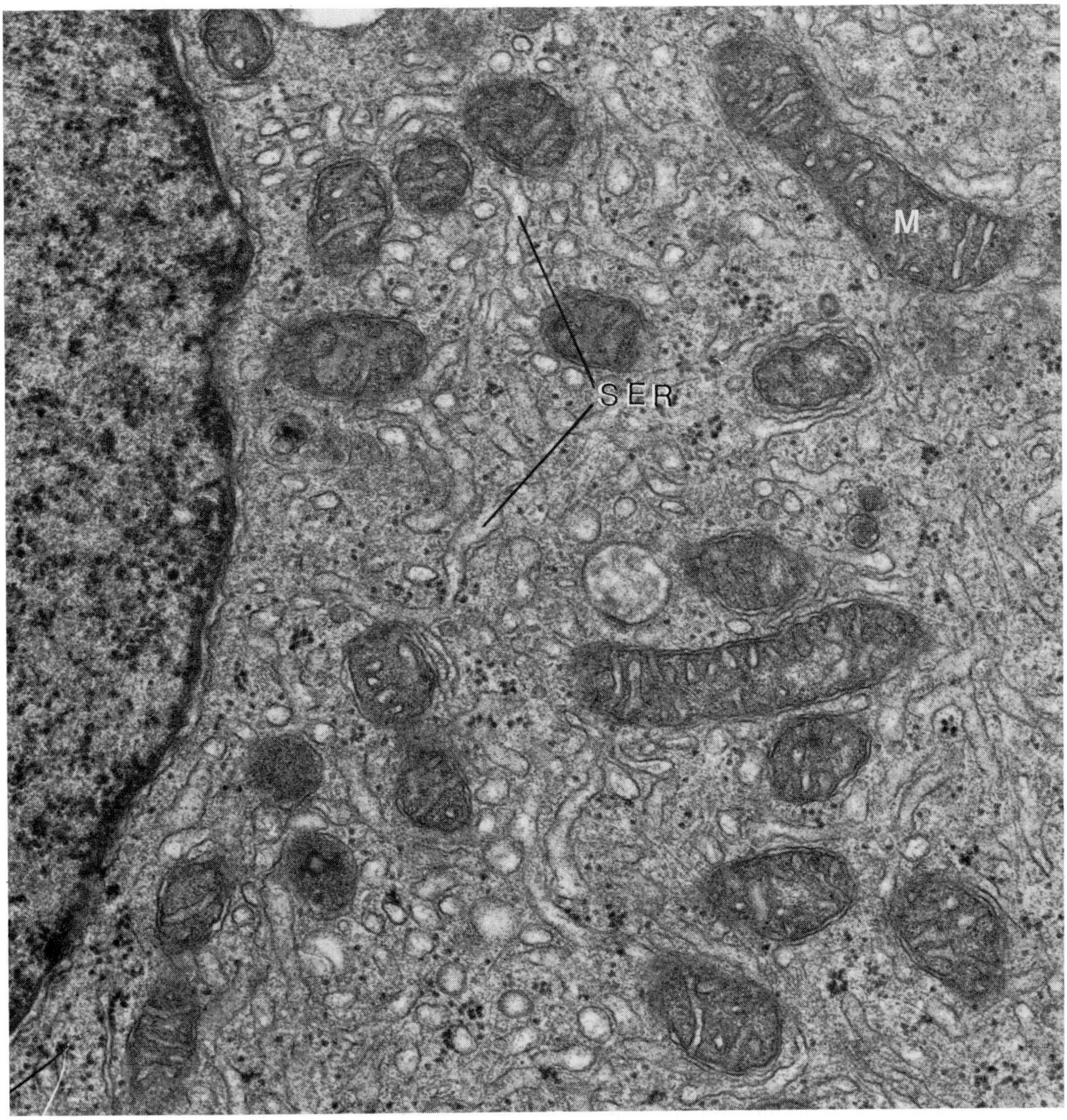

FIGURE 9–30 Part of a Leydig cell from the interstitial tissue of the human testis. Much of the cytoplasm is occupied by interconnecting tubules of smooth endoplasmic reticulum (SER). The smooth reticulum and mitochondria (M) collaborate in the synthesis of testosterone. × 24,000. (Micrograph courtesy of A. K. Christensen. *From* Handbook of Physiology, Sec. 7, Vol. 5, pp. 57–94. Reproduced by permission of the American Physiological Society, Washington, D.C.)

absence of ribosomes. The smooth endoplasmic reticulum is usually tubular or vesicular rather than cisternal in form. The common tubular variety (Fig. 9–30) is made up of a network of interconnecting channels of roughly circular cross section measuring about 1000 Å in diameter. The amount of smooth endoplasmic reticulum varies greatly with the cell type. In some cells, such as those engaged in steroid hormone biosynthesis, it is the predominant organelle, occupying much

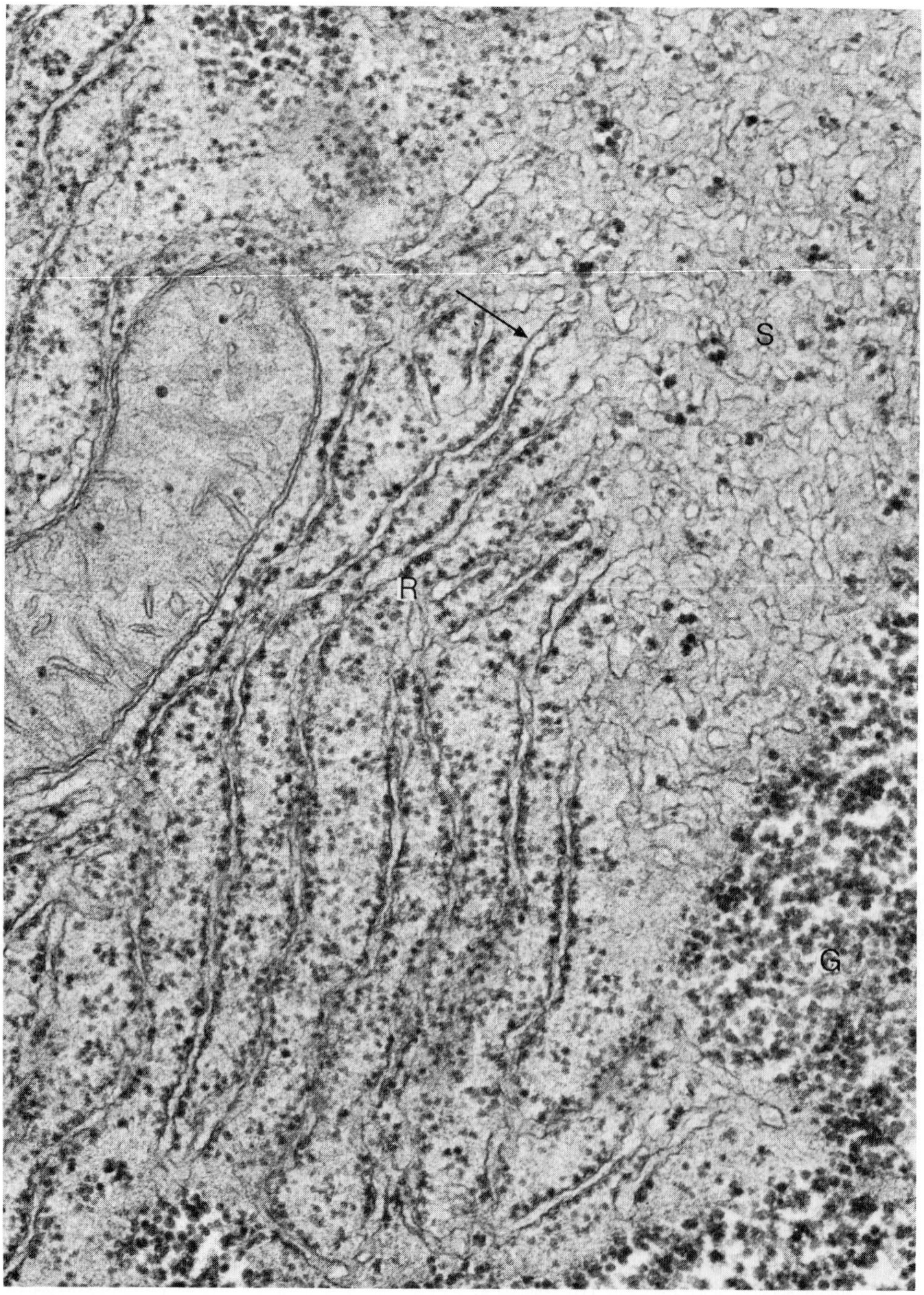

FIGURE 9–31 Cytoplasm of a rat liver cell. Note connection (*arrow*) between the rough (R) and the smooth (S) varieties of endoplasmic reticulum. G, glycogen. × 54,000. (Micrograph courtesy of R. R. Cardell.)

of the cytoplasm. In others, such as the exocrine gland cells, smooth endoplasmic reticulum is very sparse and may be virtually undetectable in electron micrographs.

The tubules of smooth endoplasmic reticulum are often observed to be connected to cisternae of the rough endoplasmic reticulum (Fig. 9–31). The presence of these connections led to the suggestion that the smooth endoplasmic reticulum is derived from the rough endoplasmic reticulum. This idea has been supported by the observation that when the smooth endoplasmic reticulum of hepatocytes is proliferating during development and in response to certain drugs, certain enzymes appear sequentially in the rough endoplasmic reticulum and in the smooth endoplasmic reticulum. This suggests that smooth endoplasmic reticulum proteins are made in the rough endoplasmic reticulum. Some recent studies, however, implicate the smooth endoplasmic reticulum itself in synthesis of phospholipids, which may include those of the lipid part of the smooth endoplasmic reticulum membranes.

In contrast to the rough endoplasmic reticulum, which has only the single major function of secretory protein synthesis, the smooth endoplasmic reticulum is known to have a great variety of functions in different cells. At first, this diversity of function seems puzzling, because the form of the organelle is similar in different cells. In an attempt to unify ideas about the activities of the smooth endoplasmic reticulum, it has been proposed that the smooth endoplasmic reticulum represents a means for the display of membrane-bound enzymes in the cytoplasm. The enzymes are part of the membranes, and since the surface area of the smooth endoplasmic reticulum membranes can be large, many enzyme molecules are placed in contact with the cytoplasmic matrix. The function of the smooth endoplasmic reticulum can vary with the nature of the enzymes that it contains, and it can be tailored to the functional role of a given cell type. Furthermore, since the functional demands of a given cell type may vary from time to time, the amount of smooth endoplasmic reticulum and its enzymatic components may be expected to change, enabling the cell to discharge its functions more efficiently under different conditions.

It is not possible to study all the functions of the smooth endoplasmic reticulum in detail, but a few of the best understood examples will be described briefly in order to illustrate the principles of smooth endoplasmic reticulum activity.

The smooth endoplasmic reticulum is known to play a role in the biosynthesis of steroid hormones. It is very abundant in cells such as the Leydig cells of the testis which synthesize testosterone (Fig. 9–30), in cells of the adrenal cortex which produce corticosteroids, and in cells of the corpus luteum of the ovary which synthesize progesterone. The cytoplasm of steroid-producing cells in most mammals is occupied mainly by an extensive network of smooth endoplasmic reticulum and numerous large mitochondria. Of the enzymes catalyzing the various steps in steroid biosynthesis, some are found in the microsomal fraction (which is derived from smooth endoplasmic reticulum) and others are in the mitochondria. Thus the smooth endoplasmic reticulum and the mitochondria, which are closely related topographically in the cell, appear to collaborate in the synthesis of steroid hormones.

Among the functions of the smooth endoplasmic reticulum in hepatocytes is the metabolism of lipid-soluble drugs, such as the barbiturates. This is carried out by a group of oxidative enzymes, which hydroxylate the drugs and render them water-soluble. Administration of phenobarbital to an experimental animal results within a few days in proliferation of large amounts of smooth endoplasmic reticulum in the hepatocytes. The newly formed smooth endoplasmic reticulum membranes are rich in the drug-hydroxylating enzymes. Thus the liver cell appears to respond to the presence of the drug by the manufacture of increased amounts of smooth endoplasmic reticulum, which contain a high proportion of drug-metabolizing enzymes. This induction of smooth endoplasmic reticulum by drugs can be of clinical importance. For example, it may explain the well-known phenomenon of tolerance to barbiturates. Progressively higher doses of the drug are required over time to produce the same effect; this may be caused by the increasing effectiveness of the drug-metabolizing system in the smooth endoplasmic reticulum. In addition, if two drugs are both metabolized by liver smooth endoplasmic reticulum, it can readily be appreciated that administration of one drug may induce smooth endoplasic reticulum proliferation and alter both the rate of metabolism and the dose required of the second drug.

Similarly, under the physiological condition of fasting, smooth endoplasmic reticulum proliferation is observed in hepatocytes, and the smooth endoplasmic reticulum is found in parts of the cytoplasm that contain glycogen. The smooth endoplasmic reticulum membranes contain the enzyme glucose-6-phosphatase, which plays a role in glycogen utilization. Thus the proliferation of smooth endoplasmic reticulum in fasting animals is thought to aid in increasing glycogen breakdown to glucose, as needed by the organism.

Another example of the functioning of the smooth endoplasmic reticulum concerns the transport of fat by the columnar absorptive cells of the intestinal epithelium. After a meal containing fat, triglycerides are broken down to monoglycerides and fatty acids by enzymes in the intestinal lumen. These molecules diffuse across the apical plasma membrane of the absorptive cells. They are then resynthesized into triglycerides before their transport out of the cells to the intestinal lymphatics. This resynthesis of triglycerides takes place in smooth endoplasmic reticulum in the apical parts of the absorptive cells, utilizing the enzyme fatty acid CoA ligase of the smooth endoplasmic reticulum. Thus, droplets of resynthesized lipids are visible in the smooth endoplasmic reticulum (Fig. 9–32). Furthermore, proliferation of smooth endoplasmic reticulum can be induced by the feeding of fat. Once again, the proliferation of smooth endoplasmic reticulum is a response by which the particular function of a cell is enhanced through the increased action of the necessary enzymes.

CYTOPLASMIC MEMBRANES AND DISEASE

The endoplasmic reticulum, Golgi apparatus, and other intracellular membranes can be involved in disease processes in many ways. Since they are essential elements in the secretory process both in exocrine protein-secreting glands and in endocrine glands that secrete either proteins or steroid hormones, alterations in the structure and function of the endoplasmic reticulum and the Golgi apparatus may be anticipated in a wide variety of glandular disorders. There may be cases of hypersecretion in which unusually large development and activity of these organelles occur, or instances of hyposecretion in which the organelles are smaller than normal and show indications of less than normal activity. Hyperthyroidism is an example of oversecretion of a hormone that has severe clinical effects on the rest of the body. Conversely, in juvenile diabetes, there may be insufficient production of insulin by the cells of the islets of Langerhans in the pancreas, with a large number of deleterious changes due to impaired entrance of glucose into cells. This is not to say that the primary defect in secretory disorders is necessarily in the cytoplasmic membranes. The alterations in the

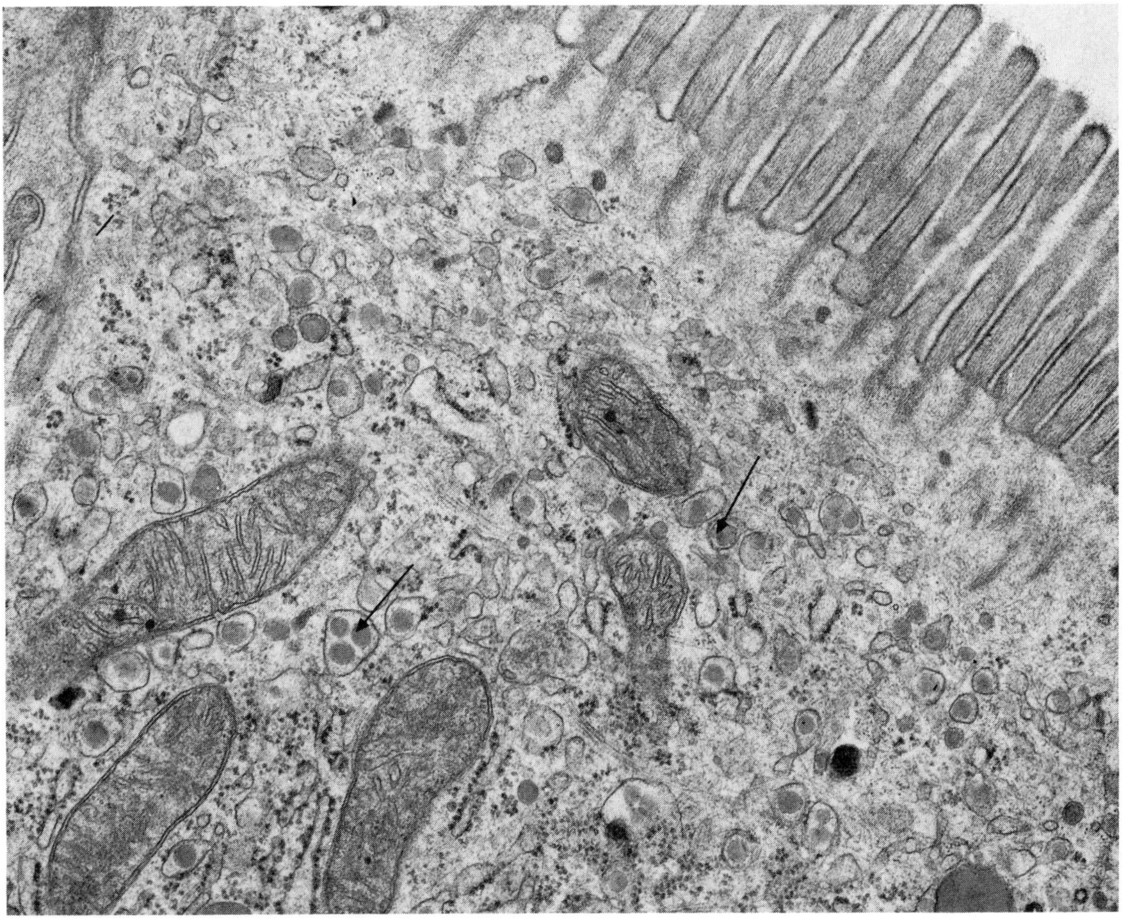

FIGURE 9–32 Apical part of an intestinal epithelial cell of the rat after a meal of fat (corn oil). Droplets of lipid (*arrows*) are present in the smooth endoplasmic reticulum following digestion, absorption, and resynthesis of the lipid. × 74,000. (Micrograph courtesy of Cardell, R. R., Badenhausen, S., and Porter, K. R.: J. Cell Biol., *34*:123, 1967. Reproduced by permission of the Rockefeller University Press.)

membranous organelles can occur in response to changes in other parts of the cell and still be instrumental in pathogenesis. Changes in nuclear activity, of course, often result in profound changes in other parts of the cell.

While alterations in the endoplasmic reticulum and Golgi apparatus are related to the secretion of various products including hormones, conversely, hormones have dramatic effects on the endoplasmic reticulum and Golgi apparatus in target cells for the hormone. For example, the development and maintenance of the cells in the male sex accessory glands are dependent on testosterone, produced by the Leydig cells of the testis. After castration, the endoplasmic reticulum and Golgi apparatus of the epithelium of the prostate gland and the seminal vesicle decline greatly. The rough endoplasmic reticulum is reduced from parallel cisternae that fill much of the cytoplasm to only a few scattered small elements. The Golgi apparatus also decreases several fold in size. Secretory vacuoles in both glands are normally very numerous, but following removal of the testes they diminish to the point of being absent from most of the cells. Similar effects can be obtained by administration of an antiandrogen such as cyproterone acetate, which is a drug that interferes with the action of testosterone by competing with it for hormone receptors in target cells. At first, alterations in the endoplasmic reticulum and Golgi apparatus following castration or antiandrogen administration might not seem relevant to clinical medicine because of the drastic nature of a procedure such as castration. Nevertheless, castration and treatment with estrogens or antiandrogens have been widely used in the therapy of disseminated cancer of the prostate gland. In a significant proportion of cases, the malignant cells derived from the prostrate retain a sensitivity to testosterone; thus a regression of the tumor can often be attained by reducing their stimulation by testosterone. This procedure has been important in reducing pain and increasing the length of survival in many patients.

A further example of a possible role of the Golgi apparatus in a disease process comes from study of the skin disease psoriasis, in which rough, raised, plaquelike lesions are distributed on the trunk and extremities. When examined with the electron microscope, an unusually large amount of a polysaccharide-rich material was observed between the cells of the epidermis, and the Golgi apparatus within the cells was distended with a similar polysaccharide-rich substance. It has been speculated that abnormal amounts of this material might prevent normal desquamation of the epidermal cells, leading to development of the lesions. This might in turn be the result of overproduction of the polysaccharide by the Golgi apparatus.

The liver normally synthesizes many proteins, including the plasma proteins albumin and fibrinogen. When liver cells are damaged owing to a wide variety of causes, including alcoholism and carbon tetrachloride poisoning, there is a dispersion of ribosomes from polysomes and a dilatation of elements of the endoplasmic reticulum and Golgi apparatus. The structural changes in these organelles are associated with diminished secretion of proteins. This deficiency is manifested clinically by a lowered concentration of serum albumin, which can lead to edema because of loss of fluid from the vascular system. Diminished protein secretion may also result in defects in the clotting of blood.

FURTHER READING

Beams, H. W., and Kessel, R. G.: The Golgi apparatus: structure and function. Int. Rev. Cytol., *23*: 209, 1968.

Christensen, A. K., and Gillim, S. W.: The correlation of fine structure and function in steroid secreting cells with emphasis on those of the gonads. *In*: McKerns, K. W. (ed.): The Gonads. New York, Appleton-Century-Crofts, 1969.

Dallner, G., and Ernster, L.: Subfractionation and composition of microsomal membranes: a review J. Histochem. Cytochem., *16*:611, 1968.

Golgi, C.: Sur la structure des cellules nerveuses des ganglions spinaux. Arch. Ital. Biol., *30*:278, 1898.

Hand, A. R., and Oliver, C.: Cytochemical studies of GERL and its role in secretory granule formation in exocrine cells. Histochem. *9*:375, 1977.

Holtzman, E., Schacher, S., Evans, J., et al.: Origin and fate of membranes of secretion granules and synaptic vesicles—membrane circulation in neurons, gland cells and retinal photoreceptors. *In*: Poste, G., and Nicolson, G. L. (eds.): Synthesis, Assembly and Turnover of Cell Surface Components. New York, Elsevier, North-Holland, 1977.

Jamieson, J. D., and Palade, G. E.: Intracellular transport of secretory proteins in the pancreatic exocrine cell. I. Role of the peripheral elements of the Golgi complex. J. Cell Biol., *34*:577, 1967. II. Transport to condensing vacuoles and zymogen granules. J. Cell Biol., *34*:597, 1967. III. Dissociation of intracellular transport from protein synthesis. J. Cell Biol., *39*:580, 1968. IV. Metabolic requirements. J. Cell Biol., *39*:589, 1968.

Kessel, R. G. Annulate lamellae. J. Ultrastruct. Res. (Suppl.), *10*:1, 1968.

Meldolesi, J.: Secretory mechanisms in pancreatic acinar cells. Role of the cytoplasmic membranes. Adv. Cytopharm., *2*:71, 1974.

Mollenhauer, H. H., and Whaley, W. G.: An observation on the functioning of the Golgi apparatus. J. Cell Biol., *17*:222, 1963.
Morré, D. J., and Ovtracht, L.: Dynamics of the Golgi apparatus: membrane differentiation and membrane flow. Int. Rev. Cytol. (Suppl.), *5*:61, 1977.
Morré, D. J., Mollenhauer, H. H., and Bracker, C. E.: Origin and continuity of Golgi apparatus. *In*: Reinert, J., and Ursprung, H. (eds.): Results and Problems in Cell Differentiation. Vol. II: Origin and Continuity of Cell Organelles. New York, Springer-Verlag, 1971, pp. 82–126.
Neutra, M., and Leblond, C. P.: Synthesis of the carbohydrate of mucus in the Golgi complex as shown by electron microscope autoradiography of goblet cells from rats injected with glucose-H^3. J. Cell Biol., *30*:119, 1966.
Neutra, M., and Leblond, C. P.: The Golgi apparatus. Sci. Am., *220*:100, 1969.
Novikoff, A. B.: The endoplasmic reticulum: a cytochemist's view (a review). Proc. Natl. Acad. Sci. U.S.A., *73*:2781, 1976.
Palade, G.: Intracellular aspects of protein synthesis. Science, *189*:347, 1975.
Porter, K. R.: Observations on a submicroscopic basophilic component of the cytoplasm. J. Exp. Med., *97*:727, 1953.
Porter, K. R.: The ground substance; observations from electron microscopy. *In*: Brachet, J., and Mirsky, A. E. (eds.): The Cell. Vol. II. New York, Academic Press, pp. 621–675.
Remmer, H., and Merker, H. J.: Effect of drugs on the formation of smooth endoplasmic reticulum and drug metabolizing enzymes. Ann. N.Y. Acad. Sci., *123*:79, 1965.
Satir, B.: The final steps in secretion. Sci. Am., *233*:29, 1975.
Siekevitz, P., and Palade, G. E.: A cytochemical study on the pancreas of the guinea pig. I. Isolation and enzymatic activities of the cell fractions. II. Functional variations in the enzymatic activity of microsomes. III. *In vivo* incorporation of leucine-1-C^{14} into the proteins of cell functions. J. Biophys. Biochem. Cytol., *4*:203, 309, 557, 1958.
Whaley, W. G.: The Golgi Apparatus. New York, Springer-Verlag, 1975.

10

GLYCOLYSIS AND ENERGY PRODUCTION

Examination of the intact cell, the cellular organelles, and the macromolecules that make up the organelles reveals a high degree of organization at every level. The cell can attain and maintain this high degree of order only through the constant expenditure of energy, which it obtains from its surroundings. If a cell is deprived of its sources of energy, it spontaneously proceeds to a state of disorder and death. A living cell is not a system at equilibrium; it is a steady-state system displaced from the point of equilibrium, and a continuous flow of energy through the system is required to maintain the steady-state position characteristic of life. Therefore, energy is of prime importance to all living cells.

Energy is the capacity to do work. The immediate source of energy for animal cells is chemical energy in the form of ingested foodstuffs. In aerobic animal cells these ingested foodstuffs are ultimately oxidized to CO_2 and H_2O by the oxygen-requiring process of respiration, and a part of the energy made available is utilized to do the work of creating and maintaining order. The ultimate source of the foodstuffs utilized by animal cells is the plant world. Plant cells utilize the radiant energy of the sun and by photosynthesis convert CO_2 and H_2O to O_2, carbohydrates, and other cellular constituents which are eventually used by animal cells as foodstuffs.

The realization of work by the expenditure of energy requires a device known as an engine. The most familiar engines are *heat* engines, such as the steam engine and the internal combustion engine. Heat engines all require a temperature differential between the intake and the exhaust. Since the cell is an isothermal device possessing no significant temperature differentials, the cell cannot be a heat engine. Instead, the cell is an example of a *chemical* engine that utilizes the *free energy* of chemical reactions to do work.

FREE ENERGY

In a system at constant temperature and pressure, the free energy is the maximum net energy available to do useful work. Hence the free energy of a chemical reaction is the maximum energy available from the reaction that the cell, as a chemical engine, can utilize to do useful work. Consider the following chemical reaction:

$$A + B \rightleftharpoons C + D \tag{1}$$

The free energy of compound A, designated as G_A, is defined as

$$G_A = G_A^{\circ} + RT \ln a_A \tag{2}$$

where G°_A is the standard free energy of compound A at temperature T, that is, the free energy of compound A at unit activity; R is the gas constant (1.987 cal/deg/mole); T is the temperature in degrees Kelvin; and a_A is the activity or "effective" concentration of compound A. Employing the approximation that the activity of a compound is equal to its concentration, equation 2 becomes

$$G_A = G^{\circ}_A + RT \ln [A] \tag{3}$$

Similarly, the free energies of compounds B, C, and D are defined as

$$G_B = G^{\circ}_B + RT \ln [B] \tag{4}$$

$$G_C = G^{\circ}_C + RT \ln [C] \tag{5}$$

$$G_D = G^{\circ}_D + RT \ln [D] \tag{6}$$

The free energy available from a chemical reaction is the free energy change, ΔG, that occurs when reactants undergo a conversion to products. Since energy is a state variable, the energy change between two states is independent of the path taken between those states. Hence, ΔG, the free energy change of the reaction, is given by

$$\Delta G = G_{products} - G_{reactants} = G_C + G_D - G_A - G_B \tag{7}$$

or

$$\Delta G = (G^{\circ}_C + RT \ln [C]) + (G^{\circ}_D + RT \ln [D]) - (G^{\circ}_A + RT \ln [A]) - (G^{\circ}_B + RT \ln [B]) \tag{8}$$

Therefore

$$\Delta G = G^{\circ}_C + G^{\circ}_D - G^{\circ}_A - G^{\circ}_B + RT \ln \frac{[C]\,[D]}{[A]\,[B]} \tag{9}$$

Since $G^{\circ}_C + G^{\circ}_D - G^{\circ}_A - G^{\circ}_B$ is the free energy change of the reaction in the standard state, that is, the free energy change when reactants at unit activity form products at unit activity, this term is designated ΔG°, the standard free energy change of the reaction. Hence, ΔG is given by equation 10

$$\Delta G = \Delta G^{\circ} + RT \ln \frac{[C]\,[D]}{[A]\,[B]} \tag{10}$$

A reaction system at equilibrium will not spontaneously undergo a change; hence, no work can be obtained from a system at equilibrium, and ΔG must equal zero. Therefore, at equilibrium, equation 10 becomes

$$\Delta G = 0 = \Delta G^\circ + RT \ln \frac{[C]_{eq} [D]_{eq}}{[A]_{eq} [B]_{eq}} \tag{11}$$

$$\Delta G^\circ = -RT \ln \frac{[C]_{eq} [D]_{eq}}{[A]_{eq} [B]_{eq}} \tag{12}$$

where $[X]_{eq}$ is the concentration of compound X at equilibrium. However, since

$$\frac{[C]_{eq} [D]_{eq}}{[A]_{eq} [B]_{eq}} = K$$

where *K* is the equilibrium constant for the reaction, equation 12 becomes

$$\Delta G^\circ = -RT \ln K \tag{13}$$

Hence, ΔG°, the standard free energy change of a reaction at a given temperature, can be calculated from the equilibrium constant for the reaction at that temperature, as shown in Table 10–1.

For a more general reaction, such as

$$\alpha A + \beta B \rightleftharpoons \gamma C + \delta D$$

the free energy change is given by

$$\Delta G = \Delta G^\circ + RT \ln \frac{[C]^\gamma [D]^\delta}{[A]^\alpha [B]^\beta}$$

Setting the activities, or concentrations, of all reactants at unity, we see that ΔG° is the free energy available when α moles of A and β moles of B are converted to γ moles of C and δ moles of D with the activities, or concentrations, of reactants and products maintained at unity. While ΔG° is a constant for any given reaction at a given temperature, the value of the actual free energy change for

TABLE 10–1 ΔG° CALCULATED FROM EQUILIBRIUM CONSTANTS AT 25°C

K	ΔG°, Calories
10,000	−5454
1000	−4090
100	−2727
10	−1363
1	0
0.1	+1363

the reaction, ΔG, will be dependent upon the activities, or concentrations, of both products and reactants.

Since many reactions of biochemical interest occur at or near pH 7, the standard free energy change at pH 7.0, designated $\Delta G^{\circ\prime}$, is often encountered in the biochemical literature. Also, if the solvent H_2O is a reactant or product, its activity is arbitrarily taken as unity.

As stated earlier, if $\Delta G = 0$, the reaction is at equilibrium. If the sign of ΔG is negative, the reaction is spontaneous and can be made to do work. A reaction with a negative ΔG is known as an exergonic reaction. If the sign of ΔG is positive, the reaction will not proceed spontaneously in the direction indicated. A reaction with a positive ΔG is designated as an endergonic reaction. Consider the reaction

$$A + B \rightarrow C + D \quad \Delta G = +5 \text{ kcal (nonspontaneous)} \tag{14}$$

Since ΔG is positive, reaction 14 is nonspontaneous and cannot be made to do work. In fact, reaction 14 requires that work be done upon it if it is to proceed to the right. However, such a reaction will proceed spontaneously in the opposite or reverse direction, as written in equation 15, and can be made to do work.

$$C + D \rightarrow A + B \quad \Delta G = -5 \text{ kcal (spontaneous)} \tag{15}$$

Note that it is the sign of ΔG (the free energy change for the reaction), and *not* $\Delta G^{\circ\prime}$ (the standard free energy change, pH 7), that determines whether a reaction will proceed spontaneously.

COUPLED REACTIONS

Being a chemical engine, the cell must use the energy derived from the exergonic reactions that take place during the metabolism of foodstuffs to do the work essential to its existence. This work includes synthetic work, mechanical work, electrical work, regulatory work, and osmotic work. The one way in which chemical energy can be transferred from one reaction to another, under isothermal conditions, is by coupling the two reactions via a common intermediate. Consider reactions 16 and 17:

$$
\begin{array}{llll}
(16) & A + B \rightleftharpoons C + D & & \Delta G^{\circ\prime} = -7 \text{ kcal} \\
(17) & C + R \rightleftharpoons P + Q & & \Delta G^{\circ\prime} = +3 \text{ kcal} \\
\hline
\textit{Sum of 16 and 17} & A+B+R \rightleftharpoons D+P+Q & & \Delta G^{\circ\prime} = -4 \text{ kcal}
\end{array}
$$

Since C is a product in reaction 16 and a reactant in reaction 17, the two reactions are coupled, and the overall reaction is the sum of the two. Likewise, the standard free energy change for the overall reaction is the sum of the standard free energy changes of the individual reactions. In the example given, the overall reaction is thermodynamically favorable even though reaction 17 is thermodynamically unfavorable. Thus reaction 16, which is thermodynamically favorable, is able to drive reaction 17, which is thermodynamically unfavorable, because they are coupled through the common intermediate, C. There are many examples of coupled reactions of this type in intermediary metabolism.

THE ROLE OF ATP

Adenosine triphosphate (ATP) plays a key role in the transfer of free energy resulting from the oxidation of foodstuffs to those cellular processes that require the input of free energy. As can be seen in Figure 10–1, ATP consists of adenine, ribose, and a triphosphate group which is attached to the ribose by a phosphoester bond. The structural features that enable ATP to function in energy transfer reactions in the cell are the two phosphoanhydride bonds that link the three phosphate groups. All phosphoanhydride bonds are said to be energy-rich bonds, or high-energy bonds, because the standard free energy change, $\Delta G^{\circ\prime}$, for the hydrolysis of a phosphoanhydride bond has a large negative value. For example,

$$ATP^{4-} + H_2O \rightleftharpoons ADP^{3-} + P_i^{2-} + H^+ \qquad \Delta G^{\circ\prime} = -7.3 \text{ kcal/mole}$$

$$ADP^{3-} + H_2O \rightleftharpoons AMP^{2-} + P_i^{2-} + H^+ \qquad \Delta G^{\circ\prime} = -7.3 \text{ kcal/mole}$$

$$PP_i^{3-} + H_2O \rightleftharpoons 2P_i^{2-} + H^+ \qquad \Delta G^{\circ\prime} = -8.0 \text{ kcal/mole}$$

where ADP is adenosine diphosphate, AMP is adenosine monophosphate, P_i is inorganic phosphate, and PP_i is inorganic pyrophosphate. The charge indicated is the charge on the species that predominates at pH 7. The phosphoester bond linking the phosphate to ribose in AMP is not considered to be energy-rich.

$$AMP^{2-} + H_2O \rightleftharpoons \text{adenosine} + P_i^{2-} \qquad \Delta G^{\circ\prime} = -3.4 \text{ kcal/mole}$$

In Table 10–2 the standard free energy changes for the hydrolysis of some biologically important phosphate compounds are listed. The $\Delta G^{\circ\prime}$ of −7.3 kcal/mole for the hydrolysis of ATP places it in an intermediate position between the very energy-rich compounds, such as phosphoenolpyruvate, and the lower energy phosphate esters. It is this intermediate position that enables ATP to function as a transfer agent of free energy in the cell. For example, the transfer of phosphate

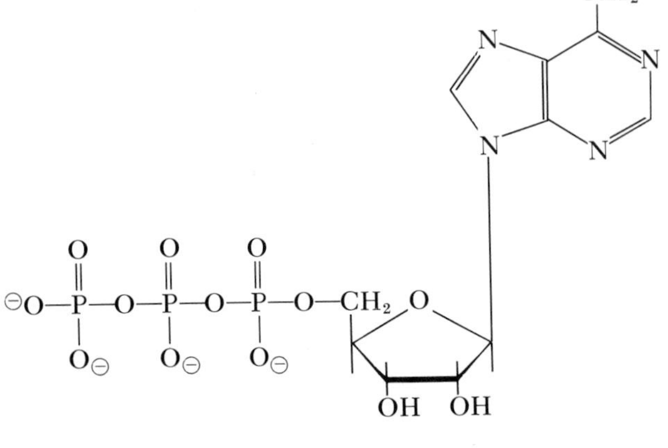

FIGURE 10–1 Adenosine triphosphate (ATP).

STANDARD FREE ENERGY CHANGES FOR THE HYDROLYSIS OF PHOSPHATES — **TABLE 10–2**

Compound	$\Delta G^{\circ\prime}$ (kcal/mole)
Phosphoenolpyruvate	−14.8
Creatine phosphate	−10.3
Acetyl phosphate	−10.3
Pyrophosphate	− 8.0
Adenosine triphosphate (ATP)	− 7.3
Adenosine diphosphate (ADP)	− 7.3
Glucose-1-phosphate	− 5.0
Adenosine monophosphate (AMP)	− 3.4
Glucose-6-phosphate	− 3.3
α-Glycerophosphate	− 2.2

from phosphoenolpyruvate to ADP is a thermodynamically favorable reaction, as shown by

$$\text{phosphoenolpyruvate} + H_2O \rightleftharpoons \text{pyruvate} + P_i \qquad \Delta G^{\circ\prime} = -14.8 \text{ kcal/mole}$$
$$\text{ADP} + P_i \rightleftharpoons \text{ATP} + H_2O \qquad \Delta G^{\circ\prime} = +7.3 \text{ kcal/mole}$$

Sum:

$$\text{phosphoenolpyruvate} + \text{ADP} \rightleftharpoons \text{pyruvate} + \text{ATP} \qquad \Delta G^{\circ\prime} = -7.5 \text{ kcal/mole}$$

If a reaction requires the input of more than 7.3 kcal of free energy to make it thermodynamically favorable, the reaction can be coupled to the hydrolysis of both high-energy phosphoanhydride bonds of ATP. An example of the way in which this is accomplished is found in the synthesis of aminoacyl-tRNA, which was discussed in Chapter 7. The synthesis is catalyzed by aminoacyl-tRNA synthetase and occurs in two steps:

$$(18) \quad \text{amino acid} + \text{ATP} \rightleftharpoons \text{aminoacyl-adenylate} + PP_i \qquad \Delta G^{\circ\prime} \sim 0 \text{ kcal/mole}$$

$$(19) \quad \text{aminoacyl-adenylate} + \text{tRNA} \rightleftharpoons \text{AMP} + \text{aminoacyl-tRNA} \qquad \Delta G^{\circ\prime} \sim 0 \text{ kcal/mole}$$

Sum of 18 and 19:

$$\text{amino acid} + \text{ATP} + \text{tRNA} \rightleftharpoons \text{aminoacyl-tRNA} + \text{AMP} + PP_i \qquad \Delta G^{\circ\prime} \sim 0 \text{ kcal/mole}$$

The equilibrium constant for the reaction represented by the sum of 18 and 19 is approximately unity; however, this reaction is driven to the right by being coupled to another reaction catalyzed by the ubiquitous enzyme pyrophosphatase.

$$(20) \qquad PP_i + H_2O \xrightleftharpoons{\text{pyrophosphatase}} 2P_i \qquad \Delta G^{\circ\prime} = -8.0 \text{ kcal/mole}$$

Thus the overall reaction is the sum of 18, 19, and 20.

Sum of 18, 19, and 20:

$$\text{amino acid} + \text{ATP} + \text{tRNA} + H_2O \rightleftharpoons \text{aminoacyl-tRNA} + \text{AMP} + 2P_i \qquad \Delta G^{\circ} \simeq -8.0 \text{ kcal/mole}$$

It is thermodynamically very favorable, being driven by the hydrolysis of both phosphoanhydride bonds of ATP.

The remainder of this chapter and much of Chapter 11 will be devoted to a consideration of the way in which the cell traps an appreciable part of the free energy change resulting from the oxidation of glucose, one of the major foodstuffs of humans, by coupling the oxidation of glucose to the formation of ATP.

GLYCOLYSIS

Glucose is one of the major fuel molecules. The standard free energy change for the oxidation of glucose to CO_2 and H_2O by molecular oxygen is large and negative.

$$(21)\quad C_6H_{12}O_6 + 6O_2 \rightarrow 6CO_2 + 6H_2O \qquad \Delta G^\circ = -686 \text{ kcal/mole glucose}$$

Aerobic organisms conserve a large fraction of this free energy change by coupling the oxidation of glucose to the generation of ATP from ADP and P_i, via the net reaction represented by equation 22.

$$(22)\quad C_6H_{12}O_6 + 6O_2 + 36ADP + 36P_i \rightarrow 6CO_2 + 36ATP + 42H_2O$$

Glucose can also be partially degraded in cells via pathways that do not require molecular oxygen. Glycolysis, or homolactic fermentation, is one of these anaerobic pathways. In glycolysis, glucose is broken down in a series of reactions to two moles of lactic acid.

$$(23)\quad \underset{\text{D-Glucose}}{C_6H_{12}O_6} \longrightarrow \underset{\text{L-Lactic acid}}{2\,CH_3-\underset{\displaystyle OH}{\underset{|}{CH}}-COOH}$$

This anaerobic fermentation is also coupled to the generation of ATP from ADP and P_i,

$$(24)\quad C_6H_{12}O_6 + 2\,ADP + 2P_i \longrightarrow 2\,CH_3-\underset{\displaystyle OH}{\underset{|}{CH}}-COOH + 2\,ATP + 2H_2O$$

and a fraction of the free energy change is trapped in the form of ATP. The enzymes that catalyze the reactions of the glycolytic pathway are easily extracted in soluble form from the cell; therefore, glycolysis appears to take place in the cytosol or soluble portion of the cytoplasm, which corresponds morphologically to the cytoplasmic matrix.

Anaerobic fermentations, such as glycolysis, are primitive biological mechanisms for obtaining energy from foodstuffs. Although glycolysis is a primitive pathway, yielding only two moles of ATP per mole of glucose, most aerobic organisms retain the ability to obtain energy from glucose through glycolysis and in fact use a part of the glycolytic pathway as a preparatory step for the aerobic metabolism of glucose. Therefore, the glycolytic pathway is important in both the anaerobic and aerobic metabolism of glucose.

Before glucose can be metabolized by a cell, it must be taken into the cell. In some cells the uptake of glucose is an active process (see Chapter 15) in which glucose can be transported into the cell even if the concentration of glucose is greater inside than outside the cell. For example, the uptake of glucose from the lumen by the mucosal cells of the small intestine involves the cotransport of sodium ion and glucose. It is an active transport process, driven by the sodium ion concentration gradient that exists across the plasma membrane. This sodium ion gradient is in turn maintained by the Na^+-K^+ pump, which is driven by the hydrolysis of ATP.

In contrast, the uptake of glucose by most human cells is a passive process which is driven only by the glucose concentration gradient across the plasma membrane. Although the driving force for the uptake of glucose is the glucose concentration gradient, the translocation of glucose across the plasma membrane appears to be mediated in some way by a protein component of the membrane. Furthermore, in some tissues, such as skeletal muscle and adipose tissue, the rate of glucose entry into the cells is modulated by the hormone insulin.

Once inside the cell, the glucose may be converted to glycogen for storage through a multireaction pathway, or it may be metabolized through other multireaction pathways, one of which is the glycolytic pathway.

The Glycolytic Pathway

PHOSPHORYLATION OF GLUCOSE

Upon entering the cell, glucose is phosphorylated to yield glucose-6-phosphate via reaction 25.

(25)

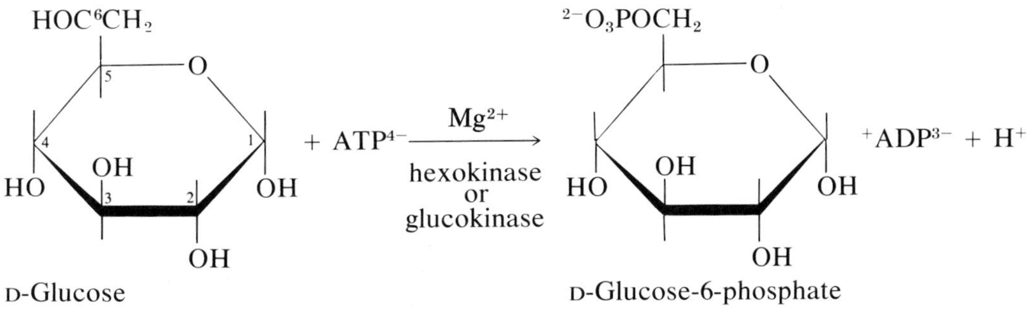

$\Delta G^{\circ\prime} = -4$ kcal/mole

This is the first step in the metabolism of glucose for all metabolic pathways that utilize glucose. As indicated, two different enzymes, hexokinase and glucokinase, are capable of catalyzing the reaction. Since ATP is utilized in this reaction, the reaction represents an investment of energy on the part of the cell. Phosphorylated compounds, such as glucose-6-phosphate, cannot permeate the plasma membrane. Therefore, an immediate return on the investment of the ATP molecule is realized, because the glucose is now trapped within the cell as glucose-6-phosphate. Additional dividends will be reaped by the cell later, since the overall pathway of glycolysis generates more ATP than it consumes. The $\Delta G^{\circ\prime}$ value of -4 kcal/mole indicates that the equilibrium constant for reaction 25 is large and that the position of the equilibrium lies far to the right. As a result, the concentration of free glucose within the cell is very small, and the reaction is essentially unidirectional, as indicated by the single arrow.

Enzymes that transfer the terminal phosphate of ATP to other organic molecules are denoted by the suffix-*kinase*. The enzyme that catalyzes the transfer of the terminal phosphate of ATP to glucose is glucokinase and is specific for glucose. Hexokinase is less specific and catalyzes the phosphorylation of glucose, fructose, and mannose, all of which are hexoses. Both enzymes require Mg^{++} for activity. There are actually three isozymes of hexokinase, each with a characteristic K_m for glucose and ATP. The three isozymes of hexokinase all have a relatively low K_m for glucose, i.e., in the range of 7×10^{-6}M to 5×10^{-5}M. Hence the hexokinases will function at near maximum capacity even when the glucose concentration is relatively low. Glucokinase has a K_m for glucose of 2×10^{-2}M; therefore, glucokinase will function at an appreciable fraction of its maximum capacity only when the concentration of glucose is high, as after a carbohydrate-rich meal. All human cells contain hexokinase. Hepatic cells contain both glucokinase and all three isozymes of hexokinase. The presence of glucokinase in the liver enables the liver to rapidly phosphorylate glucose when glucose is plentiful. The large quantities of G-6-P formed in the liver during times of glucose excess are then converted to glycogen and fat for storage.

ISOMERIZATION OF GLUCOSE-6-PHOSPHATE

The second reaction in the glycolytic pathway is the isomerization of glucose-6-phosphate to fructose-6-phosphate, a readily reversible reaction ($\Delta G^{\circ\prime} = +0.4$ kcal/mole) catalyzed by the enzyme phosphoglucoisomerase.

(26)

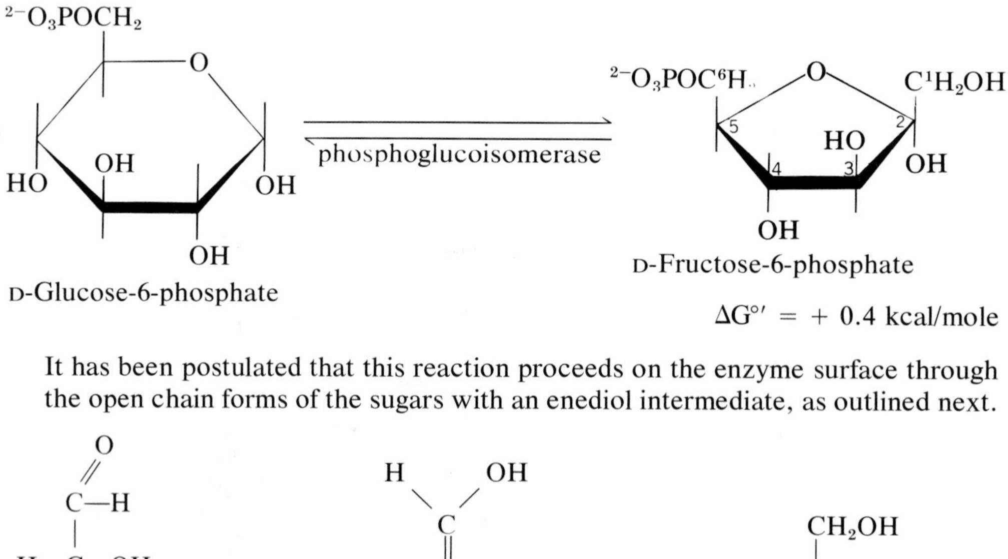

$\Delta G^{\circ\prime} = +0.4$ kcal/mole

It has been postulated that this reaction proceeds on the enzyme surface through the open chain forms of the sugars with an enediol intermediate, as outlined next.

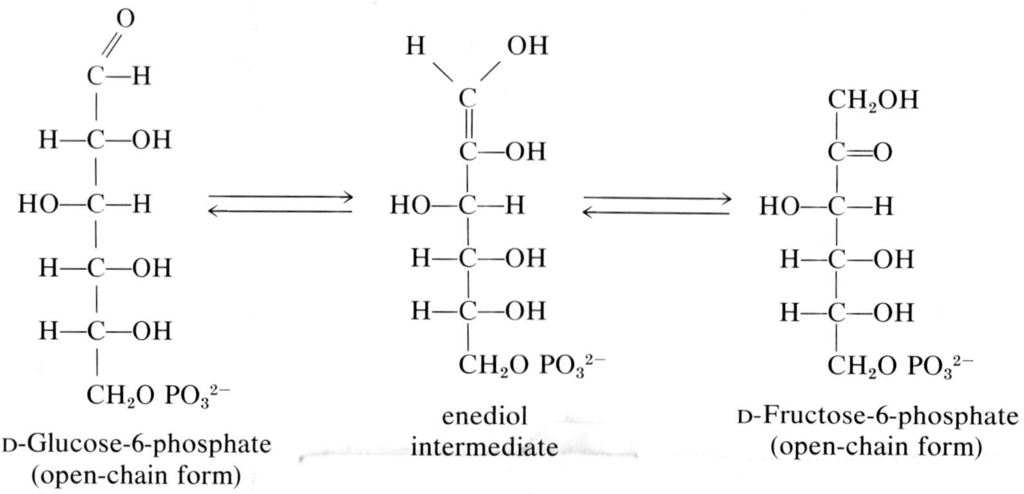

PHOSPHORYLATION OF FRUCTOSE-6-PHOSPHATE

The third reaction in the glycolytic pathway is the phosphorylation of fructose-6-phosphate to yield fructose-1, 6-diphosphate. This reaction is catalyzed by phosphofructokinase. It requires the investment of another mole of ATP and is essentially irreversible or unidirectional under the conditions existing in the cell.

(27)

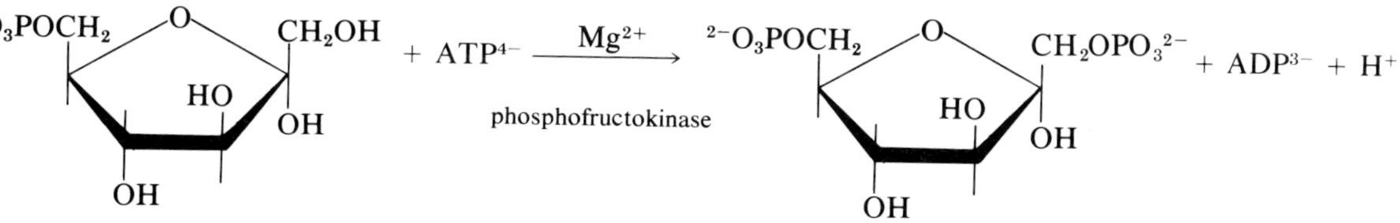

D-Fructose-6-phosphate — D-Fructose-1,6-diphosphate

$\Delta G^{\circ\prime} = -3.4$ kcal/mole

As will be discussed later, the phosphofructokinase-catalyzed reaction is an important reaction from the standpoint of the regulation of the rate of glycolysis.

CLEAVAGE OF FRUCTOSE-1,6-DIPHOSPHATE

The next reaction in the glycolytic pathway is the cleavage of fructose-1,6-diphosphate between C-3 and C-4 to form two trioses, dihydroxyacetone phosphate and glyceraldehyde-3-phosphate.

(28)

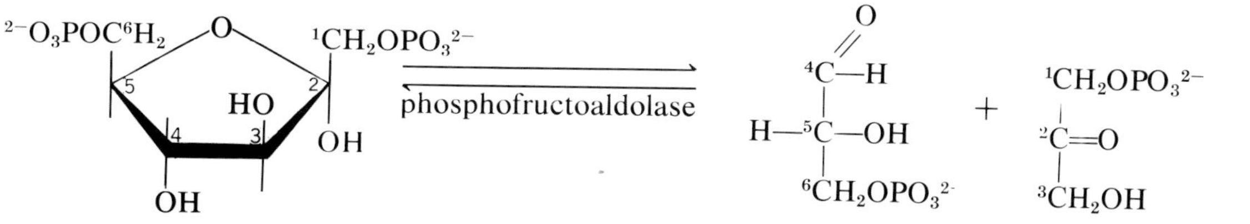

D-Fructose-1,6-diphosphate — D-Glyceraldehyde-3-phosphate — Dihydroxyacetone phosphate

$\Delta G^{\circ\prime} = +5.7$ kcal/mole

The enzyme catalyzing this reaction is phosphofructoaldolase, and it bears the suffix *-aldolase* because the reaction is analogous to the classic aldol condensation reactions of organic chemistry.

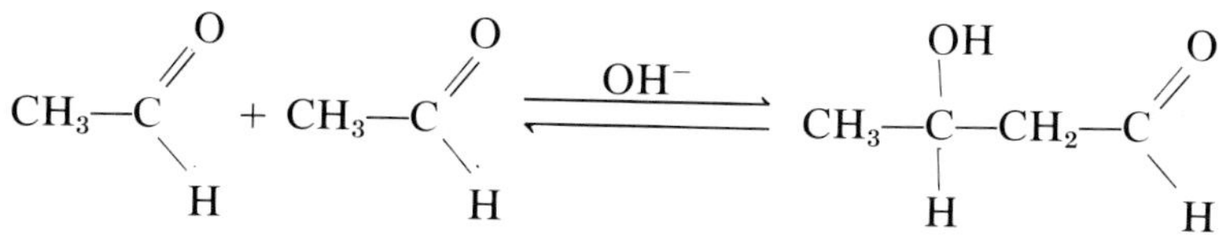

Aldol Condensation Reaction

From the standard free energy change of +5.7 kcal/mole for the phosphofructoaldolase reaction, an equilibrium constant of 6.6×10^{-5} can be calculated. This might lead one to conclude that the reaction is essentially unidirectional in the direction of fructose-1,6-diphosphate formation. However, the equilibrium constant has the dimensions of concentration.

$$K = \frac{[\text{glyceraldehyde-3-phosphate}]\,[\text{dihydroxyacetone phosphate}]}{[\text{fructose-1,6-diphosphate}]} = 6.6 \times 10^{-5}\text{M}$$

and as a result, the ratio of the concentration of products to the concentration of reactant at equilibrium will be very dependent upon the overall concentration. For example, if we consider reaction 28 only, a simple calculation will show that if the equilibrium concentration of fructose-1,6-diphosphate is 0.1M, the equilibrium concentrations of dihydroxyacetone phosphate and glyceraldehyde-3-phosphate are each approximately 2.6×10^{-3}M. Hence the equilibrium concentration of the reactant is approximately 40-fold greater than the equilibrium concentration of either product.

On the other hand, if the equilibrium concentration of fructose-1,6-diphosphate is 3×10^{-5}M, a concentration in the range of that existing in the cell, the equilibrium concentrations of dihydroxyacetone phosphate and glyceraldehyde-3-phosphate are each approximately 4.4×10^{-5}M. Thus, under the conditions existing in the cell, the equilibrium concentration of each product is very similar to that of the reactant.

ISOMERIZATION OF DIHYDROXYACETONE PHOSPHATE

The sum of reactions 25 through 28 utilizes two moles of ATP and converts one mole of glucose to one mole of glyceraldehyde-3-phosphate plus one mole of dihydroxyacetone phosphate. A second mole of glyceraldehyde-3-phosphate is produced by the isomerization of the dihydroxyacetone phosphate to glyceraldehyde-3-phosphate, a reaction catalyzed by triose phosphate isomerase.

(29)

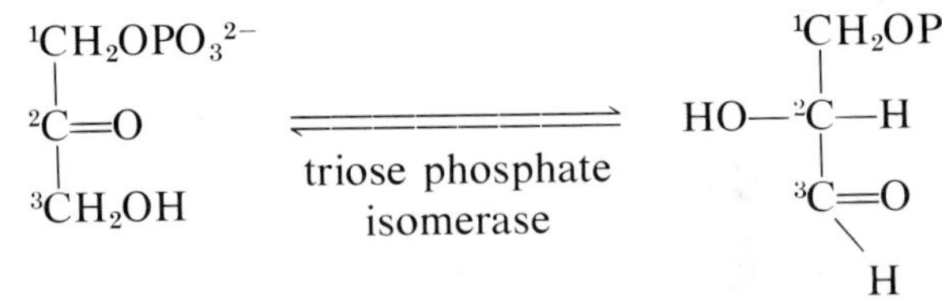

$\Delta G^{\circ\prime} = {}^{+}1.8$ kcal/mole

Dihydroxyacetone phosphate

D-Glyceraldehyde-3-phosphate

Of the two moles of glyceraldehyde-3-phosphate derived from the original mole of glucose, one mole contains a carbonyl carbon derived from C-3 of glucose (reaction 29), and the other mole contains a carbonyl carbon derived from C-4 of glucose (reaction 28). The remainder of the glycolytic pathway is concerned with the fate of these two moles of glyceraldehyde-3-phosphate.

OXIDATION OF GLYCERALDEHYDE-3-PHOSPHATE

Before the oxidation of glyceraldehyde-3-phosphate is discussed, biological oxidation-reduction reactions in general will be considered briefly. Oxidation is the removal of electrons from a compound; reduction is the addition of electrons to a compound. When one compound loses electrons and becomes oxidized, another compound must simultaneously accept electrons and become reduced; therefore, these reactions are named oxidation-reduction reactions. A very common electron acceptor in biochemical reactions is nicotinamide adenine dinucleotide, NAD^+, whose structure is shown in Figure 10–2. NAD^+ is converted to its reduced form, NADH, by the addition of two electrons and a proton (the equivalent of a hydride ion, $H:^-$). Enzymes that catalyze the transfer of a hydride ion from a substrate to NAD^+ are classified as dehydrogenases. The part of NAD^+ that accepts the hydride ion and becomes reduced is the pyridine ring system of the nicotinamide group. Because of the presence of the pyridine ring system, NAD^+ is often referred to as a pyridine nucleotide. Since humans cannot synthesize *de novo* the nicotinamide required for NAD^+ synthesis, they are dependent upon dietary sources. A diet deficient in a source of nicotinamide leads to the disease pellagra. One dietary source is the B vitamin nicotinamide, or niacin,

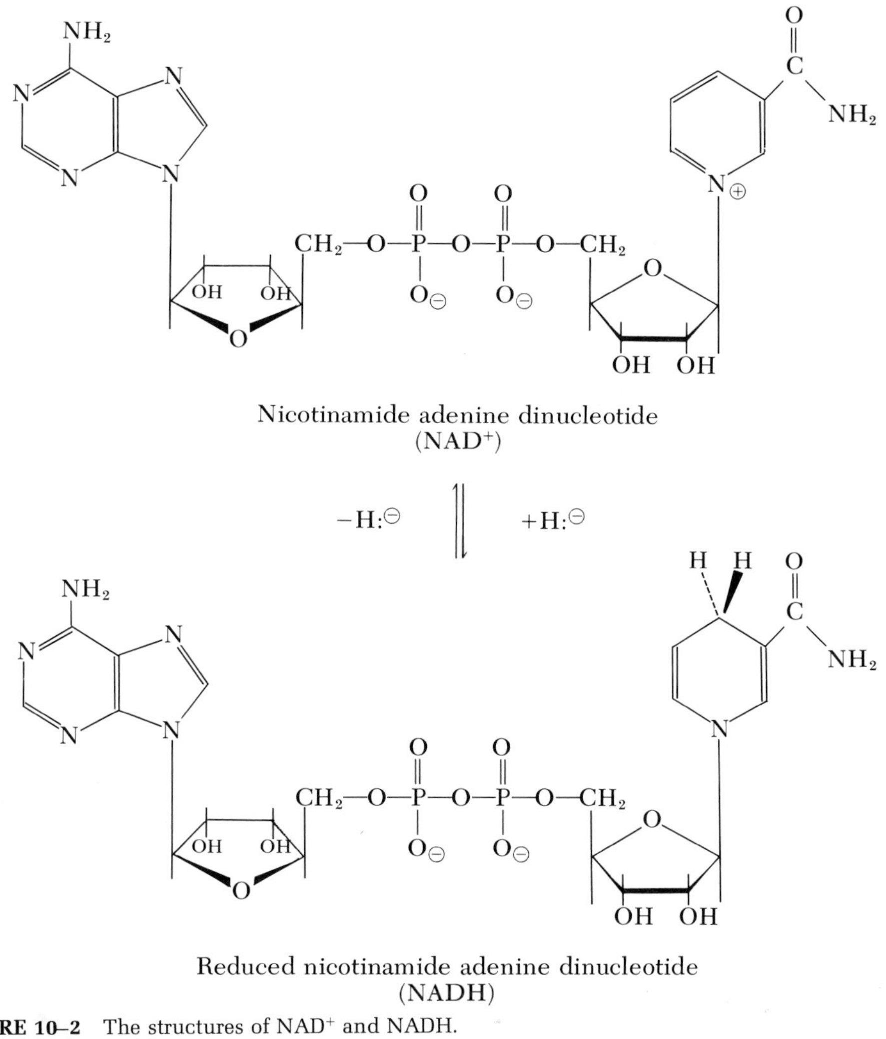

FIGURE 10–2 The structures of NAD^+ and NADH.

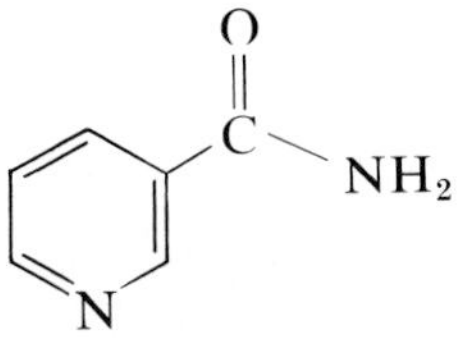

Nicotinamide

and another is the essential amino acid tryptophan, which can be converted in low yield to nicotinate mononucleotide in humans.

To return to the metabolic fate of glyceraldehyde-3-phosphate, the two moles of glyceraldehyde-3-phosphate are oxidized to two moles of 1,3-diphospho-D-glycerate in a reaction catalyzed by the enzyme glyceraldehyde-3-phosphate dehydrogenase. The electron acceptor in this oxidation-reduction reaction is nicotinamide adenine dinucleotide.

(30) 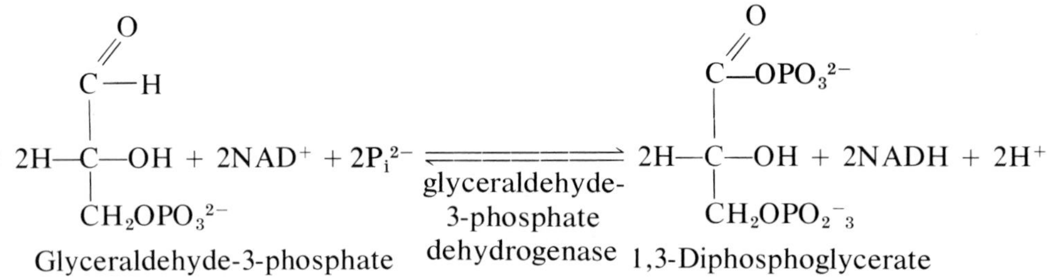

$\Delta G^{\circ\prime} = +1.5$ kcal/mole

The net result of this oxidation-reduction reaction is that a hydride ion, $H:^-$ (the equivalent of a proton plus two electrons), is transferred from glyceraldehyde-3-phosphate to the electron acceptor NAD^+, converting the aldehyde group to the oxidation state of a carboxyl group. The free energy change for this oxidation-reduction reaction is conserved through the formation of a mixed anhydride linkage between the newly formed carboxyl group and the inorganic phosphate ion. As shown in Figure 10–3, the proposed mechanism for the glyceraldehyde-3-phosphate dehydrogenase reaction involves the following steps: (1) the formation of a thiohemiacetal between the substrate and a sulfhydryl group of a cysteinyl residue at the "active site" of the enzyme; (2) the generation of an energy-rich thioester linkage by transfer of a hydride ion from the thiohemiacetal to NAD^+; and (3) the formation of 1,3-diphosphoglycerate by the phosphorolytic cleavage of the thioester by inorganic phosphate.

GENERATION OF ATP BY THE TRANSFER OF A PHOSPHORYL GROUP FROM 1,3-DIPHOSPHOGLYCERATE

The $\Delta G^{\circ\prime}$ for the hydrolysis of the mixed anhydride linkage of 1,3-diphosphoglycerate is -11.8 kcal/mole. Therefore, the transfer of the phosphoryl group from diphosphoglycerate to ADP is thermodynamically favorable.

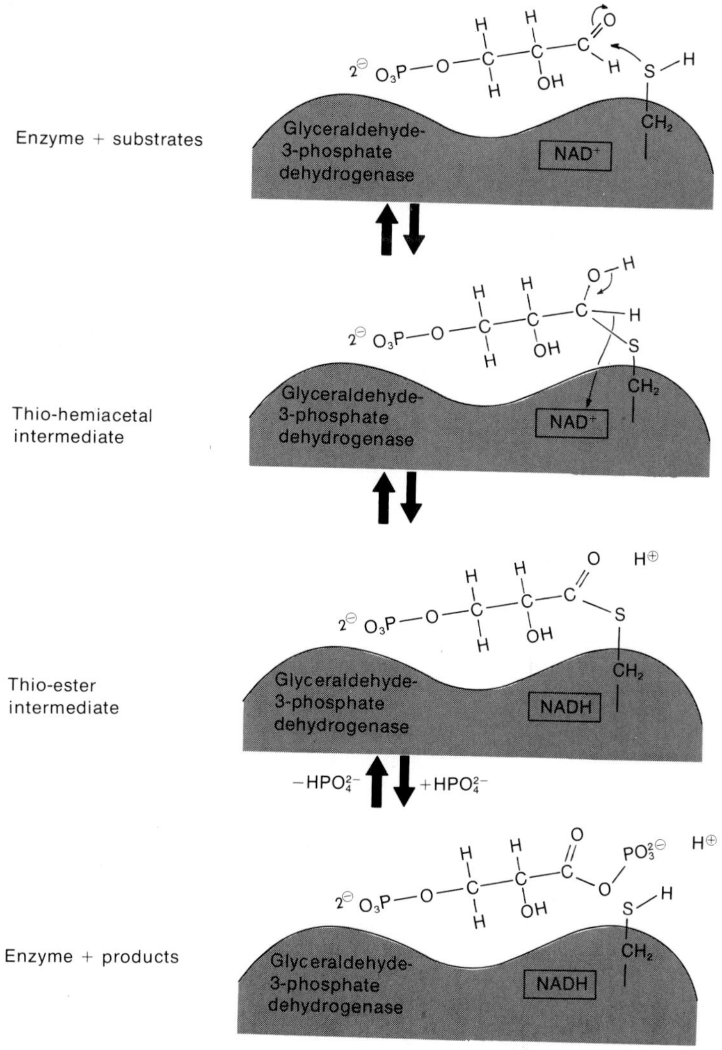

FIGURE 10–3 Proposed mechanism of action of glyceraldehyde-3-phosphate dehydrogenase.

(31) 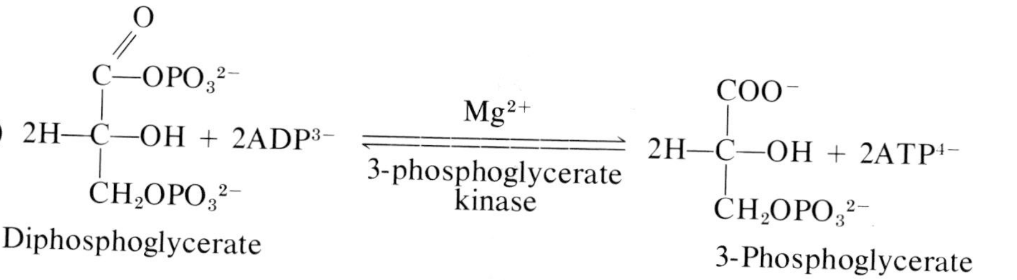

$\Delta G^{\circ\prime} = -4.5$ kcal/mole

The enzyme catalyzing reaction 31 is 3-phosphoglycerate kinase. The two molecules of ATP generated in reaction 31 cancel out the ATP debt resulting from the cell's previous expenditure of one molecule of ATP in reaction 25 and one molecule of ATP in reaction 27. The generation of ATP from ADP in reaction 31 is an example of *substrate level phosphorylation*.

CONVERSION OF 3-PHOSPHOGLYCERATE TO 2-PHOSPHOGLYCERATE

Enzymes that catalyze the apparent transfer of a phosphate group from one hydroxyl group to another hydroxyl group in the same molecule are called mutases. The next reaction in glycolysis involves a phosphate group migration of this type and is catalyzed by phosphoglyceromutase.

(32)

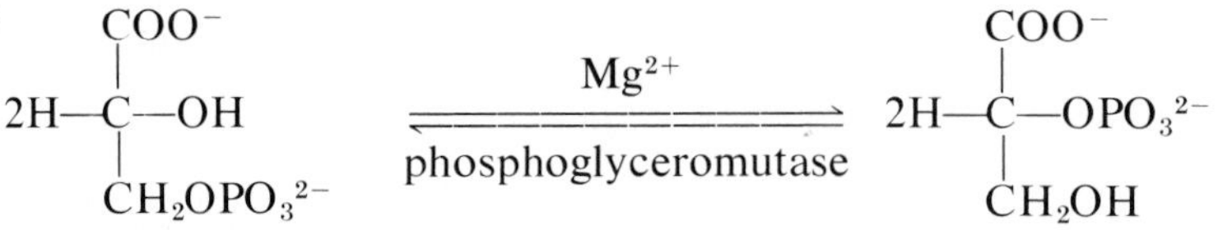

3-Phosphoglycerate 2-Phosphoglycerate

$\Delta G^{\circ\prime} = +1.1$ kcal/mole

DEHYDRATION OF 2-PHOSPHOGLYCERATE

Elimination of the elements of H_2O from 2-phosphoglycerate results in the formation of phosphoenolpyruvate (PEP).

(33)

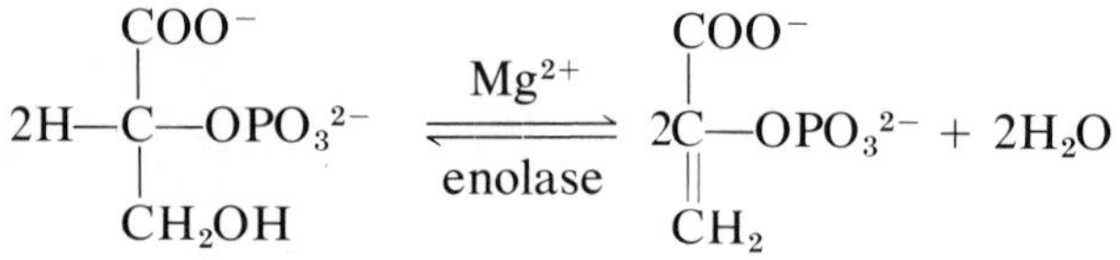

2-Phosphoglycerate Phosphoenolpyruvate (PEP)

$\Delta G^{\circ\prime} = +0.4$ kcal/mole

Enolase, the enzyme catalyzing this reaction, is strongly inhibited by fluoride ion. As shown in Table 10–2, the $\Delta G^{\circ\prime}$ for the hydrolysis of PEP is -14.8 kcal/mole. If PEP is hydrolyzed, enolpyruvate is formed, which spontaneously tautomerizes to form pyruvate.

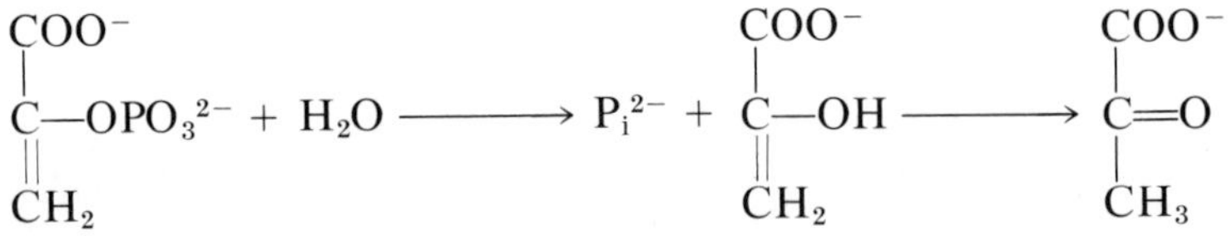

PEP Enolpyruvate Pyruvate

However, in the glycolytic pathway PEP is not hydrolyzed to produce inorganic phosphate; instead it participates in the following reaction, which generates ATP.

GENERATION OF ATP BY THE TRANSFER OF THE PHOSPHORYL GROUP OF PHOSPHOENOLPYRUVATE

The enzyme pyruvate kinase catalyzes the next reaction of glycolysis, a substrate-level phosphorylation, in which the phosphoryl group of PEP is transferred to ADP.

(34) 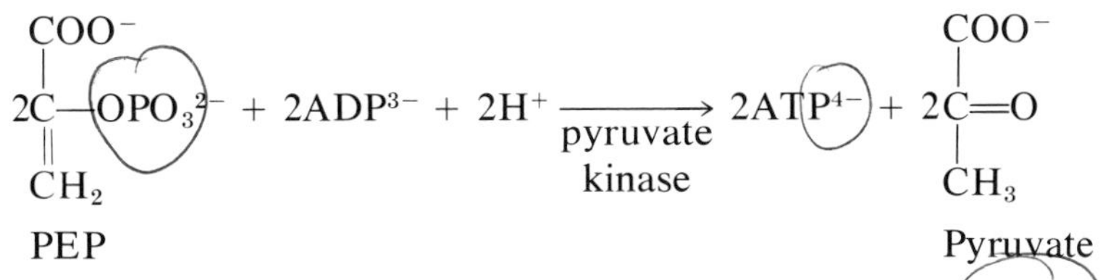

$$\Delta G^{\circ\prime} = -7.5 \text{ kcal/mole}$$

This is a very exergonic reaction ($\Delta G^{\circ\prime} = -7.5$ kcal/mole) and is an essentially irreversible, or unidirectional, reaction under the conditions existing in the cell. The two molecules of ATP generated in this reaction now raise the net ATP balance to +2 molecules of ATP per molecule of glucose entering the glycolytic pathway.

Summing up the glycolytic pathway to this point gives the following:

Sum of 25–34

$$C_6H_{12}O_6 + 2P_i^{2-} + 2NAD^+ + 2ADP^{3-} \rightarrow 2NADH + 2H^+ + 2H_2O + 2ATP^{4-} + 2CH_3\text{-}\overset{\overset{\displaystyle O}{\|}}{C}\text{-}COO^-$$

Reactions 25 through 34 constitute that part of the glycolytic pathway that is used as a preparatory step for the aerobic metabolism of glucose; it is referred to as the *Embden-Meyerhof pathway*. In addition to being a preparatory step for aerobic metabolism, the Embden-Meyerhof pathway results in the net production of two moles of ATP per mole of glucose entering the pathway. From the overall equation for the Embden-Meyerhof pathway (sum of reactions 25 through 34), it is apparent that the pathway can proceed only if ADP, P_i, and NAD^+ are present in the cell. When the cell is doing work, the energy for this work is being derived from the cleavage of ATP to form ADP + P_i. Hence when there is a demand for energy in the cell, the availability of ADP and P_i will not be limiting for the Embden-Meyerhof pathway because they are being generated by the cleavage of ATP.

However, the pathway also converts NAD^+ to NADH and can proceed only so long as NAD^+ is available in the cell. The amount of NAD^+ plus NADH within the cell is small relative to the amount of glucose and glycogen; therefore, for the metabolism of glucose to continue via the Embden-Meyerhof pathway, the NADH formed from NAD^+ must be oxidized back to NAD^+. In the *aerobic* metabolism of glucose the NADH is converted back to NAD^+ by passing the electrons into the mitochondria via a pathway that will be discussed in Chapter 11. In the aerobic metabolism of glucose the pyruvate formed by the Embden-Meyerhof pathway also enters the mitochondria where it is further metabolized. However, in glycolysis, which is *anaerobic,* NAD^+ is regenerated from NADH by the next and final reaction of the glycolytic pathway.

REDUCTION OF THE PYRUVATE BY NADH

In the absence of O_2, or when the anaerobic pathway is followed even in the presence of O_2, NAD^+ can be regenerated from NADH by employing pyruvate as an electron acceptor in a reaction catalyzed by lactate dehydrogenase.

$$\text{(35)}\quad \underset{\text{Pyruvate}}{2\begin{array}{c}COO^-\\ |\\ C{=}O\\ |\\ CH_3\end{array}} + 2NADH + 2H^+ \underset{\text{lactate dehydrogenase}}{\rightleftharpoons} \underset{\text{L-Lactate}}{2\begin{array}{c}COO^-\\ |\\ HO{-}C{-}H\\ |\\ CH_3\end{array}} + 2NAD^+$$

$$\Delta G^{\circ\prime} = -6 \text{ kcal/mole}$$

This is the final reaction of the glycolytic pathway. Thus the glycolytic pathway consists of the 11 enzyme-catalyzed reactions (reactions 25 through 35) which are summarized in Figure 10–4. The net equation for glycolysis can be arrived at by summing these eleven reactions.

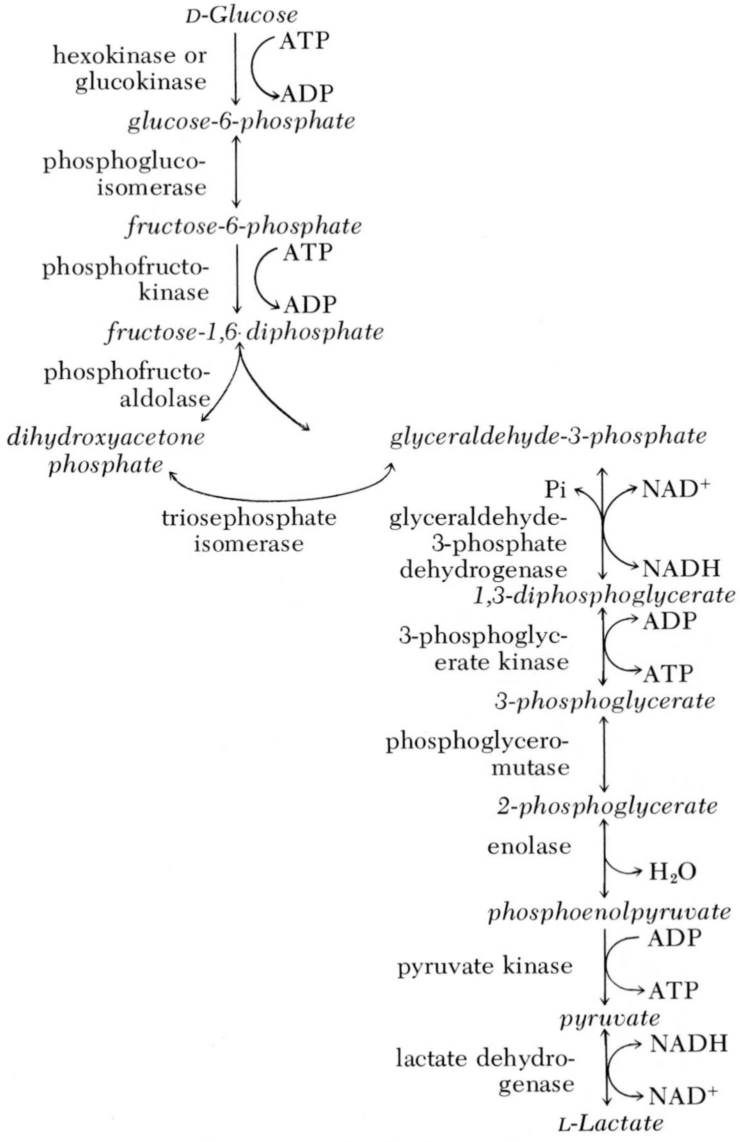

FIGURE 10–4 The glycolytic pathway.

Glycolysis: the sum of reactions 25 through 35.

$$(36) \quad C_6H_{12}O_6 + 2ADP^{3-} + 2P_i^{2-} \rightarrow 2CH_3\underset{\displaystyle OH}{\underset{|}{C}H}\text{—}COO^- + 2ATP^{4-} + 2H_2O$$

$$\Delta G^{\circ\prime} \approx -29.5 \text{ kcal/mole}$$

It is important to note that the ATP generated by glycolysis results from substrate-level phosphorylations. This does not require molecular oxygen, and it occurs in the cytosol, or cytoplasmic matrix.

Significance of the Glycolytic Pathway

Many microorganisms can live under anerobic conditions by using fermentation processes, such as glycolysis, as their sole source of energy (ATP). The only mammalian cell that is totally dependent upon glycolysis as the sole source of energy is the red blood cell, or erythrocyte, which cannot utilize the aerobic pathway because it has no mitochondria. The energy requirements of the erythrocyte are not great, and the two moles of ATP generated per mole of glucose metabolized by the glycolytic pathway are sufficient to meet its energy needs.

Other mammalian cells generate most of their ATP in mitochondria by the aerobic process that utilizes molecular oxygen as the ultimate electron acceptor. This important mitochondrial process is known as *oxidative phosphorylation* and will be discussed in Chapter 11. However, glycolysis can be an important rapid source of ATP for at least one predominantly aerobic organ, skeletal muscle. Skeletal muscle derives its energy from the metabolism of fat and carbohydrate. Under most conditions, it depends primarily upon ATP generated by oxidative phosphorylation as its major energy source for muscular contraction. However, when working at maximum rate, skeletal muscle cells cleave ATP to ADP and P_i more rapidly than the muscle mitochondria can regenerate the ATP from P_i and ADP by oxidative phosphorylation. During these brief periods of maximum muscular effort, glycolysis can rapidly supply the additional ATP that is required. A maximum muscular effort can be sustained for only a brief time because the total amount of glucose stored as glycogen in the skeletal muscle of the average 70 kg man is only enough to generate a few moles of ATP by glycolysis.

Aspects of the Control of the Rate of Glycolysis

The rate at which the Embden-Meyerhof or glycolytic pathway functions to generate ATP and pyruvate or lactate in the muscle cell is geared to the rate of utilization of ATP by the cell. When ATP is being utilized at a rapid rate, the Embden-Meyerhof or glycolytic pathway functions at a rapid rate. An important control point in the regulation of this pathway is the phosphofructokinase-catalyzed reaction (27). One interesting set of controls on phosphofructokinase activity involves ATP, ADP, and AMP. ATP, a substrate for phosphofructokinase, is essential for phosphofructokinase activity. At high concentrations ATP is an allosteric inhibitor of phosphofructokinase. However, the allosteric inhibition of phosphofructokinase by high concentrations of ATP is relieved by ADP and AMP, as illustrated in Figure 10–5.

−	ATP
+	ADP
+	AMP

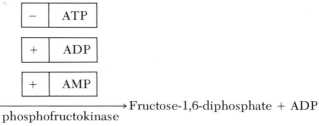

$$\text{Fructose-6-phosphate} + \text{ATP} \xrightarrow[\text{phosphofructokinase}]{} \text{Fructose-1,6-diphosphate} + \text{ADP}$$

FIGURE 10–5 Metabolite control of phosphofructokinase by ATP, ADP, and AMP. Activators are indicated by ⊞ ; inhibitors are indicated by ⊟.

This set of controls on phosphofructokinase activity can be rationalized as follows. When the cell is not utilizing ATP rapidly, the ATP concentration will be high, the AMP and ADP concentrations low, and the pathway will be blocked by the ATP inhibition of phosphofructokinase. In contrast, when ATP is being utilized rapidly, the AMP concentration as well as the ADP concentration will increase. The increased concentrations of AMP and ADP override the ATP inhibition of the enzyme and thereby turn on the Embden-Meyerhof, or glycolytic, pathway. Thus the rate at which the glycolytic pathway functions to generate ATP is geared to the rate of utilization of ATP by the cell.

The AMP concentration as well as the ADP concentration increases when ATP is being utilized owing to the presence of adenylate kinase, which catalyzes the following reaction:

(37) $$2\text{ADP} \underset{}{\overset{\text{adenylate kinase}}{\rightleftharpoons}} \text{ATP} + \text{AMP}$$

Thus, as ADP levels increase as a result of ATP utilization, AMP levels also increase. In fact, the relative change in AMP concentration with ATP utilization is much greater than the relative change in ADP levels.

The control of phosphofructokinase activity by ATP, ADP, and AMP concentrations appears to be an important part of the control mechanism that leads to a phenomenon first recognized by Pasteur and accordingly known as the Pasteur effect. It consists of the inhibition of glycolysis and lactate production by molecular oxygen, and it is observed in cells capable of metabolizing glucose by both an aerobic and an anaerobic pathway. It is apparent from equations 22 and 24 that the aerobic metabolism of glucose produces much more ATP per mole of glucose consumed than does glycolysis, the anaerobic pathway. In the presence of molecular oxygen the aerobic pathway becomes operative, leading to increased ATP concentration and decreased ADP and AMP concentrations, which in turn lead to an inhibition of phosphofructokinase. This inhibition of phosphofructokinase results in decreased glucose utilization and inhibition of glycolysis. Interestingly, most malignant tumor cells show a diminished Pasteur effect and obtain a large fraction of their energy from glycolysis even in the presence of molecular oxygen.

The control of glycolysis by ATP, ADP, and AMP is an example of *metabolite control* of a metabolic pathway. ATP, ADP, and AMP are not the only metabolites that play a role in the regulation of glycolysis, nor is phosphofructokinase the only control point in the metabolic pathway.

FURTHER READING

Axelrod, B.: Glycolysis. *In*: Greenberg, D. E. (ed): Metabolic Pathways. Vol. 1. New York, Academic Press, 1967, pp. 112–145.

Clark, M. G., and Lardy, H. A.: Regulation of intermediary carbohydrate metabolism. *In*: Whelan, W. J. (ed.): MTP International Review of Science-Biochemistry Series 1. Vol. 5. Baltimore, University Park Press, 1975, pp. 223–266.

Dickens, F., Randle, P. J., and Whelan, J. W. (eds.): Carbohydrate Metabolism and its Disorders. Vol. 1, Ch. 1–3. New York, Academic Press, 1968.

Harris, J. I., and Waters, M.: Glyceraldehyde-3-phosphate dehydrogenase. *In*: The Enzymes, 3rd ed., Vol. XIII, P. D. Boyer, (ed.), 1–49, 1976.

Krebs, H. A.: The Pasteur effect and phosphofructokinase. *In*: Horecker, B. L., and Stadtman, E. R. (eds.): Current Topics in Cellular Regulation. Vol. 8. New York, Academic Press, 1974, pp. 297–345.

Lehninger, A. L.: Bioenergetics: The Molecular Basis of Biological Energy Transformations, 2nd Ed. Menlo Park, Calif., W. A. Benjamin, 1971.

Peusner, L.: Concepts in Bioenergetics. Englewood Cliffs, N. J., Prentice-Hall, N.J., 1974.

Rose, I. A.: Mechanism of the aldose-ketose isomerase reaction. Adv. Enzymol., *43*:491, 1975.

Rose, I. A., and Rose, Z. B.: Glycolysis: regulation and mechanisms of the enzymes. *In*: Florkin, M., and Stotz, E. H. (eds.): Comprehensive Biochemistry. Vol. 17. Amsterdam, Elsevier, 1969, pp. 93–161.

11

MITOCHONDRIA

Mitochondria are membranous organelles of great importance in energy metabolism because they are the source of most of the cell's ATP and the site of many metabolic reactions. Thus, much of this chapter will deal with the fundamental biochemical processes that take place in mitochondria. The discussion will center on the citric acid cycle and oxidative phosphorylation and on the locations of these reactions with respect to mitochondrial ultrastructure. Mitochondria illustrate well how related biochemical reactions are carried out in proximity to one another within a membranous organelle.

Mitochondria were first observed in the late 19th century by Altmann, and the name mitochondria was applied a few years later by Benda. The association of respiratory enzymes with a cytoplasmic particle was reported by Warburg in the early part of this century. Many studies on mitochondria became possible following their successful isolation by Bensley and Hoerr in 1934.

STRUCTURE

In living cells viewed by phase contrast microscopy, mitochondria appear as spheres, rods, or filamentous bodies, which occupy the cytoplasm (Fig. 11–1), and undergo repeated fissions and fusions. In favorable instances, mitochondria are visible in fixed and sectioned preparations viewed with the light microscope (Fig. 11–2); they stain with acid dyes owing to their high protein content. Mitochondria can be more convincingly demonstrated in light microscope preparations with cytochemical methods, such as the DPN reaction. They can be stained supravitally with the dye Janus green B because they are capable of converting the colorless reduced form of the dye to the colored oxidized state.

In the electron microscope, mitochondria are seen to be bounded by two membranes, the inner of which is folded to form cristae (Figs. 11–3 and 11–4). Thus two compartments are defined: an outer compartment lying between the two membranes, and an inner compartment or matrix within the inner membrane. The inner compartment contains a high concentration of proteins, including many enzymes, as well as mitochondrial DNA and RNA. The inner compartment also contains the matrix granules, descriptively termed intramitochondrial dense granules, which are electron-opaque and contain divalent cations.

The abundance of mitochondria varies and is related to the size and metabolic

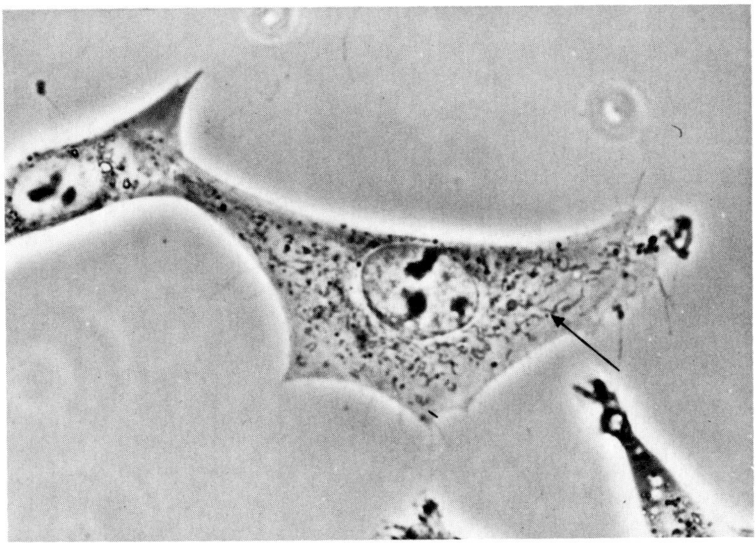

FIGURE 11–1 Phase contrast light micrograph of a living baby hamster kidney cell in tissue culture. Mitochondria are visible as filamentous structures in the cytoplasm (*arrow*). × 830. (Micrograph courtesy of R. L. Goldman, B. Tandler, and C. L. Hoppel. From Tandler, B., and Hoppel, C. L.; Mitochondria. New York, Academic Press, 1972. Reproduced by permission of the publisher.)

activity of the cell. Their number ranges from only one or a few in some algae and protozoa up to about 500,000 in large cells, such as the giant amoebae. A hepatocyte is estimated to contain about 1000. In general, the number of mitochondria is greater in cells with high energy requirements, and the size and number of their cristae are larger as well. Mitochondria are frequently about 0.5 μm in diameter, and their length ranges from a similar dimension up to 5 to 7 μm. The morphology of the mitochondrial cristae varies greatly from one cell to another (Figs. 11–4 to 11–7), but despite many detailed descriptions little is known about the functional significance of these variations. Cristae commonly appear as plates oriented transversely to the long axis of the mitochondrion, but they are sometimes longitudinally oriented. Tubular cristae, rather than the platelike variety, are seen in cells such as steroid-secreting cells and protozoa.

The functions of mitochondria, which are considered in detail in the following sections, are precisely localized within the organelle. The locations of these biochemical functions are reflected in the enzymatic compositions of the different

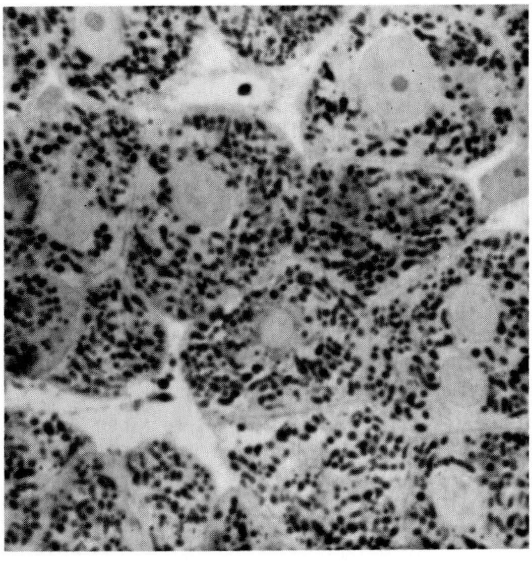

FIGURE 11–2 Light micrograph of a plastic section of rat liver cells. The spherical mitochondria are evident as dense granules in the cytoplasm of these cells. × 1290. (From Tandler, B., and Hoppel, C. L.: Mitochondria. New York, Academic Press, 1972. Reproduced by permission of the publisher.)

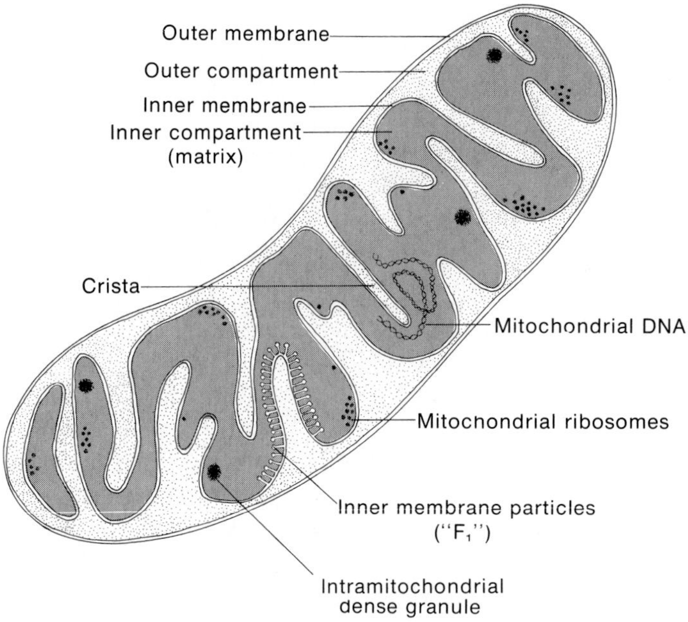

FIGURE 11–3 A diagram of the structural components of a mitochondrion.

parts of mitochondria (Table 11–1). The outer mitochondrial membrane contains monoamine oxidase, which is sometimes used as a marker for this structure, as well as other enzymes. However, oxidation of reduced coenzymes coupled with the formation of ATP, known as oxidative phosphorylation, is carried out by components of the inner membrane. That is, the molecules of the electron transport chain that are responsible for oxidative phosphorylation are integral parts of the inner membrane itself. The reactions of the citric acid cycle take place in the matrix.

Isolated mitochondria can be separated into inner and outer membrane fractions, permitting study of the structure, composition, and function of the two membranes. When inner mitochondrial membranes are negatively stained and viewed in the electron microscope, a large number of spheres, ~80 Å in diameter, are seen to be attached by means of a stalk to one surface of the inner membrane (Fig. 11–8). They are so positioned that in intact mitochondria the particles project into the matrix. Upon their discovery, these "elementary particles" were thought

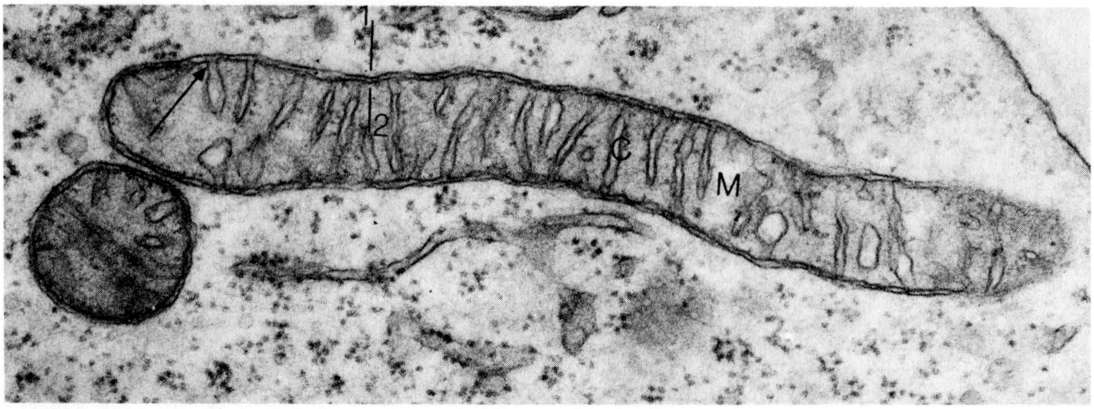

FIGURE 11–4 Wolffian duct of a fetal rat. The mitochondria are bounded by outer (1) and inner (2) membranes. Folding of the inner membrane (*arrow*) forms mitochondrial cristae (C). The inner membrane encloses the inner compartment or matrix (M). × 53,000.

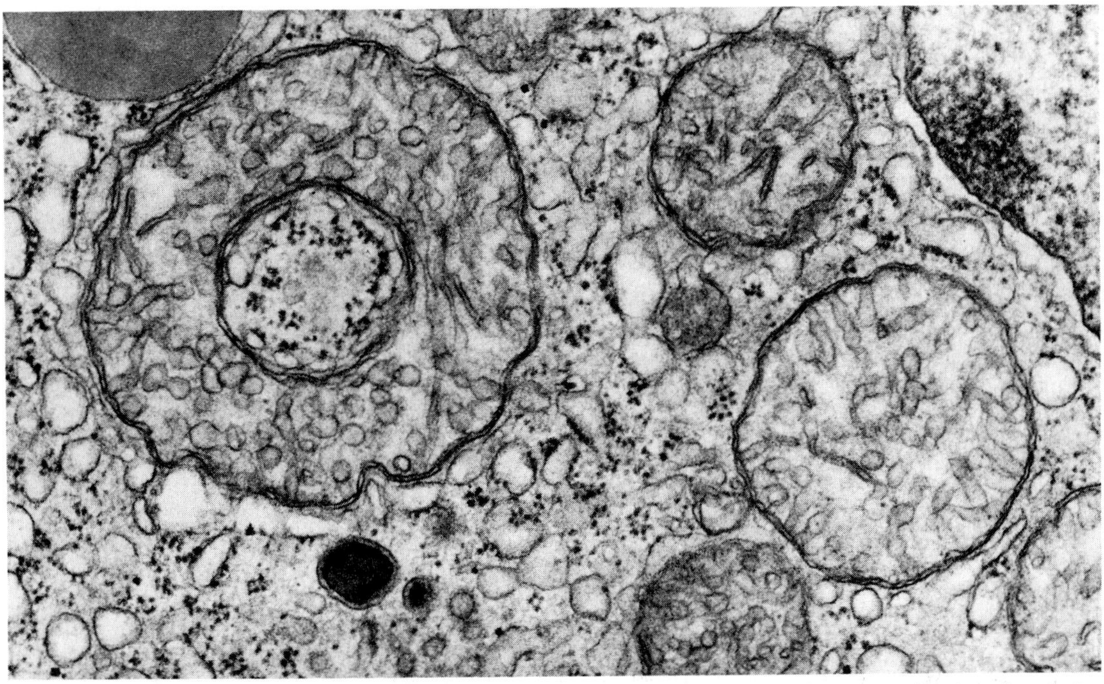

FIGURE 11–5 Interstitial cell of developing rat testis. The mitochondrial cristae have a predominantly tubular structure. × 45,000.

to contain units of the electron transport chain. However, subsequent calculations showed that they are not large enough to contain all these molecules, and the stalked spheres are now known to contain the enzyme referred to as the F_1 coupling factor (see further on).

Mitochondria in living cells undergo swelling and contraction. These relatively small changes in volume, termed low-amplitude changes, contrast with much larger volume changes designated high-amplitude swellings, which take

ENZYME DISTRIBUTION IN MITOCHONDRIA* **TABLE 11–1**

Outer Membrane
Monoamine oxidase
Rotenone-insensitive NADH-cytochrome c reductase
Kynurenine hydroxylase
Fatty acid CoA ligase
Space Between Outer and Inner Membranes
Adenylate kinase
Nucleoside diphosphokinase
Inner Membrane
Respiratory chain enzymes
ATP synthetase
Succinate dehydrogenase
β-Hydroxybutyrate dehydrogenase
Carnitine fatty acid acyl transferase
Matrix
Malate and isocitrate dehydrogenases
Fumarase and aconitase
Citrate synthetase
α-Keto acid dehydrogenases
β-Oxidation enzymes

*Courtesy of DeRobertis, E. D. P., Saez, F. A., and DeRobertis, E. M. F., Cell Biology. 6th ed. Philadelphia, W. B. Saunders, 1975. Data from A. L. Lehninger.

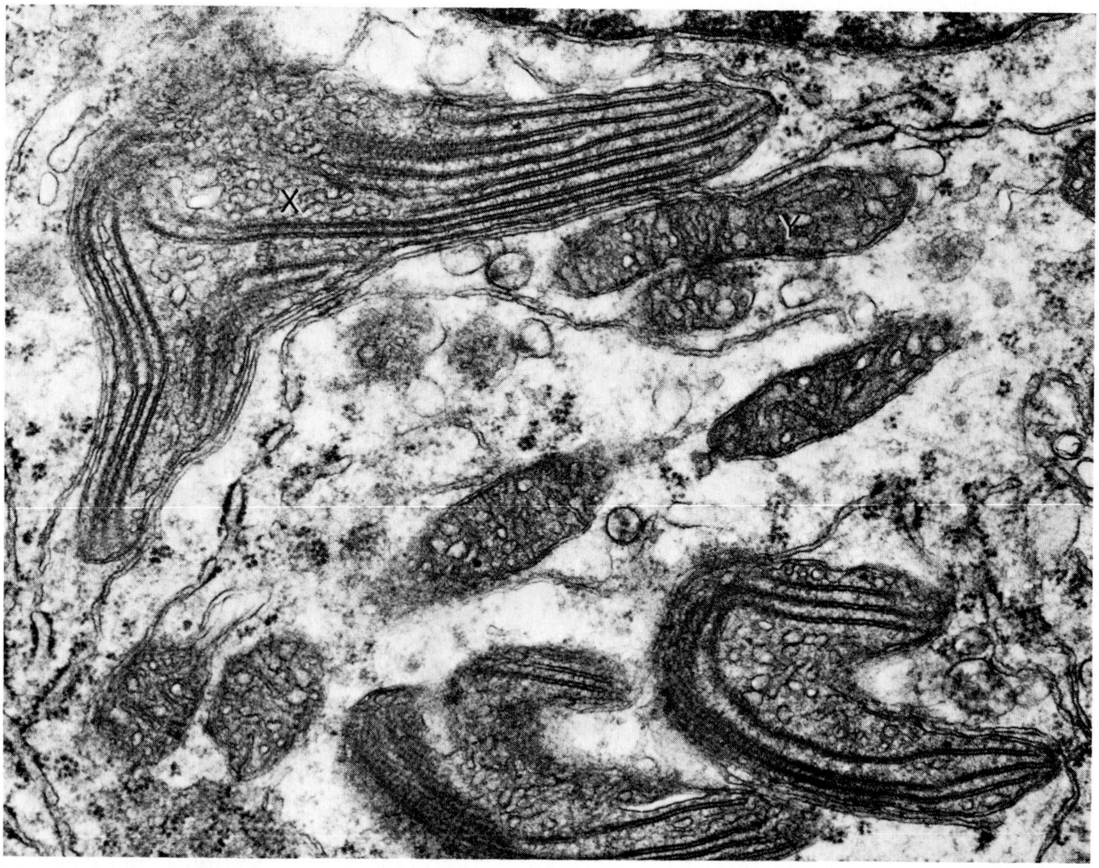

FIGURE 11–6 "Light cell" of the rat epididymis. Some of the mitochondria (X) have cristae that are oriented longitudinally with respect to the long axis of the organelle. These cristae have a periodic substructure. Note that other mitochondria (Y) have a more conventional appearance. × 39,000.

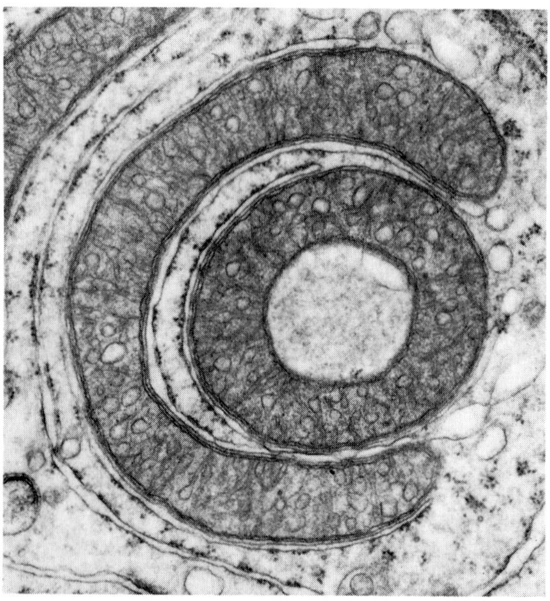

FIGURE 11–7 Sertoli cell of rat testis. The mitochondria in this field are cup-shaped. The shape of their profiles varies according to the plane of section. × 23,000.

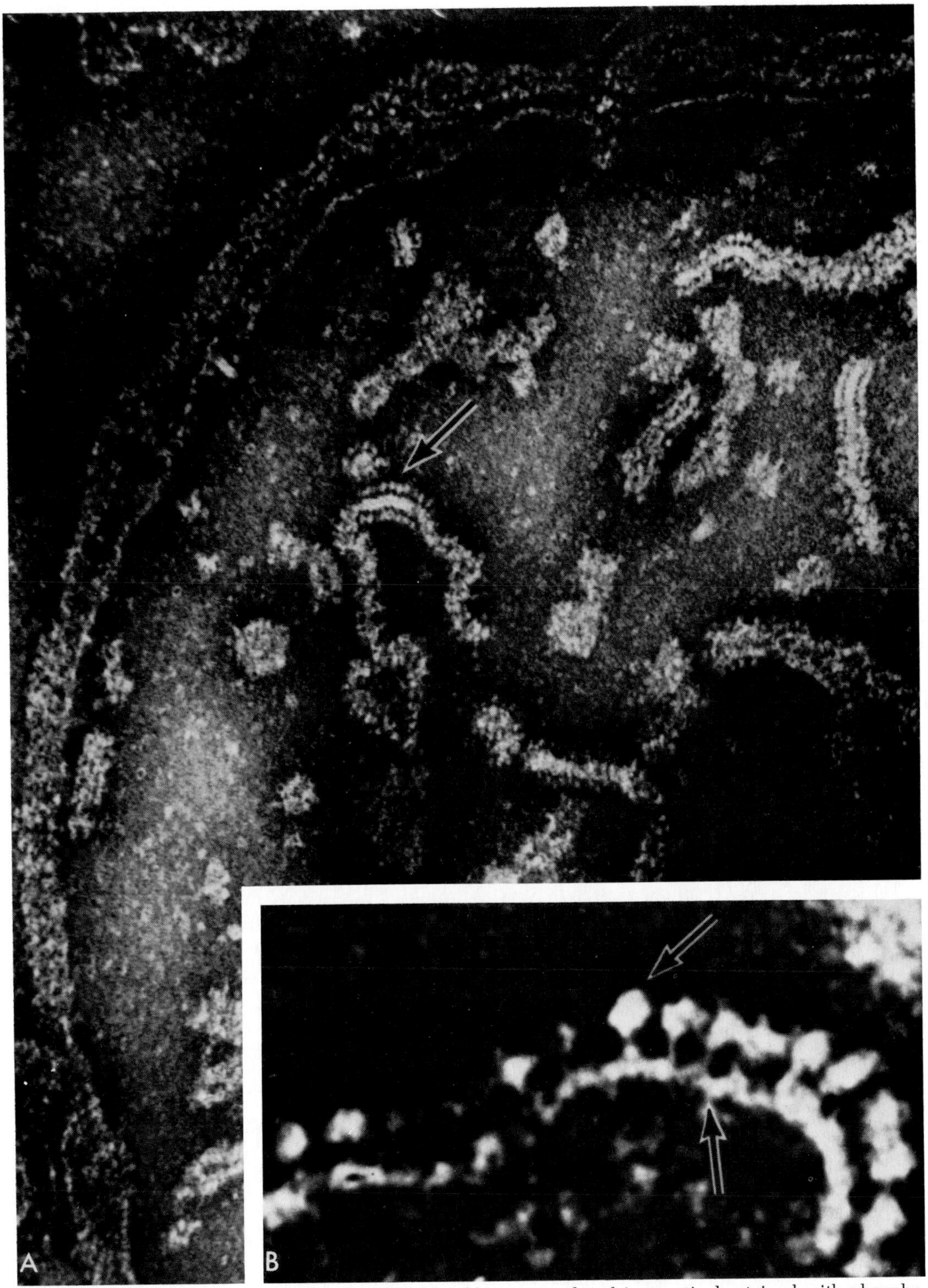

FIGURE 11–8 Electron micrographs of isolated beef heart mitochondria negatively stained with phosphotungstate. The structure of the mitochondria appears light against the dark background of the electron-opaque stain. *A*. Rows of particles are visible on the surfaces of the mitochondrial cristae (*arrow*). *In vivo* the particles are in contact with the mitochondrial matrix. *B*. Part of a crista at higher magnification. The particles are composed of a spherical head, a stalk, and a base piece. *A*. × 165,000. *B*. × 600,000. (Micrographs courtesy of Fernandez-Moran, H., Oda, T., Blair, P. V., et al.: J. Cell Biol., *22*:63, 1964. Reproduced by permission of the Rockefeller University Press.)

place in pathological processes. The mitochondrial swelling and contraction that occur normally in intact cells have been extensively studied in isolated mitochondria by electron microscopy (Fig. 11–9) and by the monitoring of change in the light-scattering ability of the preparations. The low-amplitude physiological swelling and contraction are energy-linked processes, and mitochondria vary in their internal conformation in different respiratory states. The configuration of mitochondria most familiar from electron micrographs of intact cells is known as the *orthodox conformation* (Fig. 11–9*A*). The inner membrane forms typical cristae, and the matrix occupies most of the volume of the mitochondrion, while the outer compartment comprises only a small fraction of the volume. In contrast, in the *condensed conformation* (Fig. 11–9*B*), the volume of the outer compartment is greatly increased at the expense of the matrix as the result of movement of water from the inner to the outer compartment. Thus the expanded outer compartment appears to contain a structureless fluid, while the density of the matrix is increased and the inner membrane is folded apparently at random.

The ratio between ADP and ATP seems particularly important in influencing these conformational changes. In isolated mitochondria, the orthodox state is produced when the amount of ADP is low and there is consequently little oxidative phosphorylation. If ADP is added, respiration increases and the mitochondria assume the condensed form. It is believed that active contraction of proteins in the inner membrane has a role in the conformational changes in mitochondria. Conformational changes, such as the transformation from condensed to orthodox, depend on operation of the electron transport chain, since they are prevented by inhibitors of oxidative phosphorylation, such as cyanide. The fact that in most electron micrographs of intact cells mitochondria usually appear in the orthodox conformation should not be taken to indicate that they are inactive in oxidative phosphorylation. The conditions in the intact cell that can influence the conformation of mitochondria are complex and are not as easily studied as in the case of isolated mitochondria. Perhaps the normal state of mitochondria *in vivo* is intermediate between the extremes seen *in vitro*. Another possibility is that changes in conformation take place while the cells are being fixed for microscopy.

The most important biochemical functions of mitochondria will now be described in some detail. It should be kept in mind that the reactions of the citric acid cycle take place in the matrix, or inner compartment, while electron transport and oxidative phosphorylation are carried out by components of the inner mitochondrial membrane, which is in contact with the matrix. Finally, mitochondrial nucleic acids are found in the inner compartment.

THE CITRIC ACID CYCLE

In Chapter 10 it was indicated that the Embden-Meyerhof pathway serves as a preparatory step in the aerobic metabolism of glucose. The net reaction for the metabolism of glucose, via the Embden-Meyerhof pathway, is

$$\underset{\text{Glucose}}{C_6H_{12}O_6} + 2P_i^{2-} + 2NAD^+ + 2ADP^{3-} \rightarrow 2NADH + 2H^+ + 2H_2O + 2ATP^{4-} + \underset{\text{Pyruvate}}{2CH_3{-}\overset{\overset{\displaystyle O}{\|}}{C}{-}COO^-} \qquad (1)$$

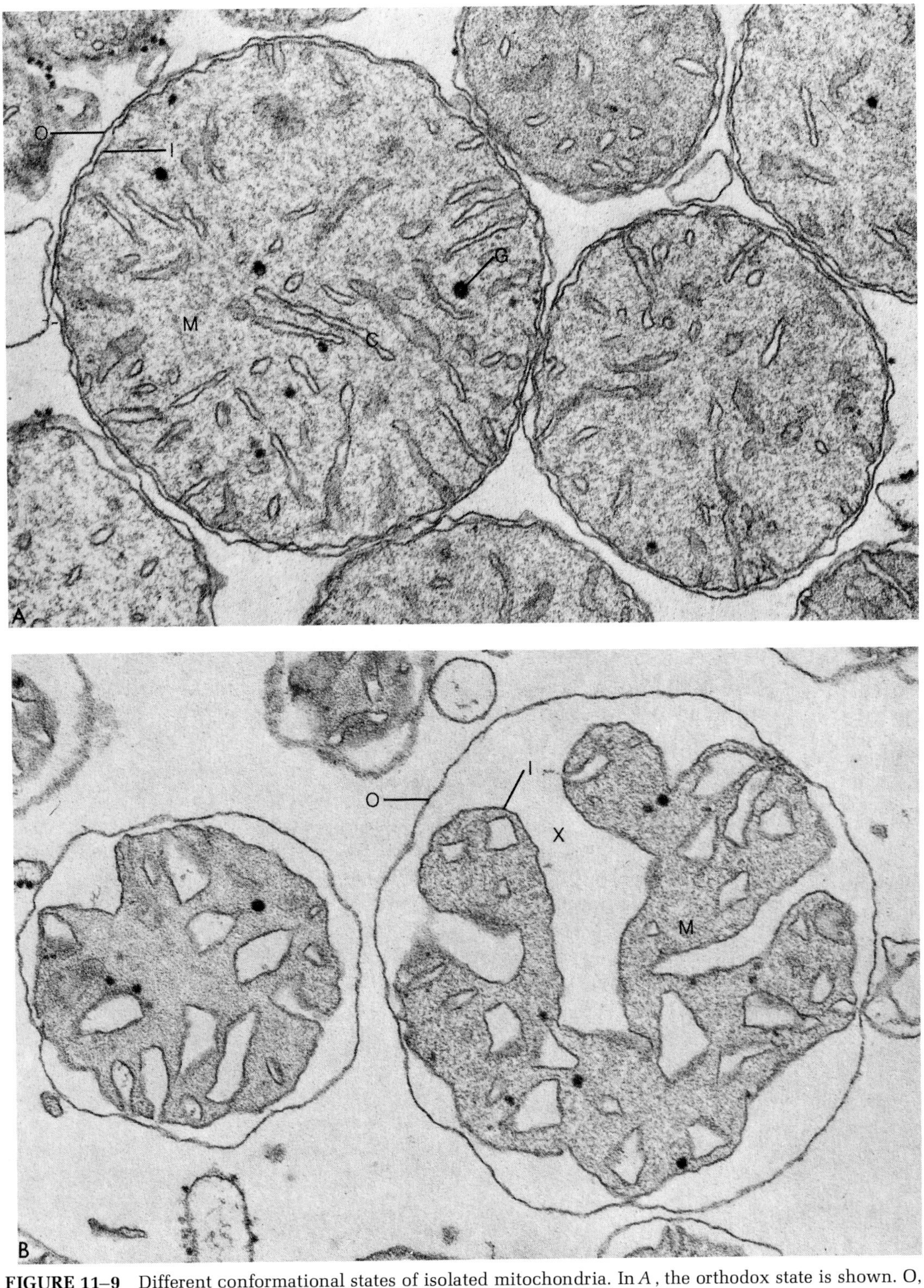

FIGURE 11–9 Different conformational states of isolated mitochondria. In *A*, the orthodox state is shown. O, outer membrane; I, inner membrane; C, crista; M, inner compartment or matrix; G, matrix granule. The outer compartment lies between the outer and inner membranes. *B* illustrates the condensed state. The volume of the outer compartment (X) is greatly expanded relative to that in *A*, with a corresponding condensation of the inner compartment or matrix (M). O, outer membrane; I, inner membrane. × 110,000. (Micrographs courtesy of Hackenbrock, C. R.: J. Cell Biol., *37*:345, 1968. Reproduced by permission of the Rockefeller University Press.)

Since the Embden-Meyerhof pathway is confined to the cytosol (or cytoplasmic matrix), the products of the pathway appear in the cytosol. Under *anaerobic* conditions the pyruvate acts as an acceptor of electrons from NADH, thereby regenerating NAD^+. In the *aerobic* pathway for the metabolism of glucose, the pyruvate and NADH generated in the cytosol by the Embden-Meyerhof pathway follow pathways that differ from those followed in glycolysis. The fate of pyruvate in the aerobic pathway will be considered first.

The Oxidative Decarboxylation of Pyruvate

In the aerobic metabolism of glucose, the pyruvate generated in the cytosol by the Embden-Meyerhof pathway enters the mitochondria where it is converted to acetyl coenzyme A (acetyl CoA) by an oxidative decarboxylation reaction that is catalyzed by the pyruvate dehydrogenase complex. The overall reaction catalyzed by the pyruvate dehydrogenase complex is

$$\underset{\text{Pyruvate}}{CH_3{-}\overset{\overset{\displaystyle O}{\|}}{C}{-}COO^-} + \text{Coenzyme A} + NAD^+ \xrightarrow[\text{pyruvate dehydrogenase complex}]{} CO_2 + NADH + \underset{\text{Acetyl CoA}}{CH_3{-}\overset{\overset{\displaystyle O}{\|}}{C}{-}SCoA} \qquad (2)$$

The pyruvate dehydrogenase complex is inhibited by acetyl CoA and NADH, two products of the reaction. This reaction is irreversible under the conditions existing in the cell. Since the reaction forms a key link between the Embden-Meyerhof pathway of the cytosol and the citric acid cycle in the mitochondria, a closer examination of the reaction and the enzyme complex is warranted.

The pyruvate dehydrogenase complex is very large and contains three different enzymes held together by noncovalent bonds. The three enzymes are pyruvate dehydrogenase, dihydrolipoyl transacetylase, and dihydrolipoyl dehydrogenase. The molecular weight for the isolated enzyme complex ranges between 4 and 9 million, depending upon the tissue from which the enzyme is isolated. Pig heart pyruvate dehydrogenase complex, with a molecular weight of approximately 9 million, appears to contain the following: about 29 molecules of pyruvate dehydrogenase, each with a molecular weight of 194,000; 1 molecule of dihydrolipoyl transacetylase with a molecular weight of 2.5 million; and 8 molecules of dihydrolipoyl dehydrogenase, each with a molecular weight of 112,000. Electron microscopy studies suggest that the 2.5 million-molecular weight transacetylase component consists of a number of subunits and forms the central core of the complex to which the molecules of the two dehydrogenases are attached. In addition, the complex contains the following three prosthetic groups: thiamine pyrophosphate on pyruvate dehydrogenase; lipoic acid covalently bound to the ϵ-amino group of a lysyl residue of transacetylase; and flavin adenine dinucleotide (FAD) associated with dihydrolipoyl dehydrogenase. The structures of these three prosthetic groups are depicted in Figure 11–10.

The thiamine pyrophosphate is synthesized in the cell by the transfer of a pyrophosphate group from ATP to thiamine, a vitamin that is obtained from the diet. A chronic dietary deficiency of this vitamin leads to the disease beriberi. The more acute thiamine deficiency observed in alcoholics who ingest little other than alcohol leads to Wernicke's encephalopathy. Both of these deficiency con-

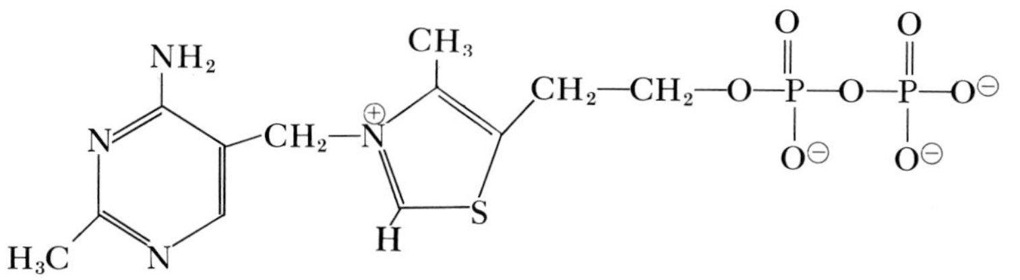

Thiamine pyrophosphate

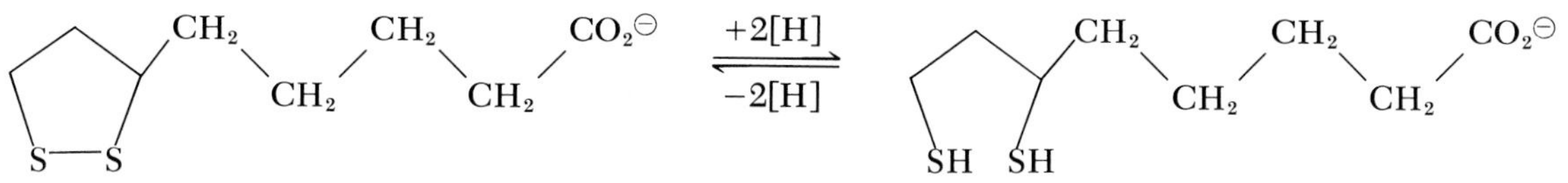

Lipoate

Dihydrolipoate

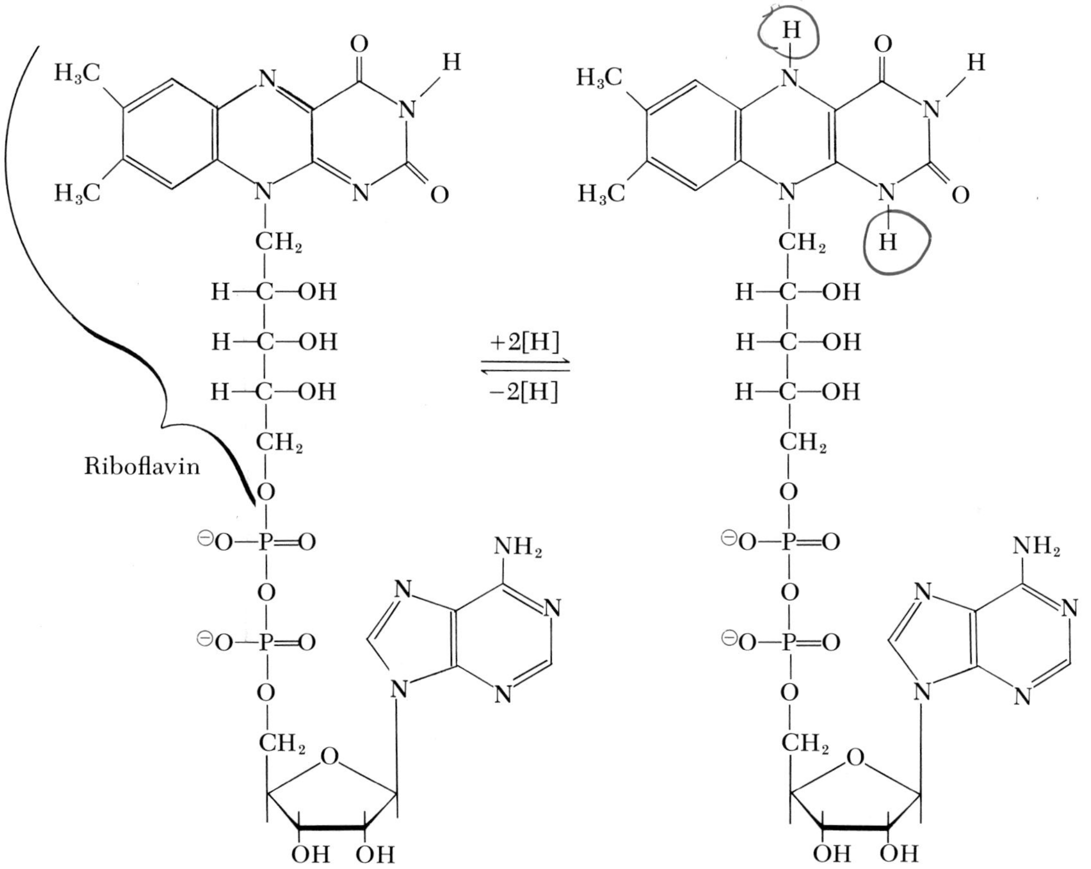

Flavin adenine dinucleotide (FAD)

Reduced flavin adenine dinucleotide ($FADH_2$)

FIGURE 11–10 The structures of the prosthetic groups associated with the pyruvate dehydrogenase complex.

ditions, particularly the latter, involve disorders of the nervous system. Impairment of the nervous system in thiamine deficiencies is not surprising when one considers the great dependence of the nervous system on glucose metabolism for energy and the key role that the pyruvate dehydrogenase complex plays in carbohydrate metabolism.

The synthesis of FAD in the cell requires ATP and riboflavin, another vitamin obtained from the diet. Proteins containing bound flavins, such as FAD, are known as flavoproteins. As shown in Figure 11–10, lipoic acid and FAD can be reduced to form dihydrolipoic acid and reduced FAD ($FADH_2$), respectively.

In addition to the three prosthetic groups listed, oxidative decarboxylation of pyruvate by the pyruvate dehydrogenase complex also requires NAD^+ and coenzyme A. The structures of coenzyme A and acetyl coenzyme A are given in Figure 11–11. The synthesis of coenzyme A in the cell requires ATP, cysteine, and another vitamin, pantothenic acid, which is also a dietary requirement. The acetyl group of acetyl CoA is attached to coenzyme A by a thioester linkage. Acetyl CoA is an energy-rich compound, as indicated by the large negative standard free energy change upon hydrolysis.

$$\underset{\text{Acetyl CoA}}{CH_3{-}\overset{\overset{\displaystyle O}{\|}}{C}{-}SCoA} + H_2O \longrightarrow \underset{\text{Acetate}}{CH_3{-}COO^-} + H^+ + \underset{\text{Coenzyme A}}{CoASH}$$

$$\Delta G^{\circ\prime} = -7.5 \text{ kcal/mole}$$

The oxidative decarboxylation of pyruvate involves four consecutive reactions, all of which take place on the multienzyme pyruvate dehydrogenase complex (Fig. 11–12). The first reaction is catalyzed by the pyruvate dehydrogenase component of the complex and results in the decarboxylation of the pyruvate and the formation of the hydroxyethyl derivative of the bound thiamine pyrophosphate. The second reaction is the oxidation of the hydroxyethyl group to an acetyl group by the lipoate of the transacetylase component and the concomitant transfer of the acetyl group from the thiamine pyrophosphate to the lipoate group. The third reaction is the transfer of the acetyl group from the acetyllipoate moiety of the transacetylase to coenzyme A to form acetyl CoA. The final reaction is catalyzed by dihydrolipoyl dehydrogenase. In this reaction the dihydrolipoyl moiety of the transacetylase is oxidized back to a lipoyl moiety by the flavoprotein which then transfers the electrons to NAD^+, forming NADH.

As depicted schematically in Figure 11–12, the lipoate group and the side chain of the lysyl residue to which it is attached form a long flexible arm. Rotation of this flexible arm may provide the necessary sequential contact between the prosthetic groups of the three enzymes in the pyruvate dehydrogenase complex.

Oxidation of Acetyl Coenzyme A Via the Citric Acid Cycle

The acetyl CoA produced by the oxidative decarboxylation of pyruvate (reaction 2) enters the citric acid cycle, which is shown in Figure 11–13. The main function of the cycle is to generate ATP by oxidizing the acetyl group of acetyl CoA. This is accomplished by one substrate-level phosphorylation and by generation of reduced coenzymes that are used to produce ATP through oxidative

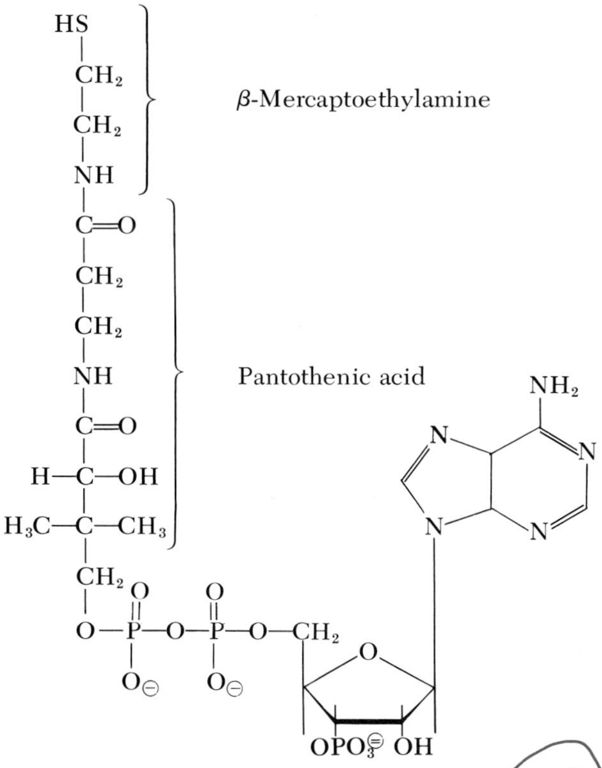

Coenzyme A (CoA—SH)

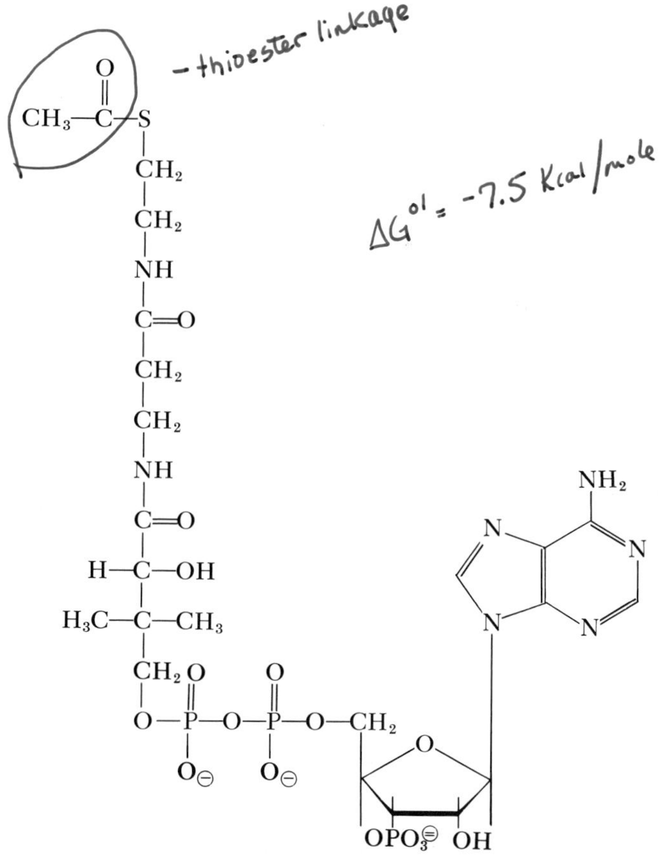

Acetyl coenzyme A (Acetyl CoA)

FIGURE 11–11 The structures of coenzyme A and acetyl coenzyme A.

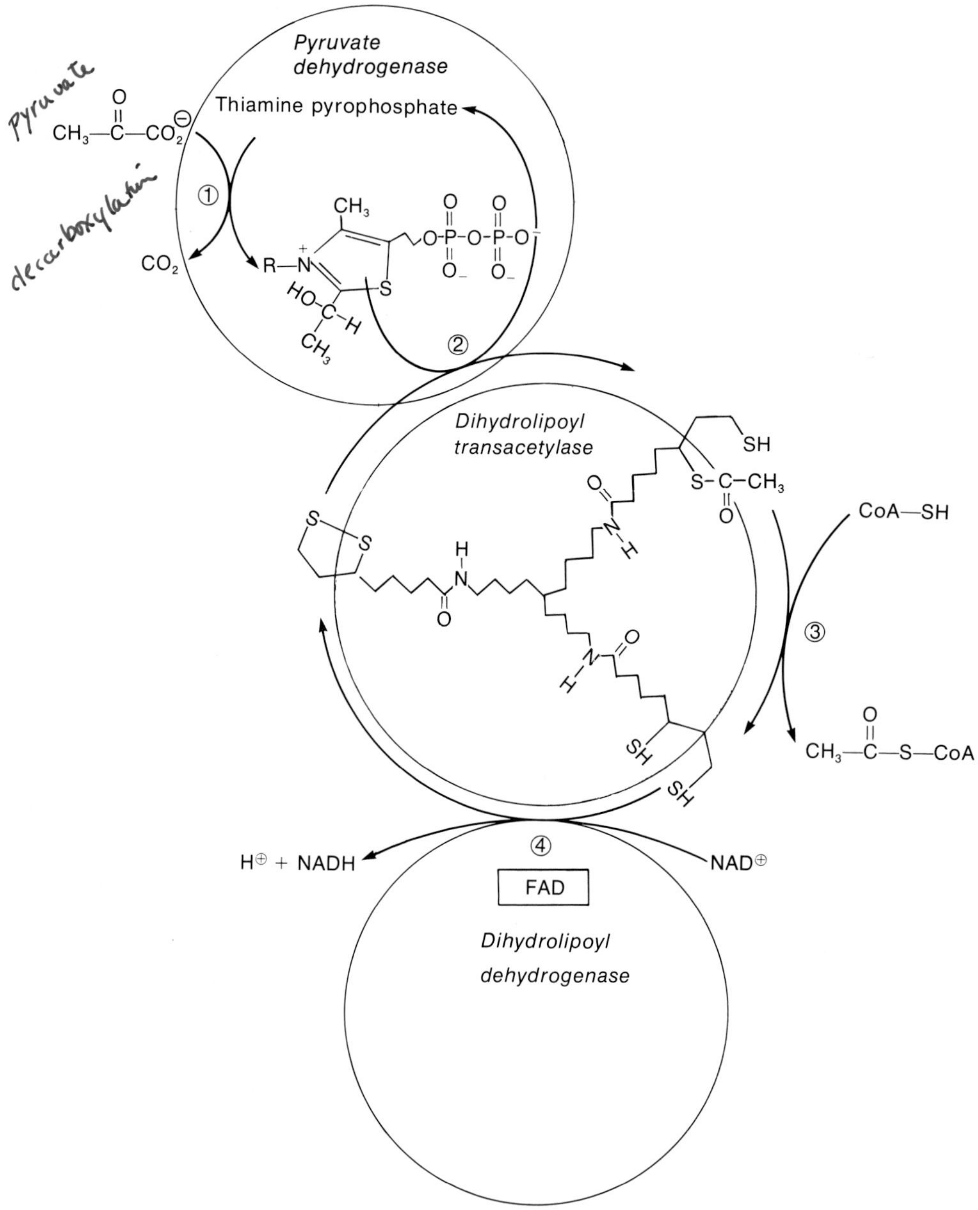

FIGURE 11–12 A schematic representation of the reactions involved in the oxidative decarboxylation of pyruvate.

phosphorylation, as explained later in this chapter. The separate reactions that comprise the citric acid cycle are considered in the following sections.

CONDENSATION OF ACETYL CoA AND OXALOACETATE

Acetyl CoA enters the citric acid cycle by condensing with oxaloacetate to form citrate, a tricarboxylic acid.

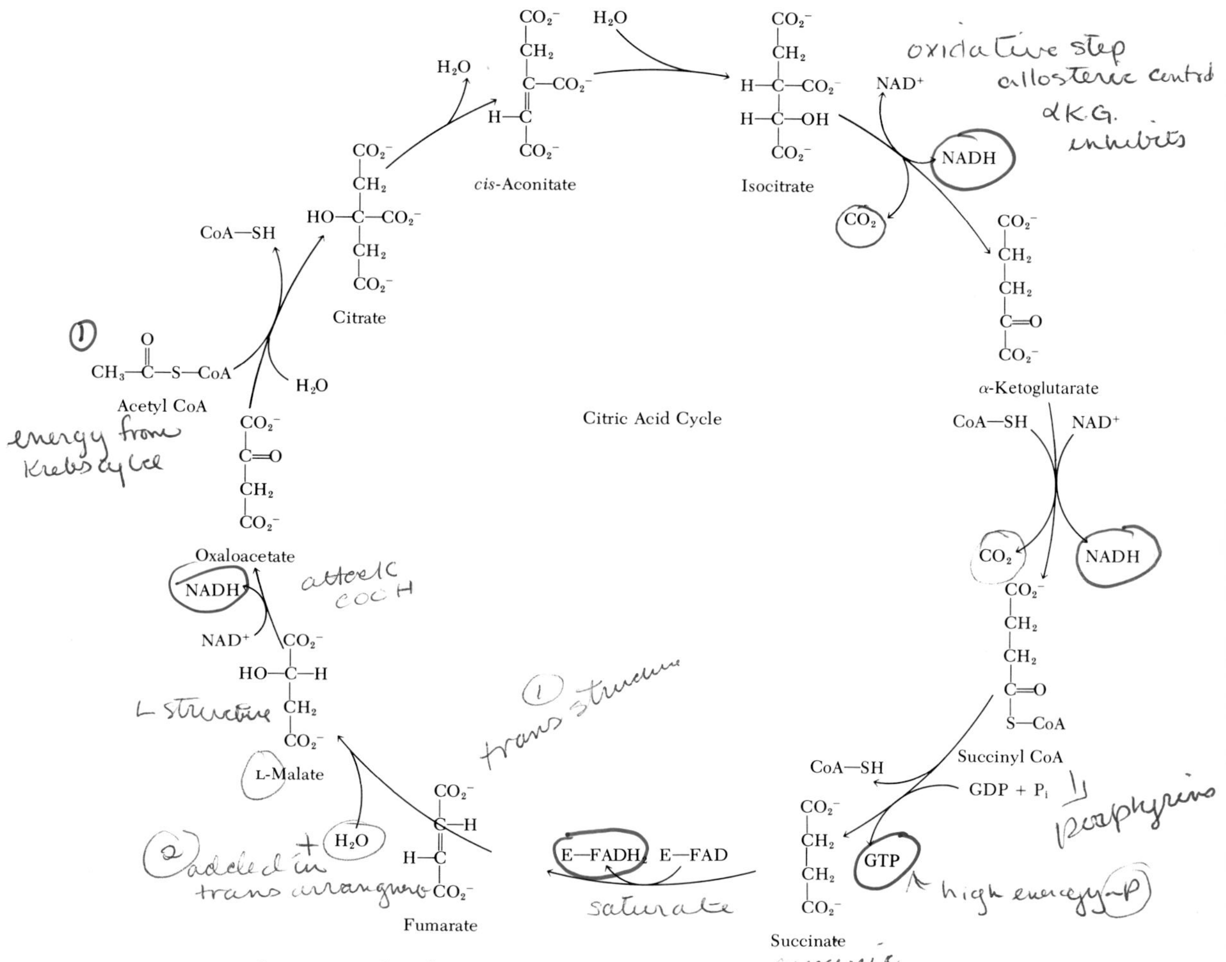

FIGURE 11–13 The citric acid cycle.

(3) $$\underset{\text{Acetyl CoA}}{{}^*CH_3\text{—}{}^*\overset{\overset{O}{\|}}{C}\text{—S—CoA}} + \underset{\text{Oxaloacetate}}{C(=O)(COO^-)(CH_2COO^-)} \xrightarrow[\text{H}_2\text{O}\quad \text{citrate synthetase}]{\boxed{-\ |\ \text{ATP}}} \underset{\text{Citrate}}{HO\text{—}C(COO^-)({}^*CH_2{}^*COO^-)(CH_2COO^-)} + H^+ + CoASH$$

This condensation reaction is catalyzed by the enzyme citrate synthetase and is essentially irreversible owing to the cleavage of the energy-rich thioester linkage in acetyl CoA. The asterisks in reaction 3 and subsequent reactions indicate the two carbons derived from acetyl CoA. This enzyme-catalyzed reaction is stereospecific, and the substituent groups of the citrate formed have the steric arrangement indicated in reaction 3.

The primary function of the citric acid cycle is to make possible the generation of ATP from acetyl CoA. Citrate synthetase, which is allosterically inhibited by ATP, plays an important role in the regulation of the cycle. When ATP concentration is high, citrate synthetase becomes less active and the rate of ATP production from acetyl CoA decreases.

ISOMERIZATION OF CITRATE TO ISOCITRATE

The isomerization of citrate to isocitrate is a two-step process, both parts of which are catalyzed by the enzyme aconitase. In the first step, aconitase catalyzes the dehydration of citrate to form *cis*-aconitate (reaction 4), and in the second step, aconitase catalyzes the hydration of *cis*-aconitate to form isocitrate (reaction 5).

(4)

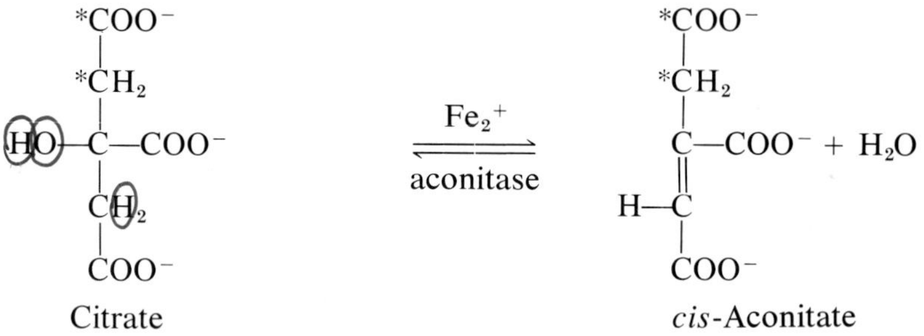

(5)

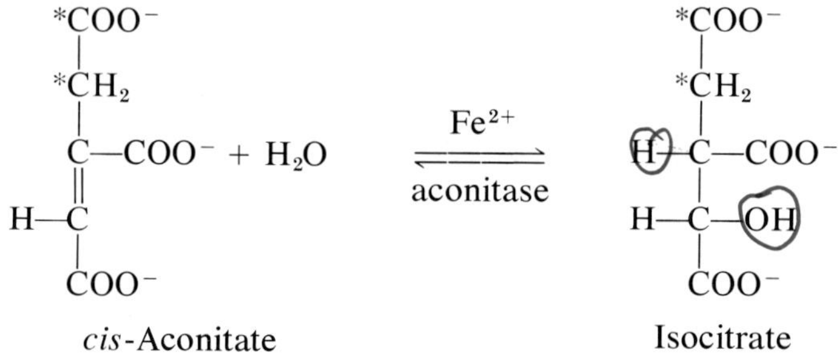

Both of these reactions are readily reversible. Although citrate is a symmetrical molecule, and the —$^*CH_2$—$^*COO^-$ group derived from acetyl CoA and the —CH_2COO^- group derived from oxaloacetate are chemically identical, aconitase can distinguish between these two chemically identical groups and always introduces the double bond as shown in reaction 4. This specificity results from the asymmetry of the aconitase molecule. A simple way to visualize how asymmetry in the enzyme might lead to this specificity is presented in Figure 11–14. If the binding of citrate to aconitase to form the enzyme-substrate complex involves interactions between three groups of the substrate and three specific sites on the enzyme, then the two —CH_2—COO^- groups of the enzyme-bound citrate molecule clearly have different spatial relationships to the enzyme and are sterically distinguishable.

CONVERSION OF ISOCITRATE TO α-KETOGLUTARATE

Isocitrate dehydrogenase catalyzes the oxidation and decarboxylation of isocitrate, which results in the formation of α-ketoglutarate.

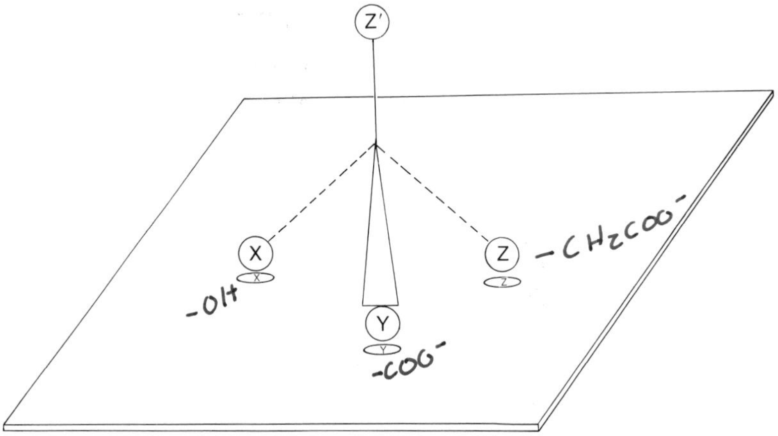

FIGURE 11–14 If x, y, and z represent the binding sites on the enzyme that are specific for the —OH,—COO^- and —CH_2COO^- or —*CH_2—*COO^- groups, respectively, and X, Y, Z, and Z′ represent the —OH, —COO^-, —CH_2—COO^- and —*CH_2—*COO^- groups of citrate, respectively, the binding of citrate to the enzyme by specific interactions at all three sites requires that Z′, the —*CH_2—*COO^- group, be directed away from the enzyme surface.

(6)

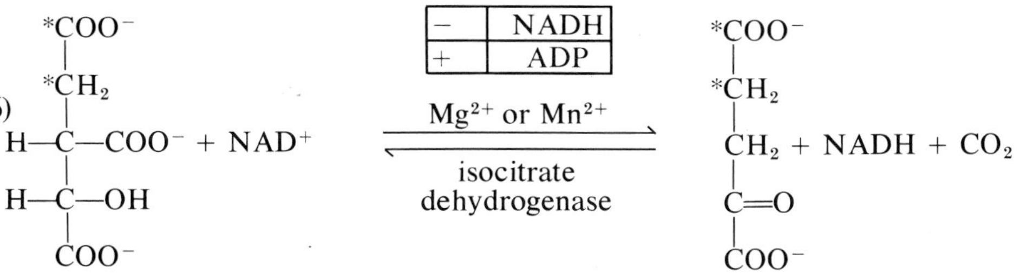

Isocitrate α-Ketoglutarate

This reaction proceeds through the formation of an intermediate, oxalosuccinate, which remains bound to the enzyme.

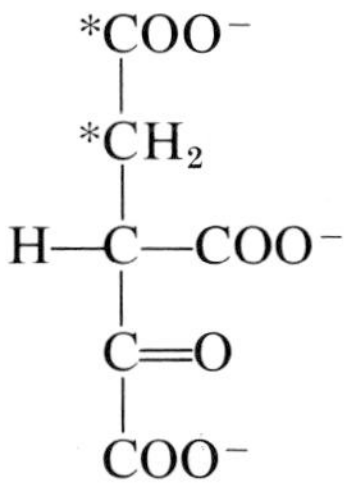

Oxalosuccinate

This oxalosuccinate intermediate results from the oxidation of the hydroxyl group of isocitrate by NAD^+, and it rapidly undergoes decarboxylation.

Isocitrate dehydrogenase is another important control point in the cycle. This enzyme is allosterically activated by ADP and inhibited by NADH. As described later, each NADH in the mitochondria leads to the generation of three molecules of ATP. Thus a high concentration of NADH represents a large potential for ATP generation without the further functioning of the citric acid cycle.

OXIDATIVE DECARBOXYLATION OF α-KETOGLUTARATE

The next reaction of the cycle is the oxidative decarboxylation of α-ketoglutarate to form succinyl coenzyme A.

(7) 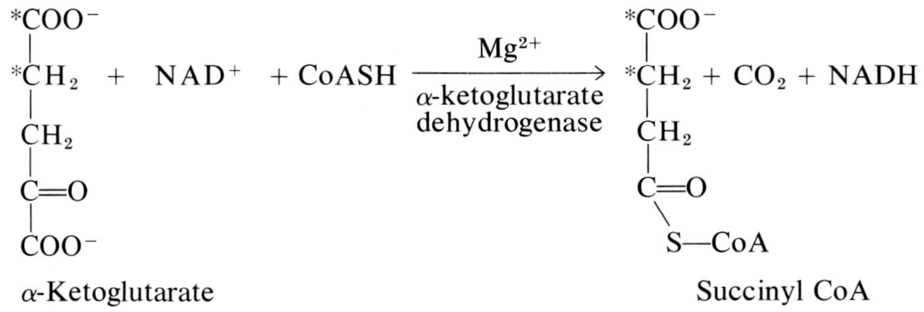

This reaction is essentially irreversible, and the mechanism and structure of the α-ketoglutarate dehydrogenase catalyzing the reaction are very reminiscent of the pyruvate dehydrogenase complex previously discussed.

A SUBSTRATE-LEVEL PHOSPHORYLATION OF GDP

Succinyl CoA contains an energy-rich thioester linkage, the hydrolysis of which is coupled to the generation of GTP in the next reaction of the cycle.

(8) 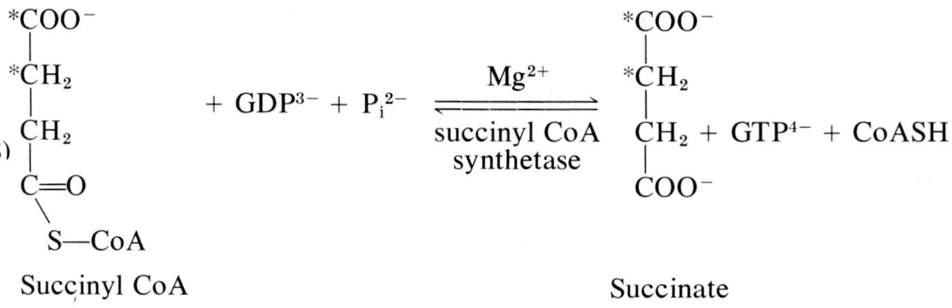

This substrate-level phosphorylation, leading to the generation of GTP, is catalyzed by succinyl CoA synthetase. The generation of a GTP is equivalent to the generation of an ATP, since they are in equilibrium via a reaction catalyzed by nucleoside diphosphokinase.

$$\text{GTP} + \text{ADP} \xrightleftharpoons[]{\text{nucleoside diphosphokinase}} \text{GDP} + \text{ATP}$$

Although the primary purpose of the citric acid cycle in the cell is to make possible the generation of ATP, this is the only reaction in the cycle leading *directly* to a nucleoside triphosphate.

OXIDATION OF SUCCINATE

Succinate is oxidized by a flavoprotein, succinate dehydrogenase, which contains a covalently bound FAD molecule that acts as the electron acceptor. Representing the enzyme containing the covalently bound FAD by E—FAD, the reaction is

(9)

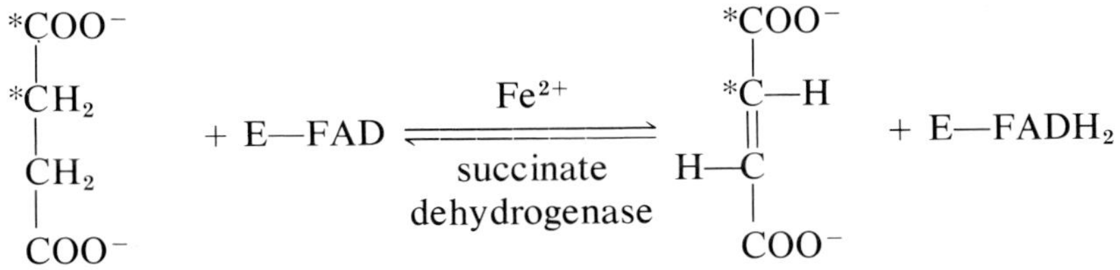

$$\text{Succinate} \begin{array}{c} *\text{COO}^- \\ | \\ *\text{CH}_2 \\ | \\ \text{CH}_2 \\ | \\ \text{COO}^- \end{array} + \text{E—FAD} \underset{\substack{\text{succinate} \\ \text{dehydrogenase}}}{\overset{\text{Fe}^{2+}}{\rightleftharpoons}} \begin{array}{c} *\text{COO}^- \\ | \\ *\text{C—H} \\ \| \\ \text{H—C} \\ | \\ \text{COO}^- \end{array} + \text{E—FADH}_2 \quad \text{Fumarate}$$

Malonate, a dicarboxylic acid whose structure closely resembles succinate, is a competitive inhibitor of succinate dehydrogenase.

$$\begin{array}{c} \text{COO}^- \\ | \\ \text{CH}_2 \\ | \\ \text{COO}^- \end{array}$$

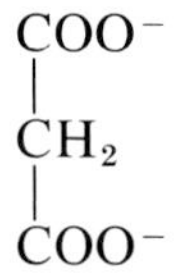

Malonate

As high concentrations, malonate poisons respiration, that is, it stops the uptake of oxygen by a tissue such as muscle. The observation that fumarate, citrate, and other intermediates in the pathway each led to a stoichiometric accumulation of succinate in malonate-poisoned muscle played a key role in the elucidation of the cyclical nature of this pathway.

THE HYDRATION OF FUMARATE

Fumarate is a symmetrical molecule. Fumarase, the enzyme catalyzing the reversible addition of H_2O to the double bond of fumarate, cannot distinguish between the two ends of the molecule when it binds fumarate. As a result, the two carbons that were marked with asterisks to indicate their origin from acetyl CoA become randomized in malate, the product of the next reaction of the cycle.

(10)

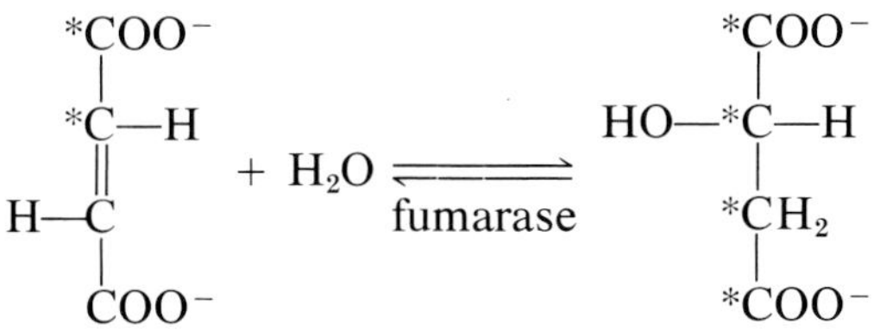

$$\text{Fumarate} \begin{array}{c} *\text{COO}^- \\ | \\ *\text{C—H} \\ \| \\ \text{H—C} \\ | \\ \text{COO}^- \end{array} + H_2O \underset{\text{fumarase}}{\rightleftharpoons} \begin{array}{c} *\text{COO}^- \\ | \\ \text{HO—}*\text{C—H} \\ | \\ *\text{CH}_2 \\ | \\ *\text{COO}^- \end{array} \quad \text{L-malate}$$

The four asterisks on L-malate indicate that the malate produced is an equimolar mixture of the following:

$$\begin{array}{c} *\text{COO}^- \\ | \\ \text{HO—}*\text{C—H} \\ | \\ \text{CH}_2 \\ | \\ \text{COO}^- \end{array} \qquad \text{and} \qquad \begin{array}{c} \text{COO}^- \\ | \\ \text{HO—C—H} \\ | \\ *\text{CH}_2 \\ | \\ *\text{COO}^- \end{array}$$

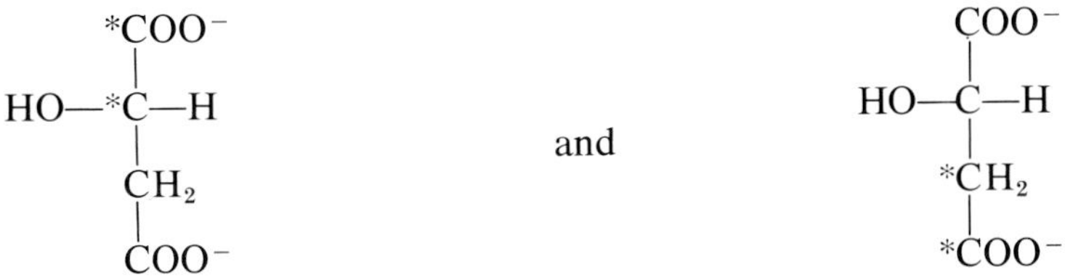

COMPLETION OF THE CYCLE—THE REGENERATION OF OXALOACETATE

The final step of the cycle is the oxidation of malate, a reaction that is catalyzed by the enzyme malate dehydrogenase.

(11) 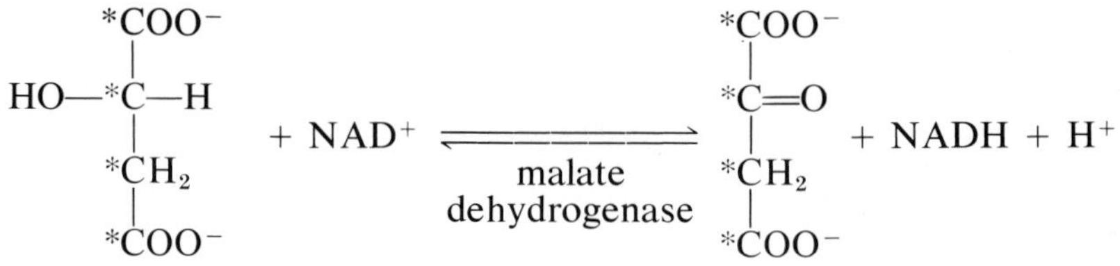

L-malate — Oxaloacetate

Since the oxaloacetate that was used in the initial reaction of the cycle has been regenerated, the cycle is now complete. Again, the four asterisks on oxaloacetate indicate that the two carbons derived from acetate are randomized, and that the oxaloacetate consists of an equimolar mixture of the following:

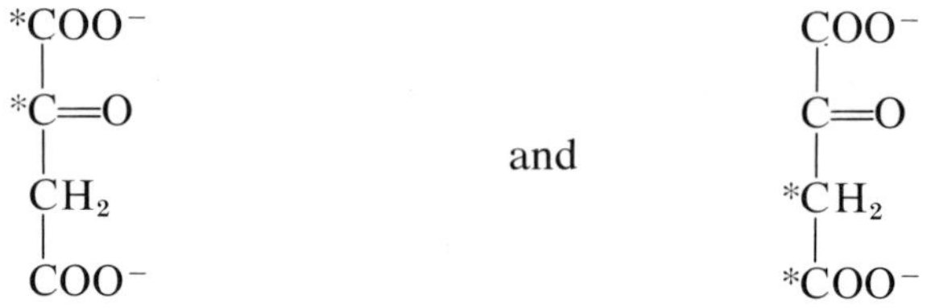

The Net Reaction of the Citric Acid Cycle

The net reaction of the citric acid cycle can be obtained by summing up reactions 3 through 11, the nine reactions that constitute the cycle.

(12)
$$\text{Acetyl Co A} + 3NAD^+ + E\text{—}FAD + GDP^{3-} + P_i^{2-} + 2H_2O \xrightarrow{\text{citric acid cycle}} 2CO_2 + CoASH + 3NADH + E\text{—}FADH_2 + 2H^+ + GTP^{4-}$$

From equation 12 it is apparent that the net effect of the cycle is to oxidize the acetyl group of acetyl CoA to CO_2 with the concomitant reduction of NAD^+ and FAD. Molecular O_2 is not involved in the cycle. Although the net effect is the conversion of the acetyl group to $2CO_2$, inspection of reactions 3 through 11 reveals that the two carbon atoms appearing as CO_2 in the cycle are not the two carbons put into the cycle as acetyl CoA. The two carbons put into the cycle as acetyl coenzyme A appear in oxaloacetate after one turn of the cycle and will appear as CO_2 only on subsequent turns of the cycle.

With the exception of succinate dehydrogenase, the enzymes of the citric acid cycle are located in the inner compartment, or matrix, of the mitochondria. Succinate dehydrogenase appears to be associated with the inner mitochondrial membrane.

Significance of the Citric Acid Cycle

The citric acid cycle is known also as the tricarboxylic acid cycle and as the Krebs cycle, since Hans Krebs was the first to show that the reactions formed a cycle. If any pathway can be said to form the central core of intermediary metabolism, it is the citric acid cycle. We have considered how glucose is converted to acetyl CoA, which is then fed into the cycle. In addition, metabolism of amino acids derived from proteins and metabolism of fatty acids derived from lipids also lead to the generation of acetyl CoA, which is then fed into the citric acid cycle. Therefore, the cycle is of central importance in the metabolism of all the major foodstuffs: carbohydrates, proteins, and lipids.

The ultimate function of the citric acid cycle is to make possible the generation of energy from the acetyl CoA derived from foodstuffs. The net reaction given in equation 12 indicates that each molecule of acetyl CoA entering the cycle gives rise to 1 molecule of GTP, 3 molecules of NADH and 1 molecule of E—$FADH_2$. When the electrons from these reduced coenzymes are passed via the electron transport chain to molecular oxygen, each molecule of NADH generated by the cycle leads to the generation of 3 molecules of ATP and each molecule of E—$FADH_2$ leads to the generation of 2 molecules of ATP.

The *cyclical* nature of the citric acid cycle (Fig. 11–13) is very important because it enables the intermediates of the cycle to have a catalytic role in the conversion of acetyl CoA to CO_2. If there are no cycle intermediates, the cycle cannot function and acetyl CoA cannot be converted to CO_2. On the other hand, a single molecule of any of the cycle intermediates can lead to the conversion of many molecules of acetyl CoA to CO_2 because the intermediates are regenerated with each turn of the cycle.

ELECTRON TRANSPORT CHAIN AND OXIDATIVE PHOSPHORYLATION

In the previous section it was shown that the conversion of a molecule of pyruvate to acetyl CoA by the pyruvate dehydrogenase complex results in the generation of a molecule of NADH in the mitochondria. In addition, the oxidation of the acetyl group of acetyl CoA via the citric acid cycle leads to the generation of one E—$FADH_2$ molecule and three molecules of NADH. In the present section, the fate of the reduced coenzymes E—$FADH_2$ and NADH will be examined, and their role in the generation of ATP by the process of oxidative phosphorylation will be described.

The Respiratory Chain

In aerobic metabolism, mitochondrial NADH and E—$FADH_2$ are oxidized to NAD^+ and E—FAD, respectively. The electrons removed from them are transferred through a complex electron transport chain to molecular oxygen, which acts as the ultimate electron acceptor. This electron transport chain is known as the respiratory chain, and it exists as assemblies of electron carriers in the inner membrane of the mitochondrion. The sequence of carriers in each of these respiratory assemblies is

$$NADH \rightarrow FP \rightarrow CoQ \rightarrow cyt.b \rightarrow cyt.c_1 \rightarrow cyt.c \rightarrow cyt.a + a_3 \rightarrow O_2$$
$$\uparrow$$
$$E\text{—}FADH_2$$

The electrons from NADH feed into the respiratory chain at its upper end, reducing first a flavoprotein (FP), NADH dehydrogenase. The prosthetic group of NADH dehydrogenase is flavin mononucleotide (FMN) (Fig. 11–15), and the reduction of FMN requires two electrons. The electrons from reduced NADH dehydrogenase (FPH_2) then pass to coenzyme Q (CoQ), forming reduced coenzyme Q ($CoQH_2$). Reduced succinate dehydrogenase (E—$FADH_2$) feeds its electrons into the electron transport chain at the level of CoQ, as indicated. Coenzyme Q, also known as ubiquinone, is a substituted benzoquinone with a very large hydrophobic isoprenoid side chain (Fig. 11–15). Reduction of CoQ to $CoQH_2$ also requires two electrons. Thus all of the reactions in the respiratory chain through the level of CoQ consist of two electron steps.

The electrons from $CoQH_2$ are transferred to the next carrier in the electron transport chain, cytochrome *b*, which is the first of five cytochromes in the chain. The cytochromes are all proteins that contain a *heme* prosthetic group, an iron

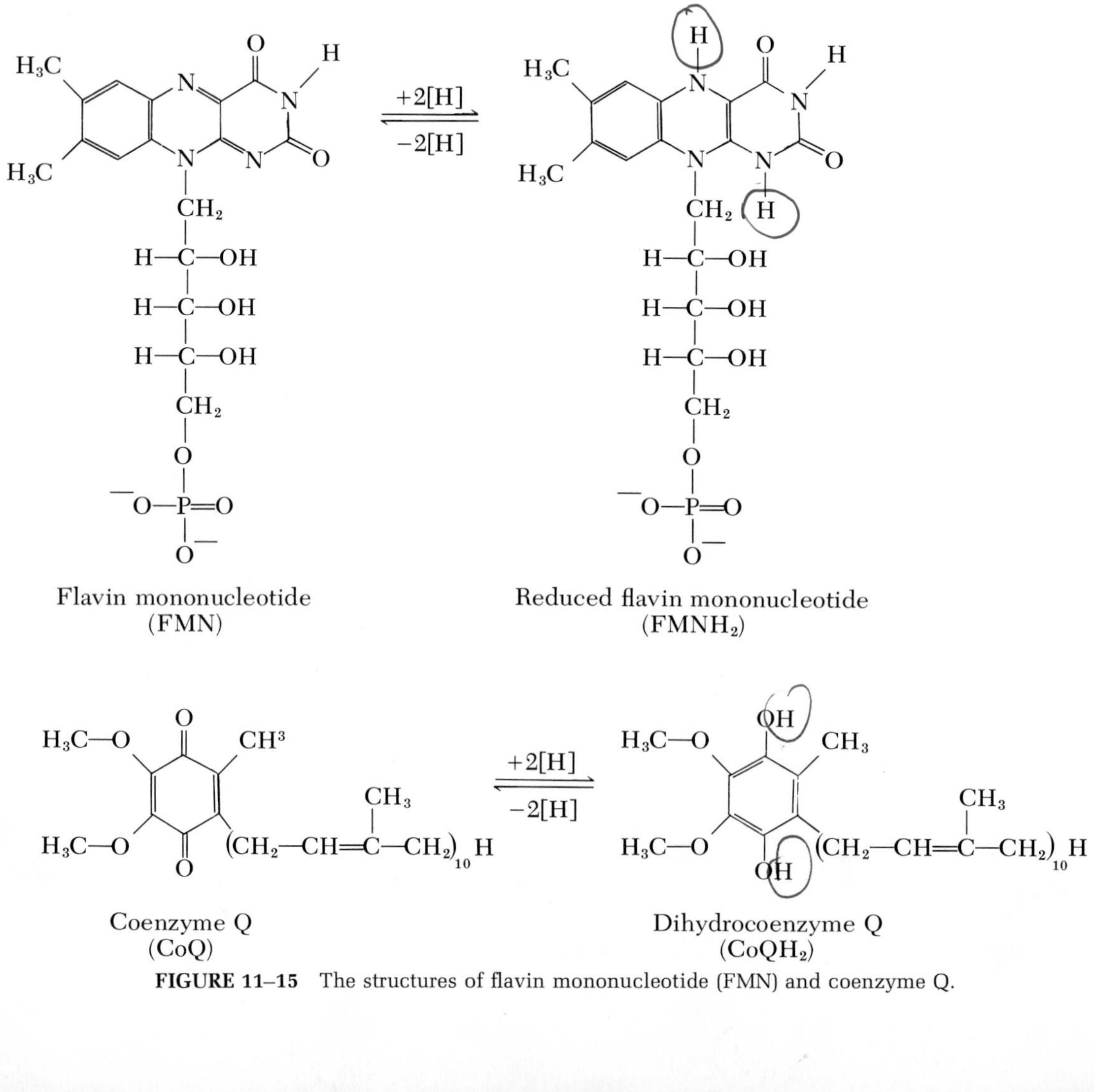

FIGURE 11–15 The structures of flavin mononucleotide (FMN) and coenzyme Q.

atom chelated by a porphyrin. The heme group of cytochrome *b* is identical in structure to the heme groups of hemoglobin and myoglobin (see Chapter 2), and it is bound noncovalently to the protein of cytochrome *b*. The heme groups in cytochrome c_1 and cytochrome *c* are also identical in structure to the heme in hemoglobin and myoglobin, but the binding of heme to the proteins in cytochromes c_1 and *c* is by covalent thioether bonds. These bonds are formed by the addition of the —SH groups of two cysteinyl residues of the proteins to the two vinyl groups, $-CH=CH_2$, of the heme, as depicted in Figure 11–16*A*. Cytochromes *a* and a_3 exist together in a complex known as *cytochrome oxidase*. The heme groups of cytochromes *a* and a_3 are known as heme A and have a structure that differs from the hemes of the other cytochromes. In heme A a formyl group,

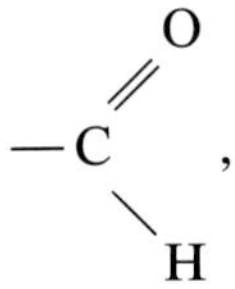

, replaces one of the $-CH_3$ groups, and a branched-chain hydrocarbon group containing an —OH group replaces one of the $-CH=CH_2$ groups, as shown in Figure 11–16*B*.

Even though the cytochromes possess heme groups, their properties are very different from hemoglobin and myoglobin. For example, they do not bind O_2 reversibly. Instead, the cytochromes function as electron carriers by undergoing reversible oxidation-reduction reactions in which the iron atom cycles between the ferric state (Fe^{3+}) and the ferrous state (Fe^{2+}). Since this is a one-electron change, the electron transport chain involves one-electron steps after the electrons have passed the point of CoQ.

The final step in the respiratory chain is the transfer of electrons from the reduced cytochrome oxidase complex to molecular oxygen, forming H_2O. This reduction of oxygen to form H_2O requires four electrons.

$$O_2 + 4H^+ + 4e^- \rightarrow 2\,H_2O$$

In addition to cytochromes *a* and a_3, the cytochrome oxidase complex contains two copper atoms, which cycle between the cuprous state (Cu^+) and the cupric state (Cu^{2+}) when electrons are transferred from cytochromes *a* and a_3 to molecular oxygen. It is thought that the flow of electrons is from cytochrome *a* to cytochrome a_3 to Cu^{++} to molecular oxygen, but the details of the process are not well understood at the present time.

From the foregoing discussion, it is apparent that the electron transport chain functions as a bucket brigade for electrons, with each carrier cycling between its oxidized and reduced states as the electrons are passed down the chain. This cycling of the carriers between their oxidized and reduced states is more clearly seen in the following representation of the respiratory chain

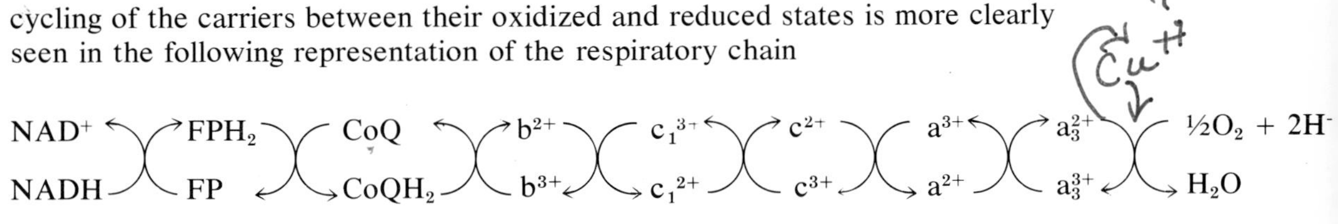

These oxidation-reduction reactions involving the carriers in the respiratory chain are accompanied by free energy changes that the cell utilizes to drive the formation of ATP from ADP. Before the process of ATP formation, known as *oxidative phosphorylation,* is considered, oxidation-reduction equilibria and the free energy changes involved will be discussed briefly.

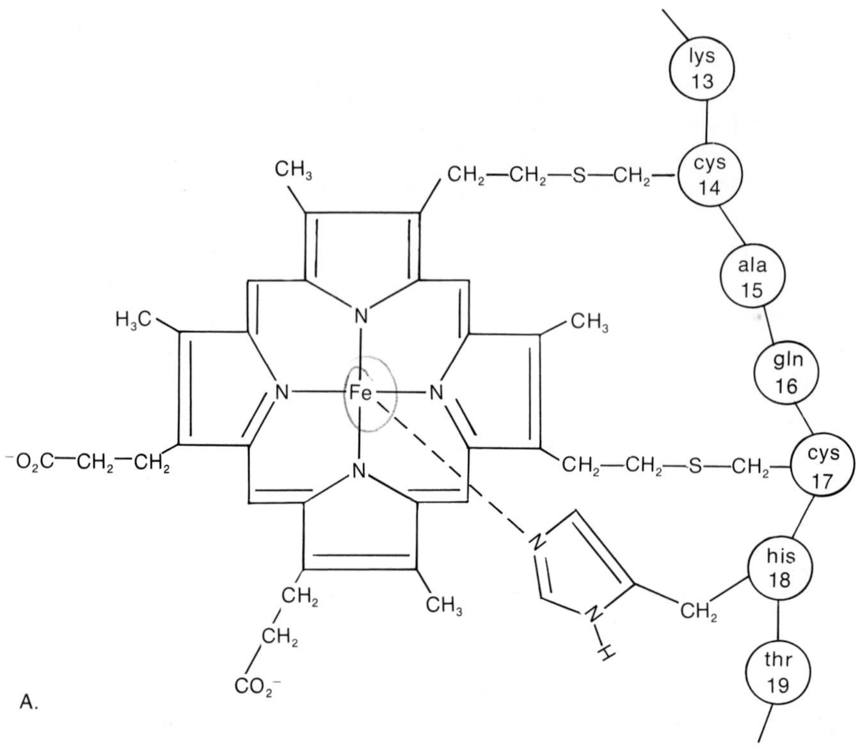

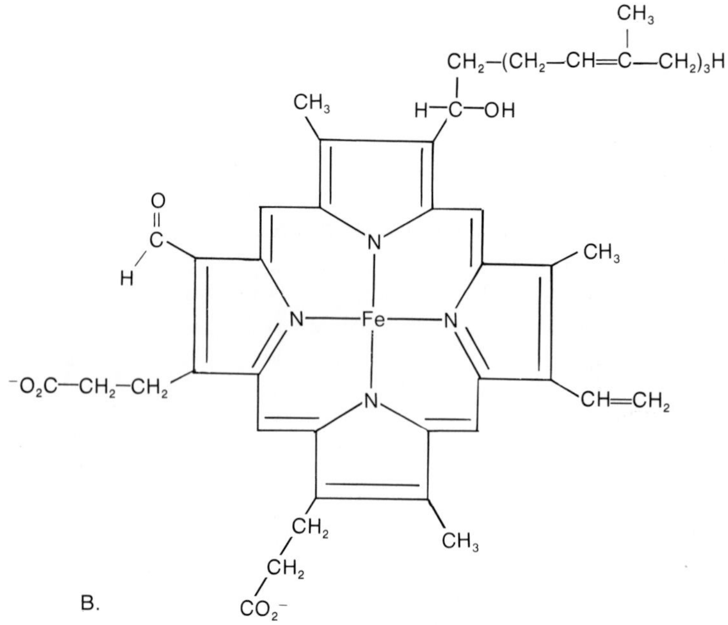

FIGURE 11–16 Heme groups of cytochromes. *A*. The heme group in cytochromes c and c_1. *B*. The heme A, in cytochromes a and a_3.

Oxidation-Reduction Reactions

A reducing agent, or reductant, is an electron donor. An oxidizing agent, or oxidant, is an electron acceptor. When a reductant transfers electrons, a conjugate oxidant or electron acceptor is formed.

$$\text{Reductant} \rightleftharpoons \text{ne}^- + \text{oxidant}$$

where *n* is the number of equivalents of electrons transferred per mole of reductant. A reductant and its conjugate oxidant constitute a redox couple. In this redox couple, *n* is two

$$\text{NADH} \rightleftharpoons \text{NAD}^+ + \text{H}^+ + 2\text{e}^-$$

and in the following couple it is one.

$$\text{cytochrome b (Fe}^{2+}) \rightleftharpoons \text{cytochrome b (Fe}^{3+}) + \text{e}^-$$

The oxidation-reduction potential, or redox potential, of a redox couple is a measure of the electron pressure or tendency of the reductant to donate electrons. The redox potential of a redox couple is determined by measuring the electromotive force of an electrochemical cell in which the redox couple constitutes one half-cell, and a reference half-cell of known redox potential constitutes the other. The ultimate standard of reference is the hydrogen couple

$$\tfrac{1}{2}\,\text{H}_2 \rightleftharpoons \text{H}^+ + \text{e}^-$$

whose redox potential is defined as 0 volts when the conditions are as follows: $[H^+] = 1M$; pressure of $H_2 = 1$ atmosphere; and $T = 25°C$.

The relationship between the redox potential and the concentrations of reductant and oxidant is given by the Nernst equation

$$E = E_o + \frac{RT}{nF}\ln\frac{[\text{oxidant}]}{[\text{reductant}]}$$

where E is the redox potential of the couple, E_o is the standard redox potential of the couple (i.e., the observed potential at 25°C when both oxidant and reductant are 1M), R is the gas constant (8.315 joules degree^{-1} mole^{-1}), T is the temperature in degrees Kelvin, n is the number of electrons transferred, and F is the faraday (96,500 coulombs mole^{-1}). Since biological scientists are particularly interested in redox reactions near pH 7, they employ standard redox potentials determined at pH 7.0, which are designated E'_o. The E'_o for the hydrogen couple is -0.42 volts.

Standard redox potentials at pH 7.0 for several redox couples of biological interest are listed in Table 11–2. The more negative the value of E'_o, the stronger the reducing powers of the reductant of the redox couple. The more positive the value of E'_o, the stronger the oxidizing powers of the oxidant of the redox couple. The reductant of a redox couple will reduce the oxidants of couples that possess more positive redox potentials.

The standard free energy change of a redox reaction is given by equation 13 (see Appendix G for the derivation of this equation).

TABLE 11–2 STANDARD REDOX POTENTIALS AT pH 7.0 FOR SOME REDOX COUPLES

Redox Couple (Reductant/Oxidant)	E'^{o}, volts
α-Ketoglutarate/succinate + CO_2	−0.67
Acetaldehyde/acetate	−0.60
H_2/H^+	−0.42
Isocitrate/α-ketoglutarate + CO_2	−0.38
NADH + H^+/NAD^+	−0.32
NADPH + H^+/$NADP^+$	−0.32
Dihydrolipoate/lipoate	−0.29
Lactate/pyruvate	−0.19
Cytochrome b (Fe^{2+})/cytochrome b (Fe^{3+})	+0.07
$CoQH_2$/CoQ	+0.10
Cytochrome c (Fe^{2+})/cytochrome c (Fe^{3+})	+0.22
H_2O/½O_2 + $2H^+$	+0.82

(13) $$\Delta G^{\circ\prime} = -nF\Delta E'_0$$

where $\Delta E'_0$ is the difference in the standard redox potentials of the two redox couples. To obtain $\Delta G^{\circ\prime}$ in calories, the caloric equivalent of the faraday (F = 23,062 calories $volt^{-1}$ $mole^{-1}$) is employed.

The total standard free energy change for the transfer of electrons from NADH to molecular oxygen, via the respiratory chain, can be calculated from the standard redox potentials of −0.32 volts for the NADH:NAD^+ couple and +0.82 volts for the H_2O:O_2 couple.

$$NADH + \tfrac{1}{2}O_2 + H^+ \rightleftharpoons H_2O + NAD^+$$

$$\Delta G^{\circ\prime} = -nF\Delta E'_0 = -2 \times 23{,}062 \times [0.82 - (-0.32)]$$

$$\Delta G^{\circ\prime} = -52.6 \text{ kcal/mole NADH}$$

Thus, for each mole of NADH oxidized by molecular oxygen via the respiratory chain, there is a large free-energy change. The cell conserves a fraction of this free energy by the process of oxidative phosphorylation.

Oxidative Phosphorylation

GENERATION OF ATP

Oxidative phosphorylation in the mitochondria is the process by which the cell couples the phosphorylation of ADP to electron transport by the respiratory chain, thereby conserving a part of the free-energy change resulting from the oxidation of NADH and E—$FADH_2$. Under most physiological conditions, electron transport via the respiratory chain is *tightly coupled* to the phosphorylation of ADP. That is, electrons can flow down the chain to molecular oxygen only when ADP and P_i are available for the formation of ATP. The availability of ADP is usually the limiting factor controlling the rate of electron transport in the respiratory chain. Since ADP and P_i are formed when ATP is being utilized to do work, the rates of substrate utilization, electron transport, and oxygen consumption automatically adjust to the demand for ATP. This control of respiration by the concentration of ADP is known as *respiratory control* or *acceptor control.*

For each NADH oxidized via the respiratory chain, three molecules of ATP are generated. As indicated in Figure 11–17, the three ATP molecules appear to be generated at three different sites in the respiratory chain. One phosphorylation site is between NADH and CoQ; another site is between cytochromes *b* and *c*; and a third site is between cytochrome *c* and molecular oxygen. Reduced succinate dehydrogenase, E—$FADH_2$, feeds its electrons into the respiratory chain at the level of CoQ, bypassing the first phosphorylation site. As a result, the oxidation of E—$FADH_2$ by the respiratory chain leads to the generation of only two moles of ATP per mole of E—$FADH_2$ oxidized.

There are a number of specific inhibitors of electron transport by the respiratory chain. The cyanide ion, CN^-, a very potent inhibitor of respiration, blocks electron flow between cytochrome ($a + a_3$) and oxygen. Antimycin A is an antibiotic that inhibits the transfer of electrons from cytochrome *b* to c_1. The insecticide rotenone blocks electron transfer between NADH and CoQ. Rotenone does not block the oxidation of succinate, because flavoproteins, such as succinate dehydrogenase, feed their electrons into the respiratory chain at CoQ, thereby bypassing the rotenone block. The antibiotic oligomycin inhibits the synthesis of ATP from ADP, and as a result inhibits electron transport by the respiratory chain in tightly coupled mitochondria.

In tightly coupled mitochondria, electrons can flow down the chain only when ADP and P_i are available for the formation of ATP. The oxidation of a mole of NADH by the respiratory chain in tightly coupled mitochondria utilizes 1 gram atom of oxygen (0.5 moles O_2) and generates 3 moles of ATP by oxidative phosphorylation. This stoichiometry is often expressed as a P:O ratio, the moles of P_i converted to ATP per gram atom of oxygen consumed. The P:O ratio for the oxidation of NADH by the respiratory chain in tightly coupled mitochondria is 3.0. The oxidation of succinate to fumarate by succinate dehydrogenase generates E—$FADH_2$. The P:O ratio for the oxidation of E—$FADH_2$ is 2.0, because E—$FADH_2$ feeds into the respiratory chain at CoQ, thereby generating 2 moles of ATP for each gram atom of oxygen consumed.

There are some compounds known to uncouple electron transport and oxidative phosphorylation. In the presence of these uncoupling agents, electrons are transferred down the respiratory chain without a concomitant phosphorylation of ADP to form ATP. All respiratory control by ADP is lost in uncoupled mitochondria, and the P:O ratio decreases and approaches zero. Uncoupling agents have no effect on substrate-level phosphorylation. These uncoupling agents have been useful tools in investigating the mechanism of oxidative phosphorylation. One uncoupling agent is 2,4-dinitrophenol (DNP).

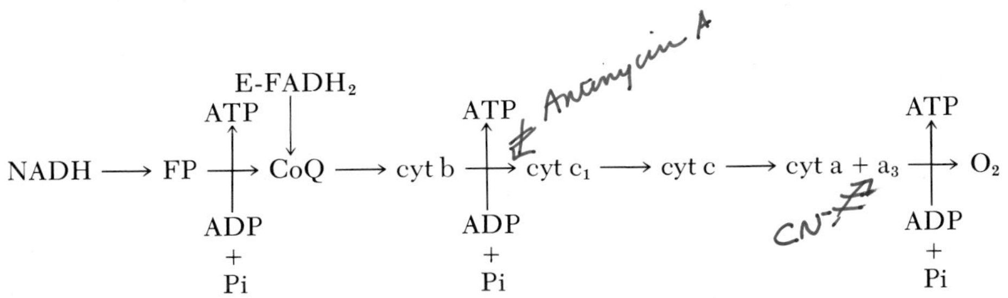

FIGURE 11–17 Generation of ATP at three sites in the respiratory chain.

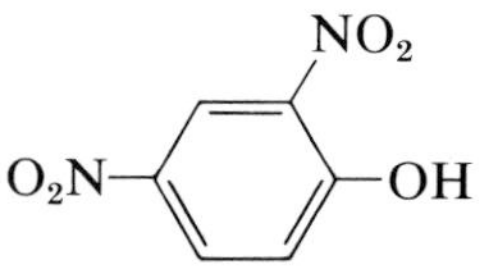

2,4-dinitrophenol

Most uncoupling agents are soluble in lipids and possess an ionizable group, such as a phenolic —OH group. The possible mechanisms by which these agents uncouple oxidative phosphorylation will be discussed later in this chapter.

THE YIELD OF ATP FROM THE OXIDATION OF PYRUVATE

The oxidation of a mole of pyruvate to CO_2 in the mitochondrion generates 1 mole of NADH by the pyruvate dehydrogenase reaction, and 3 moles of NADH, 1 mole of GTP, and 1 mole of reduced succinate dehydrogenase (E—$FADH_2$) via the citric acid cycle. Each of these 4 moles of NADH generated in the mitochondria gives rise to 3 moles of ATP via oxidative phosphorylation. Thus a total of 12 moles of ATP is formed from the NADH produced in the conversion of a mole of pyruvate to CO_2. The oxidation of E—$FADH_2$ by the respiratory chain leads to the generation of 2 moles of ATP by oxidative phosphorylation, bringing the total to 14 moles of ATP. These 14 moles of ATP plus the 1 mole of GTP generated by the substrate-level phosphorylation in the citric acid cycle make a total of 15 moles of ATP (or its equivalent) generated per mole of pyruvate converted to CO_2 and H_2O by the mitochondria.

THE OXIDATION OF NADH GENERATED IN THE CYTOSOL

NADH generated in the cytosol cannot permeate the mitochondrial membrane to be oxidized by the respiratory chain directly. In skeletal muscle the electrons from NADH in the cytosol are transported into the mitochondria through the *glycerol phosphate shuttle*. As outlined in Figure 11–18, the cytosol contains an NAD-linked glycerol phosphate dehydrogenase, which catalyzes the transfer of reducing equivalents from NADH to dihydroxyacetone phosphate, forming NAD^+ and glycerol phosphate in the cytosol. Glycerol phosphate is capable of entering the mitochondria, where it is oxidized by an FAD-linked glycerol phosphate dehydrogenase designated E_{GPD}—FAD. Oxidation of glycerol phosphate in the mitochondria results in the formation of dihydroxyacetone phosphate, which can pass back out into the cytosol, and the reduced flavoenzyme E_{GPD}—$FADH_2$, which can transfer its electrons to CoQ in the respiratory chain. Two ATP's are generated by oxidative phosphorylation as the electrons are transported through the respiratory chain from CoQ to molecular oxygen.

Thus each mole of cytoplasmic NADH gives rise to 2 moles of ATP in the mitochondria. The fact that the mitochondrial glycerol phosphate dehydrogenase is linked to FAD, a stronger oxidizing agent than NAD^+, insures that the shuttle always operates to transport reducing equivalents into, rather than out of, the mitochondria. The price that is paid for this insured unidirectionality of the shuttle is a lower yield of ATP from cytoplasmic NADH than from mitochondrial NADH.

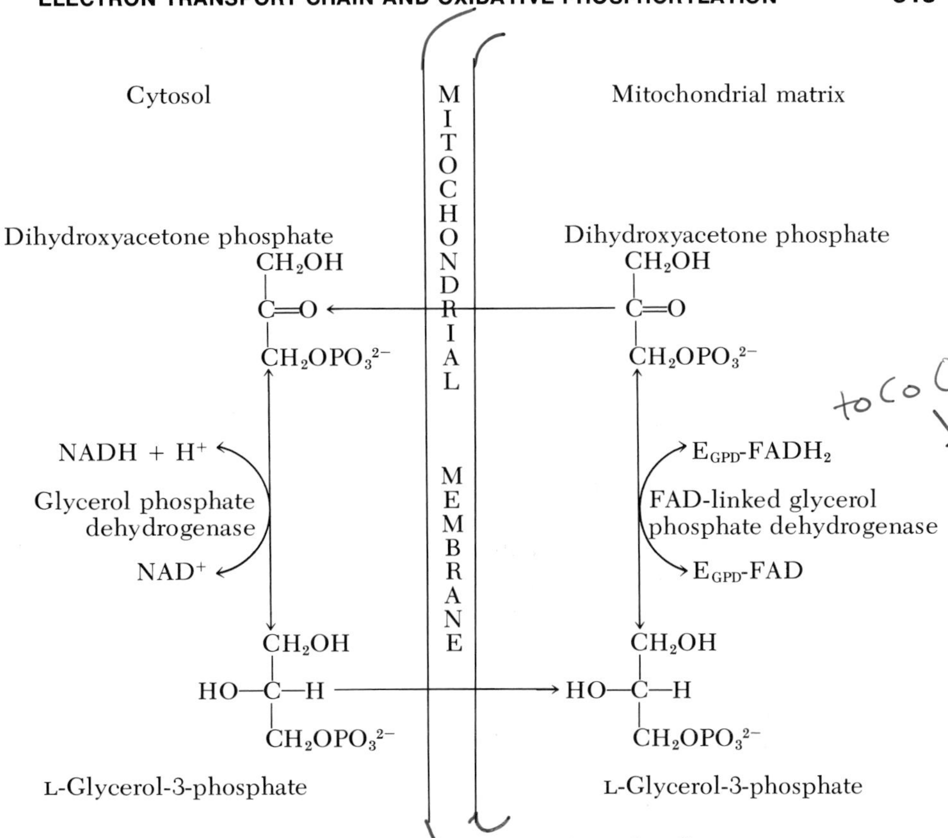

FIGURE 11–18 The glycerol phosphate shuttle in skeletal muscle cells.

A SUMMARY OF THE NET YIELD OF ATP FROM THE CONVERSION OF GLUCOSE TO CO_2 AND H_2O

The net reaction for the aerobic metabolism of glucose is

$$C_6H_{12}O_6 + 6O_2 + 36ADP^{3-} + 36P_i^{2-} + 36H^+ \rightarrow 6CO_2 + 36ATP^{4-} + 42H_2O$$

It is now possible to account for all of the ATP generated in the aerobic metabolism of glucose. The conversion of 1 mole of glucose to 2 moles of pyruvate and 2 moles of NADH through the Embden-Meyerhof pathway requires 2 moles of ATP and yields 4 moles of ATP from substrate-level phosphorylations, for a net yield of 2 moles of ATP per mole of glucose in the cytoplasm. The electrons from the 2 moles of NADH generated in the cytoplasm enter the mitochondria via the glycerol phosphate shuttle and give rise to 4 moles of ATP through oxidative phosphorylation, bringing the net production of ATP up to 6 moles of ATP per mole of glucose at this point in the pathway. In addition, each of the 2 moles of pyruvate will generate 15 moles of ATP (14 moles of ATP, plus 1 mole of GTP) in the mitochondria when converted to CO_2 and H_2O through the combined action of the pyruvate dehydrogenase complex, the citric acid cycle, and oxidative phosphorylation.

Thus, the 30 moles of ATP generated from the metabolism of the 2 moles of pyruvate to CO_2 and H_2O, plus the net 6 moles of ATP generated by the conversion of a mole of glucose to 2 moles of pyruvate, makes the total net yield 36 moles of ATP per mole of glucose. Based on standard free energy changes, the formation of 36 moles of ATP per mole of glucose oxidized represents a thermodynamic efficiency of 38 per cent $\left(\frac{36 \times 7.3 \text{ kcal}}{686 \text{ kcal}}\right)$ in energy conservation.

TRANSPORT OF ATP AND ADP ACROSS THE INNER MITOCHONDRIAL MEMBRANE

Most of the cell's ATP is generated by oxidative phosphorylation in the mitochondria and appears initially in the inner compartment or matrix of the mitochondria. However, most of the ATP-requiring processes take place outside the mitochondria. This physical separation of the major site of ATP generation from the major sites of ATP utilization implies the existence of a mechanism for transporting ATP out of the matrix of the mitochondria to the sites of utilization, as well as the existence of mechanisms for transporting ADP and P_i into the matrix for the generation of ATP.

The outer mitochondrial membrane appears to be quite permeable to most small molecules and ions, such as ATP, ADP, and P_i. In contrast, the transport of ATP, ADP, P_i and most other small molecules and ions across the inner mitochondrial membrane requires specific carriers to be present in the inner membrane. The mitochondrial inner membrane carrier system for the transport of ATP is coupled to the transport of ADP in the opposite direction; that is, the transport of an ATP molecule out of the mitochondrial matrix is obligatorily accompanied by the transport of an ADP molecule into the mitochondrial matrix. This coupled process is an example of *facilitated exchange diffusion.* Atractyloside, a plant glycoside, inhibits oxidative phosphorylation of extramitochondrial ADP by inhibiting this ATP-ADP carrier system, thereby preventing extramitochondrial ADP from entering the matrix where it can be phosphorylated.

Mechanism of Oxidative Phosphorylation

The mechanism by which phosphorylation is coupled to electron transport in mitochondria poses a question that has been intensively investigated for the past quarter century. Although considerable progress has been made, the problem is still not totally resolved. The results of these investigations have led to several hypotheses, two of which will now be considered.

THE CHEMICAL COUPLING HYPOTHESIS

The oldest of the hypotheses to explain oxidative phosphorylation is the chemical coupling hypothesis, a version of which is presented in Figure 11–19. This hypothesis is based on the effects of inhibitors and uncouplers of oxidative phosphorylation. According to this theory, the oxidation of a reduced electron carrier, $A_{red.}$, by the next carrier in the respiratory chain, B_{ox}, requires a third component, X. As illustrated in Figure 11–19 (reaction *a*) it is proposed that the oxidation of $A_{red.}$ by B_{ox} results in the formation of an energy-rich intermediate, $A_{ox} \sim X$, which traps some of the free energy change of the oxidation-reduction reaction. The next step, represented by reaction *b*, involves the displacement of A_{ox} by an enzyme, E, to form another energy-rich intermediate, $E \sim X$. In reaction *c* inorganic phosphate, P_i, displaces X, forming an energy-rich phosphoenzyme, $E \sim P$, which transfers its phosphoryl group to ADP in reaction *d*, forming ATP. According to the chemical coupling hypothesis, dinitrophenol uncouples

$$\text{a) } A_{red} + B_{ox} + X \rightleftharpoons A_{ox} \sim X + B_{red}$$

$$\text{b) } A_{ox} \sim X + E \rightleftharpoons E \sim X + A_{ox}$$

$$\text{c) } E \sim X + P_i \rightleftharpoons E \sim P + X$$

$$\text{d) } E \sim P + ADP \rightleftharpoons ATP + E$$

FIGURE 11–19 The chemical coupling hypothesis for oxidative phosphorylation.

oxidative phosphorylation from electron transfer by catalyzing the breakdown of $A_{ox} \sim X$ or $E \sim X$.

Although the chemical coupling hypothesis for oxidative phosphorylation is an attractive one, no one to date has been able to isolate or obtain strong evidence supporting the existence of the energy-rich intermediates that are postulated.

THE CHEMIOSMOTIC HYPOTHESIS

In its simplest form, the chemiosmotic hypothesis of Mitchell and others assumes that the electron carriers of the respiratory chain are arranged in the inner membrane in such a way as to translocate protons from the matrix side to the outer side of the inner membrane when electrons flow down the respiratory chain. In this way a part of the free energy change of the oxidation-reduction reactions of the carriers is conserved in the form of an energy-rich proton gradient across the inner membrane. Indeed, the generation of a pH or proton gradient across the inner mitochondrial membrane can be observed experimentally when electrons are flowing down the respiratory chain.

It is proposed that the energy stored in this energy-rich proton gradient is then utilized to drive the synthesis of ATP from ADP and P_i by a mechanism that is poorly delineated at the present time (Fig. 11–20). According to this hypothesis,

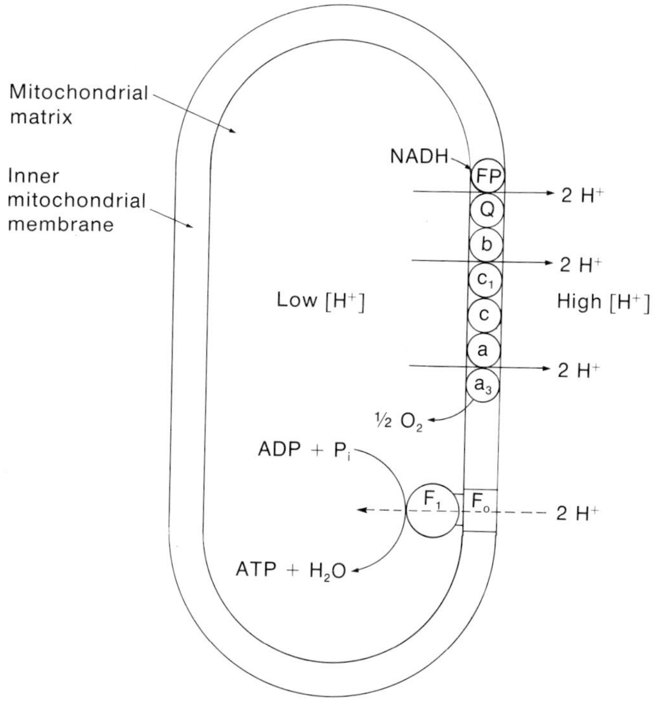

FIGURE 11–20 The chemiosmotic hypothesis for oxidative phosphorylation.

dinitrophenol uncouples oxidative phosphorylation by acting as a proton carrier across the inner membrane, thereby increasing the membrane's permeability to protons and preventing the formation of the energy-rich proton gradient. Dinitrophenol and other uncouplers of oxidative phosphorylation do in fact increase the permeability of membranes to protons.

Oxidative phosphorylation occurs in fragmented mitochondrial inner membrane preparations as well as in intact mitochondria. Electron micrographs of these fragmented inner membrane preparations indicate that they consist of small membrane-bound vesicles formed from fragments of the inner membrane. Thus far all submitochondrial preparations capable of carrying out oxidative phosphorylation have been found to contain membrane-bound vesicles. As described earlier in this chapter, the inner mitochondrial membrane contains many spherical projections known as the F_1 *coupling factor,* which are approximately 80 Å in diameter and extend into the matrix. The spheres have a molecular weight of approximately 360,000 and are made up of at least five different subunits. As shown in Figure 11–20, the F_1 factor is attached to F_0, another protein factor in the membrane. In the membrane-bound vesicles formed from the inner mitochondrial membrane by sonication, the spherical projections are attached to the outer surface of the vesicles.

Racker and his coworkers have shown that removal of the F_1 coupling factor from the surface of these vesicles destroys the ability of the vesicles to carry out oxidative phosphorylation but does not interfere with their ability to transport electrons to O_2 via the respiratory chain. When F_1 is added back to these F_1-deficient vesicles, oxidative phosphorylation is restored, indicating an essential role of F_1 in the oxidative phosphorylation mechanism.

The isolated F_1 coupling factor is a water-soluble protein possessing an ATPase activity. That is, it catalyzes reaction 14.

$$\text{ATP} + H_2O \underset{F_1}{\rightleftharpoons} \text{ADP} + P_i \tag{14}$$

The equilibrium position for this reaction lies far to the right.

According to the chemiosmotic hypothesis, migration of H^+ down its concentration gradient, which was created by electron transport, is coupled to reaction 14 and drives reaction 14 to the left, resulting in the synthesis of ATP (Fig. 11–20). It has now been demonstrated experimentally that a pH gradient imposed across the inner mitochondrial membrane can indeed drive ATP synthesis. However, the mechanism by which ATP synthesis is coupled to the migration of protons down their concentration gradient remains unclear. Racker and coworkers have successfully reconstituted active oxidative phosphorylating systems by incorporating purified electron carriers and coupling factors into synthetic phospholipid vesicles. Such studies may lead to a resolution of the important and long-standing problem of the mechanism of oxidative phosphorylation.

NUCLEIC ACIDS AND MITOCHONDRIAL REPLICATION

In 1890 when Altmann observed mitochondria with the light microscope, he gave them the name bioblasts, because he speculated that mitochondria and chloroplasts might be intracellular symbionts that arose from bacteria and algae, respectively. This idea was in disrepute with most cytologists until the discovery of mitochondrial nucleic acids in the past decade.

Prior to the early 1960's it was widely believed that all of the DNA of a cell was located in the nucleus. It then was gradually realized that mitochondria and chloroplasts contain small amounts of DNA. Several lines of evidence led to this discovery. Mitochondria stained with the Feulgen reaction in exceptional cases, such as the large mitochondrial derivatives, the kinetoplasts, of flagellated protozoa. Some investigators were able to demonstrate the presence of filaments digestible with deoxyribonuclease in the matrix of mitochondria. A small but significant amount of thymidine-^{3}H was incorporated into mitochondria and labeled them in radioautographs. Finally, DNA was isolated from many kinds of mitochondria (Fig. 11–21). The isolated DNA is usually in the form of a circular double helix, which is about 5 μm in circumference in many cells. It differs from nuclear DNA in its density and denaturation temperature, being richer in guanine and cytosine. A given mitochondrion contains one or more DNA molecules; it is thought that the multiple molecules in one mitochondrion are copies of a single type. A variation in size among mitochondrial DNA molecules is sometimes observed, the larger ones being multiples of the smaller ones in their circumference.

In addition to possessing DNA, mitochondria contain their own type of ribosomes and are capable of protein synthesis. The mitochondrial ribosomes are found in the matrix, where they are visible with the electron microscope in favorable preparations as small dense granules about 120 Å in diameter. As noted in Chapter 7, isolated mitochondrial ribosomes are found to be smaller than the cytoplasmic (extramitochondrial) ribosomes of eukaryotes and to be similar in their sedimentation properties to the ribosomes of prokaryotes. Thus mitochondrial ribosomes characteristically are 70S particles, with 50S and 30S subunits that contain 23S and 16S rRNA, respectively. Mitochondrial ribosomes also resemble those of prokaryotes in their sensitivity to inhibitors of bacterial protein synthesis, such as chloramphenicol. Conversely, they are relatively insensitive to inhibitors of cytoplasmic protein synthesis in eukaryotes, such as cycloheximide. Despite the similarities between mitochondrial and prokaryotic ribosomes, it should be noted that the two are not identical, since the RNA's differ in their base sequences and the ribosomal proteins are different.

Mitochondrial DNA is not large enough to code for all mitochondrial RNA's and proteins. It does code for some, and this is apparently the genetic basis of

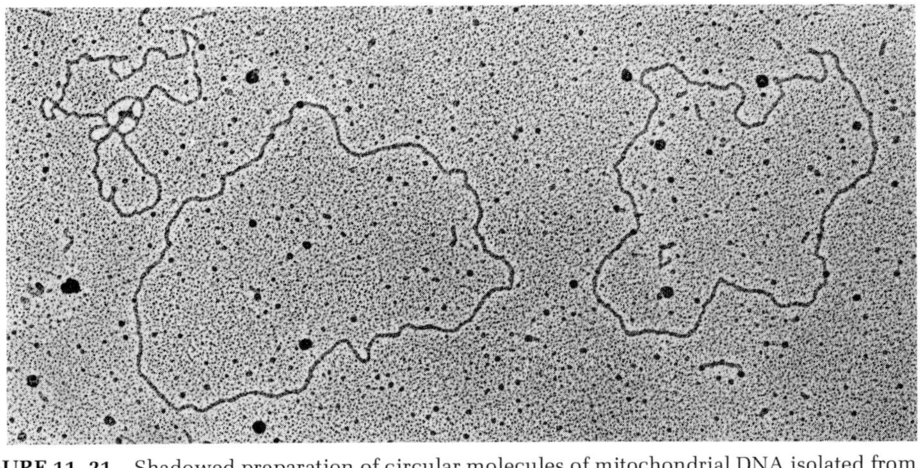

FIGURE 11–21 Shadowed preparation of circular molecules of mitochondrial DNA isolated from rat liver. The contour length of the molecules is approximately 5 μm. × 30,000. (Micrograph courtesy of Nass, M. M. K., and Buck, C. A.: J. Mol. Biol., 54:187, 1970 with permission from Journal of Molecular Biology. Copyright by Academic Press, Ltd., London.)

some instances of cytoplasmic inheritance, as the inheritance of traits independent of nuclear genes has been termed. Mitochondrial ribosomal RNA is coded for by mitochondrial DNA, as are at least some and possibly all mitochondrial transfer RNA's. Mitochondrial ribosomal proteins, however, are coded by nuclear genes. Cytochrome *c* is an example of a protein coded by nuclear genes and synthesized on cytoplasmic ribosomes before entering the mitochondria. The question of whether messages transcribed in the nucleus are translated in mitochondria is under investigation. Thus mitochondria have a complex relation to the nucleus and the remainder of the cytoplasm. In the case of cytochrome oxidase, for example, some of the subunits of the molecule apparently are coded by nuclear genes and others by mitochondrial genes.

The presence of circular DNA lacking histones, the size of the ribosome, and the sensitivities of the mitochondrial protein synthesis have recently lent credence to Altmann's hypothesis of the prokaryotic origin of mitochondria. If a prokaryote were incorporated into a eukaryotic cell by phagocytosis, the prokaryote would be surrounded by a membrane derived from the host cell. In the case of a mitochondrion, this would correspond to the outer membrane, while the inner mitochondrial membrane corresponds to the original plasma membrane of the prokaryote. It may be noted, in accord with these speculations, that both the plasma membrane of bacteria and the inner mitochondrial membrane contain respiratory enzymes. In addition, the composition of the outer mitochondrial membrane resembles that of endoplasmic reticulum, while that of the inner mitochondrial membrane has features similar to those of bacterial plasma membranes.

The mechanism of formation of mitochondria has been debated for many years, but the discovery of mitochondrial nucleic acids has led to a general acceptance of the idea that mitochondria reproduce by growth and division. An experiment that earlier led to a similar conclusion was carried out by Luck (Fig. 11–22). Growing cells of the mold *Neurospora* were exposed to choline-^{3}H, which is a precursor of phospholipids and is incorporated into membranes. The cells were then allowed to grow for several generations in a nonradioactive medium, and the distribution of the label among the mitochondria was determined by using radioautography. As the cells grew and divided, the number of mitochondria increased proportionally. It was found that all the mitochondria were labeled and that the amount of labeling per mitochondrion declined with time. This result suggested that mitochondria were formed by growth and division of the labeled mitochondria present at the beginning of the experiment, because the label was diluted and distributed among all the progeny mitochondria. New mitochondria were not formed *de novo,* because if this had occurred during growth in nonradioactive medium, two populations of mitochondria would have been observed: the original labeled mitochondria and the new unlabeled mitochondria.

More recently, different stages of partitions extending across mitochondria have been described (Fig. 11–23). Since these are particularly abundant when mitochondria are known to be increasing in number, the images have been interpreted as representing stages in mitochondrial division.

PATHOLOGICAL CHANGES IN MITOCHONDRIA

Alterations in mitochondria are particularly important in cellular reactions to injury, including the responses to ischemia and to toxic agents. Ischemia refers to a reduced blood supply to cells and tissues, and thus it is involved in the

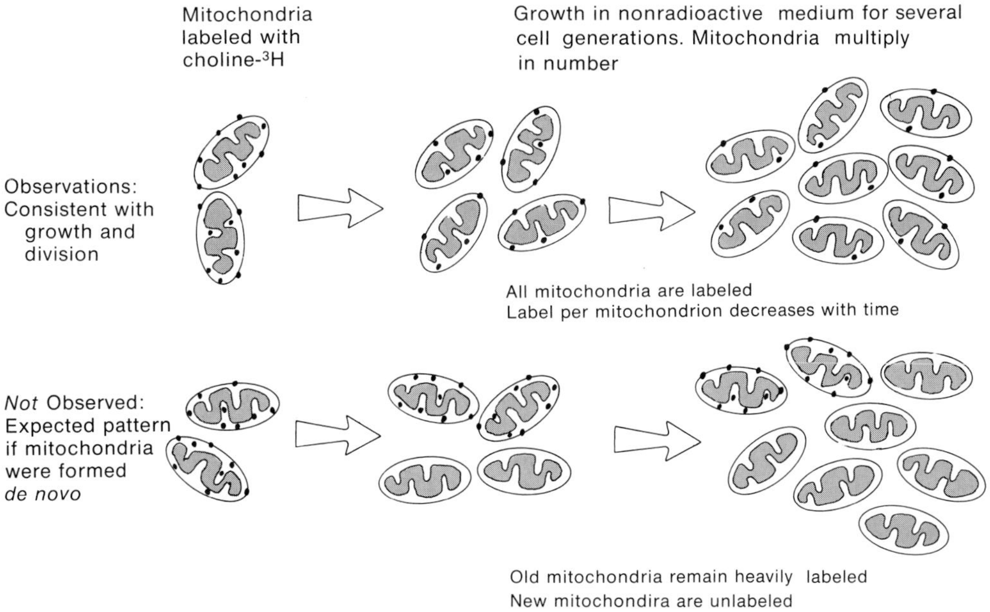
Mitochondria
labeled with
choline-³H
Growth in nonradioactive medium for several
cell generations. Mitochondria multiply
in number
Observations:
Consistent with
growth and
division
All mitochondria are labeled
Label per mitochondrion decreases with time
Not Observed:
Expected pattern
if mitochondria
were formed
de novo
Old mitochondria remain heavily labeled
New mitochondira are unlabeled

FIGURE 11–22 Luck's experiment on the formation of mitochondria. Growing cells of the mold Neurospora were labeled with choline-^{3}H, which is incorporated into the phospholipids of mitochondrial membranes. The cells were then grown for several cell generations in nonradioactive medium. During the growth in nonradioactive medium, the mitochondria increased in number. The diagram shows two different results that might have been obtained, depending on the mode of formation of mitochondria. The result depicted at the top of the figure was actually observed and is consistent with multiplication of mitochondria by growth and division. The pattern at the bottom would be anticipated if new mitochondria were formed *de novo* during growth in nonradioactive medium, but this was not observed.

pathogenesis of stroke and heart disease and is also important in injury due to trauma. If blood supply to cells is reduced, as occurs in myocardial infarction owing to occlusion of one of the coronary arteries, one of the first changes is a dilatation of the endoplasmic reticulum. This is followed rapidly by a swelling of mitochondria.

Alterations in the volume of mitochondria, both contraction and expansion, were discussed earlier in this chapter; they are thought to occur normally in living cells. These normal low-amplitude changes in volume, however, are exceeded by the high-amplitude, irreversible swelling seen in pathological conditions, such as ischemia. The first ultrastructural changes in mitochondria (Fig. 11–24) include decrease in the density of the matrix, loss of matrix granules, and appearance of masses of a dense flocculent material in the matrix. These are followed by the appearance of further densities in the matrix, and the outer mitochondrial membrane is broken as the inner compartment swells. In late stages the inner membrane also degenerates. The swollen mitochondria in the cytoplasm have a granular appearance in light microscope preparations and thus may help account for the microscopically visible changes in injured cells known in pathology as cloudy swelling or hydropic degeneration. It should be noted that during the progressive development of cellular alterations due to ischemia, the point of no return, at which changes become irreversible, is related to the initiation of high amplitude swelling in mitochondria.

Various toxic agents are known to produce changes in mitochondrial growth and replication, leading to the formation of very large bizarre mitochondria termed megamitochondria. These unusual organelles have been observed in the

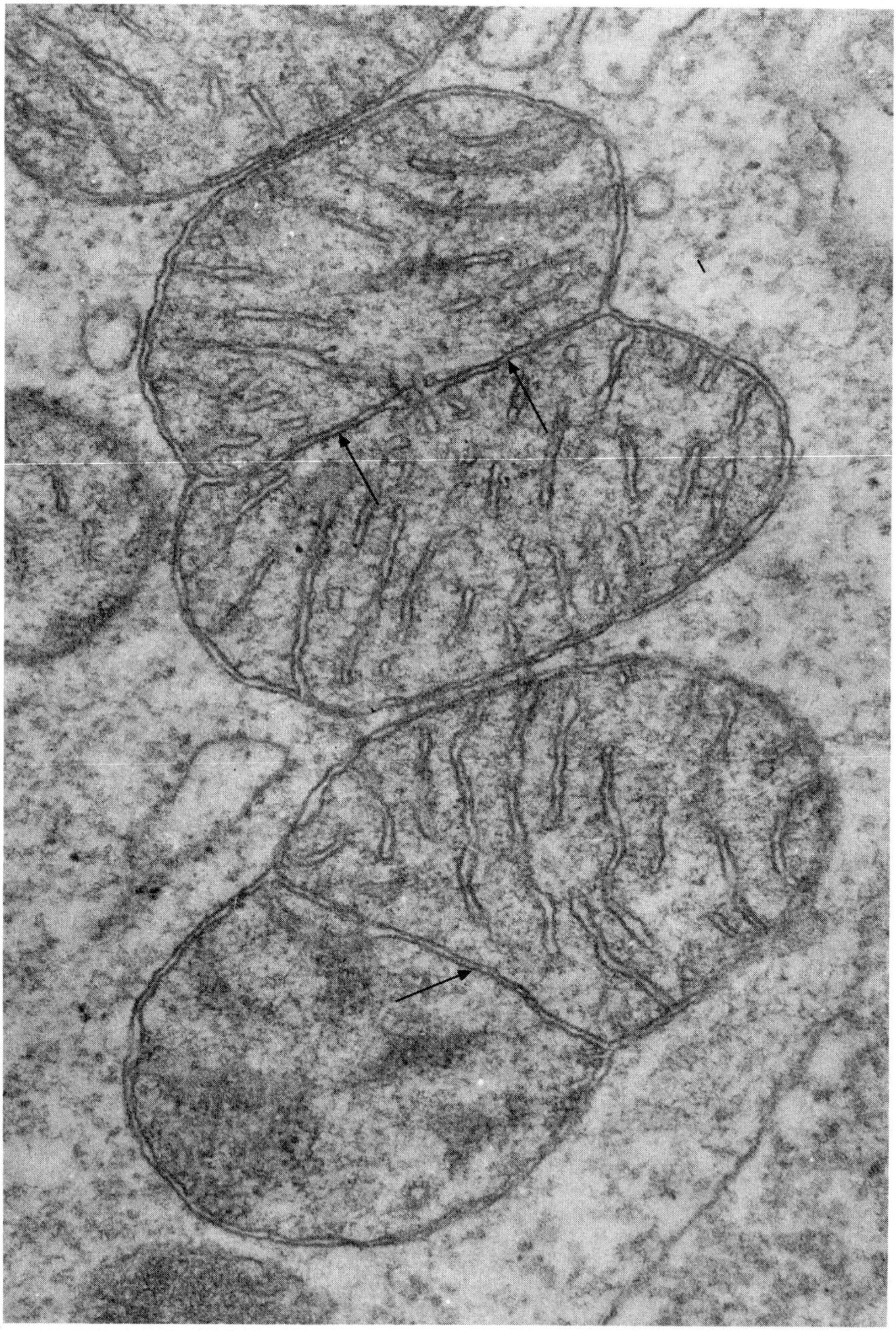

FIGURE 11–23 Mitochondria in the fat body of an insect. A partition extends across each mitochondrion (*arrows*). These are thought to represent the beginning of division of mitochondria. × 81,000. (Micrograph courtesy of Larsen, W. J.: J. Cell Biol., *47*:373, 1970. Reproduced by permission of the Rockefeller University Press.)

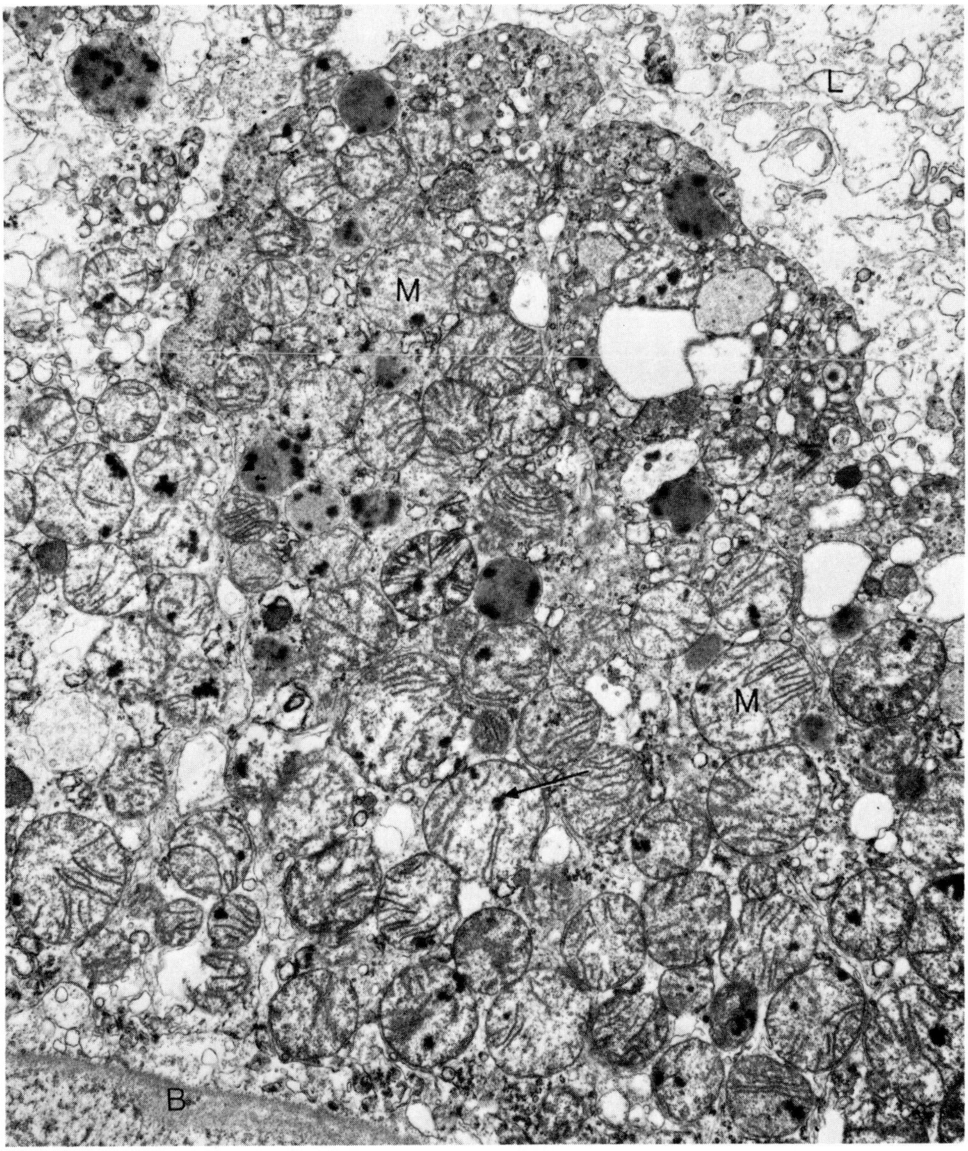

FIGURE 11–24 Proximal tubule of kidney from a patient in acute shock. The basal lamina (B) of the epithelium is the lower part of the micrograph, and the lumen of the tubule (L) is at the top. The mitochondria (M) are very swollen and contain flocculent densities (*arrow*) in the matrix. The endoplasmic reticulum is also dilated. Other cells in the vicinity are disrupted, as indicated by the presence of cellular debris in the lumen (L). (Micrograph courtesy of B. F. Trump and A. U. Arstilia. From La Via, M. F., and Hill, R. B., (eds.) Principles of Pathobiology pp. 9–96. Oxford University press, 1975. Reproduced by permission of the publisher.)

liver cells of chronic alcoholics, after administration of copper-chelating compounds, and in experimental riboflavin deficiency. In the latter instance, administration of riboflavin restores the mitochondrial population to normal, apparently by division of the megamitochondria to form larger numbers of normal-sized organelles.

FURTHER READING

Boyer, P. D., Chance, B., Ernster, L., et al.: Oxidative phosphorylation and photophosphorylation. Annu. Rev. Biochem., *46*:955, 1977.

Fernández-Morán, H., Oda, T., Blair, P. V., et al.: A macromolecular repeating unit of mitochondrial structure and function. J. Cell Biol., *22*:63, 1964.

Gibor, A., and Granick, S.: Plastids and mitochondria: inheritable systems. Science, *145*:890, 1964.

Hackenbrock, C. R.: Ultrastructural bases for metabolically linked mechanical activity in mitochondria. II. Electron transport-linked ultrastructural transformations in mitochondria. J. Cell Biol., *37*:345, 1968.

Lehninger, A. L. (ed.): The Mitochondrion. Menlo Park, Calif., W. A. Benjamin, 1964.

Lehninger, A. L.: The molecular organization of mitochondrial membranes. Adv. Cytopharm., *1*:199. Raven Press, New York, 1971.

Lowenstein, J. M.: The pyruvate dehydrogenase complex and the citric acid cycle. *In*: Florkin, M., and Stotz, E. H. (eds.): Comprehensive Biochemistry. Vol. 18S. New York, Elsevier, 1971, pp. 1–55.

Luck, D. J. L.: Formation of mitochondria in *Neurospora crassa*. J. Cell Biol., *24*:461, 1965.

Munn, E. A. (ed.): The Structure of Mitochondria. New York, Academic Press, 1975.

Nass, M. M. K.: Mitochondrial DNA: advances, problems, and goals. Science, *165*:25, 1969.

Rabinowitz, M., and Swift, H.: Mitochondrial nucleic acids and their relation to the biogenesis of mitochondria. Physiol. Rev., *50*:376, 1970.

Racker, E.: A New Look at Mechanisms in Bioenergetics. New York, Academic Press, 1976.

Racker, E.: Inner mitochondrial membranes: basic and applied aspects. Hosp. Pract., *9*:87, 1974.

Racker, E.: The membrane of the mitochondrion. Sci. Am., *218*:32, 1968.

Reed, L. J.: Pyruvate dehydrogenase complex. *In*: Horecker, B. L., and Stadtman, E. R. (eds.): Current Topics in Cellular Regulation. Vol. 1, New York, Academic Press, 1969, pp. 233–251.

Roodyn, D. B. and Wilkie, D. (eds.): The Biogenesis of Mitochondria. London, Methuen and Co., 1968.

Tandler, B., and Hoppel, C. L. (eds.): Mitochondria. New York, Academic Press, 1972.

Tedeschi, H.: Mitochondria: Structure, Biogenesis and Transducing Functions. New York, Springer-Verlag, 1976.

Trump, B. F., and Arstila, A. U.: Cellular reaction to injury. *In*: La Via, M. F. and Hill, R. B. (eds.): Principles of Pathobiology. 2nd ed. New York, Oxford University Press, 1975, pp. 9–96.

12

LYSOSOMES AND PEROXISOMES

STRUCTURE, COMPOSITION, AND TYPES OF LYSOSOMES

Lysosomes are defined cytologically as structures that are bounded by a membrane and that contain a variety of acid hydrolases (Table 12–1). Their main function is in intracellular digestion. Lysosomes were discovered by deDuve and his associates during the 1950's as contaminants of the mitochondrial fraction of liver homogenates. Differential centrifugation was used to separate what was initially referred to as a light mitochondrial fraction and was later termed the lysosomal fraction. The lysosomes were found to be particles 0.2 to 0.4 μm in diameter which contained various hydrolytic enzymes most active at acid pH. It was observed that the lysosomes displayed relatively little enzymatic activity if handled carefully. However, with certain treatments, such as sonication, osmotic shock, freezing and thawing, or exposure to detergents or lecithinase, all of which disrupt membranes, the enzymatic activity of the preparations increased dramatically. This latency of enzyme activity led to the inference that the hydrolases were contained within a membrane-bound organelle, even before lysosomes were directly visualized in the electron microscope.

As can be seen in Table 12–1, lysosomes contain a large number of different kinds of hydrolytic enzymes. They have in common an optimal activity at an acid pH and thus are known as acid hydrolases. The lysosomal enzymes are capable of digesting virtually all cellular constituents, including proteins, nucleic acids, polysaccharides, lipids, organic phosphates, and organic sulfates.

Morphologically, lysosomes have a very variable appearance, depending on their type and the stage of the digestion process in a given organelle. When viewed in the electron microscope, they often have a dense polymorphic interior, reflecting the presence of different sorts of materials undergoing digestion (Figs. 12–1 and 12–2). Lysosomes can be identified in both the light microscope and the electron microscope by cytochemical staining for one or more of the hydrolases. The acid phosphatase reaction is most commonly used. There are many variations of this technique, but most are based on the Gomori method, in which cells are incubated in a medium containing β-glycerolphosphate, lead nitrate, and a buffer. The lysosomal acid phosphatase splits the substrate, β-glycerolphosphate, and the resulting inorganic phosphate is precipitated by lead ions. Lead phosphate is

TABLE 12–1 SOME LYSOSOMAL ENZYMES AND THEIR SUBSTRATES*

Enzyme	Natural Substrate	Source of Lysosomes
Phosphatases		
Acid phosphatase	Most phosphomonoesters	Many tissues of animals and plants; protists
Acid phosphodiesterase	Oligonucleotides and phosphodiesters	
Nucleases		
Acid ribonuclease	RNA	Many tissues of animals and plants; protists
Acid deoxyribonuclease	DNA	
Polysaccharide and Mucopolysaccharide Hydrolyzing Enzymes		
β-Galactosidase	Galactosides	Animals, plants, protists
α-Glucosidase	Glycogen	Animals
α-Mannosidase	Mannosides	Animals
β-Glucuronidase	Polysaccharides and mucopolysaccharides	Animals; plants
Lysozyme	Bacterial cell walls and mucopolysaccharides	Kidney
Hyaluronidase	Hyaluronic acids; chondroitin sulphates	Liver
Arylsulphatase	Organic sulphates	Liver; plants
Proteases		
Cathepsin(s)	Proteins	Animals
Collagenase	Collagen	Bones
Peptidases	Peptides	Animals; plants; protists
Lipid Degrading Enzymes		
Esterase(s)	Fatty acid esters	Animals; plants; protists
Phospholipase(s)	Phospholipids	Animals; plants?

*From D. Pitt, Lysosomes and Cell Function, New York, Longman, 1975. Reproduced by permission of the publisher.

electron-opaque and can be visualized as a dense precipitate in the electron microscope (Fig. 12–3).

$$\beta\text{-glycerolphosphate} \xrightarrow[\text{in lysosomes}]{\text{acid phosphatase}} \text{glycerol} + \text{inorganic phosphate}$$

$$\text{inorganic phosphate} \xrightarrow{\text{lead nitrate in incubation medium}} \text{lead phosphate reaction product—electron-opaque}$$

Alternatively, the lead phosphate can be converted by treatment with ammonium sulfide to lead sulfide, which is dark-colored and visible in the light microscope. Methods are also available for certain of the other enzymes, such as aryl sulfatase and β-glucuronidase.

Some terms that are commonly used in speaking about lysosomes follow. A *primary* lysosome is one whose enzymes have not yet been engaged in a digestive process. In contrast, a *secondary* lysosome is an organelle in which active digestion is taking place. A *phagosome* is a membrane-bound structure containing material to be digested, but in which hydrolases are not yet present.

To understand how lysosomes function and how different types of lysosomes relate to one another it is necessary to consider the following phases of the

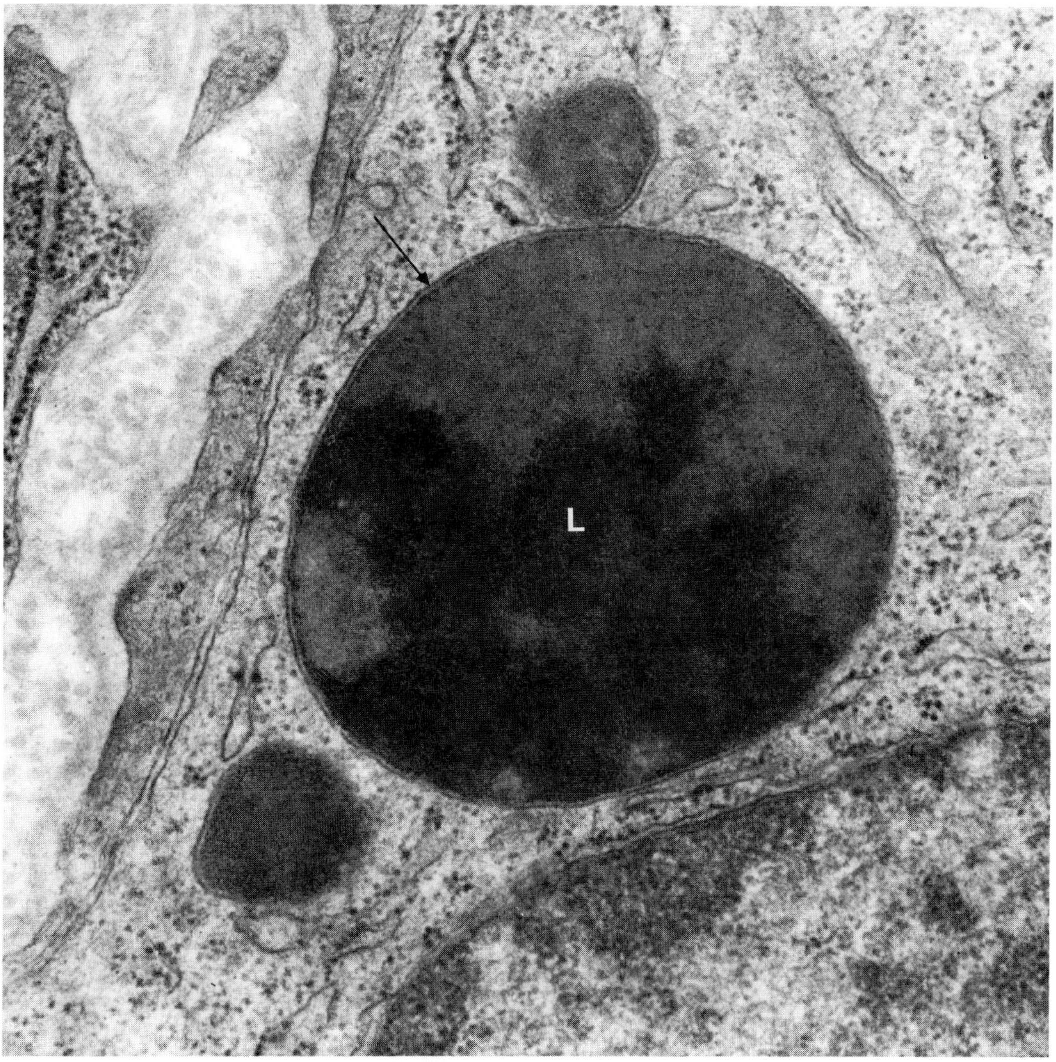

FIGURE 12–1 Rat ventral prostate. A lysosome (L) is bounded by a single membrane (*arrow*). In this preparation a thin light space separates the membrane from the dense interior. × 50,000.

process of intracellular digestion: (1) the mechanism by which material to be digested becomes enclosed in a membrane; (2) the way in which hydrolytic enzymes gain contact with the material to be digested; and (3) the fate of the digested material. The initial stage of the process—the way in which material becomes enclosed by a membrane—varies with the origin of the substance to be digested. Thus two fundamental types of digestive processes are distinguished. If the material to be digested arises from outside the cell, the process is called heterophagy; if the digested material is derived from inside the cell itself, the process is referred to as autophagy.

HETEROPHAGY

In heterophagy (Fig. 12–4), material from outside is taken into the cell by endocytosis. This consists of an infolding of the plasma membrane and a pinching-

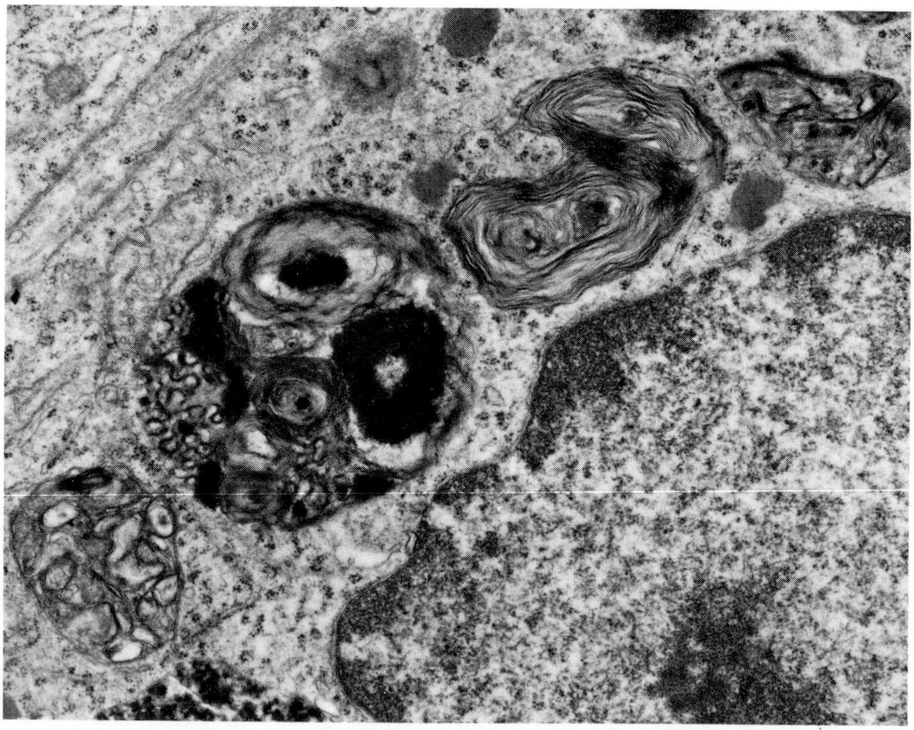

FIGURE 12–2 Seminal vesicle of a newborn rat. Several lysosomes have a heterogeneous content consisting of whorls of a membranous material and a dense substance. × 19,000. (From Flickinger, C.; Z. Zellforsch, *109*:1, 1970. Reproduced by permission of Springer-Verlag.)

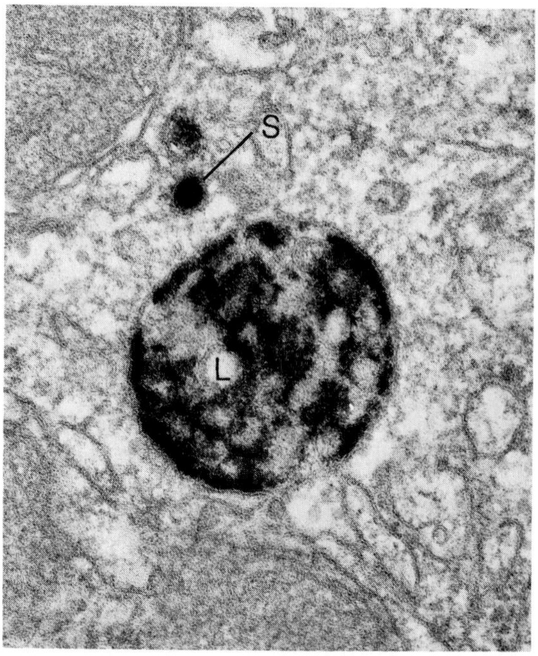

FIGURE 12–3 Lysosomes in the epithelium of the rat vas deferens. In this preparation the lysosomes contain electron-dense reaction product for acid phosphatase. The smaller lysosome (S) is a coated vesicle produced in the Golgi apparatus and represents a primary lysosome that conveys the acid hydrolases to their sites of action. The larger structure (L) is a multivesicular body, a type of secondary lysosome in which absorbed protein is digested in these cells. × 93,000. (Micrograph courtesy of Friend, D. S., and Farquhar, M. G.: J. Cell Biol., *35*:357, 1967. Reproduced by permission of the Rockefeller University Press.)

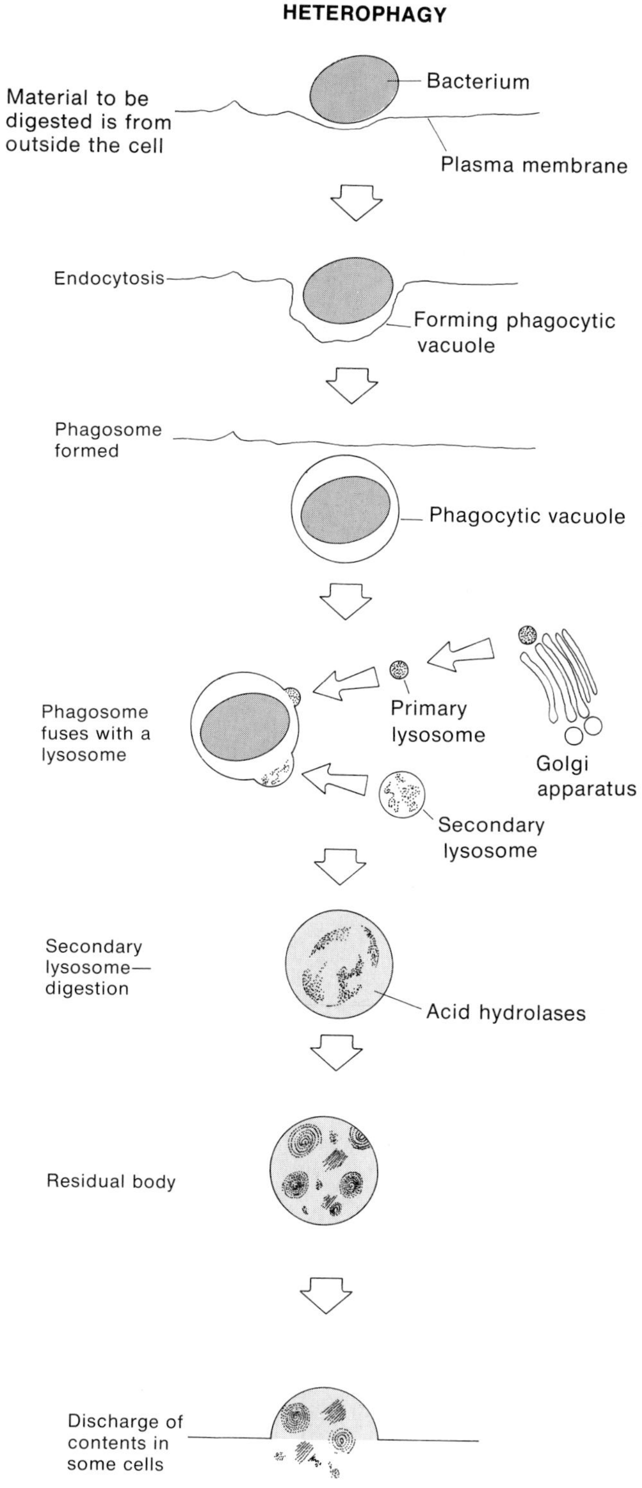

FIGURE 12–4 The process of heterophagy, in which material to be digested is obtained from outside the cell.

off of a part of the invagination to form a vesicle or vacuole containing the material from outside the cell. If a relatively large solid object, such as a microorganism, is taken up, the process is called phagocytosis and the membrane-bound structure that is formed is a phagocytic vacuole. If fluid with substances in solution or in suspension is taken into the cell, the process is termed pinocytosis and small pinocytic vesicles are formed (Fig. 12–5). Thus phagocytosis and pinocytosis are specific types of the general process of endocytosis. The size and nature of the ingested material differ, but it is to be emphasized that the basic process is similar. In both cases the result is internalization of material from outside the cell within a membrane-bound structure referred to as a phagosome. More specifically, since the material they contain is from outside the cell, pinocytic vesicles and phagocytic vacuoles are heterophagosomes.

After the material to be digested has been enclosed in a membrane, it is necessary for the hydrolases to contact the material. This is accomplished by fusion of the phagosome with a lysosome. The lysosomes participating in this fusion may be either primary or secondary. Primary lysosomes, which have not previously been involved in digestion, are often formed in the Golgi apparatus in a manner similar to the formation of secretory vacuoles. It has been suggested

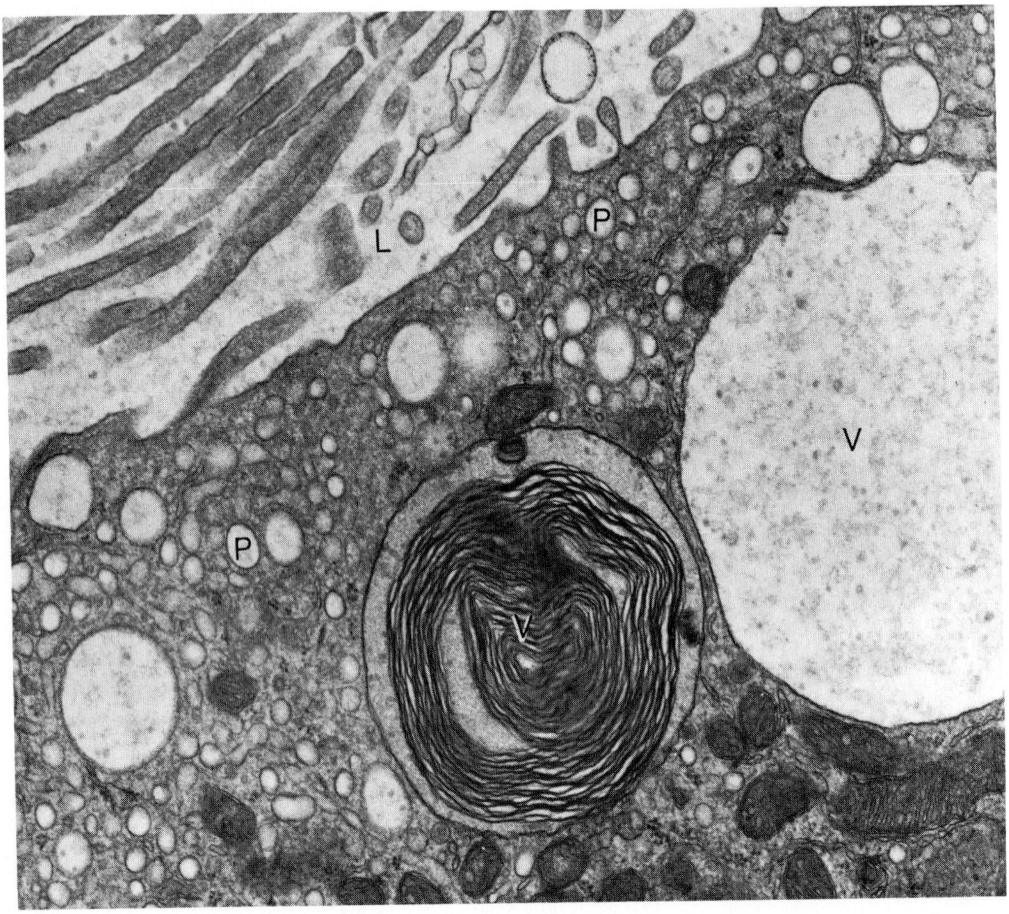

FIGURE 12–5 Apical part of a cell in the rat epididymal epithelium. Many small pinocytic vesicles (P) are present. These are apparently formed by pinching off from the plasma membrane and contain a small amount of fluid from the lumen (L). Two large vacuoles (V), one containing some membranous material, are also visible. × 28,000.

that in some cases, primary lysosomes can be formed directly by budding from the rough endoplasmic reticulum or from the smooth endoplasmic reticulum that is designated GERL and is found near the Golgi apparatus (Chapter 9).

Primary lysosomes may be difficult to identify on the basis of their morphology alone. Since they lack material to digest, primary lysosomes are frequently indistinguishable in routine electron microscope preparations from other smooth-surfaced or coated vesicles of a similar size. Identification of primary lysosomes is aided by cytochemical staining methods, such as the acid phosphatase reaction (Fig. 12–3). Fusion of a secondary lysosome with a phagosome not only brings hydrolases in contact with newly ingested material but also leads to a mixing or sharing of contents among the secondary lysosomes in a cell.

Whether the hydrolases are obtained from a primary or a secondary lysosome, the membranes of the phagosome and the lysosome become fused and form a single continuous membrane. Once a phagosome has fused with a lysosome and has acquired enzymatic activity, the resulting structure is termed a secondary lysosome, since digestion of its contents now proceeds (Fig. 12–6). Active digestion within a secondary lysosome results in the contents being broken down into small molecules, and these are presumably absorbed into the cytoplasm across the lysosome membrane. One of the puzzling features of lysosomal function is how the membrane bounding the organelle is protected from digestion, since membranes included in the contents of the lysosomes are digested.

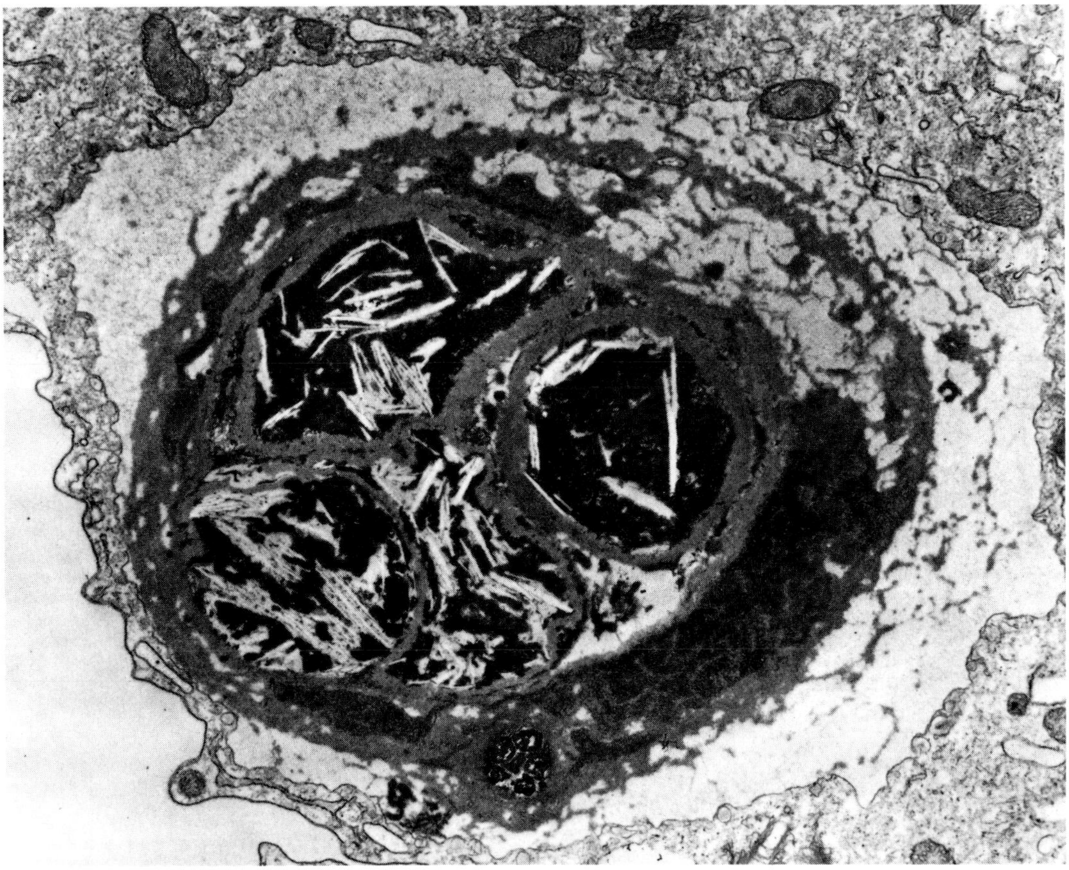

FIGURE 12–6 Cytoplasm of *Amoeba proteus*. The food vacuoles of protozoa are large secondary lysosomes in which digestion of food organisms takes place. × 8600.

Ultimately, only indigestible or slowly digestible material remains within the lysosome. At this stage, the structure is referred to as a residual body, and enzymatic activity is low or even absent. Morphologically, residual bodies often contain whorls of membranous material (Fig. 12–2). In protozoa, the contents of residual bodies are expelled from the cell by exocytosis, a process sometimes termed, in this context, cell defecation. The extent to which residual bodies are discharged from mammalian cells is uncertain. It is claimed to occur in some cells, but in others residual bodies appear to persist in the cytoplasm. Residual bodies tend to accumulate lipid which is oxidized to the pigmented material, lipofuscin. It has long been known that lipofuscin pigment accumulates with age in certain cells, such as neurons and heart muscle, and it has been speculated that this might play some role in aging of these tissues.

An example of heterophagy can be seen in the activities of the neutrophilic leukocytes of the blood and the macrophages of connective tissue. Both are significant in medicine because their phagocytosis of bacteria and other materials is an important defense mechanism. Neutrophils (Fig. 12–7) contain some granules that are primary lysosomes (azurophilic granules) and other granules that contain bacteriocidal substances (specific granules).

The primary lysosomes, or azurophilic granules, are formed at one pole of the Golgi apparatus during the development of neutrophils in the bone marrow. Interestingly, the specific granules are formed from the opposite pole of the Golgi apparatus at a later stage in development. Neutrophils can ingest bacteria by endocytosis, forming a phagosome that contains the bacteria. The membranes of both types of granule fuse with the membrane bounding the phagosome, bringing their contents into contact with the bacteria. This results in the death and digestion of the bacteria. Certain types of virulent bacteria have properties that make them resistant to phagocytosis or digestion. Heterophagy is also important in the uptake and digestion of proteins in the cells of many tissues.

AUTOPHAGY

The process of autophagy (Fig. 12–8) consists of the segregation and digestion of part of a cell's own cytoplasm. As a first step, it is necessary to separate the portion to be digested from the rest of the cell by surrounding it with a membrane. There have been many suggestions as to the source of this membrane, including *de novo* membrane formation and, at one time or another, virtually every membranous organelle in the cell. Currently the endoplasmic reticulum is favored, as the result of studies on autophagy in liver cells that is induced by the hormone glucagon. Under these conditions, elements of endoplasmic reticulum appear to surround part of the cytoplasm and fuse with one another. The result is that the portion to be digested is initially enclosed within two membranes, and then one of these, probably the innermost, is thought to degenerate.

Whatever the source of the membrane, the result is that part of the cytoplasm is surrounded by a membrane and the structure thus delineated is called an autophagosome. Once the autophagosome is formed, subsequent steps in the digestion process are the same as those described for heterophagy. That is, the phagosome fuses with a primary or secondary lysosome, acquires a content of hydrolytic enzymes, and becomes an autolysosome. Digestion proceeds, and eventually a residual body is formed.

Morphologically, the early stages of autophagy are readily identified by the presence of recognizable cell organelles enclosed within a membrane (Fig. 12–9). When digestion has proceeded to the point at which the structure of the enclosed

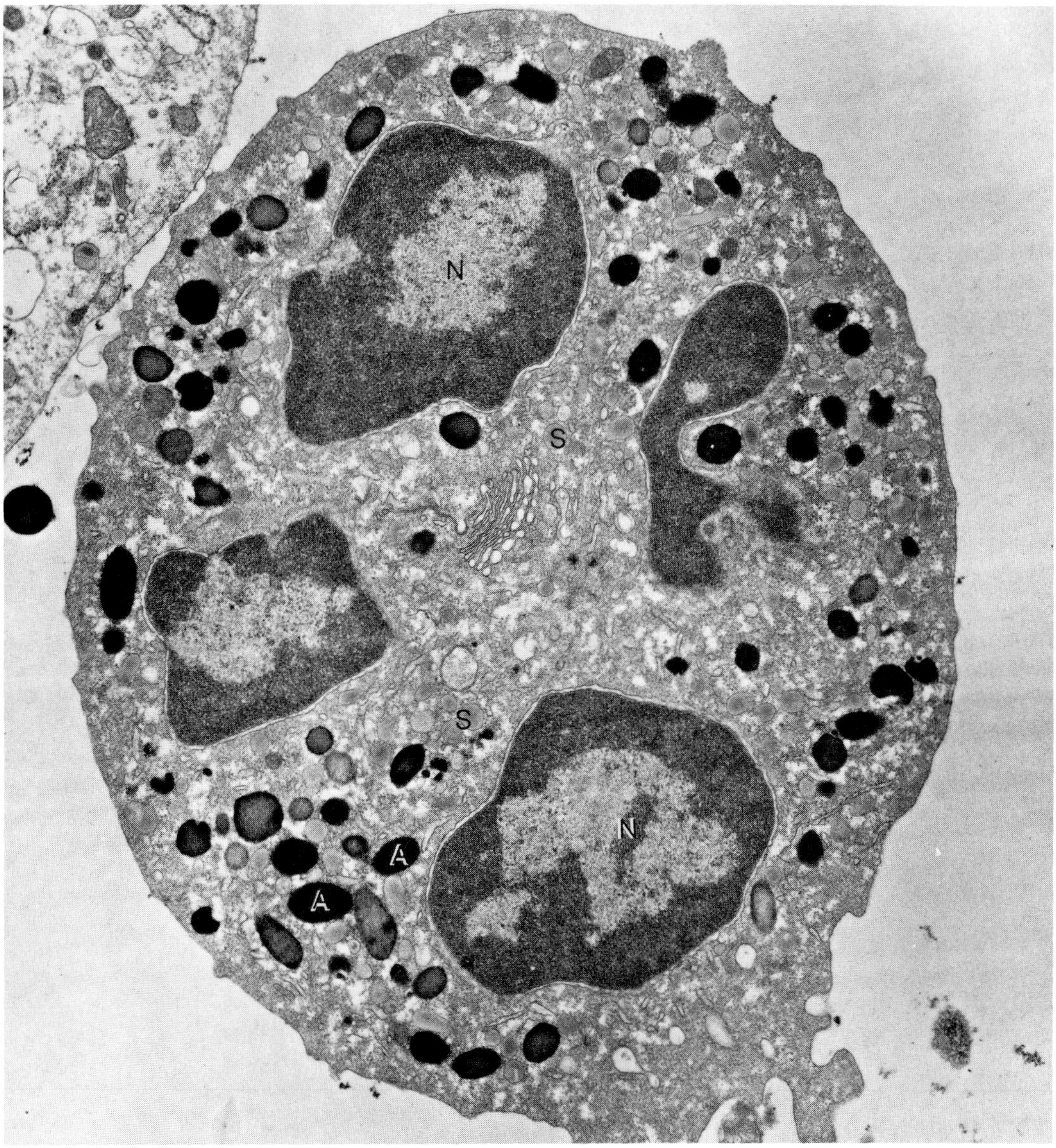

FIGURE 12–7 Human neutrophilic polymorphonuclear leukocyte, reacted for peroxidase. The cytoplasm contains two types of membrane-bound granules. The azurophilic granules (A), which are a type of primary lysosome, contain dense peroxidase reaction product. The smaller, less dense specific granules (S) lack lysosomal enzymes and peroxidase but probably contain antibacterial substances. N, lobes of the nucleus. × 21,000. (Micrograph courtesy of Bainton, D. F., Ullyot, J. L., and Farquhar, M. G.: J. Exp. Med., *134*:907, 1971. Reproduced by permission of the Rockefeller University Press.)

organelles is broken down, however, autolysosomes may be morphologically indistinguishable from heterolysosomes. The distinction may be artificial at this stage in any event, because secondary lysosomes are known to fuse with one another and exchange contents.

Autophagy is not an unusual process. It is observed in degenerating cells, such as the smooth muscle cells of the postpartum uterus, as well as in cell death during remodeling of embryonic tissues. Autophagy is also seen, however, in active cells, such as liver and kidney parenchyma. The significance of autophagy in nondegenerating cells is poorly understood, although it has been suggested that autophagy might play a role in turnover of organelles. Also, it is not clear whether

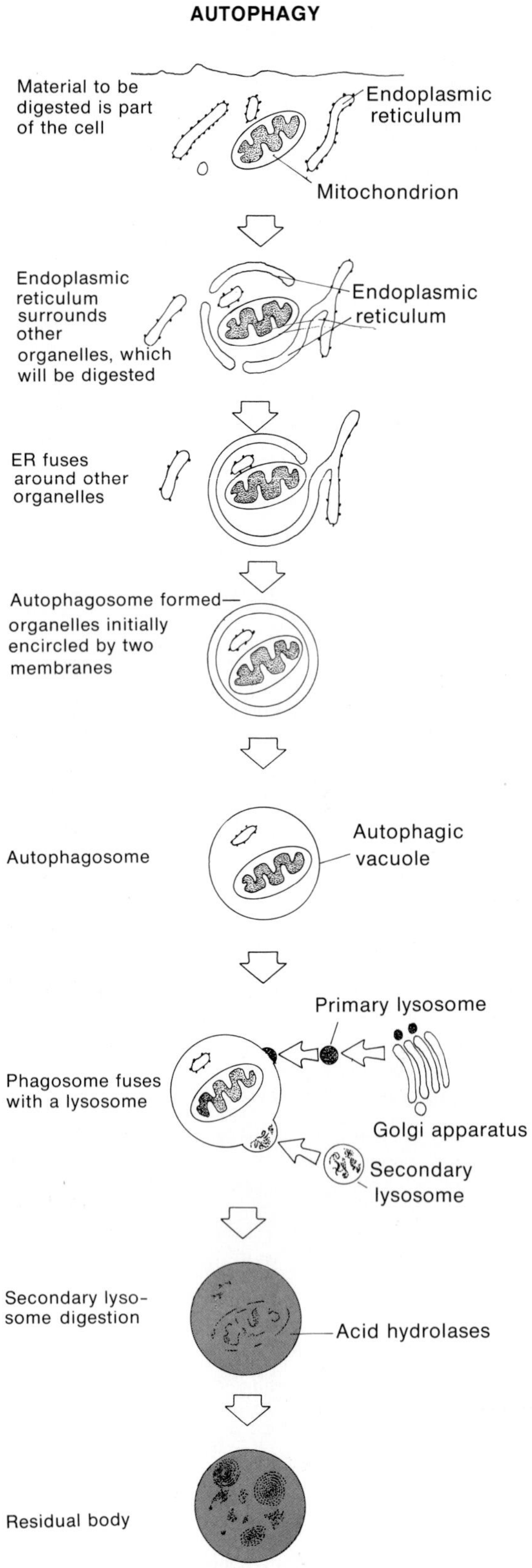
AUTOPHAGY
Material to be digested is part of the cell
Endoplasmic reticulum
Mitochondrion
Endoplasmic reticulum surrounds other organelles, which will be digested
Endoplasmic reticulum
ER fuses around other organelles
Autophagosome formed—organelles initially encircled by two membranes
Autophagosome
Autophagic vacuole
Primary lysosome
Phagosome fuses with a lysosome
Golgi apparatus
Secondary lysosome
Secondary lysosome digestion
Acid hydrolases
Residual body

FIGURE 12–8 The process of autophagy. Material to be digested is part of the cell itself. Note that once a phagosome is formed, the steps are the same as in the heterophagic process.

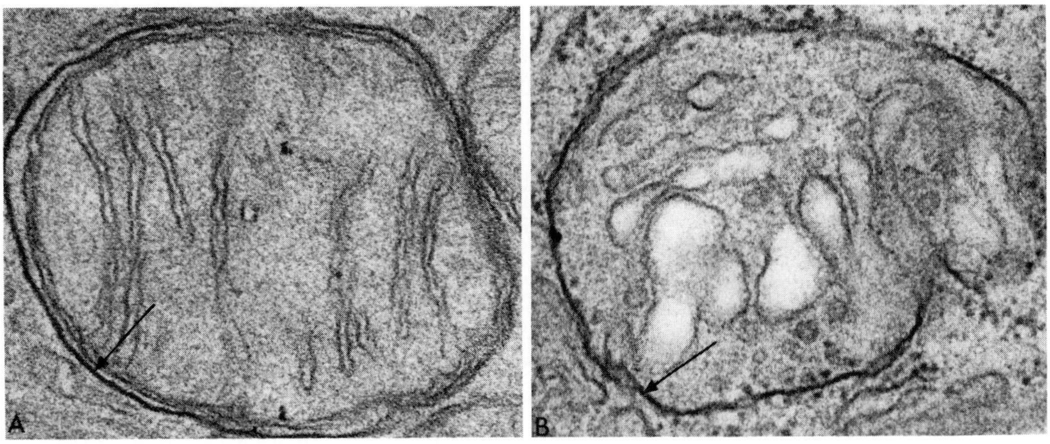

FIGURE 12–9 Autophagic vacuoles in the kidney of the rat. The membrane of the vacuole (*arrow*) encloses a mitochondrion in *A* and membranes of the Golgi apparatus or endoplasmic reticulum in *B*. × 80,000. Micrographs courtesy of Pfeiffer, U., and Scheller, H.: J. Cell Biol., 64:608, 1975. Reproduced by permission of the Rockefeller University Press.)

autophagic processes can discriminate between organelles and selectively destroy those that are defective or worn-out, or whether autophagic processes simply involve organelles at random.

LYSOSOMES AND SECRETION

Lysosomes appear to be involved in the regulation of secretion in some cells. The mammotrophs of the anterior lobe of the pituitary gland secrete the polypeptide hormone prolactin, which is synthesized in the rough endoplasmic reticulum, transported to the Golgi apparatus, packaged into secretory granules, and released by exocytosis, as in other protein-secreting cells (Chapter 9). When secretion of the hormone is diminished, as occurs when suckling young are removed from a lactating animal, excess secretory material is disposed of by fusion of secretory vacuoles with lysosomes and digestion of the secretory product. This process is referred to as crinophagy.

Lysosomes are also involved in secretion in the thyroid gland. The product of the follicular cells of the thyroid is thyroglobulin, which consists of the thyroid hormones throxine and triiodothyronine linked to protein. It is secreted from the apical ends of the cells and stored as colloid in the lumen of the follicle. Under appropriate stimulation, colloid droplets are taken up into the follicular cells by endocytosis and fuse with lysosomes. Hydrolysis of the thyroglobulin results in freeing of the thyroid hormones, which are then released to the blood from the base of the follicular cells.

In still other instances, lysosomal enzymes are secreted by exocytosis in a manner similar to secretion in exocrine gland cells. Release of lysosomal enzymes in this way occurs in connective tissue cells, resulting in digestion of components of connective tissue matrix, and in the prostate gland.

LYSOSOMES IN DISEASE

Lysosomes are important in the pathogenesis of numerous diseases, and they can be related to disease processes in several ways. The lysosomes themselves may have a normal composition, and they may participate in the pathogenesis of a disease by exercising the normal functions discussed earlier. The phagocytosis of bacteria by neutrophils and macrophages is one such example. In some diseases, lysosomes may be normal, but a pathological process may result from their inability to digest nonbiological material. For example, under certain environmental conditions, individuals are exposed to and inhale dust containing substances such as coal, asbestos, or silicates. These particles are phagocytosed by macrophages in the alveoli of the lung. They cannot be digested by the lysosomes of the phagocytes, however, and this may be important in their accumulation in the lungs and in the ensuing tissue reactions. In other instances there may be an unusual release of lysosomal enzymes. Lysosomes are thought to be involved in this way in a large number of diseases, including allergic reactions, arthritis, and certain muscle diseases. The variety of diseases included in these examples probably reflects the universal distribution of lysosomes and their importance in fundamental cell function.

In other diseases the lysosomes themselves are defective. The lysosomal storage diseases are due to genetic abnormalities, and in each instance the deficiency is usually that of a single lysosomal hydrolase. Lack of sufficient enzyme results in the accumulation of the substrate of the enzyme within lysosomes, which increase in size and become greatly distended with the indigestible substance (Fig. 12–10). These large lysosomes may occupy so much volume in a cell that normal cellular functions are impaired. Thus an important common characteristic of the storage diseases is the presence in electron microscope preparations of numerous large vacuolelike lysosomes limited by a single membrane and distended with some material. Lipids and mucopolysaccharides are often involved, perhaps because of the multiplicity of enzymes for their breakdown found in lysosomes, and many different clinical syndromes have been described.

In the lipidoses, such as Tay-Sachs disease, lipid accumulates in the brain as well as in other organs, leading to blindness and mental retardation. In the glycogen storage diseases, glycogen accumulates, particularly in liver and muscle cells. The Hurler syndrome, or gargoylism, is an example of an mucopolysaccharidosis. It is seen in several forms that include in their symptomatology dwarfism, enlarged liver and spleen, skeletal deformities, and mental retardation.

PEROXISOMES

Peroxisomes are small spherical or oval organelles, 0.3 to 1.5μm in diameter, which are bounded by a single membrane. These features lend peroxisomes some morphological similarity to lysosomes, but they are distinct from lysosomes in both structure and chemical composition. Peroxisomes have a finely granular, moderately dense interior, which is homogeneous except for the presence in some species of a crystalline inclusion known as the nucleoid (Figs. 12–11 and 12–12). Peroxisomes are found in the liver, kidney, and most other cells of vertebrates, and they are widely distributed in protozoa, plants, and other eukaryotes. The descriptive morphological term microbody has also been applied to them, but "peroxisome" is currently favored because it conveys more information about their chemical composition. The small peroxisomes found in many cells are often referred to as microperoxisomes.

Peroxisomes contain several enzymes that are related to the production or destruction of hydrogen peroxide. These enzymes include urate oxidase, D-amino acid oxidase, and α-hydroxy acid oxidase, which form hydrogen peroxide, and

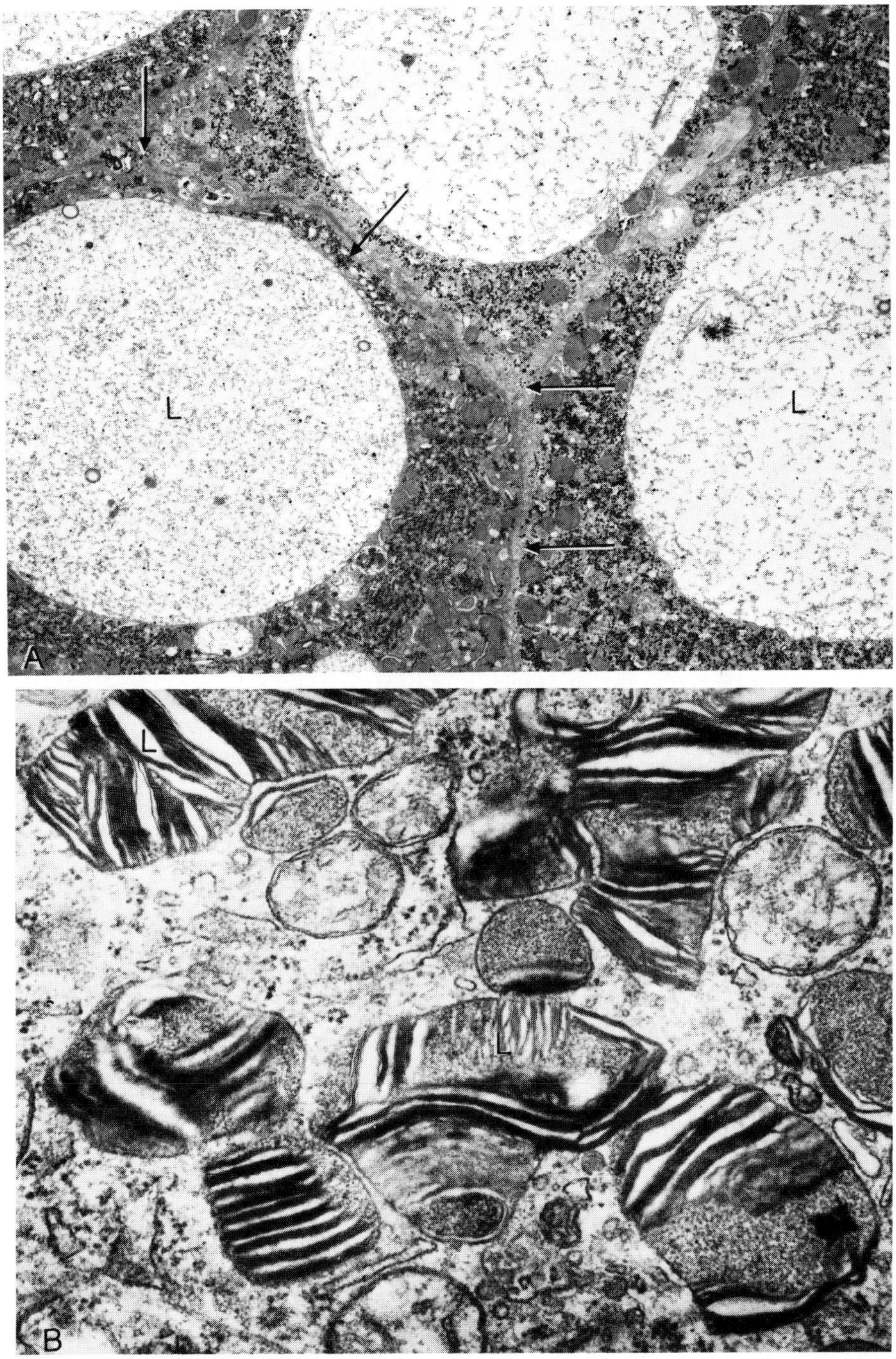

FIGURE 12–10 Electron micrographs of tissues from patients with storage diseases due to deficiency of lysosomal enzymes. *A*. Liver of a patient with G_{M1}-gangliosidosis, who lacked acid β-galactosidase. Oligo- and polysaccharides have accumulated in large vacuole-like lysosomes (L) in the cytoplasm of the hepatocytes. These lysosomes were calculated to have a volume 180,000 times that of lysosomes in a normal liver. Their large size can be appreciated by comparison with an individual liver cell (*outlined by arrows*). The cytoplasm of the hepatocyte is compressed to a rim surrounding the large vacuole. × 6000. *B*. Portion of a cerebral neuron from a patient with Sanfilippo (type A) disease. Accumulation of lipid-containing material has led to a lamellated appearance in the numerous lysosomes (L), which for this reason are known as "zebra bodies." × 47,000. (Micrographs courtesy of F. Van Hoof, A. Resibois, and H. Hers. From Dingle, J., and Fell, H. B. (eds.): Lysosomes in Biology and Pathology. Vol. 2. 1969. Reproduced by permission of Exerpta Medica, Amsterdam.)

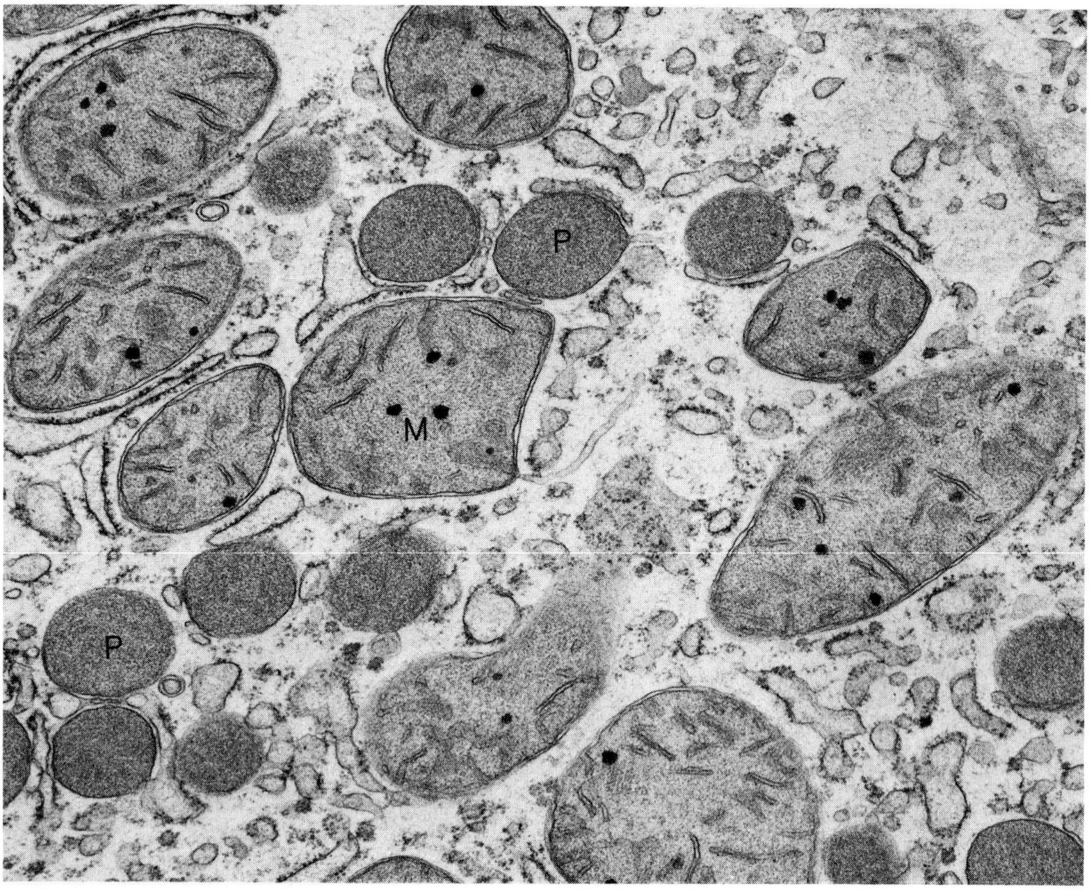

FIGURE 12–11 Peroxisomes (P) in human liver are bounded by a single membrane and have a homogeneous finely granular interior. No nucleoid is present. M, mitochondrion. × 34,000. (Micrograph courtesy of Shnitka, T. K.: J. Ultrastruct. Res., *16*:598, 1966. Reproduced by permission of Academic Press.)

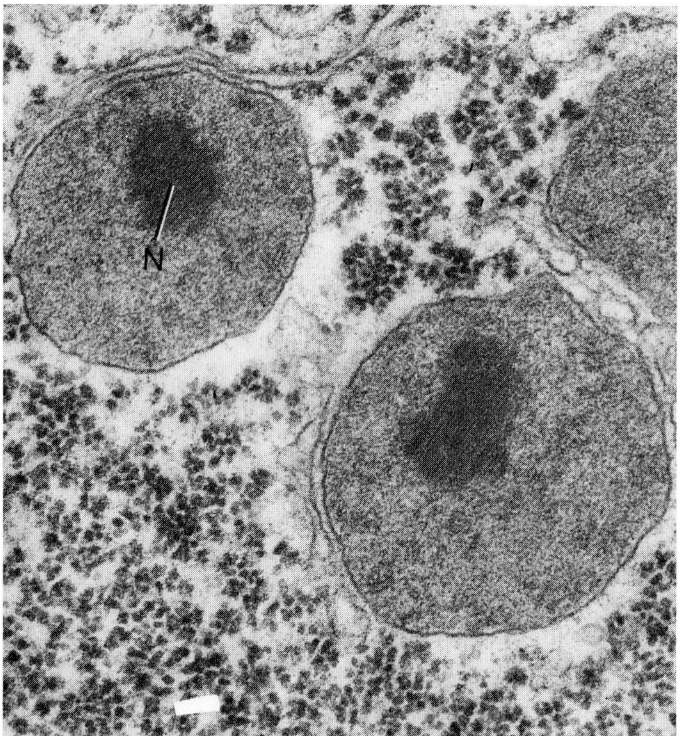

FIGURE 12–12 Peroxisomes in rat liver contain a crystalline nucleoid (N). × 77,000. (Micrograph courtesy of Shnitka, T. K.: J. Ultrastruct. Res., *16*: 598, 1966. Reproduced by permission of Academic Press.)

catalase, which destroys hydrogen peroxide. Similar organelles in plants, termed glyoxysomes, possess additional enzymes and have additional functions. Not all species possess urate oxidase, and the presence of this enzyme has been correlated with the presence of a nucleoid (Fig. 12–12). When present, urate oxidase is involved in the terminal metabolism of purines.

Although the enzymes of mammalian peroxisomes have in common some involvement with hydrogen peroxide, the functional significance of the enclosure of these particular enzymes within a single organelle is not well understood. It has been speculated that the proximity of hydrogen peroxide-producing and hydrogen peroxide-destroying enzymes may prevent the release of this toxic compound within the cell. Peroxisomes are closely related topographically to the endoplasmic reticulum, and many investigators believe that peroxisomes are formed from the endoplasmic reticulum or that they may represent local dilatations of the endoplasmic reticulum.

FURTHER READING

Cohn, Z. A., and Fedorko, M. E.: The formation and fate of lysosomes. *In*: Dingle, J. T., and Fell, H. B. (eds.): Lysosomes in Biology and Pathology. Amsterdam, North-Holland, 1969.
deDuve, C.: Lysosomes. Sci. Am., *208*: 5, 1963.
deDuve, C., and Baudhuin, P.: Peroxisomes (microbodies and related particles). Physiol. Rev., *46*: 323, 1966.
deDuve, C., and Wattiaux, R.: Functions of lysosomes. Annu. Rev. Physiol., *28*:435, 1966.
Dingle, J., and Fell, H. B. (eds.): Lysosomes in Biology and Pathology, Vols. 1–3. Amsterdam, North-Holland, 1969.
Ericsson, J. L. E.: Studies on induced cellular autophagy. II. Characteriziation of the membranes bordering autophagosomes in parenchymal liver cells. Exp. Cell Res., *56*:393, 1969.
Friend, D. S., and Farquhar, M. G.: Functions of coated vesicles during protein absorption in the rat vas deferens. J. Cell Biol., *35*:357, 1967.
Hers, H. G., and Van Hoof, F. (eds.): Lysosomes and Storage Diseases. New York, Academic Press, 1973.
Holter, H.: Pinocytosis. Int. Rev. Cytol., *8*:481, 1960.
Holtzman, E. (ed.): Lysosomes, a Survey. Vienna, Springer-Verlag, 1975.
Hruban, Z., and Rechcigl, M.: Microbodies and related particles. Int. Rev. Cytol. (Suppl.), *1*:1, 1969.
Kolodny, E. H.: Lysosomal storage diseases. N. Engl. J. Med., *294*:1217, 1976.
Masters, C., and Holmes, R.: Peroxisomes: new aspects of cell physiology and biochemistry. Physiol. Rev., *57*:816, 1977.
Novikoff, A. B., Essner, E., and Quintana, N.: Golgi apparatus and lysosomes. Fed. Proc., *23*:1010, 1964.
Pitt, D. (ed.): Lysosomes and Cell Function. London, Longman, 1975.
Shnitka, T. K.: Comparative ultrastructure of hepatic microbodies in some mammals and birds in relation to species differences in uricase activity. J. Ultrastruct. Res., *16*:598, 1966.
Silverstein, S. C., Steinman, R. M., and Cohn, Z. A.: Endocytosis. Annu. Rev. Biochem., *46*:669, 1977.
Smith, R. E., and Farquhar, M. G.: Lysosome function in the regulation of the secretory process in cells of the anterior pituitary gland. J. Cell Biol., *31*:319, 1966.

13

MICROTUBULES AND FILAMENTS

Microtubules are long, slender, hollow cylinders about 250 Å in diameter. In contrast, cytoplasmic filaments are thin threadlike structures. While microtubules in different locations and in different kinds of cells are believed to have a similar composition and basic structure, there are several types of filaments, which are thought to have distinct functions. Both microtubules and filaments are found in most eukaryotic cells. Although ubiquitous, microtubules are particularly prominent in the motile cell processes, cilia and flagella, and in the mitotic spindle. Filaments are most abundant and most highly organized in muscle cells, but it has become clear that similar structures are important in nonmuscle cells as well. Microtubules and filaments have long been thought to function in a cytoskeletal way and to play a role in cell motility, but the details of how these functions are accomplished are still emerging. Recent observations suggest that although microtubules and filaments are distinct structures, they may interact with one another in performing some of their functions.

MICROTUBULES

Structure and Chemistry of Microtubules

Although microtubules are widely distributed in different cell types, recent progress has been made by isolating components of microtubules from sources such as brain or the mitotic spindle and studying their structure, composition, and behavior *in vitro*. The long cylindrical shape of microtubules is readily appreciated in the isolated state (Fig. 13–1) as well as in sections of cells. If isolated microtubules are negatively stained before being viewed in the electron microscope, their cylindrical wall can be seen to have a globular substructure (Fig. 13–2). The globules are about 40 × 50 Å, and they usually are disposed in 13 longitudinal rows termed protofilaments. These 13 components can also be seen in the circular profile of the wall of microtubules cut in cross section in particularly favorable thin sections (Fig. 13–3). However, instances are known in which cross sections of the walls of microtubules display more or less than 13 components.

When isolated microtubules are analyzed chemically, they are found to be

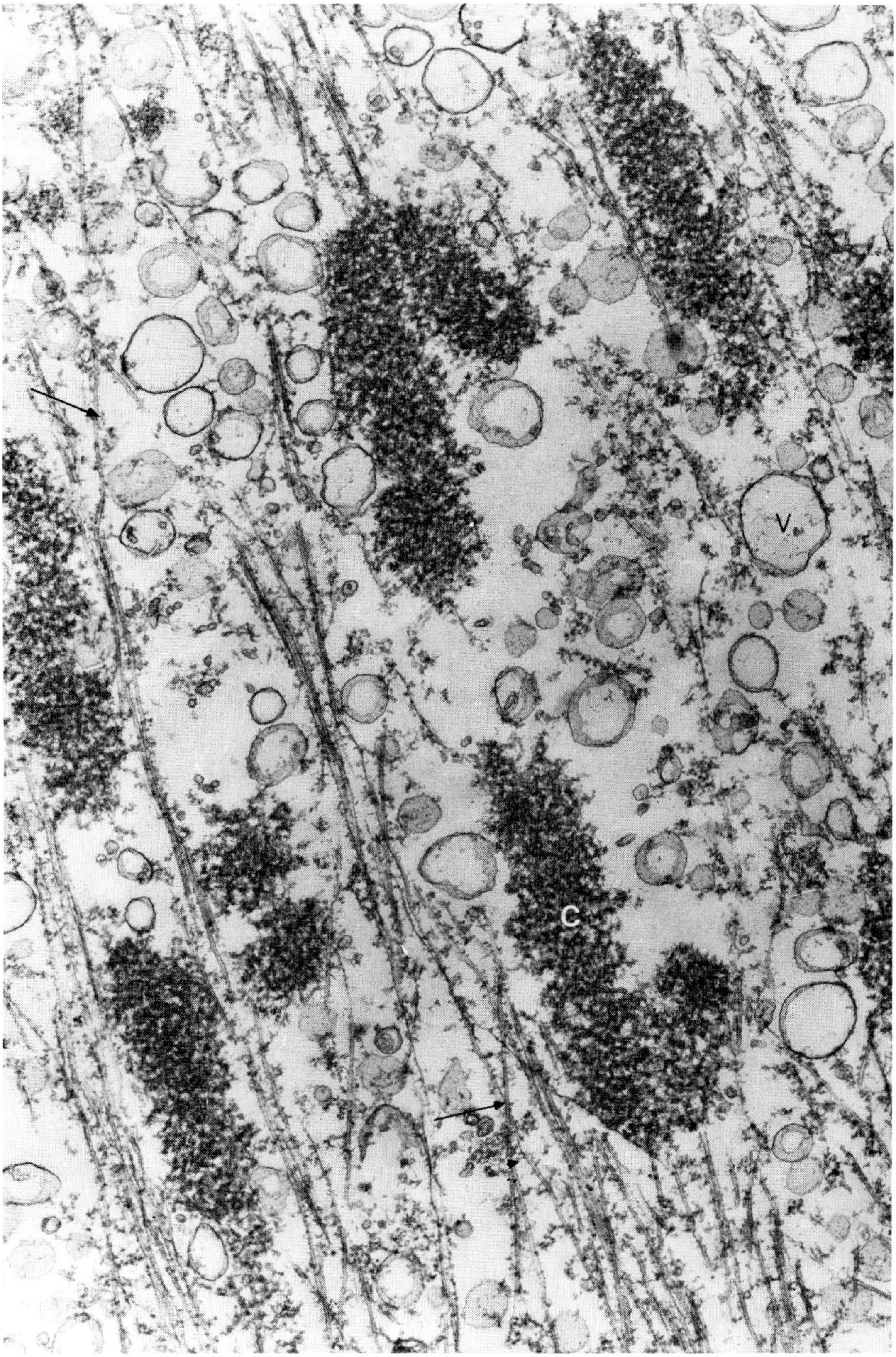

FIGURE 13–1 Isolated spindle from a sea urchin egg in early anaphase. The background is less dense than in sections of intact cells, and the long, cylindrical structure of the spindle microtubules (*arrows*) is evident. The dense masses of material represent mitotic chromosomes (C), and some microtubules appear to attach to the chromosomes. Somes membranous vesicles (V) are also present in the field. × 30,000. (Micrograph courtesy of L. Rebhun.)

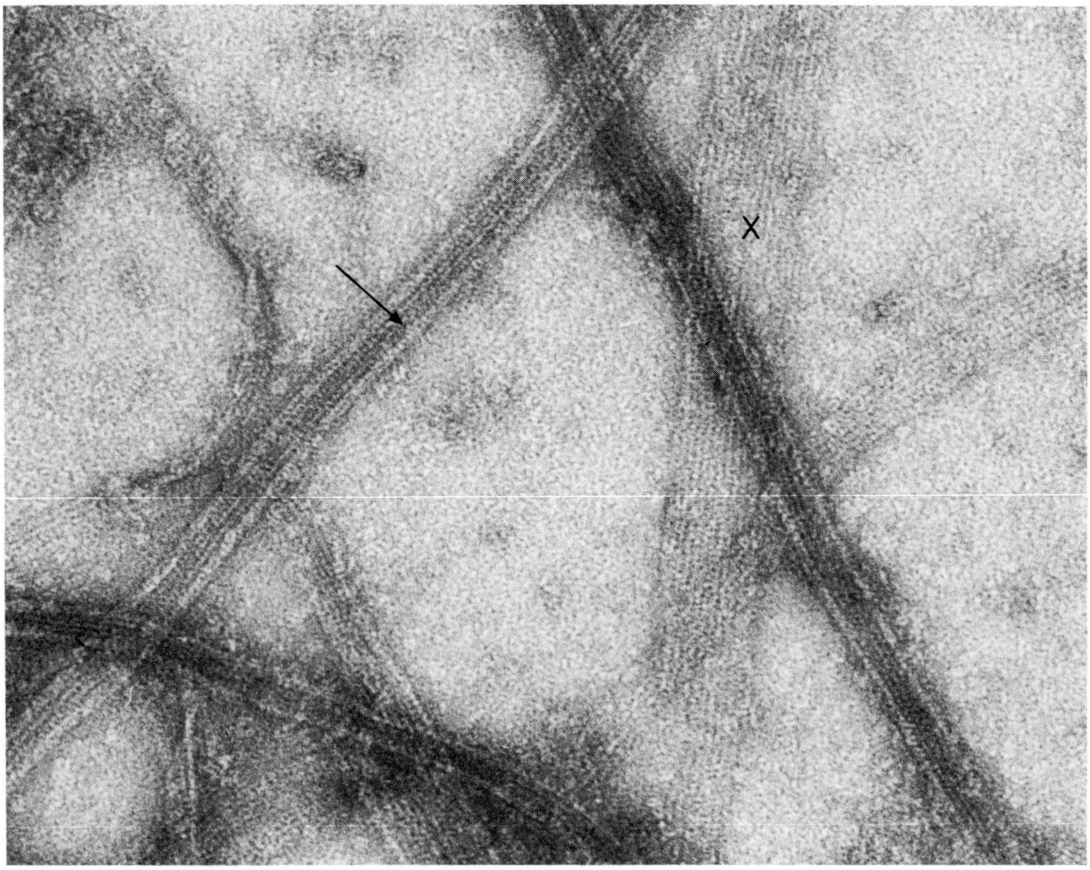

FIGURE 13–2 Isolated, negatively stained microtubules, repolymerized *in vitro* from bovine brain tubulin. The walls of the microtubules are composed of longitudinally disposed protofilaments (*arrow*), which in turn appear to be composed of a row of globules. At some places in the preparation, the microtubules have "opened up" to form flat sheets of material (X) in which their components of protofilaments and globules is especially evident. × 280,000. (Micrograph courtesy of L. Rebhun.)

composed mainly of the protein tubulin. Tubulins from different sources are very closely related proteins having a basic similarity and differing only slightly in microtubules from different organisms. Tubulin occurs as a 6S protein dimer of approximately 115,000 molecular weight. There are two main types of monomers, tubulins α and β, which have molecular weights of about 55,000, and each dimer is believed to contain one α and one β monomer. As described in the section on mitosis (Chapter 5), each tubulin dimer binds one molecule each of colchicine and of the *Vinca* alkaloid vinblastine, although the binding site for colchicine

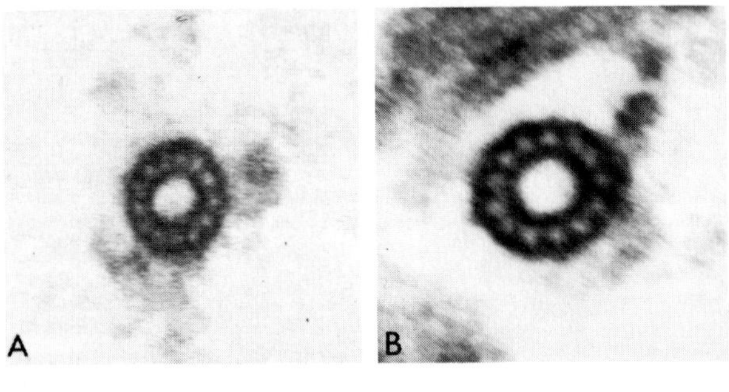

FIGURE 13–3 Cross section of microtubules fixed with glutaraldehyde and tannic acid. In material prepared in this way the protofilaments comprising the wall of the microtubule are outlined by a dense substance. Microtubules from many sources are composed of 13 protofilaments, but in some cases, as in certain cells of the crayfish, microtubules having more or fewer than 13 have been observed. *A*. × 385,000. *B*. × 450,000. (Micrographs courtesy of P. R. Burton and G. B. Pierson.)

appears to be different from that for vinblastine. These drugs inhibit mitosis by binding to tubulin dimers and reducing the polymerization of microtubules, and thus they have been extensively studied and used in the therapy of some forms of cancer. The tubulin dimer also binds two guanine nucleotides, and one mole of tubulin dimer binds about one mole of Mg^{++}; these associations are of interest because they may be important in the regulation of polymerization of tubulin into microtubules.

The correspondence between isolated tubulin molecules and the globular subunits seen in microtubules with the electron microscope is under investigation, but probably the ~40 Å globules represent tubulin monomers, since the size of the isolated tubulin dimers is roughly 40 × 80 Å.

In addition to tubulin, a number of microtubule-associated proteins have been found in preparations of isolated microtubules. These proteins are under intensive investigation and may prove to be of great functional importance. For example, some of the microtubule-associated proteins could represent "bridges" that attach to binding sites periodically disposed on the surfaces of microtubules and permit microtubules to relate to other cell organelles. Other microtubule-associated proteins may be concerned with the regulation of tubule assembly and function.

Cilia and Flagella

Microtubules are organized in a distinctive way in the core of cilia and flagella. They run longitudinally in the center of each of these processes, and when cut in cross section can usually be seen to have the 9 + 2 arrangement that is characteristic of most cilia and flagella of eukaryotes (Fig. 13–4). This pattern is so called because a central pair of individual microtubules is surrounded by a ring of nine doublet tubules. The 9 + 2 constellation of microtubules is termed the axoneme of a cilium. Each outer doublet, or outer fiber as it is sometimes called, is composed of an A and a B tubule or "subfiber." The C-shaped subfiber B has an incomplete wall, but it is fused to subfiber A in such a way that the defect in the wall of tubule B is filled by part of the wall of tubule A. Thus the two subfibers hold a portion of their wall in common. The two subfibers have different solubilities, suggesting that there is some subtle difference in their composition. The nine outer doublets are assigned numbers, as shown in Figure 13–4.

The outer doublets are canted at an angle to the circumference of the cilium, with subfiber A closer to the center of the cilium. From the surface of each subfiber A a pair of "arms" projects toward the next doublet. The geometry of the axoneme is such that the arms would be directed in a clockwise direction if the cilium were viewed from its base, looking toward its tip. The arms have been shown to contain a protein, dynein, which is significant because it has ATPase activity and thus may be involved in generating energy for ciliary movement. An inborn disease has been described which is associated with the absence of dynein arms on cilia and flagella (Fig. 13–5). Men with this disease produced sperm that were alive but immotile. Some of them also had chronic bronchitis and sinusitis along with frequent acute respiratory infections, including colds and pneumonia. These symptoms are believed to be related to the lack of ciliary motion by the cells of ciliated epithelia, leading to diminished transport of mucus in the respiratory tract.

Additional less clearly defined proteinaceous material has been described which links the peripheral doublets with another and connects with a sheath

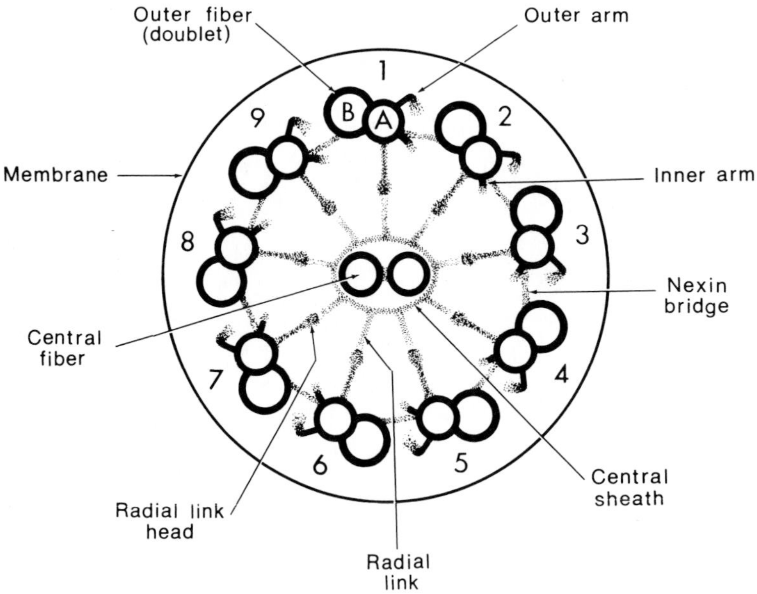

FIGURE 13–4 Diagram of the cross-sectional appearance of the components in the axoneme of a typical cilium. (Redrawn after R. Linck. Modified from Bloom, W., and Fawcett, D. W.: A Textbook of Histology. 10th ed. Philadelphia, W. B. Saunders, 1975.)

around the central pair of tubules by a series of radial projections. The protein connecting adjacent outer doublets has been termed nexin, and it is thought to play a role in binding the components of the axoneme (Fig. 13–4). A different protein forms the spokes that connect doublets to the central sheath.

Cilia and flagella are usually considered together because they are long, thin, motile processes and have in common the structural features of the axoneme. The two types of processes have differences, however, and they can be distinguished in the following way. Flagella usually occur singly on a given cell, while cilia are frequently multiple (Fig. 13–6). The longer flagella may contain structures in addition to the axoneme, such as the outer coarse fibers of mammalian sperm flagella (Fig. 13–5). The movement of flagella occurs in a more complex wavelike pattern than that of cilia, which more nearly resembles a simple bending.

When cilia move, they first bend at the base in the effective forward stroke. Recovery occurs by a wave of bending from the base toward the tip, as shown in Figure 13–7. The problem of the mechanism of ciliary movement has fascinated observers of cells for many years. Currently it is thought that ciliary movement occurs by a sliding of the tubules in the axoneme in relation to one another. If this is correct, one would expect the level of termination of the different doublets at the tip of the cilium to vary with different stages in the ciliary beat (Fig. 13–8). For example, those doublets on the side facing the direction of bending will extend farthest toward the tip during the effective stroke, while those on the opposite side will not extend as far into the tip. Painstaking analysis of electron micrographs of cross sections of cilia fixed at different stages of their bending have disclosed a pattern of termination of the components of the axoneme that is consistent with such a sliding mechanism. Recent studies indicate that dynein is involved in the sliding process.

It should be recalled that although the 9 + 2 arrangement is typical of the axonemes of cilia and flagella of mammalian cells, numerous examples of a different geometry in these processes can be found, particularly among the lower vertebrates and invertebrates. The tails of spermatozoa of insects, for example, show a remarkable variety of patterns of microtubule organization. Thus the

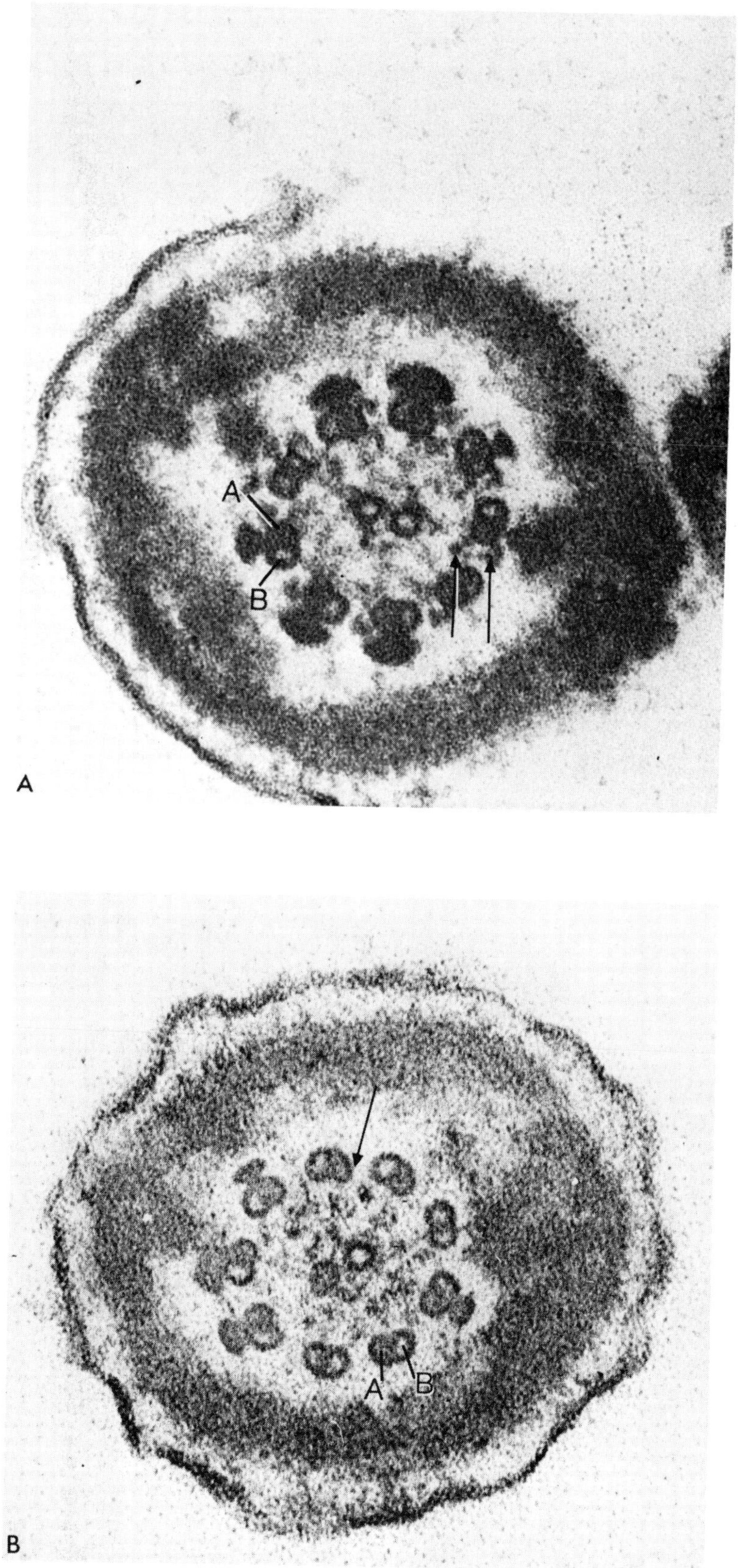

FIGURE 13–5 Cross sections of human sperm tails. *A*. From a man with normal sperm motility. Each of the nine doublet tubules of the axoneme consists of A and B tubules, and two dynein arms (*arrows*) are present on each A tubule. *B*. Immotile sperm from a patient with the "immotile cilia syndrome." Dynein arms are lacking from the A tubules, and the space they normally occupy appears empty (*arrow*). *A* and *B* × 150,000. (Micrograph courtesy of Afzelius, B. A., Eliasson, R., Johnsen, Ø, et al.: J. Cell Biol., *66*:225, 1975. Reproduced by permission of the Rockefeller University Press.)

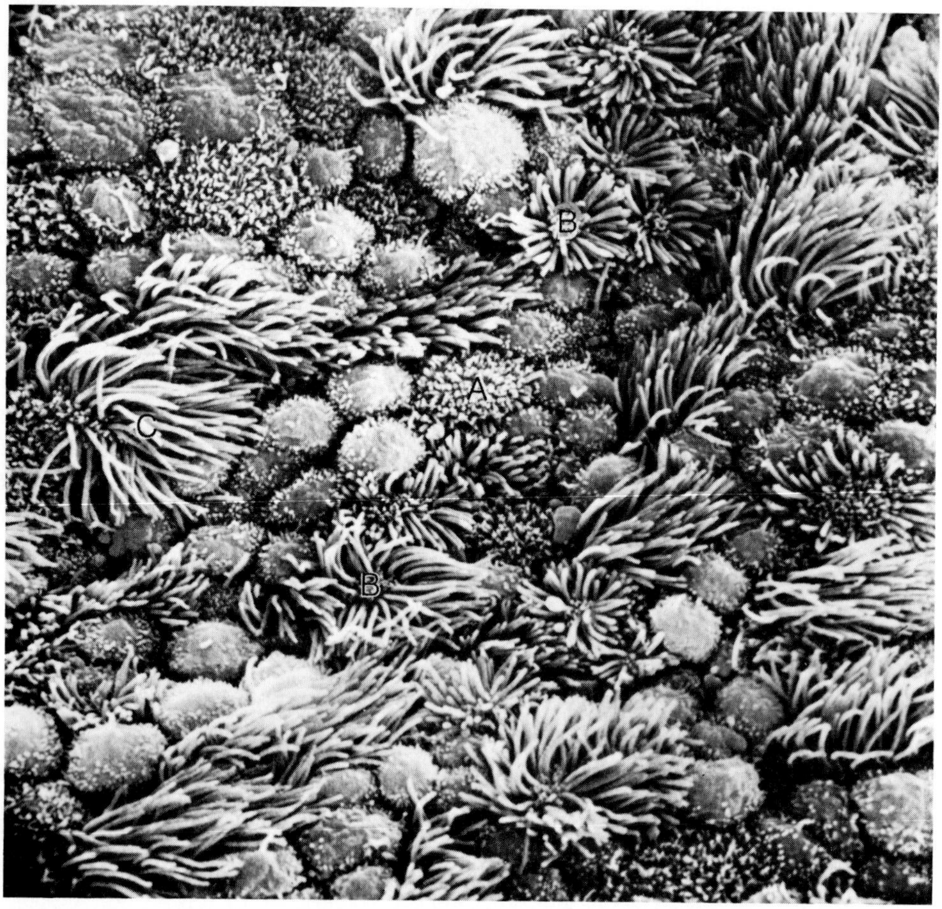

FIGURE 13–6 Scanning electron micrograph of surfaces of epithelial cells in the oviduct of a five-day-old mouse. Some cells (A) have short processes, microvilli, on their surfaces, while others have the longer motile cilia (C). Some cells (B) are in the process of forming cilia. × 3,500. (Micrograph courtesy of Dirksen, E. R.: J. Cell Biol., 62:899, 1974. Reproduced by permission of the Rockefeller University Press.)

common 9 + 2 axoneme configuration is not a necessary prerequisite for ciliary and flagellar motion.

The motions of the multiple cilia within a given cell relate to one another in two important ways. In isochronal movement the cilia all beat together. In metachronal rhythm the cilia beat in sequence, as if a wave of contraction passes through the population of cilia from one end of the cell to the other. The metachronal pattern of movement is particularly useful in moving particulate materials along the surface of the cell. The existence of isochronal or metachronal rhythms implies that there is some mechanism for coordination of ciliary contractions, but little is known about how this is accomplished. Extending into

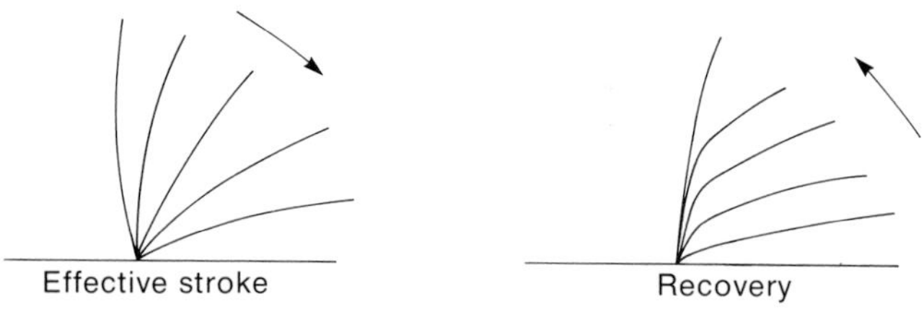

FIGURE 13–7 Diagram of the bending of a cilium during the effective stroke and in recovery.

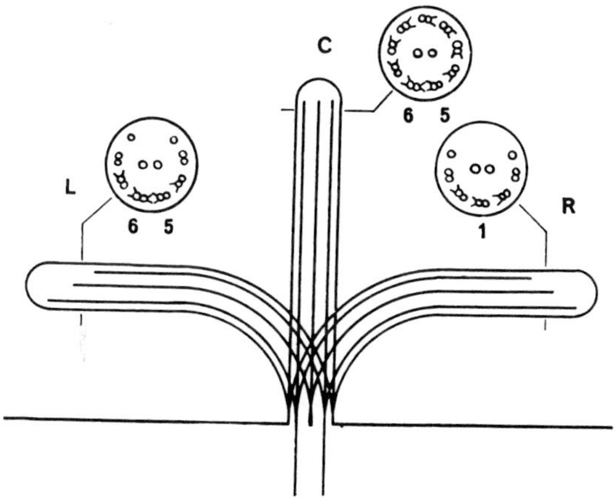

FIGURE 13–8 A diagram illustrating the sliding filament hypothesis of ciliary binding. At the center, cilium C is straight, and the doublets of the axoneme terminate at the same level. A transverse section near the tip would show all peripheral doublets. At the left, cilium L is bent toward peripheral doublets 5 and 6, which project farthest toward the tip. Doublet 1 terminates first and is missing from the cross section. Doublets become single near their termination; hence 9 and 2 are shown as single. At the right, cilium R is bent toward doublet 1, which now projects farthest. Doublets on this side of the cilium are present in the transverse section, while doublets 5 and 6 are missing. (Figure after M. A. Sleigh from Endeavour, 30:11, 1971. *From*: Bloom, W., and Fawcett, D. W.: A Textbook of Hisology. 10th ed. Philadelphia, W. B. Saunders, 1975.)

the underlying cytoplasm from the basal bodies of many cilia is a striated band, the ciliary rootlet. It is thought possibly to play a structural role by anchoring the cilia in the cytoplasm. In the ciliated protozoa, there is a very complex pattern of microtubules, fibers, and membranous elements in the cortex, and it has been speculated that these may be involved in the coordination of ciliary movement.

The microtubules of the peripheral doublets in cilia are continuous at the base of the shaft of the cilium with two of the three tubules of the triplets that comprise the wall of the basal body (Fig. 13–9). The central pair of tubules terminates in contact with a dense granule at the base of the ciliary shaft. Thus the basal body has a precise relation to the axoneme. It appears also to have a role in determining the location and orientation of the cilium, because during the formation of a cilium the ciliary microtubules extend progressively from the basal body as the shaft lengthens. Nevertheless, the basal body does not seen to play a direct role in the synthesis of protein components of the cilium, which presumably are synthesized by unbound cytoplasmic polysomes. Furthermore, it has been shown with radioautographic methods that new proteins are added at the tip of a growing cilium rather than at its base next to the basal body.

At least a limited amount of growth of a cilium is possible even in the absence of protein synthesis. If the two flagella of the alga *Chlamydomonas* are removed experimentally, inhibition of protein synthesis does not block their regrowth, although they may not attain a normal size. This behavior is attributed to the presence within the cytoplasm of a pool of free microtubule subunits, consisting of tubulin molecules and possibly other components, which need only to be polymerized to form microtubules to regenerate a cilium. Interestingly, if one of the two flagella of this organism is removed, during its regeneration the intact flagellum shortens as if it were contributing components to the pool for reconstruction of its companion.

Microtubules were readily observed in the earliest electron microscope studies of the axonemes of cilia and flagella. However, their widespread distribution in the spindle and cytoplasm of almost all eukaryotes was slower to be appreciated. At least in part, this was because microtubules in cilia and flagella seem to have a greater stability than those in most locations elsewhere in the general cytoplasm. Thus microtubules in axonemes were preserved by the OsO_4 commonly used as a primary fixative in the early days of electron microscopy, while their cytoplasmic congeners were not. It was not until the introduction of glutaraldehyde-containing fixatives in the 1960's that the ubiquitous nature of cyto-

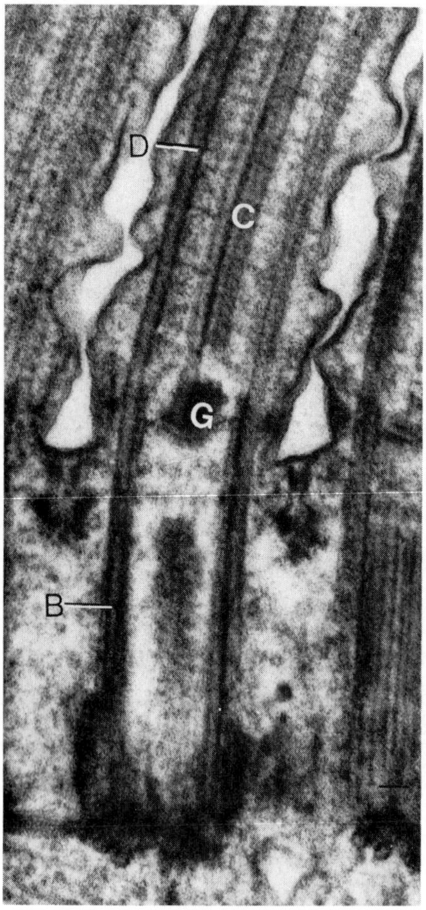

FIGURE 13–9 Electron micrograph of the base of a cilium in the protozoan Tetrahymena. The outer doublet microtubules (D) of the cilium are continuous with the tubules that comprise the wall of the basal body (B), while the central microtubules of the cilium (C) end near a dense granule (G) in the distal part of the basal body. × 94,000. (Micrograph courtesy of Sattler, C. A., and Staehelin, L. A.: J. Cell Biol., 62:473, 1974. Reproduced by permission of the Rockefeller University Press.)

plasmic microtubules as common cell organelles became known. This difference in the ease of preservation of flagellar and cytoplasmic microtubules may reflect some difference in tubulins of microtubules in various locations.

Functions of Cytoplasmic Microtubules

The many functions proposed for cytoplasmic microtubules at first seem very diverse. Possibly this is because our knowledge of the structure and composition of microtubules, their dynamic behavior, and their interaction with other parts of the cell is still developing. In any event, microtubule functions seem to lie in two main groups: (1) a cytoskeletal function in the development of cell shape; and (2) a role in certain forms of cell movement. It is not yet clear whether the behavior of the microtubules is fundamentally different in these two areas.

The role of microtubules in determining cell shape was first suggested by their prominence in certain locations, such as in long, asymmetrical cell processes. Observations that experimental destruction of microtubules resulted in changes in cell shape supported this idea. One of the most dramatic examples occurs in the protozoan *Actinosphærium*. This organism has a central cell body and numerous spikelike processes, each of which contains a bundle of longitudinally oriented microtubules in its core. Cooling the animal results in destruction of the microtubules and retraction of the processes. Upon rewarming, microtubules and processes reappear simultaneously. Microtubules are also present in the processes of nerve cells. They are prominent in the tip, or growth cone, of growing axons, and agents that affect microtubules are able to alter the growth of axons. It is noteworthy that brain is rich in tubulin, and many studies on *in vitro* polymerization of microtubules have been carried out with tubulin from this source. Certain red blood cells contain a marginal band of microtubules that encircles the cell like a hoop and may be related to the disclike shape of the erythrocyte. These and other examples imply a relationship between microtubules and the shape of cells. It has been pointed out, however, that while microtubules may have a role in the development of cell shape they are probably less important in the maintenance of cell shape in differentiated tissues of multicellular organisms, because contacts between cells most likely become prominent.

The participation of microtubules in ciliary motion and chromosome movement in the mitotic spindle have already been considered. Microtubules have also been implicated in other forms of cell motility. This is not to say that microtubules themselves are contractile or generate a motile force. Rather it appears that microtubules interact with other elements, such as filaments, and may determine the pattern or direction of movement rather than generate movement themselves. If this is correct, the the role of microtubules in motility may not be far removed from their general cytoskeletal function.

For example, cells in culture are able to move about on their substrate. Treatment with colchicine, which destroys microtubules, does not prevent the cells from moving, but it does alter the character of the motion from a gliding to an amoeboid type. It also abolishes the ability of cells to move in a particular direction in response to a chemical stimulus. Similarly, cell organelles and inclusions such as lysosomes and pigment granules undergo rapid directional movements within the cytoplasm, the so-called long saltatory movements (LSM). These movements lie in a direction parallel to the orientation of microtubules, and they can be abolished by colchicine, even though streaming of the cytoplasm continues in the presence of the drug.

In most cases, when microtubules participate either in development of cell shape or in motility, they can be seen to be plastic in their behavior, being formed temporarily in one location and then dismantled, only to reappear at another point. The formation and disappearance of the mitotic spindle is one of the most readily recognizable examples of this characteristic. Accordingly, there has been a great deal of recent investigation into the regulation of microtubule polymerization and depolymerization.

Study of this problem has been greatly aided by the ability to isolate microtubules in quantity. In addition, *in vitro* means have been discovered to depolymerize microtubules to free tubulin molecules and to repolymerize the tubulin once again to form microtubules. The details of the conditions necessary for these transformations are being determined. It appears that GTP is involved and that concentrations of Ca^{++} and Mg^{++} are particularly important. Mg^{++} is necessary for polymerization, while the concentration of Ca^{++} must be very low or polymerization is prevented. Thus nucleotides and divalent cations may play a role in the *in vivo* regulation of microtubule formation. Cyclic AMP also affects microtubules of cells in culture, suggesting that it may be a regulatory compound in microtubule formation.

Certain structures within the cell bear a close relation to the ends of microtubules, and it has been widely supposed that they may play a role in the formation of microtubules by serving as organizing or nucleation sites for the polymerization of microtubules. The basal bodies of cilia are one well-defined example, since the ciliary microtubules extend directly from the tubules of the basal body. The pericentriolar satellites of dense material furnish another example, and morphologically similar dense material has been detected elsewhere in the cell where it could serve a microtubule-organizing function (Fig. 13–10). The kinetochores of chromosomes had long been thought to be a site of organization of spindle fibers, and recent work has shown that microtubules can polymerize *in vitro* in relation to the kinetochores of isolated chromosomes.

FILAMENTS

Cytoplasmic Actin and Myosin

The distribution of the proteins actin and myosin in muscle cells has been studied intensively for many years, and as described in Chapter 19 their interactions in muscle contraction are understood at ever-increasing levels of molecular detail. One of the most interesting recent advances in cell biology has been the discovery that actin, myosin, and other muscle proteins are present not only in muscle but in almost all cells, where they are thought to have a role in many forms of cell motility.

Much of the earlier work on muscle proteins in nonmuscle cells was done on the slime mold *Physarum,* which exhibits prominent cytoplasmic streaming as a form of cell motility. Actin and myosin were isolated from this organism, and

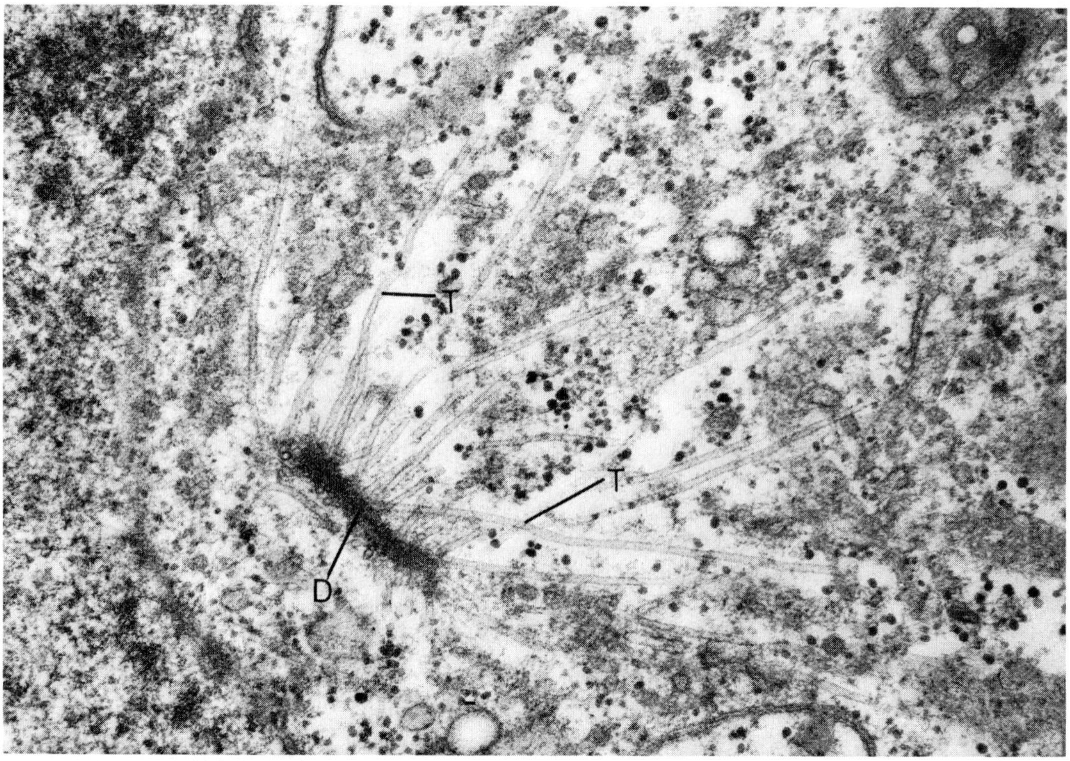

FIGURE 13–10 The soil amoeba *Acanthamoeba castellani*. Cytoplasmic microtubules (T) terminate in a mass of dense amorphous material (D). × 48,000. (Micrograph courtesy of Bowers, B., and Korn, E. D.: J. Cell Biol., 39:95, 1968. Reproduced by permission of the Rockefeller University Press.)

subsequently other proteins believed to be involved in motility were found, including additional muscle proteins and some unique proteins. Subsequently, actin and myosin were found in many other kinds of cells, and it is thought that they may be universally distributed in eukaryotic cells. Not only are actin and myosin present in virtually all vertebrate cell types that have been studied, but they have been found in invertebrates, protozoa, and even plant cells.

The composition of actin from various sources differs only slightly; it is one of the most highly conserved proteins known. Actin is present in large quantities in some nonmuscle cells. In amoebae, for example, actin exceeds any other single protein in amount. On the other hand, while myosin is present in nonmuscle cells, it is not as concentrated relative to actin as in muscle, so the ratio of myosin to actin in nonmuscle cells is lower than in muscle. Myosin from nonmuscle cells forms filaments *in vitro* that resemble the myosin filaments of muscle but are not as large. Despite this behavior *in vitro,* the state of polymerization of myosin and its morphological representation *in vivo* in nonmuscle cells is uncertain.

A major difference between the actin-myosin motile systems in muscle and nonmuscle cells is that the system is stable in muscle while it is labile in many types of nonmuscle cells. Thus mechanisms must exist for controlling the assembly and disassembly of the contractile apparatus in different places in the cell at appropriate times. One way that this might be accomplished, as suggested by certain invertebrate systems that are particularly amenable to study, is that other proteins of high molecular weight may complex with actin and keep it in storage form until it is needed.

Ultrastructure and Function of Cytoplasmic Filaments

Filaments are visible in the electron microscope in many locations in various kinds of cells. It is important to realize that all these filaments are not identical in structure or function. Although the chemical composition and functional roles of all the cytoplasmic filaments visible ultrastructurally is incompletely understood, several different types will be distinguished to illustrate possible differences in structure and function.

The thinnest filaments, also referred to as microfilaments, are about 50 Å (40 to 70 Å) in diameter and are present in a great number of cells (Figs. 13–11 and 13–12). Microfilaments such as these bind heavy meromyosin fragments in a characteristic way to form complexes that resemble a row of arrowheads. Thus the microfilaments have been identified as actin.

A second type of filament is about 100 Å thick and also is widespread in its distribution, being found in neurons, in cells in culture, and in other locations. The 100 Å filaments sometimes appear to have a light core, and they do not bind heavy meromysin. Their function is under study, and some investigators believe they participate in motility.

A third variety of filament, about 100 to 150 Å thick, is sometimes observed (Fig. 13–13). In amoebae, for example, these thick filaments are present along with microfilaments of actin, and it is thought that the thick filaments represent myosin. However, myosin has been demonstrated by extraction and by fluorescent antibody methods in many kinds of nonmuscle cells that do not contain thick filaments demonstrable in the electron microscope. Thus, as stated earlier, a major unsolved problem is the state of polymerization of myosin in nonmuscle cells.

Some cells contain conspicuous aggregations of ~70 to 80 Å filaments termed

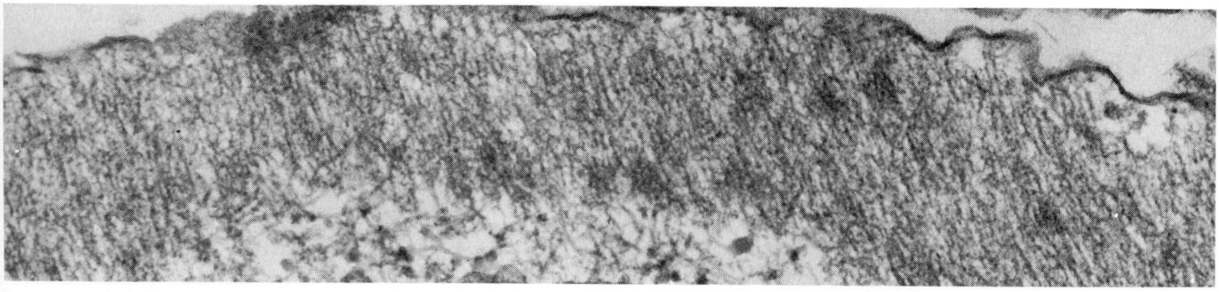

FIGURE 13–11 Microfilaments in the cortical region of a cell in a primary culture of adult human skin fibroblasts. × 60,000. (Micrograph courtesy of A. F. Miranda, G. C. Godman, and M. Schoenebeck.)

tonofilaments (Fig. 13–14). Bundles are visible with the light microscope in certain epithelia and have long been known as tonofibrils. In locations such as squamous epithelial cells, bundles of tonofilaments extend into the cytoplasm immediately underlying desmosomes and may serve to anchor the desmosomal plaques to the rest of the cytoplasm.

Some examples of the occurrence of microfilaments may serve to illustrate their distribution and possible roles. Cells growing in culture were noted by cytologists many years ago to contain bundles of birefringent material termed stress fibers. Birefringence in the polarizing microscope indicates the presence of highly ordered material. In one of the earliest demonstrations of muscle proteins in mammalian nonmuscle cells, the stress fibers were shown to be composed of bundles of microfilaments that bind heavy meromyosin and thus are actin. With fluorescent antibodies it has been discovered that stress fibers contain not only actin but also myosin, tropomyosin, and α-actinin.

Two examples of the possible functions of filaments have been discussed in relation to cell division. During cytokinesis a ring of microfilaments that bind heavy meromyosin is present underlying the plasma membrane of the cleavage furrow (Fig. 13–12), and contraction of this ring appears to pinch the cell in two. The possible role of actin and other muscle proteins in chromosome movement during mitosis is controversial but remains an intriguing possibility. Some investigators have reported that actin is present in the form of thin filaments lying between the microtubules of the spindle. Thus actin might play an important role in chromosome movement, possibly by providing a mechanism for movement of microtubules in relation to one another.

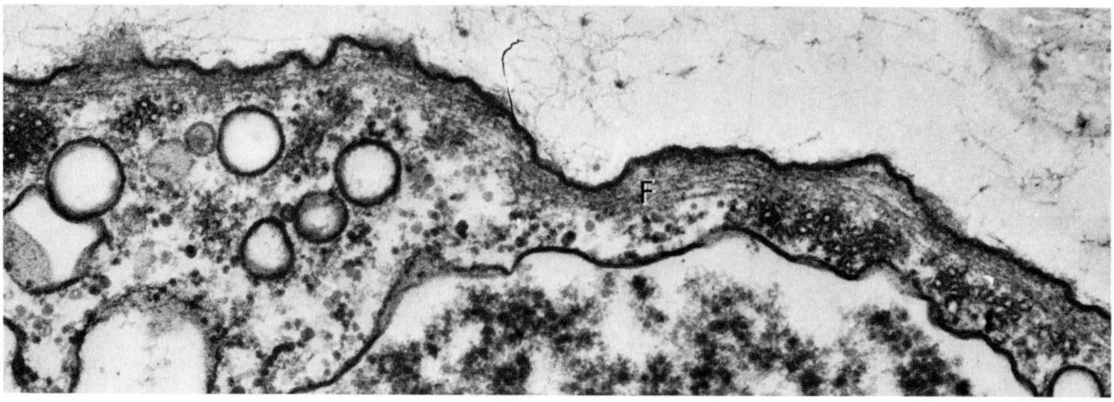

FIGURE 13–12 Microfilaments (F) in the cortex of a cleaving egg. × 65,000. (Micrograph courtesy of D. Szollosi.)

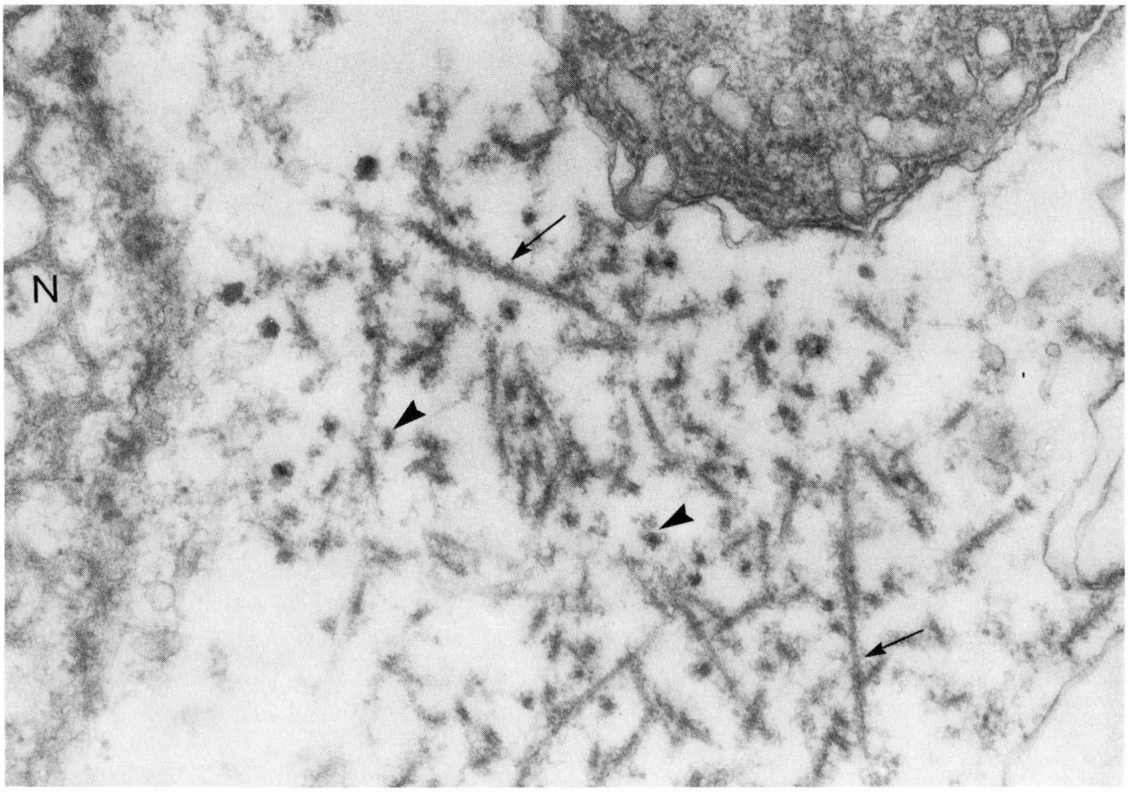

FIGURE 13–13 An aggregate of approximately 150 Å filaments in the cytoplasm near the nucleus (N) of *Amoeba proteus*. Filaments are visible in longitudinal section (*arrows*) and in cross section (◀). × 77,000. (Micrograph courtesy of G. A. Read.)

As an additional example, microvilli in the brush border of intestinal epithelial cells contain a core of actin microfilaments. If isolated preparations of the brush border are exposed to the proper amounts of ATP and calcium, the filaments of the microvilli move into the underlying terminal web. A dense material near the tip of the microvilli binds antibody to α-actinin, a component of the Z-line in muscle, and it has been speculated that a material similar to that of the Z-line may serve to anchor actin filaments to the plasma membrane in the brush border. Furthermore, associations of microfilaments and the plasma membrane have been described in other cells and may be of general significance in various forms of cell motility.

Cell Motility

Cell motility is difficult to study *in vivo,* but extensive observations have been made on cells moving on a solid substrate *in vitro*. The motility of amoebae has been studied for many years, and the mechanism of amoeboid movement has been the subject of much speculation. A long-standing hypothesis is that contraction of gelated proteins in the tail of the amoeba forces a flow of solated endoplasm forward. According to this classic view, when the sol reaches the anterior end of the advancing pseudopod, an alteration in its state from a sol to a gel occurs and it flows backward as gelated ectoplasm. At the tail, the cytoplasm is converted to a sol and flows forward again.

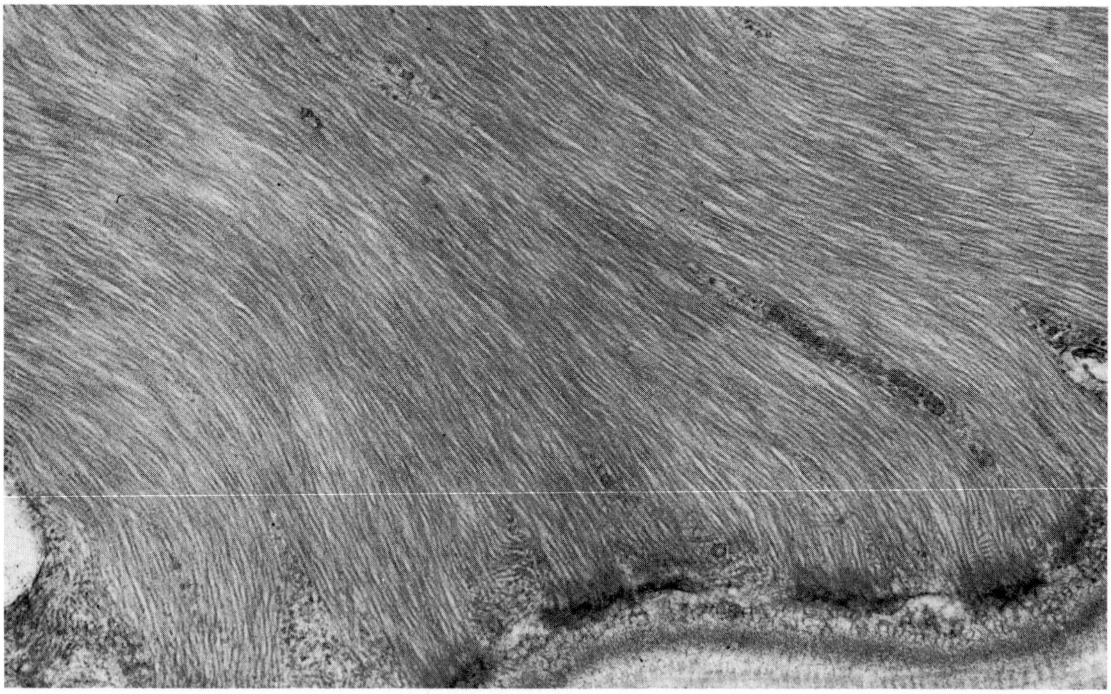

FIGURE 13–14 Tonofilaments in the epidermis of a larval lamprey. The filaments are particularly well displayed in these cells because of their great abundance; they are much less numerous in most other cells. × 52,000. (Micrograph courtesy of Fawcett, D. W.: The Cell. Philadelphia, W. B. Saunders, 1966, p. 245.)

An alternative hypothesis places the contractile force in the forward part of the amoeba. It is noted that the endoplasm of advancing pseudopods is birefringent, and with the electron microscope both thick and thin (actin) filaments are visible. Thus a current idea about amoeboid movement is that the filaments provide contractile force in this region and pull cytoplasm forward into the advancing pseudopod.

Mammalian cells, such as fibroblasts growing in culture, appear to move on a surface in a different way, by gliding rather than by amoeboid movement. The leading edge of these cells has a fan-shaped projection, termed the ruffled membrane because it exhibits wavelike undulations. Thus the cells appear to glide along the surface rather than flowing into advancing pseudopods.

The mechanism of movement of tissue cells in culture is the subject of much investigation and speculation. It has been claimed that microfilaments are involved because the drug cytochalasin, which breaks down microfilaments, arrests the activity of the ruffled membrane and stops cell movement. Interpretations of the results of experiments with cytochalasin have been criticized, however, on the grounds that the drug is known to affect other cellular activities as well as microfilaments.

Investigators have described the arrangement of microfilaments in some cell types in great detail and have attempted to relate these configurations to a mechanism of cell motility, but the subject is still under investigation, with many details to be determined. Other authors have implicated movements of the cell membrane in the gliding movements of tissue cells. In all these speculations, an additional important factor is that of the strength, extent, and localization of points of adherence between the cell and its substrate.

A feature of the behavior of cells in culture that has attracted a good deal of attention is the phenomenon of contact inhibition, because it is absent in many cancer cells and may bear a relation to important biological properties of cancer. The term contact inhibition is used in two ways: in reference to contact inhibition of cell *movement,* and in relation to contact inhibition of cell *proliferation.*

Normal fibroblasts growing on a surface in culture respond to contact with one another in the following way. If the ruffled membrane at the leading edge of an advancing cell contacts another cell, the ruffled membrane stops its undulations, the cell contracts, and movement ceases. If some other part of the cell is free of cellular contact, a new ruffled membrane forms and the cell moves off in a new direction. Normal fibroblasts will not, however, crawl over one another, and thus they form a confluent monolayer in culture. This paralysis of movement upon touching another cell is contact inhibition of cell movement. In addition to this, contact with other cells somehow inhibits DNA synthesis and mitosis in normal cells in culture, and this behavior is termed contact inhibition of cell division.

Although there are some exceptions, cancer cells, either those explanted into culture from a tumor or those transformed *in vitro* by viruses, generally do not display contact inhibition of movement or of cell division. Upon contact they crawl over one another and continue to grow and divide. This lack of contact inhibition has excited interest because of its obvious potential relation to unrestrained proliferation, invasiveness, and ability to metastasize exhibited by many cancer cells.

FURTHER READING

Afzelius, B. A.: A human syndrome caused by immotile cilia. Science, *193*:317, 1976.

Clarke, M., and Spudich, J.A.: Nonmuscle contractile proteins: the role of actin and myosin in cell motility and shape determinations. Annu. Rev. Biochem., *46*:797, 1977.

Dustin, P. (ed.): Microtubules. New York, Springer-Verlag, 1978.

Goldman, R. D., Pollard, T. D., and Rosenbaum, J. L. (eds.): Cell Motility. Vol. 3, Books A, B, and C. Cold Spring Harbor Conferences on Cell Proliferation, 1976.

Korn, E. D.: Cytoplasmic actin and myosin. Trends Biochem. Sci., *1*:55, 1976.

Locomotion of tissue Cells. Ciba Foundation Symposium 14 (new series). Amsterdam, Exerpta Medica, Elsevier, 1973.

McIntosh, J. R.: Bridges between microtubules. J. Cell Biol., *61*:166, 1974.

Pollard, T. D.: Cytoskeletal functions of cytoplasmic contractile proteins. J. Supramol. Struct., *5*: 317, 1976.

Pollard, T. D., and Weihing, R. R.: Cytoplasmic actin and myosin and cell movement. CRC Crit. Rev. Biochem., *2*:1, 1974.

Satir, P.: How cilia move. Sci. Am., *231*:44, 1974.

Schroeder, T. E.: Dynamics of the contractile ring. *In*: Inoué, S., and Stephens, R. E. (eds.): Cell Movement. New York, Abelard-Schuman, 1975.

Soifer, D. (ed.): The biology of cytoplasmic mictotubules. Ann. N.Y. Acad. Sci., *253*:1, 1975.

Szollosi, D.: Cortical cytoplasmic filaments of cleaving eggs: a structural element corresponding to the contractile ring. J. Cell Biol., *44*:192, 1970.

Wessells, N. K., Spooner, B. S., Ash, J. F., et al.: Microfilaments in cellular and developmental processes. Science, *171*:135, 1971.

14

MEMBRANES AND CELL SURFACE SPECIALIZATIONS

MEMBRANE STRUCTURE AND COMPOSITION

Membranes are sheetlike structures that form boundaries between a cell and its environment and between two cell compartments of different composition. Although the plasma membrane is only about 100 Å thick, its existence and many of its properties were known from biochemical and physical studies long before it was directly visualized. However, the membranous nature of many of the intracellular organelles was elucidated only when it became possible to study their structure with the electron microscope.

In general terms, membranes have two main biological functions. First, they serve as barriers, between the cell and its environment in the case of the plasma membrane, and between two compartments within the cell in the case of the intracellular membranes. As barriers, they determine which molecules can pass. This property varies from one membrane to another, and membranes can be very selective. Second, membranes serve as the structural base for enzymes that form part of the membrane itself. Examples include some of the enzymes of the smooth endoplasmic reticulum, such as the drug-hydroxylating enzymes and the components of the electron transport chain in the inner mitochondrial membrane.

Ultrastructure

Cellular membranes almost always appear trilaminar in the electron microscope when viewed in the proper orientation at sufficiently high magnification and resolution. In sections passing perpendicular to the surface, the membrane can be seen to consist of two electron-dense layers flanking a central light lamina (Fig. 14–1). As will be explained later, there is some information on the arrangement of various molecules in membranes, but the precise chemical mechanism of "staining" that produces the three-layered appearance is not well understood. This deficiency is not peculiar to the study of membranes, since in general the way in which various fixatives and stains combine with cellular components to

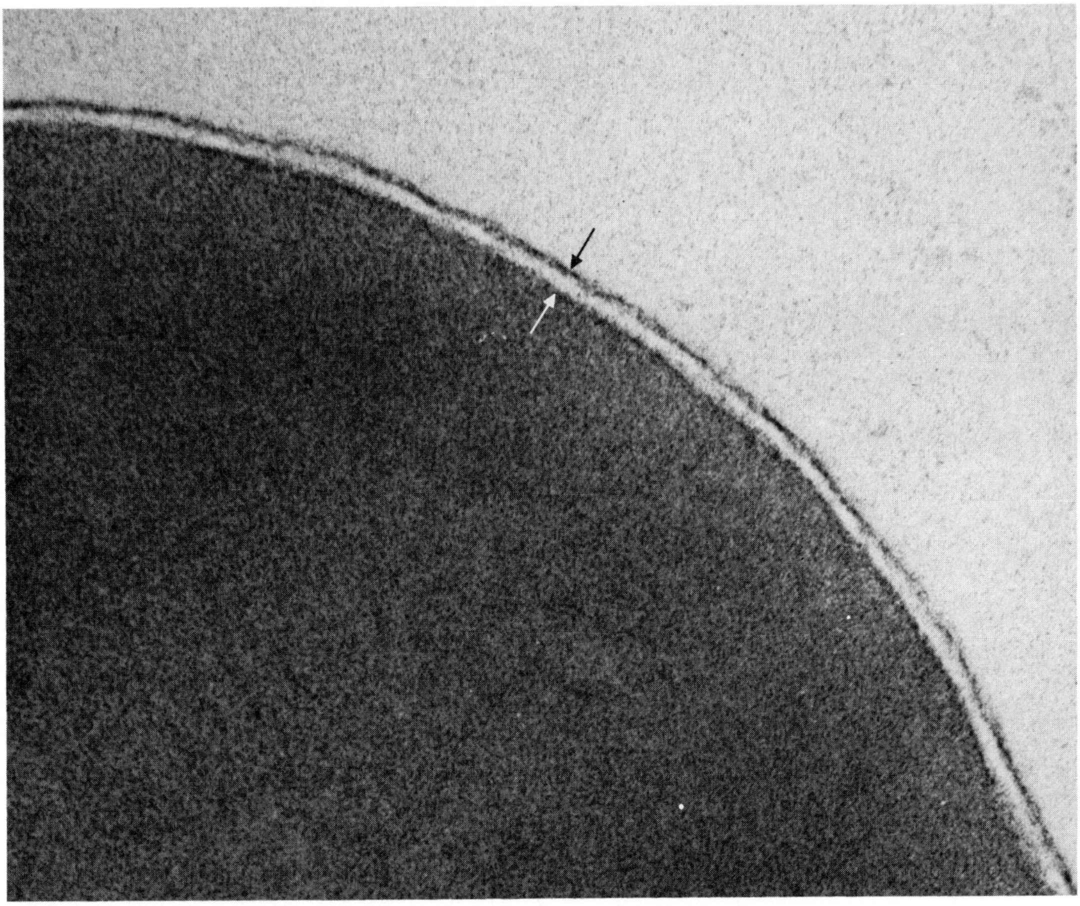

FIGURE 14–1 Very high power electron micrograph illustrating the trilaminar appearance of the plasma membrane bounding a red blood cell. The "unit membrane," between the arrows, is composed of two dark layers separated by a light lamina. × 652,000. (Micrograph courtesy of J. D. Robertson.)

produce their characteristic images in the electron microscope is poorly understood. It is to be emphasized that the proper conditions of high magnification, good resolution, and precise orientation of the membrane perpendicular to the plane of section must be met before the trilaminar character of the membrane is visible. In routine electron micrographs at low to medium magnification, a membrane usually appears as a single dense line.

Membrane thickness varies from approximately 70 to 100 Å, with the intracellular membranes usually being thinner than the plasma membrane. As seen in the electron microscope, membranes may be asymmetrical, that is, one of the dense layers may be thicker than the other. In specimens fixed in OsO_4, the inner dense line of the plasma membrane, which faces the cytoplasm, is often thicker than the outer dense line. Although the trilaminar appearance of cellular membranes may reflect some common underlying structural features, it is to a certain extent artefactual. It does not render a complete understanding of the molecular structure of membranes, which show a remarkable diversity in their biological function despite their uniformity in electron microscope images.

Cellular membranes have a polarity in their orientation. One surface always faces the cytoplasmic matrix, while the opposite surface faces either the outside of the cell or the lumen of an organelle. As a corollary of this, there are no "free ends'' of membranes in the cytoplasm, since this would bring both membrane surfaces in contact with the cytoplasmic matrix. Furthermore, this orientation is maintained during membrane fission and fusion processes.

An example is the process of exocytosis, in which the content of secretory vacuoles is discharged. The surface of the vacuole membrane that faces the interior of the vacuole comes to lie in contact with the extracellular environment after the vacuole membrane is incorporated into the plasma membrane, while the surface that faces the cytoplasmic matrix retains this orientation. The inner mitochondrial membrane might be thought to be an exception to this generalization. However, the mitochondrial matrix may be considered analogous to the cytoplasmic matrix, and the outer mitochondrial compartment to the lumen of an organelle. (See the discussion of the possible prokaryotic origin of mitochondria in Chapter 11.)

Composition

Membranes are composed mainly of lipid and protein along with a small amount of carbohydrate. The ratio of lipid to protein varies in different membranes from 0.23 in myelin to 3.0 in certain bacteria. Except for adipose cells, which store large amounts of fat in the cytoplasm, most of the lipid in eukaryotic cells is in their membranes. Many of the lipids present are phospholipids. These molecules have the form shown in Figure 14–2. They contain two fatty acids, which have an even number of carbon atoms linked to glycerol. At the remaining position, a base, which is often an amine such as choline or ethanolamine, is linked through a phosphate group to the glycerol. An important characteristic of the phospholipid molecules is that each has a polar and a nonpolar end. That is, they are amphipathic, and the polar ends tend to associate with water while the nonpolar portions are hydrophobic. If mixed with water, phospholipids form micelles, with the hydrocarbon chains of the lipids in the interior and the polar groups directed outward in contact with the water. Membranes of eukaryotes also contain cholesterol. The rings and hydrocarbon chains are nonpolar; the hydroxyl group constitutes the only polar portion.

The lipid composition of membranes varies from one cell type to another. The classes of cellular membranes in a given cell, such as endoplasmic reticulum, Golgi apparatus, and plasma membrane, also differ from one another. The lipid composition of a particular membrane of a cell can even vary within limits ac-

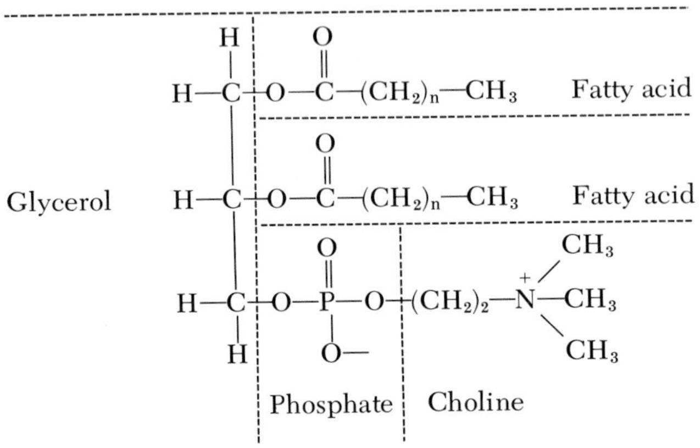

FIGURE 14–2 The chemical structure of a glycerophosphatide. Glycerophosphatides are phosphate esters of glycerol, with fatty acids linked to the glycerol at two positions and phosphate at the third. The glycerophosphatide shown here is a lecithin, because the aminoalcohol choline is joined to the phosphate group. In the cephalins, ethanolamine or serine is present in place of choline, and in the inosides, inositol is found in this location. A variety of compounds are possible within the major groups by substitution of different fatty acids at the other two positions on glycerol.

cording to the nutrients available to the cell. For example, the fatty acid composition of bacterial membranes can be altered by restricting the types of fatty acids available to the cells.

A great variety of different proteins form part of biological membranes. The membrane proteins are of two general types, integral and peripheral. The integral proteins are separated from the lipids only with difficulty. When isolated they often retain some bound lipid, and they become insoluble in water at neutral pH if the lipids are removed. Because of these properties the integral proteins are believed to represent those located in the interior of the membrane in association with the lipids. In contrast, peripheral proteins are thought to be associated with the membrane surface, since they can be removed from the membrane with mild procedures, and in the isolated state they are relatively soluble in water.

The membrane proteins are particularly important because the functional properties of the membranes are associated in large part with its proteins. These include enzymatic activity and the capacity for selective transport of ions and other molecules. Some proteins, such as the spectrin found in the membrane of red blood cells, may serve a structural purpose by forming a meshwork, while others resemble actin and may furnish contractile properties. The carbohydrate component of membranes consists of linear or branched polymers of sugars which are linked to protein to form glycoproteins, or are linked to lipid to form glycolipids.

The barrier property of membranes is associated with the membrane lipids. This was indicated in 1895 by the studies of Overton, which showed that lipid-soluble substances cross the plasma membrane of cells more readily than polar substances do. It was suggested that they do so by dissolving in the membrane. Electrical measurements subsequently showed that the capacitance of the plasma membrane is very high ($\sim 1 \mu F/cm^2$), and calculations using this information indicated that the lipid layer of the plasma membrane should be about 100 Å thick. Studies of electrical properties also showed that there is a dielectric component everywhere on the surface of the cell. Thus cells behaved as if they were bounded by a continuous 100 Å-thick layer of lipid.

The consequences of alterations in the barrier properties of membranes due to deleterious changes in membrane lipids are dramatically illustrated by the disease gas gangrene. This is caused by the bacillus *Clostridium perfringens,* which produces a variety of toxins. Among them is phospholipase C, which alters membranes by breaking down phospholipids to diglycerides. Thus the toxin rapidly causes many deleterious changes in cells, including loss of ions and intracellular enzymes, uptake of water with cell swelling, and decrease in the ability of mitochondria to carry out oxidative phosphorylation. Cell death and necrosis occur very rapidly and are widespread.

Molecular Structure

Phospholipids placed in water spontaneously form a bilayer, with their polar groups directed outward in contact with the water and their hydrocarbon chains directed inward. These experimentally produced phospholipid bilayers have been studied by forming them across a small aperture, or by producing small vesicular structures known as liposomes. In either case, artificial phospholipid bilayers have a thickness of about 60 Å, and their properties correspond closely to the barrier properties of biological membranes. During the 1920's, Gorter and Grendel extracted lipid from a known number of red blood cells, spread it on water as a monomolecular film, and calculated its surface area. Although some of their assumptions may be questioned in the light of current knowledge, their calculation that the area of lipid obtained equaled twice the surface area of the cells from which it was extracted lent credibility to the idea that plasma membranes were composed in part of a lipid bilayer.

Observations of this sort led in the 1930's to the well-known Davson-Danielli model for the molecular structure of membranes (Fig. 14–3). The membrane was visualized as being composed of a bilayer of lipid, with the hydrocarbon chains of the lipid molecules directed inward and the polar groups outward. Both surfaces of the lipid bilayer were thought to be covered with protein. The Davson-Danielli model dominated thinking about membranes for more than 30 years. It appeared to receive dramatic confirmation in the 1950's when it was observed that virtually all cellular membranes appeared trilaminar in the electron microscope, as predicted by the model.

This led to Robertson's formulation of the "unit membrane" hypothesis, which suggested that all cellular membranes had a similar basic trilaminar structure that conformed to the Davson-Danielli model. The term unit membrane is still retained by electron microscopists to designate a single membrane that appears trilaminar in the electron microscope. However, the deficiencies of the Davson-Danielli model in accounting for important biological properties of membranes, such as the transport function, along with a better knowledge of the structure and behavior of proteins has led to its replacement with more current models. The idea of a lipid bilayer has been retained, however, and is widely accepted. It is supported by many different lines of evidence, including observations on model systems and results of physical studies employing methods such as x-ray diffraction, spin-labeling, and nuclear magnetic resonance.

At present the fluid mosaic model proposed by Singer and Nicolson is regarded by many as best representing current knowledge of the molecular structure of cellular membranes (Fig. 14–4). The membrane is visualized as a fluid bilayer of lipid, with protein molecules able to move around within it. The model stresses the importance of thermodynamic factors in determining the disposition of lipid and protein, since hydrophobic and hydrophilic interactions are maximized. Thus, in the lipid bilayer the phospholipid molecules have their polar groups directed outward in contact with water, and their nonpolar parts facing inward.

The disposition of different proteins is also determined by their amphipathic character. The hydrophobic parts of the proteins are found in the interior of the membrane in association with the hydrocarbon chains of the lipid. The ionic hydrophilic parts of the proteins, on the other hand, protrude from the surface of the membrane and come in contact with water. Thus, depending on its size

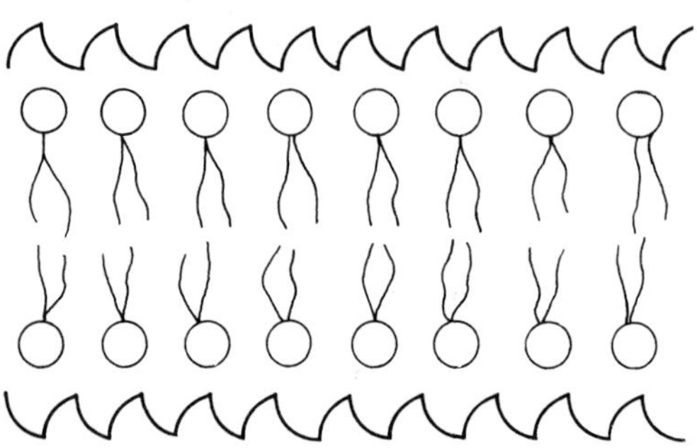

FIGURE 14–3 A diagram of the Davson-Danielli model of membrane structure. The membrane is composed of a bilayer of lipid molecules with their hydrocarbon chains (*lines*) directed inward and their polar groups (*circles*) directed outward. The polar groups are associated with the layers of protein (*zig-zag lines*) that cover both surfaces of the membrane.

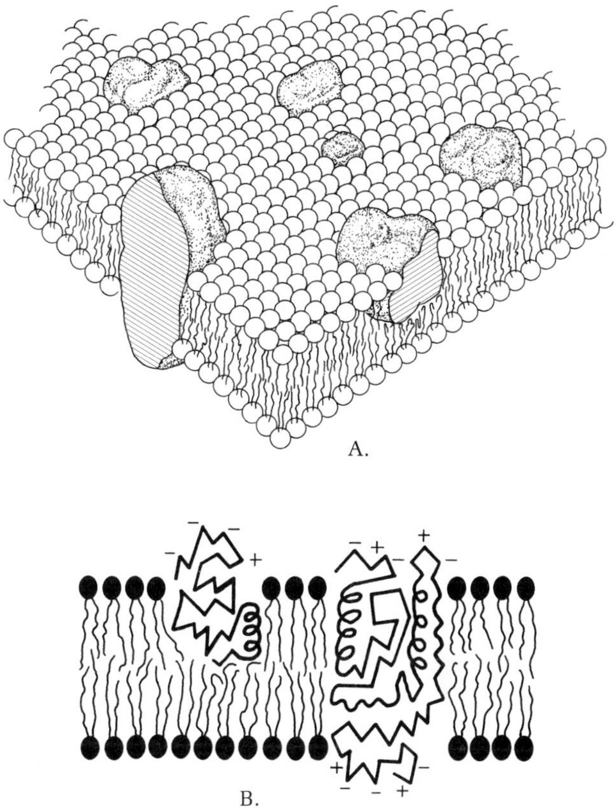

FIGURE 14–4 Diagrams of the fluid mosaic model of the structure of membranes proposed by Singer and Nicolson. *A*. Three dimensional view. *B*. cross section of a membrane. The circles represent the ionic and polar head groups of the phospholipid molecules, while the wavy lines represent the fatty acid chains. The phospholipid molecules form a bilayer with their ionic and polar groups in contact with water. The proteins are represented by globules in *A* and are shown as folded polypeptide chains in *B*. The nonpolar portions of the proteins are embedded in the membrane, while the parts that contain ionic groups protrude from one or both surfaces of the membrane. (After Singer, S. J., and Nicolson, G. L.: Science, *175*:720, 1972. Copyright 1972 by the American Association for the Advancement of Science. Reproduced by permission.)

and configuration as determined by its amino acid sequence, a given protein may be exposed on one surface or it may extend entirely through the membrane.

The carbohydrate chains of glycoproteins, being polar, project from the surface of the membrane into water. Glycophorin, the major glycoprotein of the red blood cell plasma membrane, has a molecular weight of 55,000 and is an example of a protein that spans the membrane. A region toward the carboxyl end of the molecule is hydrophobic and is associated with the lipid part of the membrane. A hydrophilic portion containing the attached carbohydrate and the amino end project from the external surface, while the carboxyl end is exposed on the inner surface of the membrane.

In general, the carbohydrate parts of glycoproteins are found only on the external surface of the plasma membrane and contribute to the negative charge of cell surfaces, to which positively charged substances can bind. This disposition of carbohydrate is one manifestation of the asymmetry of cellular membranes. Differences in the phospholipid composition of the two leaflets of the lipid bilayer may also contribute to membrane asymmetry.

In the conventional trilaminar electron microscope images of membranes, the central light lamina seems to represent the hydrocarbon chains of the membrane lipids, while the two dense lines apparently correspond to the polar groups of the phospholipids along with parts of proteins that lie on the membrane surfaces. The morphological asymmetry of membranes in the electron microscope may be due in part to the asymmetrical distribution of the carbohydrate components; the asymmetry of membrane lipids may contribute as well.

It is not possible to adequately summarize in this account the observations that led to the formulation of the fluid mosaic model. However, it can be noted that an arrangement such as this is thermodynamically favorable; it also takes into account the fact that membrane proteins are globular rather than sheetlike as proposed in the older Davson-Danielli model. The fluid mosaic model is consistent with the observation that certain protein antigens known to be part of the plasma membrane are randomly dispersed, as would be expected if proteins are able to move about in a fluid lipid bilayer.

Similarly, the model accounts for the observation that when two cells with different surface antigens are fused together to form one hybrid cell, as occurs in the presence of certain viruses, the antigens initially appear segregated in the different halves of the hybrid. However, after about 40 minutes the two antigens become intermixed over the cell surface, again suggesting that proteins are able to move about in a fluid membrane.

Finally, although understanding of the biogenesis of membranes is fragmentary, it is known that lipid and protein components of membranes can be incorporated independently into membranes and that they turn over independently of one another and have different half-lives. These observations are readily understandable on the basis of the fluid mosaic model, since individual components could easily be removed and inserted without interrupting the integrity of the membrane. The images obtained from freeze-cleave preparations for electron microscopy are also consistent with the model and will be described in the following section.

FREEZE-CLEAVE PREPARATIONS

The freeze-cleave method (also called freeze-fracture or freeze-etch) is a means of preparing specimens for electron microscopy that differs from thin sectioning. It has proved very useful in the study of membrane structure and cell contact specializations. Cells or pieces of tissue are frozen in liquid nitrogen and then cleaved with a microtome knife. This produces a fracture that goes through cells along natural planes of weakness. A small amount of ice can then be evaporated from the fracture surface in a vacuum; this step is termed etching, and it gives the method its alternative name of freeze-etching. A metal such as gold or palladium is deposited on the fracture surface by evaporation in a vacuum to form a metal replica of the surface. The cells are then discarded, and the replica is viewed in the transmission electron microscope. Replicas of this sort provide a view in relief of a plane through a cell.

Freeze-cleave preparations are of particular interest in the study of membranes because the cleavage plane generally splits a membrane in its center and follows the course of the membrane for some distance (Fig. 14–5). Thus it is possible to obtain face views of the interior of the split membrane. Cleavage through the central lamina is thought to occur because this region is occupied by the hydrocarbon chains of the membrane lipids, while the polar ends of the lipid molecules are firmly embedded in the surrounding ice.

It can be readily appreciated that if any membrane is split down the center two different faces will be obtained. One will represent that half of the membrane that was in contact with the cytoplasmic matrix. It is designated the P face in the current terminology (it was formerly called the A face). The other face is designated the E face (formerly the B face) and represents that half of the membrane that was in contact with the extracellular space, in the case of the plasma membrane, or was next to an interior space, in the case of an intracellular membrane.

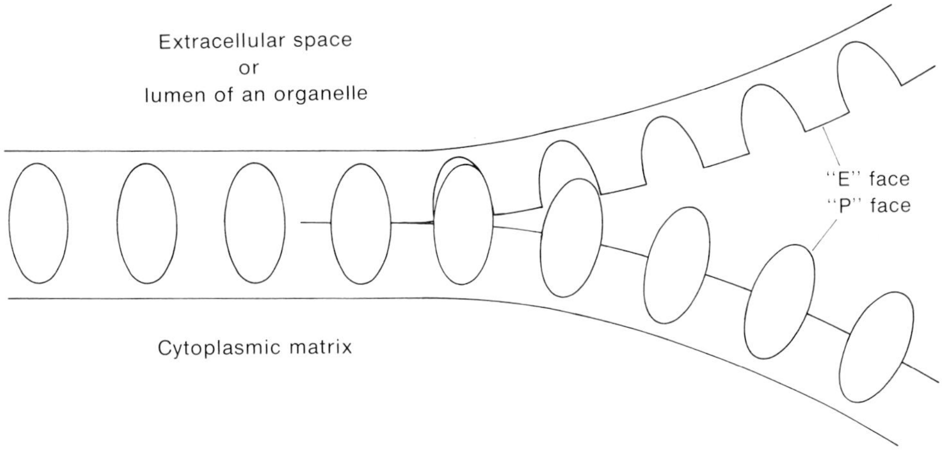

FIGURE 14–5 A diagram of the splitting of a membrane in its center during the freeze-cleave method of specimen preparation. Particles within the membrane adhered to one half of the membrane when it was split. Thus the character of the two faces differs: One is studded with projecting particles, while the other shows a corresponding array of pits. The particle-studded face usually belongs to that half of the membrane that was next to the cytoplasmic matrix.

When freeze-cleave preparations were first viewed, it was apparent that the two faces were not identical, but that one, usually the P face, contained many particles. The other had few particles and appeared relatively smooth except for the presence of numerous pits (Fig. 14–6). After several years of controversy over the nature of the particles, it now appears that they represent membrane proteins. Thus the disposition of the particles corresponds to that predicted for intrinsic membrane proteins by the fluid mosaic model of membrane structure. In addition, it has been possible experimentally to alter the pattern of membrane particles, thus showing that some proteins are able to move about in the plane of the membrane.

CELL CONTACT SPECIALIZATIONS

Definitions

Contact specializations are modifications of the surfaces of cells that permit them to relate functionally in different ways. Different specializations hold cells together firmly, seal off the extracellular space in certain regions, and conduct electrical impulses. By no means the least of the problems in understanding cell contact specializations is learning the use of the unwieldy terminology. In an attempt to clarify this, some definitions will be presented first, and then the common contact specializations will be described in more detail.

Contact specializations are often named according to the *size* and shape of the area specialized (Fig. 14–7). A *macula* is a small punctate or spotlike specialization. A *zonula* is a band that extends around a cell like a hoop. The term *fascia* is used to designate a large sheetlike area of specialization.

A second term is then applied to describe the *nature* of the specialization, particularly, with respect to the distance between the two apposed plasma membranes (Fig. 14–8). In epithelia, the plasma membranes of adjacent cells are nor-

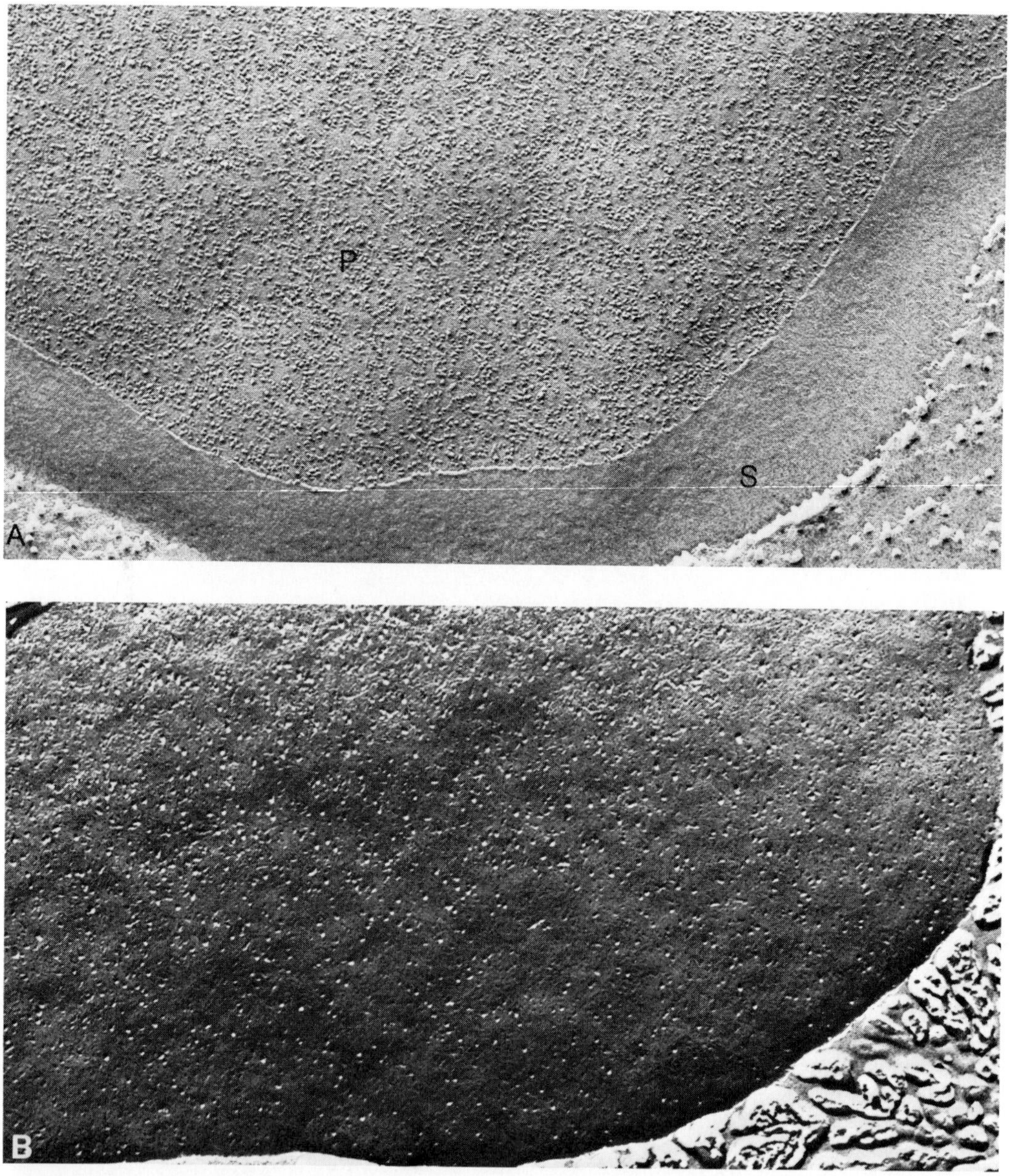

FIGURE 14–6 The appearance of the plasma membrane in freeze-cleave preparations of red blood cell ghosts. *A*. On the P face (P) many particles are visible. This specimen has been deep-etched, revealing some of the smooth external surface of the membrane (S). *B*. The E face reveals small pits but few particles. *A*. × 80,000. *B*. × 67,000. (Micrographs courtesy of T. W. Tillack. *A* is from Tillack, T. W., and Marchesi, V. T.: J. Cell Biol., *45*:649, 1970. Reproduced by permission of the Rockefeller University Press.)

mally about 150 to 200 Å apart. The term *adhaerens* (plural *adhaerentes*) describes a contact specialization in which the plasma membranes of the two cells course parallel to one another at a distance of 200 to 250 Å. The plasma membrane is underlain by a dense region of specialized cytoplasm, and the extracellular space between the cells contains specialized material that is thought to aid the cells in adhering to one another.

The word *occludens* (plural *occludentes*) is applied when the plasma membranes of two cells are fused together, obliterating the extracellular space. This

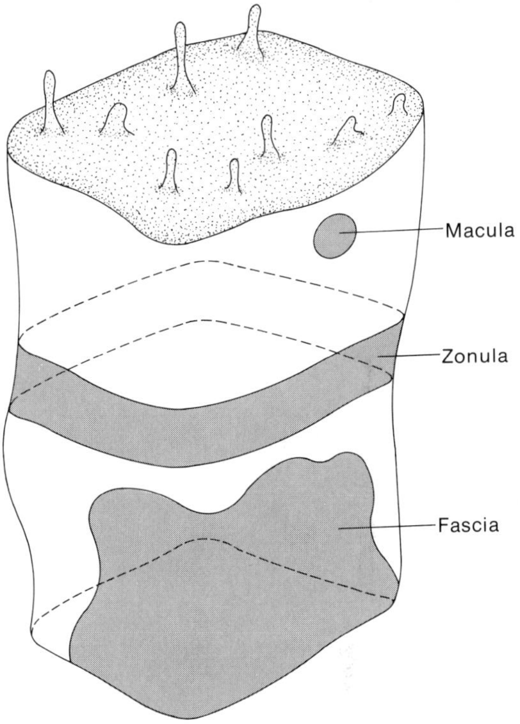

FIGURE 14–7 Contact specializations are named for their size and shape. A macula is a small spotlike specialization, a zonula is a band, and a fascia is a larger, sheetlike region of specialization. The diagram shows the specialized area on the plasma membrane of one cell; the adjacent cells have been omitted for clarity, but they would display a complementary region of specialization where they lay in contact with the cell shown.

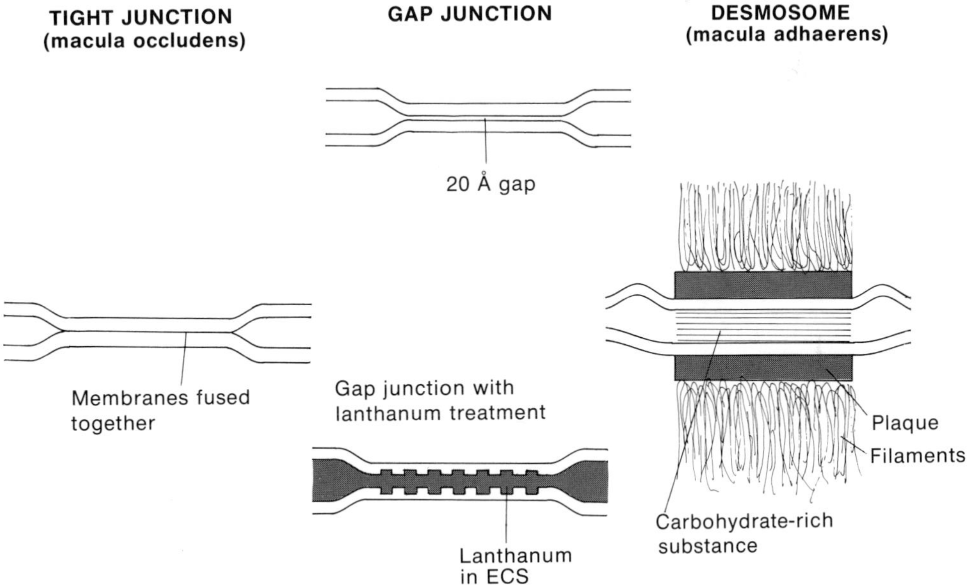

FIGURE 14–8 A diagram of the relation between the plasma membranes of apposed cells at common types of contact specializations, as seen in sections perpendicular to the surfaces of the membranes. A macula, or spotlike, specialization is shown in each case. Note that similar relationships between the membranes would cover a larger area in the case of a zonula or a fascia (see Figure 14–7).

type of contact specialization is also commonly called a tight junction, or pentalaminar junction, because the two outer dense lines of the apposed plasma membranes appear to merge with one another. Thus it is possible to characterize a contact specialization as a macula adhaerens, a zonula occludens, and so on.

The term gap junction refers to a specialization in which the plasma membranes of two cells approach one another closely, leaving an apparent "gap" of extracellular space about 20 Å wide. Since gap junctions were discovered after the introduction of the previous names, they are not readily incorporated into the terminology, although occasionally phrases such as "macular gap junction" are used to describe the size of the area specialized. Some of the common types of cell contact specializations will now be considered.

Desmosome (Macula Adhaerens)

Punctate regions of cell contact in stratified squamous epithelia were known for many years as "intercellular bridges," because it was thought by some that protoplasmic continuity between the cells occurred at these points. With the aid of the electron microscope it became clear that there is no actual continuity between the cells in these regions; instead the surfaces of the cells are modified at these locations.

Each specialization is designated a macula adhaerens or, more simply, a desmosome. As the term implies, a desmosome (Fig. 14–9) is a roughly oval or circular specialization at which two cells appear to adhere to each another. The plasma membranes of the two cells are parallel at a distance of about 200 to 250 Å. There is a plaque of electron-dense material in the cytoplasm applied to the inner aspect of the plasma membrane of both cells. Subjacent to the plaque is a somewhat less dense region containing many filaments.

Stereoscopic images of electron micrographs have shown that many tonofilaments in the epithelial cells enter these regions, form loops, and extend into the surrounding cytoplasm once again. Apparently this device provides structural support and anchors the specialized region to the remainder of the cytoplasm. The extracellular space between the cells contains a material that stains for the presence of glycoprotein. A central dense lamina has been observed in the extracellular space midway between the two apposed plasma membranes, and fine connections between the central lamina and the membranes have occasionally been visualized. The nature of this extracellular material is not definitely established, but it is widely presumed to have some adhesive property.

The function of the desmosome thus appears to be the mechanical attachment of two cells. This can be appreciated by micromanipulation studies and from electron microscope observations that cells remain attached to one another at desmosomes under conditions that lead to dilatation of the remainder of the extracellular space. Desmosomes are very abundant in epithelia such as the epidermis and the cervix of the uterus, which are subject to stress. Although first detected in stratified epithelia, desmosomes are known to be present in many different types of cells. They are probably the most common of the cell contact specializations.

Unusual forms of desmosomes have been described in certain cancers arising from epithelial tissues. These desmosomes are located not on the cell surface but within the cytoplasm in association with cytoplasmic vesicles, which are thought to form by a pinching-off of invaginations of the cell surface. An overall decrease in the number of desmosomes has been found on the surfaces of cells in some

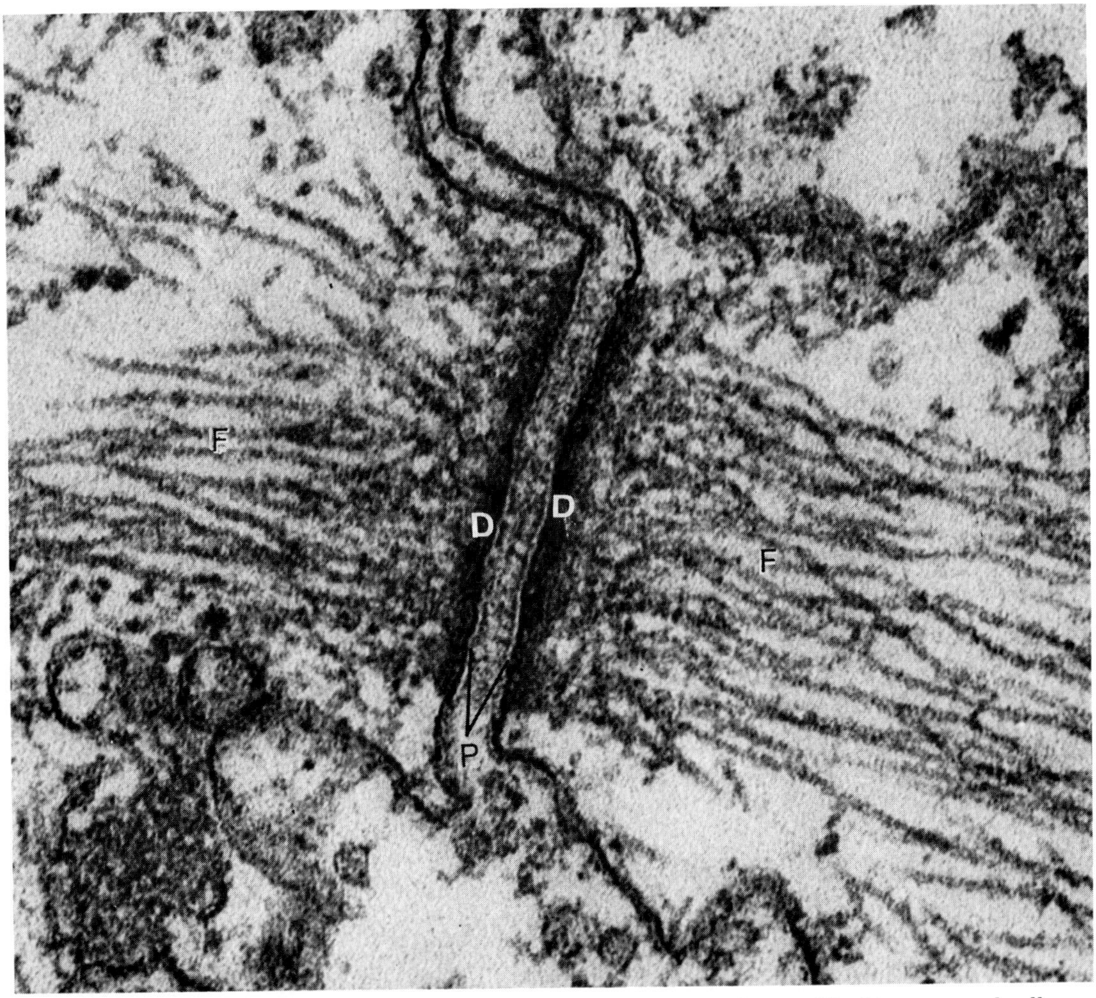

FIGURE 14–9 Desmosome in the epidermis of a newt. The plasma membranes (P) of two apposed cells are parallel to one another. A dense plaque (D) underlies both membranes, and tonofilaments (F) from the cytoplasm converge on the plaques. The density of the extracellular space between the apposed plasma membranes has been increased by staining with the dye ruthenium red, which binds to carbohydrate-rich materials. × 30,000. (Micrograph courtesy of D. E. Kelly.)

malignant invasive epithelial tumors. This may reflect differences in the degree to which cells of the tumor are attached to one another. Similarly, a decrease in the number of desmosomes in the epidermis has been found in some skin diseases. This has been related to lowered adhesion of the cells to one another and thus to premature shedding of the cells from the surface of the skin.

The desmosome is usually a mirrorlike specialization. That is, when two cells are closely apposed and this specialization is seen in one cell, the corresponding mirror-image specialization is present in the neighboring cell. At the base of some epithelia, however, half-desmosomes, or hemidesmosomes, are observed (Fig. 14–10). These are morphologically identical to one half of a normal desmosome, but since no second cell is nearby the corresponding half is absent. Extracellular specializations, consisting of fine filaments and an increased density of the basal lamina near hemidesmosomes, have been described. It is thought that hemidesmosomes may serve to attach epithelial cells more firmly to the extracellular materials of the underlying connective tissue.

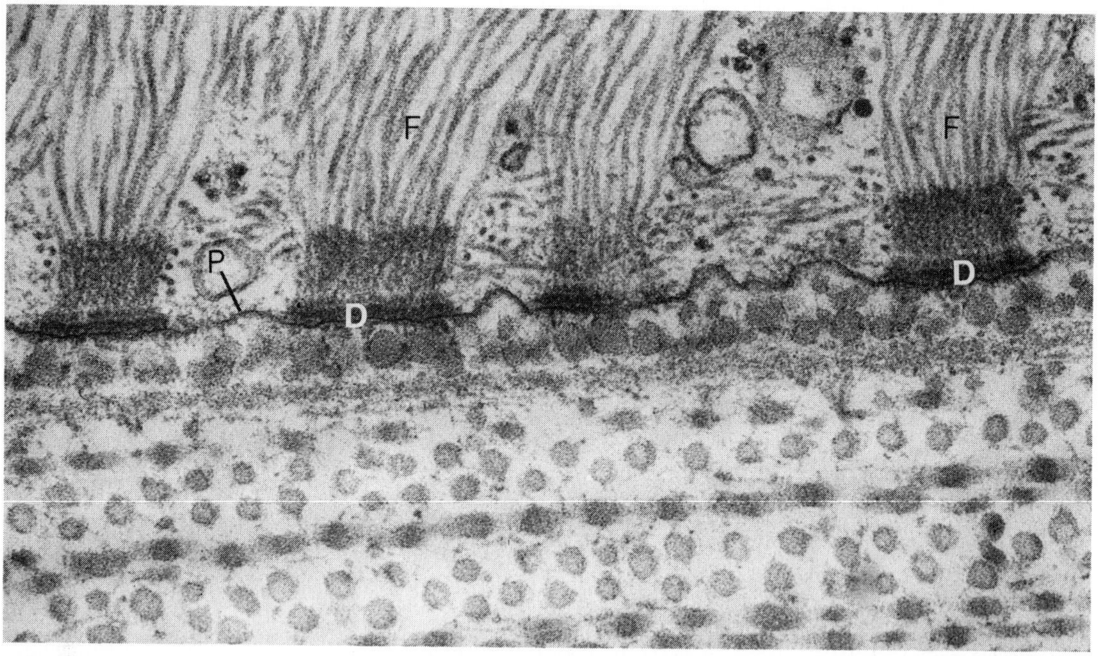

FIGURE 14–10 Hemidesmosomes at the basal surface of a cell in the newt epidermis. Each hemidesmosome resembles one half of a complete desmosome. Underlying the plasma membrane (P) are dense plaques (D) and tightly packed bundles of tonofilaments (F). × 80,000. (Micrograph courtesy of Kelly, D. E.: J. Cell Biol., *28*:51, 1966. Reproduced by permission of the Rockefeller University Press.)

Terminal Bar (Junctional Complex of Epithelia)

With the light microscope a region of increased density was observed next to the lumen between the cells of the columnar epithelia, such as that of the intestine. For many years cytologists believed these "terminal bars" represented an accumulation of intercellular cement. Although from a functional point of view this may not be far from the truth, electron microscopy revealed that the terminal bar is actually made up of a series of cell contact specializations (Figs. 14–11 and 14–12). Proceeding from the lumen toward the base of the cells, it comprises a zonula occludens, a zonula adhaerens, and a series of desmosomes.

Thin sections of the zonula occludens, or tight junction, show that the outer dense lines of the plasma membranes of the two apposed cells are fused together to form a five-layered junction, thus obliterating the extracellular space. Tight junctions of the sort found in the terminal bar have a characteristic appearance in freeze-cleave preparations (Fig. 14–13). They present a pattern of linear ridges that anastomose and form a network in the plane of the membrane.

By their distinctive appearance the ridges provide a means for readily identifying true tight junctions, although the manner in which the membranes are cleaved to produce this picture is still under investigation. The ridges visualized in freeze-cleave preparations are believed to represent the locations of actual fusion between the plasma membranes of the two cells. Thus the membranes of the two cells are attached along a series of lines rather than in large bands.

The zonula adhaerens is commonly described as having ultrastructural features similar to those of the macula adhaerens or spot desmosome, except that the zonula adhaerens of course extends as a band around the entire circumference

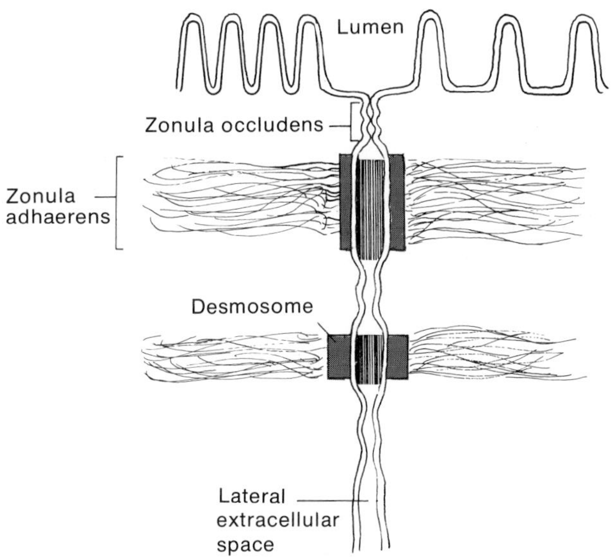

FIGURE 14–11 A diagram of the cell contact specializations that comprise the terminal bars present in epithelia, such as that of the intestine. The zonula occludens and the zonula adherens extend in bands around the entire circumference of the cells, while the desmosomes are spotlike specializations.

of the cell. Recent observations, however, have indicated that there are differences in the detailed structure of the macula adhaerens and zonula adhaerens that are probably of functional significance.

In the zonula adhaerens, the plasma membranes of the two cells are parallel and are separated by ~150 to 200 Å. Although the extracellular space is slightly more dense than in nonjunctional areas, it lacks the central dense lamina of the desmosome. The underlying cytoplasm of the zonula adhaerens also lacks the dense plaque that is so conspicuous in the macula adhaerens. More important, the cytoplasmic filaments associated with the cytoplasmic surface of the membranes in the zonula are part of the terminal web and are reported to be ~70 Å in diameter and composed of actin. This is in contrast to the ~100 Å, nonactin, filaments of the macula adhaerens.

The desmosomes of the terminal bar structurally resemble those found elsewhere. Several desmosomes are disposed in a ring around each epithelial cell immediately toward the base from the zonula adhaerens.

The function of the zonula occludens appears to be to separate the lumen from the lateral extracellular space between the cells. If this is the case, then solutes transported across the epithelium cannot pass between the cells but are forced to go through them. One model for the transport of salt and water by epithelia such as those of the intestine and gall bladder suggests that ion pumps on the lateral plasma membranes of the epithelial cells participate in this process. The sealing capacity of tight junctions varies from one tissue to another, and the permeability of the zonula occludens has been found to be related to the number of ridges visible in the electron microscope. Thus epithelia across which there is a steep ionic and osmotic gradient exhibit many ridges (Fig. 14–13), while relatively "leaky" epithelia have few ridges.

The function of the zonula adhaerens is in mechanical attachment between cells in epithelia and in other tissues such as heart muscle. The association of actin with the zonula adhaerens has led to the suggestion that these specializations transmit active forces generated within cells, while the maculae adhaerentes with their tonofilaments transmit passive stresses.

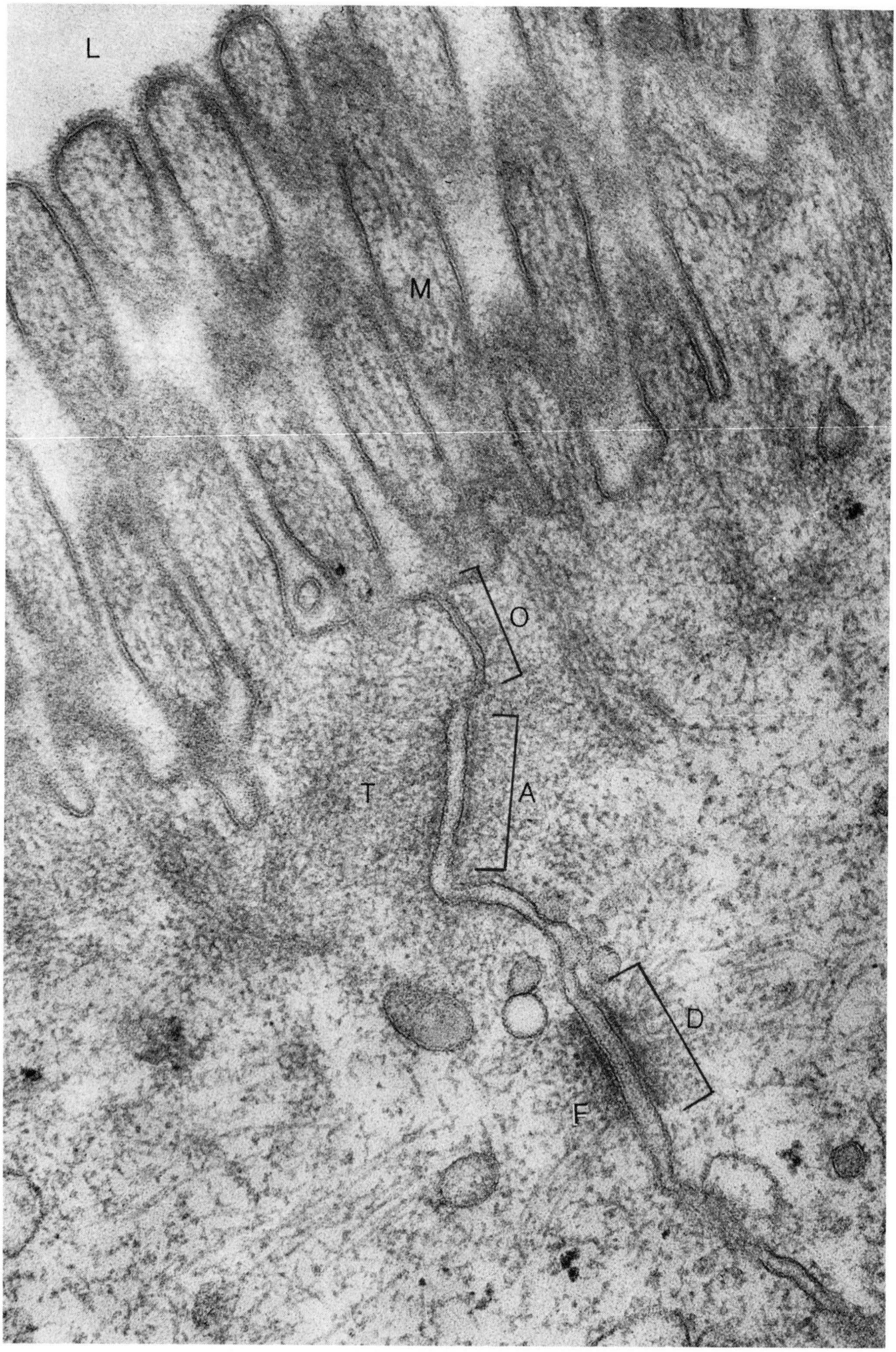

FIGURE 14–12 High-power electron micrograph of a junctional complex between two cells in the rat intestinal epithelium. The lumen of the intestine (L) is toward the top of the field. Three types of contact specializations constitute the junctional complex: a zonula occludens (O), a zonula adherens (A), and a series of desmosomes (D). Note the fine filaments continuous with those of the terminal web (T) that are associated with the zonula adherens, and the coarser tonofilaments (F) that converge on the desmosome. M, oblique sections of microvilli, × 107,000. (Micrograph courtesy of Farquhar, M. G., and Palade, G. E.: J. Cell Biol., *17*:375, 1963. Reproduced by permission of the Rockefeller University Press.)

FIGURE 14–13 Freeze-fracture preparation of frog urinary bladder. The zonula occludens appears as a series of interconnected ridges (*arrow*) on this, the P fracture face. Corresponding grooves would appear on the E face (not shown). This is a very "tight" epithelium, across which steep ionic and osmotic gradients can be maintained, and correspondingly the zonula occludens is wide, consisting of five or more ridges between the lumen (L) and the lateral intercellular space. × 45,500. (Micrograph courtesy of Claude, P., and Goodenough, D. A.: J. Cell Biol., *58*:390, 1973. Reproduced by permission of the Rockefeller University Press.)

Gap Junction

In thin sections of gap junctions, the plasma membranes of the two apposed cells approach one another very closely, but they appear to remain separated by a gap approximately 20 Å wide (Fig. 14–14). The space between the cells has been demonstrated more clearly with the use of lanthanum nitrate as a tracer, because this electron-dense colloidal substance penetrates narrow extensions of the extracellular space but does not enter intact cells.

With lanthanum, a space between the two cells 55 Å wide is outlined in sections perpendicular to the plane of the membrane. Since this is wider than the gap that appears in sections of untreated tissue, it has been presumed that lanthanum also stains part of the outer dense laminae of the two apposed plasma membranes. Sections that pass tangential to the plane of the plasma membranes of specimens treated with lanthanum permit visualization of a number of electron-lucent prisms or particles surrounded by electron-dense lanthanum. The prisms are often arranged in a hexagonal pattern with a center-to-center spacing of about 90 Å. They evidently occupy the extracellular space since they are outlined by lanthanum, and they are thought to represent projections from the surfaces of the apposed membranes.

Gap junctions have a characteristic appearance in freeze-cleave preparations, and they are most easily identified and studied by this method (Fig. 14–15). The plasma membrane at a gap junction displays a hexagonal pattern of particles, with a 90 Å center-to-center spacing on the P face of the membrane (in vertebrates) and a corresponding pattern of pits in the E face. The intramembranous particles seen in freeze-cleave preparations and the prisms outlined by lanthanum show a similarity in size and pattern. This resemblance suggests that the prisms displayed

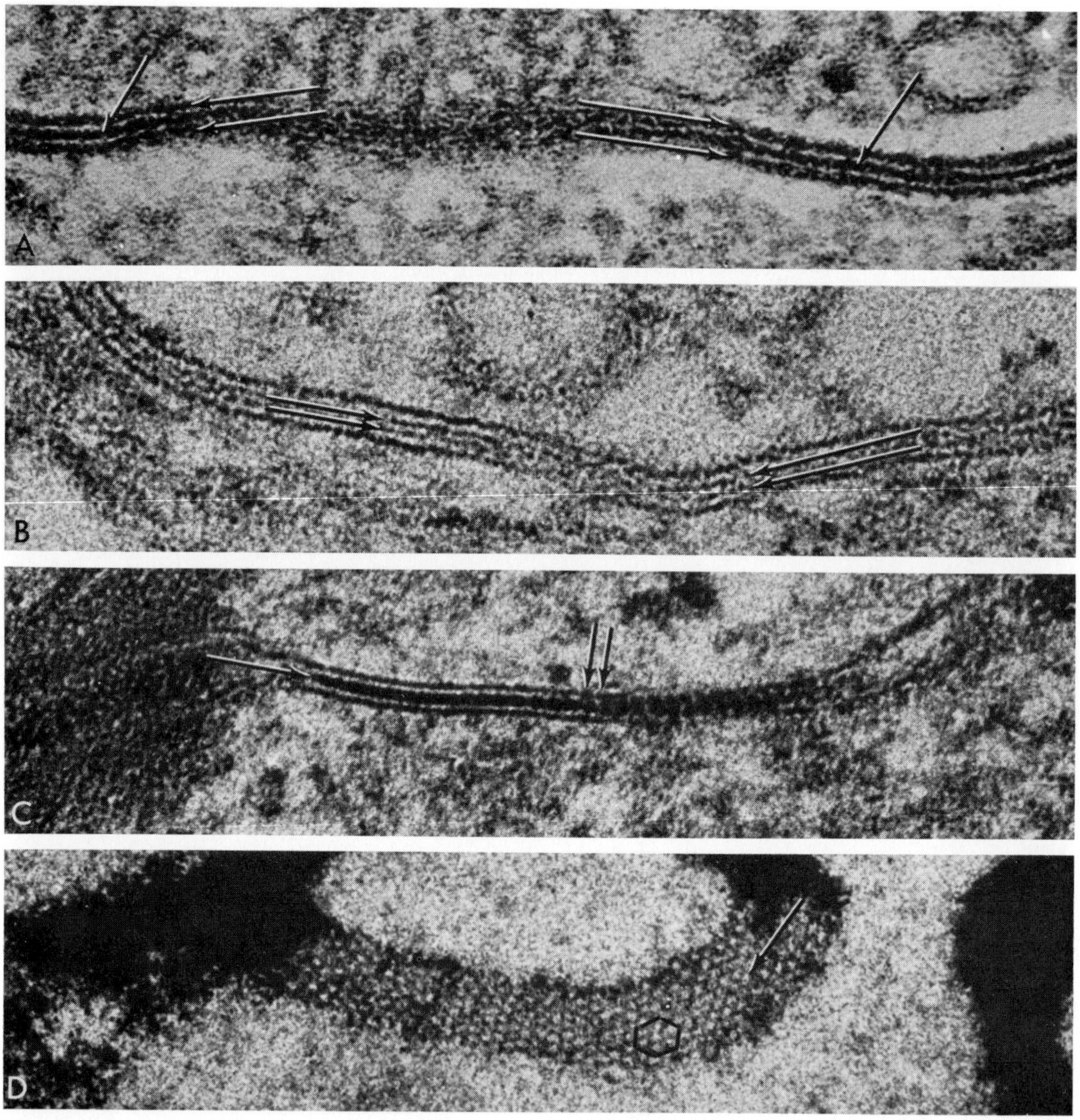

FIGURE 14–14 The appearance of gap junctions in thin sections. A and B are routine preparations stained *en bloc* with uranyl acetate; C and D show tissue that was treated with lanthanum hydroxide, which serves as an electron-dense tracer of the extracellular space between the cells. *A*. In many electron micrographs it is difficult to resolve a gap between the membranes of the apposed cells, and the junction appears to have five alternating dark (*arrows*) and light layers. Myocardium × 250,000. *B*. At high magnification in favorable preparations, a 30 to 40 Å gap between the apposed membranes is visible (*arrows*). Cervical epithelium × 260,000. *C*. The gap between the cells is demonstrated by the electron-dense lanthanum, which has penetrated the space between the cells and part of the outer dense layer of the plasma membranes. Cervical epithelium × 200,000. *D*. A gap junction in a preparation treated with lanthanum is shown in face view in a section that has passed tangential to the membranes of the apposed cells. The dense lanthanum fills the area between the cells and outlines electron-lucent subunits (*arrow*), which are 60Å in diameter and often are hexagonally packed. Cervical epithelium × 235,000. (Micrographs courtesy of McNutt, N. S., and Weinstein, R. S.: J. Cell Biol., *47*:666, 1970. Reproduced by permission of the Rockefeller University Press.)

by lanthanum are projections of intramembranous particles from the surface of the membrane (Fig. 14–16).

The appearance of gap junctions has been described in some detail because they are thought to be of physiological importance and are currently the subject of intensive investigation. The presence of gap junctions has been correlated with

FIGURE 14–15 Electron micrograph of a gap junction in a freeze-cleave preparation. The P face of the membrane (P) shows the closely packed 80 to 90 Å particles that are characteristic of most gap junctions. A corresponding array of pits is visible on the E face of the membrane (E). × 113,250. (Micrograph courtesy of C. Peracchia.)

electrical coupling between many different kinds of cells and has been demonstrated by neurophysiological methods. Gap junctions are thought to represent low-resistance pathways through which ions and possibly other molecules can move between cells, allowing the direct passage of electrical current.

The way in which this physiological behavior is related to the detailed ultrastructure of the gap junction is not yet known. Images of some specimens have led to the suggestion that the membrane particles of the gap junction have a fine central channel, ~15 Å in diameter, which courses in a direction perpendicular to the plane of the membrane. Since the particles of apposed membranes meet one another in the extracellular space, it is speculated that these channels provide a route for direct transmission of ions and other substances from one cell to another.

Gap junctions and the associated electrical coupling are found between cardiac and smooth muscle cells and between certain neurons. That is, they are

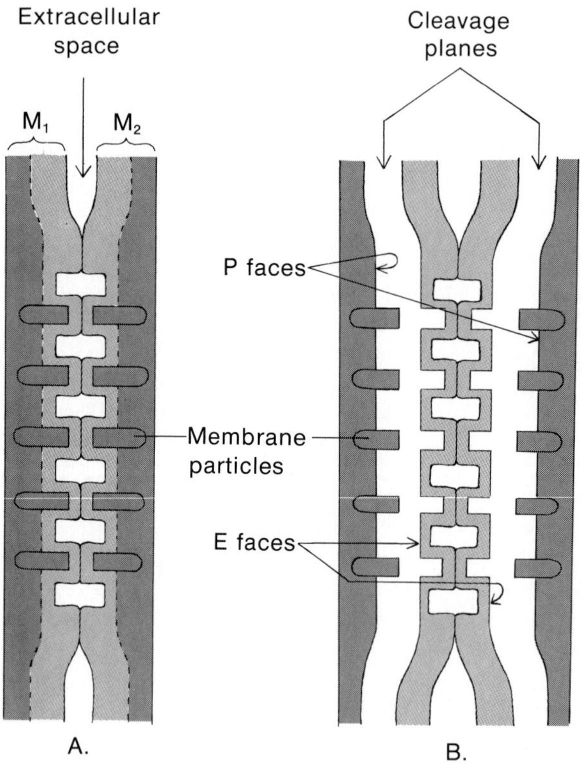

FIGURE 14–16 Diagram of a gap junction. *A*. The gap junction is formed by apposition of two membranes M_1 and M_2. Note the cylinders projecting into the extracellular space where they would be outlined in lanthanum preparations. *B*. Splitting of the membranes in a freeze-cleave preparation reveals particle-rich P faces and particle-poor E faces of the membranes. (Modified from McNutt, N. S., and Weinstein, R. S.: J. Cell Biol., *47*:666, 1970.)

present between certain cells that are known to transmit impulses as part of their normal physiological activity. An intriguing feature of gap junctions, however, is that they are also found between cells that are not known to transmit impulses. For example, the cells of early squid and chick embryos have gap junctions and are extensively coupled, and with development this coupling changes in its pattern.

The significance of coupling in cells that do not normally transmit is not entirely clear, but it is thought that the junctions may provide a pathway for exchange of nutrients or metabolites, play some regulatory role in development, or be involved in cell motility and migration patterns. Some cancer cells have been reported to have reduced coupling compared with normal cells, and there may be functional significance in the reduced opportunities for intercellular communication in cancerous tissues.

Gap junctions are now recognized as a type of contact specialization that is structurally and functionally distinct from true "tight" or "occluding" junctions. However, in some references the use of the term tight junction may be confusing for an historical reason. Initially, tight junctions were described in numerous locations, such as in terminal bars, between smooth muscle cells, and so on. In 1967, with use-improved techniques and better resolution in electron micrographs, it was discovered that *some* of the junctions previously referred to as tight junctions actually displayed a gap between the apposed plasma membrane; subsequently other characteristics of these gap junctions were revealed.

Thus those contact specializations that up to 1967 were called tight junctions were later seen to be of two types. First, the *true tight junctions,* to which the term occludens is properly applied, are those in which the extracellular space is obliterated by formation of a pentalaminar junction. They show a pattern of ridges in freeze-cleave preparations. An example is the zonula occludens of the terminal bar. Second, *gap junctions* are those that demonstrate a small space in lanthanum-stained preparations and exhibit the characteristic hexagonal array of particles in freeze-cleave preparations. Examples are present between smooth muscle cells and in the intercalated discs of cardiac muscle. One should be aware of the fact, however, that true tight junctions and gap junctions are both referred to as "tight" or "occluding" junctions in the literature prior to the late 1960's and even today in some nonmorphological texts and articles.

SPECIALIZATIONS OF THE FREE SURFACES OF CELLS

Cell Coats

Since cell coats are found on the external surface of the plasma membrane of virtually all eukaryotic cells, it is probably appropriate to consider them as fundamental parts of all cells. However, cell coats are much more highly developed than usual in some locations, such as on the apical surface of certain epithelia. These cell coats probably have a protective function and may facilitate absorption by enhancing the attachment of materials to be absorbed, as has been demonstrated for the thick cell coat of the amoeba (Fig. 14–17).

A highly developed cell coat, such as that on the surface of the amoeba or the intestinal epithelium (Fig. 14–18), is readily visible in the electron microscope as a mat of fine filamentous material. It is detectable with the light microscope by staining with PAS. Cell coats are glycoprotein in composition. It has been noted previously (Chapter 9) that radioautographic studies have shown that the Golgi apparatus plays a role in the assembly of the carbohydrate part of this glycoprotein. Cell coats have proved very difficult to separate from the plasma membrane, since they are resistant to a variety of mechanical and chemical pro-

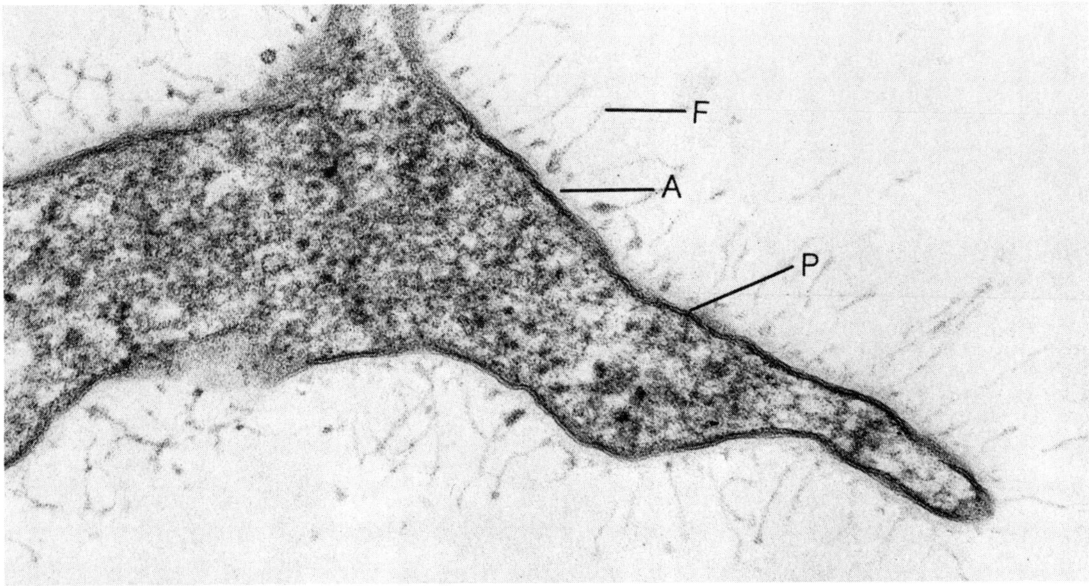

FIGURE 14–17 Cell surface of an amoeba. The glycoprotein cell coat is highly developed in these cells. It comprises an amorphous layer (A) next to the plasma membrane and long filaments (F) which extend outward from the plasma membrane (P). × 93,000.

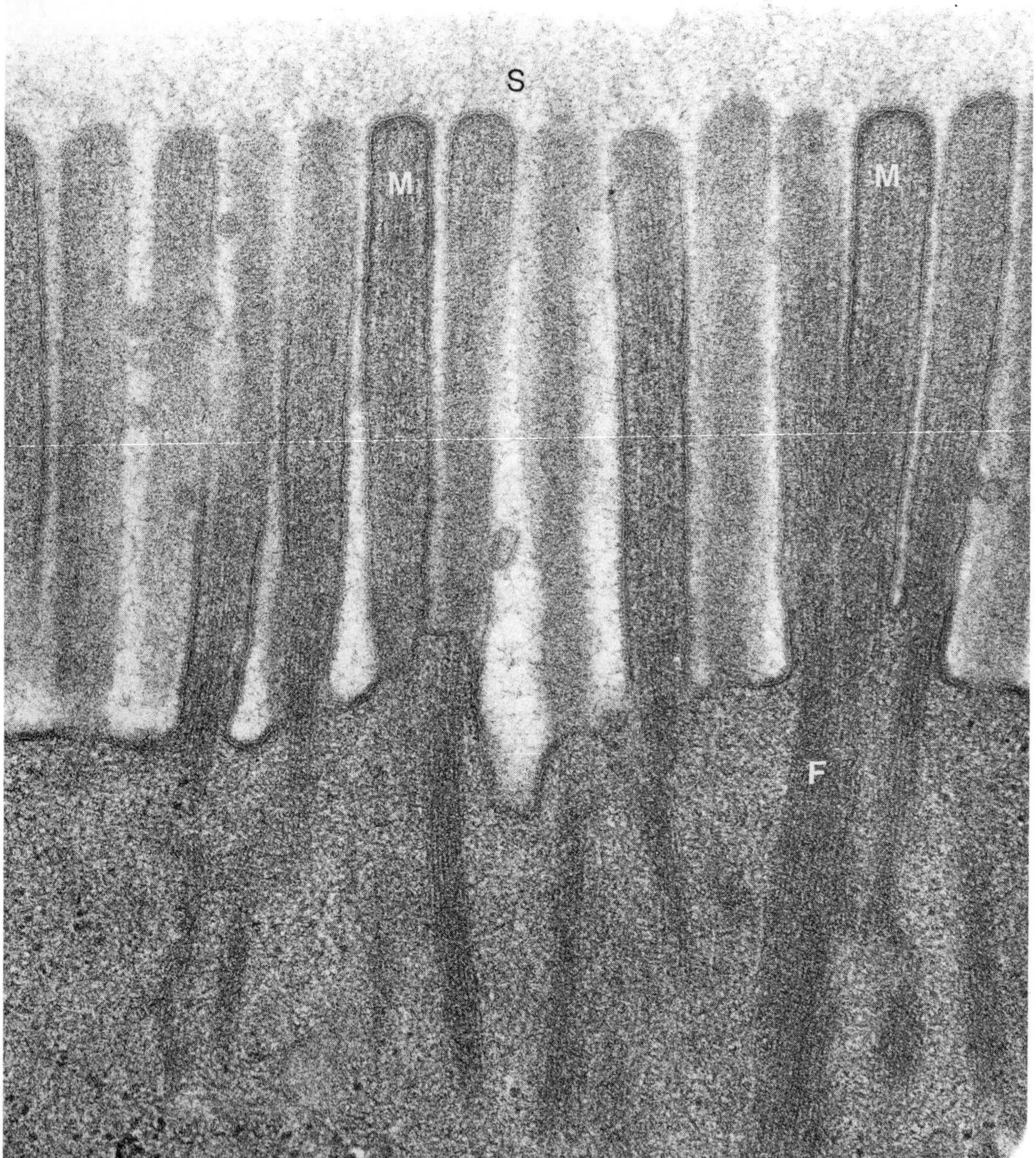

FIGURE 14–18 High-power electron micrograph of the apical surface of an absorptive cell in the cat intestine. The surface exhibits many fingerlike microvilli (M). Note in the core of each microvillus many fine filaments (F) which extend into the underlying apical cytoplasm. The fuzzy cell coat (S) forms a thick mat on the surface of the microvilli. × 100,000. (Micrograph courtesy of S. Ito.)

cedures. This has led to their being regarded as an integral part of the plasma membrane rather than as an extraneous layer applied to its surface.

According to the fluid mosaic model of membrane structure, the relation of the cell coat to the rest of the plasma membrane can be visualized as follows. The hydrophilic portions of membrane proteins project from the membrane into the surrounding water. In the case of glycoproteins, the hydrophilic portions include the carbohydrate chains, and the aggregate of carbohydrate chains on the external surface of the plasma membrane forms a layer that is sufficiently thick to be visualized as a cell coat.

Although the cell coat is most easily visualized and studied in cells such as the intestinal epithelium, in which it is very thick, a similar but thinner layer can be demonstrated on most other eukaryotic cells. In some instances it is visible with the electron microscope, while in others special staining methods are required. By virtue of their location and universal presence, cell surface coats are thought to be important in interactions between cells. Thus they may play a role in the development and functioning of tissues. Alterations in cell surface coats may be involved in the ability of cancer cells to invade and metastasize.

The lectins, substances of plant origin, have been widely used in recent years to study properties of cell surfaces. These are proteins or glycoproteins, and some of the most common examples are concanavalin A, wheat germ agglutinin, ricin, and phytohemagglutinin. Their usefulness lies in the fact that they bind specifically to certain saccharides that are found on the surfaces of cells. Concanavalin A (Con A), for example, binds to D-mannose and D-glucose residues. Thus lectins have been used as markers for certain constituents of the cell surface.

The lectins can be coupled to molecules such as ferritin or horseradish peroxidase, which enables them to be visualized in the electron microscope. In this way it has been possible to follow changes in lectin binding (and thus in the constituents of the cell surface) under different experimental conditions, as well as changes in certain cells during their development. It has been possible also to assess the mobility of components of the cell surface by determining the distribution of lectin-binding sites under different conditions. Lectins have been used in the study of malignant cells because such transformed cells in culture are more readily agglutinated by lectins than are their normal counterparts. This is apparently due to some difference in the composition of the cell surface, and the basis of this difference is under investigation.

Microvilli

Small fingerlike projections known as microvilli (Figs. 14–18 and 14–19) are among the most common specializations of the free surfaces of cells. They are seen in variable numbers on the surfaces of many cells both *in vivo* and in cell culture. Microvilli are highly developed on absorptive epithelia, such as those of the renal tubules and the intestine.

A specialization of the apical surface of the absorptive epithelium in the intestine was recognized many years ago by light microscopy. A refractile, vertically striated band was seen on the luminal surface of the cells and for this reason was given the descriptive name of striated, or brush, border. With the electron microscope, it was discovered that the striated border consists of a large number of parallel, fingerlike microvilli (Figs. 14–18 and 14–19). The apical surface of the epithelium of the epididymis bears extraordinarily long processes of this sort, which were thought by light microscopists to bear a resemblance to cilia, and since they were not observed to be motile they were called stereocilia. Electron microscopy has shown that stereocilia resemble very long microvilli.

A microvillus is about 800 Å in diameter and 1 μm long (Fig. 14–18), and in the case of the epithelium of the duodenum there are thousands of them per cell. Underlying the striated border in the apical cytoplasm of intestinal epithelial cells there is a sheet or mat of fine filaments referred to as the terminal web. A core

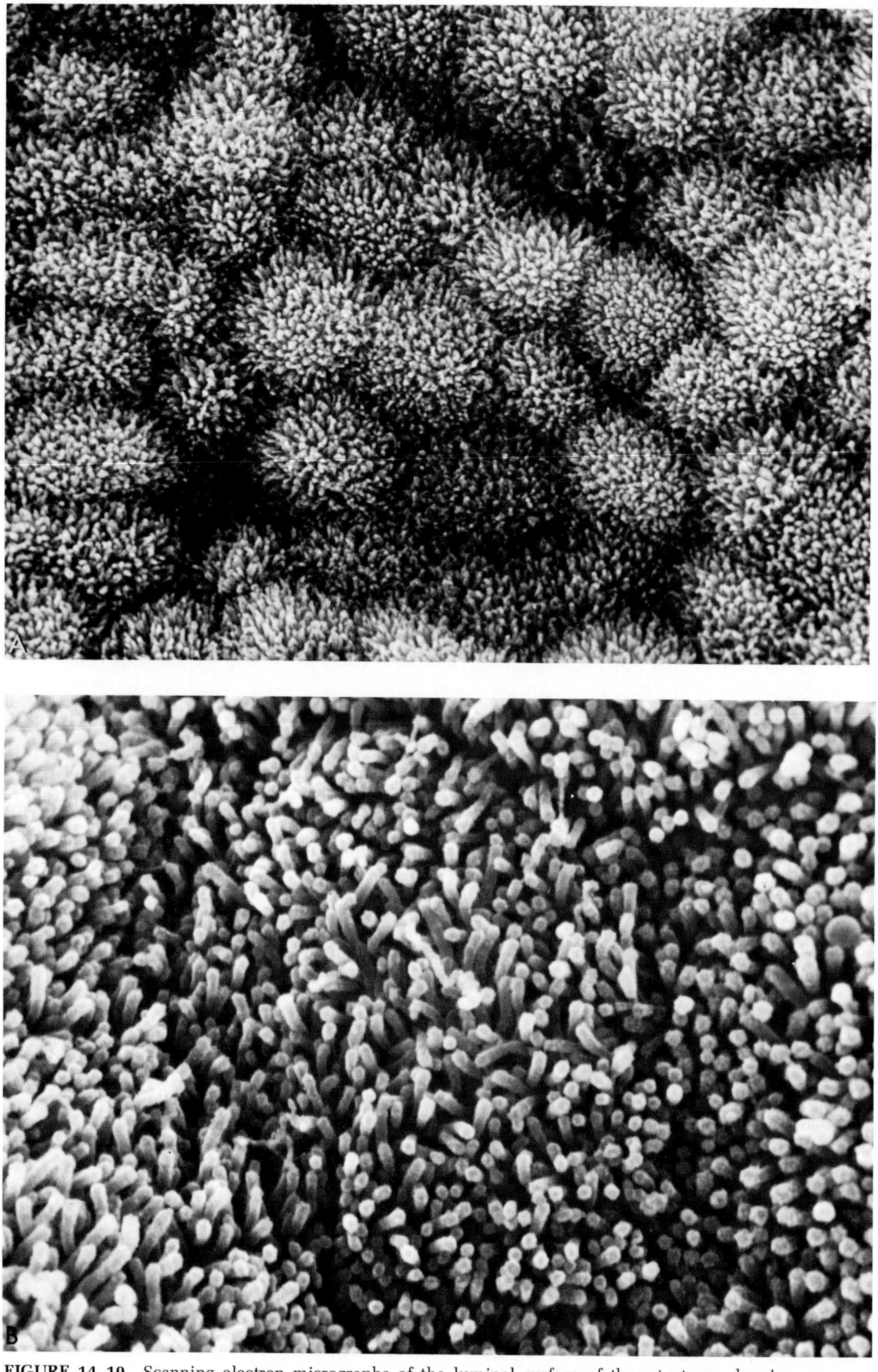

FIGURE 14–19 Scanning electron micrographs of the luminal surface of the rat uterus showing numerous microvilli. *A*. The dark bands outline a number of individual cells, all of which have many microvilli on their surfaces. × 4300. *B*. At higher magnification, the slender, fingerlike character of the microvilli can readily be appreciated. The animals were experimentally treated, as explained in the reference. (Micrographs courtesy of Anderson, W. A., Kang, Y.-H., and De Sombre, E. R.: J. Cell Biol., *64*:692, 1975. Reproduced by permission of the Rockefeller University Press.)

of fine filaments runs longitudinally in the microvillus, being anchored to a patch of dense material underlying the plasma membrane at the apex of the process and extending from the base of the microvillus to interdigitate with the filaments of the terminal web. The filaments of the terminal web appear to be composed of actin (see Chapter 13). They are thought to furnish support and possibly to provide for movement of the microvilli. The presence of numerous microvilli greatly increases the surface area of the cell, and thus they are thought to enhance the absorptive capacity of the cell.

Absorptive epithelia, such as that of the intestine, renal tubules, and part of the male duct system, often have a system of apical pits and canaliculi (Fig. 14–20), which are small invaginations of the plasma membrane extending into the apical cytoplasm from between the bases of the microvilli. These pits and canaliculi can often be observed to have a coating of fine spikes or bristles on the inner surface of the plasma membrane, facing the cytoplasmic matrix. Sometimes a coat is also visible on the external surface of the membrane. Coated vesicles, so called because similar coats are present on the surface of their membranes, are found in the apical cytoplasm. The use of tracer materials has shown that the

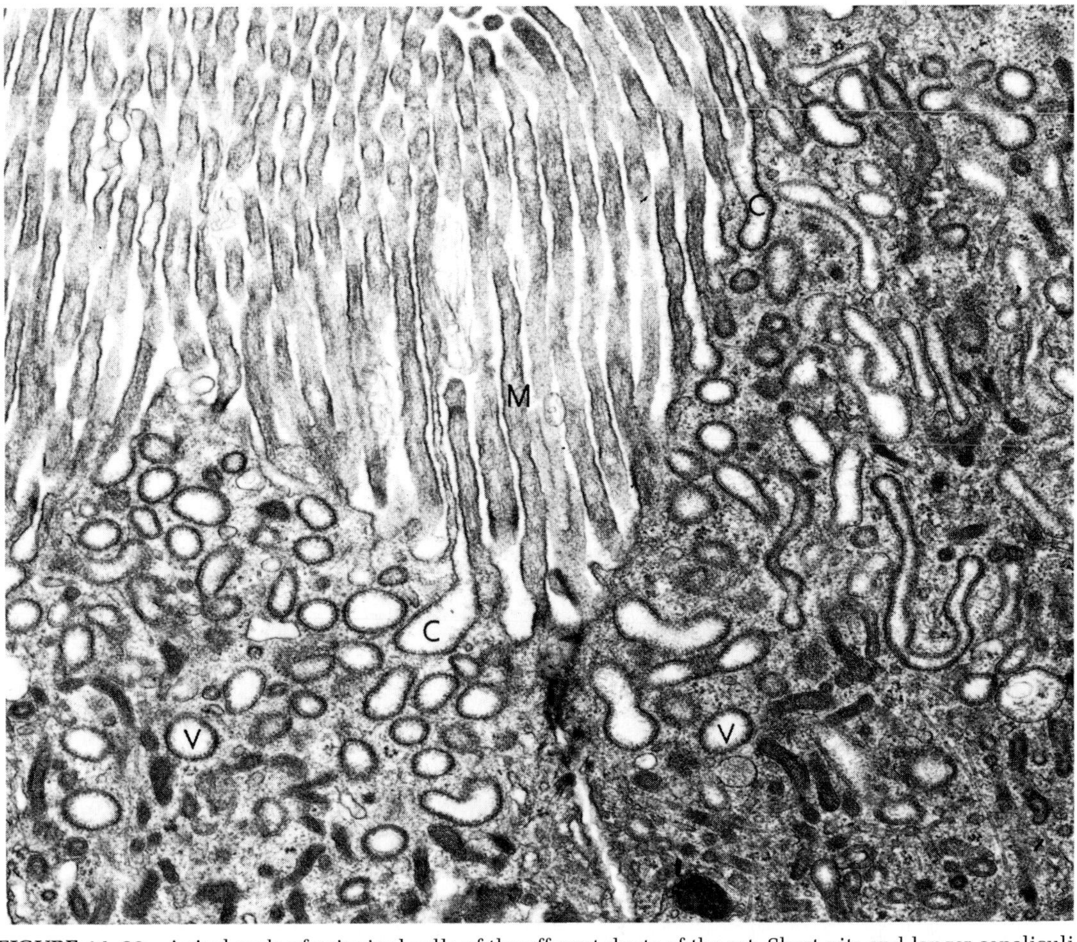

FIGURE 14–20 Apical ends of principal cells of the efferent ducts of the rat. Short pits and longer canaliculi (C) are invaginations of the apical plasma membrane between the microvilli (M). Coated vesicles (V) are formed by pinching off of the ends of the canaliculi. × 19,000. (Micrograph courtesy of H. English.)

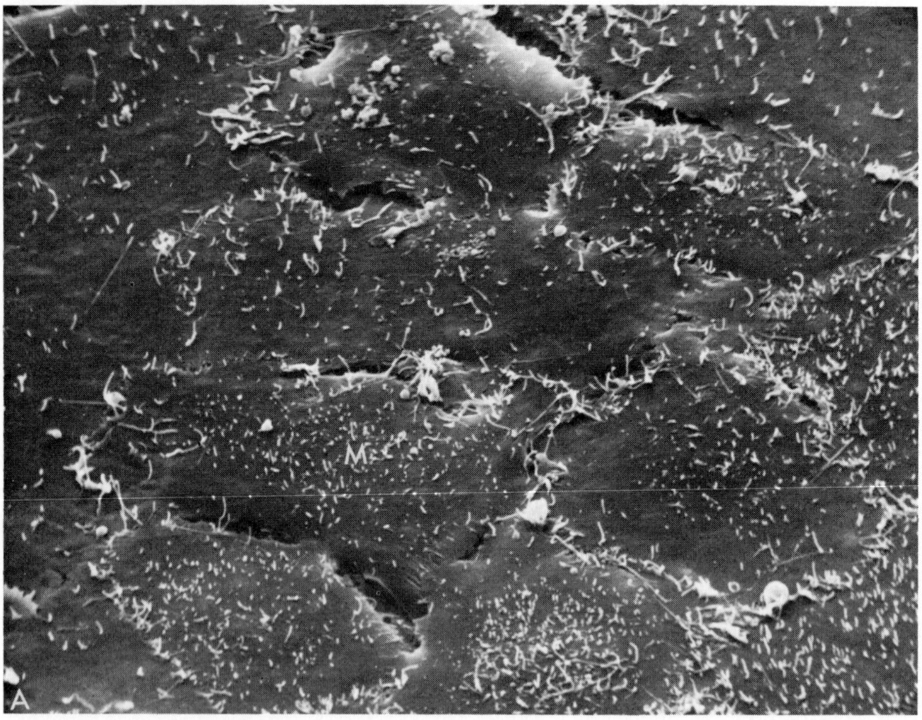
M
A

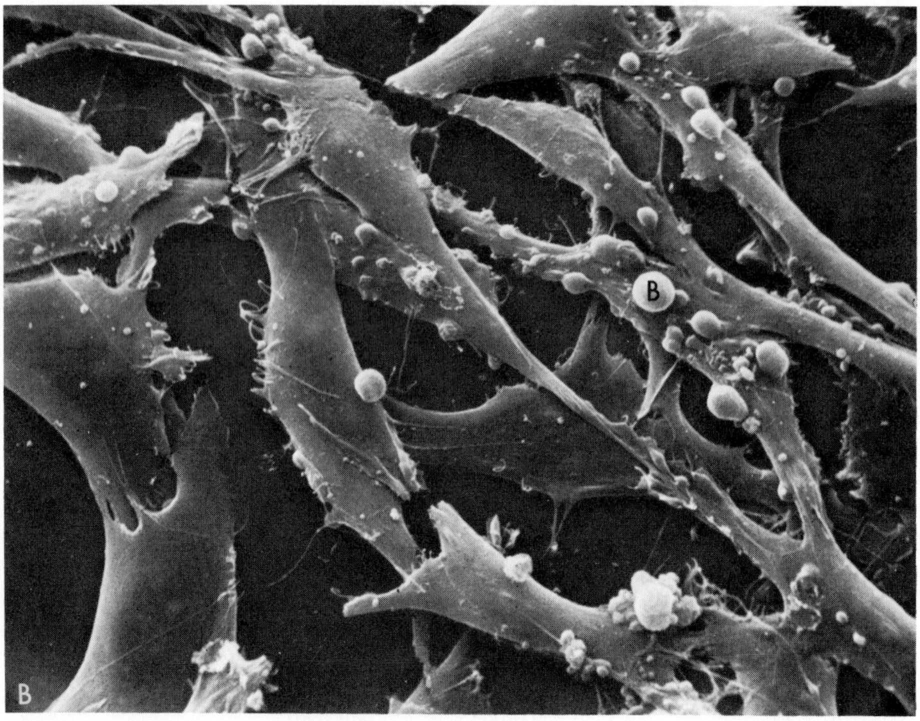
B
B

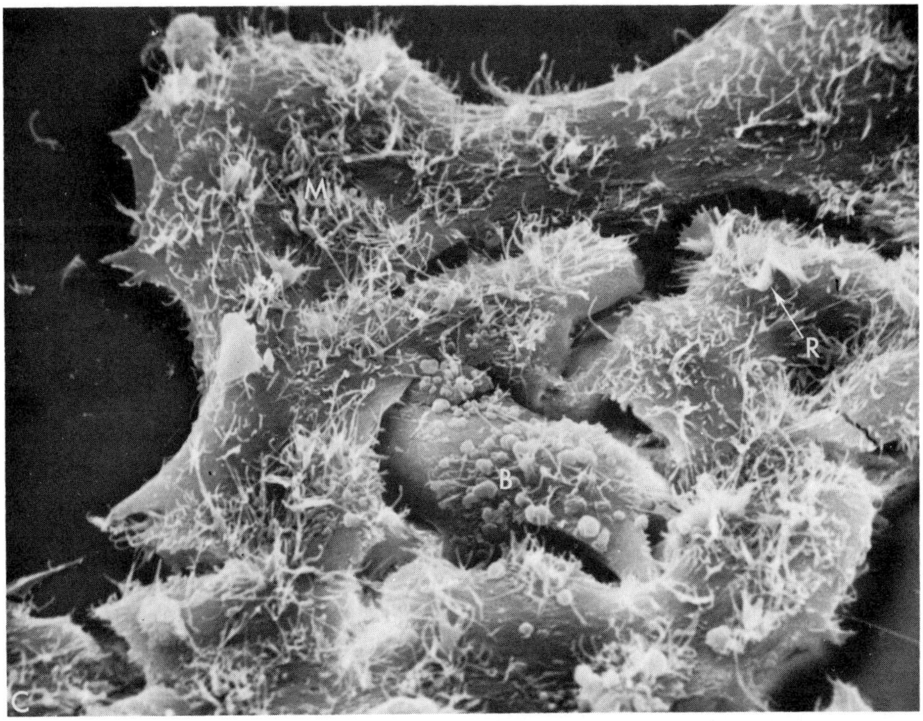

FIGURE 14–21 Scanning electron micrographs of normal and transformed mouse cells grown in culture (Balb/373, clone A 31). *A*. Normal cells form an epithelioid sheet. Small microvilli (M) are concentrated over the centers of the cells. *B*. Cells transformed by virus. The cells have lost their contact inhibition and have assumed an elongated shape. Microvilli are almost absent, but blebs (B) are numerous on their surfaces. *C*. Spontaneously transformed cells. These cells also have lost contact inhibition and differ in shape from normal cells. Compared with the normal cells (A), these cells have more long microvilli (M), "ruffling" (R), and blebbing (B). × 1000. (Micrographs courtesy of Porter, K. R., Todaro, G. J., and Fonte, V.: J. Cell Biol., 59:633, 1973. Reproduced by permission of the Rockefeller University Press.)

coated vesicles are formed by a pinching-off from the pits and canaliculi and that some substances are absorbed by this form of pinocytosis.

A variety of structures can be observed on the surfaces of cells with the scanning electron microscope (Fig. 14–21). These include microvilli, spherical bulbous projections known as blebs, and folds or "ruffles." Under certain conditions in cell cultures, changes in surface morphology at different stages in the cell cycle have been observed. For example, during G_1, cells show many surface specializations, including microvilli, blebs, and ruffles. During the S phase, however, the surface becomes relatively smooth, while microvilli and some ruffles return during G_2. Cells in mitosis are spherical and display a large number of long thin processes referred to as filopodia.

Interestingly, changes on the cell surface during interphase seem to depend upon contact between the cells, since these changes are observed in confluent cultures but not in similar cells growing in low-density cultures. In addition, tumor-producing varieties of cultured cells, transformed to this state either spontaneously or by certain viruses, show a variety of strain-specific surface abnormalities visible with the scanning electron microscope.

Basal Folds

The base of most epithelia is smoothly contoured. However, in some that are engaged in active transport of ions and water, the basal plasma membrane is highly folded, and numerous mitochondria lie in the cytoplasm within the folds (Fig. 14–22). This has been widely regarded as an adaptation for increasing the surface area of the membrane available for transport. The proximity of many mitochondria is interpreted as providing a ready source of energy for ion pumps located in the plasma membrane.

Specializations of this sort are seen in the base of the epithelium of renal tubules. They are highly developed in the salt glands of marine birds, which can excrete 1N sodium chloride, enabling the birds to drink sea water.

Basal Lamina

The basal lamina is not truly a cell surface specialization, but since it is almost universally associated with epithelial cells and is produced by them, it is considered briefly here. The basal lamina is a sheet of material that underlies epithelial cells. In sections perpendicular to the base of the epithelium (Fig. 14–23) it is visualized as a moderately dense band about 500 to 800 Å thick, which is separated from the basal plasma membrane of the epithelial cells by a lighter region.

The basal lamina has an amorphous or finely filamentous texture. It is composed of a type of nonbanded collagen peculiar to basal lamina ($\alpha 1$, type IV collagen) and a carbohydrate component, which is responsible for its staining with PAS. The basal lamina and epithelial cells appear to be attached, possibly by means of fine filaments that have been described as extending between the two.

Production of basement membrane material is altered in some diseases. For example, in diabetes mellitus the cells of the renal glomerulus produce much more material than normal, and the glomerular basal lamina becomes highly thickened. Since urine is formed by filtration across this basal lamina, changes in kidney function can result, and renal failure is one of the complications that develops in patients with severe diabetes.

Some confusion exists in the terminology used to designate this structure. Basal lamina is currently the preferred term, but "basement lamina" and "basement membrane" are sometimes seen. The term basement membrane has been used for years by light microscopists and thus more correctly refers to a structure visible in the light microscope. The PAS-positive basement membrane of light microscopy represents the basal lamina (visible only in the electron microscope) and some of the underlying connective tissue. It is evident that neither the basal lamina nor the "basement membrane" represents a true lipoprotein membrane.

Motile Cell Processes

Cilia and flagella are considered in Chapter 13 under microtubules and filaments.

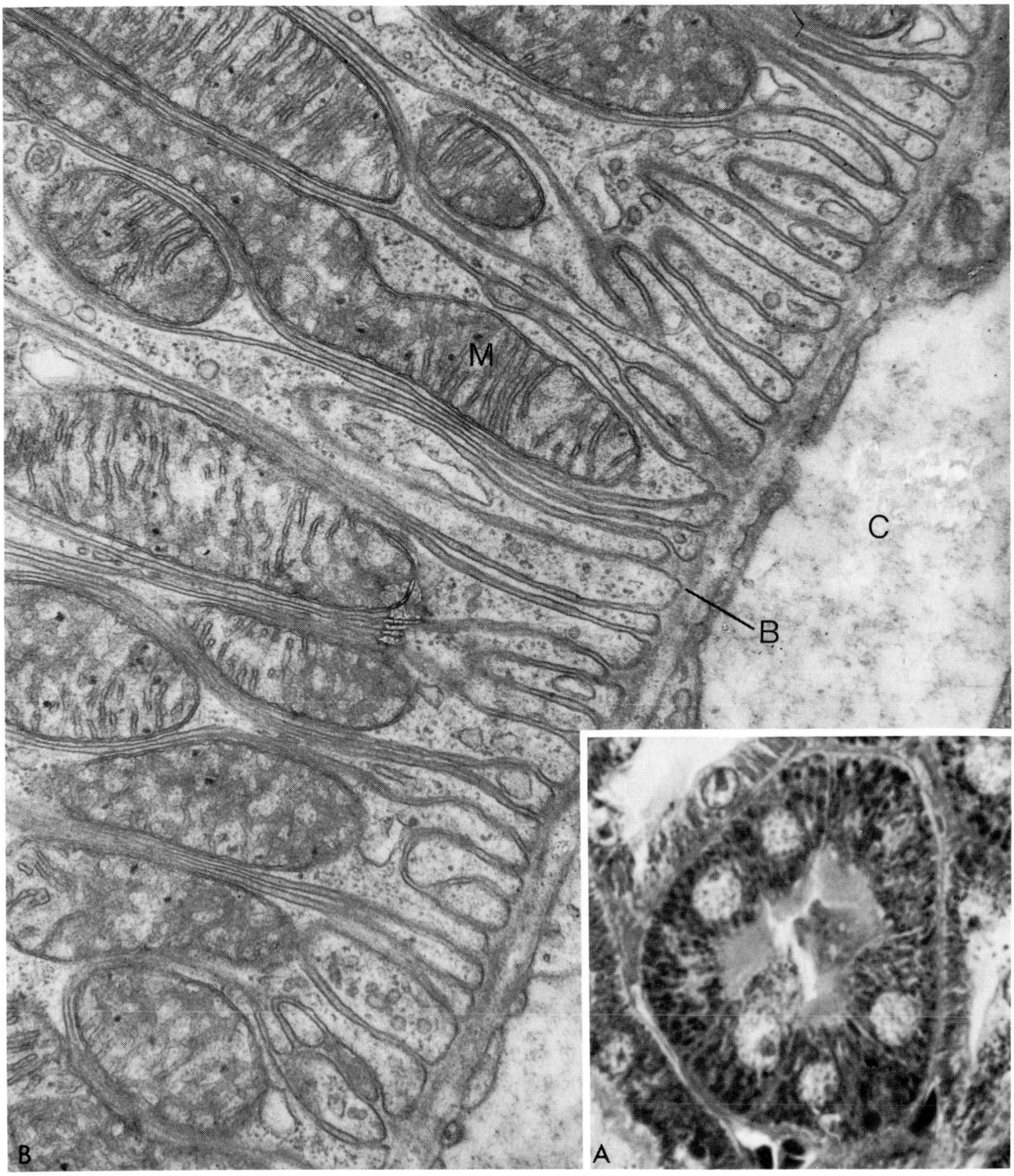

FIGURE 14–22 *A*. Light micrograph of a paraffin section of the proximal convoluted tubule of the kidney. The orientation of many mitochondria along basal folds of the plasma membrane lends a pattern of vertical striations to these epithelial cells. The circular, lightly staining areas represent nuclei. Altmann stain. × 1350. (Micrograph courtesy of B. Tandler and C. L. Hoppel. Reproduced by permission of Academic Press.) *B*. Electron micrograph of the base of the epithelium of the distal convoluted tubule in the guinea pig kidney. The plasma membrane at the base of the epithelial cells is highly folded, forming narrow compartments that interdigitate in a complicated way with similar processes of neighboring cells. Many mitochondria (M) lie in the cytoplasm near the plasma membrane. The basal lamina (B) of the epithelium appears as a moderately dense band underlying the base of the cell. *C*. Lumen of a capillary. × 32,000. (Micrograph courtesy of Fawcett, D. W.: The Cell. Philadelphia, W. B. Saunders, 1966. Micrograph by A. Ichikawa.)

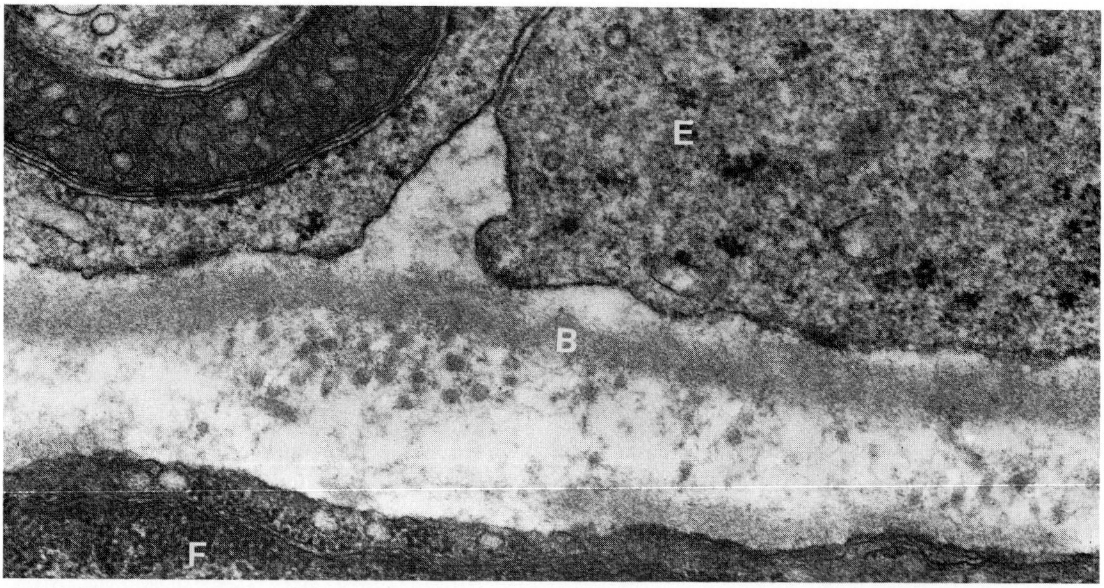

FIGURE 14–23 Base of the seminiferous epithelium of the rat testis. The basal lamina (B) appears as a moderately dense band underlying the epithelium (E). Part of a cell (F) of the interstitial tissue is visible at the bottom of the field. × 43,000.

HORMONES AND THE CELL SURFACE

Properties of the cell surface are important in the response of cells to hormones of the polypeptide and catecholamine varieties. Specific receptors for the hormone are located on the external surface of the plasma membrane of "target cells," which are cells responsive to the hormone. Binding of the hormone to the surface receptors results in increased activity of the enzyme adenyl cyclase, which is associated with the inner aspect of the plasma membrane. Adenyl cyclase catalyzes the conversion of ATP to the cyclic nucleotide, 3′, 5′-cyclic adenosine monophosphate (cAMP).

According to the model proposed by Sutherland, the discoverer of cAMP, the hormone is regarded as a "first messenger," while the cAMP is considered to be a "second messenger" in the response to a hormone. The hormone-induced increase in cAMP initiates a train of events leading to metabolic alterations in the target cell. The precise nature of the ensuing changes depends upon the cell type. For example, the pituitary hormone ACTH, acting on cells of the adrenal cortex, increases intracellular levels of cAMP, which enhances synthesis of adrenal cortical steroids. The hormone glucagon interacts with the surface of liver cells and increases cAMP levels, which leads through a sequence of events to glycogen breakdown.

The chain of intracellular events following an increase in cAMP may be complex and may consist of several sequential stages (Fig. 14–24). In the case of the liver cell's response to glucagon or epinephrine, for example, the hormone-induced elevation of cAMP leads to phosphorylation of the enzyme protein kinase. In turn, protein kinase phosphorylates the enzyme phosphorylase b kinase. Phosphorylase b kinase then converts the relatively inactive phosphorylase b to the more active form, phosphorylase a, which breaks down glycogen to glucose.

The important point to bear in mind in the context of the cell surface is that specific receptors for a hormone are located on the surface of target cells. Their

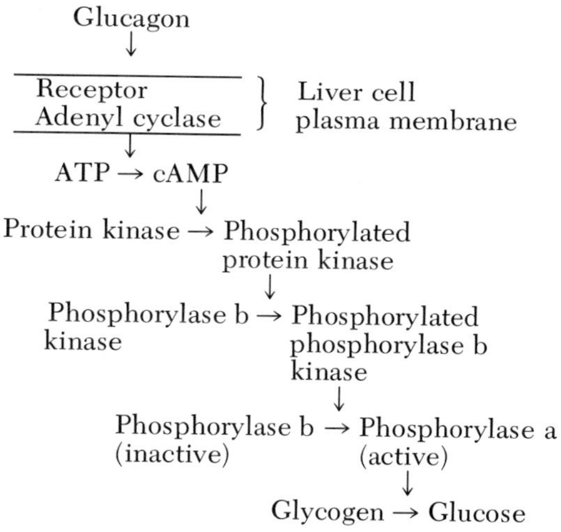

FIGURE 14–24 A diagram of steps in the response of liver cells to the hormone glucagon, leading to the breakdown of glycogen to glucose.

presence determines whether a given cell is able to respond to a particular hormone. The nature of the response, however, depends upon the metabolic machinery present in the stimulated cell, with cAMP serving as a common intermediary in many hormonally responsive cells. Hormones known to work by this sort of mechanism include adrenocorticotropic hormone (ACTH), luteinizing hormone (LH), thyroid-stimulating hormone (TSH), parathyroid hormone, thyroxine, vasopressin, glucagon, epinephrine, and norepinephrine.

It should be noted that the steroid hormones, such as estrogen and testosterone, act by a different kind of mechanism. The steroid enters a target cell and binds to a cytoplasmic receptor protein. This hormone-receptor complex then enters the nucleus and interacts with DNA. The result is increased RNA synthesis, through which the characteristic changes in the target cell are mediated.

THE CELL SURFACE AND CANCER

The study of properties of cell surfaces is important in medicine not only because characteristics of the cell surface determine how cells relate to one another in normal tissues, but also because they bear on altered relationships between cells in cancer. Alterations in the cell surface may be related to the relatively unrestrained cell proliferation in tumors, which is not regulated normally by cell contacts, hormones, or other substances that interact with the cell surface. Furthermore, cancers are characterized by the ability to spread to other locations by invasion and frequently by metastasis. Cell surface changes are probably important in metastasis because the cells must separate from the primary tumor and relate to their surroundings in the new location. In addition, changes in the cell surface may help to explain alterations in the relation of cancer cells to the body's immune apparatus.

Much research has been conducted on the surfaces of tumor cells, and many differences from the normal cell have been found. Discussion of this complicated and interesting field is well beyond the scope of this text. It may be noted, however, that one current problem is the difficulty of relating these observed surface changes to the biological properties of tumors, such as proliferation, invasion, and metastasis. The mechanisms by which the surface changes are produced also are poorly understood. Recent studies suggest that alterations in the mobility and distribution of surface components may be important changes in cancer cells.

FURTHER READING

Bennett, H. S.: Morphological aspects of extracellular polysaccharides. J. Histochem. Cytochem., *11*:14, 1963.

Bolis, L., Hoffman, J. F., and Leaf, A. (eds.): Membranes and Disease. New York, Raven Press, 1976.

Branton, D.: Freeze-etching studies of membrane structure. Philos. Trans. R. Soc. Lond., *261*:133, 1971.

Claude, P., and Goodenough, D. A.: Fracture faces of zonulae occuludentes from "tight" and "leaky" epithelia. J. Cell Biol., *58*:390, 1973.

Farquhar, M. G., and Palade, G. E.: Junctional complexes in various epithelia. J. Cell Biol., *17*:375, 1963.

Fawcett, D. W.: Physiologically significant specializations of the cell surface. Circulation, *26*:1105, 1962.

Friend, D. S., and Gilula, N. B.: Variations in tight and gap junctions in mammalian tissues. J. Cell Biol., *53*:758, 1972.

Gilula, N. B.: Junctions between cells. *In*: Cox, R. P. (ed.): Cell Communcation. New York, John Wiley and Sons, 1974, pp. 1–29.

Ito, S.: Structure and function of the glycocalyx. Fed. Proc., *28*:12, 1969.

Kelly, D. E.: Fine structure of desmosomes, hemidesmosomes and an adepidermal globular layer in developing newt epidermis. J. Cell Biol., *28*:51, 1966.

Leblond, C. P., and Bennett, G.: Elaboration and turnover of cell coat glycoproteins. *In*: Moscona, A. A. (ed.): Cell Surface in Development. New York, John Wiley and Sons, 1974, pp. 29-49.

Luft, J. H.: The structure and properties of the cell surface coat. Int. Rev. Cytol., *45*:291, 1976.

McNutt, N. S., and Weinstein, R. S.: The ultrastructure of the nexus. J. Cell Biol., *47*:666, 1970.

Nicolson, G. L., and Poste, G.: The cancer cell: dynamic aspects and modifications in cell-surface organization. N. Engl. J. Med., *295*:197, 253, 1976.

Porter, K. R., Todaro, G. J., and Fonte, V.: A scanning electron microscope study of surface features of viral and spontaneous transformants of mouse Balb/3T3 cells. J. Cell Biol., *59*:633, 1973.

Revel, J. P., and Karnovsky, M. J.: Hexagonal array of subunits in intercellular junctions of the mouse heart and liver. J. Cell Biol., *33*:C7–C12, 1967.

Robertson, J. D.: Unit membranes: a review with recent new studies of experimental alterations and a new subunit in synaptic membranes. *In*: Cell Membranes in Development. New York, Ronald Press, 1964.

Singer, S. J., and Nicolson, G. L.: The fluid mosaic model of the structure of cell membranes. Science, *175*:720, 1972.

Sjöstrand, F. S.: The structure of cellular membranes. Protoplasma, *63*:248, 1967.

Staehelin, L. A.: Structure and function of intercellular junctions. Int. Rev. Cytol., *39*:191, 1974.

Whaley, W. G., Danwalder, M., and Kephart, J. E.: Golgi apparatus: influence on cell surfaces. Science, *175*:596, 1972.

15

TRANSPORT PROCESSES: DIFFUSION, MEMBRANE PERMEABILITY, AND MEDIATED TRANSPORT

The discussion of cellular membranes is continued in this and succeeding chapters by considering some of their physiological properties. Some of the ways in which various substances are able to cross biological membranes are discussed in this chapter. In Chapter 16 the movement of water across membranes by the process of osmosis is considered, along with the distribution of ions across membranes according to their electrochemical potentials. This information is the basis for the study in Chapter 17 of the resting potential, the electrical potential across the plasma membrane of all cells, and the mechanism of the action potential displayed by excitable cells. Some of the physiological properties of nerve and muscle cells, which are specialized for excitation, conduction, and contraction, are considered in Chapters 18 and 19.

TYPES OF TRANSPORT PROCESSES

The processes that contribute to movement of molecules from one place to another in biological systems are known as *transport processes*. The exchanges of matter between an organism and its environment as well as the movement of substances within the organism are mediated by one or more transport processes. The transport processes considered here are defined as follows.

1. *Diffusion*. Diffusion is the process by which atoms or molecules inter-

mingle owing to their thermal motion. It results in the net movement of substances from a place of higher to a place of lower concentration.

2. *Mediated transport.* This term refers to transport of molecules across a barrier separating two compartments by means of a specific channel or *carrier molecule* that resides in certain cell membranes. If the transport can occur against a concentration difference (more properly, against an electrochemical potential difference), it is called *active transport;* otherwise it is called *facilitated* transport.

3. *Osmosis.* The passage of water through a semipermeable membrane separating solutions with different concentrations of solutes is called osmosis. An ideal semipermeable membrane is permeable to water, but impermeable to all solutes.

4. *Flow of ions (electrodiffusion).* Electrodiffusion is the movement of ions in response to a concentration difference, an electrical potential difference, or both.

The following are some examples of physiological processes in which transport plays a central role.

In the mammalian lung, air from the environment exchanges with the gas in the pulmonary alveoli. The walls of the alveoli are traversed by numerous blood capillaries, so that the gas in the alveoli and the blood in the pulmonary capillaries are brought into close approximation. The transfer of oxygen from alveolar gas to capillary blood and the movement of carbon dioxide from blood to alveolar gas occur by the *diffusion* of these gases across the alveolocapillary membrane.

In times of water deprivation, the mammalian kidney can produce a urine that is much more concentrated in solutes than the blood from which it is derived. This permits the organism to conserve water. Urine in the renal collecting ducts passes through the medulla of the kidney. Since the concentration of solutes in the interstitial fluid (fluid between cells) of the medulla is higher than that in the collecting duct urine, water is drawn by *osmosis* from the urine to the interstitium, thereby concentrating the solutes in the urine.

Most cells possess an electrical potential (voltage) difference across their plasma membranes. That is, the cytoplasm is at a different electrical potential than the extracellular fluid surrounding the cells. This so-called resting membrane potential is especially pronounced in nerve and muscle cells, both of which are electrically excitable. A later chapter will show how the resting membrane potential results from the *diffusion of ions* across the plasma membrane under the influence of differences in concentration and electrical potential. The action potential of nerve and muscle cells is essential for communication and movement and is also the result of the *flow of ions* across the plasma membrane.

As the contents of a meal traverse the alimentary canal, enzymes act on foodstuffs to convert them to smaller, less complex molecules. Starches and sugars are broken down to monosaccharides; proteins and peptides are split to amino acids. As the digested food passes through the small intestine, sugars and amino acids are removed from the intestinal contents and transported to the blood by specific *active transport systems* which are capable of reducing the concentrations of sugars and amino acids in the gut contents to levels far below those in circulating blood.

There are a number of disease states in which defects in transport processes are involved. Diabetes mellitus (sugar diabetes) is among the most common of these diseases. In the normal human an increase in the level of blood glucose triggers the pancreas to release insulin into the bloodstream. Insulin then acts on various cells, most notably those of muscle, fat, and liver tissue, to increase the rate of transport of glucose into the cytoplasm by *facilitated transport* and to increase the rate at which glucose is metabolized. In the diabetic the amount of insulin released by the pancreas is below normal. The result is that a high level of blood sugar is maintained and glucose cannot be utilized at normal rates. Many symptoms of diabetes are thus results of defective transport and utilization of glucose.

In a different disease, diabetes insipidus (tasteless urine), which is unrelated to diabetes mellitus, there is a defect of osmotic water reabsorption in the kidney. The permeability to water of the collecting duct epithelium is under the influence of antidiuretic hormone (ADH), which is produced in the hypothalamus and released by the pituitary gland. The effect of ADH is to render the collecting duct epithelium permeable to water so that the *osmotic* process can function to concentrate the urine. In patients with diabetes insipidus there is insufficient release of ADH so that the collecting duct epithelium is relatively impermeable to water. Thus when urine passes down the collecting duct, little water can be extracted from it by *osmosis* into the medullary interstitium. As a result, such patients produce large volumes of very dilute urine, which may reach 20 liters per day in severe cases.

Hereditary spherocytosis is a disease that involves defective ion transport in red blood cells. In this disease the red blood cell membrane has an abnormally high permeability to sodium ion. Normal red blood cells have a membrane-bound enzyme, the sodium-potassium ATPase, which is responsible for extruding Na^+ from the cell by *active transport*. The concentration of this enzyme is increased in hereditary spherocytosis, and if provided sufficient nutrients the red cells are able to pump Na^+ out as fast as it *diffuses* in, despite its abnormally high rate of entry. In the venous sinuses of the spleen, however, the levels of nutrients are quite low, so that red cells in the splenic sinuses are not able to produce ATP fast enough to pump Na^+ out as rapidly as it enters. The increased cellular content of Na^+ causes water to enter the red cells by *osmosis*, so that the cells swell. The swelling predisposes the cells to be destroyed by the spleen, and anemia results.

These examples illustrate that an understanding of transport processes is useful in understanding both the function of the normal organism and the pathophysiology of certain disease states. Transport processes will now be considered in detail, beginning with diffusion.

DIFFUSION

Diffusion was defined earlier as the process by which atoms or molecules intermingle owing to their random thermal motion. Consider a box divided in two by a removable partition.

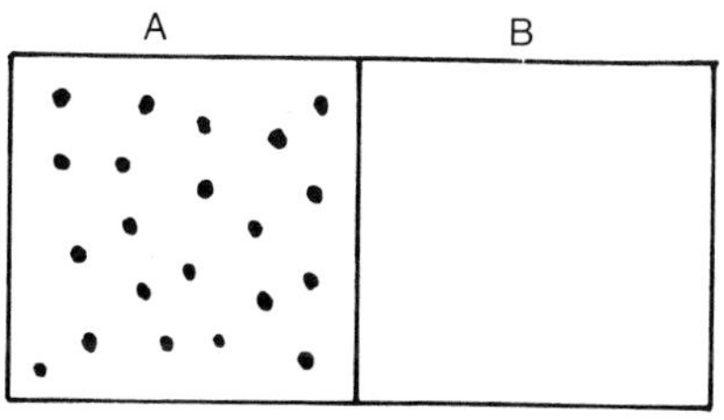

Suppose that a much larger number of molecules of a particular compound is placed on side A than on side B, and then the partition is removed. Each molecule is in random thermal motion. It is equally likely that a molecule that starts out on side A will move to side B in a given time period as it is that a molecule starting on side B will end up on side A. Since there are many more molecules present on side A, however, the total number of molecules moving from side A to side B will be greater than the number moving from side B to side A. As a result, the number of molecules on side A will decrease while the number of molecules on side B increases.

This process of net diffusion of molecules will continue until the number of molecules on side A equals that on side B. At this point the rate of diffusion of molecules from A to B will just equal that from B to A, so no net flux will occur. A state of dynamic *equilibrium* will thus result when concentrations on side A and side B are equal. Diffusion across cellular membranes always tends to equalize the concentrations on the two sides of the membrane, since net diffusion always occurs from the side of higher to the side of lower concentration.

The rules that govern diffusion processes were first enunciated in the 19th century by Adolf Fick, a German physician, physiologist, and physical scientist. He saw that the diffusion rate across a particular plane surface must be proportional to the area of that plane and to the difference in concentration of the diffusion substance on the two sides of the plane. Expressed mathematically, *Fick's first law of diffusion is*

$$J = -DA\frac{dc}{dx} \tag{1}$$

where J is the net diffusional flow in moles or grams per unit time, A is the area of the plane, dc/dx is the *concentration gradient* across the plane, and D is a constant of proportionality called the diffusion coefficient.

The nature of the concentration gradient, dc/dx, in Fick's first law requires further explanation. The concentration profile of the diffusing substance, i.e., the plot of concentration versus position, may take many forms. In Figure 15–1, two possible forms of the concentration profile are shown, and the positions of the planes across which the rates of diffusion are to be determined are represented by X_1, X_2, and so on. In panel A the concentration profile is a straight line. In this case the value of the concentration gradient is simply the slope of the line $\Delta c/\Delta x$, and the rate of flux is the same at all three x values, namely,

$$J = -DA\frac{dc}{dx} = -DA\frac{\Delta c}{\Delta x} \tag{2}$$

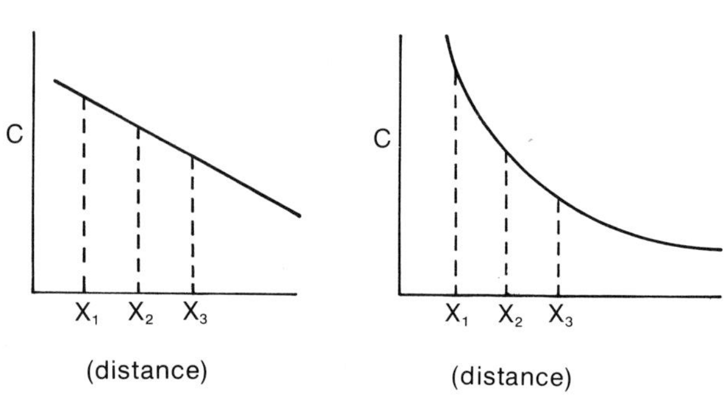

FIGURE 15–1 Examples of concentration profiles.

In panel B, however, the concentration profile is curvilinear. At each point, X_1, X_2, and X_3, the value of dc/dx is equal to the slope of the tangent to the curve at that point. Since the slopes of those three tangents are all different, the rates of diffusion across planes placed at X_1, X_2, and X_3 will be different.

Note also that the equation for Fick's first law contains a minus sign. The significance of the minus sign is that it indicates the *direction* of diffusion. In the diagram shown on page 388, the slope of the concentration profile is *negative*, but the direction of diffusion is in the *positive* X-direction. The minus sign at the beginning of Fick's first law times the minus concentration gradient gives a positive sign to the flux; this is appropriate since the flux is in the positive X-direction. Thus, the need for the minus sign in the equation arises because molecules flow down a concentration gradient, i.e., from higher to lower concentration.

The Diffusion Coefficient

The diffusion coefficient, D, has units of cm^2/sec. This can be seen by solving equation 1 for D and expressing all the variables on the right hand side of the resulting equation in c.g.s. units.

$$D = \frac{J}{A\dfrac{dc}{dx}} = \frac{\text{moles/sec}}{cm^2\dfrac{\text{moles/cm}^3}{cm}} = cm^2/sec$$

D can be thought of as a factor proportional to the ease with which the diffusing molecule can move in the medium in which it is diffusing. It will be smaller the larger the molecule and the more viscous the medium.

For large spherical solute molecules (much larger than the solvent molecules), Albert Einstein derived the following relation by analogy to Stokes' law for a spherical object falling through a viscous medium

(3) $$D = kT/(6\ \pi r \eta)$$

where k is Boltzmann's constant, T is the absolute temperature (kT is proportional to the average kinetic energy of a solute molecule), r is the molecular radius, and η is the viscosity of the medium. This is known as the Stokes-Einstein relation, and the molecular radius defined by this expression is called the Stokes-Einstein radius.

Note that for large molecules equation 3 predicts that D will be inversely related to the molecular radius of the diffusing species. Since the molecular weight (MW) of the molecule should be proportional to r^3, D should be inversely related to $(MW)^{1/3}$, so that a molecule A which is eight times lighter than molecule B will have a diffusion coefficient only twice as large as molecule B. For smaller solutes, with a molecular weight less than about 300, D is inversely proportional to $(MW)^{1/2}$ rather than to $(MW)^{1/3}$.

Diffusion is quite a rapid process when the distance over which it must take place is small. This can be seen from another relation determined by Einstein. Consider the random movements of molecules that are originally located at $X = 0$. Since a given molecule is equally likely to diffuse in the $+X$ or $-X$ direction, the *average* displacement of all the molecules that begin at $X = 0$ will be zero.

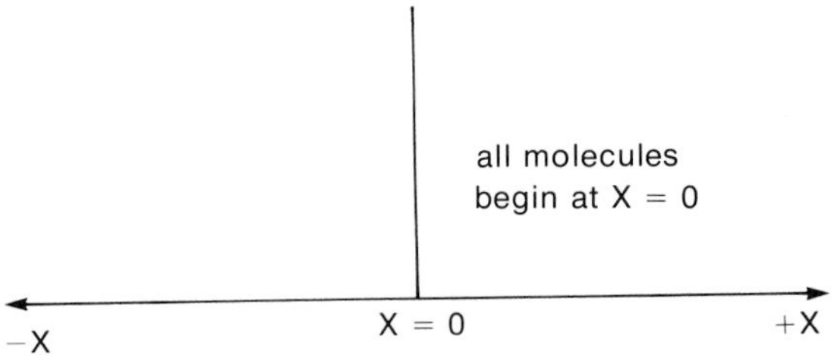

The average displacement squared $\overline{(\Delta X)^2}$, which of course is a positive quantity, is represented by

(4) $$\overline{(\Delta X)^2} = 2\,Dt$$

where *t* is the time elapsed since the molecules started diffusing. The Einstein relation (equation 4) tells us how far the average molecule will diffuse in time t, and it is a useful rule of thumb to estimate the time scale of a particular diffusion process.

A useful application of the Einstein relation is to compute how long it will take an average molecule, such as a small solute with $D = 1 \times 10^{-5} cm^2/sec$, to diffuse 100 μm (10^{-2}cm). Solving equation 4 for time, it can be seen that only 5 seconds are required.

$$t = \frac{\overline{(\Delta X)^2}}{2D} = \frac{(10^{-2}cm)^2}{2 \times 10^{-5}cm^2/sec} = 5\ sec$$

This is a relatively short period with respect to the time scale of metabolic events. Most cells in the body are within 100 μm of a blood capillary and thus are only 5 seconds away from a source of small nutrient molecules. By contrast, consider how long it would take a small molecule to diffuse 1 cm, a distance that is still small compared to the size of the human body. Using equation 4 it can be seen that the average molecule requires 14 hours to diffuse 1 cm. The important point of these calculations is that while diffusion can occur quite rapidly over quite small distances, it cannot serve as an effective means of biological transport over distances as large as a centimeter. It is not surprising then that as larger multicellular organisms evolved, various forms of circulatory systems were developed, which bring molecules within reasonable diffusion range of the cells that require them.

PERMEABILITY OF CELL MEMBRANE

Permeability of Lipid-Soluble Molecules

The plasma membrane serves as a barrier between the cytoplasm and the extracellular fluid, enabling the cell to maintain cytoplasmic concentrations of many substances that differ markedly from those in the extracellular fluid. As early as the turn of the century the relative impermeability of the plasma membrane to most water-soluble substances was attributed to its "lipoid" nature.

As noted in Chapter 14, the hypothesis that the plasma membrane has a "lipoid" character arose from experiments showing that compounds that are soluble in nonpolar solvents (such as ether, benzene, or olive oil) enter cells more readily than water-soluble substances. Figure 15–2 shows the relationship between membrane permeability and lipid solvent solubility for a number of substances. As a measure of the lipid solvent solubility, the ratio of the solubility of the solute in olive oil to its solubility in water is used. This ratio is called the olive oil/water partition coefficient. Note that the permeability of the plasma membrane increases with the "lipid solubility" of the substance. For compounds with the same oil/water partition coefficient there is decreasing permeability with increasing molecular weight. As described in Chapter 14, the fluid mosaic model of membrane structure depicts the plasma membrane as a lipid bilayer with proteins embedded in it. The data of Figure 15–2 support the idea that the lipid bilayer part of the membrane is the principal barrier to substances that permeate by simple diffusion.

The relation between lipid solvent solubility and membrane permeability also suggests that "lipid-soluble" molecules can enter or leave cells simply by dissolving in the plasma membrane and diffusing across it. Imagine a substance that dissolves in the lipid bilayer and then passively diffuses across the plasma membrane to the other side (Fig. 15–3). If the substance equilibrates with the lipid bilayer, its concentration at the outer face of the bilayer, $C_m(o)$, will be $\beta \cdot C_0$: β the partition coefficient (β) times the concentration in the extracellular medium. Its concentration at the inner face of the membrane, $C_m(i)$, will be $\beta \cdot C_i$: times the concentration in the cytoplasm. The concentration difference within the membrane is not $C_o - C$ but $C_m(o) - C_m(i)$. Since $C_m(o) - C_m(i) = \beta[C_o - C_i]$, the concentration gradient relevant to diffusion across the membrane is

$$\beta \frac{C_o - C_i}{\Delta X}$$

Thus, the more lipid-soluble the substance, the larger is β, and the larger the concentration gradient that causes it to diffuse across the membrane. According to Fick's first law, then, the flux of the substance across the membrane is

$$J = -DA\frac{dc}{dx} = -DA\beta\frac{C_o - C_i}{\Delta X} = -\frac{D\beta}{\Delta X}A[C_o - C_i]$$

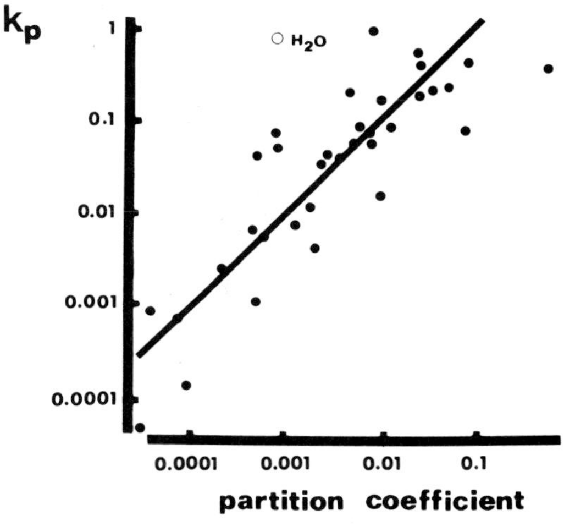

FIGURE 15–2 The permeability of the plasma membrane of the alga *Chara ceratophylla* as a function of the lipoid solubility of the permeating solute. The permeability coefficient (k_p) in moles/($cm^2 \cdot hr \cdot M$) is plotted versus the olive oil:water partition coefficient for a number of organic nonelectrolytes and water. (From data of Collander, R., and Barlund, H.: Acta Bot. Fenn., *11*:1, 1933.)

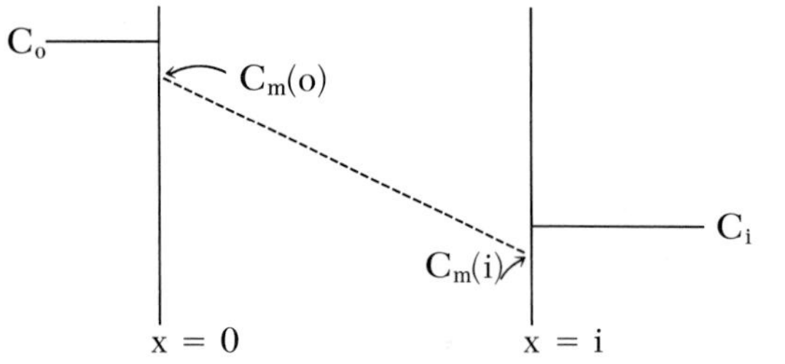

FIGURE 15–3 Concentration profile inside a cell membrane.

Since $D\beta/\Delta X$ is a constant for a particular substance and a particular membrane, it is often called the permeability coefficient (k_p), so that

$$J = -kp\ A\ [C_o - C_i]$$

The units of k_p are cm/sec since $k_p = \frac{D\beta}{\Delta X}$, β is dimensionless, and the units of D are cm^2/sec.

Permeability of Water-Soluble Molecules

Very small water-soluble substances pass cell membranes at a much faster rate than that predicted by their "lipid solubility." For example, Figure 15–2 shows that water permeates the cell membrane much more readily than larger molecules with a similar olive oil/water partition coefficient. The basis for this anomalously high permeability of water is not known.

Some investigators think that cell membranes are traversed by small polar or water-filled pores that allow very small water-soluble molecules to pass. Others believe that very small water-soluble molecules can pass between adjacent phospholipid molecules without actually *dissolving* in the phase made up by the fatty acid side chains. As the size of water-soluble molecules increases, their membrane permeability decreases. The cell membranes of most cells are essentially impermeable to the simple diffusion of water-soluble molecules whose molecular weight exceeds about 200.

Ions, because they are charged, are generally insoluble in lipid solvents and thus usually have low permeability to cell membranes. Most ionic flow that occurs across membranes is believed to occur through "channels" that span the membrane. Some channels are highly specific with respect to the ions that can pass, while others may permit the passage of all ions below a certain size limit. The flows of specific ions that cause the action potentials of nerve and muscle cells will be discussed in Chapters 17 through 19.

Certain other water-soluble molecules that cannot cross the plasma membrane by simple diffusion are essential to the survival of cells. Among these are sugars, amino acids, and nucleotides. Many cells have specific mechanisms in the plasma membrane to allow the passage of vital metabolites into (or out of) the cell by mediated transport. Such mechanisms are often termed *transport systems,* and certain of their characteristics will be discussed in the following section.

MEDIATED TRANSPORT

Properties of Mediated Transport

Some molecules enter or leave certain cells by way of specific carriers or channels in the plasma membrane. Transport via such carriers or channels is termed *mediated transport*. Mediated transport systems include both *active transport* systems and *facilitated transport* systems. The basic properties of mediated transport are the following:

1. The substance is transported more rapidly than other molecules of similar molecular weight and lipid solubility which simply diffuse across the membrane.

2. The rate of transport shows saturation kinetics. That is, as the concentration of the transported compounds is increased,

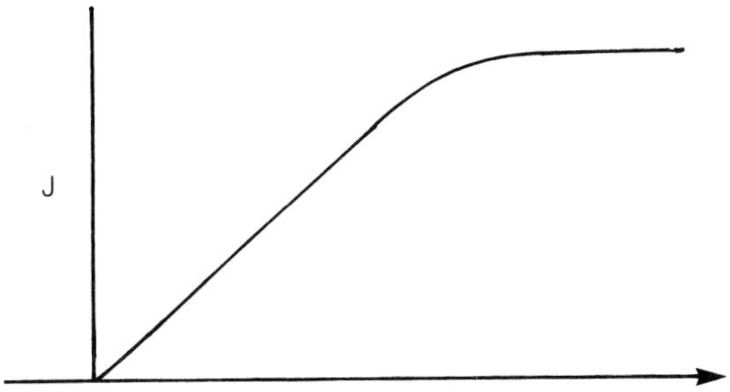

a point is reached after which the transport rate increases no further. At this point, the transport system is said to be saturated.

3. The carrier or channel has chemical specificity, so that only molecules with the requisite structure are transported. The specificity of most transport systems is not absolute, however, and in general it is broader than the specificity of most enzymes. The transport of glucose into red blood cells, for example, is a mediated process. Glucose is the preferred substrate, but mannose, galactose, xylose, and certain other sugars are also taken up by the same transport system. D-arabinose, on the other hand, is a poor substrate for the glucose carrier, and sorbitol and mannitol do not enter red cells at all. Mediated processes usually have some degree of stereospecificity as well. In the case of the red cell sugar transport system, D-glucose is transported, but L-glucose is transported only poorly.

4. Structurally similar compounds may compete for transport. In most cases, this means that the presence of one transported substance will decrease the rate of transport of a second substance by competing for the binding site on the carrier or channel. The competition is often regarded as analogous to enzyme inhibition. Inhibition in transport process is characterized as competitive or noncompetitive by the same criteria used to classify enzyme inhibition.

5. Transport may be inhibited by compounds apparently unrelated to transport substrates. This may occur with an inhibitor that binds to the carrier or channel in such a way as to decrease its affinity for transport substrates. For example, phloretin does not resemble a sugar molecule, yet it is a fairly potent inhibitor of sugar transport systems. In the case of active transport systems, which require a link to cellular metabolism, transport may be inhibited by sub-

stances that block various metabolic processes. For example, the rate of Na^+ extrusion from cells by the sodium pump is decreased by any substance that interferes with ATP generation.

Facilitated transport, sometimes called facilitated diffusion, is the transport via a carrier or channel that has no link to energy metabolism. Facilitated transport has all of the characteristics described previously except that it is not depressed by inhibitors of energy metabolism. Since they have no link to energy metabolism, facilitated diffusion processes cannot transport uncharged substances against concentration gradients or ions against electrochemical potential gradients. Facilitated transport systems that reside in the plasma membrane act to equalize the concentrations in the cytoplasm and extracellular fluid of the substances for which they are specific.

For example, monosaccharides enter muscle cells by a facilitated transport process. Several sugars, such as glucose, galactose, arabinose, and 3-O-methylglucose, compete for the same carrier. The rate of transport shows saturation kinetics. The nonphysiological stereoisomer L-glucose enters the cells very slowly, and nontransported sugars, such as mannitol or sorbose, enter muscle cells very slowly if at all. Phloretin inhibits uptake of sugars. This transport system is stimulated by insulin. Since in the absence of insulin sugar transport is rate-limiting for muscle sugar metabolism, insulin is an important regulator of muscle metabolism.

Active transport processes have all the properties of facilitated transport but can also concentrate their substrates against electrochemical potential gradients. This requires energy, and therefore active transport must be linked to energy metabolism in some way. Some active transport systems utilize ATP directly, while others are linked indirectly to metabolism. Because of their metabolic dependence, active transport processes are likely to be inhibited by any of the substances that interfere with energy metabolism.

A Model of Mediated Transport

Figure 15–4 depicts the operation of a hypothetical mobile carrier. The carrier (C) binds substrate (S) at the outer surface of the membrane to form a carrier-substrate complex (CS). CS can be translocated across the membrane to the inner surface of the membrane. There it dissociates, leaving S free to diffuse into the cytoplasm. Note that the same processes can serve to transport S from the cytoplasm to the extracellular fluid. The K_m values define the strength of binding of the substrate to the carrier at the two faces of the membrane, the K_m being

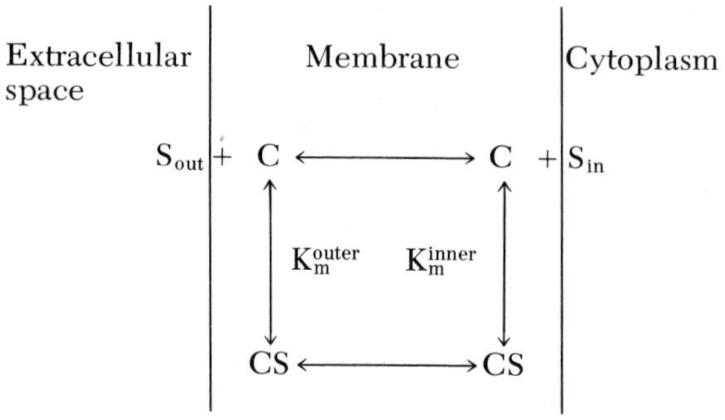

FIGURE 15–4 Model of mediated transport. C, carrier; S, substrate.

equal to the concentration of substrate required to occupy one half of the total number of carriers. If the K_m values for binding of substrate by the carrier are the same at the two faces of the membrane, and if the ease of translocation of C and CS are the same in the two directions within the membrane, the system will function as a facilitated transport system.

To see the relation between the K_m's and S inside and outside the cell in a different way, consider influx (entry of substrate into the cell) and efflux (exit of substrate from the cell) as separate processes. Since this is a carrier-mediated system, both influx and efflux show saturation kinetics and can be described in terms of the Michaelis-Menten formalism. That is,

(1) $$\textit{rate of influx} = \frac{V_{max}\, S_{out}}{K_m^{outer} + S_{out}}$$

(2) $$\textit{rate of efflux} = \frac{V_{max}\, S_{in}}{K_m^{inner} + S_{in}}$$

Assume for this example that V_{max} for influx and efflux are equal, although this is not necessarily true for all transport systems.

In a steady state, the cellular and extracellular concentrations are constant in time. Thus when a steady state is reached in this system, the rates of influx and efflux must be equal. Setting equations 1 and 2 equal to one another

$$\frac{V_{max}\, S_{out}}{K_m^{outer} + S_{out}} = \frac{V_{max}\, S_{in}}{K_m^{inner} + S_{in}}$$

Dividing both sides by V_{max}, cross-multiplying, and then subtracting $S_{out}\, S_{in}$ from both sides gives

$$S_{out} K_m^{inner} = S_{in} K_m^{outer}$$

so that

$$\frac{S_{in}}{S_{out}} = \frac{K_m^{inner}}{K_m^{outer}}$$

Note that if the K_m values for the binding of substrate by carrier are the same at both faces of the membrane, then $S_{in}/S_{out} = 1$. This verifies the contention that a completely symmetrical model corresponds to facilitated transport and serves to equalize the concentation of substrate on the two sides of the membrane.

If, on the other hand, K_m^{inner} is twice as large as K_m^{outer}, then $S_{in}/S_{out} = 2$, and S_{in} is twice as large as S_{out}. In this case the transport system serves to concentrate the substrate in the cell, and the model corresponds to an active transport system.

This is perhaps the simplest way in which an active transport system might work. Note that energy is needed to convert the carrier cyclically, back and forth, between its two forms. In the absence of an energy input the distribution of the carrier between the two will remain stable in time, with the proportions of carrier in the two forms dependent on the potential energy of the two forms of the carrier.

In the examples of active transport systems described further on, two ways in which a carrier can be converted to a form with different affinity for substrate are illustrated. One way is by phosphorylation of the carrier protein by ATP, catalyzed by a protein kinase. This is a rather direct link to metabolism since it depends directly on ATP.

A second means is by alteration in the affinity of the carrier depending on the concentration of another substance that is itself actively transported by a different carrier. If there are different concentrations of the effector substance at the inner and outer faces of the membrane, different carrier forms with different substrate-binding affinities will predominate at the two faces. This type of linkage to metabolism is less direct, depending only on the concentration difference of another substance that is actively transported. Thus this type of active transport is sometimes termed secondary active transport.

Examples of Active Transport Processes

The sodium-potassium pump, which is present in the plasma membrane of most mammalian cells, is linked quite directly to a supply of metabolic energy, since ATP itself is involved in conversion of the carrier from one affinity state to the other. As can be seen in Figure 15–5, when the carrier (E) is phosphorylated by ATP at the inner surface of the membrane, it is converted from a form that prefers K^+ to a form that prefers to bind Na^+. At the outer surface of the membrane, the binding of K^+ promotes hydrolysis of E~P by a mechanism that involves a phosphoprotein phosphatase activity. Removal of the P converts the carrier back to E, which preferentially binds K^+.

Since the sodium-potassium pump splits ATP, the pump is also called the sodium-potassium ATPase. The sequence of changes in the pump results in transport of K^+ into the cell and Na^+ out of the cell. The rate of splitting of ATP and of pump activity is stimulated by increasing intracellular (Na^+) and extracellular (K^+). In human erythrocytes the sodium-potassium pump extrudes 3 Na^+ ions for every 2 K^+ taken up. It is not known, however, whether this 3/2 coupling ratio applies to all other cells and under all conditions.

The neutral amino acid transport system of the small intestine is an example of an active transport system that is believed to use the energy present in the gradient of another actively transported species. Because of the action of the sodium-potassium pump in the epithelial cells of the small intestine, Na^+ is present in the cells at much lower concentration than in the contents of the intestinal lumen. It is believed that some or all of the energy required for the active uptake

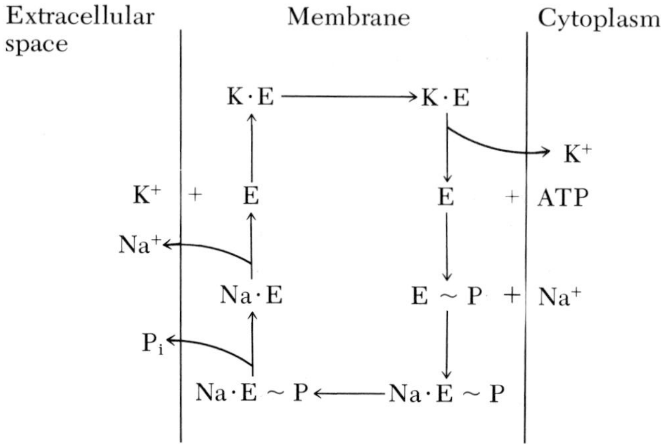

FIGURE 15–5 Simplified model of the Na^+-K^+ pump. E, Na^+,K^+-ATPase; E~P, phophorylated enzyme.

of the neutral amino acids is derived from the electrochemical potential of the Na^+ gradient in the manner described in the following section.

For the simple carrier system diagrammed in Figure 15–4, active transport will occur across the plasma membrane if the affinity of the carrier for the transported substance is increased at the outer surface or decreased at the inner surface of the membrane. It appears that in the case of the neutral amino acid transport system the high [Na^+] at the outer surface has the effect of causing the carrier to bind amino acid with greater affinity. In accord with this, the apparent K_m for alanine uptake is lower in the presence of high external Na^+ than in the absence of Na^+ (Fig. 15–6), while the apparent V_{max} of the transport process does not depend on the external Na^+ concentration.

The transport of Na^+ into the intestinal epithelial cell is also enhanced in the presence of neutral amino acids in the lumen, so it is believed that Na^+ and amino acid are transported into the cell on the same carrier. Subsequently amino acids are transported across the serosal surface of the epithelial cell presumably by a facilitated transport mechanism, while Na^+ is extruded from the cell by the Na^+ pump.

The model shown in Figure 15–7 has been proposed to account for the active transport of alanine into intestinal epithelial cells. It is proposed that the binding of Na^+ to the carrier (X), either before or after binding of the amino acid (A), results in an increase in the affinity of the carrier for the amino acid. When the complex XANa is translocated to the intracellular face of the membrane, the [Na^+] it encounters is quite low. Therefore, Na^+ tends to dissociate from the carrier, which converts the carrier back to the form with a lower affinity for the amino acid. This promotes the dissociation of A from the carrier and thus the entry of amino acid into the cell.

Since the active transport of alanine depends on the presence of the Na^+ gradient, but does not depend directly on ATP or some other high-energy metabolic intermediate, the active transport of alanine is termed secondary active transport.

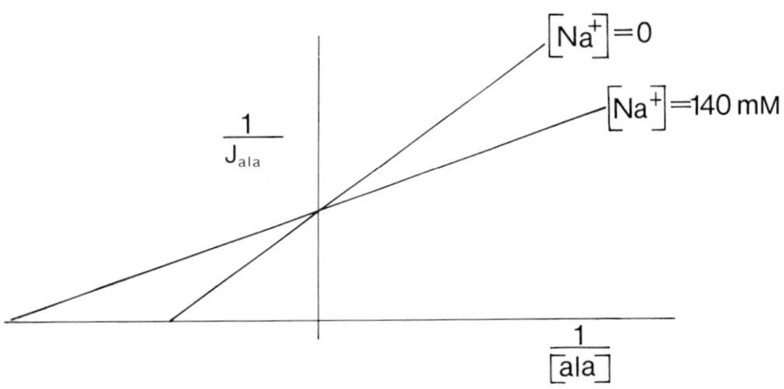

FIGURE 15–6 Dependence of alanine entry into the intestinal epithelial cell on [Na^+] in the gut lumen.

Diseases Involving Transport Defects

Diabetes mellitus, diabetes insipidus, and hereditary spherocytosis have been mentioned as examples of diseases in which defects in transport processes are involved. Some additional examples of disease states that involve malfunctioning transport systems are listed below.

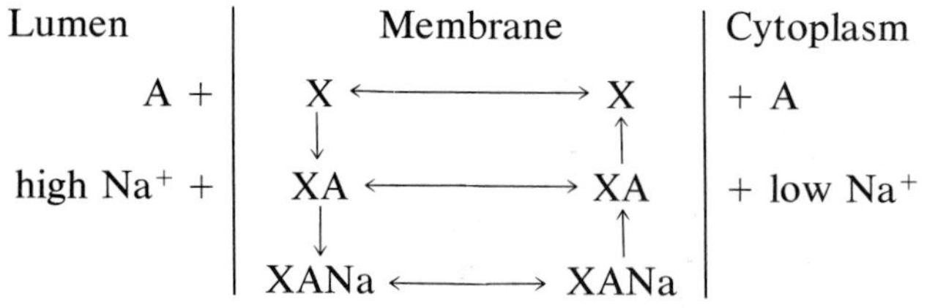

FIGURE 15–7 Simplified model of alanine entry into intestinal epithelial cells.

The disease commonly known as *rickets* results from dietary deficiency of vitamin D. In this disease there is defective absorption of Ca^{++} from the intestine, and this results in defective growth of the skeleton. It has been shown that vitamin D induces the intestinal epithelial cells to produce a calcium-binding protein that is necessary for Ca^{++} absorption. In the absence of vitamin D, the levels of the Ca^{++} binding protein fall and Ca^{++} absorption diminishes.

Pernicious anemia is due to congenital lack of a gastric intrinsic factor (IF) that is required for the intestinal absorption of vitamin B_{12}. IF binds vitamin B_{12}, and the complex is recognized by receptors on the ileal mucosa. Subsequently B_{12} is taken up by the epithelial cells and transferred to the blood. In the absence of IF only a small fraction of an oral dose of B_{12} is absorbed. Since vitamin B_{12} is required for erythrocyte maturation, anemia results from B_{12} deficiency.

Gout involves excessive reabsorption by the renal tubular epithelium of uric acid from the urine. This leads to higher blood levels of uric acid and ultimately to the deposition of uric acid crystals in the synovial membranes that line the joint capsules, causing swelling and great pain. Gout is often treated with drugs, such as probenecid, which decrease renal reabsorption of uric acid.

Cholera is caused by toxins produced by certain species of bacteria. Cholera toxin causes large amounts of salt and water to be lost from the intestinal mucosa. Recently it has been shown that the absorption of salts and water from the intestine to the blood is normal in cholera but that there is a large *secretion* of salts and water from the blood to the intestinal lumen. The mechanism that causes the increased secretion of salts and water is not known. However, since cholera toxin causes increased levels of cyclic AMP in intestinal epithelial cells, it is suspected that cyclic AMP is somehow involved in causing the increased secretion.

Cystinuria is a defect in the renal reabsorption of cystine and certain other amino acids. This is an hereditary disorder, and certain forms of the disease involve defective intestinal absorption of various amino acids.

There are many other diseases that involve transport defects. The point here is not to make a comprehensive list but to assert that the understanding of transport mechanisms has led to a deeper knowledge of certain disease states. This has in turn permitted more rational therapy for these diseases. For example, the current therapy for cholera involves replacement of the fluid that the patient loses in the stool. It is likely that when the mechanism of the increased salt and water secretion by the intestine in cholera is understood, it will become possible to prevent the occurrence of the disease or to greatly diminish the severity of its symptoms.

FURTHER READING

Christensen, H. N.: Biological Transport. 2nd ed. Reading, Mass., W. A. Benjamin, 1975.

Curran, P. F., Schultz, S. G., Chez, R. A., et al.: Kinetic relations of the Na-amino acid interaction at the mucosal border of the intestine. J. Gen. Physiol., *50*:1261, 1967.

Davson, H. A: Textbook of General Physiology. 4th ed. Baltimore, Williams & Wilkins, 1970, pp. 395–507.
Davson, H., and Danielli, J. F.: The Permeability of Natural Membranes. 2nd ed. New York, Cambridge University Press, 1952.
Dowben, R. M.: General Physiology. New York, Harper & Row, 1969, pp. 427–473.
Einstein, A.: Investigations on the Theory of the Brownian Movement. New York, Dover, 1956.
Glynn, I. M., and Karlish, S. J. D.: The sodium pump. Annu. Rev. Physiol., *37*:13, 1975.
Goldin, S. M.: Active transport of sodium and potassium ions by the sodium and potassium ion-dependent adenosine triphosphatase from renal medulla. Reconstitution of the purified enzyme into a well-defined *in vitro* transport system. J. Biol. Chem., *252*:5630, 1977.
Harris, E. J.: Transport and Accumulation in Biological Systems. 3rd ed. London, Butterworth, 1972, pp. 1–92.
Hill, A. V.: The diffusion of oxygen through tissues. *In*: Trails and Trials in Physiology. London, Edward Arnold, 1965, pp. 208–241.
Jacobs, M. H.: Diffusion processes. Ergebnisse der Biologie. Vol. 12, 1935, pp. 1–160.
Kotyk, A., and Janacek, K.: Cell Membrane Transport: Principles and Techniques. New York, Plenum Press, 1970.
Stein, W. D.: The Movement of Molecules Across Cell Membranes. New York, Academic Press, 1967, pp. 36–241.
Van Holde, K. E.: Physical Biochemistry. Englewood Cliffs, N.J., Prentice-Hall, 1971, pp. 47–49, 85–90.
Whittam, R.: Transport and Diffusion in Red Blood Cells. London, Edward Arnold, 1964.

16

OSMOSIS AND IONIC EQUILIBRIA

OSMOSIS

Osmosis is the flow of water across a semipermeable membrane from a compartment in which the *solute* concentration is lower to one in which the solute concentration is greater. A *semipermeable membrane* is one permeable to water but impermeable to solutes. Osmosis occurs because the presence of solute causes a reduction of the *chemical potential* of water, and water will flow from where its chemical potential is higher to where it is lower if it is allowed to do so. (Chemical potential will be discussed further on in this chapter.) Other effects caused by the reduction of the chemical potential of water by solute include reduced vapor pressure, lower freezing point, and higher boiling point of the solution as compared with pure water. Because these properties and osmotic pressure all depend on the *concentration* of the solute present rather than on its chemical nature, they are called *colligative properties*.

Osmotic Pressure

Consider the situation diagrammed in Figure 16–1. The membrane separates a solution from pure water. Water flow from side A to side B by osmosis tends to occur because the presence of solute on side B lowers the chemical potential of the water in the solution. Pushing on the piston will increase the chemical potential of the water in the solution and slow down the net rate of osmosis. If the force on the piston is increased, eventually a point is reached at which net water flow ceases. Application of still more pressure will cause water to flow in the opposite direction, from the solution to pure water. The pressure on side B that is just sufficient to prevent water entry by osmosis is called the *osmotic pressure* of the solution on side B.

Since the osmotic pressure of a solution depends on the number of particles in solution, the degree of ionization of the solute must be taken into account. Thus a 1M solution of glucose, a 0.5M solution of NaCl, and a 0.333 . . . M solution of $CaCl_2$ have theoretically the same osmotic pressure (in actuality they will have somewhat different osmotic pressures owing to the deviations of actual

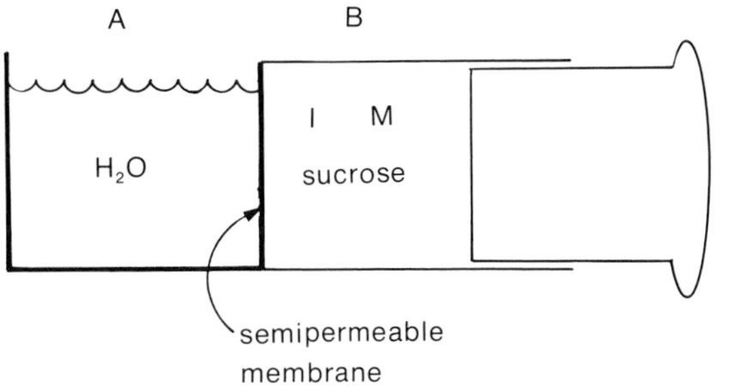

FIGURE 16–1 Schematic illustration of osmosis.

solutions from ideal solution theory). Many of the laws that pertain to osmotic pressure and related colligative properties were formulated by the Dutch chemist van't Hoff in the 19th century. The most frequently used form of *van't Hoff's law* for calculation of osmotic pressure is

(1) $$\pi = iRTm$$

where π is osmotic pressure, i is the number of ions formed by dissociation of a solute molecule, R is the ideal gas constant, T is the absolute temperature, and m is the molal concentration of solute (moles of solute per kg water). This relation applies more and more exactly as the solution becomes more and more dilute.

In the biological sciences, molar concentrations are more frequently used than molal concentrations, and van't Hoff's Law can be expressed as

(2) $$\pi = iRTc$$

where c is the molar concentration of solute (moles of solute per liter of solution). The discrepancy between theory and reality is somewhat greater for this form of van't Hoff's law than for equation 1.

Equations 1 and 2 do not apply exactly to real solutions, and at the concentrations of many substances in body fluids the deviations may be substantial. For example, sodium is the principal cation of the extracellular fluids and chloride is the main anion, Na^+ being present at about 150 mEq/L and Cl^- at about 120 mEq/L. Table 16–1 shows that NaCl solutions in this concentration range differ considerably in their osmotic pressures (and in all the other colligative properties as

TABLE 16–1

OSMOTIC PRESSURES OF NaCl SOLUTIONS OF VARIOUS CONCENTRATIONS*

[NaCl] mole/l	Theoretical Osmotic pressure at 0°C (Atmospheres)	Actual Osmotic Pressure at 0°C (Atmospheres)	Actual / Theoretical
0.010	0.448	0.448	1.000
0.020	0.896	0.874	0.975
0.050	2.24	2.13	0.950
0.100	4.48	4.19	0.935
0.200	8.96	8.29	0.925
0.500	22.4	20.59	0.919
1.00	44.8	41.62	0.929

**From:* Weast, R. C. (ed.): Handbook of Chemistry and Physics. 53rd ed. Cleveland, The Chemical Rubber Co., 1972. Reproduced by permission of the publisher.

well) from the predictions of ideal solution theory for dilute solutions as embodied in van't Hoff's law.

One way of correcting for the deviations of the real world from the theoretical predictions of equation 2 is to use a correction factor called the *osmotic coefficient* (ϕ). Including the osmotic coefficient, equation 2 becomes

$$\pi = \phi iRTc \tag{3}$$

The osmotic coefficient may be greater or less than 1 (it is less than 1 for electrolytes of physiological importance), and it approaches 1 as the solution becomes more and more dilute. Since ϕiC can be thought of as the osmotically effective concentration, ϕiC is often referred to as the *osmolar concentration,* with units in osmoles/liter. Values of the osmotic coefficient depend on the nature of the solute and on its concentration. Table 16–2 lists, for a number of solutes, osmotic coefficients that may be expected to apply fairly well in the ranges of concentrations in which these solutes are present in the body fluids of mammals. Solutions of proteins deviate markedly from van't Hoff's law, and different proteins deviate to different extents. The deviation from ideality may be even more concentration-dependent for proteins than for smaller solutes.

The following are some examples of calculations involving osmotic pressure.

What is the osmotic pressure (at 0°C) of a 150 mM NaCl solution?

$$\pi = \phi iRTc$$

Taking ϕ for NaCl from Table 16–2, we obtain

$$\pi = 0.93 \times 2 \times 22.4 \text{ l-atm/mole} \times 0.150 \text{ mole/l} = 6.24 \text{ atm}$$

TABLE 16–2 OSMOTIC COEFFICIENTS (ϕ) OF VARIOUS SOLUTES OF PHYSIOLOGICAL INTEREST*

Substance	i	Molecular Weight	ϕ
NaCl	2	58.5	0.93
KCl	2	74.6	0.92
HCl	2	36.6	0.95
NH_4Cl	2	53.5	0.92
$NaHCO_3$	2	84.0	0.96
$NaNO_3$	2	85.0	0.90
KSCN	2	97.2	0.91
KH_2PO_4	2	136.0	0.87
$CaCl_2$	3	111.0	0.86
$MgCl_2$	3	95.2	0.89
Na_2SO_4	3	142.0	0.74
K_2SO_4	3	174.0	0.74
$MgSO_4$	2	120.0	0.58
Glucose	1	180.0	1.01
Sucrose	1	342.0	1.02
Maltose	1	342.0	1.01
Lactose	1	342.0	1.01

*From: Glasser, O. (ed.): Medical Physics. Chicago, Year Book Publishers, 1944. Reproduced by permission of the publisher.

What is the osmolarity of this solution?

$$\text{Osmolarity} = \phi ic = 0.93 \times 2 \times 0.150 = 0.279 \text{ osmole/l} = 279 \text{ mosmolar}$$

Measurement of Osmotic Pressure

Measurement of osmotic pressure by determining the pressure that one must exert to prevent water from entering a solution by osmosis across a semipermeable membrane is time-consuming and technically difficult. Therefore, the osmotic pressure often is estimated from another colligative property, depression of the freezing point. The relation that describes the freezing point depression effect is

(4) $$\Delta T_f = 1.86\ \phi ic$$

where ΔT_f is the freezing point depression in °C. Thus, the effective osmotic concentration (in osmoles/liter) is

(5) $$\phi ic = \Delta T_f/1.86$$

When the freezing-point depression of a multicomponent solution is determined, the effective osmolar concentration (in osmoles/liter) of the solution as a whole is obtained; this can then be used to calculate the total osmotic pressure. Because of the substantial deviations of real solutions from van't Hoff's law, as expressed in equations 1 and 2, any exact calculations of the osmotic pressure from the concentrations of the solutes present should use the osmotic coefficients (ϕ) discussed previously.

If the total osmotic pressure of solution A (as calculated or measured by freezing point depression or by the osmotic pressure developed versus pure water across a true semipermeable membrane) is the same as that of solution B, then solutions A and B are said to be *isosmotic* (or iso-osmotic). If solution A has greater osmotic pressure than solution B, A is said to be *hyperosmotic* with respect to B. If solution A has less total osmotic pressure than solution B, A is said to be *hypo-osmotic* to B.

Osmotic Swelling and Shrinking of Cells

The plasma membranes of body cells are effectively impermeable to many of the solutes of the interstitial fluid but are highly permeable to water. Therefore, when the osmotic pressure of the interstitial fluid is increased, water leaves the cells by osmosis and the cells shrink until the effective osmotic pressure of the cytoplasm is equal to that of the interstitial fluid. Conversely, if the osmotic pressure of the extracellular fluid decreases, water enters the cells, causing them to swell.

Red blood cells are often used to illustrate the osmotic properties of cells because they are readily obtained and easily studied. Within a certain range of external solute concentrations the red cell behaves as a "perfect osmometer," that is, its volume is directly related to the solute concentration in the extracellular medium. In Figure 16–2 the red cell volume, as a fraction of its normal volume

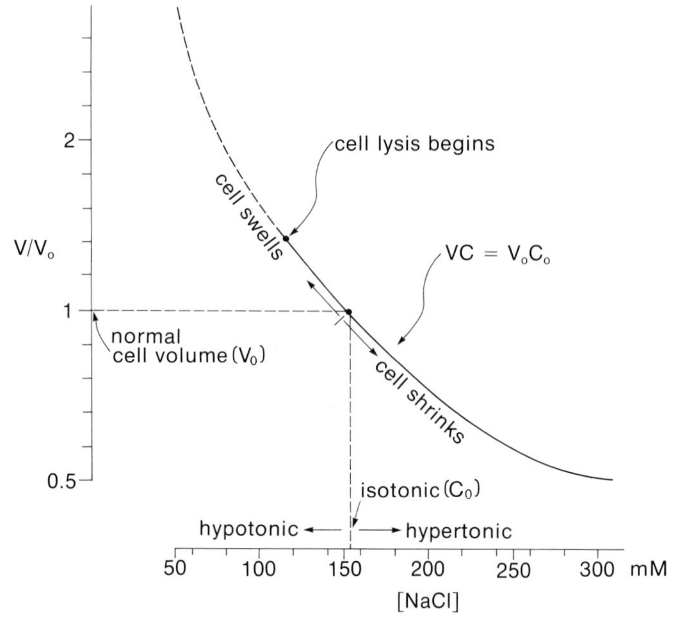

FIGURE 16–2 The red cell as a "perfect osmometer." The red cell volume (V) as a fraction of its volume in normal plasma (Vo) is plotted versus the concentration (mM) of a solution of NaCl in which the cells are suspended.

in plasma, is plotted versus the concentration of NaCl solution in which the red cells are suspended.

Note that at an NaCl concentration of 154 mM (308 mM particles) the volume of the cells is the same as their volume in plasma, that is, in this concentration of NaCl the cells neither shrink nor swell. For this reason 154 mM NaCl is called *isotonic* to the red cell. A concentration greater than 154 mM is called *hypertonic* (cells shrink) and a solution less concentrated than 154 mM is termed *hypotonic* (cells swell). When the red cell has swollen to about 1.4 times its original volume, the cell begins to lyse. At this volume the properties of the red cell membrane abruptly change, and hemoglobin leaks out of the cell. Other large molecules also become permeable at this point.

The moieties inside the red blood cell that produce an osmotic pressure that just balances that of 154 mM NaCl include hemoglobin, Na^+, organic phosphates such as ATP, 2,3-diphosphoglycerate, and other glycolytic intermediates. Regardless of the chemical nature of these moieties, the red cell behaves as if it were filled with a solution of 308 mM *impermeant* molecules.

The behavior of the cell becomes more complicated, however, if it is exposed to *permeant* molecules as well as impermeants, as shown in the following examples. NaCl will be considered impermeant in these illustrations.

Example 1. Red cells are suspended in a very large volume of 154 mM NaCl plus 30.8 mM glycerol. Glycerol slowly permeates the red cell membrane. Initially the concentration of osmotically active particles outside the cells is 308 mM from NaCl plus 30.8 mM glycerol. The total is 10 per cent greater than 308 mM, so the cell volume will diminish by 10 per cent. But then the glycerol will begin to diffuse into the red cell. As it does so, osmotically active molecules are transferred from outside to inside the red cell, and the cell will increase in volume. At equilibrium the concentration of glycerol will be 30.8 mM both within the cell and without, so that the original concentration difference that caused the shrinking is completely dissipated. At this time the red cell will have returned to its original volume. These changes in cell volume can be illustrated in graphical terms:

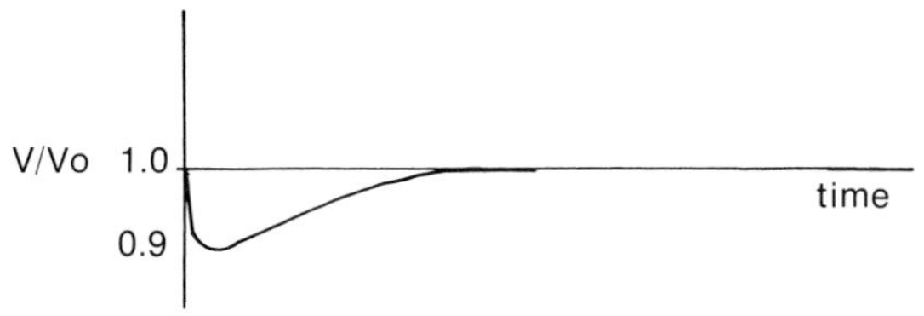

If the same experiment is performed with 30.8 mM urea, which permeates the red cell membrane very rapidly, in place of the glycerol, the final result will be the same but the swelling phase will occur much more rapidly owing to the rapid entry of urea.

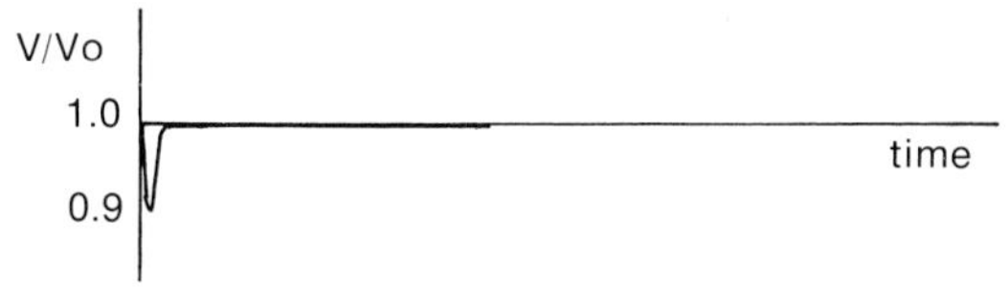

This example shows that the presence of permeant molecules in the extracellular medium has only a *transient* effect on the cell volume. The greater the permeability of the permeant, the more rapidly is the steady-state volume reached. However, the *steady-state volume* is determined only by the concentration of impermeant molecules in the extracellular fluid (Fig. 16–2).

This information can be expressed in the form of rules for predicting the volume changes that occur when the extracellular fluid composition is modified.

1. The steady-state volume of the cell is determined by the concentration of impermeant particles in the extracellular fluid.

2. Permeant particles cause only transient changes in cell volume.

3. The time course of the transient changes will be more rapid the greater the permeability of the permeant molecule.

In the examples that follow, these rules are applied in different situations. In each case red blood cells are placed in an infinite volume of a particular solution, and changes in the red cell volume are studied.

Example 2. Red cells are suspended in a solution of 129 mM NaCl and 50 mM urea. Urea is a permeant; NaCl is considered impermeant. The concentration of impermeant particles is 129 × 2 = 258 mM, which is less than the 308 mM impermeant particles inside the cell. Therefore, the cell will swell, and the steady-state volume will be (308/258) V_o (*Rule 1*).

Initially, however, the total concentration of particles in the extracellular medium is 258 mM for Na^+ + Cl^- plus 50 mM urea = 308 mM. Thus at first there is no tendency for water to move in either direction. Urea permeates rapidly, however, and as it does so it generates an osmotic force that moves water into the red cell (*Rule 2*). Since urea is very permeable, the cell attains its steady-state volume rapidly (*Rule 3*). If the permeating solute were glycerol, which permeates much more slowly than urea, then the time course over which the volume changes would be slower, as shown by the dotted line.

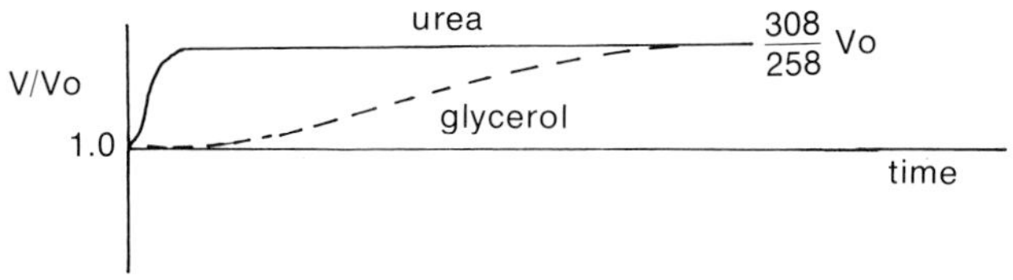

Since the total concentration of particles in the external solution is 308 mM (NaCl + urea), its freezing point depression or its total osmotic pressure across a membrane impermeable to all solutes would be the same as that of the red cell interior (ignoring the osmotic coefficients ϕ). For this reason it is said that the external solution is *isosmotic* to the red cell. However, since the steady-state volume of the red cell in that solution is larger than V_0, the solution is *hypotonic* to the red cell.

Note that the first term refers to the total concentration of solute particles in the external solution relative to the interior of the unswollen red cell. The second term refers to the steady-state volume of the red cell in that solution.

Example 3. Red cells are suspended in a solution of 154 mM NaCl and 75 mM glycerol. The steady-state volume of the red cell will be determined only by the concentration of impermeants (NaCl), which is 154 × 2 = 308 mM. Thus the steady-state volume will be the same as the initial volume, V_0 (*Rule 1*). At first, however, the cell will shrink because the total concentration of osmotically active particles in the external solution is 308 mM (Na^+ + Cl^-) + 75 mM (glycerol) = 383 mM (*Rule 2*). Since glycerol is a slow permeant, the transient shrinkage will have a long time course (*Rule 3*).

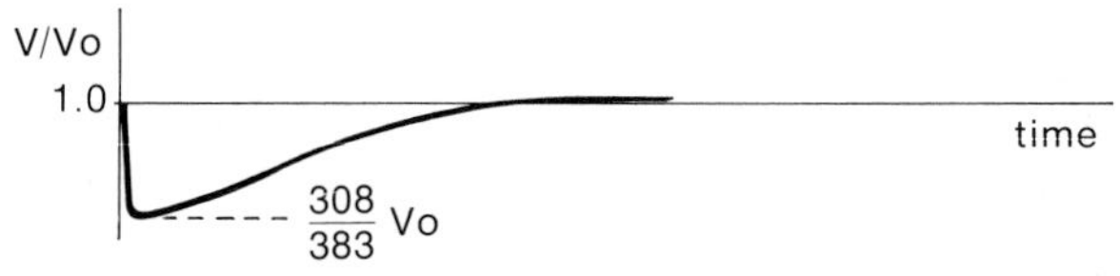

The solution is *hyperosmotic* and the cell originally shrinks, but it is *isotonic* since the steady-state volume equals V_0.

Example 4. The cells are suspended in 308 mM urea. The concentration of impermeants in the extracellular medium is zero. Thus the cell will swell until it lyses (*Rule 1*). Initially, the concentration of particles outside the cell is 308 mM, just as it is inside the cell, and the red cell will neither swell nor shrink (*Rule 2*). Since urea enters the cell quite rapidly, it will do so generating an osmotic force for water to enter and the cell will rapidly lyse (*Rule 3*).

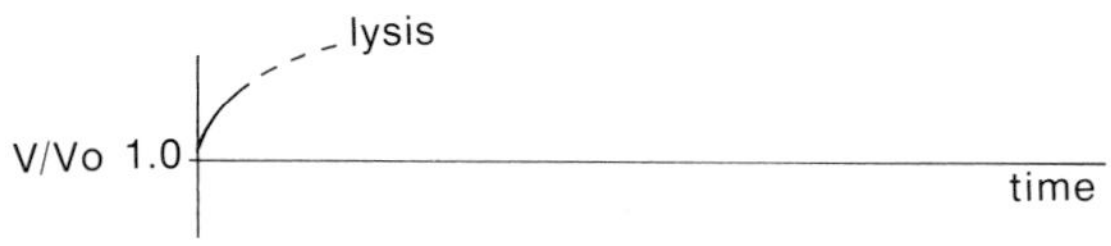

The solution is isosmotic and hypotonic.

Example 5. The cells are placed in 350 mM glycerol. Again the concentration of impermeants is zero, so the cell must eventually lyse (*Rule 1*). At first, however, there are 350 mM particles outside, compared with 308 mM particles inside, the red cell, so the cell will initially shrink, and the smallest volume it can achieve is $V_0 \times 308/350$ (*Rule 2*). Since glycerol is a slow permeant the cell will slowly swell as glycerol enters until it lyses (*Rule 3*).

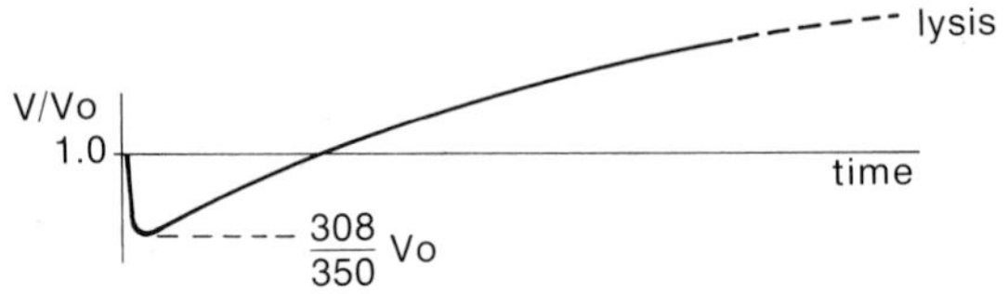

This solution is *hyperosmotic* (the total concentration of particles exceeds that of the normal red cell interior) but *hypotonic*, since the cell swells and ultimately lyses.

As these illustrations show, only *impermeable* solutes can have a long-lasting osmotic effect across a membrane. This is of practical importance in understanding how best to restore the blood volume of a patient who has hemorrhaged. Isotonic NaCl is the cheapest infusion and may be most readily available, but since NaCl quite readily crosses the capillary wall, the solution will distribute in all the extracellular space of the body. Since the plasma volume is about 8 per cent of the body water, while the interstitial water is about 25 per cent of the body water, only 32 per cent of the isotonic NaCl administered to the patient will remain in the vasculature.

Blood plasma is more efficient at restoring lost blood volume because it contains the plasma proteins, which are impermeable to the capillary wall. As a result, much of the fluid volume of plasma that is administered remains in the patient's blood vessels for a long time. In some cases, use is made of solutions of synthetic macromolecular "volume expanders," such as the large polysaccharide dextran, which are large enough to make the capillary wall impermeable to them. Such macromolecules can substitute for plasma proteins in expanding a patient's blood volume.

The Magnitude of the Osmotic Effects of Permeating Molecules

In the examples given of red cells placed in different solutions, it was demonstrated that permeants such as glycerol exert a transient osmotic effect. In some situations it is useful to determine the size of the osmotic effect that a permeant has.

The flow of water across a membrane is directly proportional to the osmotic pressure difference of the solutions on the two sides

(6) $$J = K_f \Delta\pi$$

where J is water flow, $\Delta\pi$ is the osmotic pressure difference, and K_f is a proportionality constant called the filtration coefficient. Equation 6 holds for osmosis caused by impermeants, but with permeants less osmotic flow is achieved. The greater the permeability of the solute, the less the osmotic flow. Table 16–3 shows the osmotic water flow induced across a porous membrane by solutions of solutes of different molecular size; the solutions have identical freezing points, so their total osmotic pressures are the same. Note that the larger the solute molecule, and thus the more impermeable it is across the membrane, the greater the osmotic water flow.

TABLE 16–3

OSMOTIC WATER FLOW ACROSS A POROUS MEMBRANE (DIALYSIS MEMBRANE) INDUCED BY VARIOUS SOLUTES*†

Water Flow Produced by a Gradient of	Net Volume Flow (μl M^{-1} min^{-1})	Solute Radius (Å)	Reflection Coefficient (σ)
D_2O	0.06	1.9	0.0024
Urea	0.6	2.7	0.024
Glucose	5.1	4.4	0.205
Sucrose	9.2	5.3	0.368
Raffinose	11	6.1	0.440
Insulin	19	12	0.760
Bovine serum albumin	25.5	37	1.02
Hydrostatic pressure	25		

*Flow is expressed as μl/min caused by a 1-molar concentration difference of solute across the membrane. The flows are compared to the flow caused by a theoretically equivalent hydrostatic pressure.

†Data of Durbin, R. P.: J. Gen. Physiol., *44*:315, 1960.

Equation 6 can be rewritten to take this into account by including σ, the reflection coefficient.

$$J = \sigma K_f \Delta\pi \tag{7}$$

σ is a dimensionless quantity that ranges from one for completely impermeable solutes down to zero for very permeable solutes. σ is a property of a particular solute and a particular membrane, and it is the fraction of the theoretical maximum osmotic flow the solute could induce that the solute actually causes (see Table 16–3). In kidney physiology the reflection coefficients of the renal tubular epithelium for certain solutes are used to determine the effect of these solutes on water movement across the renal tubule.

IONIC EQUILIBRIA

Electrochemical Potentials of Ions: The Nernst Equation

Consider a membrane separating two aqueous solutions containing Cl^- ions,

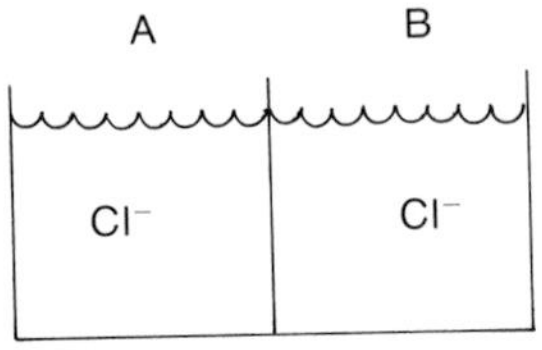

and assume that Cl^- is at higher concentration on side A than on side B. If there is no electrical potential difference between side A and side B, Cl^- will tend to diffuse from side A to side B, just as if it were an uncharged molecule. If, however, side A is electrically positive with respect to side B, the situation is more complex.

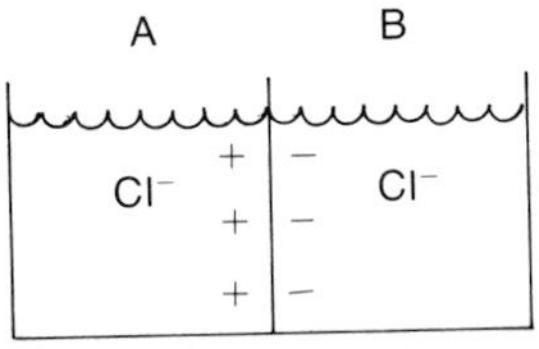

The tendency remains for Cl^- to diffuse from side A to side B because of the concentration difference, but now there is also a tendency for Cl^- to move in the reverse direction from B to A as a result of the electrical potential difference between the two sides, since opposite signs attract. The direction in which Cl^- will move depends on whether the concentration tendency or the electrical potential difference tendency is stronger. By comparing the two tendencies, concentration and electrical, it is possible to predict which way the ion will tend to flow, provided the concentrations of Cl^- on sides A and B and the magnitude of the electrical potential difference are known.

The electrochemical potential difference of an ion is a quantity that includes its tendency to move as a result of a concentration difference and an electrical potential difference. The electrochemical potential difference ($\Delta\mu$) of an ion between side A and side B separated by a membrane is

(1) $$\Delta\mu = RT\ 1n\left(\frac{C_a}{C_b}\right) + zF\,(E_a - E_b)$$

where *C* is concentration, *E* is electrical potential, *F* is Faraday's number (96,500 coulombs/mole), and *z* is the charge number of the ion (e.g., +1 for Na^+, 2 for Ca^{++}, −2 for SO_4^{--}, and so on).

The first term on the right-hand side is the tendency for the ion to move from A to B because of a concentration difference, and the second term quantifies the tendency of the ion to move from A to B owing to an electrical potential difference. The units of RT are energy/mole, and the RT 1n (C_a/C_b) term represents the chemical potential energy difference between a mole of ions on side A as compared to a mole of ions on side B. The second term also has units of energy/mole: since F has units of charge/mole and E is potential difference, their product is energy/mole. Thus $\Delta\mu$ describes the difference in electrochemical potential between a mole of ions on side A and a mole of ions on side B resulting from both concentration differences and electrical potential differences.

An ion will tend to move from where its electrochemical potential is higher to where it is lower. Since $\Delta\mu$ is defined as the electrochemical potential of the ion on side A minus that on side B, note that if $\Delta\mu$ is positive the electrochemical potential on side A is greater than that on side B. Therefore, if $\Delta\mu$ is positive, the ion will tend to move from A to B; if $\Delta\mu$ is zero, there is no net tendency for the ion to move at all; if $\Delta\mu$ is negative, the electrochemical potential is higher on side B and the ion will tend to move from side B to side A.

$\Delta\mu$ may be thought of as the *net* force on the ion, while RT in (C_a/C_b) is the force due to the concentration difference, and zF ($E_a - E_b$) is the force due to the electrical potential difference. When the two forces are equal and opposite, $\Delta\mu = 0$ and there is no net force on the ion. When there is no net force on the ion, there is no net movement of the ion and the ion is said to be in *equilibrium* across the membrane. At equilibrium $\Delta\mu = 0$.

(2) $$\text{Since } \Delta\mu = RT1n\left(\frac{C_a}{C_b}\right) + zF\,(E_a - E_b)$$

at equilibrium $0 = RT1n\,(C_a/C_b) + zF\,(E_a - E_b)$

(3) $$E_a - E_b = \frac{-RT}{zF}\,1n\,\frac{C_a}{C_b} = \frac{RT}{zF}\,1n\,\frac{C_b}{C_a}$$

Equation 3 is called the *Nernst equation* after the famous 19th-century physical chemist who derived it. Since the condition of equilibrium was assumed in its derivation, the Nernst equation holds *only* for ions at equilibrium. It describes the electrical potential difference, $E_a - E_b$, required to produce an electrical force, zF ($E_a - E_b$), that is exactly equal and opposite to the concentration force tending to move the ion from A to B, RT 1n (C_a/C_b).

The following examples illustrate some uses of the Nernst equation.

Example 1. In the situation diagrammed, Na^+ is 10 times more concentrated in chamber A than in chamber B. What electrical potential difference between the chambers is required so that Na^+ will be in equilibrium?

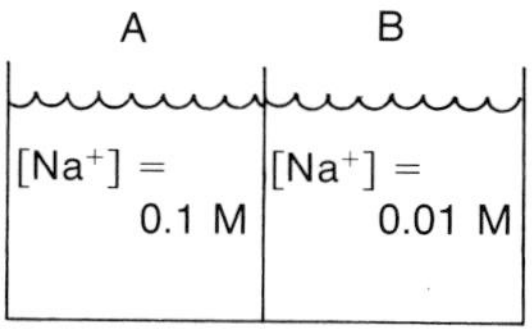

Since we have specified that Na^+ should be in equilibrium, the Nernst equation will hold

$$E_a - E_b = (RT/zF)\ln(C_b/C_a)$$

Since $\ln(x) = 2.303 \log_{10}(x)$

$$E_a - E_b = (2.303RT/zF)\log_{10}(C_b/C_a)$$

If we choose units so that E is expressed in millivolts (mV), then at 37°C (2.303RT/F) is approximately equal to 60 mV, so

(4) $$E_a - E_b = \frac{60}{z}\,\text{mV}\log_{10}(C_b/C_a)$$

Equation 4 is a convenient form of the Nernst equation which we will use from this point on.

Then for this situation

$$E_a - E_b = 60\ \text{mV}\log_{10}(0.01/0.1) = -60\ \text{mV}\log_{10}(10)$$

Since $\log_{10}(10) = 1$

$$E_a - E_b = -60\ \text{mV}$$

That is, at equilibrium side A must be 60 mV negative with respect to side B. We can see intuitively that this polarity is correct since this will cause Na^+ to tend to move from B to A owing to the electrical force, which will counteract the tendency for it to move from A to B as a result of the concentration difference.

It can be seen from this example that an electrical potential difference of about 60 mV is required to balance a 10-fold concentration difference.

Example 2. In the situation diagrammed here, is Cl^- in equilibrium? If Cl^- is not in equilibrium, in which direction will the net flow of Cl^- be?

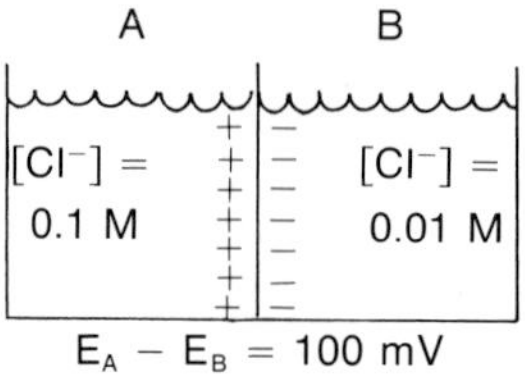

The Nernst equation will tell us the electrical potential difference, $E_a - E_b$, that will just balance the concentration difference of Cl^- across the membrane.

$$E_a - E_b = (60\ \text{mV}/-1)\log_{10}(0.01/0.1)$$
$$= 60\ \text{mV}\log_{10}(0.1/0.01) = +60\ \text{mV}$$

So a potential difference of +60 mV between A and B would just balance the tendency of Cl^- to move because of its concentration difference. Since $E_a - E_b$ is actually +100 mV, the electrical potential is of the correct sign to counterbalance the concentration force, but it is 40 mV larger than it needs to be. Since the electrical force on Cl^- is larger than the concentration force, it will determine the direction of Cl^- movement. Cl^- will thus tend to move from B to A.

In summary, if the potential difference across a membrane, as shown in the diagram, is equal to the potential difference calculated from the Nernst equation, using the concentrations for X on side A and side B, then X is in equilibrium across the membrane.

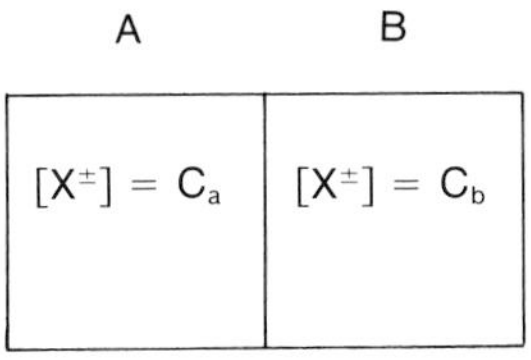

If the electrical potential is of the same sign as predicted by the Nernst equation but larger in magnitude, then the electrical force is larger than the concentration force and net movement of the ion will tend to occur in the direction determined by the electrical force. When the electrical potential difference is of the same sign but is numerically less than that calculated from the Nernst equation, then the concentration force is larger than the electrical force and net movement of the ion tends to occur in the direction determined by the concentration difference.

The electrical activity of nerve and muscle cells will be considered in Chapter 17. The concepts of what it means for an ion to be at equilibrium and how the balance of electrical and concentration forces determines the direction of ionic flow will form the basis for a discussion of the ionic mechanisms of the resting membrane potential and the action potential.

The Gibbs-Donnan Equilibrium

The cytoplasm of all cells contains proteins, organic polyphosphates, and other ionized substances that cannot permeate the plasma membrane, as well as Na^+, K^+, Cl^-, and other ions to which the plasma membrane is somewhat permeable. This mixture of permeant and impermeant ions has interesting steady-state properties that are described by the term Gibbs-Donnan equilibrium.

Consider as an example a membrane separating a solution of KCl from a solution of KD, where D^- is an anion to which the plasma membrane is completely impermeable. The membrane is permeable to water, K^+, and Cl^-.

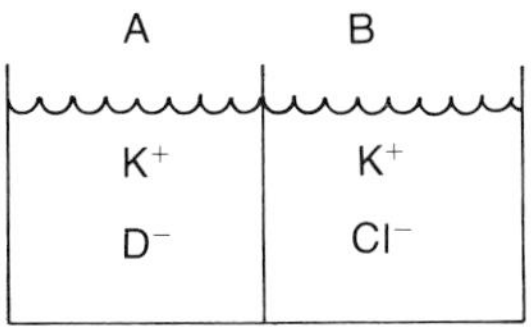

Suppose that on side A there is a 0.1 M solution of KD, and an equal volume of 0.1 M KCl is on side B. Because $[Cl^-]_B$ exceeds $[Cl^-]_A$, Cl^- will flow from chamber B to chamber A. Negatively charged Cl^- ions flowing from side B to side A will create an electrical potential difference, side A negative, that will cause K^+ also to flow from side B to side A. In fact, the same number of K^+ ions as Cl^- ions will flow from side B to side A in order to preserve *electroneutrality*. The principle of electroneutrality states that any macroscopic region of a solution must have an equal number of positive and negative charges on its ionic constituents. In actuality, slight separation of charges does occur in certain situations, but in chemical terms the imbalance between positive and negative changes is infinitesimal.

Given sufficient time, those components of the system that can permeate the membrane, K^+ and Cl^- in this example, will come to equilibrium. At equilibrium $\Delta\mu_{K^+} = 0$ and $\Delta\mu_{Cl^-} = 0$. Using the definitions of $\Delta\mu_{K^+}$ and $\Delta\mu_{Cl^-}$ from equation 2

$$RT\ 1n\frac{[K^+]_a}{[K^+]_b} + F\,(E_a - E_b) = 0$$

$$RT\ 1n\frac{[Cl^-]_a}{[Cl^-]_b} - F\,(E_a - E_b) = 0$$

Adding these two equations and dividing each side by RT gives

$$1n\frac{[K^+]_a}{[K^+]_b} + 1n\frac{[Cl^-]_a}{[Cl^-]_b} = 0$$

This yields

$$1n\frac{[K^+]_a}{[K^+]_b} = -1n\frac{[Cl^-]_a}{[Cl^-]_b} = 1n\frac{[Cl^-]_b}{[Cl^-]_a}$$

so that

$$\frac{[K^+]_a}{[K^+]_b} = \frac{[Cl^-]_b}{[Cl^-]_a}$$

Cross-multiplying,

$$[K^+]_a\,[Cl^-]_a = [K^+]_b\,[Cl^-]_b \qquad (3)$$

Equation 3 is called the Donnan relation and holds for any univalent cation-anion pair in equilibrium between the two compartments. If there were other ions present that could attain an equilibrium distribution, the same reasoning and an equation similar to equation 3 would apply to them as well.

By using the Donnan relation and the principle of electoneutrality it is possible to determine the equilibrium concentrations of the components in the problem posed at the beginning of this section. In the initial situation

A	B
$[K^+] = 0.1M$ $[D^-] = 0.1M$	$[K^+] = 0.1M$ $[Cl^-] = 0.1M$

However, since Cl^- moves from B to A, the equilibrium value of $[Cl^-]_b$ can be denoted as 0.1−X. By electroneutrality principle $[K^+]_b = [Cl^-]_b = 0.1 - X$. Since the volumes of A and B are the same at equilibrium, $[Cl^-]_a$ = X and $[K^+]_a$ = 0.1+X. Substituting these concentrations into the Donnan relation

$$[K^+]_a\,[Cl^-]_a = [K^+]_b\,[Cl^-]_b$$

$$(0.1 + X)\,(X) = (0.1 - X)\,(0.1 - X)$$

Solving this equation for X gives X = 0.03333. . . ., so that at equilibrium

A	B
$[K^+]$ = 0.1333 . . M $[Cl^-]$ = 0.0333 . . M $[D^-]$ = 0.1M	$[K^+]$ = 0.0666 . . . M $[Cl^-]$ = 0.0666 . . . M

In a Gibbs-Donnan equilibrium only the permeable ionic species achieve equilibrium. The impermeable ions, D^- in the example given, cannot reach an equilibrium distribution. Although it is not quite so obvious, water also will not achieve equilibrium, unless provision is made for that to occur. Note that the total number of K^+ and Cl^- ions on side A in this example exceeds that on side B. This is a general property of Gibbs-Donnan equilibria.* Taking the protein into account as well, it can be seen that the total number of osmotically active particles on side A is considerably greater than on side B. Since the ions K^+ and Cl^- are present in an equilibrium condition, they will generate their full osmotic force even though the membrane is quite permeable to them.

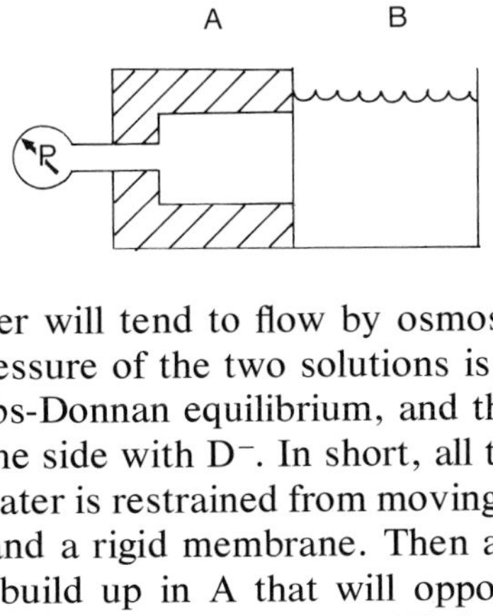

In this example, water will tend to flow by osmosis from side B to side A until the total osmotic pressure of the two solutions is equal. But then ions will flow to set up a new Gibbs-Donnan equilibrium, and that requires more osmotically active particles on the side with D^-. In short, all the water from side B will end up on side A unless water is restrained from moving by enclosing the solution of side A by rigid walls and a rigid membrane. Then as fluid flows from side B to side A, pressure will build up in A that will oppose further osmotic water movement. The pressure that exists in A at equilibrium is the difference in the total osmotic pressures of the solutions in A and B. In this example the approximate pressure in chamber A at equilibrium is

$$
\begin{aligned}
\text{total } \Delta\pi &= \Delta\pi_{K^+} + \Delta\pi_{Cl^-} + \Delta\pi_{D^-} \\
&= RT\,(\Delta C_{K^+} + \Delta C_{Cl^-} + \Delta C_{D^-}) \\
&= RT\,(0.06667 - 0.0333 - 0.1) \\
&= 22.4 \text{ atm X } (.13334) = 2.99 \text{ atm}
\end{aligned}
$$

*If $a \cdot b = c^2$, then $a + b > c + c$

Regulation of Cell Volume

K^+ and Cl^- are approximately in equilibrium across most cell membranes, and their distribution is influenced by the predominantly negatively charged impermeant ions inside the cell, such as proteins and nucleotides. Thus, K^+ and Cl^- can be thought of as satisfying the conditions for the Gibbs-Donnan equilibrium. That being the case, why does not the osmotic imbalance just discussed cause the cell to swell and finally burst? One reason is that cells actively pump Na^+ out of the cytoplasm to the extracellular fluid, decreasing the osmotic pressure of the cell sap and increasing that of the extracellular fluid. The Na^+ pumping is done by the enzyme Na^+,K^+-ATPase, which resides in the plasma membrane. The Na^+,K^+-ATPase hydrolyzes ATP and uses some of the energy released to extrude Na^+ from the cell and to actively pump K^+ into the cell. While K^+ is only slightly removed from an equilibrium distribution, Na^+ is pumped out against a large electrochemical potential difference.

When the ATP production of the cell is compromised (in the presence of low O_2 levels, for example) or when the Na^+,K^+-ATPase is specifically inhibited, cells swell. Some authorities contend that the irreversible damage to the brain that results from acute oxygen deprivation is due to cellular injury that results from cell swelling. As mentioned in the previous chapter, in hereditary spherocytosis a high permeability of the red cell membrane to Na^+ results in cell swelling when the Na^+ pump is metabolically compromised and cannot keep up with the Na^+ influx into the cell.

FURTHER READING

Christensen, H. N.: Biological Transport. 2nd ed. Reading, Mass., W. A. Benjamin, 1975, pp. 69–96.
Davson, H.: A Textbook of General Physiology. 4th ed. Baltimore, Williams & Wilkins, 1970, pp. 368–394, 508–713.
Dick, D. A. T.: Cell Water. London, Butterworth, 1965.
Dowben, R. M.: General Physiology. New York, Harper & Row, 1969, pp. 456–478.
Durbin, R. P.: Osmotic flow of water across permeable cellulose membranes. J. Gen. Physiol. *44*:315, 1960.
Harris, E. J.: Transport and Accumulation in Biological Systems. 3rd ed. London, Butterworth, 1972, pp. 73–146.
Kotyk, A., and Janacek, K.: Cell Membrane Transport. Principles and Techniques. New York, Plenum Press, 1970, pp. 91–130.
Landis, E. M., and Pappenheimer, J. R.: Exchange of substances through the capillary walls. *In*: Handbook of Physiology, Section 2: Circulation, Volume 2. Washington, D.C., The American Physiological Society, 1963, pp. 961–1034.
Moore, W. J.: Physical Chemistry. 4th ed. Englewood Cliffs, N.J., Prentice-Hall, 1972, pp. 250–254, 309.
Pappenheimer, J. R.: Passage of molecules through capillary walls. Physiol. Rev., *33*:387, 1953.
Ponder, E.: Hemolysis and Related Phenomena. New York, Grune & Stratton, 1948.
Renkin, E. M. Filtration, diffusion, and molecular sieving through porous cellulose membranes. J. Gen. Physiol., *38*:225, 1954.
Stein, W. D.: The Movement of Molecules Across Biological Membranes. New York, Academic Press, 1967, pp. 242–253.
Whittam, R.: Transport and Diffusion in Red Blood Cells. London, Edward Arnold, 1964, pp. 45–75.
Van Holde, K. E.: Physical Biochemistry. Englewood Cliffs, N.J., Prentice-Hall, 1971, pp. 39–47.

17

MEMBRANE POTENTIALS

The cells of the nervous system communicate with each other by means of a propagated electrical disturbance in the plasma membrane. This propagated electrical disturbance is called an *action potential*. Similarly, in skeletal muscle tissue an action potential serves to spread the contractile activity rapidly over the entire muscle cell. Later in this chapter, the action potential and the ionic mechanisms that account for its properties will be described. All cells that are able to produce an action potential have a sizeable resting electrical polarization across the plasma membrane. Thus it is necessary first to consider the electrical potential difference across the plasma membrane known as the *resting membrane potential*.

THE RESTING MEMBRANE POTENTIAL

The resting membrane potential can be measured by means of a glass microelectrode that has a tip diameter of about 0.1 μm and can puncture the plasma membrane without injuring the cell greatly. For a muscle cell the electrical potential difference between the tip of the microelectrode inside the cell and a reference electrode in the extracellular fluid is about −90 millivolts (mV). Thus, the myoplasm is almost 0.1 volt electrically negative with respect to the cell's surroundings. This resting membrane potential is absolutely necessary in order for the cell to produce an action potential. If the resting membrane potential is significantly decreased, for example, to −50 mV or less, the cell is no longer able to produce an action potential.

Ionic Equilibria

Since familiarity with ionic equilibria is essential to understand the generation of the resting potential, some salient points about this subject will be given here. Additional background can be obtained by reviewing the discussion of ionic equilibria in Chapter 16.

Recall that if a chamber, A, is separated by a membrane from a second chamber, B, the tendency for an ionic species to flow from side A to side B is given by the *electrochemical potential* of the ion on side A minus that on side B. The electrochemical potential difference ($\Delta\mu$) between sides A and B is

$$\Delta\mu = RT \ln \frac{C_A}{C_B} + zF(E_A - E_B) \tag{1}$$

The first term on the right-hand side of equation 1 can be thought of as the tendency for the ion to move from A to B because of a concentration difference (the "concentration force"), and the second term can be regarded as the tendency for the ion to move because of the electric field (the "electrical force"). Which way the ion will move depends on the balance between the concentration force and the electrical force. If $\Delta\mu$ is positive, μ_A is greater than μ_B, so the ion will tend to flow from side A to side B. If $\Delta\mu$ is negative, the ion will tend to flow from B to A. If $\Delta\mu$ is 0, there is no net force acting on the ion and thus no tendency for the ion to move in either direction; in this case the ion is in *equilibrium* between chambers A and B.

For an equilibrium situation, $\Delta\mu = 0$ in equation 1. Solving for $E_A - E_B$

$$E_A - E_B = -\frac{RT}{zF} \ln \frac{C_A}{C_B} = \frac{RT}{zF} \ln \frac{C_B}{C_A} = \frac{2.303\,RT}{zF} \log_{10} \frac{C_B}{C_A} \tag{2}$$

Equation 2 is the Nernst equation. It permits calculation of the electrical potential difference ($E_A - E_B$) that will create an electrical force that is equal, but opposite in direction, to the concentration force. An ion that satisfies the Nernst equation is by definition in equilibrium. That is, suppose the actual values of $E_A - E_B$, C_A, and C_B are determined empirically and substituted into equation 2. If *equality* of the right- and left-hand sides is obtained, then the ion in question is in equilibrium between A and B. At equilibrium there is no net force on the ion, and no net flow (or current) of the ion will occur.

In contrast, if $E_A - E_B$ is *not* equal to the right-hand side of equation 2, then the concentration force and the electrical force are *not* equal and the ion is not in equilibrium. If the measured $E_A - E_B$ has the *same* sign predicted by the right-hand side of the Nernst equation, then the electrical force is *opposite* in direction to the concentration force. However, if the measured $E_A - E_B$ is *smaller* in magnitude than predicted by the right-hand side, the electrical force is too small to balance the concentration force; if the measured $E_A - E_B$ is *larger* in magnitude than the calculated value, the electrical force is larger than the concentration force. On the other hand, if the measured $E_A - E_B$ has the *opposite* sign from that predicted by the right-hand side of the Nernst equation, this means that the electrical force and the concentration force are in the *same* direction, so there is no possibility for the ion to be in equilibrium. The larger the discrepancy between the measured $E_A - E_B$ and the calculated $E_A - E_B$, the larger the imbalance between electrical and concentration forces, and the greater the net tendency for ion flow to occur.

The Concentration Cell

Consider a situation such as the one shown in this diagram.

A	B
1 M KCl	.1 M KCl

Imagine that the membrane separating chambers A and B is permeable to cations, but not to anions, and that initially there is no electrical potential difference across the membrane. K^+ will flow from A to B owing to the concentration force acting on it. Cl^- has the same force on it, but cannot flow from A to B because the membrane will not allow it to pass. The flow of K^+ from A to B, the K^+ current, will transfer net positive charge to side B and leave an excess of negative charges behind on side A, causing A to become electrically negative to B.

A			B
	−	+	
1 M	−	+	.1 M
	−	+	
KCl	−	+	KCl

This electrical force is oppositely directed to the concentration force on K^+. The more K^+ that flows, the larger the opposing electrical force. K^+ flow will cease when the electrical force just balances the concentration force, that is, when the electrical potential difference is equal to the equilibrium potential for K^+, or when

$$E_A - E_B = \frac{60 \text{ mV}}{+1} \log_{10} \frac{0.1}{1} = -60 \text{ mV} \log_{10} 10 = -60 \text{ mV}$$

It should be noted that only a very small amount of K^+ flows from A to B before equilibrium is reached. This is because the separation of positive and negative charges requires a large amount of work, and the potential difference that builds up to oppose further K^+ movement is a manifestation of that work.

A very important point to appreciate is that the K^+ concentration difference in this example acted much like a *battery*. Because the natural tendency for any ion that can flow is to seek equilibrium, K^+ tended to flow until its equilibrium potential difference was established. As we will see, in a system such as a cell membrane, with more than one permeable ion, each ion "strives" to make the potential difference equal to its equilibrium potential. The more permeable the ion is, the more successful it is in forcing the electrical potential difference toward its equilibrium potential.

The Distribution of Ions across Plasma Membranes

In most tissues certain ions are not in equilibrium between the extracellular fluid and the cytoplasm of the cells, and in excitable tissues this is always the case. Table 17–1 gives values for the concentrations of Na^+, K^+, and Cl^- in the

TABLE 17–1 DISTRIBUTION OF Na^+, K^+, AND Cl^- ACROSS THE PLASMA MEMBRANES OF FROG MUSCLE AND SQUID AXON.*

		Extracellular Fluid (mM)	Cytoplasm (mM)	Approximate Equilibrium Potential (mV)	Actual Resting Potential (mV)
Frog Muscle	$[Na^+]$	120	9.2	+67	
	$[K^+]$	2.5	140	−105	−90
	$[Cl^-]$	120	3–4	−89 to −96	
Squid Axon	$[Na^+]$	460	50	+58	
	$[K^+]$	10	400	−96	−70
	$[Cl^-]$	540	about 40	about −68	

*From Katz, B., Nerve, Muscle, and Synapse. New York, McGraw-Hill, 1966. Reproduced by permission of the publisher.

extracellular fluid and in the cytoplasmic water for frog skeletal muscle and for squid giant axon, two tissues that have been well characterized in this regard. Values for mammalian muscle are similar to those for the more readily studied frog.

Chloride is approximately in equilibrium across the plasma membrane of both frog muscle and squid axon since its potential difference for equilibrium, as calculated from the Nernst equations, is about equal to the measured membrane potential difference. In both tissues K^+ has a concentration force tending to make it flow out of the cell. In frog muscle the electrical force on K^+ is oppositely directed to the concentration force, and if the $E_{in} - E_{out}$ were −105 mV the forces would exactly balance. However, since $E_{in} - E_{out}$ is only −90 mV, the concentration force is stronger than the electrical force and K^+ has a net tendency to flow out of the cell. In frog muscle and squid axon Na^+ has *both* the concentration and electrical forces tending to make it flow into the cell. Thus Na^+ is the ion farthest from an equilibrium distribution, because the larger the difference between the measured membrane potential and the equilibrium potential for a particular ion, the larger the net force tending to make that ion flow.

Active Ion Pumping

In Chapter 15 the Na^+, K^+-ATPase, also called the Na^+-K^+ pump, was discussed. This enzyme is located in the plasma membrane, and it uses the energy of the terminal phosphate bond in ATP to actively extrude Na^+ from the cell and take K^+ into the cell. This ion pump is responsible for the large intracellular K^+ concentration and the small intracellular Na^+ concentration. Since the pump apparently moves a larger number of Na^+ ions out than K^+ ions in (the ratio usually cited is 3 Na^+ to 2 K^+), it also causes a net transfer of positive charge out of the cell and thus contributes to the resting membrane potential. Because of this contribution to the generation of the membrane potential, the pump is termed *electrogenic*.

The size of the contribution of the pump to the resting potential can be determined in experiments in which the pump is completely inhibited by a cardiac glycoside, such as ouabain. These studies show that in some cells the electrogenic pump is responsible for a large fraction of the resting potential. In most vertebrate nerve and muscle cells, however, the contribution of the pump to the resting potential is minor under most circumstances, being less than 5 mV. Thus it is necessary to look further for the major factors that generate the resting membrane potential in nerve and muscle.

Generation of the Resting Membrane Potential by the Ion Gradients

In the section on concentration cells, it was shown that an ion gradient can function as a battery. When several ions are distributed across a membrane so that all are removed from equilibrium, each ion will tend to force the transmembrane potential toward its own equilibrium potential, as calculated by the Nernst equation. The more permeable the membrane to a particular ion, the greater success that ion will have in forcing the membrane potential toward its equilibrium potential. In the case of the concentration differences across the plasma membrane of frog muscle (Table 17–1), the Na^+ concentration difference can be regarded as a battery that tends to make $E_{in} - E_{out}$ equal to +67 mV. The K^+ concentration difference is a battery that attempts to make $E_{in} - E_{out} = -105$ mV, and the Cl^- concentration difference resembles a battery trying to make $E_{in} - E_{out} = -90$ mV. This situation is often represented in terms of an electrical equivalent circuit for the plasma membrane.

Extracellular fluid

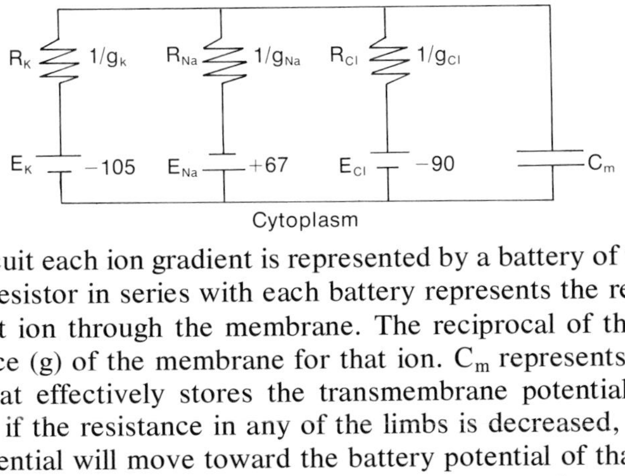

Cytoplasm

In this circuit each ion gradient is represented by a battery of the appropriate polarity. The resistor in series with each battery represents the resistance to the passage of that ion through the membrane. The reciprocal of this resistance is the conductance (g) of the membrane for that ion. C_m represents the membrane capacitance that effectively stores the transmembrane potential difference. In such a circuit, if the resistance in any of the limbs is decreased, the total transmembrane potential will move toward the battery potential of that limb.

The way in which the interplay of ion gradients creates the resting membrane potential can be seen in relation to this equivalent circuit. If the transmembrane potential is equal to the equilibrium potential for a particular ion, there is no net force on the ion and there will be no net flow of that ion. However, if the membrane potential is *not* equal to the equilibrium potential for a given ion, then the *difference* between the membrane potential and the equilibrium potential for that ion can be regarded as the *driving force* for the ion. Since ions bear charge, ionic flow is equivalent to electrical current. The net current (I) of an ion across the membrane is the product of the driving force and the conductance of the membrane for that ion. For the ions in the membrane equivalent circuit

$$I_K = g_K (E_m - E_K)$$
$$I_{Na} = g_{Na} (E_m - E_{Na})$$
$$I_{Cl} = g_{Cl} (E_m - E_{Cl})$$

In the steady state, when the transmembrane potential is constant, the sum of all the ionic currents across the membrane must be zero. This is so because if there were net current across the membrane, the membrane capacitor would be charging or discharging and the membrane potential would be changing, since any net charge transfer across the membrane leads to a change in the degree of charge separation and hence in the membrane potential. If K^+, Na^+, and Cl^- are considered to be the only important ions, then the algebraic sum of their currents must be zero in the steady state.

(4) $$I_K + I_{Na} + I_{Cl} = 0$$

Substituting from equation 3 for the currents

(5) $$g_K(E_m - E_K) + g_{Na}(E_m - E_{Na}) + g_{Cl}(E_m - E_{Cl}) = 0$$

For a cell membrane across which chloride is in equilibrium $E_m = E_{Cl}$ and $E_m - E_{Cl} = 0$, so the third term can be dropped. However, it should be kept in mind that Cl^- is not in equilibrium for all excitable cells; whenever E_m is changed away from E_{Cl}, Cl^- will exert a restoring force to bring E_m back toward E_{Cl}, just as any ion tries to force E_m toward its equilibrium potential. Dropping the third term, equation 5 becomes

(6) $$g_K(E_m - E_K) + g_{Na}(E_m - E_{Na}) = 0$$

Multiplying through and solving for E_m gives

(7) $$E_m = \frac{g_K}{g_K + g_{Na}} E_K + \frac{g_{Na}}{g_K + g_{Na}} E_{Na}$$

Equation 7 is known as the *chord conductance equation.* It shows that the membrane potential across the cell membrane is a weighted average of the equilibrium potentials for K^+ and Na^+, with the weighting factor for each ion being that fraction of the total ionic conductance attributable to that particular ion. To consider more ions, it is necessary to take only their conductances into account by adding on similar terms. For example, in a cell in which Cl^- is not in equilibrium and Ca^{++} plays an important role, the chord conductance equation becomes

(8) $$E_m = \frac{g_K}{g_T} E_K + \frac{g_{Na}}{g_T} E_{Na} + \frac{g_{Cl}}{g_T} E_{Cl} + \frac{g_{Ca}}{g_T} E_{Ca}$$

where $g_T = g_K + g_{Na} + g_{Cl} + g_{Ca}$

For a cell in which the important ions are Na^+ and K^+, and Cl^- is in equilibrium, the simpler form of the chord conductance equation (equation 7) applies.

Since the frog muscle fiber discussed earlier has $E_{in} - E_{out} = -90$ mV, the membrane potential is much closer to E_K than to E_{Na}. The prediction of the chord conductance equation is that in frog muscle g_K is 10.5 times larger than g_{Na} in the resting state. This has been confirmed by measurements using radioactive tracers for K^+ and Na^+, which have shown that in resting frog sartorius muscle g_K is about 10 times g_{Na}. In resting squid axon the chord conductance equation predicts that g_K is 4.9 times larger than g_{Na}. In other types of excitable cells the relationship between g_K and g_{Na} may be somewhat different, or other ions may play a role in the resting membrane potential. Thus resting membrane potentials vary from −30 mV in some types of smooth muscle up to −90 mV in vertebrate muscle and in cardiac ventricular cells.

The chord conductance equation also predicts that if g_{Na} were suddenly increased, the membrane potential would move toward E_{Na} (i.e., +67 in frog muscle). This is actually what occurs during an action potential when there is a transient increase in g_{Na}. The ionic mechanism of the action potential will be considered further in the next section.

In summary, the Na^+-K^+ pump sets up gradients of Na^+ and K^+ across the plasma membranes of cells. Since a larger amount of Na^+ is pumped out than K^+ is pumped in, the pump transfers net charge across the membrane and contributes to the membrane potential; thus the pump is electrogenic. In most cases, however, the electrogenic activity of the pump is responsible for only a small fraction of the resting potential. The major portion of the resting membrane potential is the result of the interplay between Na^+ and K^+, with each diffusing down its electrochemical potential gradient—Na^+ flowing into the cell and K^+ flowing out. The main role of the pump is then to maintain the ion gradients it has established. Both K^+ and Na^+ tend to force the transmembrane potential toward their respective equilibrium potentials. The resulting E_m is a weighted average of E_K and E_{Na}, with the weighting factors being that fraction of the total ionic conductance contributed by K^+ and Na^+. While the resting membrane potential is directly due to the diffusion of K^+ and Na^+ down their respective electrochemical potential gradients, it is indirectly due to the Na^+-K^+ pump which maintains the gradients.

THE ACTION POTENTIAL

The *action potential* is a rapid change in the membrane potential followed by a return to the resting membrane potential (Fig. 17–1). The form of the action potential differs somewhat from one excitable tissue to another. The action potential is propagated in unchanged form and magnitude along the whole length of a nerve axon or muscle fiber, and it is the basis of the signal-carrying ability of nerve cells and the capacity of all parts of a long muscle cell to contract almost

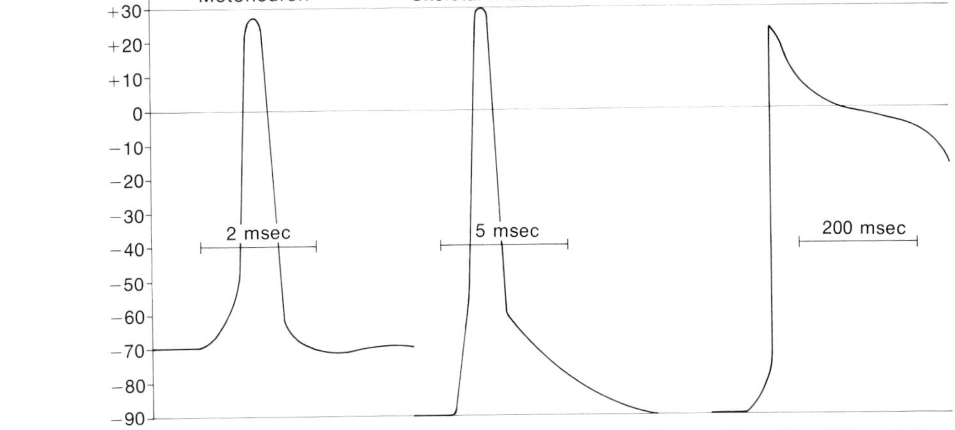

FIGURE 17–1 Action potentials in some excitable cells of vertebrates. Note the different time scales.

simultaneously. The ionic currents that cause the action potential to occur and the mechanism by which it is propagated rapidly along the length of a cell will now be considered.

Observations of Membrane Potentials

Much of our knowledge of the ionic mechanism of the action potential was obtained first in experiments on the giant axon of the squid and confirmed later in other preparations. The squid giant axon has been so widely studied because its large diameter, which reaches up to 0.5 mm, makes it a convenient object for electrophysiological investigations. The frog sartorius muscle is another extremely useful electrophysiological preparation. It can be easily removed from the frog without damaging the surface muscle cells, which may then be visualized with a microscope and penetrated with one or more microelectrodes.

Consider a frog sartorius fiber that has been removed from the animal and placed in a solution of similar composition to the frog's extracellular fluid. Two microelectrodes and instruments to record the potential difference between them are used to perform the following experiment (Fig. 17–2).

The Resting Potential. When both microelectrodes are placed in the bath, no potential difference between them is observed. Electrode #2 is then advanced toward a muscle cell until it penetrates the plasma membrane. At the moment the electrode penetrates the membrane an immediate change of the potential difference between the two electrodes is observed, with the intracellular electrode (#2) becoming about 90 mV negative with respect to the external electrode (#1). This is the *resting membrane potential* of the muscle fiber. If the plasma membrane is penetrated with another microelectrode (#3), electrode #3 is also −90 mV relative to the external solution. At rest there is no net internal current in the axon and no potential difference between the two intracellular electrodes.

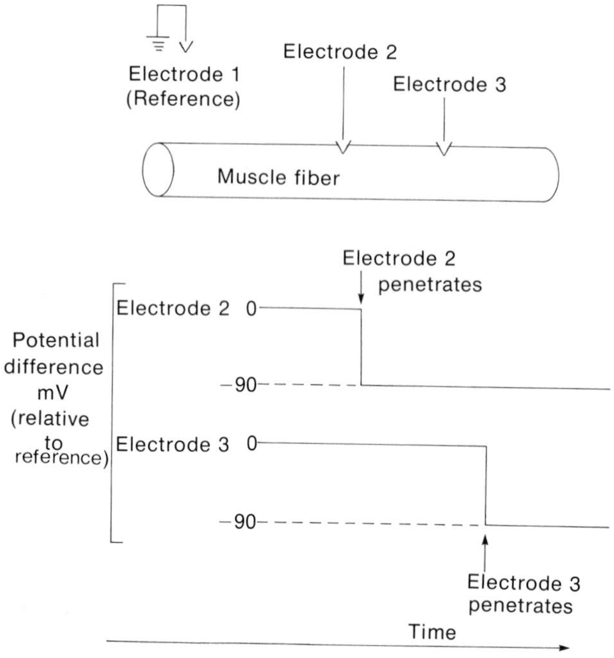

FIGURE 17–2 The resting membrane potential. A muscle cell is penetrated first with one microelectrode [2] then with another [3].

SUBTHRESHOLD BEHAVIOR—THE LOCAL RESPONSE. An intracellular electrode is now placed in the cell near the entry point of electrode #2 for the purpose of passing current across the plasma membrane. If a small pulse of current is passed into the cell (positive charges move into the cell), this will diminish the polarization of the membrane, or *depolarize* the membrane, at electrode #2, as shown in Figure 17–3. No change in polarization is observed at electrode #3, if electrode #3 is more than a few mm from the site of current injection.

Similarly, a pulse of outward current (positive charges move out of the cell) will increase the polarization across the plasma membrane, or *hyperpolarize* the cell, at electrode #2 (Fig. 17–3). Again no change will be seen at electrode #3 unless it is within a few mm of the site of current passage.

Sometimes the words depolarization and hyperpolarization cause confusion. If the membrane potential changes from −90 mV to −70 mV, this is depolarization, since it lessens the potential difference or polarity across the cell membrane. If the membrane potential goes from −90 mV to −100 mV, the polarity across the membrane has increased; hence this is hyperpolarization. Polarization refers to the absolute magnitude of the difference in electrical potential between the cytoplasm and the extracellular fluid.

To the same experimental preparation, add another intracellular electrode (#4) that penetrates the cell membrane 2 to 3 mm from electrode #2 (Fig. 17–4). Now successively stronger hyperpolarizing current pulses are applied via the current electrode, and recordings at electrodes #2, #3, and #4 (all relative to the extracellular reference electrode) are taken. Figure 17–4 shows that the extent of hyperpolarization observed at electrode #2 is directly related to the strength of the current pulse applied there. At electrode #4 (2 to 3 mm away), smaller hyperpolarizing responses are observed. At electrode #3 no change is observed,

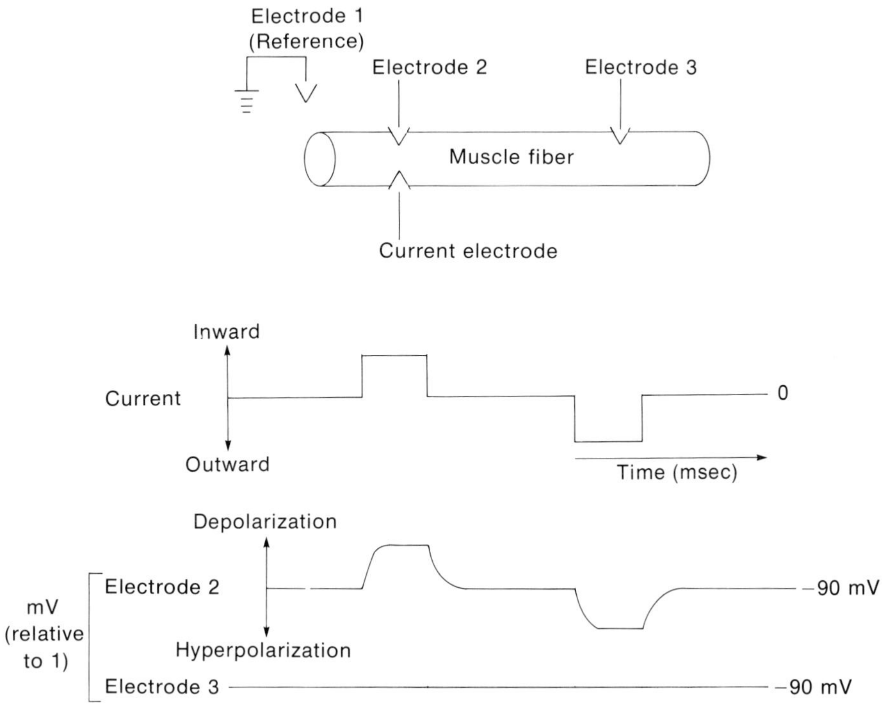

FIGURE 17–3 Response of the muscle fiber membrane potential to small local currents. Electrode 3 is more than 10 mm from the current electrode.

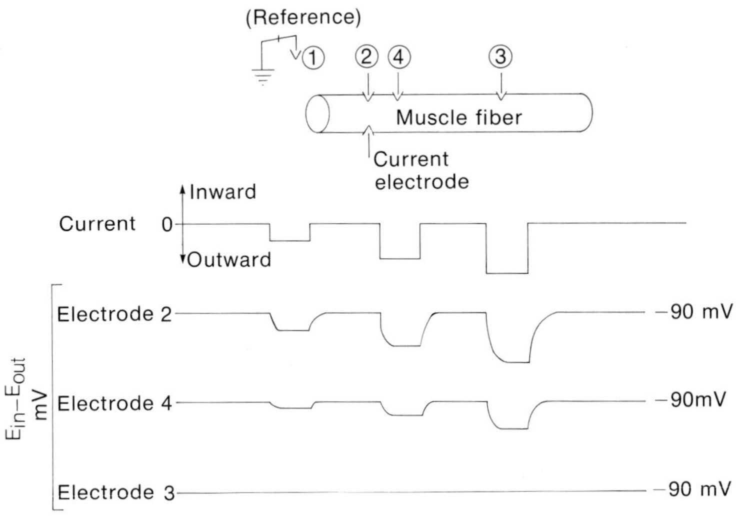

FIGURE 17–4 Hyperpolarization of the muscle fiber membrane potential in response to small local currents as recorded by electrodes that are: [2] very close to the site of current passage; [4] 2–3 mm from the site of current passage; and (3) greater than 10 mm from the site of current passage.

since it is too far away from the site of current withdrawal. Since this hyperpolarizing response is observed only near the site of current withdrawal, it is called the *local response* (or the *electrotonic response*). If electrode #4 were moved a bit farther from the current withdrawal site, the responses would be still smaller. The magnitude of the local response declines exponentially with distance from the site of current passage.

If we apply *inward* pulses of current, transient *depolarizations* will be observed. The features described earlier for hyperpolarizations apply to the depolarizations observed at the various electrodes in response to small inward currents. The decay of the magnitude of the electrotonic, or local, response with distance will be shown in Figure 17–5.

The Action Potential. If progressively larger inward current pulses are applied, a point is reached at which a different sort of response, the *action potential,* occurs (Fig. 17–6). In order to trigger an action potential the depolarization caused by the inward current must be sufficient for the membrane potential to reach a threshold value, which is about −60 mV for frog sartorius muscle.

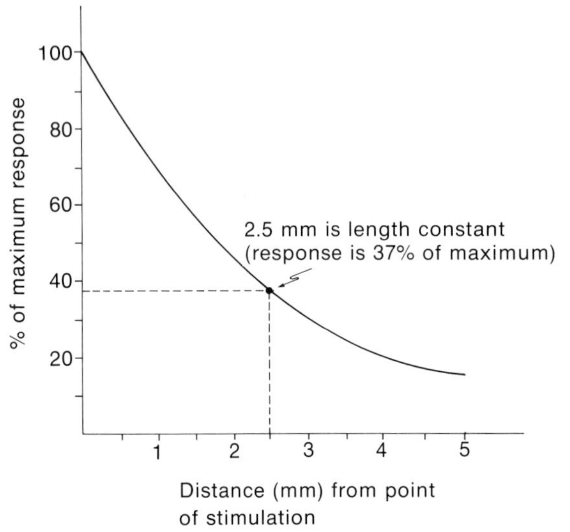

FIGURE 17–5 The magnitude of the local response declines exponentially with distance along the axon or muscle fiber.

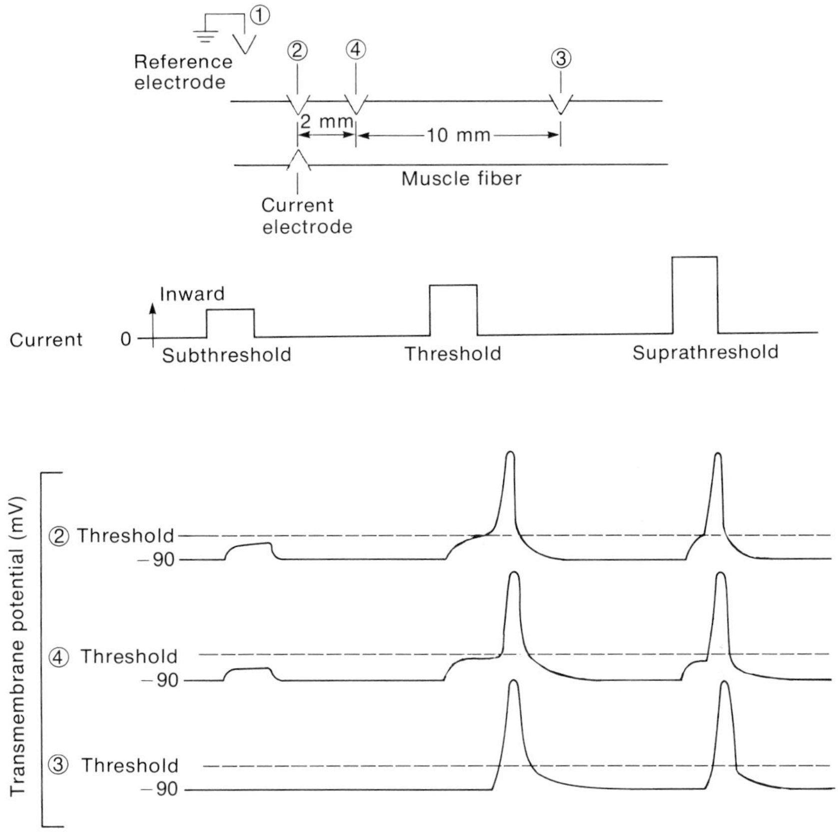

FIGURE 17–6 Response of muscle fiber membrane potential to currents that are subthreshold, threshold, and suprathreshold.

The action potential differs from the local depolarizing response in two important ways. First, it is usually a much larger response. The membrane polarity actually reverses, the cell interior becoming positive with respect to the exterior. Second, the action potential is *propagated without decrement* down the length of the fiber. This means that the height and form of the action potential response remain the same as it travels along the fiber, and it does not decay exponentially with distance like the local response.

Note in Figure 17–6 that when the threshold is reached, an action potential is fired and propagated along the fiber. The size and form of the action potential at electrodes #2 and #3 are the same as at electrode #1. Even when a stimulus larger than the threshold stimulus is applied, the size and form of the action potential do not change; that is, the action potential does not grow with increased stimulus strength. A stimulus either fails to elicit an action potential and is called a subthreshold stimulus, or it produces a full-strength action potential and is referred to as a threshold or suprathreshold stimulus. For this reason the action potential is often called an *all-or-none response*.

THE FORM OF THE ACTION POTENTIAL. The form of the action potential of a squid giant axon is shown in Figure 17–7. Once the membrane is depolarized sufficiently to reach the threshold, an explosive depolarization occurs, which completely depolarizes the membrane and then *overshoots*, so that the membrane acquires the reverse polarity. The action potential peaks at around +50 mV and returns toward the resting membrane potential almost as rapidly as it had pre-

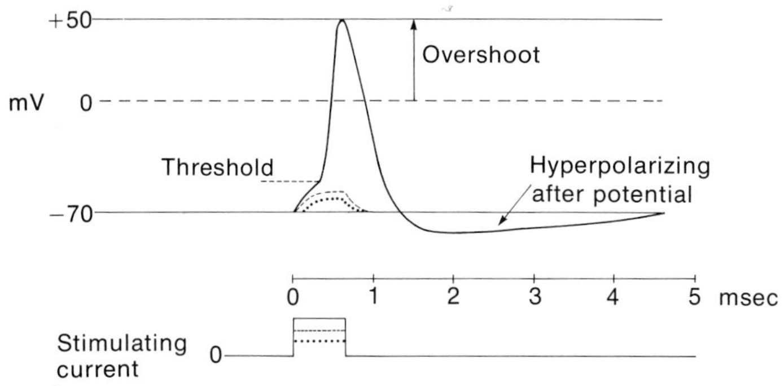

FIGURE 17–7 An action potential in the giant axon of the squid as recorded with an intracellular microelectrode.

viously depolarized. A transient hyperpolarization occurs after repolarization which is known as the *hyperpolarizing afterpotential* and persists for about 4 msec.

Ionic Mechanism of the Action Potential

Earlier in this chapter, the membrane potential was described as the weighted sum of the equilibrium potentials for Na^+, K^+, and Cl^-, the weighting factor for each ion being the fraction it contributes to the total ionic conductance of the membrane. These relationships were defined by the chord conductance equation. Since E_K is about −100 mV in squid axon, an increase in g_K would hyperpolarize the membrane. Since E_{Cl} is about −70 mV, an increase in g_{Cl} would stabilize E_m at −70. An increase in g_{Na} would cause depolarization and reversal of the membrane polarity, since E_{Na} is about +60 mV in squid axon. A decrease in g_K would also have the effect of depolarizing the membrane.

In fact, it has been known for a long time that Na^+ is involved in the action potential. Overton showed in 1902 that Na^+ in the extracellular fluid is required for excitability. Also in 1902, Bernstein proposed that the action potential was the result of a transient breakdown of the barrier properties of the plasma membrane, so that the conductances to all ions increased transiently. About 60 years later A. L. Hodgkin and A. F. Huxley were awarded the Nobel Prize for showing that the action potential of squid giant axon is due to a pattern of conductance increases to sodium and potassium ions. Hodgkin and Huxley found (Fig. 17–8) that the conductance to Na^+, g_{Na}, increases rapidly and markedly during the first part of the action potential, peaking approximately when the action potential peaks and then decreasing rather rapidly. A delayed increase in the K^+ conductance, g_K, occurs somewhat more slowly, peaks at about the middle of the repolarization phase, and rather slowly returns to baseline values.

As already described, the chord conductance equation shows that the membrane potential is a result of the opposing tendencies of the K^+ gradient to pull E_m toward the equilibrium potential for K^+ and of the Na^+ gradient to pull E_m toward the equilibrium potential for Na^+. An increase in conductance of either ion will increase its power to pull E_m toward its equilibrium potential. For this reason the rapid increase in g_{Na} during the action potential will cause the membrane potential to move toward the equilibrium potential for Na^+ (+58 mV). Actually the peak of the action potential never reaches +58 mV, but only about +50 mV. This is because the conductance to K^+ also increases, albeit more

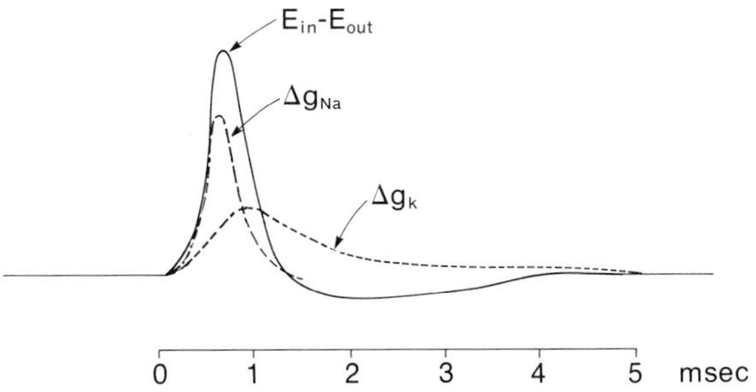

FIGURE 17–8 changes in g_{Na} and g_K that occur during the action potential in squid axon. (Redrawn from Hodgkin and Huxley.)

slowly, providing an opposing tendency, and also because the increase in g_{Na} tends to inactivate itself after a certain time.

The rapid return of the membrane potential toward the resting potential is then due to the rapid decrease of g_{Na} and the continued increase in g_K, both of which diminish the size of the Na^+ term in the chord conductance equation and increase the K^+ term. During the hyperpolarizing afterpotential, when the membrane potential is actually below the resting potential, g_{Na} has returned to resting levels but g_K is above resting levels, so that E_m is pulled closer to the K^+ equilibrium potential (−100 mV).

Evidence for this Understanding of the Action Potential

It was suspected in the first years of this century that the action potential was caused by a transient increase of the permeability of the plasma membrane to ions. Subsequently, evidence began to accumulate that the entry of Na^+ into the cell plays a central role in the action potential. It was found that removal of Na^+ from the extracellular fluid abolishes the action potential. The height of the action potential and the maximum rate of rise of the action potential both depend on the extracellular concentration of Na^+, increasing with increasing external Na^+.

By repetitively stimulating the squid giant axon in the presence of radioactive $^{22}Na^+$, it was possible to determine that 3 to 4 $\times$ 10^{-12} moles of Na^+ enter the axon through each cm^2 of surface area for each impulse. This is a very small amount of Na^+, but the measurement is in reasonable agreement with calculations of the amount of Na^+ that must enter, based on electrical considerations.

It is known that the capacitance of 1 cm^2 of squid axon membrane is about 1 microfarad. The amount of charge that must flow to discharge the membrane capacitor by 100 mV, which is the approximate height of the action potential, is then C times V or 10^{-6} farad $\times$ 10^{-1} volt $=$ 10^{-7} coulombs. Dividing this by the amount of charge on 1 mole of Na^+ ions, 96,500 coulombs/mole, gives about 10^{-12} moles of Na^+ ions that must enter.

If only Na^+ and K^+ are involved, the amount of K^+ that leaves the cell during the repolarization phase must be equivalent to the amount of Na^+ that enters to depolarize the cell. The Na^+-K^+ pump is thought to pump out the Na^+ that enters and to reaccumulate the lost K^+, but because of the extremely small number of ions that cross the membrane during each impulse the Na^+-K^+ pump is not required in the short run. For example, a squid axon that has been poisoned with CN^- or ouabain so that ion pumping has virtually ceased can conduct nearly 100,000 action potentials before it fails. Smaller axons, because their ratio of surface to volume is much larger, are not as independent of the Na^+ pump as larger axons and will fail after fewer impulses.

A more detailed knowledge of the ionic mechanism of the action potential came from application of a method known as the *voltage clamp* technique, which was developed in the 1940's by K. S. Cole and his collaborators in the United States and was applied to the squid giant axon by A. L. Hodgkin and A. F. Huxley in England. The voltage clamp technique utilizes the principle of electronic feedback to rapidly set the transmembrane potential to whatever value the experimenter desires. The voltage clamp then holds the membrane potential at this level and provides a precise measure of the net ionic current that flows across the membrane for as long as the clamp is maintained.

Hodgkin and Huxley found that when the membrane was rapidly depolarized to some point beyond the threshold there was an almost instantaneous current that corresponded to discharge of the membrane capacitance. This was followed by another pattern of currents which occurred over roughly the same time course that the action potential would have taken (Fig. 17–9). They suspected that the inward phase of the current was caused by Na^+ entering the axon and the outward current by K^+ leaving the axon. They were able to separate the Na^+ current from the K^+ current in the following ways.

1. Enough of the external Na^+ was replaced by choline$^+$, which is an impermeant cation, so that the Na^+ concentration inside the axon equalled that in the bathing medium. Then the membrane potential was clamped to zero. At this membrane potential there was no net force causing Na^+ to flow, since there was no concentration difference across the membrane and no electrical force. When this was done, the inward current after stimulation disappeared as expected (Fig. 17–10, trace *B*), indicating that it was normally caused by Na^+. In the same axon, when E_m was clamped to a positive value there was a net electrical force causing Na^+ to leave the axon, and an outward Na^+ current was observed (Fig. 17–10, trace *C*), the reverse of the normal situation.

2. The involvement of Na^+ in the inward current during the action potential was also demonstrated without changing the external $[Na^+]$. When E_m was clamped to +58 mV, the Na^+ equilibrium potential at which no net flow of Na^+ occurs, the inward current disappeared. Clamping to positive E_m greater than +58 mV caused an outward Na^+ current. The potential at which the Na^+ current reverses direction (+58 mV) is called the *reversal potential* for the Na^+ current.

It was concluded that rapidly depolarizing the membrane causes a rapid increase in the flow of Na^+ into the axon and a slower increase in the flow of K^+ out of the axon. Using the techniques described earlier to separate the Na^+ current from the K^+ current, Hodgkin and Huxley were able to determine the conductance, g, for each ion as a function of time during the voltage clamp at different

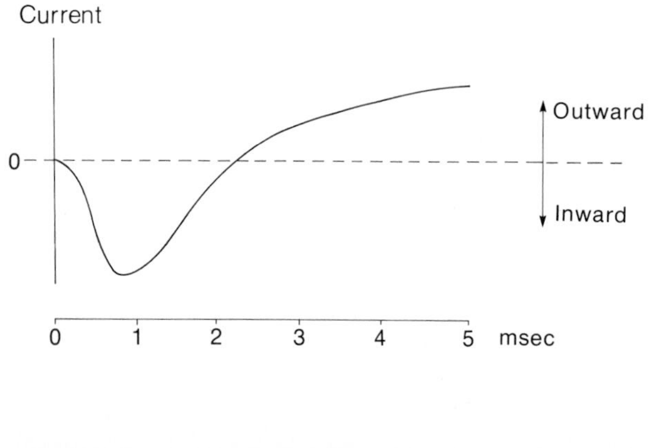

FIGURE 17–9 Net membrane current in squid giant axon following voltage clamp to O mV transmembrane potential. (Redrawn from Hodgkin and Huxley.)

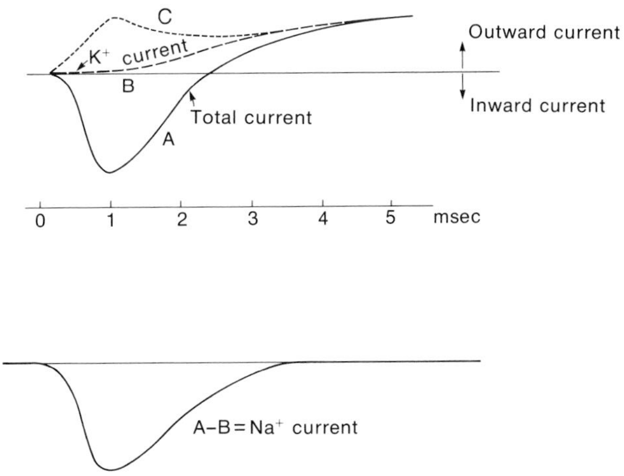

FIGURE 17–10 Net membrane currents during voltage clamp of squid giant axon. *A*. Clamp to O mV in normal seawater. *B*. Clamp to O mV when [Na^+] in medium equals that in the cytoplasm. *C*. Clamp to a positive membrane potential under same conditions as in *B*. (Redrawn from Hodgkin and Huxley.)

E_m. [If I and E are known, g can be computed, since $I_{Na} = g_{Na} (E_m - E_{Na})$ and $I_K = g_K (E_m - E_K)$]. For example, when the membrane potential was clamped to zero, the results shown in Figure 17–11 were obtained. Note that the increase in g_{Na} shuts off even though the clamp is maintained, but the g_K increase remains stable and does not decrease to basal levels until the clamp is released. Therefore, it is said that the g_{Na} increase is self-inactivating. It was found also that smaller depolarizations lead to smaller increases in g_{Na} and g_K.

In summary, these studies showed that there are three basic kinetic events in the action potential: the increase of g_{Na}, the return of g_{Na} to resting levels, and the increase of g_K. The rate and extent of each of these depends on the level of membrane potential. By performing voltage clamp experiments at many different E_m levels, Hodgkin and Huxley were able to reconstruct the time courses and magnitudes of the changes in g_{Na} and g_K that must occur during the action potential.

The Na^+ and K^+ Channels

In order to account for the changes in conductance that occur during the action potential, Hodgkin and Huxley proposed that there are separate Na^+ and K^+ channels, each with distinct characteristics, in the plasma membrane. More recent research, especially by Hille, has given us some ideas about the characteristics of these channels, as shown in Figure 17–12. The Na^+ channel is believed to have both an activation gate and an inactivation gate to account for the self-inactivation of g_{Na} during voltage clamp. It is noteworthy that digestion of the interior of the squid axon with the proteolytic enzyme pronase abolishes the self-inactivation, presumably by destroying the inactivation gate.

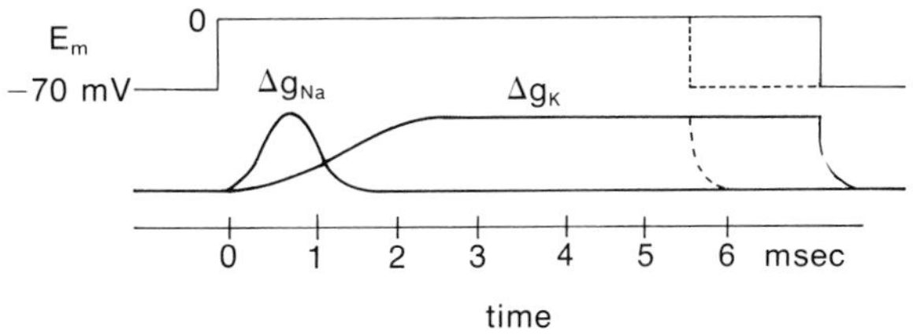

FIGURE 17–11 Changes in conductances of sodium and potassium elicited by voltage clamping the squid giant axon to O mV. (Redrawn from Hodgkin and Huxley.)

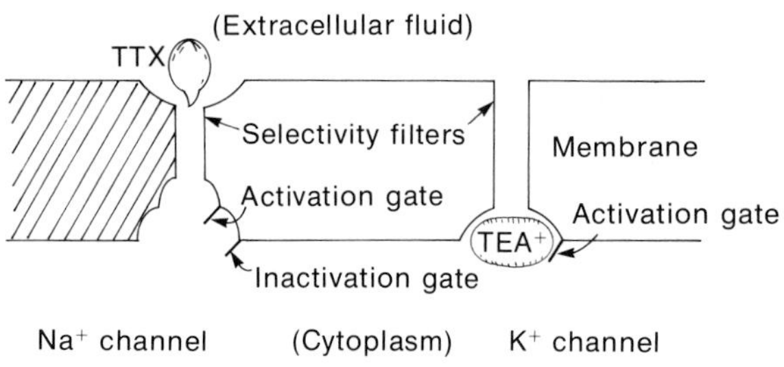

FIGURE 17–12 Schematic representation of Na^+ and K^+ channels of squid giant axon. (Based on the work of Hille and Armstrong.)

In order to enter the part of the channel known as the selectivity filter, it is believed that K^+ and Na^+ must shed most of their water of hydration. They can then form coordination bonds with oxygen atoms which line the interior of the selectivity filter. It is believed that the configuration of the oxygen atoms results in the specificity of the Na^+ and K^+ channels. Tetrodotoxin (TTX) is a specific blocker of the Na^+ channel, and tetraethylammonium ion (TEA^+) blocks the K^+ channel. The sites at which they are thought to block the channels are shown in Figure 17–12.

It is believed that the gates of the channels may be charged and perhaps are peptide chains. The opening and closing of the gates as a function of membrane potential is associated with a redistribution of charges across the membrane which are called gating currents. Currents that occur during the action potential and that have some of the properties expected of gating currents have recently been observed.

Explanation of Certain Properties of the Action Potential

Voltage Inactivation of the Action Potential. If the magnitude of the resting membrane potential is progressively decreased (made less negative), by increasing the concentration of K^+ in the extracellular fluid, for example, the action potential has a progressively slower rate of rise and a smaller overshoot. This is the result of two factors: (1) the smaller force driving Na^+ into the partly depolarized cell; and (2) *voltage inactivation* of the Na^+ channels. As already described, the increase in g_{Na} during the action potential is self-inactivating. However, once the Na^+ channels are inactivated, before they can be reopened the membrane must be repolarized toward the normal resting membrane potential. As the membrane potential is restored to normal resting levels, more and more of the Na^+ channels return to the state in which they can be activated again (Fig. 17–13). Note, however, that as a squid giant axon is depolarized from −70, fewer and fewer Na^+ channels can be activated. When O mV potential is reached only

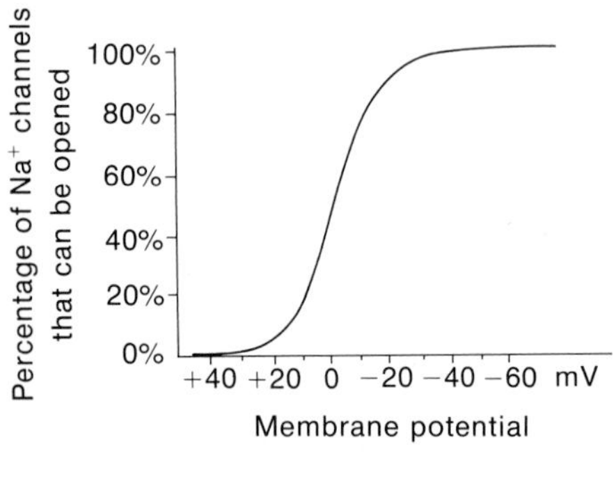

FIGURE 17–13 Voltage inactivation of sodium channels of squid giant axon. As the resting membrane potential is decreased, the number of sodium channels that remain activatable decreases. (After Hodgkin and Huxley.)

half of the Na^+ channels of a squid axon can be activated. In fact, since the action potential mechanism requires that a "critical number" of Na^+ channels be opened, it may fail even though a considerable fraction of Na^+ channels remain in a state of activation. This is a *voltage inactivation* of the action potential due to voltage inactivation of the Na^+ channels. Voltage inactivation of the Na^+ channels is involved in some important properties of excitable cells, such as refractoriness and accommodation.

THE REFRACTORY PERIODS. During most of the action potential, the membrane is refractory to further stimulation. That is, no matter how strongly the cell is stimulated it cannot fire a second action potential. This period is called the *absolute refractory period* (Fig. 17–14). The reason that the cell is refractory is that a large fraction of its Na^+ channels are voltage-inactivated, and they can be restored to the state in which they can be reopened only by repolarization of the plasma membrane.

During the latter part of the action potential, the cell can be caused to fire a second action potential, but a stronger than normal stimulus is required. This period is called the *relative refractory period*. During the early part of the relative refractory period, before the membrane potential has returned to the resting potential level, a fraction of the Na^+ channels are voltage-inactivated. Thus a stronger than normal stimulation is required to open a "critical number" of Na^+ channels to trigger the action potential. In addition, throughout the relative refractory period the conductance to K^+ is elevated, which provides an increased opposition to depolarization of the membrane and thus contributes to the refractoriness.

ACCOMMODATION TO SLOW DEPOLARIZATION. If the plasma membrane of a nerve or muscle cell is depolarized slowly, the threshold for firing may be passed without an action potential being fired. This is called *accommodation*. Both Na^+ and K^+ channels are involved in accommodation. If the cell is depolarized slowly, some of the Na^+ channels that are opened by depolarization have sufficient time to become voltage-inactivated before the threshold potential can be reached. If the depolarization is sufficiently slow, the "critical number" of open Na^+ channels required to trigger the action potential may never be achieved. In addition, the K^+ channels open in response to prolonged depolarization and the increased value of g_K tends to repolarize the membrane, making it still more refractory to stimulation. Because of accommodation, a very weak stimulus will not cause an action potential no matter how long it is applied.

The Strength-Duration Curve. An electrical stimulus depolarizes the cell membrane by causing charge to flow across it. The relevant quantity is the *total*

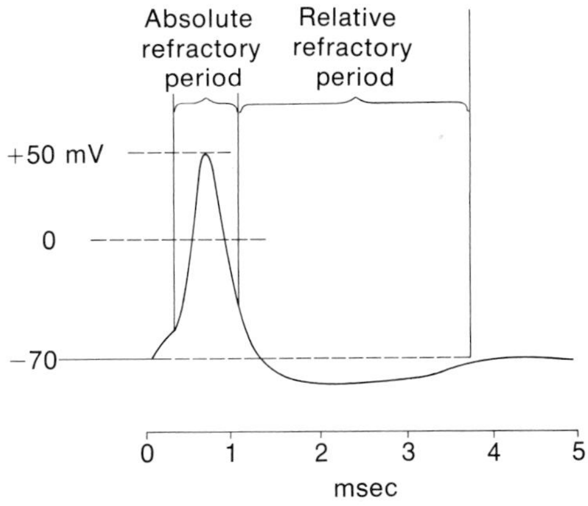

FIGURE 17–14 The refractory periods of squid giant axon are shown relative to the time course of the action potential.

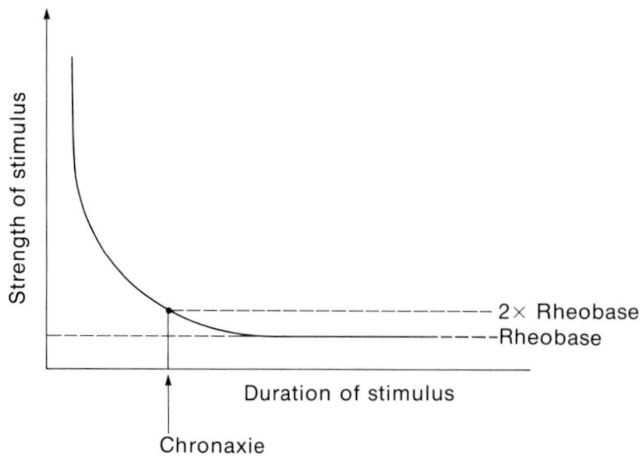

FIGURE 17–15 Schematic depiction of a strength-duration curve.

amount of charge that flows across the membrane (current × time). It follows that a strong current can depolarize the membrane to threshold quite quickly, but a weaker current must be applied longer, so that the critical amount of charge can flow across the membrane. These relations are often depicted in the *strength-duration curve* (Fig. 17–15), which plots the strength of stimuli versus the time they must be applied to elicit an action potential.

A very weak stimulus will fail to elicit an action potential even if applied for a very long time. The *minimum* stimulus strength that can elicit an action potential from a given preparation is called the *rheobase*. The time required for a stimulus strength *twice* rheobase to elicit an action potential is called the *chronaxie* of the preparation. Chronaxie is often used as an index of the excitability of a preparation. The larger the chronaxie, the less excitable the preparation.

FURTHER READING

Adelman, W. J., Jr. (ed.): Biophysics and Physiology of Excitable Membranes. New York, Van Nostrand, 1971.

Aidley, D. J.: The Physiology of Excitable Cells. New York, Cambridge University Press, 1971, pp. 1–92.

Armstrong, C., and Bezanilla, F.: Current related to the gating particles of the sodium channels. Nature, *242*:459, 1973.

Baker, P. F.: The nerve axon. Sci. Am., *214*:74, 1966.

Baker, P. J., Hodgkin, A. L., and Shaw, T. I.: The effects of changes in internal ion concentrations on the electrical properties of perfused giant axons. J. Physiol., *164*:355, 1962.

Davson, H.: A Textbook of General Physiology. 4th ed. Baltimore, Williams & Wilkins, 1970, pp. 548–713, 1031–1126.

Dowben, R. M.: General Physiology. New York, Harper & Row, 1969, pp. 210–211, 390–402, 474–480.

Hodgkin, A. L.: The Conduction of the Nervous Impulse. Springfield, Charles C Thomas, 1964, pp. 11–46, 56–85.

Hodgkin, A. L., and Huxley, A. F.: The components of membrane conductance in the giant axon of *Loligo*. J. Physiol., *116*:473, 1952.

Hodgkin, A. L., and Huxley, A. F.: Currents carried by sodium and potassium ions through the membrane of the giant axon of *Loligo*. J. Physiol., *116*:449, 1952.

Hodgkin, A. L., and Huxley, A. F.: The dual effect of membrane potential on sodium conductance in the giant axon of *Loligo*. J. Physiol., *116*:497, 1952.

Hodgkin, A. L., and Huxley, A. F.: Movement of radioactive potassium and membrane current in a giant axon. J. Physiol., *121*:403, 1953.

Hodgkin, A. L., and Huxley, A. F.: A quantitative description of membrane current and its application to conduction and excitation in nerve. J. Physiol., *117*:500, 1952.

Katz, B.: Nerve, Muscle, and Synapse. New York, McGraw-Hill, 1966, pp. 1–90.

Kotyk, A., and Janacek, K.: Cell Membrane Transport: Principles and Techniques. New York, Plenum Press, 1970, pp. 91–150, 247–268.

Kuffler, S. W., and Nicholls, J. G.: From Neuron to Brain. Sunderland, Mass. Sinauer Associates, 1976, pp. 77–131.

18

NERVE CELLS

In Chapter 17, the basic concepts of electrical activity of cells were introduced, including the way in which cells produce a resting membrane potential, and the mechanism of the action potential in excitable cells. In the present chapter, the structure and physiology of neurons, cells that are specialized for excitation and conduction of electrical impulses, will be considered. In Chapter 19, the principles of electrical activity of cells will be applied to the muscle cells, which are specialized not only for excitation but also for contraction. In both chapters, emphasis is placed on the way in which basic *cellular* processes have developed and have become specialized in the nerve and muscle cell. This can serve as the basis for consideration elsewhere of nerve and muscle *tissues*, which involve more complicated interactions between cells and are appropriate subjects for texts of nerve and muscle anatomy and physiology.

MORPHOLOGY OF NERVE CELLS

A central dogma in neuroanatomy is the neuron doctrine. This theory holds that the nervous system is composed of neurons, which are separate and distinct cellular units and are related to one another through cell contacts, or synapses, at which impulses are transmitted from one cell to another. This seems like an obvious statement, but it is often emphasized in beginning a consideration of the structure of nerve cells because historically there was much debate over whether the nervous system was composed of separate cells or was a continuous network of protoplasm. One contribution of the famous neuroanatomist Ramon y Cajal was his role in demonstrating the cellular nature of the nervous system through studies carried out around the turn of the century.

Neurons are cells specialized for excitation and for the conduction of impulses. Thus their shape is important (Figs. 18–1, 18–2), because it determines how they relate to one another. Neurons have a cell body that comprises the nucleus and the surrounding cytoplasm, or perikaryon. Two types of cell processes, dendrites and axons, permit nerve cells to relate to one another. The dendrites are multiple branched processes that together with the cell body receive impulses from other neurons. The shape of the dendritic tree varies greatly in neurons found in different locations. It reflects the neuron's receptive field or how that particular neuron receives input from other cells. The axons are rela-

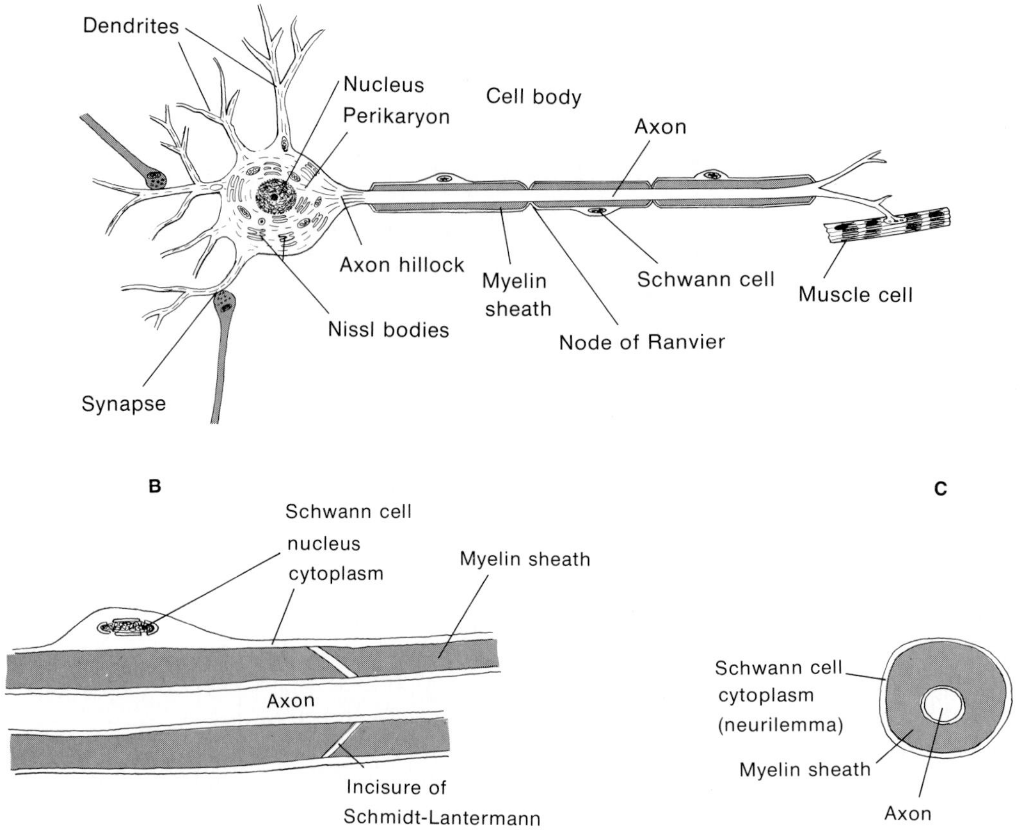

FIGURE 18–1 *A*. Diagram of parts of a neuron, such as one of the motor neurons of the ventral horn of the spinal cord. *B*. Part of the axon and its sheaths are shown diagrammatically in longitudinal section. *C*. The axon and sheaths in cross section.

tively straight, less branched processes that conduct impulses over long distances. Each neuron has only one, or a small number, of axons.

Neurons are frequently classified according to their shape into multipolar, bipolar, or unipolar cells. Multipolar cells, such as the pyramidal cells of the motor cortex, have numerous processes that arise from the cell body; usually there are multiple branched dendrites and a single relatively unbranched axon. Bipolar neurons have two processes, which arise from opposite sides of the cell body. Examples are the bipolar cells of the retina. Unipolar cells are found in the sensory ganglia. They have only a single process which divides a short distance from the cell body into two branches that proceed in opposite directions. Neurons are also distinguished by their shape as Golgi type I and Golgi type II cells. Golgi type I neurons are those with long axons that extend to some other part of the nervous system, while Golgi type II cells have short axons that remain within the region of gray matter in which the cell body lies.

Cell Body

Many neurons have a large round or oval nucleus located near the center of the cell body (Fig. 18–3). It contains much euchromatin and a large nucleolus, indicating a high level of activity. In the cytoplasm of the perikaryon are numer-

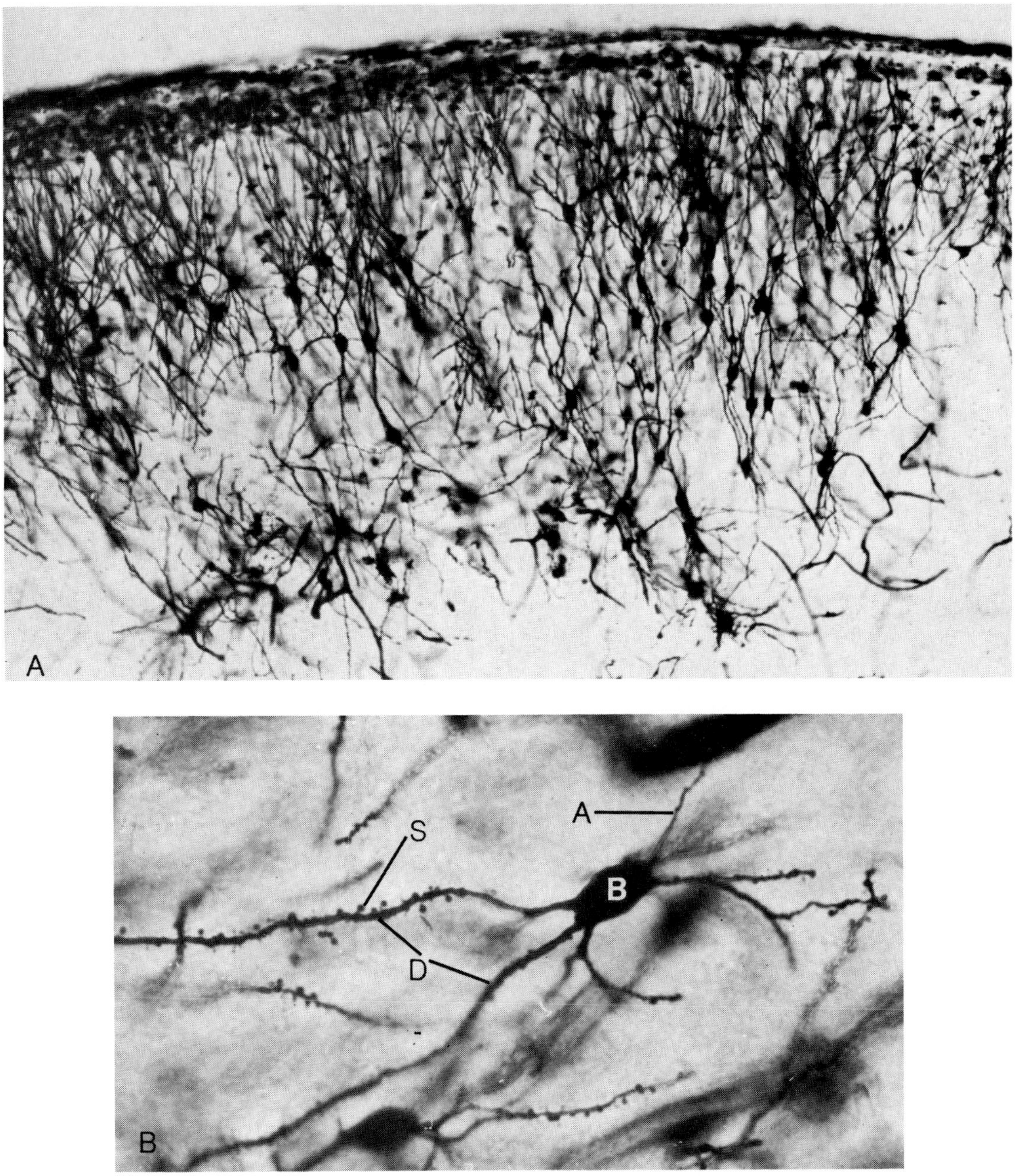

FIGURE 18–2 Light micrographs of neurons in the rat olfactory cortex, prepared by a silver impregnation method which blackens some neurons throughout their entire extent. This method is particularly useful for demonstrating the shape and orientation of various types of neurons. *A*. Processes of many neurons extend toward the surface of the brain at the top of the field. × approx. 200. *B*. Processes extending from the cell body (B) of a neuron include several dendrites (D) and a single thin axon (A). Small spines (S) are visible on the dendrite. × approx. 1,500. (Micrographs courtesy of L. Heimer.)

ous basophilic bodies termed Nissl bodies, the Nissl stain being a basic dye that is widely used in neuroanatomy. With the electron microscope (Figs. 18–4 and 18–5), the Nissl bodies are seen to be composed of cisternae of rough endoplasmic reticulum (RER) and free polysomes. The pattern of Nissl bodies varies and is characteristic of neurons in different locations. Nissl bodies are present in the

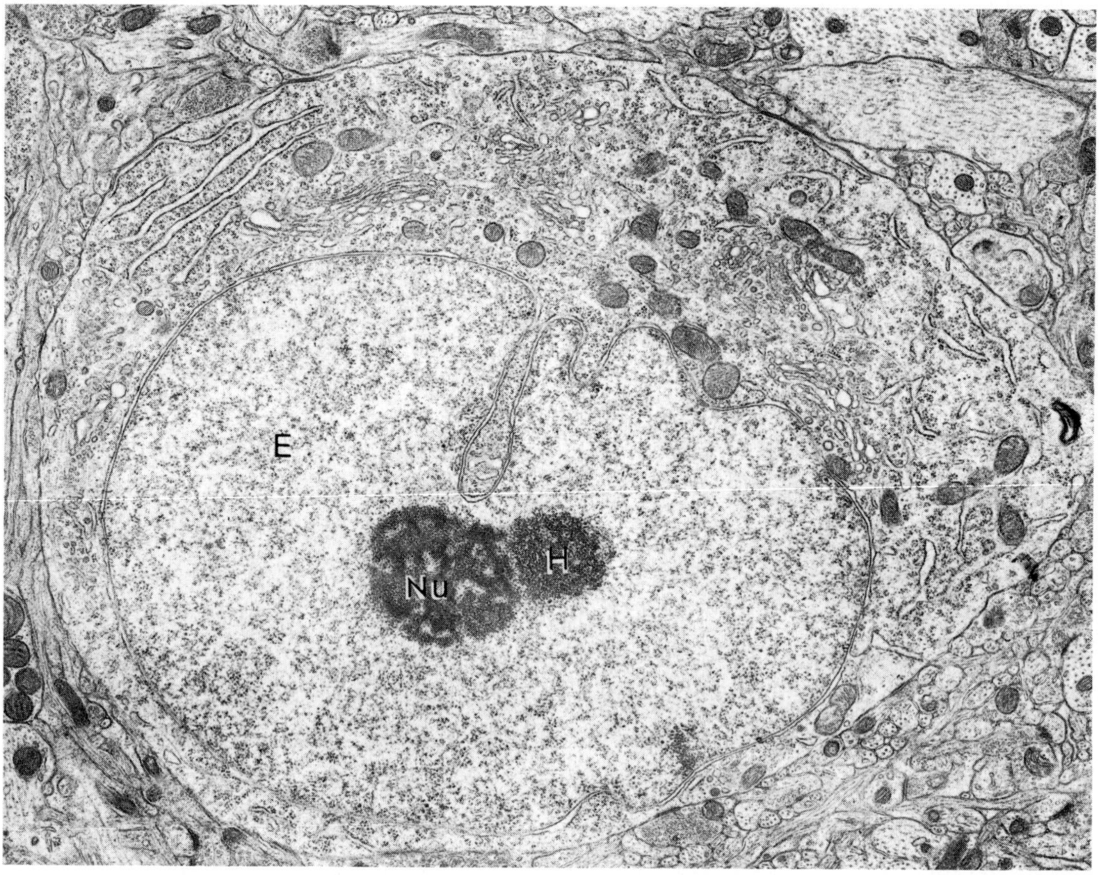

FIGURE 18–3 Cell body of a neuron in the rat olfactory cortex. The nucleus contains a prominent nucleolus (Nu) and much euchromatin (E). There is little heterochromatin except for that (H) associated with the nucleolus. × 11,000. (Micrograph courtesy of Heimer, L., and Kalil, R.: J. Comp. Neurol., *178*:559, 1978. Reproduced by permission of the Wistar Institute Press.)

perikaryon and in dendrites but are absent from axons. The perikaryon (Fig. 18–4) also contains the Golgi apparatus, which was first discovered in neurons. It forms a network in the cytoplasm around the nucleus, so that in sections through the perikaryon, the characteristic stacks of cisternae are seen to be arranged in a ring surrounding the nucleus.

Both microtubules and filaments are prominent in neurons (Fig. 18–6). The microtubules, or neurotubules, as they are sometimes called, are about 200 Å in diameter. They resemble microtubules found in other cells and are very numerous. The brain is a rich source of tubulin for biochemical studies. The neurofilaments are about 100 Å in diameter. Thus individual microtubulues and filaments are not visible in the light microscope, but bundles of the tubules and filaments stain with metal impregnation methods and were described years ago by light microscope cytologists as neurofibrils. Microtubules and filaments tend to run longitudinally in neuronal processes. They are thought to have a role in transport of materials along the processes. The microtubules participate in growth of axons and may aid in maintaining the shape of the processes.

The neuronal cytoplasm, of course, also contains the other common cell organelles, such as mitochondria and lysosomes. As will be described later, specialized vesicles play an important part in neuronal function.

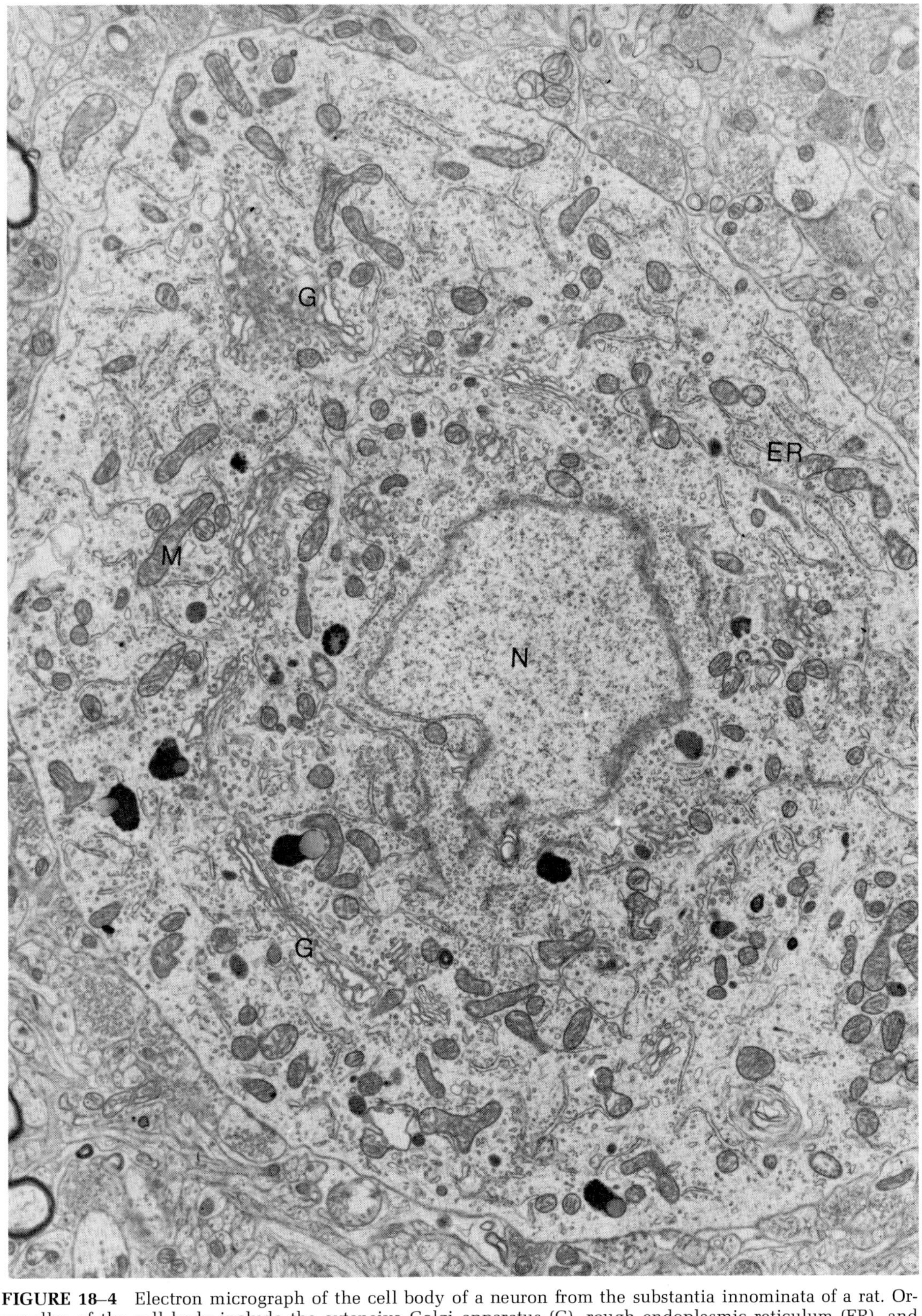

FIGURE 18–4 Electron micrograph of the cell body of a neuron from the substantia innominata of a rat. Organelles of the cell body include the extensive Golgi apparatus (G), rough endoplasmic reticulum (ER), and mitochondria (M). N, nucleus. × 11,000. (Micrograph courtesy of L. Heimer.)

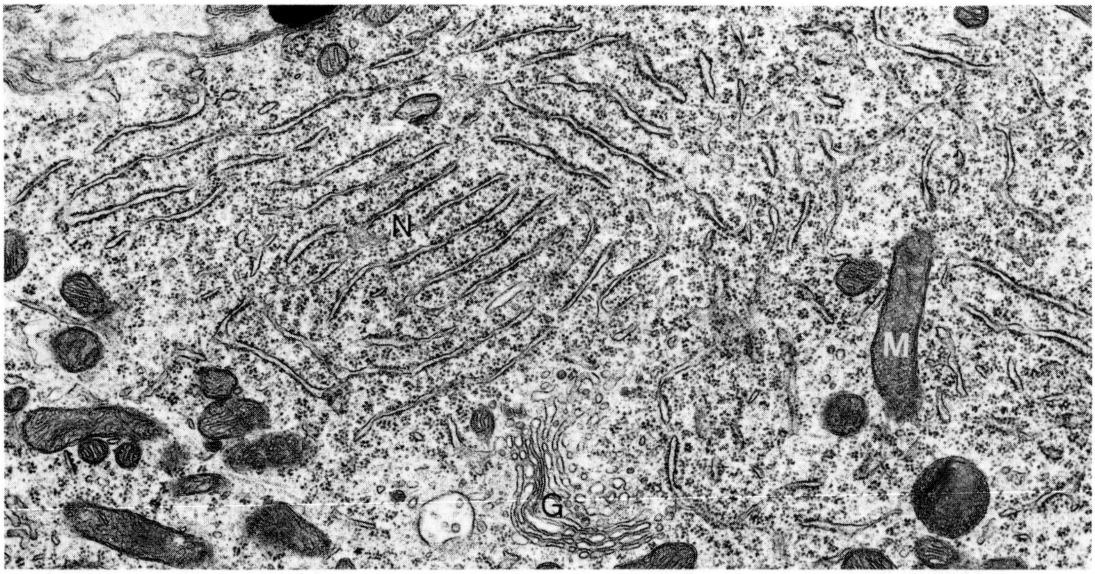

FIGURE 18–5 Part of the cytoplasm of the cell body of a neuron in the rat olfactory cortex. Nissl bodies (N) are composed of rough endoplasmic reticulum and free ribosomes. G, Golgi apparatus; M, mitochondria. × 19,000. (Micrograph courtest of L. Heimer.)

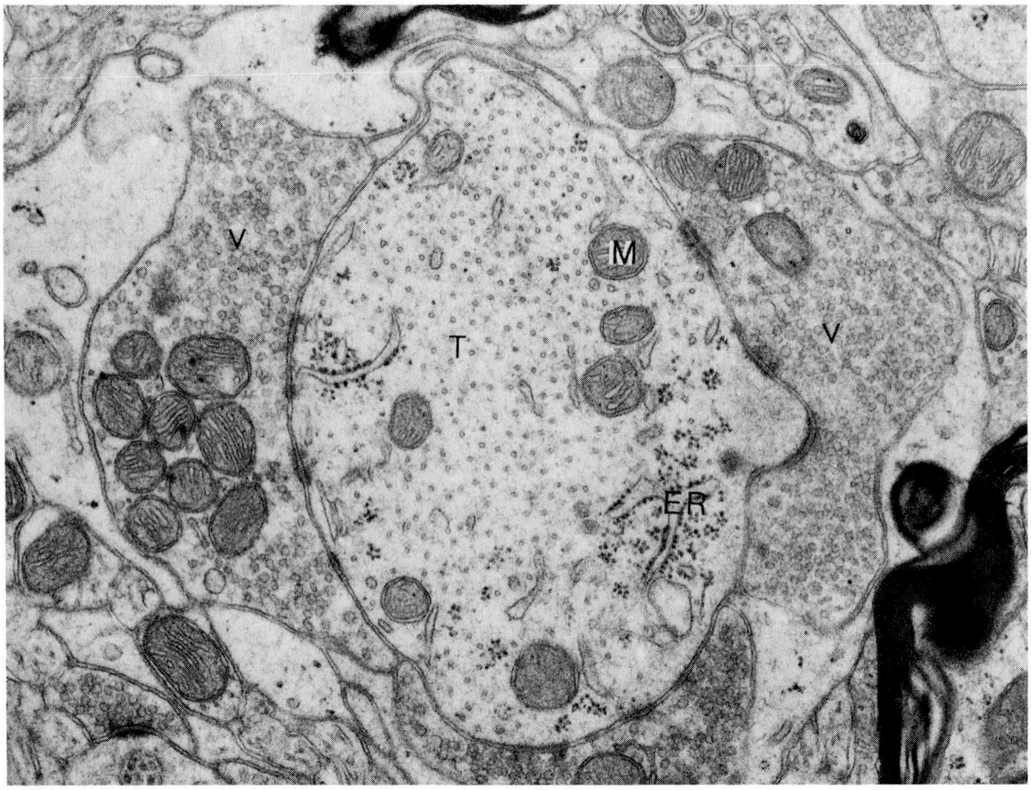

FIGURE 18–6 Cross section of a dendrite in the rat brain, showing mitochondria (M), some rough endoplasmic reticulum (ER), and cross sections of microtubules (T). Processes of other neurons containing synaptic vesicles (V) make synaptic contact with the dendrite. × 27,000. (Micrograph courtesy of L. Heimer.)

Dendrites

Cytologically and functionally, the dendrites (Figs. 18–6 and 18–7) can be regarded as extensions of the cell body. They contain the same organelles as the perikaryon, including Nissl bodies, and never have a myelin sheath. Dendrites taper, branch, and generally have irregular contours. These features serve to increase the surface area and are related to their function of receiving impulses. Most of the synapses on a neuron occur on the dendritic stems (Fig. 18–6) or on specialized projections, the dendritic spines, (Figs. 18–2 and 18–7), which are composed of a stalk and a terminal expansion or bulb.

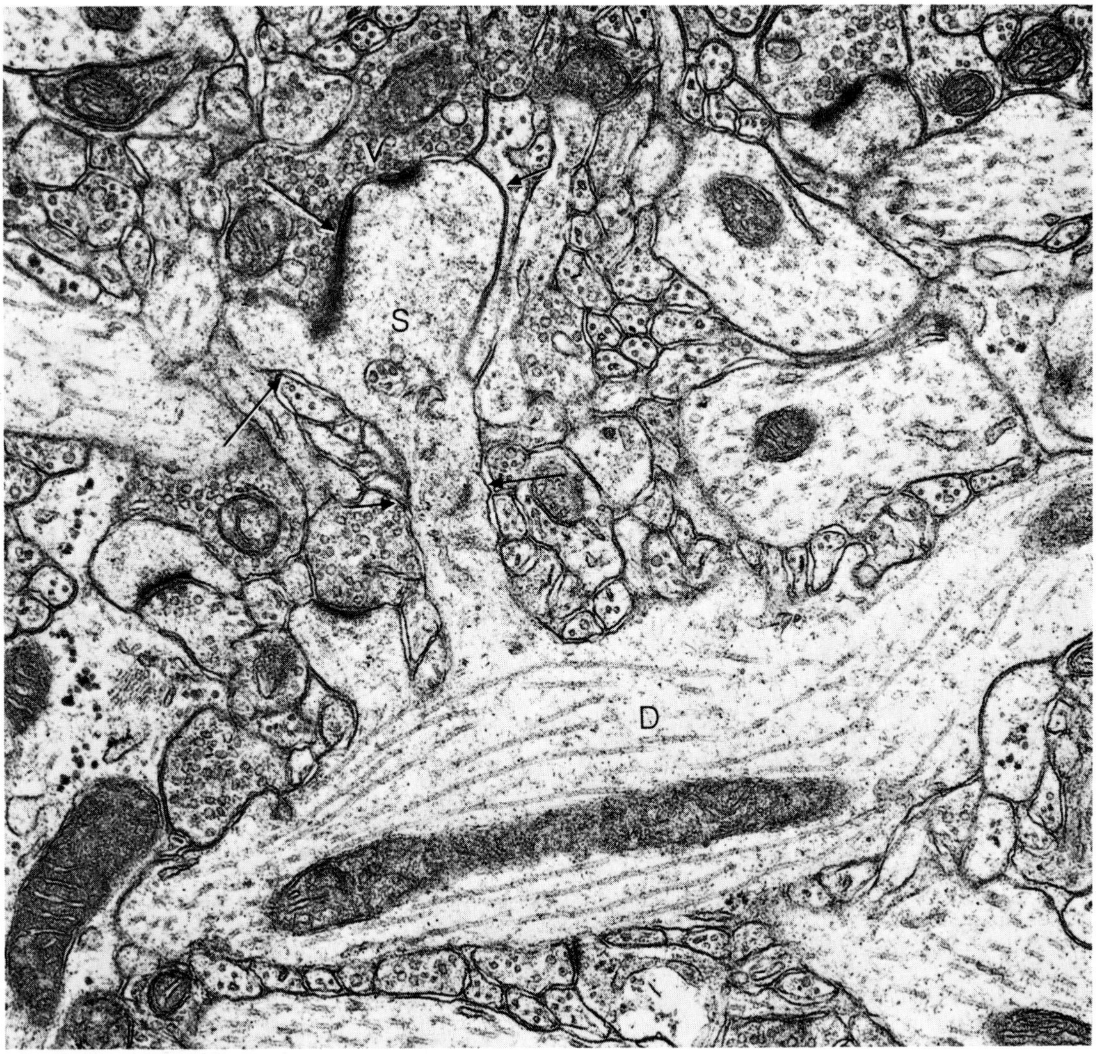

FIGURE 18–7 Portion of a dendrite in the rat olfactory cortex. A spine (S), which projects from the stem of the dendrite (D), is outlined by arrows. An axon containing synaptic vesicles (V) makes contact with the spine. × 32,500. (Micrograph courtesy of L. Heimer.)

Axons

Axons are usually long straight processes of relatively uniform caliber (Figs. 18–1 and 18–2). These features reflect their main function of transmitting impulses. They may be branched, but this usually occurs infrequently in comparison with dendrites. Axons arise from a conical elevation on the surface of the cell body, referred to as the axon hillock.

Since axons and axon hillocks lack ribosomes and rough endoplasmic reticulum, the proteins necessary for their maintenance must be transported from the cell body. There has been much investigation of transport of substances along axons. Materials are transported at different speeds, and both slow and fast axonal transport components have been described. The fast-moving proteins generally are thought to be those synthesized on ribosomes attached to the rough endoplasmic reticulum and involved in the renewal of membranous components of the axons, such as the plasma membrane and synaptic vesicles. The proteins that are transported more slowly may consist mainly of those synthesized on unattached ribosomes and subsequently found in the cytoplasmic matrix and in microtubules.

Some structural differences between dendrites and axons are summarized in Table 18–1.

Nerve Fibers

In the peripheral nervous system, axons are surrounded by sheath cells, the Schwann cells. A nerve fiber is defined as an axon along with its accompanying Schwann cells. Axons are related to Schwann cells in two main ways.

In the case of *unmyelinated nerve fibers*, the axon simply occupies an indentation in the surface of the Schwann cell (Fig. 18–8). One Schwann cell may enclose several axons in this way. However, one Schwann cell does not usually cover an axon through its entire course, and Schwann cells are found in series along the length of an axon.

The relationship between the axon and the Schwann cells is more complicated in *myelinated nerve fibers* because the Schwann cells have elaborated a myelin sheath around the axon. In contrast to unmyelinated nerves, one Schwann cell forms a myelin sheath for only one axon in the peripheral nervous system. As in the case in unmyelinated fibers, multiple Schwann cells occur in series

TABLE 18–1 SUMMARY OF STRUCTURAL DIFFERENCES BETWEEN DENDRITES AND AXONS

Dendrites	Axons
Many	Single
Irregular contour	Smooth
Taper	Uniform caliber
Twisting	Relatively straight
Highly branched	Few or no branches
Never myelinated	May be myelinated or unmyelinated
Ribosomes and RER present	No ribosomes or RER
Microtubules>neurofilaments	Neurofilaments>microtubules

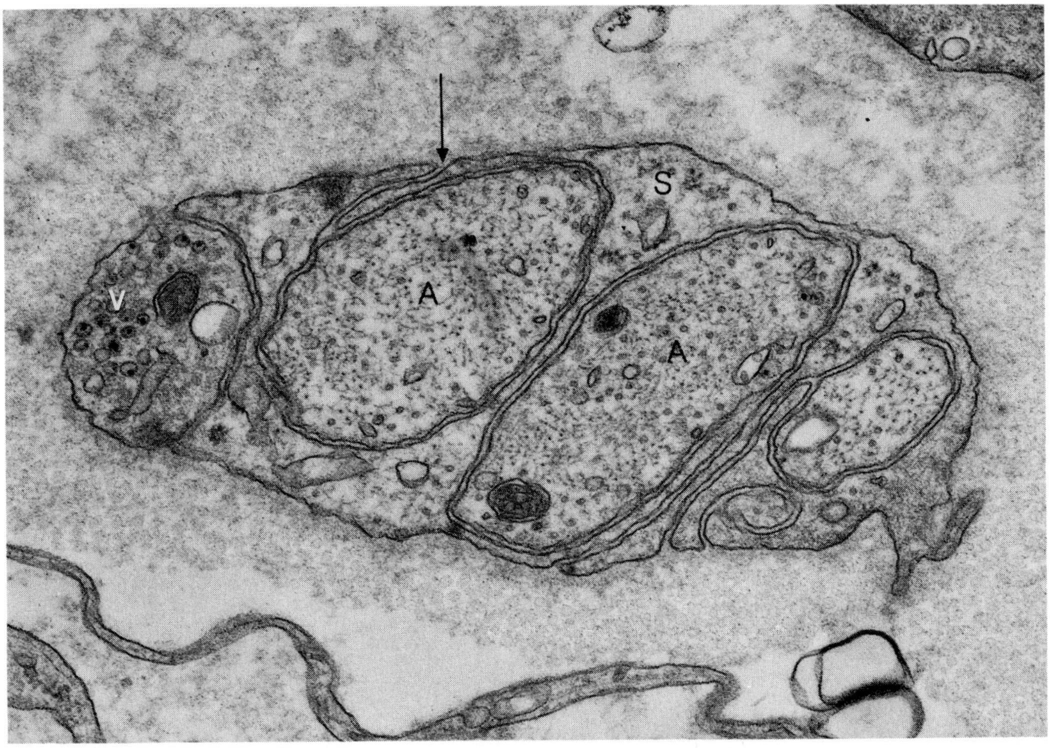

FIGURE 18–8 Cross section of a small unmyelinated nerve. Several axons (A) occupy individual recesses in the surface of a Schwann cell (S). Note that although some axons are deeply embedded in the troughs in the surface of the Schwann cell, the extracellular space surrounding the axons remains continuous with that around the nerve as a whole (*arrow*). One of the axons contains some synaptic vesicles (V). × 32,000. (*From*: Flickinger, C. J., Am. J. Anat., *134*:107, 1972. Reproduced by permission of the Wistar Institute Press.)

along the myelinated fiber. As will be described later, the myelin sheath is of great importance in the conduction properties of nerve fibers. In the fresh state, the lipid-rich myelin is glistening and white, which gives "white matter" in the central nervous system its characteristic appearance.

In light microscope preparations, the myelin sheath is found next to the axon in a location analogous to that of insulation on a wire (Fig. 18–1). Unless special methods are used, however, the lipid in the myelin is usually extracted, and all that remains is a network of precipitated protein that is sometimes termed neurokeratin. The cytoplasm of the Schwann cell forms a tenuous layer, called the neurilemma in classic histology, which in turn surrounds the myelin sheath (Fig. 18–1).

The structure of the myelin sheath has been greatly clarified by electron microscope studies. It is readily seen to be composed of a pattern of light and dark lines (Figs. 18–9 and 18–10). The thickest of the dark lines, the *major dense* lines, are spaced about 170 Å apart, and they determine a repeating period of about 170 Å in the radial direction with respect to the axis of the nerve fiber. Midway between the major dense lines there is a thinner, less dense *intraperiod line*. Light lines lie between the major dense lines and the intraperiod lines. Myelin in the peripheral nervous system is derived by wrapping of the plasma membrane of the Schwann cell many times in a spiral about the axon (Fig. 18–11). Most of the cytoplasm of the Schwann cell within the spiral wrapping is lost, so that myelin consists almost entirely of many apposed layers of Schwann cell plasma

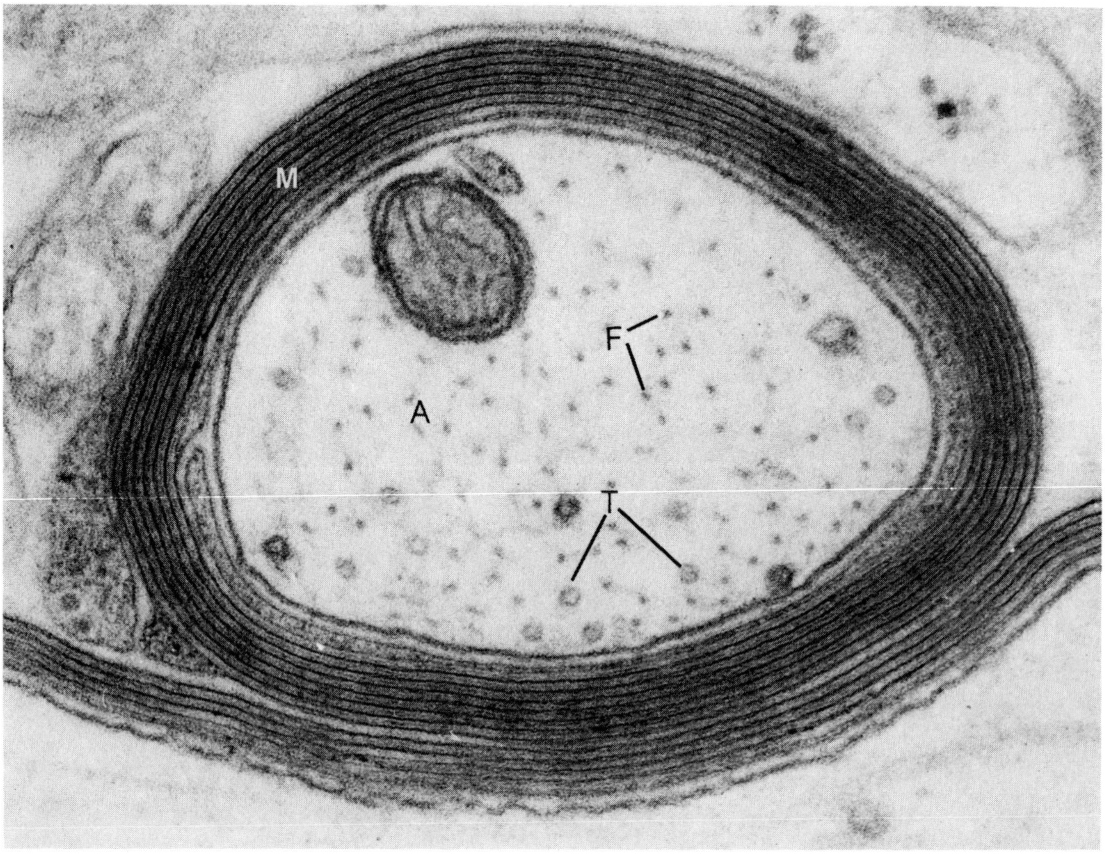

FIGURE 18–9 Cross section of a myelinated axon in the lateral hypothalamus of the rat. The myelin sheath (M) displays a pattern of alternating light and dark lines. Within the cytoplasm of the axon (A) are punctate cross-sectional profiles of neurofilaments (F) and a few microtubules (T). × 85,000. (Micrograph courtesy of L. Heimer.)

membrane. With this information, the origin of the pattern of dark and light lines in myelin can be discerned (Fig. 18–11).

The Schwann cell plasma membrane has the usual trilaminar appearance in the electron microscope, with a cytoplasmic dense line, a light line, and an external dense line. If the major dense line in myelin is traced to the Schwann cell,

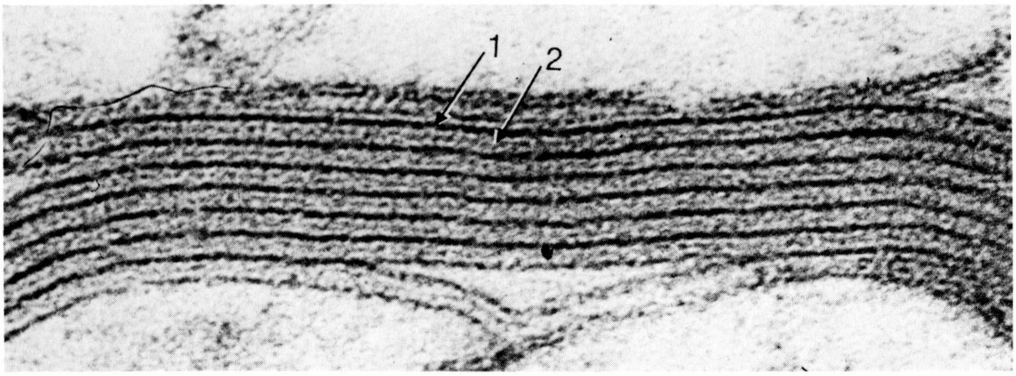

FIGURE 18–10 Part of a cross section of a myelin sheath at high magnification. The denser lines (1) represent the major period lines. Midway between them are the intraperiod lines of lesser density (2). × 175,000. (Micrograph courtesy of L. Heimer.)

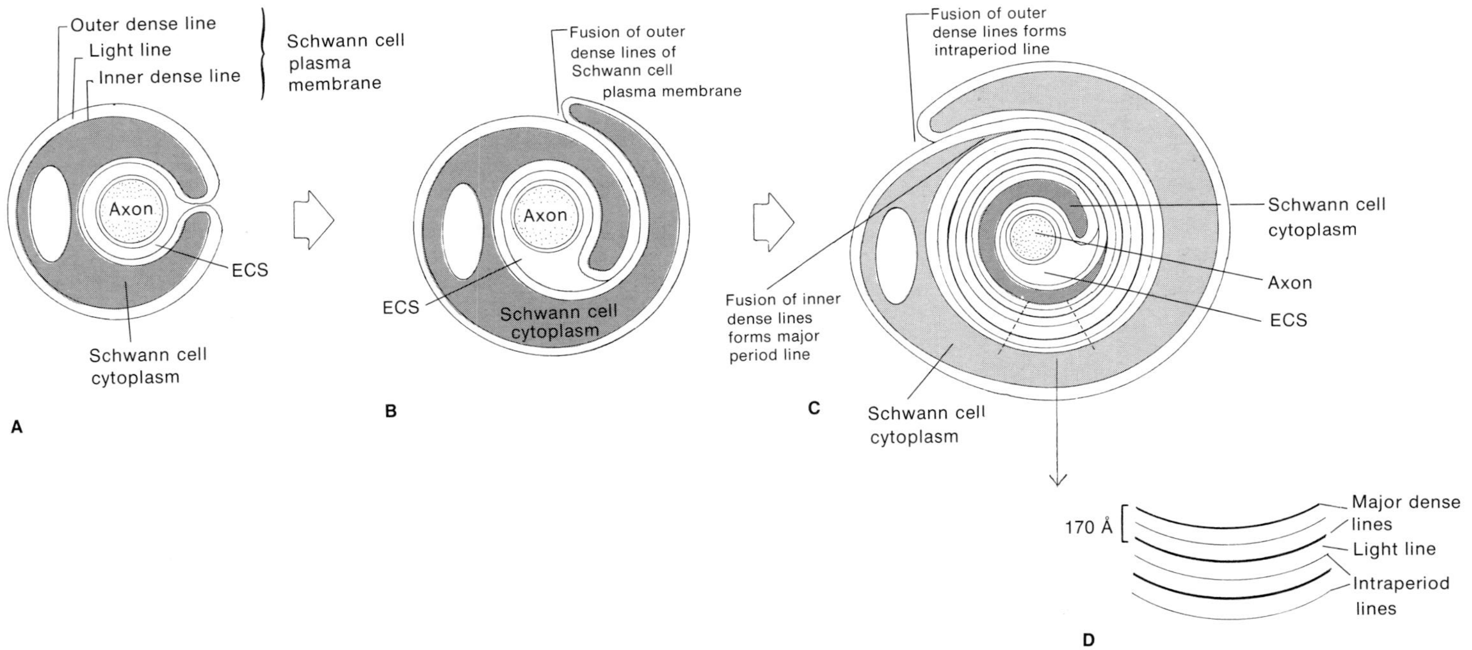

FIGURE 18–11 Diagram of the formation of myelin by wrapping of Schwann cell plasma membrane around an axon. *A*. The axon lies in an indentation in the surface of the Schwann cell. The Schwann cell plasma membrane is three-layered, consisting of an outer dense line next to the extracellular space, a central light line, and an inner dense line next to the cytoplasm. *B*. The Schwann cell has begun to wrap around the axon in a spiral. The outer dense lines of the Schwann cell plasma membrane are fused together for a short distance. *C*. Myelin has been formed by wrapping of the Schwann cell around the axon and loss of Schwann cell cytoplasm from within the wrappings, so that the layers of Schwann cell plasma membrane can fuse with one another. Note that the major dense line is formed by fusion of two inner dense lines of the plasma membrane, while the intraperiod line is formed by fusion of two outer dense lines of the plasma membrane. Some Schwann cell cytoplasm remains in the innermost portion of the spiral close to the axon. *D*. A portion of the myelin sheath. The repeating period of 170 Å in the radial direction is defined by the major dense lines.

it is seen to be formed by the apposition of two cytoplasmic dense lines of the Schwann cell plasma membrane. Similarly, the intraperiod line is formed by apposition of two external dense lines of the Schwann cell plasma membrane. The light lines in myelin correspond to the light line in the Schwann cell plasma membrane.

At intervals along the length of an axon, the myelin sheath is interrupted at nodes of Ranvier. The nodes represent the places at which two Schwann cells in series along an axon abut. Thus all of the myelin sheath between two successive nodes is formed by one Schwann cell. At a node of Ranvier, successive spirals of Schwann cell plasma membrane touch the axon, beginning with the innermost spiral and followed by the others in succession (Fig. 18–12). In longitudinal sec-

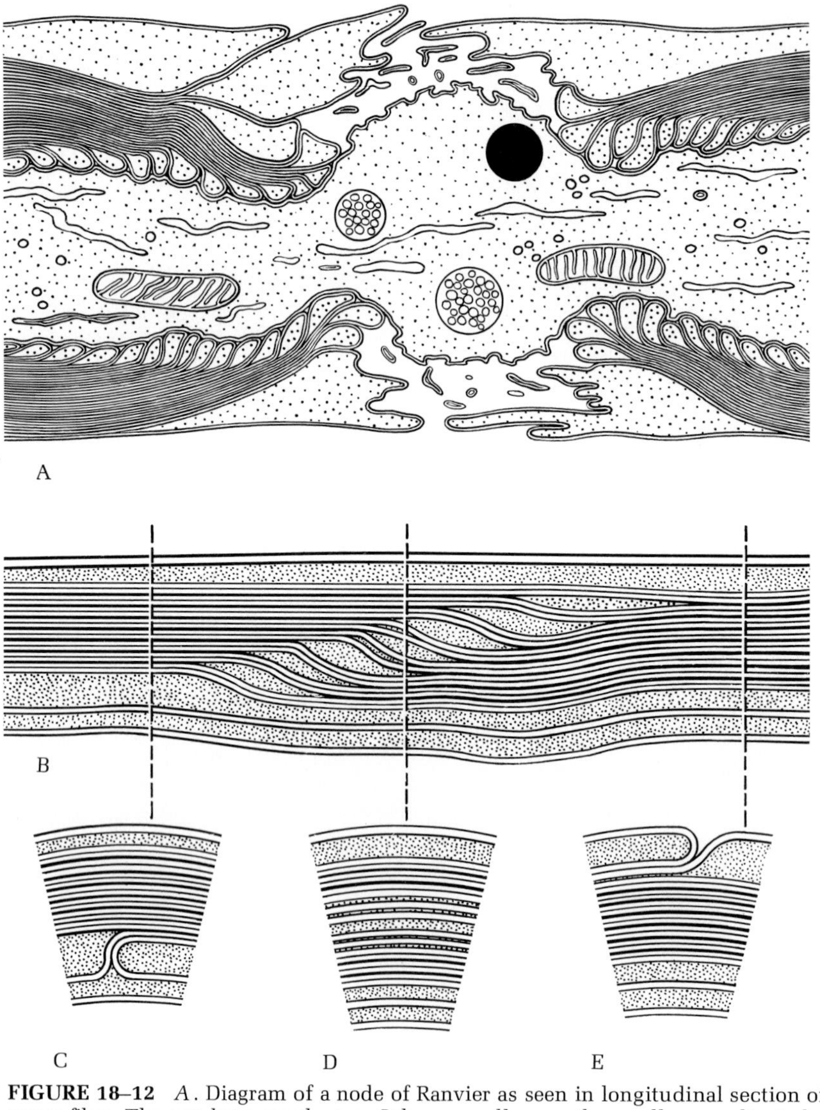

FIGURE 18–12 *A*. Diagram of a node of Ranvier as seen in longitudinal section of a myelinated nerve fiber. The gap between the two Schwann cells over the small unmyelinated segment of the axon is shown. (Redrawn from Robertson, Prog. Biophys. Mol. Biol. *10*:344, 1960.) *B–E*. Diagrams of a cleft of Schmidt-Lanterman. *B*. Cleft in longitudinal section. *C,D,E*. Transverse sections of the cleft at various points. (Redrawn from Robertson, Mol. Biol., New York, Academic Press, 1960.) *A–E*. From Bloom, W., and Fawcett, D. W.: A Textbook of Histology. 9th ed. Philadelphia, W. B. Saunders, 1968.

tions of myelinated nerves, small oblique clefts, the incisures of Schmidt-Lantermann, are seen (Fig. 18–1). These represent regions at which the myelin lamellae are buckled and a small amount of Schwann cell cytoplasm intervenes within the major dense line (Fig. 18–12). A thin tongue of Schwann cell cytoplasm thus pursues a spiral course from the bulk of the cell cytoplasm outside the myelin sheath to the inner part of the sheath close to the axon. This could provide a course for cytoplasmic transport of materials to and from the axon through the myelin sheath.

In the central nervous system, there are no Schwann cells. However, myelin is formed in a manner similar to that described by the oligodendroglial cells of the central nervous system. One difference is that a single oligodendroglial cell may form myelin sheaths for several axons in the central nervous system, whereas Schwann cells in the peripheral nervous system form myelin for only one axon.

CONDUCTION OF THE ACTION POTENTIAL

One of the main functions of neurons, carried out by their axons, is the conduction of action potentials. In the case of certain neurons, such as the motor neurons of the ventral horn of the spinal cord, an action potential is conducted from the cell body of the neuron in the spinal cord, by way of a motor axon, to a skeletal muscle fiber. The distance from the motoneuron to one of the muscle fibers it innervates may be as long as 1 meter. The mechanism by which the action potential is conducted and the factors that determine the speed of conduction are considered in the following section.

The Local Response

Figure 18–13 schematically depicts the membrane of an axon or muscle fiber that has been depolarized in one region. The external face of the membrane in the depolarized region is negative relative to the adjacent membrane, and the internal face of the membrane is positively charged relative to neighboring areas. Thus currents flow as shown in Figure 18–14. This current flow has the effect of depolarizing the areas of the membrane adjacent to the initial site of depolarization. The newly depolarized areas can then depolarize other segments of the membrane that are still further removed from the initial site of depolarization. This is the mechanism by which the *local response*, or electrotonic response, occurs (see Chapter 17).

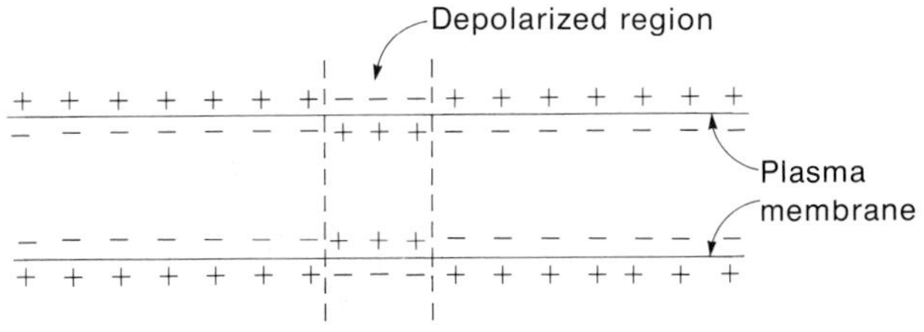

FIGURE 18–13 Diagram of an axon or muscle fiber with a localized depolarization of its plasma membrane.

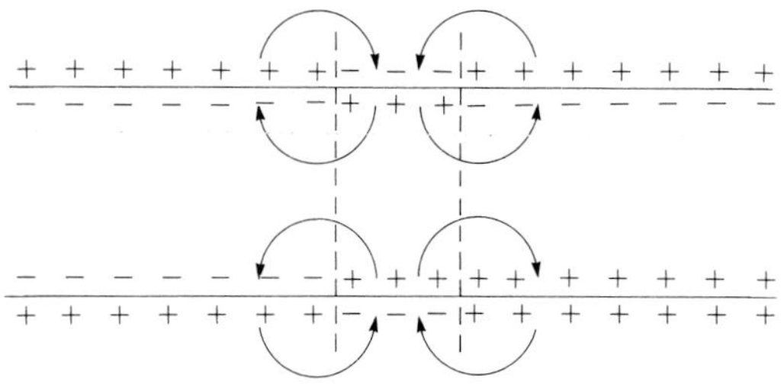

FIGURE 18–14 The spread of depolarization along an axon or muscle fiber by local current flow.

Speed of Conduction

Two main factors determine the speed with which the conduction mechanism for the local response can spread a depolarization along a nerve or muscle fiber: the membrane capacitance and the electrical resistance inside and outside the axon.

The cell membrane is a capacitor, which is an insulator separating two conductors, and the membrane potential is a manifestation of the charge stored on the membrane capacitor. The amount of charge that must flow to depolarize 1 cm^2 of membrane depends on the membrane capacitance, which is the amount of charge that is stored per volt of potential difference across the membrane. A typical value for membrane capacitance (C_m) is 10^{-6} farad/cm^2 membrane. Therefore, to depolarize the membrane from -100 mV (-0.1 v) to 0 mV requires that 10^{-6} farad/cm^2 $\times$ 10^{-1} volt $=$ 10^{-7} coulomb/cm^2 must flow.

Thus membrane capacitance determines *how much* charge must flow to depolarize the membrane. The internal and external resistance of the cell determines *how rapidly* the charge can flow. As seen in Figure 18–4, local current flow takes place within the cell and in the extracellular space surrounding the cell. In most tissues the electrical resistance of the extracellular space is quite low. Within the cell, however, the resistance to current flow may be substantial, principally because the current must flow through a small cross-sectional area. Accordingly it is the resistance to longitudinal current flow within the cell (R_{in}) that limits the rate at which current can flow to depolarize the membrane.

A larger membrane capacitance (C_m) means that more charge must flow to depolarize the membrane. Larger internal resistance (R_{in}) implies greater impedance to current flow to depolarize the membrane. Thus, the larger the product $R_{in}C_m$, the slower the spread of a localized depolarization. $R_{in}C_m$ is called the *time constant* for conduction. The larger $R_{in}C_m$, the smaller the *conduction velocity* of the nerve or muscle cell. The smaller the time constant, the faster the conduction.

The membrane capacitance depends on the thickness of the plasma membrane and on its insulating properties. The internal resistance of the interior of the cell depends on the electrical resistivity of the cytoplasm and on the diameter of the cell. Just as with a wire, the larger the cross-sectional area of the cell, the smaller its internal resistance. As a nerve or muscle fiber gets larger, the surface area and hence the capacitance per unit length of fiber increases in proportion to the radius of the fiber. The cross-sectional area, however, increases with the square of the radius. Thus while C_m per unit length increases with fiber radius, R_{in} per unit length decreases with the square of the radius, so that $R_{in}C_m$ (the time constant) decreases with increasing radius. For this reason, the larger the diameter of the nerve or muscle fiber, the larger the conduction velocity.

Conduction with Decrement: The Need for a Self-Reinforcing Signal

When the local response was discussed in Chapter 17, it was noted that over the course of several millimeters the local response dies away to almost nothing (see Figure 17–5). A nerve or muscle fiber may be thought of as an electrical cable. In a perfect cable the insulation surrounding the conductile core prevents all loss of current so that a signal is transmitted down the cable with undiminished strength. The insulation in a nerve or muscle fiber is provided by the plasma membrane, which has a resistance much higher than that of the cytoplasm. However, partly because of its extreme thinness, it is a far from perfect insulator. The ratio of R_m (the membrane resistance) to R_{in} (the internal resistance) describes the quality of a particular cell as a cable. The higher the ratio of R_m to R_{in}, the better the cable and the longer a signal can be transmitted without significant decrement. R_m/R_{in} determines the *length constant* of a fiber, which is defined as the distance over which a conducted signal falls to 37 per cent of its initial height. The length constant for most nerve and muscle fibers is about 2 to 3 mm. However, since some axons in the human body are about 1 meter long, it is clear that the local response alone will not suffice to conduct a signal over such a distance.

Many nerve and muscle fibers are substantially longer than their length constants. In order to be able to conduct an electrical impulse with undiminished strength along the full length of these fibers the action potential mechanism, which *reinforces itself* as it propagates, is required. The action potential propagates by the mechanism described in Figures 18–13 and 18–14. When the segments on either side of the depolarized region reach threshold, these segments fire an action potential, which locally reverses the polarity of the membrane potential. By local current flow the areas of the fiber adjacent to these sites are brought to threshold, and they then fire an action potential. This cycle of depolarization by local current flow, followed by generation of an action potential in a restricted region, travels along the length of the fiber. In this way the action potential is propagated over long distances without alteration in form or amplitude.

It should be noted that the information-carrying capacity of nerves is restricted by this mechanism of propagation. Since the form and size of the action potential are invariant, only variations related to the frequency of action potentials can code information for transmission along axons. The frequency is limited by the duration of the absolute refractory period (about 2 msec) to a maximum of 500 impulses/sec.

Myelination and Conduction Velocity

A squid giant axon of 500 μm diameter (which is unmyelinated) has a conduction velocity of 25 m/sec. Since conduction velocity is directly proportional to fiber radius, a human fiber of 10 μm diameter should conduct at 0.5 m/sec. With such a conduction velocity, a reflex arc from the foot would require about 4 seconds. That is, if a man stepped on a hot coal, it would be 4 seconds before he began to withdraw his foot. Of course, our reflexes are much faster than this, despite our nerve fibers being much smaller in diameter than squid giant axons. This is due to the presence of the myelin sheath in some human fibers, which results in the myelinated fibers having much faster conduction than nonmyelinated fibers of similar diameter. A 10 μm myelinated fiber actually has a conduction velocity of 50 m/sec, which is twice that of the 500 μm squid giant axon.

This high speed of conduction allows our reflexes to be fast enough to avoid dangerous stimuli. Figure 18–15 illustrates the effect of myelination in greatly increasing the conduction velocity.

The development of myelinated fibers in vertebrates is of considerable evolutionary importance. If our peripheral nerve fibers had to be as large as squid giant axons, then the peripheral nerve trunks, which contain hundreds of nerve fibers, would be so large that human anatomy would need to be considerably different just to accommodate the increased size of nerve trunks.

As described earlier, wrapping of Schwann cell plasma membrane around a nerve axon results in the formation of a myelin sheath consisting of several to 100 or more plasma membrane layers (Fig. 18–16). The gaps in the sheath every 1 to 2 mm along the axon are the nodes of Ranvier, which are about 1 μm wide and represent the spaces between different Schwann cells lined up along the axon.

The myelin sheath drastically alters the electrical properties of the axon. The multiple wrappings of membrane around the axon result in the *membrane resistance* being substantially elevated and R_m/R_{in} being much greater. This means that less of a conducted signal is lost through the electrical insulation of the membrane (see page 447), and consequently the amplitude of a conducted signal does not decline so rapidly as it is conducted along the axon.

Each of the Schwann cell membranes has a capacitance similar to that of the axolemma. The capacitances of each of the membranes in the myelin sheath are essentially connected in series. The rule for adding capacitors in series is $1/C_t = 1/C_1 + 1/C_2 + 1/C_3 + \ldots\ldots$ and so on, where C_t is the overall capacitance and C_1, C_2, C_3, and so on are the individual capacitances. If $C_1 = C_2 = C_3 = \ldots\ldots C$ and there are 50 capacitances in series (for 25 Schwann cell wraps), then $1/C_t = 50/C$ and $C_t = C/50$. Thus a myelin sheath with 50 individual membranes has the effect of lowering the *membrane capacitance* by a factor of 50. On page 446 it was shown that the time constant $R_{in}C_m$ determines the conduction velocity, the lower the time constant, the higher the conduction velocity. Since 25 Schwann cell turns have no effect on R_{in}, but lower C_m by about 50-fold, there is about a 50-fold increase in conduction velocity owing to the myelination.

There is still another characteristic of conduction in myelinated fibers that contributes to its rapidity. This is *saltatory conduction*, which means that the impulse jumps from node to node. Saltatory conduction occurs because the action potential is regenerated only at the nodes of Ranvier (1 to 2 mm apart) rather than being regenerated continuously as the impulse is propagated. The internodal plasma membrane does not fire an action potential because the depolarization it experiences is spread over 50 or so membranes in series, including the axonal and Schwann cell membranes in the myelin sheath, so that the actual depolari-

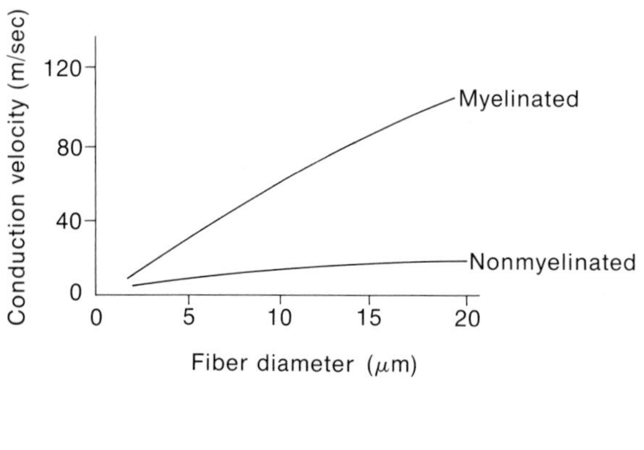

FIGURE 18–15 Conduction velocities of myelinated and nonmyelinated axons as functions of fiber diameter.

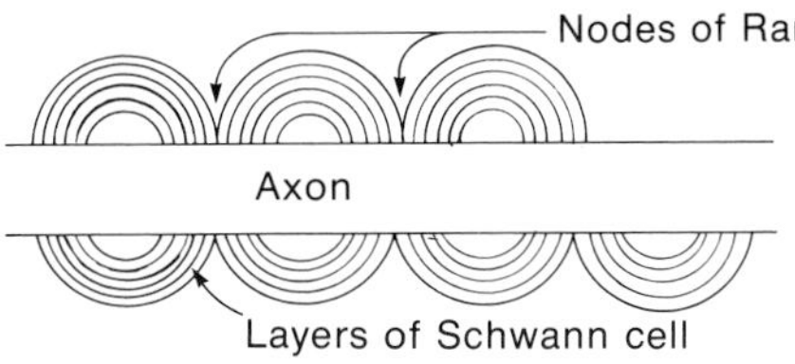

FIGURE 18–16 Schematic depiction of the myelin sheath of a myelinated axon.

zation of the neuronal plasma membrane is only 1 or 2 mV, which is not nearly sufficient to reach threshold.

Myelinated axons are also more metabolically economical than nonmyelinated axons. After the action potential has occurred, the sodium-potassium pump must extrude the sodium that entered and reaccumulate the potassium that left the axoplasm during the action potential. Restricting the ionic currents to a small fraction of the membrane surface at the nodes of Ranvier markedly decreases the ion pumping required.

A number of human neurological disorders are termed demyelinating diseases and are characterized by injury to myelin with relative preservation of axons. Among these diseases are: the lipidoses, the leukodystrophies, and those diseases in which there is abnormal breakdown of apparently normal myelin. In the lipidoses, such as Tay-Sachs and Niemann-Pick diseases, there is abnormal accumulation of particular lipids. In the leukodystrophies there is abnormal formation of myelin. Multiple sclerosis is one of the diseases in which there is abnormal breakdown of apparently normal myelin. The demyelinating diseases cause a wide variety of mental abnormalities and motor and sensory dysfunctions.

SYNAPTIC TRANSMISSION: GENERAL FEATURES

A synapse is a site of functional contact between excitable cells. It is the place at which an impulse is transmitted from one cell to another. Two general types of synapses can be distinguished, electrical synapses and chemical synapses.

At an *electrical synapse*, the two excitable cells communicate by the direct passage of electric current between them, a process known as ephaptic or electrotonic transmission. Gap junctions are present between electronically coupled cells, and they are thought to be the places where current passes directly between the cells. There are few examples of ephaptic transmission in the vertebrate nervous system; in most cases it appears that information is transferred instead by means of chemical synapses. Perhaps this is because the chemical synapse is better suited for the complex modulation of synaptic activity and the integration that occurs at synapses in the vertebrate central nervous system.

At a *chemical synapse* a transmitter substance is released from one neuron, the presynaptic cell. The transmitter diffuses across the extracellular space and interacts with receptors on the surface of the postsynaptic cell to produce a change in its electrical properties. Chemical synapses are characterized by a synaptic delay due to the time required for these events to take place. A chemical mechanism is also utilized for transmission of an impulse between nerve and muscle cells at the myoneural junction. In the peripheral nervous system the two main transmitter substances are acetylcholine (ACh) and norepinephrine. Acetylcholine is also utilized as a transmitter in the central nervous system, but other substances have been implicated as well.

Although the structure of the pre- and postsynaptic cells, the structure of the synapse, and the transmitter substance vary, certain features of chemical synapses are almost universal and are outlined in Table 18–2.

Although others before him had suggested the existence of chemical transmitters, Otto Loewi in 1921 was the first to unequivocally demonstrate the action of a transmitter substance. Electrical stimulation of the vagus nerve of the frog causes a dramatic decrease in heart rate. Loewi perfused the heart of a frog while repetitively stimulating the vagus nerve. A second frog heart was perfused with the effluent from the first heart. The second heart showed a decrease in heart rate when the vagus nerve of the first frog was stimulated, demonstrating that a substance released into the fluid perfusing the first heart when its vagus was stimulated could cause a decrease in the heart rate of the second heart. Loewi called this substance vagusstoff (vagus substance) and was able to show that acetylcholine, when infused into the frog heart, had the same effect as vagusstoff. He also found that the actions of both acetylcholine and vagusstoff were blocked by the drug atropine, which we now know blocks the postsynaptic receptors for acetylcholine, and were potentiated by eserine, which inhibits acetylcholine esterase—the enzyme that breaks down acetylcholine. Loewi concluded that stimulation of the vagus nerve of the frog resulted in the release of acetylcholine which in turn caused the decreased heart rate.

Subsequent investigation showed that acetylcholine is a transmitter at a number of vertebrate synapses, including the neuromuscular junction, in autonomic ganglia, at postsynaptic terminals of all parasympathetic neurons, and in certain sympathetic fibers.

COMMUNICATION BETWEEN NERVE AND MUSCLE CELLS

The synapse between the axons of motor neurons and skeletal muscle fibers is known as the neuromuscular junction, the myoneural junction, or the motor end plate. Because it is more accessible to direct investigation than most other synapses, the neuromuscular junction was the first vertebrate synapse to be studied in detail. Much of our knowledge of the mechanism of transmission at chemical synapses was discovered in pioneering investigations of the neuromuscular

TABLE 18–2 GENERAL CHARACTERISTICS OF TRANSMISSION AT CHEMICAL SYNAPSES

Action potential in presynaptic cell

↓

Depolarization of the plasma membrane of the presynaptic axon terminal

↓

Release by the presynaptic terminal of transmitter

↓

Chemical combination of the transmitter with specific receptors on the plasma membrane of the postsynaptic cell

↓

Transient change in the conductance of the postsynaptic plasma membrane to specific ions

↓

Transient change in the membrane potential of the postsynaptic cell junction. The neuromuscular junction thus serves as a model of the chemical synapse and provides a basis for understanding more complicated synaptic relations between neurons in the central nervous system.

It should be kept in mind that most of the following discussion of the neuromuscular junction deals specifically with the junction between nerves and *skeletal* muscle cells. The structural relationship between autonomic nerves and smooth or cardiac muscle cells differs from that between somatic motor nerves and skeletal muscle cells and is considered briefly later in the chapter.

Structure of the Neuromuscular Junction

At the neuromuscular junction in skeletal muscle (Fig. 18–17) the motor nerve loses its myelin sheath and breaks up into fine terminal branches that are covered by a Schwann cell, referred to as a teloglial cell. Each of these terminal branches of the axon lies in a shallow groove, or synaptic trough, on the surface of the muscle cell (Figs. 18–17 and 18–18). The plasma membrane, or sarcolemma, of the muscle cell lining the trough is pleated to form numerous junctional folds. The ends of the axons lying in the troughs contain many 400 Å smooth-surfaced vesicles, which have a clear interior and contain acetylcholine. The axon and the muscle cell remain separated by a carbohydrate-rich amorphous material which is a continuation of a similar layer, the external lamina, that surrounds the entire muscle cell. One motor neuron may branch and innervate more than one muscle cell; when this occurs, the nerve and the group of muscle cells that it innervates is called a motor unit.

Some electron microscopic investigations of the neuromuscular junction suggest that the acetylcholine receptor molecules are concentrated on the crests of the junctional folds, while acetylcholinesterase is evenly distributed on the surface of the postsynaptic membrane. The transmitter-containing vesicles in the nerve terminals appear to be concentrated opposite the mouths of the junctional folds. However, thickenings of the presynaptic membrane just opposite the junctional folds may prevent the release of acetylcholine directly into the depths of the folds (Fig. 18–19).

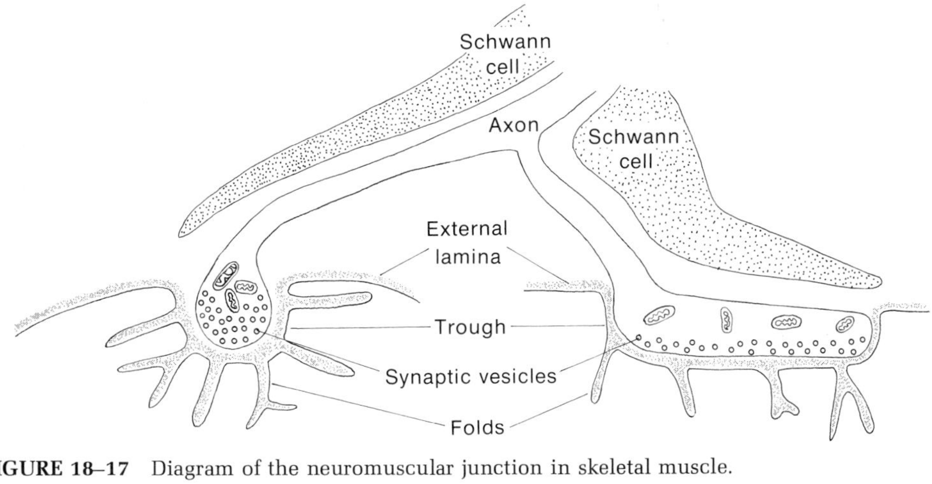

FIGURE 18–17 Diagram of the neuromuscular junction in skeletal muscle.

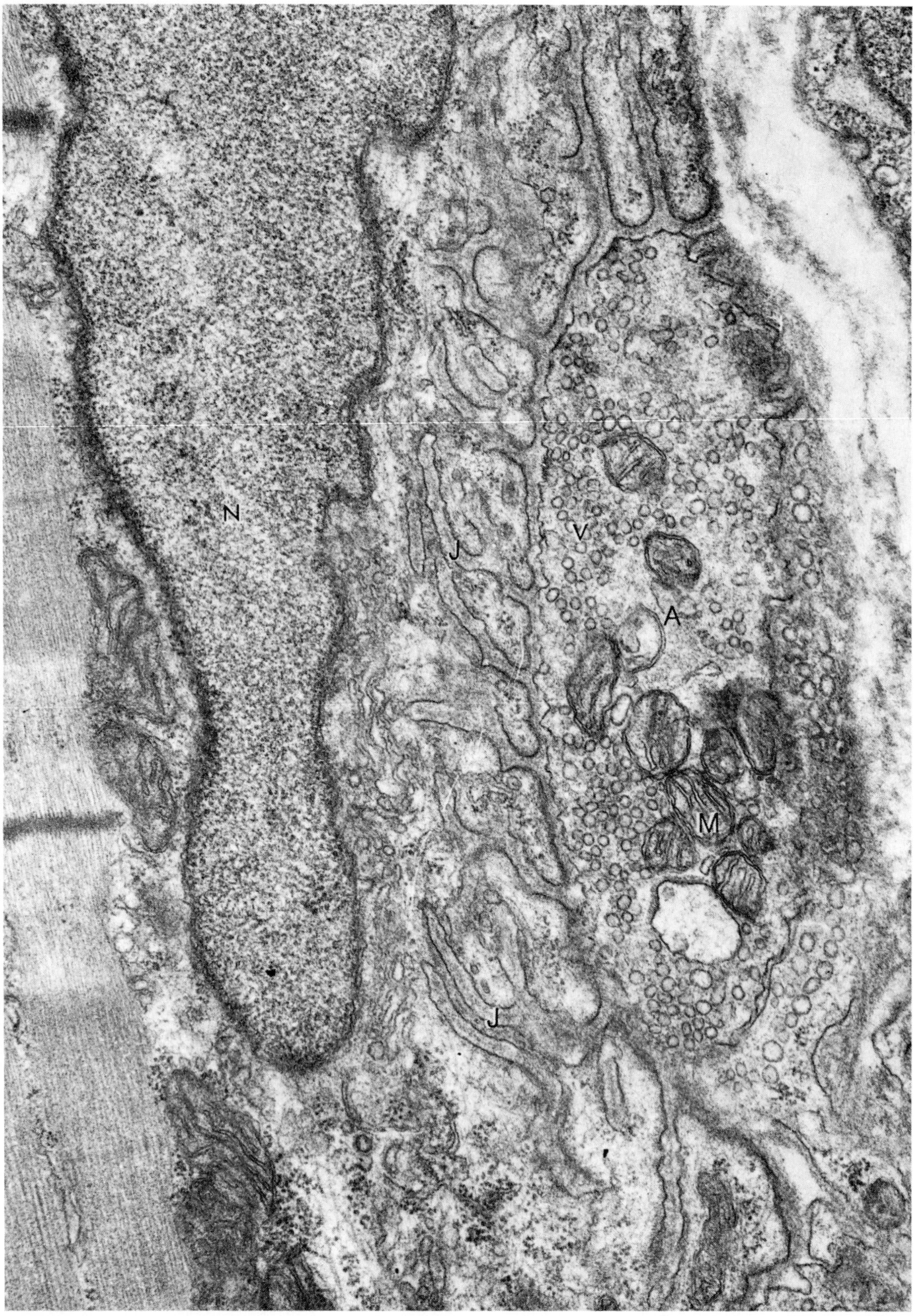

FIGURE 18–18 Electron micrograph of the neuromuscular junction at a red skeletal muscle fiber in the rat diaphragm. The end of an axon (A) lies in a depression on the surface of a muscle cell. The nerve ending contains synaptic vesicles (V) and mitochondria (M). Junctional folds (J) of the plasma membrane of the muscle cell underlie the area of contact with the axon. N, nucleus of the muscle cell. × 37,000. (Micrograph courtesy of Padykula, H. A., and Gauthier, G.: J. Cell Biol., *46*:27, 1970. Reproduced by permission of the Rockefeller University Press.)

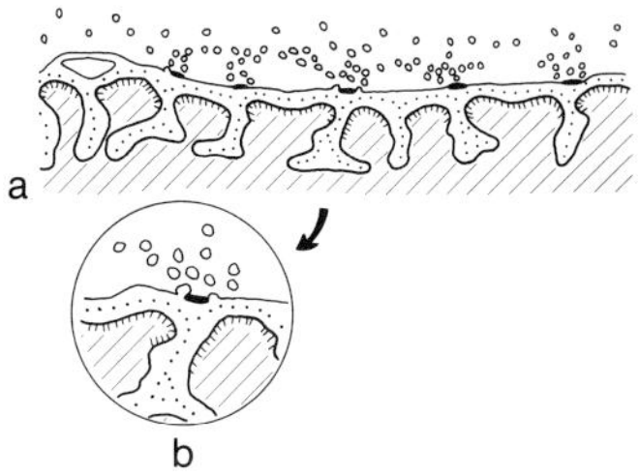

FIGURE 18–19 Highly schematic diagram of frog neuromuscular junction, showing the concentration of presynaptic vesicles opposite the mouths of the postjunctional folds and the concentration of acetylcholine receptors on the tips of the postjunctional folds. Note the thickening of the presynaptic membrane just opposite the mouths of the postjunctional folds. (From Porter, C. W., and Barnard, E. A.: J. Membrane Biol., *20*:1975. Reproduced by permission of Springer-Verlag.)

Overview of Transmission at the Neuromuscular Junction

Our current knowledge of the mechanism of transmission at the neuromuscular junction is the result of a great deal of experimental work, most notably that of Bernard Katz and his associates. The action potential is conducted down the motor axon to the presynaptic axon terminals. The depolarization of the plasma membrane of the axon terminal causes a transient increase in its permeability to Ca^{++}. Ca^{++} flows into the axon terminal, and by a mechanism that remains poorly understood induces the acetylcholine-filled synaptic vesicles to fuse with the plasma membrane and to empty their contents into the synaptic cleft by a process of exocytosis. Acetylcholine diffuses across the synaptic cleft and combines with a specific receptor protein that is embedded in the muscle plasma membrane of the motor end plate. The combination of acetylcholine with the receptor molecule causes a transient increase in the permeability of the membrane of the motor end plate to Na^+ and K^+ and thus a transient depolarization of the end plate region, which is known as the end plate potential or EPP. The EPP is a transient event because the action of ACh is terminated by its hydrolysis into choline and acetate by the enzyme acetylcholine esterase which is in high concentration on the postjunctional membrane.

The molecular details of the way in which combination of acetylcholine with the receptors causes increased Na^+ and K^+ conductance is currently the subject of intense study. The plasma membrane at the neuromuscular junction is not electrically excitable (i.e., it does not fire action potentials), but when it is depolarized, adjacent regions of the muscle cell membrane also become depolarized by electrotonic conduction. When these regions are depolarized to the threshold, an action potential is generated. The action potential is propagated along the length of the muscle fiber at high velocity and induces the muscle fiber to contract by a mechanism discussed in Chapter 19. These events are summarized in Table 18–3. Certain aspects of the neuromuscular transmission process will now be considered in more detail.

Physiological Features of Transmission at the Motor End Plate

Synthesis of Acetylcholine. Motor neurons and their axons synthesize acetylcholine, but muscle cells and most other cells do not have the ability to produce acetylcholine. The enzyme choline-o-acetyltransferase in the cytoplasm of the motor neuron catalyzes the condensation of acetyl coenzyme A and choline. Acetyl CoA is a product of the metabolism of the neuron, but choline is obtained from outside the cell. The plasma membrane of the neuron possesses a transport system capable of accumulating choline against a large concentration gradient.

TABLE 18–3 SUMMARY OF EVENTS OCCURRING DURING NEUROMUSCULAR TRANSMISSION

Action potential in presynaptic motor axon terminals
↓
Increase in Ca^{++} permeability and inrush of Ca^{++} into axon terminal
↓
Release of ACh from synaptic vesicles into synaptic cleft
↓
Diffusion of ACh to postjunctional membrane
↓
Combination of ACh with specific receptors on postjunctional membrane
↓
Increase in permeability of postjunctional membrane to Na^+ and K^+ causes EPP
↓
Depolarization of areas of muscle membrane adjacent to end plate and initiation of action potential

About half of the choline liberated in the synaptic cleft when acetylcholine is hydrolyzed by cholinesterase is taken back up into the motor neuron to be used for the resynthesis of more acetylcholine.

Quantal Nature of Transmitter Release. It was first observed by Fatt and Katz that even if the motor neuron is not stimulated, small depolarizations of the postjunctional muscle cell occur spontaneously. The spontaneous depolarizations are called miniature end plate potentials, or MEPP's, and they occur at random with a frequency that averages about 1/sec. The postjunctional membrane is depolarized only about 0.4 mV with each MEPP. However, the MEPP has the same time course as the end plate potential (EPP), evoked by arrival of an action potential in the nerve terminal, and it responds identically to the EPP in response to pharmacological agents. For example, the EPP and the MEPP are both prolonged by agents that inhibit acetylcholinesterase and are both depressed by agents that compete with acetylcholine for binding to the receptor. While the frequency of MEPP's varies widely from moment to moment, their amplitude is quite constant. The MEPP is the result of the spontaneous release of a small number of vesicles of transmitter into the synaptic cleft.

The quantal nature of transmitter release can be demonstrated in another way as well. Extracellular Ca^{++} is absolutely required for the release of transmitter. If the extracellular Ca^{++} is reduced to very low levels or if Mg^{++} (which is a competitive antagonist of Ca^{++}) is added, the size of the EPP that is evoked by stimulation of the motor neuron is dramatically reduced. Under these conditions small variations in the size of the stimulation-evoked EPP occur spontaneously. The size of the EPP does not vary continuously, but in small steps. Statistical analysis of the variations in EPP size shows each step to correspond to the size of a single MEPP.

Termination of Transmitter Action by Cholinesterase and Reuptake of Choline. The enzyme acetylcholinesterase is present in high concentration on the surface of the postjunctional membrane. The importance of cholinesterase in terminating transmitter action is illustrated by the action of agents such as eserine and edrophonium, which are inhibitors of the enzyme and thus are called anticholinesterases. In the presence of an anticholinesterase, the EPP is larger and dramatically longer in duration.

About half of the choline liberated by cleavage of the transmitter is taken back up into the presynaptic terminal to be used for transmitter resynthesis. The hemicholiniums are drugs that block the choline transport system and inhibit reuptake. Repeated stimulation of the motor neuron in the presence of hemicholinium leads to depletion of the store of transmitter and ultimately to a decrease in the acetylcholine content of the quanta.

The Ionic Mechanism of the End Plate Potential

The cation channels that open in the postjunctional membrane in response to acetylcholine are different from the cation channels of the nerve and muscle membranes in being independent of the membrane potential. They are gated by the action of transmitter rather than by the transmembrane potential. Although the cation channels of the postjunctional membrane are not permeable to anions, they are not very selective among the various small cations. Na^+, K^+, Rb^+, and NH_4^+ can move through these channels with approximately equal ease.

Recall from Chapter 17 that the membrane potential is set primarily by the membrane conductances to K^+ and Na^+ as indicated in the conductance equation

$$E_m = \frac{g_K}{g_K + g_{Na}} E_K + \frac{g_{Na}}{g_K + g_{Na}} E_{Na}$$

If g_{Na} becomes equal to g_K as the result of acetylcholine action the conductance equation becomes

$$E_m = \tfrac{1}{2}(E_K) + \tfrac{1}{2}(E_{Na})$$

Thus during an EPP, the membrane potential should tend to the average of the equilibrium potentials of K^+ and Na^+. Taking $E_K = -105$ mV and $E_{Na} = +67$mV, -19mV is obtained for the "equilibrium potential" of the EPP. That this value is approximately correct was demonstrated in experiments in which EPP's were evoked by stimulation of the motor neuron while the postjunctional level of E_m prior to stimulation was set at various levels by passing current into the postjunctional muscle cell (Fig. 19–4). For E_m values less than about -15 mV the EPP caused depolarization of the postjunctional membrane, and the depolarizations became smaller and smaller as E_m was increased (depolarized). For E_m about -15 mV the EPP produced no change in the potential of the postjunctional membrane at all, since the potential was already at the equilibrium value for the EPP. When the E_m prior to stimulation was set at values more positive than -15 mV the direction of the EPP actually reversed, with the EPP tending to return the postjunctional E_m toward -15 mV. For this reason the equilibrium value of the postjunctional E_m is more often termed the *reversal potential* of the EPP.

The Acetylcholine Receptor

In recent years the acetylcholine receptor has been the object of intense study. Progress has been aided by the development of techniques for isolation and purification of highly hydrophobic membrane proteins, such as the acetylcholine receptor, and by the availability of certain snake venom neurotoxins that bind almost irreversibly to this receptor. The most widely used toxin is α-bun-

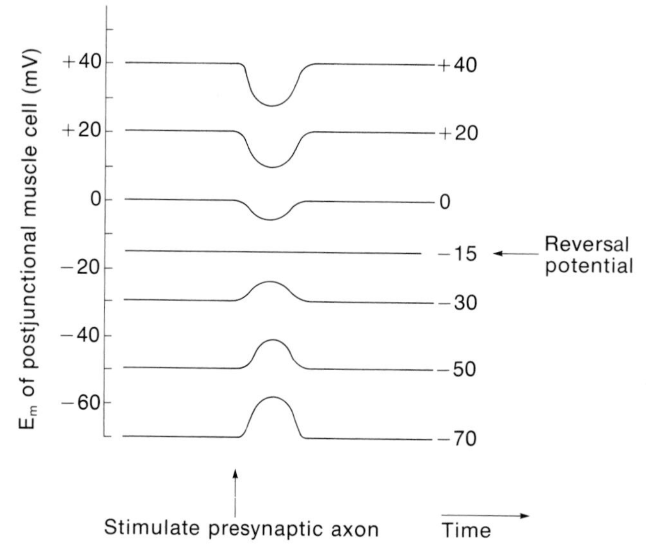

FIGURE 18–20 Schematic illustration of an experiment to determine the reversal potential of the neuromuscular junction. The resting potential of the postjunctional muscle cell is set to various levels by passing current into it. The prejunctional axon is then stimulated and the end plate potential recorded.

garotoxin, which is obtained from the venom of the Formosan krait, a relative of the cobra.

The binding of tritium-labeled α-bungarotoxin by various neuromuscular junctions has permitted quantitative studies of acetylcholine receptors. It appears that there are 10^7 to 10^8 binding sites per end plate. Evidence from mouse diaphragm neuromuscular junctions suggests that the binding sites concentrated on the crests of the postjunctional folds have a density of about 20,000 per μm^2. This means that the receptor molecules are rather closely packed, since the maximum density possible has been calculated to be 50,000 per μm^2, based on the estimates of receptor size.

The acetylcholine receptor protein is an integral membrane protein (Chapter 14), which is deeply embedded in the hydrophobic core of the postjunctional membrane. The cholinesterase, by contrast, is apparently loosely associated with the surface of the postjunctional membrane.

The acetylcholine receptor protein has been extracted from certain neuromuscular junctions and from the electroplax of electric fish, where its concentration is especially high, and it has been purified by means of affinity chromatography. The precise nature of the structure of the acetylcholine receptor protein remains controversial. Purified receptor protein has been inserted into lipid bilayer model membranes. In such systems it can be shown that the binding of acetylcholine by the receptor complex leads to an increase in the Na^+ and K^+ conductance of the lipid bilayer. One peptidic fragment of the receptor has been found to act alone to increase the cation permeability of lipid bilayers, suggesting that the cation channel that is activated in the EPP may be part of the receptor protein complex.

Myasthenia gravis is a disease that involves defective transmission at neuromuscular junctions. There is evidence that in myasthenia there is defective synthesis or packaging of acetylcholine in the presynaptic terminals. The EPP is smaller, as are the miniature end plate potentials, suggesting a decreased amount of acetylcholine in each quantum. More recent evidence based on binding of tritiated bungarotoxin suggests that there is also a markedly reduced density of acetylcholine receptors on the postjunctional membrane. The relative importance of these two factors in myasthenia gravis is not yet understood.

Autonomic Nerve Endings

The relationship between nerve and muscle in cardiac and smooth muscle tissues differs from that in skeletal muscle. Cardiac and smooth muscle cells are innervated by two types of unmyelinated autonomic nerves. Cholinergic nerves release acetylcholine, and their terminals contain agranular vesicles, so named because they have a clear interior. Adrenergic nerves release norepinephrine and contain granular vesicles that are slightly larger (500 Å) than the agranular variety (400 Å) and have a dense granule in their interior. The synaptic vesicles are found in terminal expansions of the axons and in swellings, termed varicosities, which occur along the course of the axons in muscle tissue.

In cardiac and smooth muscle, the terminal expansions and varicosities do not make the specialized contacts with muscle cells that occur in skeletal muscle, and thus there are no motor end plates in cardiac and smooth muscle. The axons usually are located at a distance up to several hundred Å from the surface of the muscle cells (Fig. 18–21). This structural relationship implies that the transmitter substance is released at a distance from the muscle cells and can diffuse to influence several muscle cells in the vicinity. Such a relationship is possible in cardiac muscle, for example, because the heart has an intrinsic rhythmic contractility independent of the nerves. By release of transmitter, however, the nerves can influence the rate of the heart beat, acetylcholine from parasympathetic nerves slowing the heart and norepinephrine from adrenergic nerves speeding contrac-

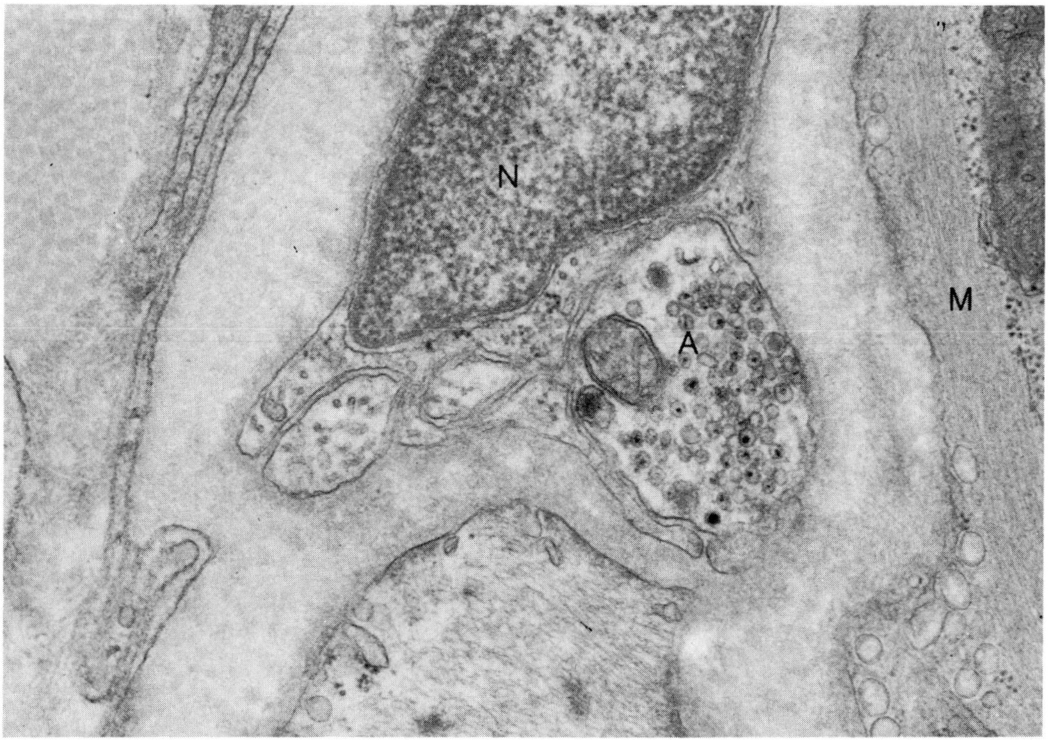

FIGURE 18–21 An autonomic nerve in smooth muscle tissue of the rat prostate. An axon (A) containing synaptic vesicles lies in a recess in the surface of a Schwann cell. At least some of the synaptic vesicles have a dense core. The vesicle-containing portion of the axon does not make close contact with a smooth muscle cell (M); instead the nerve and muscle lie at a distance of several hundred nm from each other. The nucleus (N) of the Schwann cell is included in the section. × 40,000. (*From*: Flickinger, C. J.: Am. J. Anat., *134*:107, 1972. Reproduced by permission of the Wistar Institute Press.)

tions. In some smooth muscles, axon terminals or varicosities closely approach a muscle cell and lie in an indentation in its surface, the two cells being separated by only about 100 Å space. Even in these instances, however, the surface of the muscle cell does not show the structural specializations seen in a motor end plate.

SYNAPSES BETWEEN NEURONS

Structure of Synapses

In the central nervous system, the end of an axon that makes synaptic contact with another neuron is often expanded. These expansions are blackened in preparations impregnated with silver and resemble small buttons. Hence they have been named *boutons terminaux* by neuroanatomists. The terminal boutons often contact a dendritic spine, but they may contact a neuron at any place on its surface.

When viewed in the electron microscope, a synapse in the central nervous system has certain features reminiscent of a macula adhaerens (Figs. 18–22 and 18–23). There is often an increased density in the cytoplasm underlying the plasma

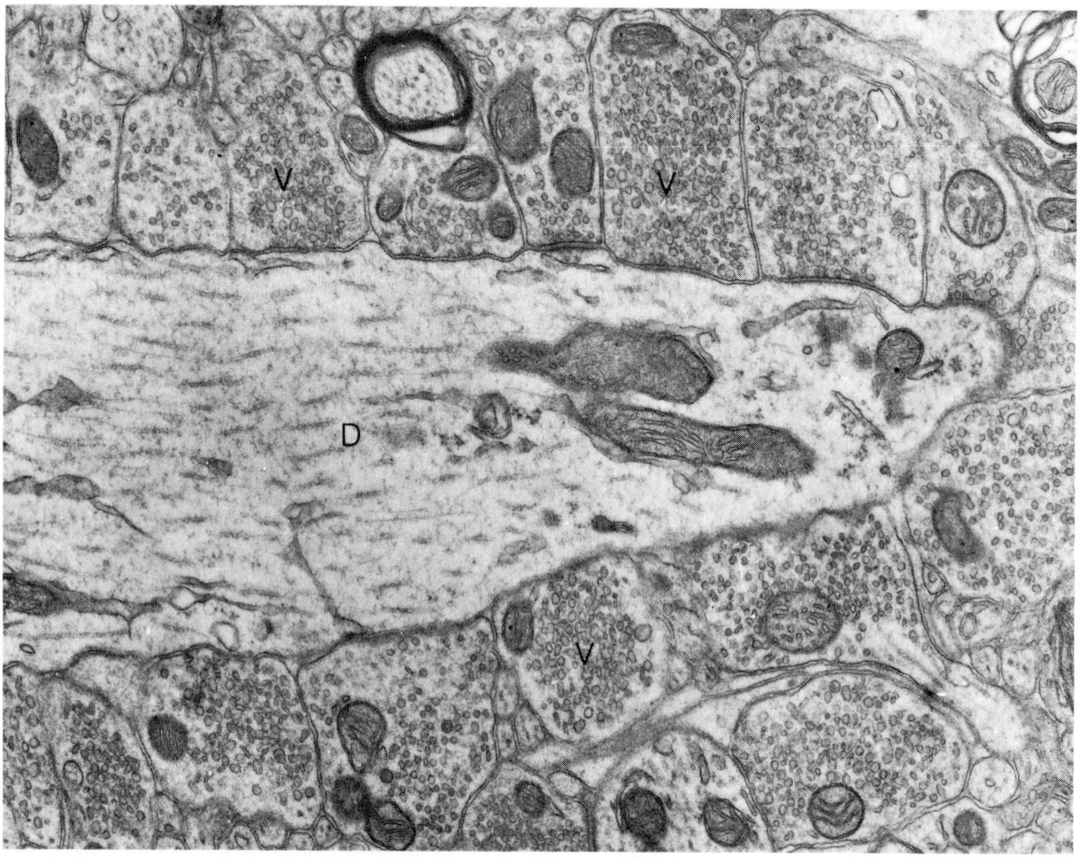

FIGURE 18–22 Synapses in the globus pallidus of a rat. Many neuronal processes containing synaptic vesicles (V) contact a dendrite (D). The postsynaptic densities are not pronounced in most of these contacts, so these synapses probably represent the relatively "symmetrical" type. Compare with Figure 18–23. × 24,000. (Micrograph courtesy of L. Heimer and R. D. Wilson. *From:* Santini, M. (ed.): Golgi Centennial Symposium Proceedings. New York, Raven Press, 1975, p. 177. Reproduced by permission of the publisher.)

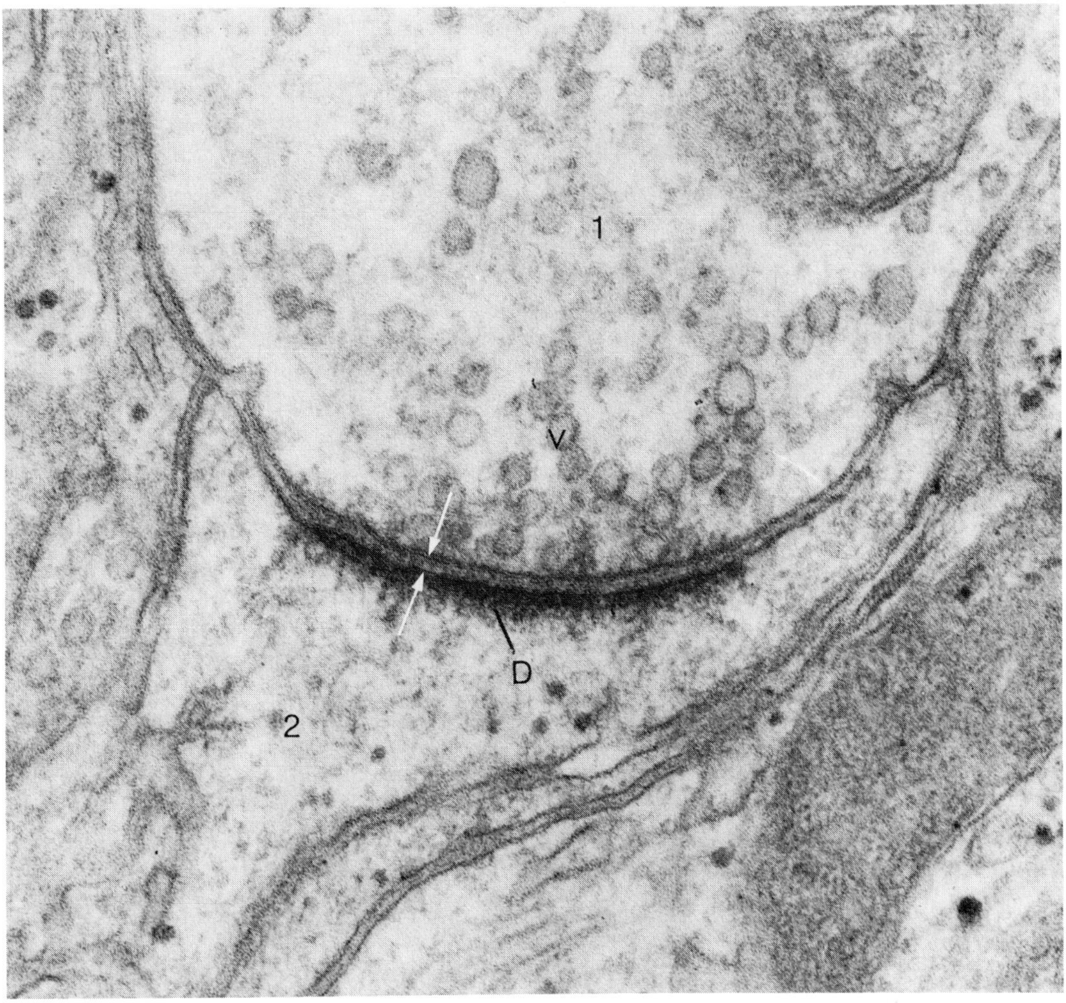

FIGURE 18–23 High magnification of a synapse in the rat olfactory cortex. The presynaptic cell (1) contains synaptic vesicles (V). In the postsynaptic neuron (2) there is a layer of dense material (D) underlying the plasma membrane. The synaptic cleft lies between the plasma membranes of the two apposed cells (*arrows*). The prominence of the postsynaptic density indicates a relatively "asymmetric" type of synapse. × 110,000. (Micrograph courtesy of L. Heimer.)

membranes in the region of contact, the two cell membranes are parallel to one another, and the membranes are actually adherent since they stay together during homogenization of the tissue. However, unlike the macula adhaerens, the synapse is polarized and it has additional structural features. The presynaptic cell contains a multitude of synaptic vesicles about 400 Å in diameter, which contain the transmitter substance. Mitochondria are numerous in the presynaptic cell. A variety of other structural specializations of the pre- and postsynaptic cells also have been described in different locations in various species.

Much effort has been directed toward identification of different morphological types of synapses and attempts have been made to correlate the various forms with physiological characteristics, such as the excitatory and inhibitory properties of synapses. Chemical synapses are of course all polarized with respect to the direction of release of transmitter. However, differences in the extent of development of the postsynaptic density, also known as the subsynaptic web, have led to the distinction of two main types of synaptic junctions: *asymmetrical synapses* (Fig. 18–22), which have a relatively prominent postsynaptic density, and so-called *symmetrical synapses* (Fig. 18–23), in which the postsynaptic density is less marked. Synapses vary continuously in form between two extremes, however, and intermediate types are abundant. These two forms of synaptic junctions also differ in the shape of their synaptic vesicles. The asymmetrical synapses usually contain round vesicles, while the symmetrical synapses contain flattened or elliptical vesicles. On the basis of the morphology of certain synapses of known function, it has been proposed that the asymmetrical synapses with round vesicles are excitatory and that the symmetrical synapses with flat vesicles are inhibitory. This hypothesis is under study, and problems remain. For example, the different shapes of vesicles imply different contents of transmitters. Yet it is known that the same transmitter can have excitatory or inhibitory functions in different locations; thus the physiological nature of the synapse also depends on the character of the postsynaptic cell.

It is evident that fusion of many synaptic vesicles with the plasma membrane would lead eventually to a depletion of synaptic vesicles and to an increase in the area of the plasma membrane. Studies on the uptake of tracer materials at the neuromuscular junction after repeated stimulation of the nerve have shown that the membranes are recycled at the nerve ending. After fusion of synaptic vesicles with the plasma membrane, the nerve ending appears swollen. Then coated vesicles form by invagination of the plasma membrane. From the membrane taken up into the cytoplasm by the coated vesicles, new synaptic vesicles are re-formed and acquire a content of transmitter substance. Membranous cisternae may be an intermediate stage between the coated vesicles and the new synaptic vesicles. Nerve endings also take up choline that is formed by the splitting of acetylcholine by acetylcholinesterase in the extracellular space, and this choline is reused in the synthesis of new transmitter.

Basic Physiological Properties of Neuronal Synapses

Much of what we know about synaptic properties was first learned in experiments on certain invertebrate preparations or in experiments conducted by John C. Eccles and his associates on spinal motor neurons of the cat. Evidence has since accumulated in studies on other synapses in the mammalian central nervous system that these are general properties of synaptic behavior.

In an electrical synapse that is made by a gap junction, conduction is generally bidirectional. In a *chemical* synapse, however, owing to the structure and organization of the synapse, conduction is necessarily one way. *One-way conduction* contributes to the order of the complex central nervous systems of vertebrates. A synaptic delay of some 0.5 msec is associated with transmission at chemical synapses. In certain polysynaptic pathways in the central nervous system, synaptic delay accounts for a significant fraction of the total conduction time.

Any change in the conductance of the postsynaptic plasma membrane to an ion that is not in equilibrium across that membrane will lead to a flow of that ion and thus to a change in the membrane potential of the postsynaptic cell. In most cases transmitters have been found to increase the conductance of the postsynaptic membrane to one or more ions, but certain invertebrate transmitters apparently act by decreasing the conductances of specific ions.

The postsynaptic membrane, that part of the membrane of the postsynaptic neuron that forms the synapse, is specialized for *chemical sensitivity* rather than for electrical sensitivity. The change in membrane potential, be it depolarization or hyperpolarization, that occurs here is conducted by the local response (electrotonically) over the surface of the postsynaptic cell, but action potentials do not originate at the synapse. Instead there is a special region of the postsynaptic neuron, the *axon hillock-initial segment* region (Fig. 18–24), which by virtue of having a lower threshold than the rest of the membrane will generate an action potential if the integrated effect of all the inputs to the cell exceeds the threshold. Once generated, the action potential is both conducted back over the surface of the presynaptic cell soma and propagated along its axon.

Input-Output Relations

In the simplest kinds of synapses a single action potential in the presynaptic cell (the input) results in a single action potential in the postsynaptic cell (the output). In other synapses the output differs from the input, but is some function of the input. At such synapses *integration* is said to take place. Thus, depending on the relationship between input and output, synapses can be classified as one-to-one, one-to-many, or many-to-one.

In a *one-to-one synapse* the input and output are the same since a single action potential in the presynaptic cell leads to a single action potential in the postsynaptic cell. Since the output is the same as the input, no integration takes place at this type of synapse. The neuromuscular junction is an example of such a synapse.

In a *one-to-many synapse* a single action potential in the presynaptic cell leads to many action potentials in the postsynaptic cell. Such synapses are relatively rare. An example is the synapse of motor neurons on the Renshaw cell in the spinal cord. A single impulse in the motor neuron causes the Renshaw cell to fire a burst of action potentials.

In a *many-to-one synaptic arrangement* a single action potential in the presynaptic cell is not sufficient to cause the postsynaptic cell to fire an action potential. The nearly simultaneous arrival of presynaptic impulses in several cells that synapse on the same postsynaptic cell is required to fire an action potential in the postsynaptic cell. The spinal motor neuron is a good example of this type of synaptic organization. On the order of 100 presynaptic neurons synapse with each spinal motor neuron (Fig. 18–24). Some are excitatory inputs that depolarize the postsynaptic cell and bring it closer to its threshold potential; others are inhibitory inputs that hyperpolarize the motor neuron and take its membrane potential farther away from the threshold. The postsynaptic potentials caused by

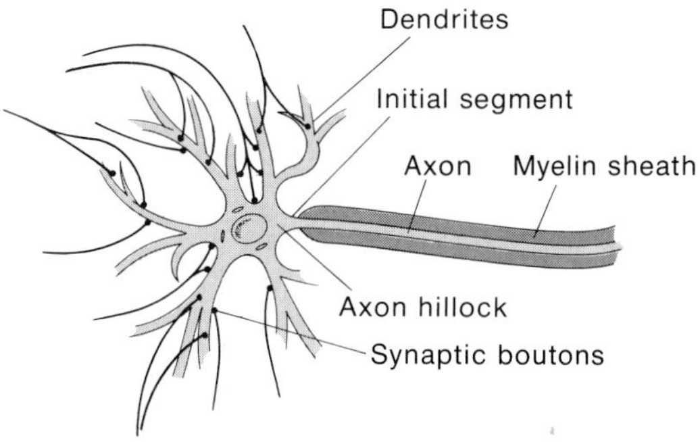

FIGURE 18–24 Semidiagrammatic representation of a spinal motor neuron with many synapses on its soma and dendrites.

a single input are about 1 to 2 mV, so no single excitatory input is capable of causing the motor neuron to fire.

The transient depolarization of the postsynaptic neuron caused by an action potential in the presynaptic cell is called an *excitatory postsynaptic potential* or *EPSP*. The transient hyperpolarization caused by the firing of an inhibitory input is called an *inhibitory postsynaptic potential* or *IPSP*. At any moment in time the cell integrates its various inputs by adding them algebraically. If the sum of the inputs is sufficient to depolarize the postsynaptic cell to its threshold potential, it fires an action potential. This is an example of integration at the synaptic level.

The summation (or integration) of multiple inputs is of two types: *spatial summation* and *temporal summation*. *Spatial summation* occurs when two inputs arrive almost simultaneously. Their postsynaptic potentials are summated, so that if both are excitatory inputs the resulting depolarization of the postsynaptic cell will be about twice as large as either input alone would cause. On the other hand, an EPSP and an IPSP occurring simultaneously might completely cancel each other. Even inputs that synapse at opposite ends of the postsynaptic cell will summate. This is because the postsynaptic potentials are rapidly spread over the cell membrane of the postsynaptic cell with very little decrement, since the cellular dimensions (about 100 μ) are much smaller than the length constant (about 1 to 2 mm) for electrotonic conduction. *Temporal summation* occurs when two or more impulses in a single presynaptic neuron are fired close enough in time so that the resulting postsynaptic potentials can summate. A train of impulses in a single presynaptic cell can cause the membrane potential of the postsynaptic cell to change in a stepwise fashion, each step resulting from the transmitter released by one of the presynaptic impulses. Integration takes place as described previously because many positive and negative inputs impinge on a single motor neuron. This allows for fine control of the firing pattern of the spinal motor neuron.

Ionic Mechanism of the Postsynaptic Potentials in Spinal Motor Neurons

The *excitatory postsynaptic potential (EPSP)* appears to be caused by a transient increase of the conductance of the postsynaptic membrane to both Na^+ and K^+. One way in which this can be demonstrated is by injection of either Na^+ or K^+ into the cell interior to raise the concentration of the ion in question. This results in a smaller EPSP, because when the conductance increase occurs there is a smaller tendency for Na^+ to flow in to depolarize the cell or a larger tendency for K^+ to flow out and oppose depolarization. Furthermore, if the postsynaptic membrane is progressively depolarized, the EPSP gets smaller and smaller owing to a decreased tendency for Na^+ to enter and an increased tendency for K^+ to leave the cell. When E_m is about zero, the EPSP vanishes, and if E_m is made positive the EPSP reverses direction. Thus zero is the *reversal potential* for the EPSP. Recall that if Na^+ were the only ion involved in the EPSP, then the reversal potential for the EPSP should equal the equilibrium potential for Na^+, which is about +65 mV. Injection of Cl^- into the cell has no effect on the EPSP, so presumably Cl^- is not involved in the EPSP.

The *inhibitory postsynaptic potential (IPSP)* is apparently due to increased chloride conductance at the postjunctional membrane. The reversal potential for the IPSP is somewhat more negative than the resting potential of the postsynaptic cell, and this corresponds to the equilibrium potential for Cl^-. At rest there must be a net tendency for Cl^- to enter the cell. When g_{Cl} is increased as the result of transmitter release at an inhibitory synapse, Cl^- enters the cell and hyperpolarizes it. Injecting Cl^- into the cell or hyperpolarizing it decreases the net tendency for Cl^- to enter and reduces the size of the IPSP, but injection of Na^+ or K^+ produces no change in the IPSP. Inhibitory synapses play a vital role in stabilizing the central nervous system. It is noteworthy that certain convulsants, such as strychnine and tetanus toxin, act by decreasing the size of the IPSP.

Transmitters in Mammals

The substances that are *known* to be transmitters at particular mammalian synapses are few. *Acetylcholine* is the transmitter at the neuromuscular junction, in sympathetic and parasympathetic ganglia, and at parasympathetic nerve endings. It is also believed to be the transmitter at the synapse made by motor neuron axon collaterals on Renshaw cells in the spinal cord. *Norepinephrine* is known to be the transmitter at most sympathetic (postganglionic) nerve endings.

A host of substances are *suspected transmitters* at various synapses, but the evidence is not yet sufficient to absolutely identify each of them as a transmitter. *Glycine* is believed to be a transmitter at some inhibitory synapses on spinal motor neurons, because iontophoresis of glycine onto the motor neuron causes hyperpolarization, and this effect is blocked by strychnine which depresses the IPSP. Certain cortical synapses respond to *acetylcholine*. *Gamma-aminobutyric acid (GABA)* is suspected to be an inhibitory transmitter at certain synapses in the brain. Glutamate, aspartate, and cysteine stimulate certain cortical neurons when applied iontophoretically. Serotonin (5-hydroxytryptamine) is present in high concentration in certain neurons and is thought to be a transmitter.

Because it is a complex task to prove that a substance is the transmitter at a particular synapse, workers in the field have enunciated criteria that a candidate substance (X) must satisfy before it can be accepted as a transmitter.

1. The presynaptic neuron must contain X and must be able to synthesize it.

2. X should be released by the presynaptic neuron on stimulation.

3. Microapplication of X to the postsynaptic membrane must mimic the effect of stimulating the presynaptic neuron.

4. The effects of presynaptic stimulation and of microapplication of X should be modified in the same way by pharmacological agents.

The list of known transmitters is small partly because of the experimental difficulty of establishing all four of these criteria at a particular synapse. However, this is an active area of research, and we can anticipate the identification of the transmitters at a number of synapses in the near future.

FURTHER READING

Aidley, D. J.: The Physiology of Excitable Cells. New York, Cambridge University Press, 1971, pp. 37–70, 93–184.

Barondes, S. H. (ed.): Cellular dynamics of the neuron. Symposia of the International Society for Cell Biology. Vol. 8, New York, Academic Press, 1969.

Bloom, W., and Fawcett, D. W., et al.: A Textbook of Histology. Philadelphia, W. B. Saunders, 1975.

Carpenter, M. B.: Human Neuroanatomy. 7th ed. Baltimore, Williams & Wilkins, 1976, pp. 49–158.
Dale, H. H., Feldberg, W., and Vogt, M.: Release of acetylcholine at voluntary motor nerve endings. J. Physiol., *86*:353, 1936.
Davson, H: A Textbook of General Physiology. 4th ed. Baltimore, Williams & Wilkins, 1970, pp. 1031–1242.
Del Castillo, J., and Katz, B.: Quantal components of the end-plate potential. J. Physiol., *124*:560, 1954.
Dowben, R. M.: General Physiology. New York, Harper & Row, 1969, pp. 413–421.
Eccles, J. C.: The Physiology of Synapses. Berlin, Springer Verlag, 1964.
Eccles, J. C.: The synapse. Sci. Am., *212*:56, 1965.
Fatt, P., and Katz, B.: Spontaneous subthreshold activity at motor nerve endings. J. Physiol., *117*: 109, 1952.
Hodgkin, A. L.: The Conduction of the Nervous Impulse. Springfield, Charles C Thomas, 1964, pp. 30–55.
Hodgkin, A. L., and Rushton, W. A. H.: The electrical constants of a crustacean nerve fibre. Proc. R. Soc. B, *133*:444, 1946.
Kandel, E. R. (ed.): Cellular Biology of Neurons. The Handbook of Physiology. Section 1, Volume 1, Baltimore, Williams & Wilkins, 1977.
Katz, B.: Nerve, Muscle, and Synapse. New York, McGraw-Hill, 1966, pp. 73–158.
Kuffler, S. W., and Nicholls, J. G.: From Neuron to Brain. Sunderland, Mass. Sinauer Associates, 1976, pp. 77–87, 132–236.
Peters, A., Palay, S. L., and Webster, H. deF.: The Fine Structure of the Nervous System: The Neurons and Supporting Cells. Philadelphia, W. B. Saunders, 1976.
Takeuchi, A., and Takeuchi, N.: On the permeability of end plate membrane during the action of transmitter. J. Physiol. *154*:53, 1960.

19

MUSCLE CELLS

Muscle cells are specialized for the function of contraction. Therefore, they show a high degree of development of the cytoplasmic contractile proteins and of other organelles associated with the contractile process. Several different types of muscle cells will be discussed in this chapter. Skeletal and cardiac muscle are known as striated muscles because the cells display transverse bands, or striations; they are distinguished from smooth muscle, which does not have such a banding pattern. Skeletal muscle cells constitute the numerous voluntary muscles responsible for conscious movements of the body. Both cardiac and smooth muscles are involuntary muscles. Cardiac muscle cells are found in the heart, of course, while smooth muscle cells are widely distributed in the walls of blood vessels and the digestive tract.

SKELETAL MUSCLE CELLS

Structure of Skeletal Muscle Cells

Skeletal muscle is composed of very large, cylindrical, multinucleated cells called muscle fibers (Fig. 19–1). The muscle cells are bound together by connective tissue to form the grossly visible "muscles." The individual muscle cells are roughly circular or polygonal in cross section (Figs. 19–2 and 19–3) and have a diameter of 10 to 100 μm. Their length is of the order of centimeters. In some cases, a muscle cell may extend the entire length of the muscle, while in others several fibers are found end-to-end. The multiple nuclei in skeletal muscle fibers are located peripherally in the cytoplasm, lying just underneath the plasma membrane, and much of the interior of the fiber is occupied by the contractile material (Figs. 19–2 to 19–4).

The most striking feature of skeletal muscle fibers is the presence of striations in both longitudinal and transverse directions (Figs. 19–1 and 19–4). The longitudinal striations are due to the presence of myofibrils, which are cylindrical bundles of the contractile elements, the thick and thin myofilaments which are composed of myosin and actin along with other proteins. The arrangement of thick and thin filaments within the myofibrils gives rise to the pattern of transverse banding (Fig. 19–6).

There are two main transverse bands, the A band and the I band, both of which are repeated many times along the length of a myofibril (Figs. 19–5 and

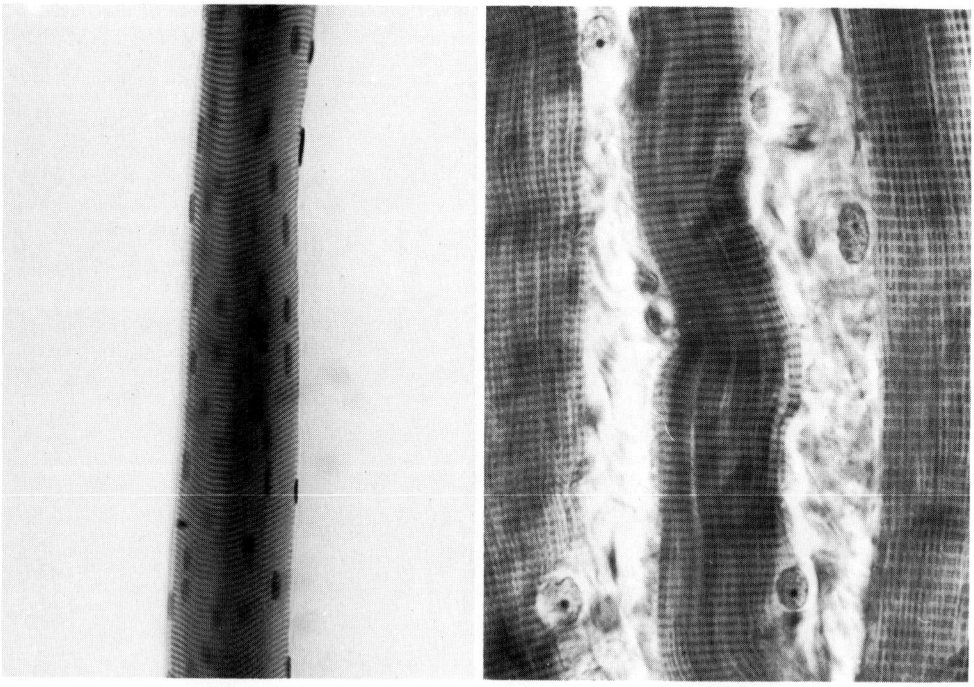

A B

FIGURE 19–1 Light micrographs of skeletal muscle cells. *A*. Part of a single skeletal muscle cell teased from the human gastrocnemius muscle. × 275. *B*. Parts of three skeletal muscle fibers with connective tissue between them, from the sketetal muscle of the lip. × 400. (*From*: Leeson, C. R., and Leeson, T. S.: Histology, Philadelphia, W. B. Saunders, 1976.)

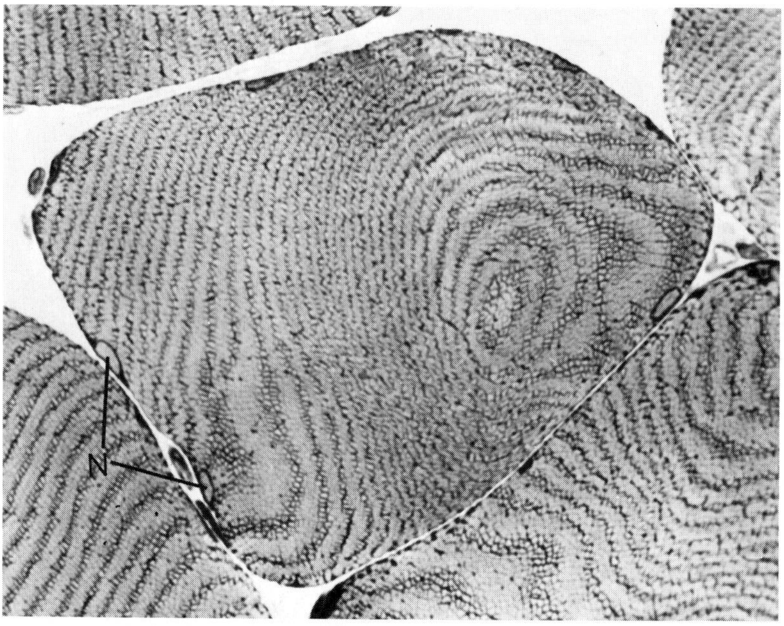

FIGURE 19–2 Light micrograph of a cross section of white skeletal muscle fibers in the rat semitendinosus muscle. Nuclei (N) are located at the periphery of the cell, while much of the interior is occupied by the lightly staining myofibrils. The darkly staining structures in the interior of the fiber are mitochondria. × 760. (Micrograph courtesy of Gauthier, G.: Z. Zellforsch, 95:462, 1969. Reproduced by permission of Springer-Verlag.)

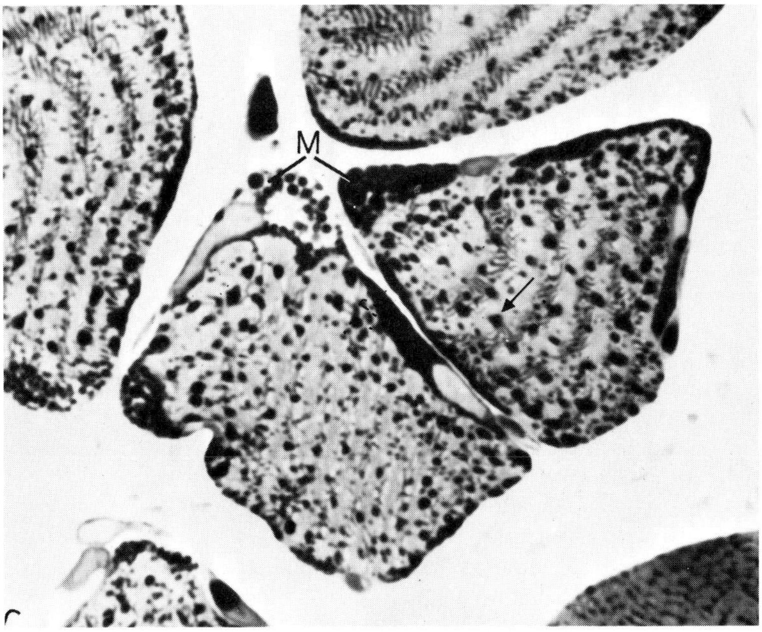

FIGURE 19–3 Light micrograph of a cross section of red skeletal muscle fibers in the rat diaphragm. Densely staining mitochondria are very numerous in this type of muscle fiber, both in aggregations at the margins of the cells (M) and in the interior of the fibers (*arrow*) between the lightly staining myofibrils. × 1,200. (Micrograph courtesy of G. Gauthier. *From* Briskey, E. J., Cassens, R. G., and Marsh, B. B. (eds.): The Physiology and Biochemistry of Muscle as a Food. Madison, University of Wisconsin Press, 1970, pp. 103–130. Reproduced by permission of the publisher.)

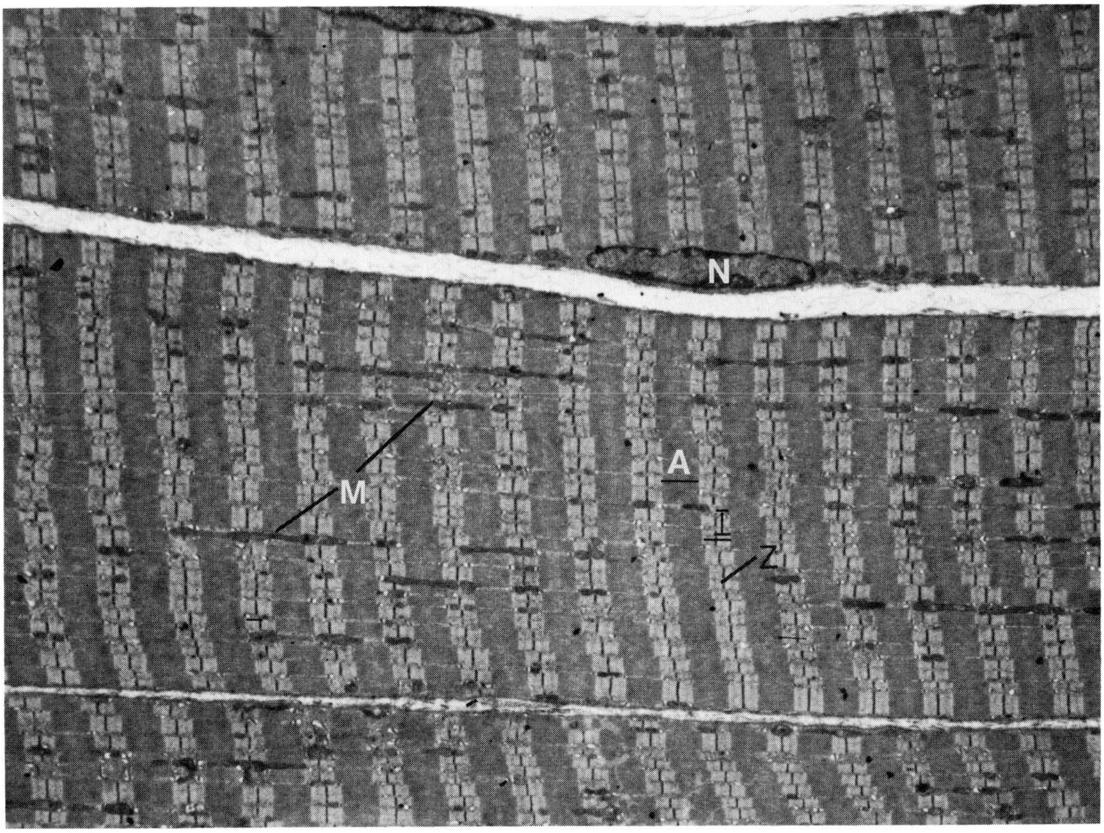

FIGURE 19–4 Low-power electron micrograph of portions of three skeletal muscle fibers from a mouse. The nuclei (N) are located at the margin of the fiber, and some cytoplasmic organelles accumulate at the poles of the nuclei. Mitochondria (M) are arranged in rows between the myofibrils, which display the characteristic pattern of transverse A, I, and Z bands. × 3,400. (Micrograph courtesy of J. H. Venable. From Bloom, W. and Fawcett, D. W.: A Textbook of Histology, Philadelphia, W. B. Saunders, 1976.)

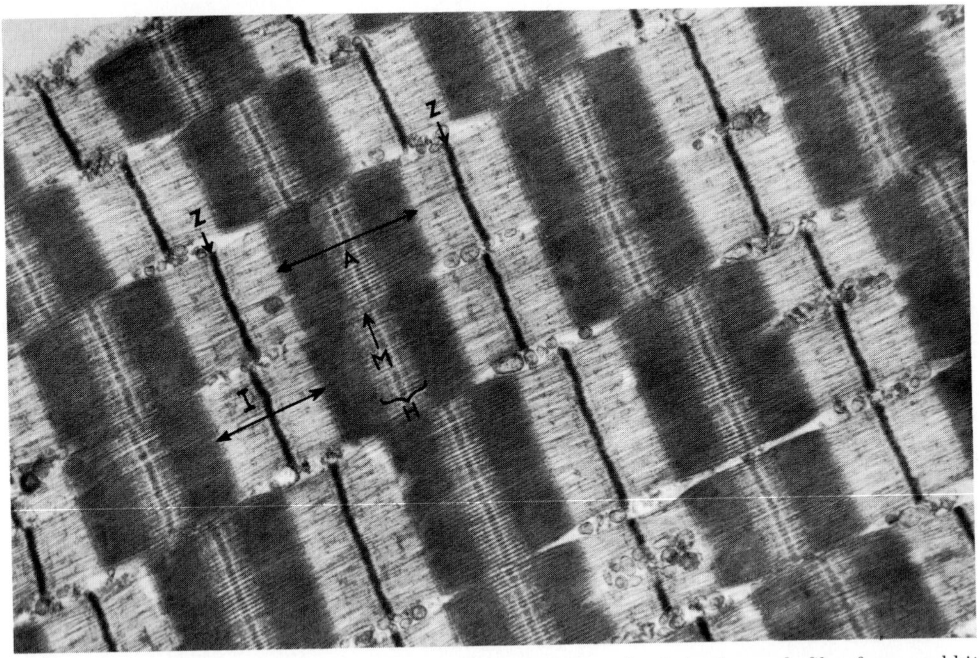

FIGURE 19–5 Electron micrograph of a longitudinal section of a skeletal muscle fiber from a rabbit. Seven myofibrils are represented. A sarcomere extends from one dark Z line to the next. Note the light I bands and the dark A bands. The A bands contain a central lighter zone, the H band, which in turn is bisected by a thin dark M band. (Micrograph by H. E. Huxley. *From* Leeson, C. R., and Leeson, T. S.: Histology. Philadelphia, W. B. Saunders, 1976.)

19–6). The A band is so named because it is relatively anisotropic in polarized light. With the electron microscope it is seen to contain filaments which are about 100 Å in diameter and are referred to as thick filaments. As will be discussed further on, each thick filament contains a number of myosin molecules, each of which is composed of a heavy meromyosin and a light meromyosin portion. The I band is so designated because it is relatively isotropic in polarized light. The I band contains filaments which are about 50 Å in diameter and are referred to morphologically as thin filaments. The thin filaments are composed mainly of actin.

The I band is bisected by a narrow dark line, the Z line (Figs. 19–5 and 19–6). The unit of the myofibril, the sarcomere, is defined as extending from one Z line to another. At the Z band, the thin actin filaments of adjacent sarcomeres interdigitate in a complex fashion, joining one sarcomere to the next. Thus the sarcomeres are cylindrical bundles of myofilaments linked end-to-end at their Z lines to form myofibrils.

The thin filaments of the I bands also penetrate part way into both ends of the A band and interdigitate with the thick myosin filaments in the A band. However, in relaxed muscle cells there is a region in the center of the A band into which the thin filaments do not penetrate (Fig. 19–5 and 19–6). As a result, the center of the A band appears lighter than the rest and is designated the H band. In the center of the H band, and thus in the center of the sarcomere, midway between two Z lines, there is a dense line called the M band, in which the thick filaments appear to be bound together in a lattice (Fig. 19–7).

Since the individual myofibrils with their transverse A, I, H, and Z bands are aligned in register with one another, the various bands appear to extend transversely across the entire muscle cell. Because of the arrangement of the

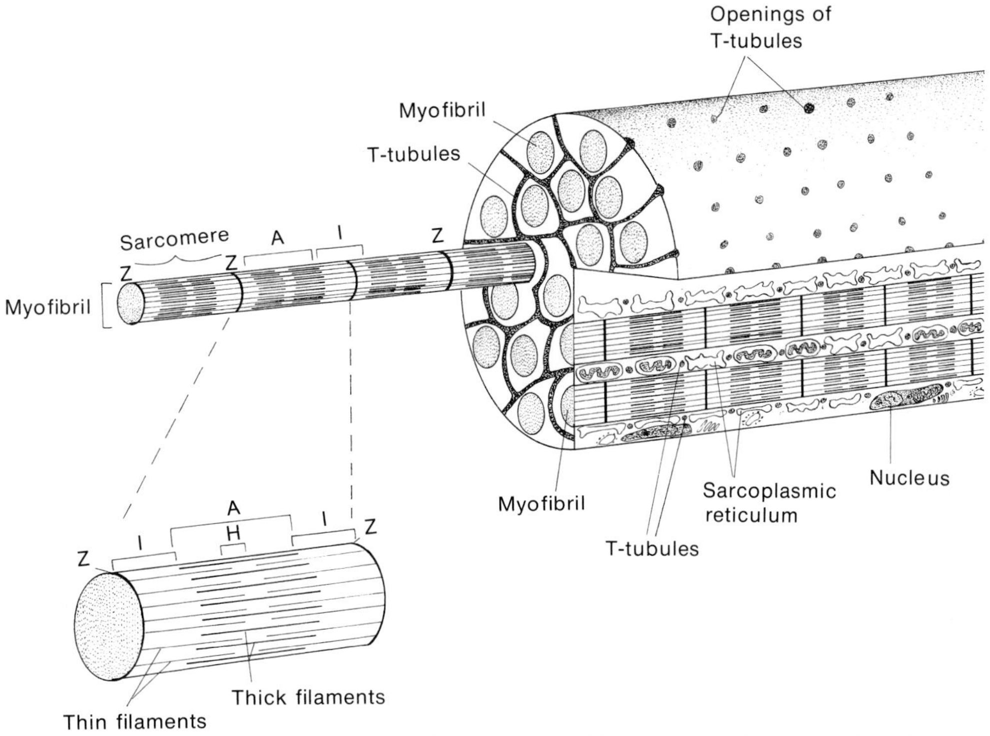

FIGURE 19–6 Diagram of a skeletal muscle cell (muscle fiber), showing the myofibrils, which are composed of thick and thin myofilaments, and the location between the myofibrils of mitochondria and sarcoplasmic reticulum. The transverse banding pattern of the myofibrils is also illustrated.

thick and thin myofilaments which produces the pattern of transverse banding, the appearance of myofibrils in cross section will vary with the level of the sarcomere at which the plane of section passes (Fig. 19–7). Thus, cross sections through the I band will reveal only thin filaments, viewed end-on. Sections through the H band contain only thick filaments. Those passing through the major part of the A band, in which thick and thin filaments overlap, contain both kinds of myofilaments. The myofilaments have a precise geometrical relation to one another in this region of interdigitation. In vertebrate skeletal muscles, six thin filaments are disposed around each thick filament. This places three thick filaments about each thin filament. Fine bridges extending from the surface of the thick filaments toward the thin filaments are visible in the electron microscope. They represent the heads of heavy meromyosin molecules and are believed to be important in the interaction between myosin and actin filaments during muscle contraction.

The morphology of the transverse bands in muscle cells is important because there is a change in the width of some of the bands during muscle contraction and relaxation (Fig. 19–8). These changes form part of the basis for the sliding filament hypothesis of muscle contraction. I bands and H bands decrease in width during muscle contraction, and they increase in width during muscle relaxation. This is thought to be the result of myosin and actin filaments sliding in relation to one another, interdigitating to a greater degree during muscle contraction and to a lesser degree during muscle relaxation and stretching. The "bridges" between thick and thin filaments in the region of interdigitation as well as certain other muscle proteins found in association with myosin and actin in the myofibrils

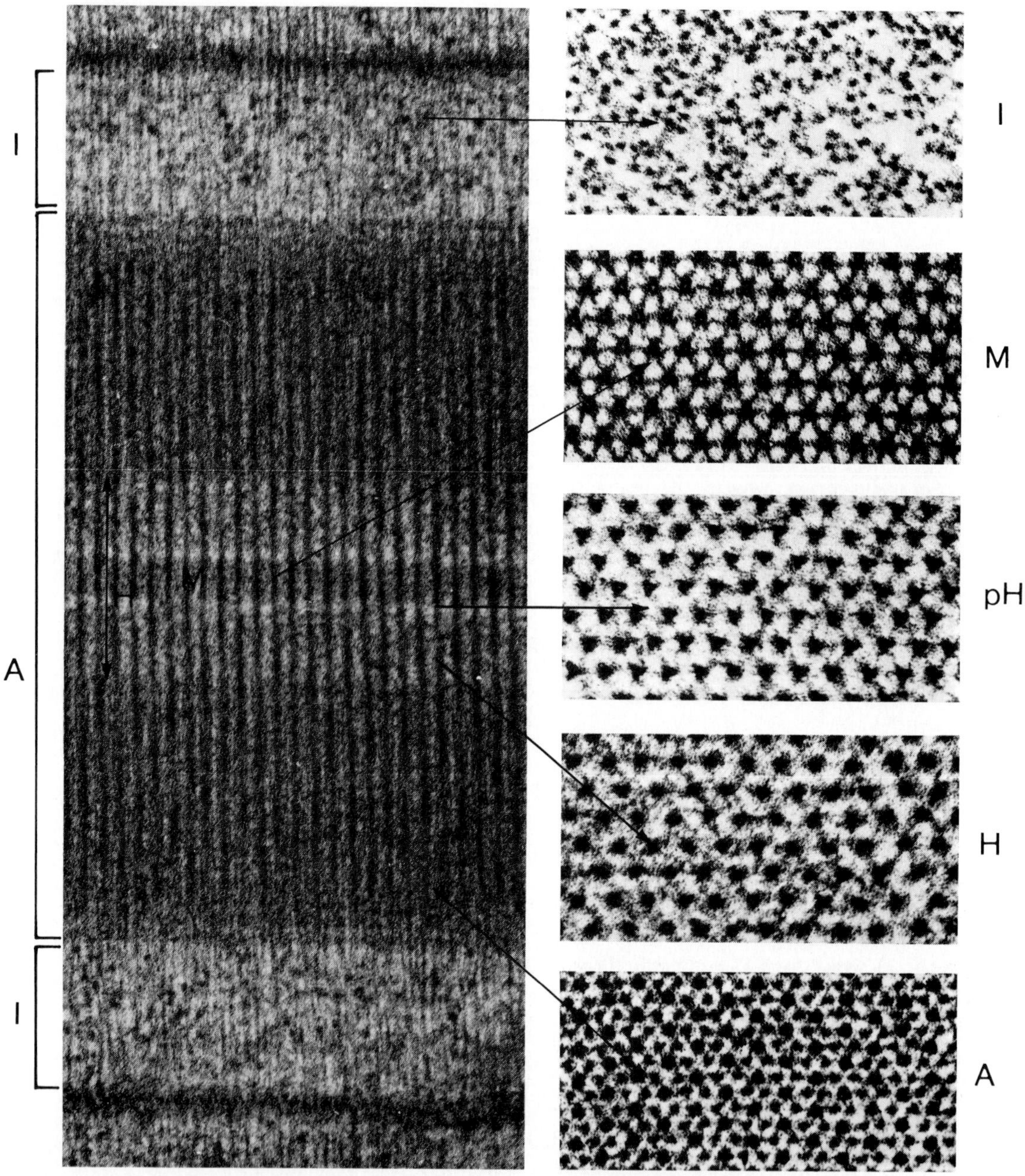

FIGURE 19–7 Electron micrographs of skeletal muscle. *Left*, longitudinal section of a sarcomere; *right*, cross sections at various levels of the sarcomere. In the I band, only thin filaments are present, while in the A band thin and thick filaments overlap. Only thick filaments are found in the H band, and the thick filaments appear connected to one another by extensions in the M band. A small portion near the center of the H band lacks the bridges found elsewhere on the thick filaments and is referred to as the pseudo-H band (pH). (Micrographs courtesy of F. A. Pepe. *From*: Timasheff, S. W., and Fasman, G. D. (eds): Biological Macromolecules V. New York, Marcel Dekker, 1971. Reproduced by permission of the publisher.)

are important in the interaction between thick and thin filaments. These will be discussed in detail later in the chapter.

The cytoplasm, or sarcoplasm as it is sometimes called, in muscle cells contains the usual cellular organelles, but their disposition is directed by the presence of the myofibrils which occupy the bulk of the cytoplasm. Mitochondria are

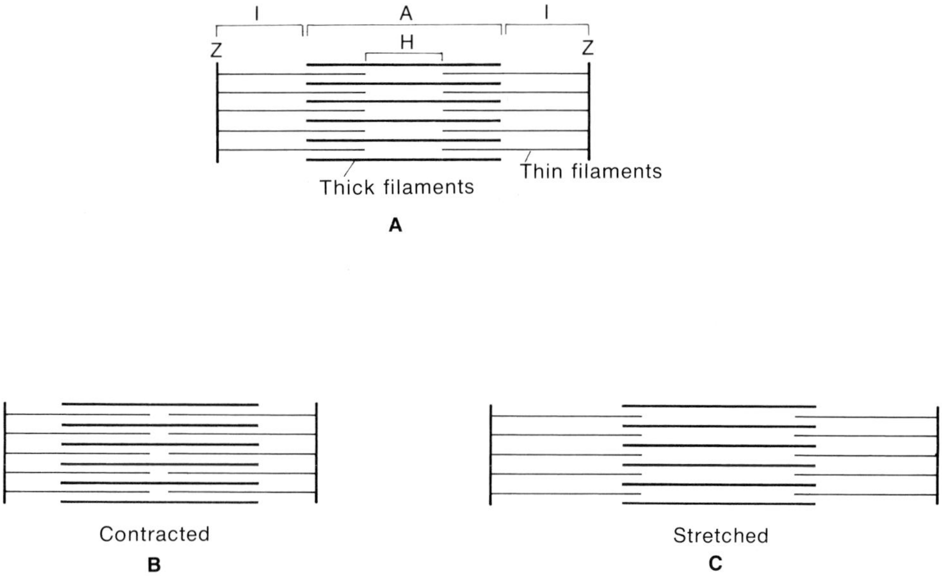

FIGURE 19–8 Diagram of a sarcomere of mammalian striated muscle, illustrating the sliding filament hypothesis of muscle contraction. *A*. The arrangment of thick and thin myofilaments gives rise to the transverse banding pattern of striated muscle. *B*. During contraction the interdigitation of thick and thin filaments increases, and the sarcomere undergoes shortening. *C*. When the muscle is stretched, the thick and thin filaments interdigitate to a lesser degree.

disposed in longitudinal rows between myofibrils (Fig. 19–4). Most of the other organelles, such as rough endoplasmic reticulum and Golgi apparatus, are restricted to small islands of cytoplasm, especially the conically shaped regions found at the poles of the nuclei (Fig. 19–4).

There are two systems of tubules present in the cytoplasm of muscle cells, and they are both very important in muscle physiology. The transverse tubule, or T-tubule, system consists of invaginations of the plasma membrane. Therefore, the lumen of the T-tubules is continuous with the extracellular space. In contrast, the sarcoplasmic reticulum lies within the cytoplasm of the muscle cell and is a differentiated form of smooth endoplasmic reticulum. Thus, although the T-tubules and the sarcoplasmic reticulum are closely related functionally, they are derived from different classes of cellular membranes.

At regular repeating intervals along the length of the muscle fiber, the T-tubules (Fig. 19–8) begin as invaginations of the plasma membrane and extend into the interior of the muscle cell, insinuated between the myofibrils. As their name indicates, their course lies in a series of planes transverse to the long axis of the muscle fiber. Within each plane, T-tubules arising from different points around the circumference of the fiber anastomose with one another and encircle the myofibrils. The T-tubules are slender, have a relatively uniform diameter, and often appear circular when cut in cross section.

The sarcoplasmic reticulum, which is specialized smooth endoplasmic reticulum, forms a network of irregular, anastomosing tubules lying between the myofibrils (Figs. 19–8 and 19–9). The sarcoplasmic reticulum surrounding a single myofibril can be visualized as a sleeve encircling the cylindrical myofibril. Generally, the tubules of the sarcoplasmic reticulum run longitudinally with respect to the myofibrils. However, when the sarcoplasmic reticulum abuts a T-tubule, the elements of the sarcoplasmic reticulum become confluent to form a sac, the terminal cisterna. Two terminal cisternae of the sarcoplasmic reticulum flank each

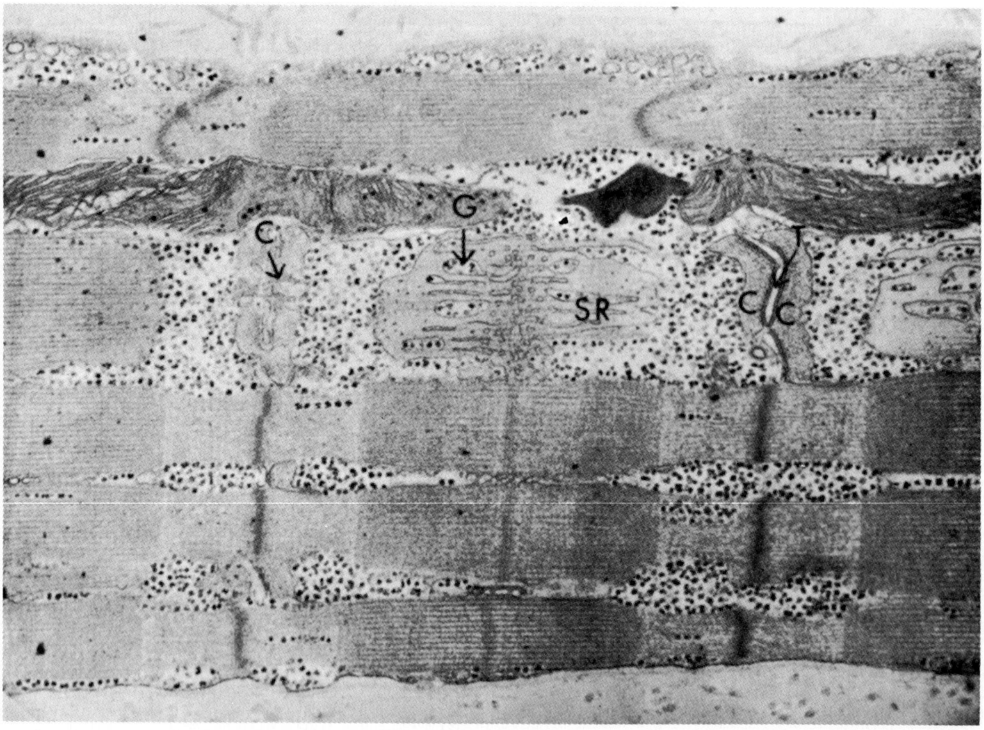

FIGURE 19–9 Electron micrograph of a longitudinal section of frog skeletal muscle. Three myofibrils are visible at the bottom of the field. Toward the top of the micrograph, the plane of section has grazed the surface of a myofibril, revealing the arrangement of the sarcoplasmic reticulum. The longitudinally oriented tubules of the sarcoplasmic reticulum (SR) are connected (in other planes of section) with the terminal cisternae (C) of the reticulum. Two terminal cisternae (C) flank a T-tubule (T) to form a triad. The dense granules are glycogen (G). Micrograph by H. E. Huxley. *From*: Leeson, C. R., and Leeson, T. S.: Histology. Philadelphia, W. B. Saunders, 1976.)

T-tubule, and this complex of three elements is known as a triad (Fig. 19–9). In some instances, a series of punctate densities has been observed to link the T-tubule with the terminal cisternae. The location of the triads with respect to the transverse banding pattern is constant for a given kind of skeletal muscle cell. In mammalian skeletal muscle, the triads are located at the level of the A-I junction, and thus there are two per sarcomere. In amphibian skeletal muscle, which has been widely studied in physiology, the triads are at the level of the Z lines, so there is only one per sarcomere. As will be discussed in more detail, the T-tubules undergo depolarization and conduct impulses rapidly from the surface of the muscle fiber into its interior. The terminal cisternae of the sarcoplasmic reticulum sequester and release calcium ions, which play a role in muscle contraction.

The Contractile Proteins

The major contractile proteins are actin and myosin. As described in the preceding section, myosin is located in the thick filaments, while actin resides in the thin filaments. This has been shown in several ways. For example, actin and myosin have been differentially extracted from a glycerol-treated muscle fiber (glycerol removes the sarcolemma), and the structural changes caused by each step of protein extraction have been monitored (Fig. 19–10). Other evidence lead-

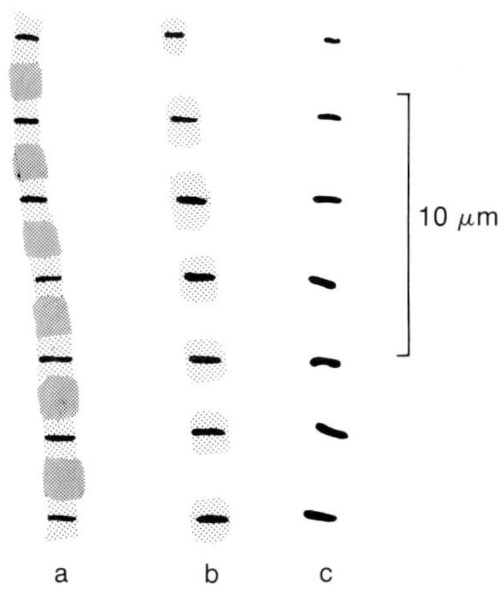

FIGURE 19–10 Sequential extraction of action and myosin from a glycerol-treated muscle fiber. The diagram shows the appearance in the phase contrast microscope of a fiber: (*a*) before protein extraction, (*b*) after extraction of myosin, and (*c*) after extraction of actin. (Drawn from a photograph in Hanson, J., and Huxley, H. E.: Symp. Soc. Exp. Biol., 9:228, 1955.)

ing to the same conclusion has come from experiments in which it was shown that antibodies against purified myosin combined with the thick filaments, while antibodies against purified actin bound to thin filaments. Purified actin and myosin monomers have been observed under appropriate conditions to polymerize into structures resembling thin and thick filaments, respectively.

THE THICK FILAMENT

Each thick filament is an assembly of many myosin molecules with a smaller number of minor protein species. A myosin molecule has the structure shown in Figure 19–11. Myosin consists of two similar major polypeptides which are intertwined to form a molecule with a long tail region consisting of the intertwined α-helical regions and a head region composed of the two globular ends of the subunits. Proteolytic enzymes cleave the myosin molecule into the parts shown in Figure 19–11. The fragment designated as LMM (light meromyosin) is quite insoluble and tends to self-aggregate. HMM (heavy meromyosin) is more soluble and retains the ATP-splitting activity of myosin as well as its ability to bind to actin. Further cleavage of HMM into the SF_1 and SF_2 subfragments shows that the globular heads (the SF_1 fragments) possess the ATPase and actin-binding activities.

Treatment of myosin with concentrated solutions of guanidine or urea dissociates it into its six constituent polypeptide chains (Fig. 19–12). There are two identical *heavy chains*, each having a molecular weight of about 200,000. The α-helical portions of the heavy chains intertwine in the intact molecule to form the

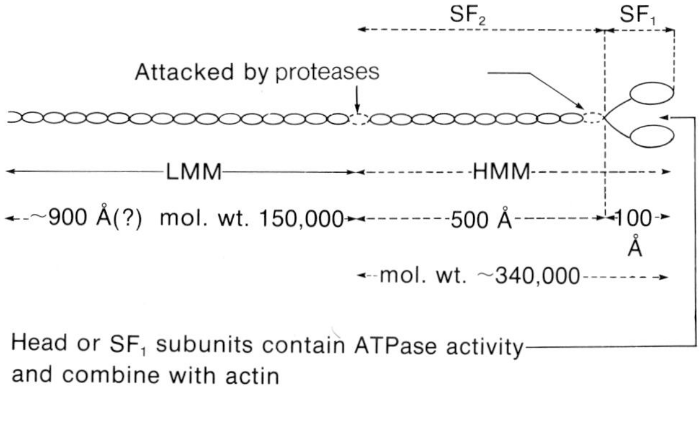

FIGURE 19–11 Diagrammatic representation of myosin molecule and its proteolytic products. (After S. Lowey.)

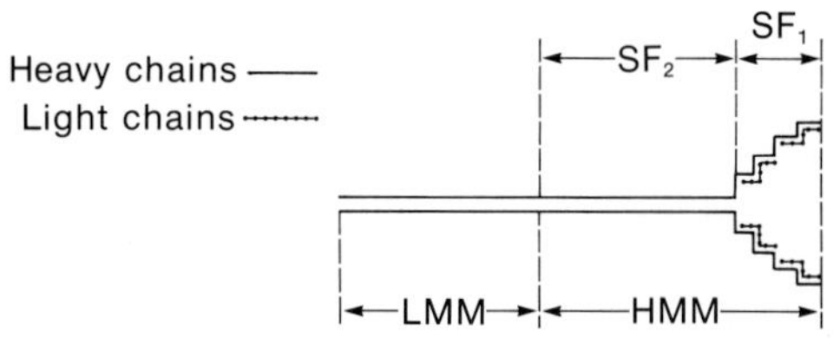

FIGURE 19–12 Diagram showing the six polypeptide chains of the myosin molecule.

tail portion shown in Figure 19–12, and the globular ends of the heavy chains form part of the myosin heads. Associated with the myosin heads are four much smaller peptides called the *light chains*. The light chains vary among myosins from different fiber types, but appear to consist of two sets of similar units in each molecule.

Under suitable conditions myosin molecules will self-aggregate to form structures that resemble thick filaments, except that they are of variable length (Fig. 19–13). Like the native thick filaments, these artificial thick filaments have projections that are believed to form cross-bridges to the thin filaments and to participate in the contraction process (Fig. 19–14). Studies by H. E. Huxley and others of thick filaments by means of x-ray diffraction suggest that the cross-bridges are arranged in a helix on the surface of the thick filament (Fig. 19–14). It appears that, at any given level, two bridges project directly opposite from each other. The next pair of cross-bridges is displaced 143 Å along the axis of the thick filament and rotated relative to the first pair by 120 degrees.

THE THIN FILAMENT

The major protein of the thin filament is actin. Thin filaments also contain tropomyosin and, in vertebrates, the Ca^{++}-binding regulatory protein troponin. The *actin* of the filaments is in the form of F-actin, the F indicating its fibrous character. F-actin is formed by the polymerization of G-actin monomers (Fig. 19–15), which are globular proteins 55 Å in diameter. In each thin filament, two chains of F-actin are intertwined to form the long two-stranded helical structure shown in Figure 19–15. The thin filament is of the order of 1 μm long in vertebrate striated muscle and contains 300 to 400 actin monomers. *Tropomyosin* is a long thin protein molecule. Tropomyosin molecules bind to actin to form filamentous strands that lie in each groove of the actin double helix. In this position, each tropomyosin molecule rests on about seven actin monomers. *Troponin* is a globular protein, and one troponin molecule apparently is associated with each tropomyosin molecule.

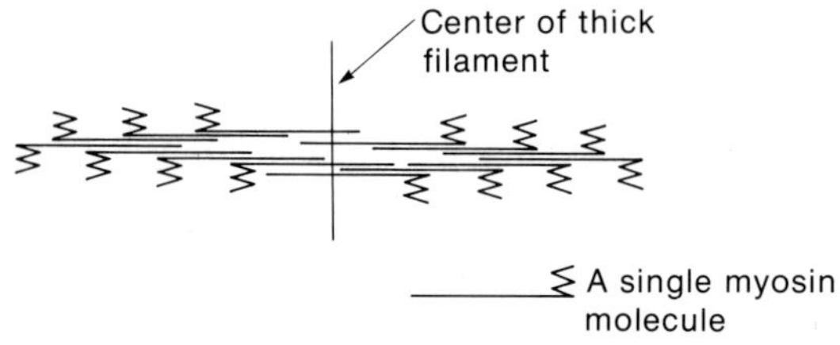

FIGURE 19–13 Diagram showing the way in which myosin molecules are supposed to aggregate to form the thick filament. Note that the myosin heads, which protrude from the surface of the thick filament, have an opposite polarity on the two sides of the center line. (Redrawn after Huxley, H. E.: J. Mol. Biol., 7:281, 1963.)

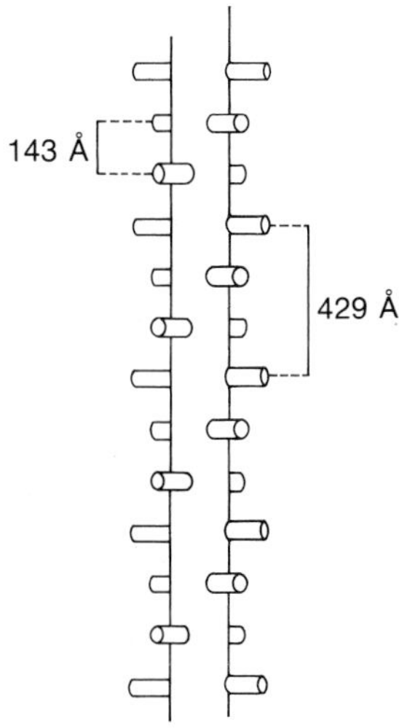

FIGURE 19–14 Diagram showing an arrangement of projections from the frog sartorius thick filament which accounts for the x-ray diffraction data obtained by H. E. Huxley. (Redrawn from Huxley, H. E.: Science, *164*:1356, 1969.)

The Sliding Filament Model of Muscle Contraction

Microscopic observation of relaxed and contracted striated muscle reveals that upon shortening, the Z lines come closer together, and thus the length of the sarcomere decreases. The shortening is due to a decrease in the width of the I bands, with the A bands remaining the same width.

However, electron microscopic observations have shown that the lengths of the thick and thin filaments themselves do not change on contraction. These observations led to the idea that when a muscle contracts the thin filaments are drawn in toward the center of the sarcomere by sliding between the thick filaments. This sliding filament model (Fig. 19–8) is almost universally accepted, but the mechanism by which the force is generated to cause the thin filaments to be pulled toward the center of the sarcomere is not so well understood and requires a more extended discussion.

Generation of the Contractile Force

The mechanism of interaction between thick and thin filaments during muscle contraction is the subject of intensive investigation. The following is a description

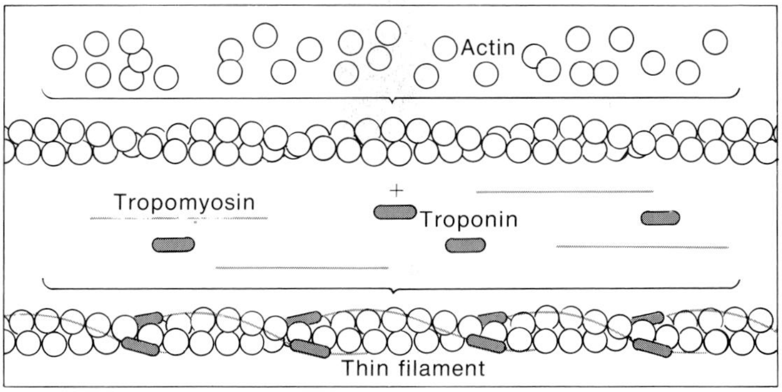

FIGURE 19–15 Diagram of the structure of the thin filament showing the two-stranded actin filament and the way in which tropomyosin and troponin are believed to be arranged in the thin filament. (Redrawn from Murray, J. M., and Weber, A.: Sci. Am., *230*(2):58, 1974.)

of a current model of how force is generated between actin and myosin to cause the thin filaments to be drawn toward the center of the sarcomere.

It has long been known that purified myosin can bind to F-actin to form a complex known as actomyosin. As noted, both the actin-binding ability of myosin and its ability to split ATP are located in the myosin head (the SF_1 subfragment of heavy meromyosin). The myosin head also has a rather high affinity site for binding ATP, so that at the ATP concentrations present in a living muscle almost every myosin head has a bound ATP. The complex between myosin and ATP, designated M·ATP, has only a low affinity for actin. However, the bound ATP is hydrolyzed by the myosin ATPase to form an "activated complex," designated $M \cdot ADP \cdot P_i^*$. In the activated complex, ATP has been split, but P_i (inorganic phosphate) and ADP have not dissociated from the complex. A current theory is that some of the energy released by hydrolysis of the high-energy phosphate bond is stored in a "high-energy configuration" of the myosin molecule. Now $M \cdot ADP \cdot P_i^*$ has a high affinity for actin, and thus the activated myosin head promptly binds to a site on the adjacent thin filament, forming a cross-bridge between the filaments.

The combination of the activated myosin head group, $M \cdot ADP \cdot P_i^*$, with actin is believed to promote two events: (1) the release of ADP and P_i, and (2) a conformational change (or swiveling) of the myosin cross-bridge which pulls the bound thin filament toward the center of the sarcomere. The molecular details of these events remain obscure.

The complex of actin and myosin that remains is stable in the absence of ATP. Indeed the formation of such complexes in dead muscle when the intracellular [ATP] falls to very low levels produces the muscular rigidity known as *rigor mortis*. In the presence of ATP in normal cells, each myosin head tends to bind an ATP, and when it does so it reverts to the M·ATP state, which has a low affinity for actin. Under these conditions, the myosin head detaches from the thin filament.

The cycle can now repeat itself. Each cycle moves the thin filament a distance of only about 100 Å relative to the thick filament, so that many repeats of the cycle are required to produce macroscopically observable shortening.

The model of the contractile cycle is summarized in Figure 19–16 in terms of the chemical moieties involved, with a schematic representation of the interaction of the myosin head groups with the thin filaments.

Control of Contractile Activity

An important consideration in the physiology of muscle cells is the necessity for muscle contraction to be switched on and off. A clue to how this might occur came from experiments in which myosin heads were allowed to interact in the presence of ATP with thin filaments from which tropomyosin and troponin had been removed. The myosin head groups proceeded to interact with the thin filaments with concomitant splitting of ATP. In fact, ATP was split until no more remained in the solution. In contrast, when the myosin heads were allowed to interact with *intact* thin filaments containing tropomyosin and troponin, ATP was not split unless Ca^{++} was provided at a concentration in excess of 10^{-7}M. If a greater concentration of Ca^{++} was provided, splitting of ATP took place at a greater rate, and maximum ATP splitting occurred at about 10^{-5}M Ca^{++}.

This experiment and many others suggest that the presence of tropomyosin

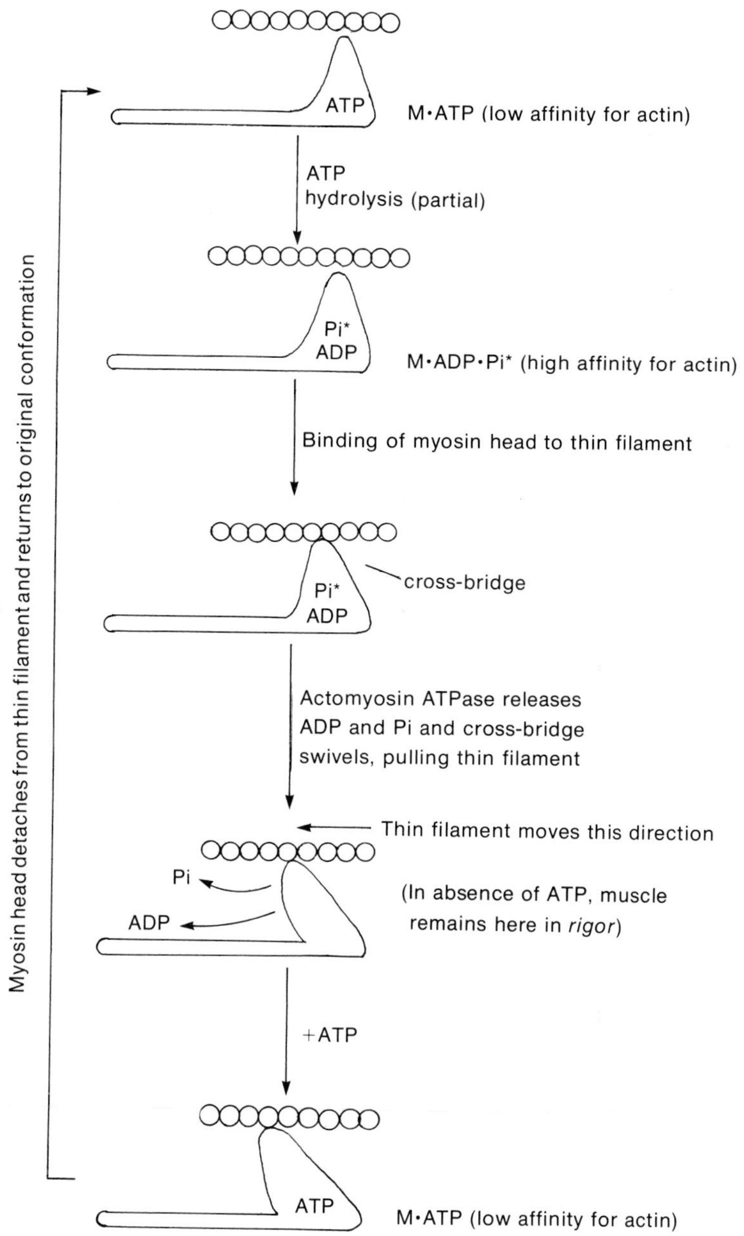

FIGURE 19–16 A current view of the contractile cycle.

and troponin on the thin filament confers on Ca^{++} the ability to regulate the interaction between the actin of the thin filament and the myosin heads (cross-bridges) of the thick filament. Troponin has a Ca^{++} binding site on one of its three subunits known as troponin C. A current notion is that when troponin binds Ca^{++}, a structural change causes tropomyosin to rotate into the groove of the F-actin double helix and out of the way of the myosin-binding site on actin. In the absence of bound Ca^{++} on troponin, the tropomyosin lies out of the groove and interferes with the binding of myosin to actin (Fig. 19–17). In this way the level of intracellular $[Ca^{++}]$ can control the activity of the contractile machinery.

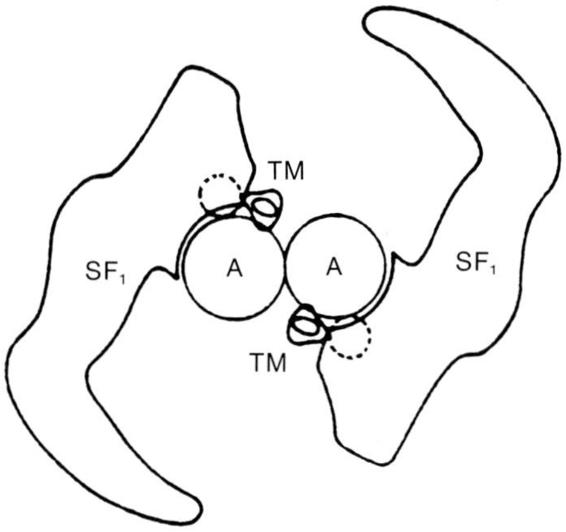

FIGURE 19–17 Composite end-on view of actin-tropomyosin-SF_1 structure. The solid contours represent the position of tropomyosin (TM) in the activated state. The dotted contours show the position of tropomyosin in relaxed muscle. In the relaxed state, tropomyosin, because of its position, inhibits the interaction of SF_1 with actin. (Redrawn from Huxley, H. E.: Cold Spring Harbor Symp. Quant. Biol., *37*:361, 1972.)

Membrane Potentials in Skeletal Muscle

As discussed in Chapter 17, the resting membrane potential in a typical skeletal muscle cell is about −90 mV (cell interior-negative). This is somewhat more polarized than a typical nerve cell, principally because the muscle cell has a lower ratio of sodium conductance to potassium conductance. The action potential of skeletal muscle is similar in certain important respects to the action potential of the squid giant axon. The rapid depolarization and overshoot of the muscle action potential are due to the inrush of Na^+ through fast sodium channels that open explosively when the threshold is reached. Repolarization is due to a spontaneous return of the sodium conductance to resting levels and a delayed opening of the potassium channels. The action potential is rapidly propagated over the surface of the muscle cell as described in Chapter 18.

Excitation-Contraction Coupling

The term excitation-contraction coupling refers to the mechanism by which an action potential in the cell membrane causes the muscle cell to contract. As noted in the section on muscle structure, the membranes of transverse tubules (T-tubules) are continuous with the plasma membrane, and the T-tubules are thus open to the extracellular space. An action potential in the muscle cell plasma membrane is believed to spread throughout the entire network of T-tubules.

The membrane of the sarcoplasmic reticulum contains a Ca^{++}-activated ATPase, also known as the Ca^{++} pump, which is capable of accumulating Ca^{++} in the sarcoplasmic reticulum against a large concentration gradient. The pumping action of the sarcoplasmic reticulum maintains the concentration of Ca^{++} in the sarcoplasm of resting muscle at less than 10^{-7}M, causing the contractile machinery to remain inactive. When an action potential spreads throughout the T-tubular system, the sarcoplasmic reticulum releases Ca^{++} into the sarcoplasm. The mechanism by which an action potential in the T-tubule causes the sarcoplasmic reticulum to release Ca^{++} remains obscure, but there is a close structural relationship between the T-tubules and the terminal cisternae of the sarcoplasmic reticulum. When the concentration of Ca^{++} in the sarcoplasm becomes high enough, Ca^{++} binds to troponin. This causes tropomyosin to move deep into the groove on the thin filament, allowing actin and myosin to interact and producing muscle contraction. When the membranes of the T-tubular system return to their resting membrane potential, the sarcoplasmic reticulum membranes again become impermeable to Ca^{++}. The Ca^{++} pump reaccumulates Ca^{++} into the interior of the sarcoplasmic reticulum, thus decreasing the Ca^{++} concentration in the sarcoplasm. When the sarcoplasmic $[Ca^{++}]$ falls sufficiently, Ca^{++} dissociates from troponin, tropomyosin rises out of the groove, blocking further myosin interaction, and muscles relax. In this way, the electrical state of the sarcolemma is believed to control the state of activation of the contractile machinery via an effect on the sarcoplasmic $[Ca^{++}]$.

The relation between the membrane potential and the degree of muscle contraction can be seen by using the voltage clamp method to maintain the plasmalemma at various potentials (Fig. 19–18). The membrane is depolarized in small steps, beginning at the resting membrane potential. Although the threshold for the action potential (which cannot occur because of the clamp) is about −60 mV, contractile force does not begin to develop until the membrane is depolarized 10 mV more to reach $E_m = -50$ mV. As the membrane is depolarized still further, a greater force of contraction is elicited. The maximum mechanical force is elicited when the membrane potential reaches − 20 mV; further depolarization does not result in greater contraction. Figure 19–19 shows a skeletal muscle action potential, the contractile threshold (−50 mV), and the potential for maximum contraction (−20 mV). Agents that alter the shape of the action potential in such a way as to increase the area of the shaded region in Figure 19–19 are potentiators of muscle contraction.

Characteristics of Muscle Contraction due to the Properties of Muscle Cells

The Twitch. In skeletal muscle, a single action potential in the plasma membrane leads to a single, all-or-none contraction called a *twitch* (Fig. 19–20). The twitch lasts from 20 to 150 msec. This is much longer than the action potential, which is over in 3 to 10 msec. This is because the release of Ca^{++} and its reuptake by the sarcoplasmic reticulum occur on a longer time scale than the electrical events at the plasma membrane.

Summation of Contractions and Tetanus. When a skeletal muscle cell or an entire muscle is stimulated at low frequency, a series of twitches is elicited that do not differ individually from the twitches elicited by a single stimulus. However, when the stimuli occur more frequently, so that each stimulus arrives before the muscle has completely relaxed from the preceding stimulus, the contractions induced by the individual stimuli summate (Fig. 19–21). If the stimulus frequency

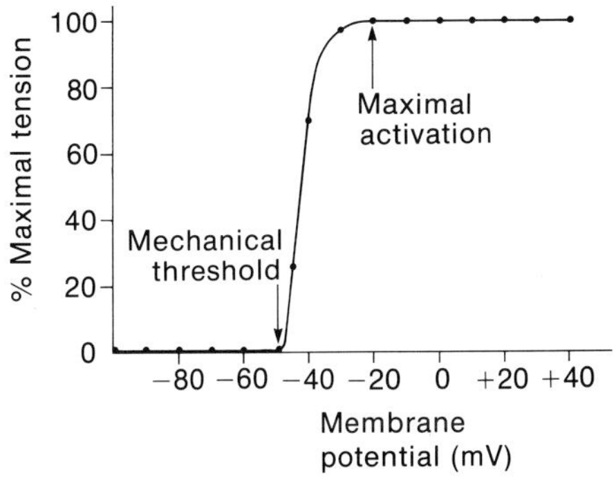

FIGURE 19–18 An idealized experiment in which a muscle plasma membrane is depolarized in a stepwise fashion, using voltage clamp techniques, and the tension developed by the muscle is recorded. No tension is produced until the membrane potential reaches about −50 mV (the mechanical threshold). As the membrane is further depolarized, the developed tension increases, reaching a maximum at a membrane potential of −20 mV. Further depolarization causes no further increase in tension.

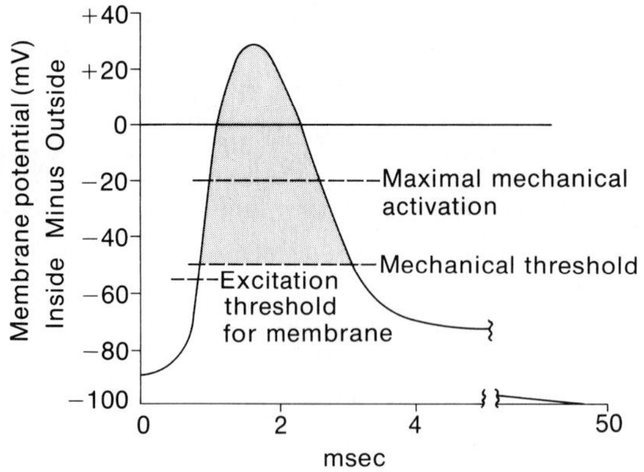

FIGURE 19–19 The information of Figure 19–18 is seen in the context of the normal skeletal muscle action potential. The stippled area is above the mechanical threshold, and the area of this section relates directly to the amount of tension developed by the muscle. (Redrawn from Sandow, A.: Arch. Phys. Med. Rehab., 45: 62, 1964.)

is sufficiently high, the contractions fuse into a single sustained contraction called a *tetanus*.

Isometric versus Isotonic Contractions. When a muscle shortens against a constant load, it is said to contract *isotonically* (constant force). When a muscle develops force but cannot shorten because the two ends of the muscle are fixed, it is said to contract isometrically (constant length).

Muscle Length and Muscle Contraction. The force that a muscle develops when it contracts depends markedly on its initial length. If the greatest isometric force a muscle can develop is determined as a function of muscle length, an optimum length, l_0, for force development will be found (Fig. 19–22). It is significant that the resting length of most muscles in the body is close to l_0. Furthermore, most skeletal muscles do not shorten a great deal in their normal function and thus remain in the range of length in which they can develop force effectively.

This behavior can be observed both at the level of entire muscles and at the level of single skeletal muscle fibers (Fig. 19–23). It is found that the sarcomere length at which maximum force development occurs permits maximum overlap of thick and thin filaments, without overlap of the thin filaments themselves in the center of the sarcomere.

The Force-Velocity Relationship. When a muscle lifts a weight in an isotonic contraction, the force that the muscle develops equals the load. The larger the weight the muscle lifts, the slower its velocity of shortening. As shown in Figure 19–24, the velocity of shortening decreases hyperbolically as the load increases.

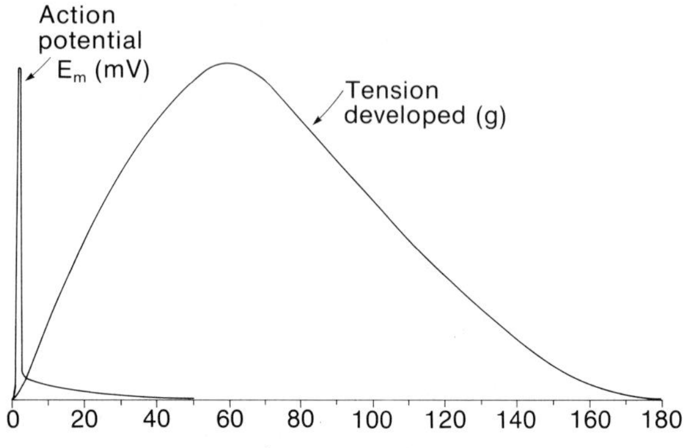

FIGURE 19–20 The action potential and the tension developed in a typical skeletal muscle fiber are shown on the same time base. Note the much more rapid time course of the action potential.

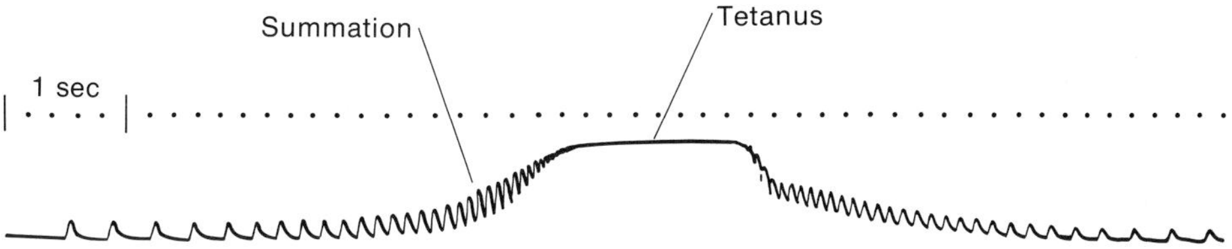

FIGURE 19–21 A record of the tension produced by a single muscle fiber, stimulated with a frequency that increases continuously until full tetanus is obtained and then decreases. Note that the maximum tetanic force is several times that of a single twitch. (Redrawn from Buchtal, Dan. Biol. Med., *17*:1, 1942.)

The maximum force that the muscle can develop (f_{max}) is the force of an isometric contraction when the muscle length is set to the optimum length for force development (1_0). If the muscle attempts to lift a weight equal to f_{max}, it will not shorten at all, i.e., the velocity of shortening will be zero. A muscle fiber shortening against no load has a maximum shortening velocity, V_{max}, which is characteristic of the fiber type.

Gradation of Skeletal Muscle Contraction In Vivo

Up to this point we have discussed properties of muscle related to the behavior of single muscle cells. In the body, however, the relevant functional unit of muscle contraction is the *motor unit.* Motor axons branch, and each motor axon innervates a number of skeletal muscle fibers. In mammals, each muscle fiber is served by one motor end plate. A motor axon and the muscle fibers it innervates are known as a motor unit. Since an action potential in the motor nerve causes contraction of each fiber it innervates, the whole motor unit is the fundamental contractile element. Motor units contain as few as two to three muscle fibers per motor unit in muscles responsible for fine movements, such as the extraocular muscles and the inner ear muscles. Large motor units containing as many as 2000 muscle fibers are present in large muscles used to generate large forces. The large leg muscles are in this category but have a considerable range in the number of fibers per motor unit. The different motor units in a muscle are activated asynchronously, with the smaller units recruited first, so the muscle contracts smoothly. Impulses in motor nerves can vary in frequency up to a maximum of 150 per second.

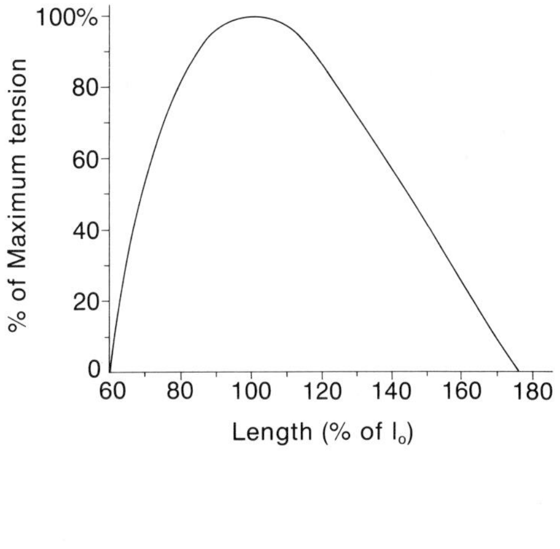

FIGURE 19–22 Dependence of the tension developed by a skeletal muscle upon its length. The isometric tension developed in response to tetanic stimulation (expressed as a percentage of maximal tension) is plotted against the length of the muscle (expressed as a percentage of the length at which tension is maximal, 1_0.)

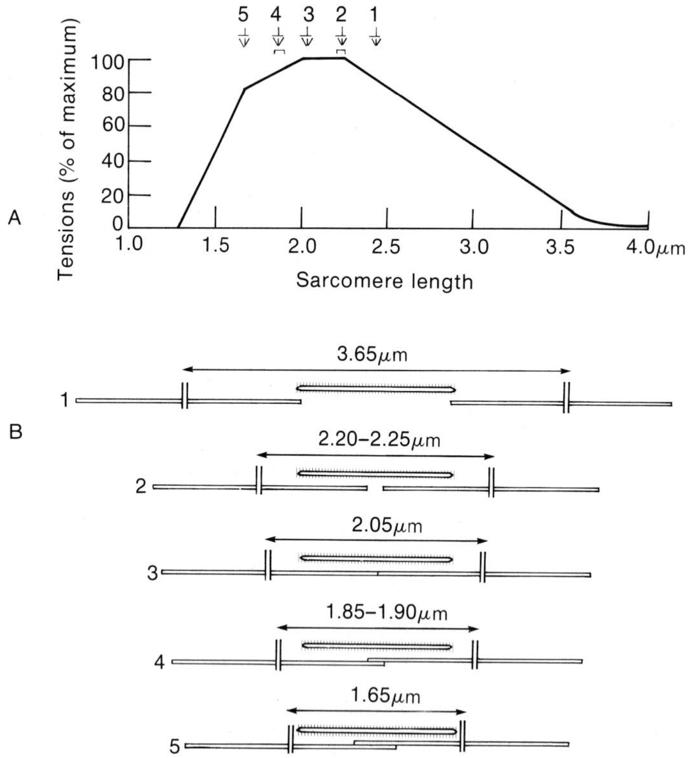

FIGURE 19–23 The relationship between sarcomere length and isometric tension developed by single fibers of frog sartorius muscle. *A*. Isometric tension as a function of sarcomere length. *B*. The arrangment of the thick and thin filaments for different sarcomere lengths. (Redrawn from Gordon, A. M., Huxley, A. F., and Julian, F. J.: J. Physiol., *184*: 170, 1966.)

There are basically two ways in which the force of contraction of a muscle can be increased: (a) by an increase in the number of active motor units, known as *recruitment,* and (b) by increasing the frequency of stimulation of the individual motor units. In the body, these two mechanisms are used to produce smoothly graded contractions over a considerable range of total muscle force developed.

The Types of Striated Muscle Cells

The two main types of striated muscle fibers are white-fast fibers and red-slow fibers, so named because of their gross appearances and speeds of contrac-

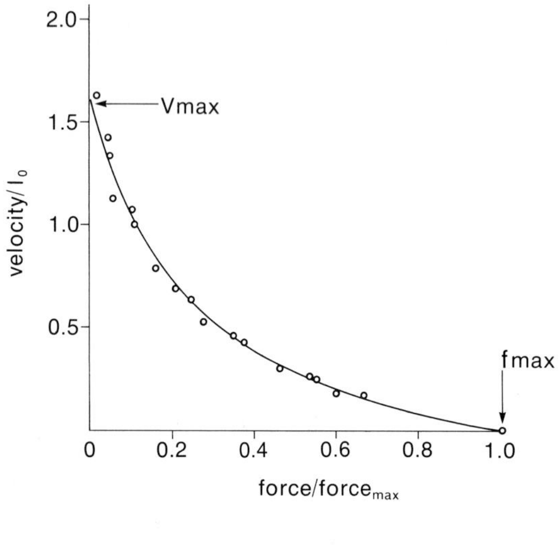

FIGURE 19–24 The force–velocity relationship for frog sartorius muscle at 0°C. The muscles were made to shorten against various loads and the speed of shortening determined. Shortening velocity is plotted as $v/1_0$, where 1_0 is the muscle length for maximum isometric force development. Force is plotted as force/maximum force. (Redrawn from Hill, A. V.: First and Last Experiments in Muscle Mechanics. New York, Cambridge University Press, 1970, p. 32).

tion. White-fast fibers are best suited for performing high intensity work for short periods of time. Red-slow fibers are specialized for prolonged activity with less force development. A third fiber type, red-fast, is less frequent in occurrence. Many of the properties of red-fast fibers are intermediate between those of the other two fiber types, with contractile properties approaching those of white-fast fibers and metabolic characteristics resembling those of red-slow fibers. Most muscles contain a mixture of fiber types, but a few muscles in some species contain largely one fiber type. All the fibers in a single motor unit are of the same type, and the differentiation of the embryonic muscle fibers into the two types reflects a neurotrophic influence of the axon type that innervates the cells.

Red muscles are red because they contain significant amounts of myoglobin, a sarcoplasmic oxygen-binding protein, and have a high capillary density. Both of these factors play a role in the highly oxidative metabolism of red-slow muscle. The slow but more frequent contractions of these fibers demand ATP at a rate which can be met efficiently by oxidative metabolism based on the circulation supplying O_2 and substrate. White-fast muscle, on the other hand, derives much of its ATP from glycolysis. This inefficient process is very fast and can supply ATP at the high rates demanded for high-velocity contractions of brief duration. Accordingly, white-fast fibers have larger stores of glycogen than are present in red muscle. Table 19–1 compares the properties of red-slow and white-fast fibers, with emphasis on their adaptations for oxidative and glycolytic metabolism, respectively.

At birth, all muscles are red-slow, and white-fast fibers develop postnatally. The motor nerve has an important influence on the development of muscle fibers. If the motor nerves to red-slow and white-fast fibers are interchanged, the red-slow fibers become whiter and faster and the white-fast fibers become redder and slower. This so-called *trophic effect* of the motor nerve on muscle development is a subject of current study.

Exercise has a significant influence on muscle development. High-intensity, short-duration exercise, such as weight-lifting, results in the hypertrophy of white

TABLE 19–1

COMPARISON OF THE PROPERTIES OF WHITE-FAST AND RED-SLOW MUSCLE FIBERS

Characteristics Related to Mechanical Performance		
	White-Fast	*Red-Slow*
Maximum contraction velocity	High	Moderate
Twitch duration	Short (25 msec)	Long (85 msec)
Frequency for fused tetanus	High (70/sec)	Lower (20/sec)
Myosin ATPase-specific activity	High	Moderate
Pumping rate of sarcoplasmic reticulum	Higher	Lower
Fiber diameter	45–70 μm	30–50 μm
Metabolic Factors		
Primary Source of ATP	*Glycolysis*	*Oxidative Phosphorylation*
Number of adjacent capillaries	Low (0–1)	High (4–8)
Myoglobin content	Low	High
Mitochondrial density	Low	High
Dehydrogenase activities	Low	High
Glycogen content	High	Low
Glycogen phosphorylase activity	High	Low
Susceptibility to fatigue	High	Low

fibers caused by the synthesis of more myofibrils. This results in a muscle with more myofibrils contracting in parallel, which thus can develop more total force. In contrast, low-intensity, long-duration training, such as distance running, increases the oxidative capacity of all fibers used for the exercise, with no increase in fiber size.

The Role of Creatine Phosphate in Muscle Metabolism

During a contraction, the rate of utilization of ATP by a muscle cell rises many fold above the basal level. Should the ATP concentration in the sarcoplasm fall to very low levels, rigor would develop. This is prevented by the presence in the muscle of a large pool of high-energy bonds in the form of creatine phosphate. Under normal circumstances, as rapidly as ATP is split by the actomyosin ATPase, the ADP released is rephosphorylated from creatine phosphate by the enzyme creatine phosphokinase. If the muscle cell continues to work at a high level, the creatine phosphate level falls steadily, but the intracellular ATP concentration is maintained at an almost constant level. Not until the pool of creatine phosphate is severely depleted does the ATP level fall markedly. Muscular fatigue intervenes before this happens, but the factors leading to fatigue remain unclear. Thus creatine phosphate plays the role of a high-energy phosphate "buffer" which maintains the level of ATP in the sarcoplasm almost constant, even in the face of drastically increased ATP utilization.

CARDIAC MUSCLE

Structure of Cardiac Muscle Cells

The pattern of striation in mammalian cardiac muscle cells (Fig. 19–25) is the same as that in skeletal muscle cells. Cardiac cells, however, display some modifications of the basic plan of striated muscles described for skeletal muscle. Unlike skeletal muscle cells, cardiac muscle cells are not simple cylinders; they branch and by doing so form a network that enables them to carry out their squeezing action when the heart contracts. Cardiac muscle fibers contain only one nucleus, or occasionally two, and the nuclei are located centrally in the fibers. The mitochondria are exceptionally large and numerous with elaborate cristae (Figs. 19–25 and 19–26). Their structure reflects the fact that cardiac muscle cells have a great need for energy derived from oxidative phosphorylation since they are continuously contracting and relaxing without rest. In cardiac muscle cells the myofilaments are not generally grouped into as discrete myofibrils as skeletal muscle; instead, the regions occupied by myofilaments have a tendency to become confluent (Fig. 19–26).

The T-tubules of cardiac muscle (Fig. 19–27) have a larger caliber than those of skeletal muscle and contain a lining of polysaccharide-rich material, which is a continuation of the PAS-positive external lamina found on the outer surface of each cell. The sarcoplasmic reticulum of cardiac muscle cells forms a network around the myofibrils, but it is not as well developed as that of skeletal muscle. Large terminal cisternae are not present, but some expanded elements of sarcoplasmic reticulum are closely apposed to T-tubules (Fig. 19–27). The T-tubules penetrate into the fiber at the level of the Z line.

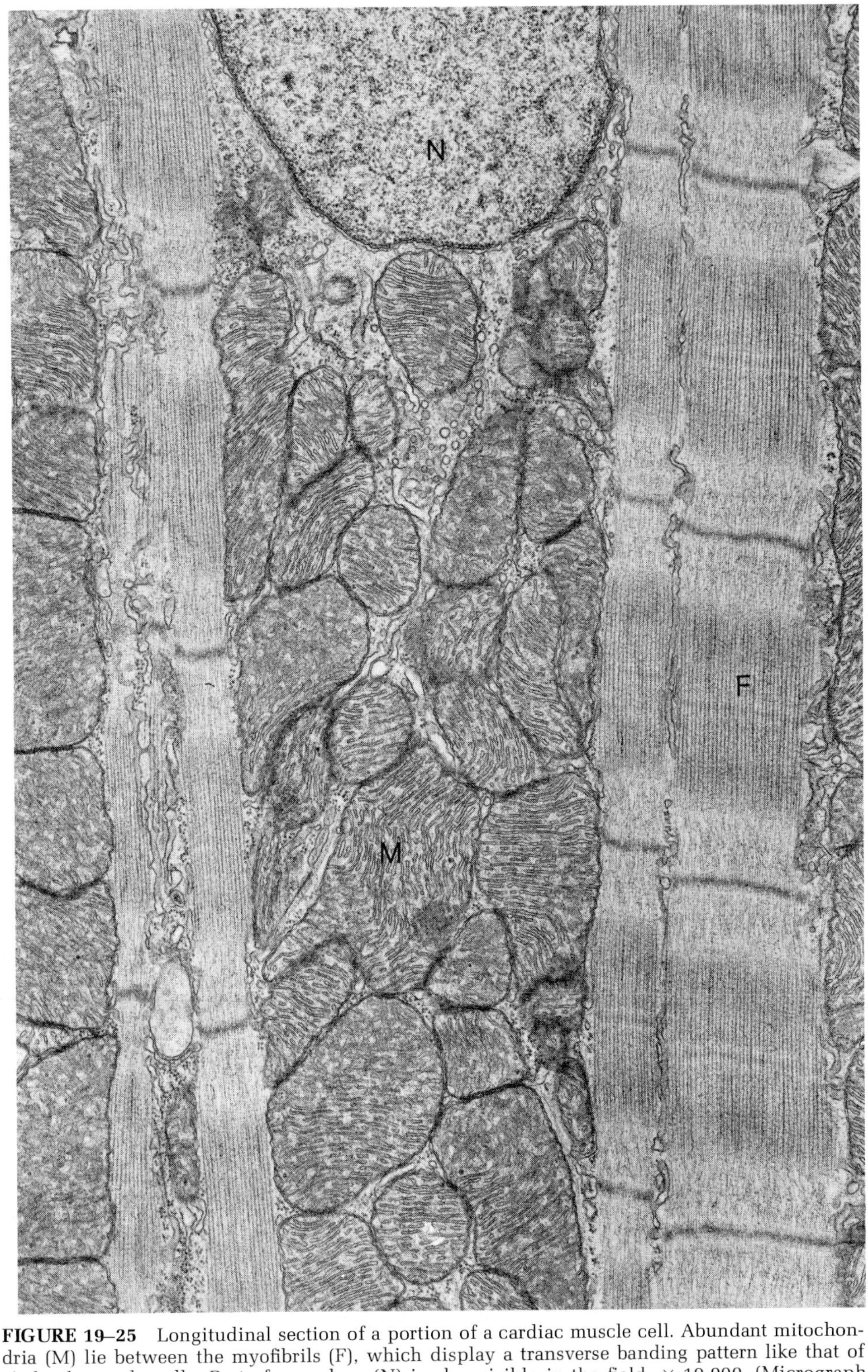

FIGURE 19–25 Longitudinal section of a portion of a cardiac muscle cell. Abundant mitochondria (M) lie between the myofibrils (F), which display a transverse banding pattern like that of skeletal muscle cells. Part of a nucleus (N) is also visible in the field. × 19,000. (Micrograph courtesy of N. S. McNutt and D. W. Fawcett. From Langer, G. A., and Brady, A. J. (eds.): Mammalian Myocardium. New York, John Wiley, 1974, pp. 1–49. Reproduced by permission of the publisher.)

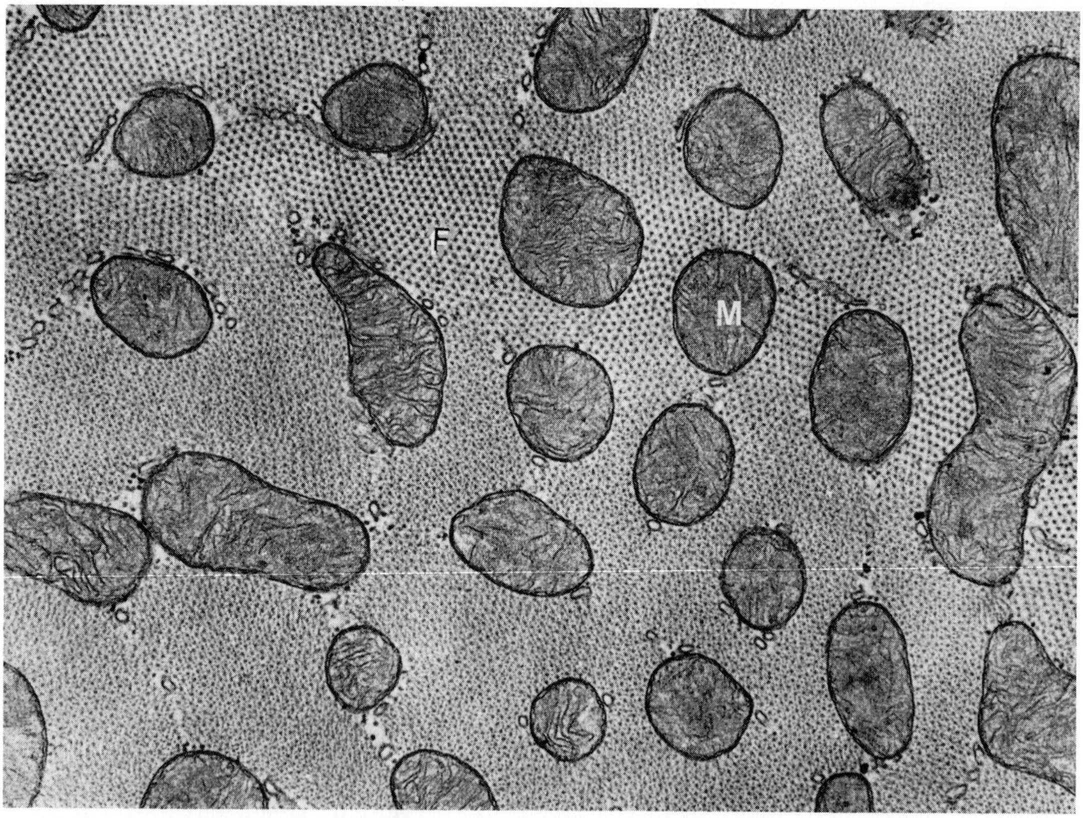

FIGURE 19–26 Cross section of a cardiac muscle cell. The myofilaments (F) are seen end-on. Note that they are not grouped in as distinct bundles to form cylindrical myofibrils as in skeletal muscle cells. Many mitochondria (M) lie between the myofilaments. × 30,000. (Micrograph courtesy of Fawcett, D. W., and McNutt, N. S.: J. Cell Biol., *42*:1, 1969. Reproduced by permission of the Rockefeller University Press.)

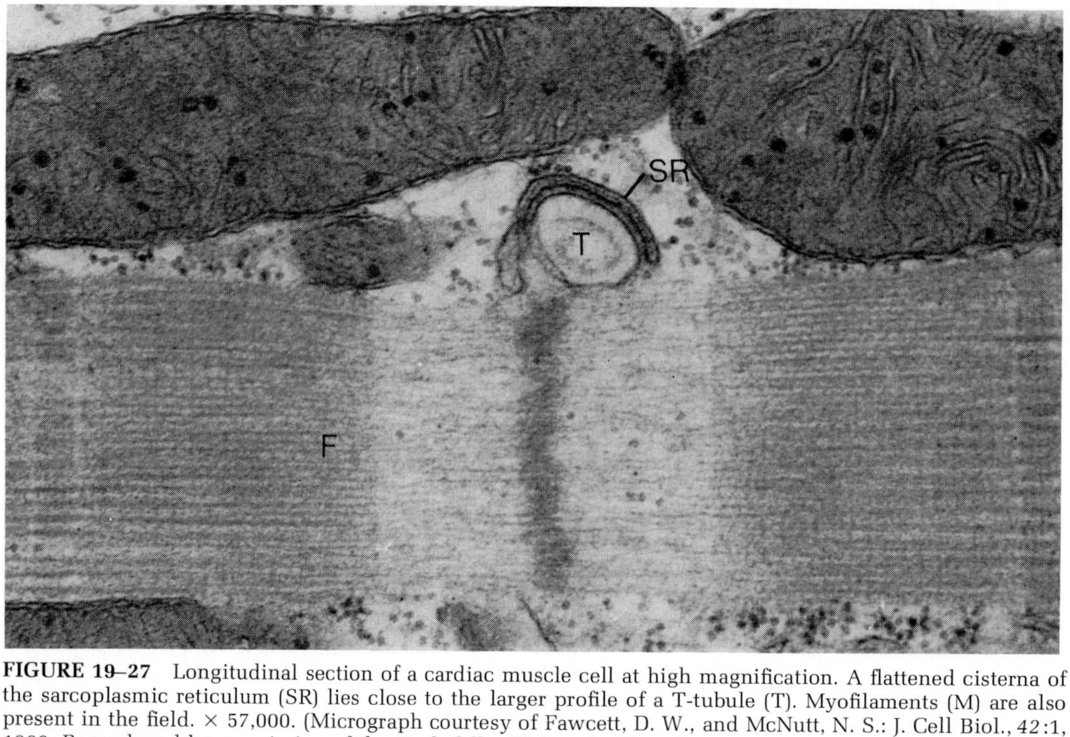

FIGURE 19–27 Longitudinal section of a cardiac muscle cell at high magnification. A flattened cisterna of the sarcoplasmic reticulum (SR) lies close to the larger profile of a T-tubule (T). Myofilaments (M) are also present in the field. × 57,000. (Micrograph courtesy of Fawcett, D. W., and McNutt, N. S.: J. Cell Biol., *42*:1, 1969. Reproduced by permission of the Rockefeller University Press.)

A unique characteristic of cardiac muscle cells is the presence of intercalated discs. These appear in the light microscope as dense bands that extend transversely across the muscle fibers. With the electron microscope it can be seen that they are actually contact specializations between the ends of adjacent muscle cells that serve to link cardiac muscle fibers together end-to-end (Figs. 19–28 and 19–29). The ends of the muscle fibers and the intercalated discs occur at the level of a Z line in each fiber. However, proceeding transversely across a fiber, the disc jumps from one sarcomere to the next, so that the intercalated disc is not actually a flat plate but has a steplike configuration. As a result, some parts of the intercalated disc are oriented transversely to the long axis of the muscle fiber, while other parts course longitudinally.

Several different type of contact specializations are present in the intercalated disc (Figs. 19–28 and 19–29). The transverse portions of the intercalated disc contain desmosomes, a wavy-appearing fascia adhaerens, and small punctate gap junctions. The longitudinal parts of the intercalated discs contain extensive sheetlike areas of gap junctions. The function of the desmosomes and the fasciae adhaerentes is to hold the cells together, while the gap junctions serve to transmit impulses from one cell to another.

Physiologically, cardiac muscle is similar in many respects to red-slow skeletal muscle. The metabolic properties of red-slow muscle listed in Table 19–1 apply equally well to cardiac muscle. Cardiac muscle contracts even more slowly than red-slow muscle, with a cardiac contraction lasting about 250 msec.

Autonomy of the Heart

The heart receives no motor input from the central nervous system, and cardiac muscle has no motor end plates. Nerve fibers from both the sympathetic and parasympathetic divisions of the autonomic nervous system are present in the heart, however, and influence the rate and force of contraction. Certain cardiac muscle cells have an unstable resting potential which tends to depolarize spontaneously. When the threshold is reached, these cells fire an action potential and repolarize, prompting the membrane potential to immediately begin to move toward threshold again. Such cardiac cells fire action potentials in a rhythmical fashion. The cells of the sinoatrial (SA) node have the fastest intrinsic rhythm of all the cardiac cells and hence serve as the *pacemaker* for cardiac contraction (Fig. 19–30). Since cardiac cells are coupled by low-resistance junctions in the intercalated discs, an action potential that originates in the SA node can rapidly propagate throughout the entire heart, causing every cardiac muscle cell to contract. For this reason the heart is often termed a *functional syncytium*.

The Cardiac Action Potential

Figure 19–31 shows a typical ventricular action potential. It differs from the action potential of skeletal muscle in being 25 times longer in duration, primarily as the result of a prolonged plateau phase. The upstroke of the cardiac action potential is due to a rapid inrush of Na^+ through *fast sodium channels,* which are apparently identical to the sodium channels involved in the action potentials in skeletal muscle and nerve. During the plateau phase of the cardiac action potential, both sodium and calcium enter the cell by way of so-called *slow channels*.

FIGURE 19–28 Longitudinal section of parts of two cardiac muscle cells from the cat atrium, showing an intercalated disc. Transverse (T) and longitudinal (L) portions of the intercalated disc are present. Most of the transverse portion is a fascia adherens, but a punctate gap junction is present at the arrow. The longitudinal portions contain large gap junctions. × 25,000. (Micrograph courtesy of Fawcett, D. W., and McNutt, N. S.: J. Cell Biol., 42:46, 1969. Reproduced by permission of the Rockefeller University Press.)

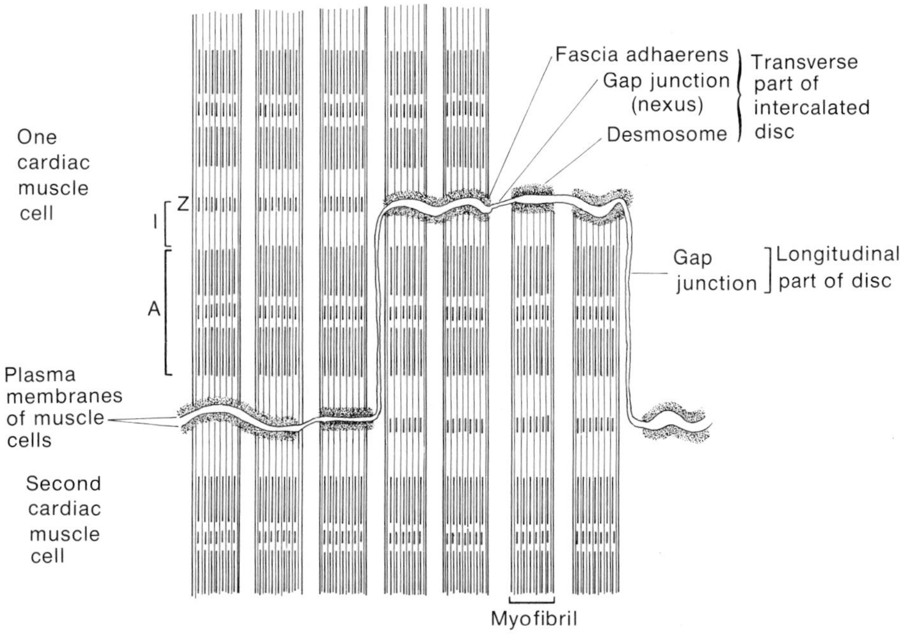

FIGURE 19–29 Diagram of the cell contact specializations in the intercalated disc.

Repolarization is caused by a closing of the slow channels and a delayed increase in the potassium conductance.

Excitation-Contraction Coupling in Cardiac Muscle

Cardiac cells have relatively less sarcoplasmic reticulum than skeletal muscle cells, and their sarcoplasmic reticulum is less active in accumulating calcium. The calcium that enters the sarcoplasm via the slow channels during the plateau phase of the cardiac action potential plays a role in activating the contractile proteins and may also serve as a signal to trigger the release of calcium from the sarcoplasmic reticulum.

Regulation of Contraction in Cardiac Muscle

The refractory period of cardiac muscle lasts almost as long as the contraction itself (Fig. 19–32). Therefore, cardiac muscle cannot be summated by repetitive stimuli. Since in a normal beat every cell contracts, the entire heart can be thought of as a single motor unit.

FIGURE 19–30 Schematic depiction of intracellular recording from a cell in the sinoatrial node of the heart. The resting membrane potential (mV) is plotted versus time (sec). The cell spontaneously fires repetitive action potentials.

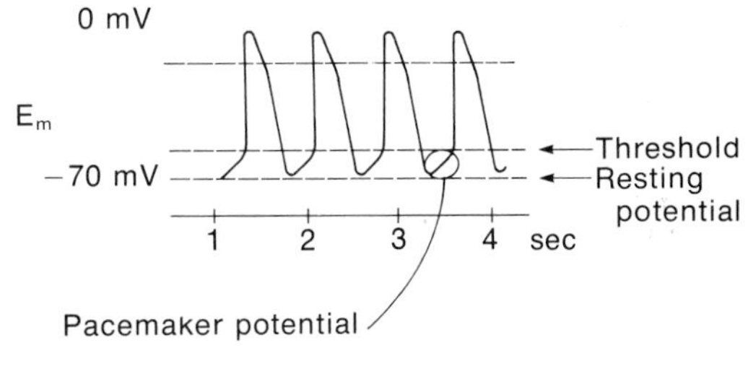

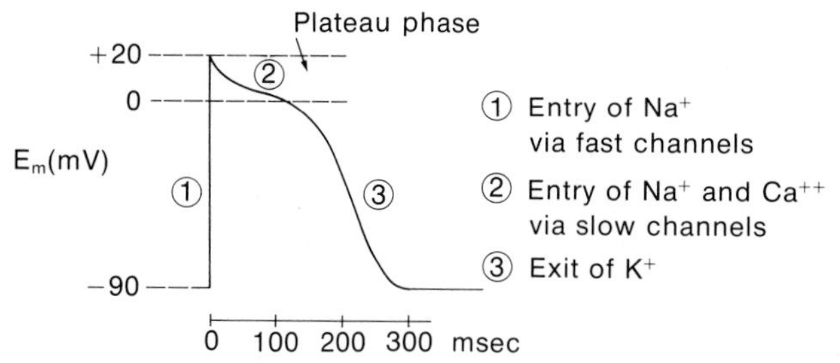

FIGURE 19–31 Schematic illustration of the action potential in a cardiac ventricular cell as recorded with an intracellular microelectrode. The transmembrane potential difference (mV) is plotted versus time (msec).

For these reasons the heart cannot increase its force of contraction by temporal summation of contractions or by recruitment of additional motor units, as skeletal muscle does. The force of cardiac contraction does vary with the sarcomere length at the beginning of contraction and is affected by cardioactive neurotransmitters and hormones. The heart increases its force of contraction in response to the norepinephrine that is released from sympathetic nerve endings in the myocardium and in response to circulating epinephrine. The acetylcholine that is released from parasympathetic nerve endings in the heart causes a decrease in the force of contraction. Increased levels of epinephrine and norepinephrine in the myocardium during exercise are said to increase the force of contraction by shifting the heart to a different length-tension curve, as illustrated in Figure 19–33.

SMOOTH MUSCLE

Smooth muscle is found in the walls of blood vessels, the gastrointestinal tract, the respiratory tree, the urogenital organs, and other tissues. Smooth muscle is more variable in its properties than skeletal or cardiac muscle. In the human, for example, the smooth muscle of the vasculature differs in important respects from the smooth muscle of the intestine, and within each system the smooth muscle cells vary considerably. Because of the variability of smooth muscle and because its molecular architecture is less orderly than that of striated muscle, our understanding of it is comparatively rudimentary.

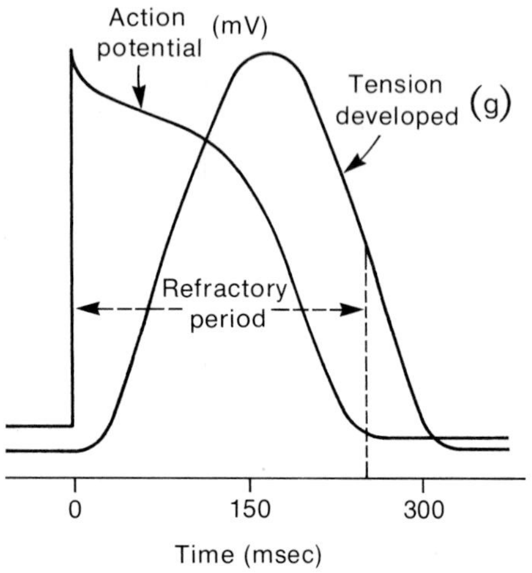

FIGURE 19–32 The time relationships between the action potential in a cardiac ventricular cell and its mechanical response. Note that the refractory period lasts almost as long as the contraction.

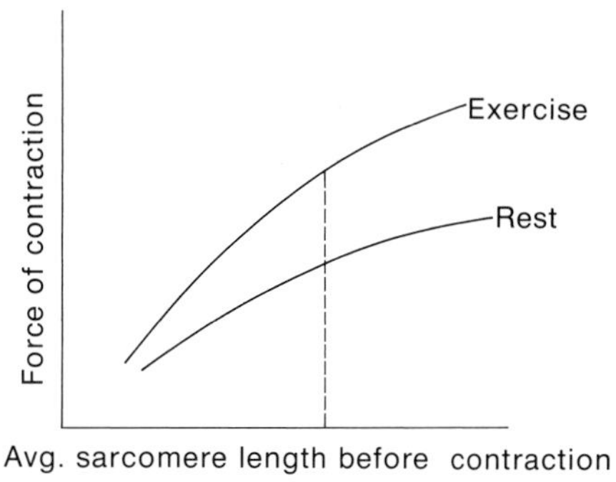

FIGURE 19–33 Schematic depiction of the length-tension relation for the heart. In exercise, the heart responds to circulating epinephrine and to norepinephrine released by cardiac sympathetic nerves by shifting to higher contractility (i.e., greater force of contraction for a given sarcomere length before contraction).

Structure of Smooth Muscle

Smooth muscle is so named because when viewed with the light microscope it lacks the cross-striations of striated muscle. The spindle-shaped smooth muscle cells (Fig. 19–34) are about 2 to 6 μm in diameter and from 50 to 400 μm long. Each contains a single central nucleus that is elongated in the long axis of the cell. Transverse tubules (T-tubules) are not present in smooth muscle cells, and the sarcoplasmic reticulum is poorly developed. Some smooth muscle types are almost totally lacking in sarcoplasmic reticulum. The usual organelles, such as mitochondria, rough endoplasmic reticulum, and Golgi apparatus, are located in conical masses of cytoplasm at the poles of the nucleus and in islands scattered through the cell (Fig. 19–34). Numerous vesicles are found in the cytoplasm near the surface of the muscle cells, and some are connected by a neck to the plasma membrane. These have long been called pinocytic vesicles, but the basis of such apparently intense pinocytic activity is not clear. It has been suggested that they might have some other function, perhaps playing a role analogous to that of short T-tubules.

Most of the cytoplasm is filled with myofilaments (Figs. 19–34 and 19–35) which course in a generally longitudinal direction. For a long time, routine preparations for electron microscopy showed all the filaments to be of a similar size, about 50 Å in diameter, resembling the thin actin filaments of skeletal muscle. More recently, thick filaments have been observed with the electron microscope in sections of smooth muscle cells under certain conditions of fixation and of cell contraction. Although there is not complete agreement on the structure of thick filaments in living smooth muscle, a current view is that both thick and thin filaments are present in smooth muscle cells and that they are disposed in a nonrandom way. Smooth muscles contain about one fifth the myosin of skeletal muscle, but their actin content is as high or higher than that of skeletal muscle. In electron micrographs of cross sections of smooth muscles in which thick filaments are visible, there are 12 to 18 thin filaments for each thick filament, compared with a ratio of two thin filaments to one thick filament in skeletal muscle. It is not clear whether smooth muscle filaments have an arrangement that is analogous to the sarcomere of skeletal muscle.

The existence in smooth muscle of a structure analogous to the Z line is uncertain, but smooth muscle cells appear to have an internal "skeleton" consisting of "intermediate" filaments and dense bodies. The intermediate filaments are so called because they are intermediate in size between thin and thick filaments. The dense bodies are electron-opaque structures which may be sites of attachment for thin filaments in the cytoplasm and to the plasma membrane.

The cells of some smooth muscles are related to one another by punctate gap

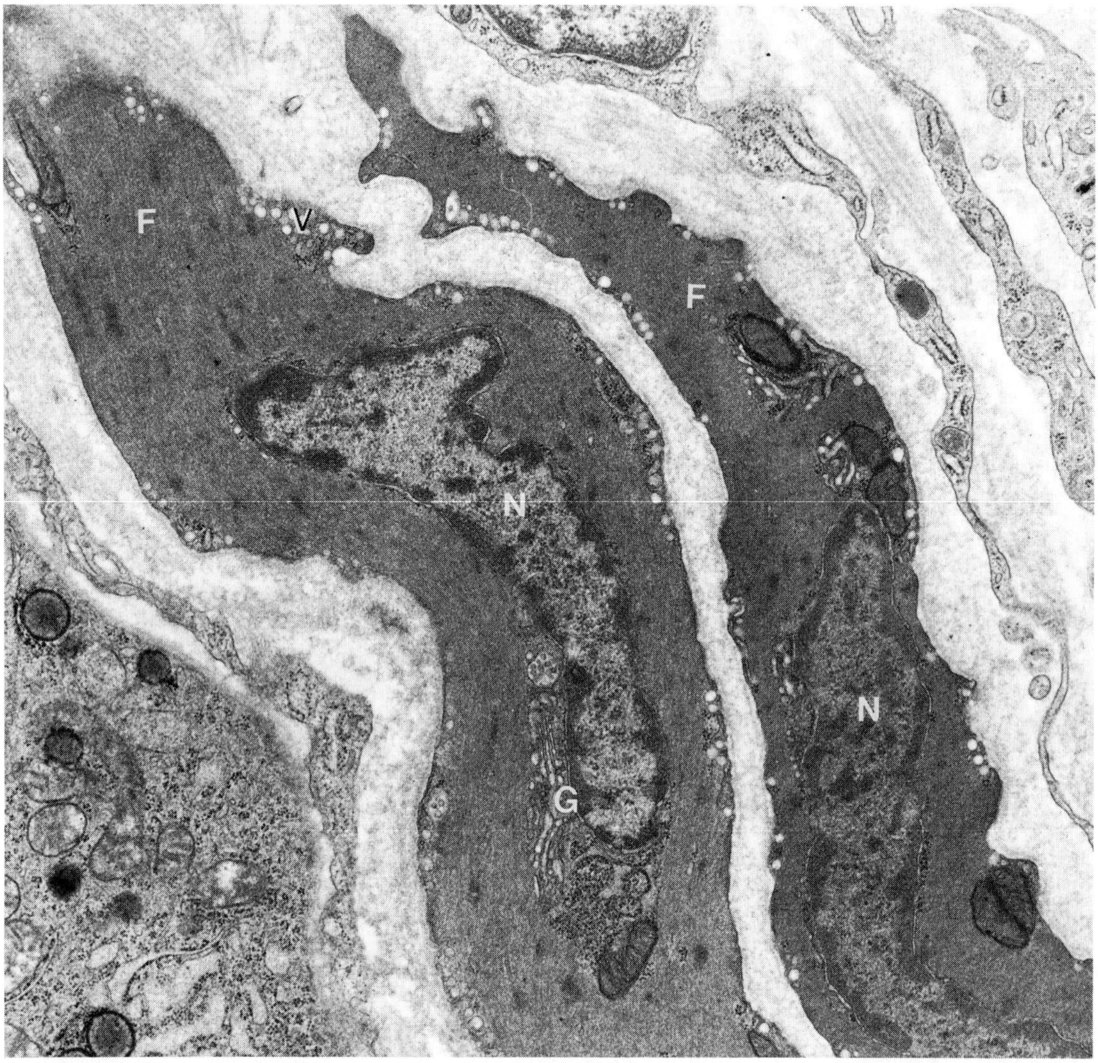

FIGURE 19–34 Portions of two smooth muscle cells from the interstitial tissue of the rat prostate. The nucleus (N) is centrally located in each fiber. Most of the cytoplasm is occupied by the myofilaments (F). A Golgi apparatus (G) and some endoplasmic reticulum and mitochondria are located at one pole of the nucleus, and vesicles (V) are found at the margins of the cells. × 17,500.

junctions, or nexuses, that are found at intervals over the surfaces of adjacent cells. The nexuses provide a connection between the cells with low electrical resistance, so that the cells of smooth muscle are electrically coupled. The degree of electrical coupling between cells varies among different types of smooth muscle.

Tropomyosin as well as actin and myosin is present in smooth muscle, and most investigators believe that the sliding filament model of contraction applies to smooth muscle. The way in which the contractile apparatus is regulated by Ca^{++} differs from striated muscle, but the mechanism(s) remains obscure.

Mechanical Properties of Smooth Muscle

Smooth muscle contracts and relaxes 10 to 1000 times more slowly than skeletal muscle. Repetitive electrical stimulation of smooth muscle leads to sum-

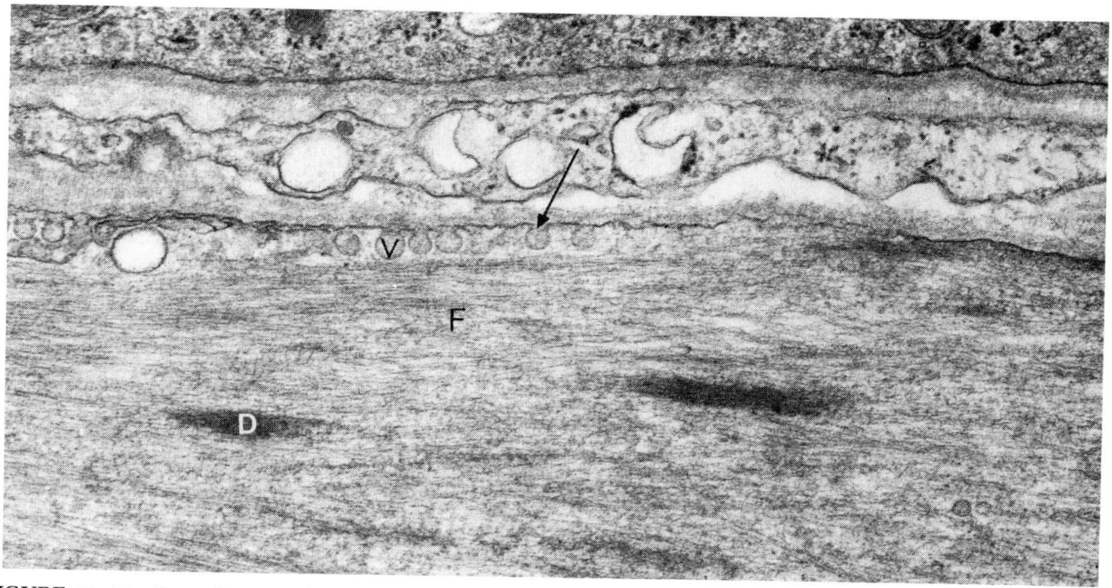

FIGURE 19–35 Part of a smooth muscle cell in the rat prostate. Most of the field is occupied by myofilaments (F). Dense bodies (D) may bind filaments together at intervals, and in other locations they appear to anchor the filaments to the plasma membrane. Vesicles (V) are numerous at the periphery of the cell, and some of them are connected to the plasma membrane (*arrow*). × 41,000.

mation of contractions, as in skeletal muscle, and at a particular frequency of stimulation results in a fused tetanus. The length-tension curve of smooth muscle is broader than that of striated muscle, so that smooth muscle can develop tension over a wider range of muscle length. However, the change in the volume of a hollow organ reflects reorientation as well as shortening of muscle cells. Although smooth muscles have a significantly lower myosin concentration than skeletal muscles, smooth muscle develops about the same maximum tension as skeletal muscle. The longer thick filaments of smooth muscle and the higher ratio of thin to thick filaments may contribute to the ability of smooth muscle to develop the same force as skeletal muscle with less myosin.

Electrophysiology of Smooth Muscle

The resting membrane potential of smooth muscle cells varies from −40 to −80 mV in different types of smooth muscle. In most smooth muscles, the resting potential is not stable in time. The membrane potential may slowly vary in a cyclical fashion, as seen in Figure 19–36*A*, or it may vary in a noncyclical way in response to the levels of norepinephrine, acetylcholine, and other effector substances. Slow cyclical variations in membrane potential (*slow waves*) with a duration of five to seven seconds are typical of intestinal smooth muscle. Note in Figure 19–36 that the slow waves are associated with rhythmical changes in the force produced by the muscle, with depolarization of the membrane causing increased contractile force. There is good evidence that the intestinal slow waves are due to rhythmic activity of the electrogenic Na^+ pump, which contributes substantially to the resting membrane potential in many types of smooth muscle (see Chapter 17). In the intestine, it appears that periods of high pumping rate alternate with periods of lower pumping rate. When the pump rate is high, the membrane becomes more polarized, and when the pump rate is lower the membrane depolarizes. Cyclical variations in pumping rate thus cause cyclical varia-

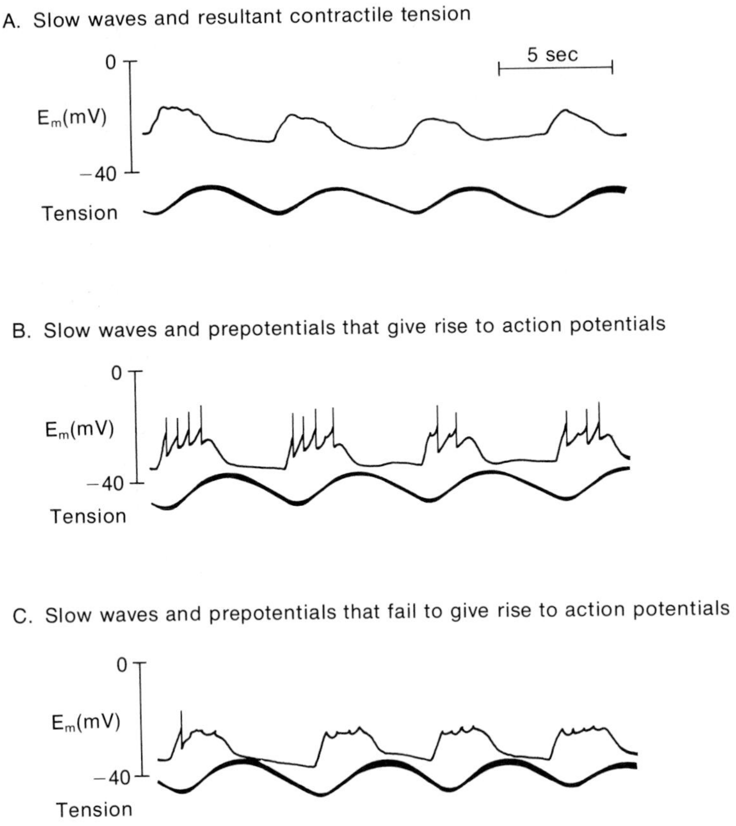

FIGURE 19–36 Electrical and mechanical activity of longitudinal smooth muscle of rabbit jejunum. The upper trace in each panel is the transmembrane electrical potential (mV), and the lower trace shows contractile tension. It is noteworthy that one tissue displays such a range of electrical activities. (Redrawn after Bortoff, A.: Am. J. Physiol., *201*: 203, 1961.)

tions in the resting membrane potential. Cyclical variations in membrane potential may also occur via various other mechanisms.

Some types of smooth muscle, for example the smooth muscle of the pulmonary artery, do not ordinarily fire action potentials, and slow variations in membrane potential are the only membrane electrical events that occur. In gastrointestinal smooth muscle, however, contraction is associated with action potentials triggered at the peaks of slow waves. In those smooth muscle cells that do fire action potentials, it is not uncommon to find each action potential spike preceded by a *prepotential* (or generator potential) which resembles the pacemaker potential of the sinoatrial node of the heart. Each prepotential brings the membrane to threshold for the action potential (Fig. 19–36*B*). As seen in Figure 19–36*C*, there may be conditions under which the prepotentials fail to elicit action potentials. Most pre- or pacemaker potentials studied are the result of an increase in the conductances of the plasma membrane to Na^{++} and Ca^{++}.

Smooth muscle action potentials differ from those of skeletal and cardiac muscle in that the rate of rise of the spike is smaller and the overshoot is small or nonexistent. The action potential of smooth muscle is not blocked by tetrodotoxin, the puffer fish poison that blocks the fast Na^+ channels which are responsible for the rapid depolarization (spike) during the action potential in nerve and striated muscle. The channels responsible for the spike of smooth muscle resemble the slow channels of the heart in that an influx of both Na^+ and Ca^{++} is involved in generating the rising phase of the action potential. It appears that in some types of smooth muscle, entry of Ca^{++} is principally responsible for generating the action potential, in other types Na^+ entry is primary, and in still other types both Na^+ and Ca^{++} are important.

Control of Smooth Muscle Contraction

Smooth muscles, unlike striated muscles, often maintain a particular level of contractile tension continuously over an extended period of time. This maintained level of smooth muscle tension is known as *tone*. Tone may be thought of as the baseline level of contractile activity of smooth muscle, upon which time-dependent variations in contraction are superimposed. As described earlier, the level of the membrane potential in most smooth muscles controls the contractile activity of the muscle. Smooth muscles are innervated by the autonomic nervous system. Both sympathetic and parasympathetic postganglionic fibers end in beaded nerve terminals among the smooth muscle cells (Chapter 18). The norepinephrine that is released from the sympathetic terminals plays an important role in the regulation of smooth muscle tone. Different types of smooth muscle vary, however, in their responses to these agents. Bloodborne substances may also be important in regulating the function of smooth muscle. Circulating epinephrine augments the effect of the norepinephrine release from nerve terminals in most smooth muscles. Vasopressin (antidiuretic hormone), oxytocin, and certain prostaglandins, among many other substances, also significantly affect smooth muscle function.

Excitation-Contraction Coupling in Smooth Muscle

While certain molecular details of the control process may turn out to be unique to smooth muscle, the level of cytoplasmic Ca^{++} controls the degree of activation of the contractile apparatus, as is the case in striated muscle. In many smooth muscles this is related to the resting membrane potential or to action potentials. Ca^{++} that enters the cell through the plasma membrane plays an important role in excitation-contraction coupling in many smooth muscles. In such tissues, contractions are diminished if the extracellular Ca^{++} concentration is lowered and are increased in the presence of elevated extracellular Ca^{++}. The more poorly developed the sarcoplasmic reticulum, the more marked the dependence of the contractile force on the level of extracellular Ca^{++}. Some smooth muscles (mostly vascular) contract in response to hormones and drugs with little measurable change in membrane potential. This is called pharmacomechanical coupling and appears to reflect Ca^{++} release from the sarcoplasmic reticulum or through the cell membrane without involving conductance changes to Na^+ or K^+.

The light chains of smooth muscle myosin are subject to phosphorylation and dephosphorylation by specific light-chain kinases and phosphatases. It is possible that these processes play an important role in excitation-contraction coupling or in modulation of contractile activity.

FURTHER READING

Aidley, D. J. The Physiology of Excitable Cells. New York, Cambridge University Press, 1971, pp. 191–294.

Bloom, W., and Fawcett, D. W. (eds.): A Textbook of Histology. Philadelphia, W. B. Saunders, 1975.

Briskey, E. J., Cassens, R. G., and Marsh, B. B. (eds.): The Physiology and Biochemistry of Muscle as a Food. Madison, University of Wisconsin Press, 1970.

Bülbring, E., and Shuba, M. F.: Physiology of Smooth Muscle. New York, Raven Press, 1976.

Burnstock, G.: Structure of smooth muscle and its innervation. *In*: Bülbring, E., Brading, A. F., Jones, A. W., et al. (eds.): Smooth Muscle. Baltimore, Williams & Wilkins, 1970.
Carlson, F. D., and Wilkie, D. R.: Muscle Physiology. Englewood Cliffs, N.J., Prentice-Hall, 1974.
Casteels, R., Gotfraind, T., and Rüegg, J. C. (eds.): Excitation-Contraction Coupling in Smooth Muscle. New York, Elsevier, 1977.
Cold Spring Harbor Symposium on Quantitative Biology, *37*. Mechanism of Muscle Contraction. New York, Cold Spring Harbor Press, 1973.
Davson, H.: A Textbook of General Physiology. 4th ed. Baltimore, Williams & Wilkins, 1970, pp. 1273–1474.
Gauthier, G. F.: The ultrastructure of three fiber types in mammalian skeletal muscle. *In*: Briskey, E. J., Cassens, R. G., and Marsh, B. B. (eds.): The Physiology and Biochemistry of Muscle as a Food. Madison, University of Wisconsin Press, 1970, pp. 103–130.
Huxley, H. E.: The mechanism of muscular contraction. Sci. Am., *213*:18, 1975.
Katz, A. M.: Physiology of the Heart. New York, Raven Press, 1977.
Katz, B.: Nerve, Muscle, and Synapse. New York, McGraw-Hill, 1966, pp. 159–168.
Langer, G. A., and Brady, A. J. (eds.): The Mammalian Myocardium. New York, John Wiley and Sons, 1974.
Leeson, C. R., and Leeson, T. S.: Histology. 3rd ed. Philadelphia, W. B. Saunders, 1976.
McNutt, N. S., and Fawcett, D. W.: Myocardial ultrastructure. *In*: Langer, G. A., and Brady, A. J. (eds.): The Mammalian Myocardium. New York, John Wiley and Sons, 1974, pp. 1–49.
Murray, J. M., and Weber, A.: The cooperative action of muscle proteins. Sci. Am. *230*:58, 1974.
Needham, D. M.: Machina Carnis: The Biochemistry of Muscular Contraction in Historical Development. New York, Cambridge University Press, 1971.
Padykula, H. A., and Gauthier, G. F.: The ultrastructure of the neuromuscular junctions of mammalian red, white, and intermediate skeletal muscle fibers. J. Cell Biol., *46*:27, 1970.
Peachey, L. D.: The sarcoplasmic reticulum and transverse tubules of the frog's sartorius. J. Cell Biol., *25*:209, 1965.
Pepe, F. A.: The structural components of the striated muscle fibril. *In*: Timasheff, S. N., and Fasman, G. D. (eds.): Biological Macromolecules Series. Vol. V. New York, Marcel Dekker, 1971, p. 247.
Rhodin, J. A. G.: Fine structure of vascular walls in mammals, with special reference to smooth muscle component. Physiol. Rev., *42*(Suppl.) *5*:48, 1962.
Somlyo, A. P., and Somlyo, A. V.: Vascular smooth muscle. I. Normal structure, pathology, biochemistry and biophysics. Pharmacol. Rev., *22*:249, 1970.
Weber, A., and Murray, J. M.: Molecular control mechanisms in muscle contraction. Physiol. Rev., *53*:612, 1973.

20

INTERACTIONS OF DRUGS WITH CELLS

Among the aims of the physician are understanding the nature of malfunctions that produce disease and learning ways in which the course of disease can be modified by appropriate treatment. Thus, beginning with cell structure and function, studies of basic medical sciences can be approached by considering how normal processes can go wrong and how these errors may be modified in the treatment of disease.

GENERAL MECHANISM OF DRUG ACTION

One of the main ways in which the physician administers therapy is through the use of drugs. It is important to understand at the outset that drugs do not *create* new responses; instead, they alter the *rate* at which *normal* cellular activities occur. Therefore, it is pertinent for the physician to study normal cell structure and function and to consider the fundamentals of how drugs interact with cells. Since there are many drugs in current use, the purpose of this chapter is not to describe the properties of specific drugs; this will be accomplished by a course in basic pharmacology. The objective here is to define the principles of interaction of drugs with cells as a foundation for later, more detailed studies of different types of drugs. Specific drugs are mentioned only as examples to illustrate these principles.

Pharmacology is defined as the study of the effects of chemicals on biological systems. The biological system can be a purified enzyme in a test tube, an intracellular organelle, an intact cell or tissue, or a complex living organism, such as a human being. *Pharmacodynamics* refers to the study of how chemicals produce biological effects at any or all levels of organization of living material. The term *pharmacokinetics* refers to the study of the factors that determine the concentration of a chemical at its site(s) of action at any given time after administration of the agent to a biological system. The discipline of *toxicology* is the study of the toxic effects of chemicals on biological systems.

The number of chemical compounds with proven and potential biological effects is staggering. Of these chemicals, more than 7000 are marketed as drugs or therapeutic entities. In addition to their use as therapeutic agents, drugs are useful research tools. Many cases exist in which the establishment of the mechanism of action of a drug was the key to understanding the physiology of a system or the pathophysiology of a disease. For example, studies with the *Rauwolfia serpentina* alkaloid, reserpine, showed that the neurotransmitter substance, norepinephrine, in postganglionic sympathetic nerve endings was synthesized and stored in vesicles in the cytoplasm and that the vesicles were involved in the release mechanism for the transmitter. The dramatic reponse of patients suffering with Parkinson's disease to L-dopa was important in understanding the nature of neuronal pathways and neurochemical transmitters in the substantia nigra of the central nervous system. The therapeutic response to L-dopa allowed clear definition of the neurotransmitter deficiency in Parkinson's disease and established dopamine as an important transmitter in the central nervous system.

In the following sections, the different ways in which drugs produce their effects will be considered. These mechanisms are divided into two main groups: responses that do not depend on an interaction of the drug with a specific receptor, and those that require a reaction of the drug with a specific receptor molecule. Since drug actions produced by reaction with specific receptors comprise the largest and most complex group, the simpler responses, which are independent of receptors will be considered first. It will be seen, however, that within this category there is a range in the specificity of action of the drugs from very nonspecific to highly specific.

RESPONSES INDEPENDENT OF SPECIFIC RECEPTORS

Effects of Drugs that are Nonspecific

Nonspecific effects of drugs are a result of the chemical or physical properties of the agents, and the site of action of the drug is usually extracellular or at the cell surface. Perhaps the simplest example is the neutralization of gastric hydrochloric acid by an antacid drug: $NaHCO_3 + HCl \rightarrow NaCl + H_2O + CO_2$. In this reaction, the base (antacid), sodium bicarbonate, reacts chemically with hydrochloric acid in the stomach. Similarly, the compound ammonium chloride, NH_4Cl, can be used to acidify the urine. It is administered orally, and after absorption from the gastrointestinal tract the NH_4^+ is metabolized to urea and hydrogen ions. The presence of excess protons and Cl^- in the blood is known as hyperchloremic acidosis, and the kidney excretes the H^+ and Cl^-, resulting in diuresis and acidification of the urine.

Nonspecific effects of drugs are also illustrated by several compounds that exert an effect in the body through an osmotic action. The osmotic diuretic mannitol (or the excess glucose in diabetes mellitus) is filtered by the renal glomerulus and is presented as additional solute in the tubular urine, resulting in increased osmolarity. The increased osmotic gradient for water reabsorption means that more water remains in the renal tubules and is excreted as urine. Osmotic action also plays a role in the use of compounds of high molecular weight, such as dextrans and polyvinyl pyrrolidone as plasma expanders or substitutes for plasma or whole blood. These compounds are administered intravenously, and because of their size, their distribution is limited to the vascular space. With added solute in the vascular tree, water is attracted into the plasma to maintain isotonicity.

Similarly, the popular antacid $Mg(OH)_2$ has a cathartic action because Mg^{++} retains water in the lumen of the colon.

Activated charcoal acts nonspecifically as an adsorbent of a wide variety of compounds in the stomach. It acts rapidly and has a broad spectrum of adsorptive activity. Thus in certain cases of drug poisoning, charcoal is considered to be the most valuable agent for emergency treatment.

A wide variety of nonspecific agents are used to destroy living matter. These nonspecific agents disperse or denature cell surface components and subsequently intracellular organelles and macromolecules. Such drugs are classified as disinfectants, antiseptics, and spermicides. Short-chain aliphatic alcohols (e.g., ethanol and isopropyl alcohol) and aldehydes (e.g., formaldehyde) are representative of desiccants and denaturants that destroy the integrity of membranes and macromolecules. Detergents destroy lipid membranes and macromolecular complexes whose structural integrity and function depend on ionic bonds. Oxidizing agents (e.g., halogens and peroxides) are popular antiseptics that drastically alter the functional capacity of cell membranes and intracellular organelles or macromolecules.

Drug Interactions with Small Molecules

In the following examples of drug interactions with small molecules, the drug effects remain limited primarily to the cell membrane or extracellular sites of action. However, specificity is increased since these drugs react only with specific ions or organic molecules.

The chelating agents are organic chemicals that combine specifically with metal ions. The chelator contains atoms of O, N, or S which form coordinate covalent bonds with the ions, yielding a stable complex. A chelating agent often encountered in laboratory experiments is ethylenediaminotetraacetic acid (EDTA). The stability constants for several ions with EDTA range from a low of 1.7 with Na^+ to 18.2 for Pb^{++} (Na^+, 1.7; Mg^{++}, 8.7; Ca^{++}, 10.6; Cd^{++}, 16.4; Pb^{++}, 18.2). Any ion that has a high stability constant with EDTA can replace a complexed ion with a lower stability value. For example, Ca^{++} can replace Mg^{++} or Na^+. The chelating agent reacts with ions from tissues and fluid to form a water-soluble chelation complex. Dimercaprol (BAL) is a chelating agent that is useful in cases of lead or arsenic poisoning because it complexes ionic lead and arsenic from body tissues and fluids and the chelation-complex is excreted by the kidney. Penicillamine is an effective chelator of copper, mercury, and zinc.

Cholestyramine is the chloride salt of a basic anion-exchange resin with a polymeric molecular weight of greater than 10^6. Cholestyramine is not absorbed following oral administration and acts in the intestine to bind bile acids, which are then excreted in the feces in greater than normal amounts. This resin is useful in the treatment of patients with high plasma cholesterol levels, patients with elevated plasma bile acids, and patients intoxicated with the cardiac glycoside, digitoxin.

Protamines are simple, low molecular weight proteins that are strongly basic molecules. Heparin, the widely used anticoagulant, is strongly acidic owing to its high degree of sulfation. When combined, heparin and protamine form a stable complex which is devoid of anticoagulant activity. Protamine is useful therefore if hemorrhage occurs in patients treated with heparin.

These examples of drug interaction with small molecules represent what is referred to as *chemical antagonism*. Each example clearly depicts a chemical reaction between two reactants that forms a stable inactive product and results in the loss of biological activity of the reactants.

Drug Incorporation into Macromolecules

Drugs can act as counterfeit parts for normal body components, and in this manner they display a high degree of specificity. The drug replaces a normal precursor in a pathway for the synthesis of an important cellular constituent. Thus the reaction product rather than the drug itself produces the ultimate pharmacological response.

One of the most carefully studied examples of this type is the substitution of 5-bromouracil (BU) in the reaction sequences that normally utilize thymine. BU is synthesized to bromodeoxyuridine triphosphate instead of the thymidine triphosphate derived from thymine. The deoxy-BU-triphosphate resembles thymidine triphosphates and is paired opposite adenine in DNA by DNA polymerase. Replication and transcription of nucleic acids can proceed normally in cells in which up to 40 per cent of the thymine has been replaced with BU. In mammalian cells that contain BU-DNA, however, there is an increased rate of mutation, frequency of chromosomal breakage, and sensitivity to x-irradiation.

In striking contrast, 5-fluorouracil (FU) is handled metabolically like uracil rather than thymine. FU is converted to the monophosphate riboside that inhibits the enzyme thymidine synthetase, resulting in the blockade of thymine synthesis. Furthermore, the FU riboside is phosphorylated successively to the nucleoside triphosphate and then incorporated into messenger RNA in place of uracil. The presence of FU in mRNA results in misreading of the genetic code by the polysomes and the assembly of polypeptide chains with incorrect amino acid sequences. The protein synthesized may be functional, partly functional, or inert, depending upon the functional importance of the amino acid residues where inappropriate amino acids were inserted through miscoding. 5-FU is used as a cancer chemotherapeutic agent because the inhibition of thymine synthesis impairs DNA synthesis and therefore cell division and growth.

Drug Interactions with Functionally Important Macromolecules

In general, interactions of drugs with macromolecules involve highly specific interactions between a drug and a single type of macromolecule. There are exceptions, however, such as the relatively nonspecific types of interactions displayed by the cytotoxic alkylating agents used in cancer chemotherapy. All alkylating agents form carbonium ions that are involved in strong electrophilic chemical reactions with cellular constituents. These reactions between cellular nucleophilic substances (compounds that contain PO_4, NH_2, SH, OH, and COOH groups) and the alkylating agent carbonium intermediate result in the formation of covalent bonds. The alklyating agents react with DNA, primarily with the strongly nucleophilic #7 nitrogen of guanine. Once alkylated, the guanine residue may base pair with thymine residues instead of cytosine, thus leading to miscoding, inhibition of cell division, and death of the cell.

As previously described, heparin is a strong acid and forms ionic bonds with basic groups. A wide variety of proteins contain basic moieties that react with heparin. The anticoagulant action of heparin is the consequence of a reaction between heparin and a protein in plasma that is required for blood coagulation. When combined with heparin, this protein (antithrombin) cannot participate in coagulation.

The antimitotic and anti-inflammatory effects of colchicine may be explained by the ability of the drug to bind to microtubular protein. Colchicine binds to the microtubular protein, tubulin, and prevents the assembly of fibrillar microtubules, which not only interferes with mitotic spindle formation but also inhibits mobilization of leukocytes during inflammation. In this example, the acceptor for colchicine is the highly specific macromolecule tubulin.

In addition, there are many drugs that exert quite specific biological effects by binding to and inhibiting a specific enzyme. Classic examples are the inhibition of acetylcholinesterase by diisopropylfluorophosphate (DFP) and inhibition of "prostaglandin synthetase" (cyclo-oxygenase) by aspirin or indomethacin.

Many antibiotics inhibit protein synthesis by binding specifically to the ribosome. The antibiotic puromycin is structurally similar to phenylalanyl ~ tRNA and binds to the ribosome. Once in place on the ribosome, puromycin accepts the peptide chain that is intended for transfer to the next aminoacyl ~ tRNA, and the peptide chain is released from the ribosome with a COOH terminal puromycin residue. Puromycin exerts its effect on protein synthesis in both prokaryotic and eukaryotic cells. Other compounds that inhibit mammalian protein synthesis have been found useful in cancer chemotherapy.

Other useful antibiotics are selective and block only microbial protein synthesis. Tetracyclines and aminoglycosides (streptomycin, gentamicin, kanamycin) bind specifically to the 30S ribosomal subunit in some bacteria and prevent access of aminoacyl ~ tRNA to the ribosomal tRNA ~ complex. The macrolide antibiotics (erythromycin) lincomycin, and chloramphenicol bind primarily to bacterial 50S ribosomal subunits suppress peptidyl transferase activity, and impair peptide bond formation. Ribosomal translocation is uncoupled from peptide bond synthesis.

RECEPTORS

The interactions of most drugs with biological systems are dependent upon the reaction of the drug with a specific receptor. A receptor is a macromolecular component of the cell with which the drug interacts to produce its characteristic biological effect. The word "receptor" is imprecise and indicates nothing about the molecular nature of a binding site or the reaction sequence imposed between the drug receptor complex and the final effect of the drug. Since the receptor usually is detected by the response elicited by the drug-receptor complex, biochemical isolation and identification of receptors are difficult because the response is often lost during manipulation of the tissue.

Specificity and Antagonism

The concept that drugs interact with a specific cellular site derives from the studies of Paul Ehrlich and J. N. Langley. Ehrlich observed that minor changes in the chemical structure of organic compounds markedly altered the effects of these compounds on bacteria. He suggested that tissues or cells contained reactive groups (receptors) with which drugs combined. From his studies of the relationship between the structure of the drug and its activity it followed that chemical modification of the drug could vastly improve or impair the ability of a drug to combine with its receptor. This is often referred to as the lock and key theory of drug action. Clearly one of the most important properties of a receptor is its specificity for certain drug molecules.

Langley carried out his classic studies on the neuromuscular junction. Previous studies by Claude Bernard had shown that the drug curare acted at the junction of a motor nerve with skeletal muscle and prevented muscle contraction induced by nerve stimulation. Langley found that nicotine acted at the neuromuscular junction to produce skeletal muscle contraction and that curare blocked the response to nicotine. He postulated a "receptive substance" or specialized material in the neuromuscular junction with which drugs interacted. The combination of nicotine with this receptive substance triggered muscle contraction, while curare combined with this material and blocked contraction. Furthermore, it was apparent that curare prevented the binding of nicotine by the receptive substance. Thus Langley advanced receptor theory by adding another key concept—receptors can be blocked. Therefore, a drug that binds to a receptor and produces a biological response is known as an *agonist*, while a drug that binds to a receptor without eliciting a response is referred to as an *antagonist*.

The receptors for the catecholamines, the adrenergic receptors, illustrate the specificity of agonists and antagonists (Table 20–1). The different responses of the heart and the blood vessels to the three catecholamines show that there are clearly different cardiac and vascular receptors for the catecholamines shown in the table. The cardiac receptor has been classified as a β-adrenergic receptor and the vascular as an α-adrenergic receptor. The α-adrenergic receptors are blocked selectively by phentolamine and are not activated by isoproterenol. β-receptors are activated best by isoproterenol and inhibited by propranolol.

TABLE 20–1 RESPONSES OF THE HEART AND BLOOD VESSELS TO CATECHOLAMINES

	Cardiac Receptor (β)	Vascular Receptor (α)
	Increased heart rate	Vasoconstriction
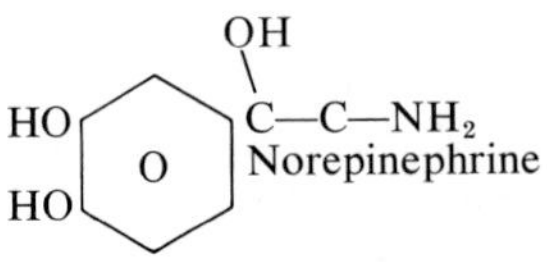	+	++++
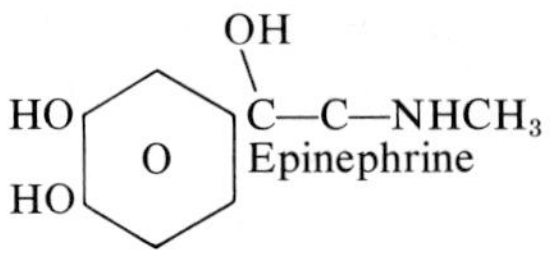	++	++
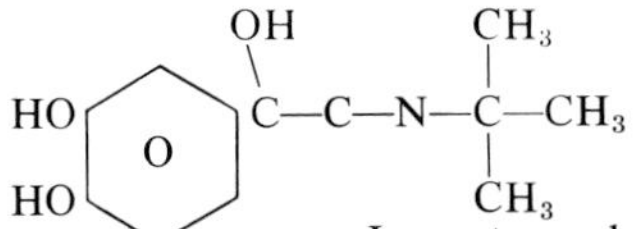 Isoproterenol	+++	0
	All heart rate responses are blocked by propranolol but are not affected by phentolamine.	Phentolamine blocks ↑ blood pressure responses to NE and Epi, but propranolol does not attenuate responses.

Types of Bonds Involved in Interactions of Drugs with Receptors

An important functional property of receptors, in addition to their specificity, is that the drug-receptor complex must be stable for that period of time required for the initiation of subsequent events that ultimately are manifested as the characteristic drug response. The stabilizing forces of this drug-receptor interaction are typical chemical bonds that hold two atoms together.

The bond formed when two atoms share a pair of electrons is a covalent bond, and the strength is about 100 kcal/mole (20 times greater than an ionic bond). When the electrons shared are contributed by only one of the atoms, a coordinate covalent bond is formed. The formation of coordinate covalent or covalent bonds between drugs and small molecules (chelators), macromolecules (alkylating agents), and enzymes (DFP) has been discussed for effects of drugs independent of receptors. However, formation of covalent bonds between drugs and receptors would not be advantageous, because the high strength of the covalent bond dictates an essentially irreversible reaction for the drug. Therefore, since covalent bonds are so stable, a covalent drug-receptor complex would result in a response of very long duration. The fact that most drug responses are readily reversible indicates that most drug-receptor complexes dissociate at body temperature. Thus the covalent binding of a drug to its receptor is rather rare. Instead, interactions between drugs and receptors more commonly utilize weaker bonds, such as ionic bonds, hydrogen bonds, and van der Waals forces.

An ionic bond is formed by the transfer of one or more electrons from one atom to another. This bond represents the well-known electrostatic attraction between two atoms of opposite charge and has a strength of about 5 kcal/mole. The strength, however, is dependent on the distance separating the two ions and decreases in proportion to the square of the interatomic distance.

When hydrogen is ionically or covalently bound to an electronegative atom, the hydrogen may coordinate two electrons contributed by another strong electronegative atom (O, N). As described in Chapter 2, the coordination of these additional electrons is referred to as hydrogen bonding and results in the bridging of two electronegative groups. The strength of a single hydrogen bond (about 2 kcal/mole) requires extreme atomic proximity and is less than that of an ionic bond. However, when multiple hydrogen bonds are formed they add remarkable stability to a complex.

At close proximity two neutral atomic groups form weak attractive forces (van der Waals forces, 0.5 kcal/mole). The strength of van der Waals forces is inversely proportional to the seventh power of the interatomic distance.

Quantitative Aspects of Drug-Receptor Interactions

BINDING OF DRUGS TO RECEPTORS

The ionic bond forms at high velocity and is of sufficient strength to overcome the random nature of the drug and attract it to the receptor. The formation of an ionic bond requires atoms of opposite charge. Most macromolecules contain exposed charged groups and most drugs have one or more charged centers, so the prerequisite for opposite charges is easily satisfied. However, drugs will form ionic bonds with many different macromolecules and not just the receptor, so an ionic interaction cannot explain the specificity of the drug-receptor interaction. In addition, although the strength of the ionic bond may be great enough to initiate the drug-receptor interaction, it is not sufficient to produce a stable complex for any period of time. Clearly, the ionic bond must be reinforced by hydrogen bonding and by van der Waals forces that form between specific atoms of the drug and the macromolecular receptor. This reinforcement adds the stability required for the drug-receptor complex to withstand the energy of thermal agitation and also is the key to specificity.

Once all these bonds are formed it seems probable that a reversible change in the receptor's molecular configuration occurs. This drug-induced change in the structure of a macromolecule is thought to be the factor that initiates the biological response characteristic for that drug and tissue. Receptor-blocking agents may bind near or at the active site and do not trigger the biological response typical of the agonist.

KINETICS OF DRUG-RECEPTOR INTERACTIONS

Many indirect inferences about how drugs and receptors interact have been made from kinetic studies of the drug-receptor interaction. The use of kinetic studies to examine these interactions will be described in this section. Recently some receptors have been isolated. This sort of work promises that more objective description of these interactions will become available in the future as additional receptors are isolated and studied.

In the description of the forces involved in drug binding, it was emphasized that this represented a reversible reaction.

$$\text{macromolecule} + \text{drug} \rightleftharpoons \text{macromolecule-drug complex}$$

A. J. Clark was responsible for applying the law of mass action to drug receptor interactions. Clark arrived at the following conclusions (which are derived mathematically further on) regarding the interaction of a drug with its receptor.

1. The law of mass action is applicable to the reversible reaction between one drug and one receptor species.

2. All receptors for a given drug are identical and readily accessible.

3. The intensity of the biological response is directly proportional to the fraction of the total receptor population that is occupied.

4. The concentration of drug greatly exceeds the receptor concentration so that the concentration of drug bound is negligible. In other words, free [drug] $\simeq$ total [drug].

The binding of a drug to its receptor induces a pharmacological response. When the response is added to the reversible binding equation, one obtains the fundamental equation for a drug-receptor interaction.

$$D_{\text{drug}} + R_{\text{receptor}} \underset{K_2}{\overset{K_1}{\rightleftharpoons}} [D \sim R]_{\text{complex}} \xrightarrow{K_3} \Delta_{\text{response}}$$

Note the similarity of this equation to that for an enzyme-substrate reaction (see Chapter 4).

$$E_{\text{enzyme}} + S_{\text{substrate}} \underset{K_2}{\overset{K_1}{\rightleftharpoons}} [E \sim S] \xrightarrow{K_3} E + \text{product}$$

Application of the law of mass action to the drug-receptor-response equation generates the equation

$$\frac{[R]\,[D]}{[DR]} = \frac{k_2}{k_1} = K_D$$

where [R] is receptor concentration, [D] is drug concentration, [DR] is drug-receptor concentraion, and K_D is the dissociation constant of the receptor-drug complex. If $[R_T]$ is the total receptor concentration,

$$[R_T] = [DR] + [R].$$

Substituting for [R] in the equilibrium equation,

$$\frac{([R_T] - [DR])\,[D]}{[DR]} = K_D.$$

By rearrangement,

$$\frac{[DR]}{[R_T]} = \frac{[D]}{K_D + [D]}$$

The response (Δ) to the drug is proportional to the concentration of the drug-receptor complex.

$$\Delta = K_3[DR]$$

The drug will achieve its maximum effect (Δ_{max}) when all receptors are occupied by the drug,

$$[DR] = [R_T]$$

and by substitution

$$\Delta_{max} = k_3\,[R_T].$$

Therefore,

$$\frac{\Delta}{\Delta_{max}} = \frac{[DR]}{[R_T]} = \frac{[D]}{K_D + [D]}.$$

When half-maximal response is obtained,

$$\frac{\Delta}{\Delta_{max}} = \frac{[D]}{K_D + [D]} = \frac{1}{2}$$

and the concentration of D required for half-maximal response is equal to K_D. Thus it is possible to determine the dissociation constant for the receptor-drug complex by measuring the pharmacological effect in response to gradually increasing concentrations of a drug. The association of the drug with its receptor is given by $1/K_D$ and is a measure of the *affinity* of the drug for the receptor.

The constant K_3 in the drug-receptor-response equation is an index of the capacity of the drug-receptor complex to induce the pharmacological response, Δ. This property is intrinsic in the drug molecule and is known as its *efficacy*. The efficacy is proportional to the maximal response obtainable with the drug, Δ_{max}.

To obtain values for the affinity and efficacy of a drug, dose-response curves are constructed. The data are obtained by adding increasing concentrations of a drug to a biological system and measuring the magnitude of each response. An example of one such biological system is an isolated strip of thoracic aorta. Its response to increasing concentrations of drugs is a graded contraction. When the response (Δ) is plotted against the log dose [D] (Fig. 20–1), the typical curve is sigmoidal in shape.

The dissociation constant (K_D) for the receptor complex formed with each drug can be read from the graph at 0.5 Δ_{max}: $K_{D_1} = 10^{-9}$; $K_{D_2} = 10^{-8}$; $K_{D_3} = 10^{-7}$. Since $1/K_D$ is a measure of affinity, clearly D_1 has the greatest and D_3 the least affinity for the receptor.

Note the similarity between this graph and the plot of the velocity of an enzyme reaction, V, against the concentration of substrate, [S] (Chapter 4). The equation:

$$\frac{\Delta}{\Delta_{max}} = \frac{[D]}{K_D + [D]}$$

can be written as the equation of a straight line:

$$\frac{1}{\Delta} = \left(\frac{K_D}{\Delta_{max}}\right)\frac{1}{[D]} + \frac{1}{\Delta_{max}}$$

This is comparable to a Lineweaver-Burk plot for enzyme-substrate interactions, and a plot of $1/\Delta$ versus $1/D$ will yield a straight line with a slope of K_D/Δ_{max} (Fig. 20–2).

In Figure 20-1, the three drugs (D_1, D_2, and D_3) produced the same maximal response. However, the Δ_{max} produced by various agonists may differ. If the Δ_{max}

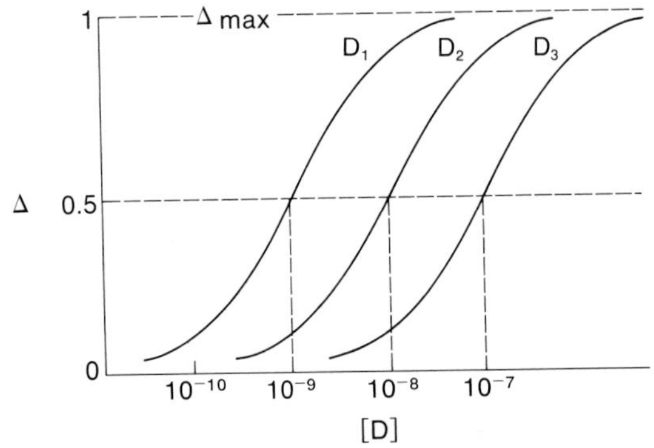

FIGURE 20–1 A plot of pharmacological response, Δ, against concentration of drug, [D], for three drugs, D_1, D_2, and D_3. At 0.5 (Δ max), [D] = K_D. Δ is in grams of tension. (Modified from Gourley, D. R. H.: Interactions of Drugs with Cells. A Topic in Cell Biology. Springfield, Charles C Thomas, 1971.)

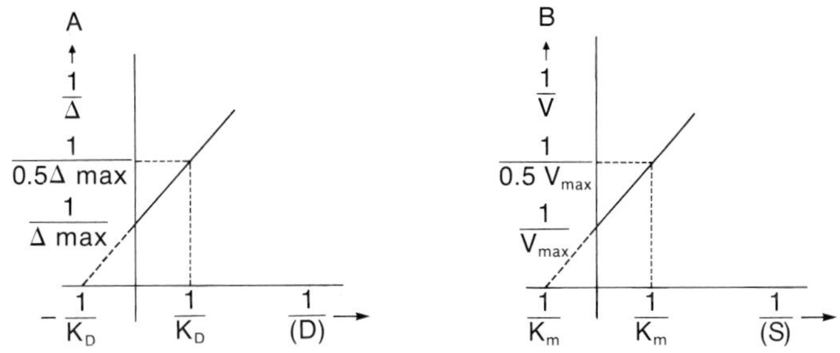

FIGURE 20–2 Comparison between drug-receptor and enzyme-substrate double reciprocal plots. (Symbols are defined in text.) *A*. In the drug-receptor double reciprocal plot, the intercept of the y axis gives the reciprocal of Δ_{max}; the intercept on the x axis gives the negative reciprocal of K_D. By extrapolation

$$\frac{1}{0.5\Delta_{max}} \simeq \frac{1}{K_D}$$

at the intercept of the x axis. *B*. In the enzyme-substrate double reciprocal plot, the intercept on the y axis gives the reciprocal of V_{max}; the intercept on the x axis gives the negative reciprocal of K_m. By extrapolation

$$\frac{1}{0.5v_{max}} \simeq \frac{1}{K_m}$$

at the intercept of the x axis. (Modified from Goldstein, A. et al.: Principles of Drug Action: The Basis of Pharmacology. 2nd ed. New York, John Wiley & Sons, 1974.)

is different among a series of structural analogues, the compounds that induce a Δ_{max} that is lower than the best response obtained are referred to as *partial agonists* (Fig. 20–3).

Figure 20–3 could depict the responses of aortic strips to norepinephrine (D_1), histamine (D_4), and angiotensin (D_5). Each compound interacts with its own specific receptor and induces a different Δ_{max}. Norepinephrine (D_1) has the greatest maximum efficacy, but angiotensin (D_5) has greater affinity for its receptor than does NE for the α-adrenergic receptor or histamine (D_4) for the histamine receptor. This figure could also represent dose-dependent responses induced by a series of chemical congeners that interact with a common receptor. Clearly, a partial agonist can have an affinity as great or greater than that of a full agonist. The defining characteristic of a partial agonist is that its efficacy once combined with the receptor is less than that of the full agonist.

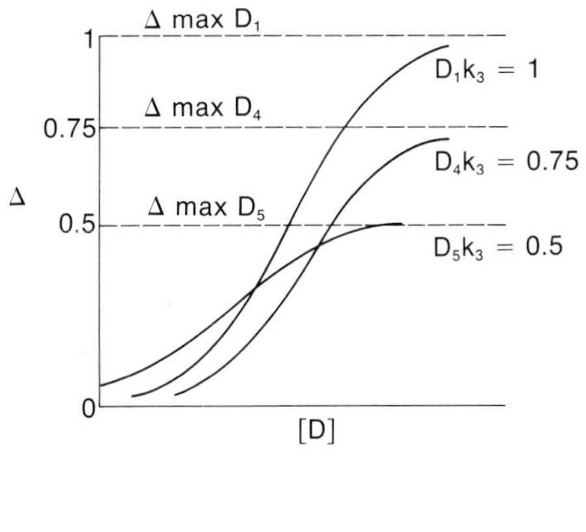

FIGURE 20–3 Responses to three different drugs. (Modified from Gourley, D. R. H.: Interactions of Drugs with Cells. A topic in Cell Biology. Springfield, Charles C Thomas, 1971.)

INHIBITION OF DRUG ACTION

In the discussion of the studies by Langley, it was pointed out that curare is an antagonist of nicotine at the neuromuscular junction in skeletal muscle. A drug that reduces the effect of another drug but has no efficacy when used alone is an antagonist. These agents are classified as noncompetitive antagonists and competitive antagonists.

A *noncompetitive antagonist* may prevent the interaction of the drug and receptor by combining with a site on a macromolecule adjacent to the receptor. By doing so, the noncompetitive antagonist might interfere with the conformational changes that normally follow the combination of the agonist with its receptor. Alternatively, the noncompetitive antagonist may induce an allosteric conformational change in the receptor so that the functional groups of the receptor site are no longer capable of an interaction with the agonist. In either case, the structure of the noncompetitive antagonist can be totally unrelated to that of the agonist.

In case of phenoxybenzamine (PBA) versus norepinephrine, the noncompetitive kinetics shown in Figure 20–4 are obtained. PBA and other haloalkylamine α-blockers are related chemically to nitrogen mustards and can alkylate various sites. The parallel shift of the dose-response curve to NE obtained a few minutes after the administration of PBA (cf., curves *control* and *1*) led investigators to postulate a number of "spare receptors." In the first curve only 10 out of every 100 receptors need to be occupied to achieve Δ_{max}. The remaining 90 per cent are spare receptors. In the second curve (*1*), PBA has alkylated 50 per cent of the total receptor population. However, 10 out of 50 receptors can be occupied and achieve 100 per cent response. By the law of mass action, twice the original NE concentration is required, and the dose response curve shifts in a parallel fashion to the right. At the time curve 2 was obtained, more than 90 per cent of the total receptor populations has been irreversibly alkylated. It is therefore impossible for NE to bind to 10 receptors, and the Δ_{max} for NE decreases. In curves 3, 4, and 5, NE receptor population has been increasingly alkylated, so that fewer and fewer receptors are available to interact with NE.

The presence of an α-agonist or competitive antagonist protects the α-adrenergic receptor from alkylation by PBA (so-called receptor protection). However, after blockade with PBA has developed fully, the inhibition is unaltered by other drugs capable of an interaction with the α-adrenergic receptor. This is referred to as nonequilibrium blockade and is the result of covalent binding of PBA to the α-receptor.

Competitive antagonism is depicted in Figure 20–5. The response to the antagonist B alone is virtually zero; when a constant amount of B is added to the tissue, the response to identical concentrations of D is less than in the absence

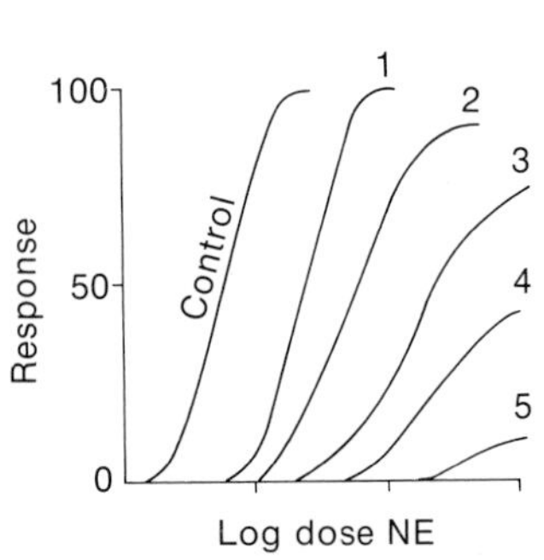

FIGURE 20–4 Curves 1 through 5 represent responses to NE at progressively later times after treatment with PBA. The peak effect of PBA usually requires at least one hour, and the half-life for reversal of blockade is about 24 hours.

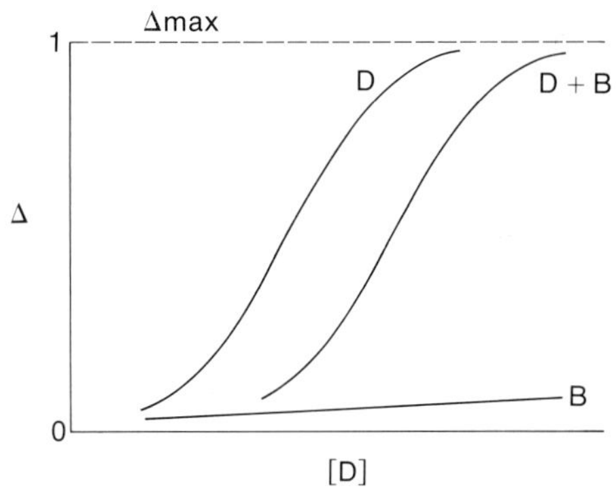

FIGURE 20–5 A plot of pharmacological response, Δ, against concentration of drug, [D], for agonist drug D, competitive antagonist drug B, and a combination of D + B. (Modified from Gourley, D. R. H.: Interactions of Drugs with Cells, A Topic in Cell Biology. Springfield, Charles C Thomas, 1971.)

of B, and the curve is shifted to the right. However, a maximal response still is obtained when sufficiently high doses of D are added to the system. B is a *competitive antagonist* and will reversibly occupy a fraction of the drug receptors, depending on its concentration.

In summary, the distinguishing difference between a noncompetitive antagonist and a competitive antagonist is in the ability of an agonist to induce the same Δ_{max} in the absence and presence of a competitive, but not a noncompetitive, antagonist.

QUANTAL RESPONSES

The previous discussion of drug-receptor kinetics has dealt with the graded response in which, once a threshold concentration of drug is exceeded, the response is proportional to the drug concentration. Comparable consideration should be given to a different type of response, the *quantal or all-or-nothing response*. In this case one simply determines the presence or absence of a definite end point. For example, with a given dose of anesthetic agent, an animal either goes to sleep or it does not. It is possible to obtain a frequency distribution for a quantal response in relation to different doses of the drug (Fig. 20–6*A*), and it usually will not approximate a normal distribution curve. Because of the variability in response from one individual to another, there is no such thing as a standard dose of a drug when a quantal response is measured. Some animals will be very sensitive to a drug, and others will be resistant to a given concentration of a drug. When one plots a frequency dose-response curve (quantal response) versus dose, the distribution curve is skewed to the right (Fig. 20–6A). However, when plotted as frequency versus log dose, the data will approximate a normal distribution (Fig. 20–6*B*). From the cumulative frequency dose-response curve (Fig. 20–7), the dose can be obtained that will induce the desired quantal response in 50 per cent of the subjects studied (ED_{50}). It is obvious that a drug may have several quantal responses. For example, consider the frequency response for animals put to sleep by an anesthetic compared with the cumulative frequency of deaths induced by the same anesthetic agent (Fig. 20–8). From the cumulative frequency distribution of deaths in these animals, we can obtain the dose of drug that will kill half the animals treated (LD_{50}). Some idea of the safety margin with a drug may be obtained by the ratio of the LD_{50}/ED_{50}. This ratio is called the *therapeutic index (TI)*. We would like this number to be as great as possible; however, a therapeutic index of about 3 is common. In practical terms, it is legally impossible of course to obtain a clinical therapeutic index, since the treated patient population would have to be killed to obtain an approximation of the

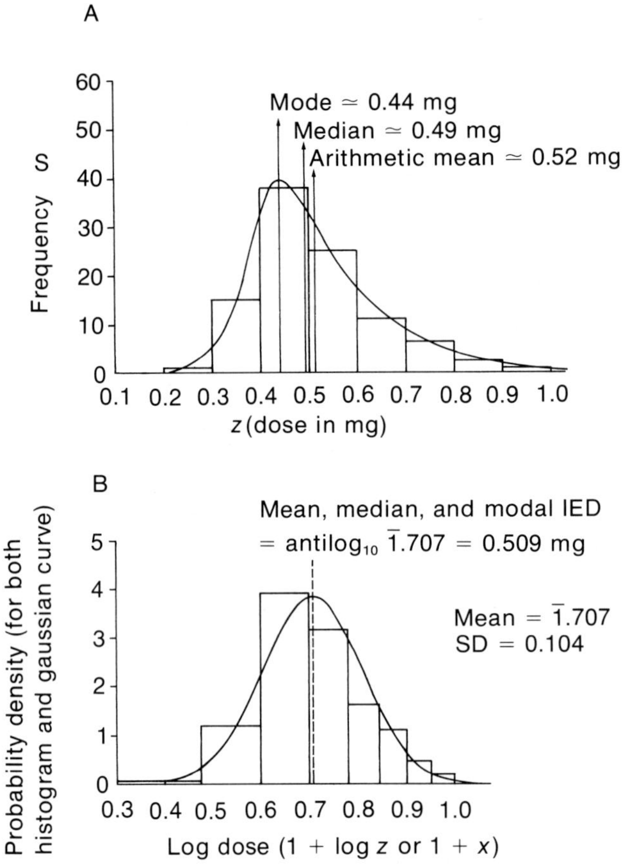

FIGURE 20–6 Frequency distribution for a quantal response. *A*. Frequency versus dose. *B*. Frequency versus log dose. (From Colquhoun, D.: Lectures on Biostatistics. London, Oxford University Press, 1971. pp 348, 350. Reproduced by permission of the publisher.)

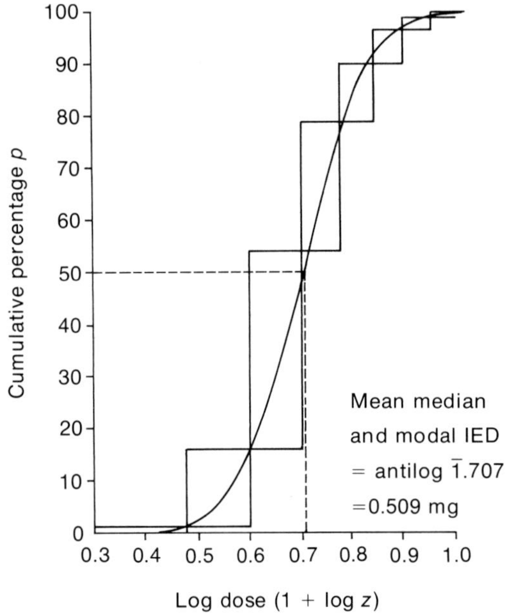

FIGURE 20–7 Cumulative frequency versus log dose for a quantal response. (From Colquhoun, D.: Lectures on Biostatistics. London, Oxford University Press, 1971. Reproduced by permission of the publisher.)

LD_{50}. What one tries to be certain of with patients is to never exceed the minimum toxic dose or lethal $dose_1$ (see Figure 20–8). If one divides the LD_1/ED_{99} the resulting value is referred to as the *certain safety factor (CSF)*. The medical goal is to determine the dosage that will treat effectively 99 per cent of patients without causing any deaths. The CSF is a more meaningful clinical value than the TI, which can be calculated only from animal studies, but the CSF value is impossible to determine accurately because the desired graphical points lie on nonlinear portions of the dose-response curves.

Families of Drugs

The actions of any drug are related intimately to its chemical composition. Relatively minor modifications in the drug molecule may result in major changes in pharmacological properties. Exploitation of the structure-activity relationship has at times led to the synthesis of valuable therapeutic agents. Since changes in molecular configuration need not alter all actions and effects of a drug equally, it is sometimes possible to develop a congener with a more favorable therapeutic index or with more acceptable secondary characteristics than those of the parent drug. In addition, effective therapeutic agents have been fashioned by developing chemically related competitive antagonists of other drugs or of endogenous substances known to be important in biochemical or physiological function.

Perhaps the best example of the development of a family of drugs is the sulfonamides. When they were first introduced for the treatment of infection, it was noticed that many patients had a diuresis or became hypoglycemic. These observations were reported and taught as the side effects of the sulfa-drugs. One patient's side effect, however, might be sound therapy for another patient. The

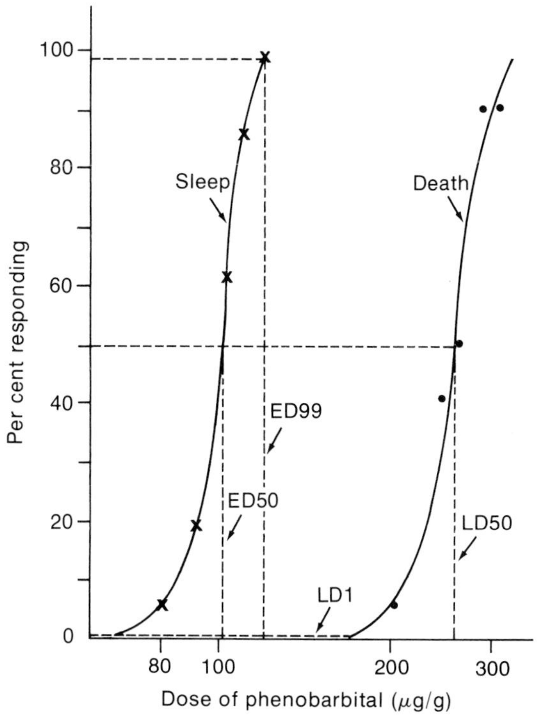

FIGURE 20–8 Quantal dose response curves representing the cumulative number of animals responding as the dose is increased. Groups of 20 mice were injected with different doses of phenobarbital. (From Levine, R. R.: Pharmacology: Drug Actions and Reactions. Boston, Little, Brown, 1970. Reproduced by permission of the publisher.)

drug manufacturers did structure-activity studies to develop more efficacious drugs to reduce blood sugar and induce diuresis. From these studies came the sulfonylureas (oral hypoglycemic agents) used in the treatment of diabetes, carbonic anhydrase inhibitors, and thiazide diuretics, used to treat edematous states. Sometimes the structure-activity relationship appears quite broad. For example, many chemically dissimilar drugs exhibit local anesthetic activity, and compounds of totally different chemical structure produce similar CNS-depressant effects. These examples do not negate the significance of the structure-activity relationship in other instances. They do, however, emphasize that much remains to be learned of the basic mechanisms of action of most drugs and that overtly similar effects may be produced by more than a single mechanism.

Receptors as Part of a Chain

The most fundamental aspect of pharmacodynamics is that which deals with the mechanisms of drug action. It is essential to emphasize the distinction between drug action and drug effect.

Although often considered synonymous, the terms action and effect have useful pharmacological connotations that should be preserved. Most drugs produce their effects by combining with enzymes, cell membranes, or other specialized functional components of cells. Drug-cell interaction alters the function of the cell component and thereby initiates the series of biochemical and physiological changes that are characteristic of the drug. Only the *initial consequence* of drug-cell combination is correctly termed the *action* of the drug; the *remaining events* properly are called drug *effects*.

The effects of the hormone glucagon offer an excellent example of the cascade of events that are initiated following the reaction of glucagon with its receptor in hepatocytes. The overall effect of glucagon on the liver is a dose-related increase in glucose output, resulting from an increse in glycogenolysis and gluconeogenesis and a decrease in glycogen synthesis. The stimulatory action of glucagon on glycogenolysis depends upon an increase in the activity of phosphorylase, the rate-limiting enzyme that catalyzes the removal of glycosyl units from the outer branches of the glycogen molecule and hence leads to the production of glucose. This activation of phosphorylase is the result of a change in the equilibrium between its inactivation by a specific phosphorylase phosphatase and its reactivation by a phosphorylase kinase. Phosphorylase kinase in turn must be activated by another specific kinase, a step that is controlled by the intracellular concentration of cAMP and therefore by the relative rate of cAMP formation by adenylate cyclase and of cAMP destruction by a specific phosphodiesterase. Glucagon functions as a "first messenger" and initiates this complex series of reactions by activation of adenyl cyclase which, being located on the plasma membrane of the hepatocyte, forms part of its "hormonal receptor" system (Fig. 20–9).

Cyclase activation increases the intracellular concentration of cAMP. This nucleotide, acting as a "second messenger," stimulates the other reactions mentioned earlier. ATP is the phosphate donor for the activation of phosphorylase and phosphorylase kinase and is the precursor of cAMP, while Mg^{++} is absolutely required. Methylxanthines, such as theophylline and caffeine, inhibit phosphodiesterase, while other agents, such as imidazole, activate it. In this example, the *action* of glucagon is the stimulation of adenylate cyclase to increase the intra-

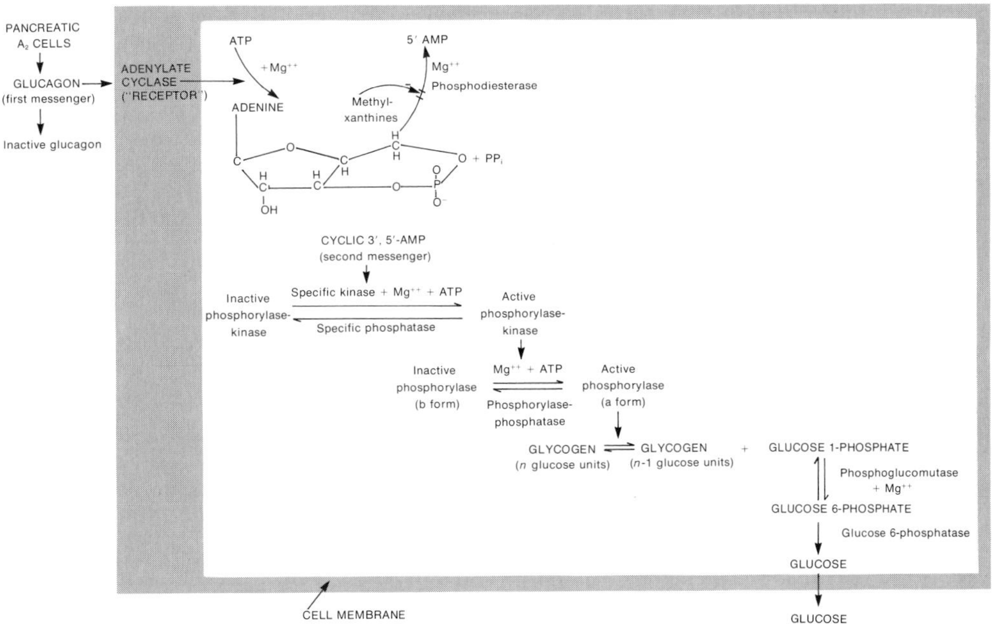

FIGURE 20–9 Glycogenolytic effect of glucagon in the hepatocyte, according to the "second messenger" theory. (From Bacq, Z. M. (ed.): Fundamentals of Biochemical Pharmacology. Oxford, Pergamon Press, 1971. Reproduced by permission of the publisher.)

cellular concentration of cAMP, while the remaining reactions are considered *effects* of glucagon.

The objectives of analysis of drug action are identification of the primary action, delineation of the details of the chemical reaction between drug and cell, and characterization of the full action-effects sequence. Only the complete analysis provides a truly satisfactory basis for the therapeutic use of the drug.

TRANSPORT OF DRUGS AND THEIR DISTRIBUTION IN THE BODY

The magnitude of response induced by a drug is determined by the concentration of drug at the receptor sites for the drug. The multiple factors influencing the drug concentration at its receptor are the subject of the following discussion.

Several processes determine the movement of a drug from the alimentary canal to plasma and to cells of an organ. In general, there are two types of barriers to be considered: the walls of blood vessels and the plasma membranes of cells. In the case of the walls of blood vessels, small molecules, such as drugs, are able to escape the vascular space by a process of ultrafiltration. These low molecular weight compounds, which are dissolved in plasma, filter through physiological "pores" in the capillary endothelium under the driving force of hydrostatic pressure. The capillary endothelium is a barrier to large molecules, and most drugs and nutrients pass as a result of diffusion and pinocytosis.

The plasma membrane of cells represents the most important barrier to the movement of drug molecules. There may be several layers of cells that oppose the penetration of a drug (e.g., the epithelial border of the intestine), but the factors governing the passage of a drug through a multicellular obstacle are remarkably similar to those involved in the movement of a compound through the plasma membrane of a single cell. Drugs may pass through the plasma membrane by (1) penetration through "pores" or channels, (2) facilitated diffusion, (3) diffusion, and (4) active transport.

Numerous cell membranes have been found to behave as if they contained pores with effective radii of approximately 4 Å, and pores with these dimensions would permit the movement of polar molecules that are three carbon atoms or less in length. However, since most drugs have radii that exceed 4 Å, the penetration of compounds through pores in membranes is of minor importance. Facilitated diffusion and endocytosis also are of theoretical interest only. Therefore, the following discussion will concentrate on the passage of drugs through the plasma membrane by diffusion and active transport.

Penetration of Semipermeable Membranes by Diffusion

There are physical-chemical properties of drugs that influence their diffusion through a plasma membrane. Most drugs are weak electrolytes and exist in a polar solvent (water) as a mixture of ionized and un-ionized molecules. The ionized form of a drug has very poor lipid solubility and does not pass through a cell membrane. In contrast, the un-ionized drug is lipid-soluble and permeates cell membranes (see Chapter 15). The driving force for diffusion is the concentration gradient of the un-ionized form of the drug across the membrane. Diffusion is a first-order process, since it is proportional to the concentration of the diffusing species (un-ionized drug). At equilibrium, the concentration of drug in the un-ionized form is identical on both sides of a semipermeable membrane.

The lipid solubility of a drug usually is determined by establishing the partitioning of a drug between water and a simple nonpolar solvent, such as chloroform. After adequate mixing, the ratio of the drug concentrations in nonpolar/polar solvent yields the partition coefficient for that compound. Any drug with a partition coefficient greater than 0.01 is considered to have appreciable lipid solubility.

Since most drugs are salts of weak acids or bases, the fraction of the total drug concentration that exists in water as the ionized and un-ionized species is a function of pH and the dissociation constant of the compound. As described in Chapter 2, the dissociation of weak acids and bases is represented as

$$HA \rightleftharpoons H^+ + A^-$$

$$BH^+ \rightleftharpoons B + H^+$$

where A is acid and B is base. The dissociation constant (K_a) of an acid is represented by the general equation

$$K_a = \frac{[H^+][A^-]}{HA}$$

By using logarithms, one can substitute pH for the $-\log$ of the H^+ and pK_a for $-\log$ of K_a and derive the Henderson-Hasselbalch equation

$$pK_a = pH + \log \frac{[\text{un-ionized}]}{[\text{ionized}]}$$

Rearrangement of the Henderson-Hasselbalch equation shows that for a weak acid, the proportion of ionized drug is determined by pH − pk_a.

$$pH - pK_a = \log \frac{[\text{ionized}]}{[\text{un-ionized}]} \text{ (weak acid)}$$

Similarly, the fraction of a weak base that is ionized at any pH is reflected by pK_a − pH, where pK_a refers to the conjugate acid of the base.

$$pK_a - pH = \log \frac{[\text{ionized}]}{[\text{un-ionized}]} \text{ (weak base)}$$

From the approximate pH of two compartments in the body and the pK_a of a drug, one can calculate the proportions of ionized and un-ionized drug in each compartment at equilibrium, as in the examples shown in Figure 20–10. Remember that only the un-ionized form of the drug can pass through a membrane and will reach equal concentrations on each side of a plasma membrane at a steady state.

Active Transport of Drugs through Membranes

Certain lipid-*in*soluble compounds pass rapidly through cell membranes. Since the membrane penetration is much too rapid to be dependent on diffusion, the concept of a membrane "carrier," or active transport, has been advanced to explain passage of these drugs that have poor lipid solubility. All these specialized transport systems are characterized as follows:

1. The compound being transported moves against a concentration gradient.
2. Energy is required for transport to proceed.
3. The number of "carrier sites" is limited and can be saturated.
4. If two compounds are transported by the same system, one compound will competitively inhibit transport of the second substance.
5. Each carrier is specific for a chemical group or structure.

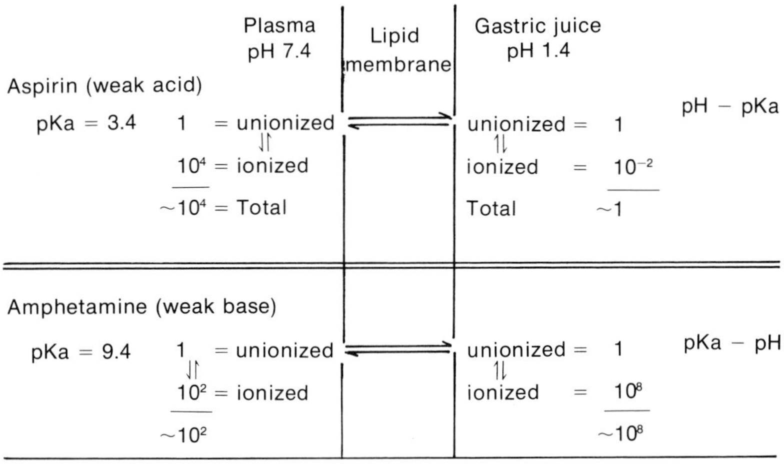

FIGURE 20–10 The distribution of aspirin, a weak acid, and amphetamine, a weak base, between plasma and gastric juice.

Certain cells, such as those of the endothelium of the capillaries surrounding the renal tubules, hepatic parenchymal cells, and cells of the blood-brain barrier, appear to have two specialized active transport mechanisms. One system transports ionized acidic drugs, and the other transports the ionized form of basic compounds. Neither transport system is very specific and keys only on the acidic or basic properties of a drug and not on its molecular structure. Thus any two acidic compounds may compete for transport by the system for acids. In each of the tissues cited, the transport process carries the drug away from a tissue and enhances its elimination from the body. The transport process in the kidney is not well developed in an infant, and renal function in general is often poor in the aged. Thus drugs that are dependent on renal function for elimination from the body will tend to accumulate in the newborn and in geriatric patients.

In addition to the active transport of dissociated acids and protonated bases, other specialized systems are important for the movement of numerous endogenous compounds through cell membranes. These transport systems for naturally occurring substances are very specific, and only drugs with structures that are similar to the endogenous material will compete with these carriers for transport. Frequently drug molecules are designed to utilize these specific transport systems to gain entry to a particular cell type or to prevent cellular accumulation of a specific endogenous compound.

Drug Elimination from the Body

Foreign molecules are eliminated from the body by a variety of mechanisms. The three most important processes are renal excretion (filtration and tubular secretion), biliary excretion, and drug biotransformation. Drugs also may be eliminated in sweat, milk, and saliva, but these represent minor routes of excretion. The primary route of excretion for volatile agents is expiration from the lungs.

In the kidney, drugs are filtered with plasma and pass into the ultrafiltrate, which becomes urine. If the drug is highly bound to plasma protein, however, the bound compound is not filtered in the renal glomerulus but is retained in the blood. In the case of unbound compounds, any drug that is in the *un-ionized* form in the kidney ultrafiltrate will tend to pass *out* of the urine through the tubular cells and into the peritubular fluid. Therefore, acidification of urine enhances the excretion of basic compounds and the reabsorption of acids. Conversely, renal excretion of a weak acid is greatest when the urine is alkaline. Compounds with high lipid solubility may be excreted by the liver into the intestine in bile. Such a drug can then be reabsorbed from the intestine and re-excreted by the liver, thus establishing an "enterohepatic" cycle. The important process of biotransformation of drugs and its role in elimination of drugs is considered in the following section.

BIOTRANSFORMATION OF DRUGS

For a very water-soluble drug eliminated totally by glomerular filtration in the kidney, the half-life ($t\frac{1}{2}$) or time required to eliminate half of the amount of drug in the body is about 4 to 5 hours. If the drug is secreted actively by the renal tubules, a shorter $t\frac{1}{2}$ of 30 to 60 minutes is possible. However, in the case of a highly lipid-soluble compound which diffuses back across the renal tubules into the extracellular fluid, about 30 days could be required to excrete half the compound from the body. Obviously, in such a case, the kidney is not a sufficient means to eliminate the nonpolar drug from the body and enzymatic mechanisms are important.

Properties of Drug-Metabolizing Enzymes

The body contains enzymes that biotransform a diverse array of foreign chemicals that enter the body. The biotransformations of drugs involve enzymes that occur in the cells of the liver, placenta, small intestine, thyroid gland, kidney, plasma, testes, skin, and lung. While the liver is the most important organ for drug biotransformation, the enzymes in other tissues and in plasma display more substrate specificity than the hepatic enzyme system. Usually, biotransformation of a drug is an inactivating reaction, but a drug metabolite may be quite active or toxic. Furthermore, the pharmacological action of an active metabolite may be totally different from that of the parent compound. Conversely, some compounds are inactive when administered to the body (prodrugs) and are converted to an active agent by the drug-metabolizing enzymes. There are four general types of reactions involved in drug metabolism: (1) oxidation, (2) reduction, (3) hydrolysis, and (4) conjugation. These reactions are summarized in Figure 20–11.

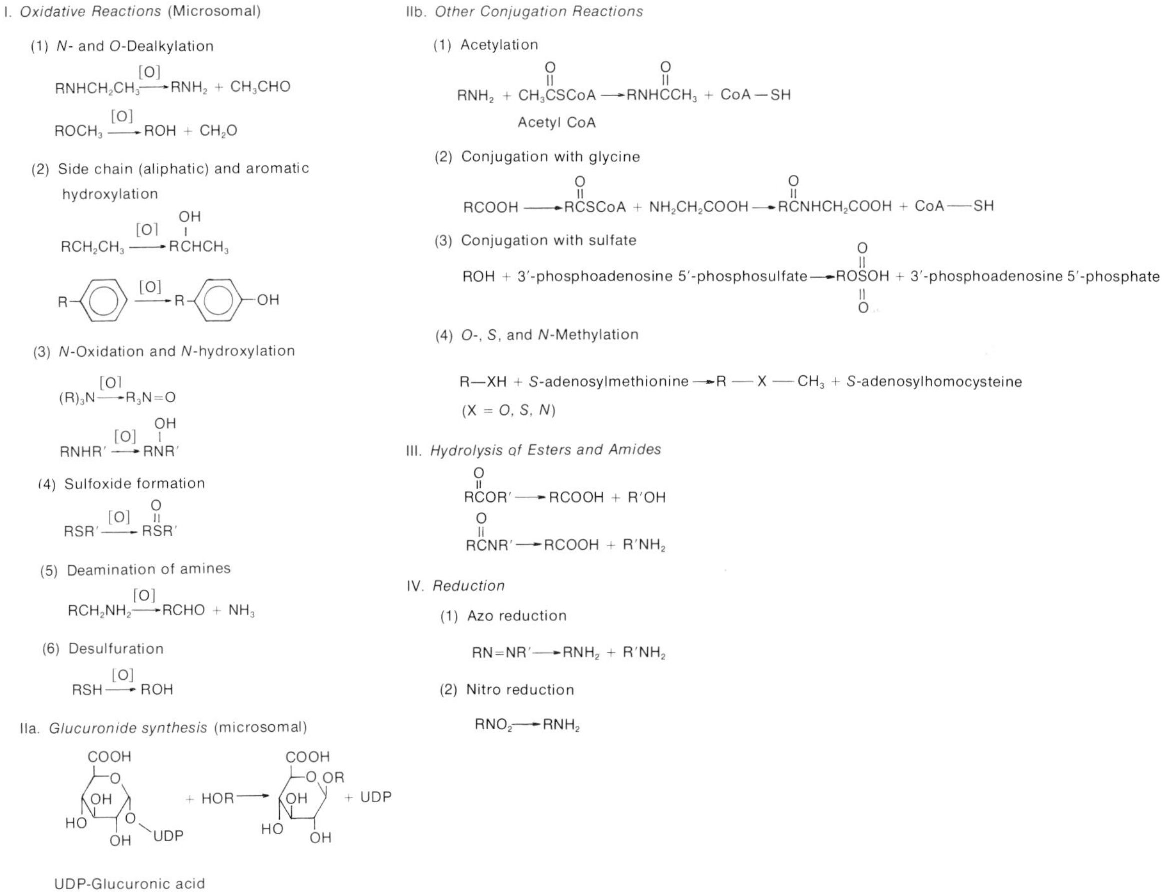

FIGURE 20–11 A summary of the main types of reactions by which drugs are metabolized: I, oxidative reactions; II, conjugations; III, hydrolyses; and IV, reductions. (*From*: Fingl, E., and Woodbury, D.: General principles. *In*: Goodman, L. S., and Gilman, A.: The Pharmacological Basis of Therapeutics. 5th ed. New York, Macmillan, 1975. Reproduced by permission of the publisher.)

Oxidative metabolism of a nonpolar drug results in a product that is more polar and thus more water-soluble and more easily excreted than the parent compound. Many of the oxidative enzymes of biotransformation are associated with the thin tubular network of smooth endoplasmic reticulum and thus are known as microsomal enzymes. The drug-metabolizing enzyme of the hepatic smooth endoplasmic reticulum is a mixed-function oxidase known as the cytochrome P-450 system. Lipid solubility is an important property of the drug molecule, since it must enter the hepatocyte and penetrate into the smooth endoplasmic reticulum and bind to cytochrome P-450. Enzyme induction or synthesis of cytochrome P-450 can be initiated by a variety of lipid-soluble compounds, of which barbiturates are the best characterized (see Chapter 9). The diagram of the electron transport sequence of the P-450 system in Figure 20–12 shows that this hepatic oxidase requires NADPH for reducing power and molecular oxygen for the final oxidation reaction; the types of reactions catalyzed by cytochrome P-450 are illustrated in Figure 20–11.

Since this is a highly nonspecific enzyme system, competitive inhibition may occur among the many substrates. However, the concentrations of most drugs in the body are far below that required to saturate the hepatic microsomal enzymes of biotransformation, and thus competition among substrates of cytochrome P-450 is seldom a problem *in vivo* in normal adults. However, the activity of drug-metabolizing enzyme systems is low in the newborn, and difficulties may be encountered in eliminating lipid-soluble drugs from a neonate. Similarly, impaired liver function can dramatically decrease the rate of drug biotransformation. The rate of biotransformation of any drug that is rapidly metabolized by the liver is also susceptible to changes because of variation in hepatic blood flow.

Other metabolic transformations of drugs are catalyzed by enzymes located in the mitochondria. These enzymes are nonmicrosomal and include the enzymes that are responsible for the oxidation of alcohols and aldehydes, the reduction of aldehydes and ketones, the hydrolysis of some esters and amides, and the oxidation of primary and secondary aromatic amines (monamine oxidase). Nonmicrosomal drug metabolism occurs in many tissues, but the liver is of primary importance.

Enzymes responsible for the biotransformation of drugs also may be soluble

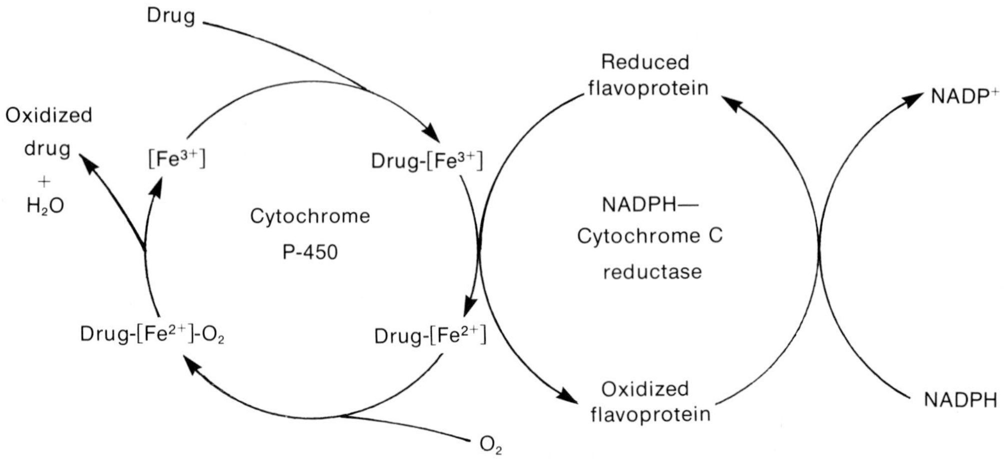

FIGURE 20–12 Electron transport sequence in the cytochrome P-450 system. (*From*: Fingl, E., and Woodbury, D., General principles. *In*: Goodman, L. S. and Gilman, A.: The Pharmacological Basis of Therapeutics, 5th ed., New York, Macmillan, 1975. Reproduced by permission of the publisher.)

and occur in plasma. Perhaps the best-characterized plasma enzyme is pseudocholinesterase, an enzyme that hydrolyzes several ester compounds.

Metabolism of a drug may be via a single enzymatic reaction or through more complicated multiple sequential reactions. In this way, a drug may be converted to a metabolite which also is biologically active but whose activity differs quantitatively or even qualitatively from the parent drug. Alternatively, a compound may be converted to one or more pharmacologically inactive metabolites. Both active and inactive metabolites may be conjugated via additional enzymatic reactions. Conjugated metabolites invariably are inactive. In some cases, a drug may be conjugated directly without intermediary steps of biotransformation.

Several drugs act by inhibiting drug-metabolizing enzymes, as shown in the following examples. The compound SKF-525A inhibits the cytochrome P-450 system and potentiates the actions of numerous other drugs. Disulfiram is an inhibitor of aldehyde dehydrogenase and blocks the metabolism of ethanol to acetate. Several agents that have been used clinically as antidepressants act primarily as inhibitors of monoamine oxidase (MAO), because inhibition of MAO interferes with the metabolism of several neurotransmitters in the central nervous system. Since allopurinol blocks the enzyme xanthine oxidase, resulting in a decrease in plasma uric acid levels, allopurinol is very useful in the treatment of chronic gouty arthritis.

Genetic Differences in Drug Metabolizing Enzymes

It is well known that the rate of drug biotransformation varies greatly in different species. The species differences may be either qualitative or quantitative. Certain metabolic reactions that are common in one species may be absent in another. With drugs that enter two or more routes of biotransformation, the relative importance of the different routes can vary among species. Such differences are particularly important if one of the metabolites is biologically active. Most species differences in response to a drug may be accounted for by variations in the rates of drug biotransformation.

In humans, genetic heterogeneity among individuals has long been recognized. The study of hereditary factors that influence drug metabolism and drug response is a new area of pharmacological interest called *pharmacogenetics*. Hereditary differences in genetic material can lead to alterations in the structure, activity, or amount of specific proteins, including enzymes involved in drug biotransformation. Since many of these enzymes have no known function in the metabolism of endogenous compounds, alterations in their activity do not become apparent unless exogenous chemicals (drugs) are administered. There are several well-authenticated cases in which genetic polymorphisms of drug-metabolizing enzymes are known to account for the occurrence of drug toxicity in humans. The best examples of pharmacogenetic study involve (1) pseudocholinesterase and the inability of certain patients to metabolize succinylcholine; and (2) the vast variability in the rates of biotransformation of isoniazide and hydralazine by acetylating enzymes.

SUMMARY

The most important mechanism by which drugs pass through cell membranes is simple diffusion. Drugs that are lipid soluble diffuse readily into cells. Most drugs are weak electrolytes, and the un-ionized form generally is more lipid soluble than the ionized form. Thus the amount of drug that passes through a membrane depends upon the dissociation constant of the drug and the pH of the aqueous environment on each side of the membrane. Drugs that are relatively insoluble in lipid can penetrate the membrane barrier only by specialized transport processes or via "pores" in the membrane. Although few drugs compete successfully for the specialized transport mechanisms that exist in cell membranes for naturally occurring substances, some cells have nonspecific active transport mechanisms for drugs that are weak acids or weak bases. Only a few pharmacologically active compounds are small enough to permit passage through pores in cell membranes, although most can pass readily through the larger openings between the cells of capillary epithelial sheets.

Biotransformation of drugs involves a relatively small number of enzymes that catalyze metabolic reactions that make the drug progressively more polar. Polar compounds are more easily excreted by the kidney. Drug-metabolizing enzymes are nonspecific and react with a wide variety of foreign compounds. The liver is the most important organ for the biotransformation of drugs. The most important enzyme systems (cytochrome P-450) are associated with smooth endoplasmic reticulum (microsomal enzymes); however, a few enzymes are found in mitochondria, plasma, and cytoplasm (nonmicrosomal enzymes). Drug metabolism is controlled genetically, and rates of biotransformation for a single agent may vary widely from patient to patient.

The biological response to a drug is determined by the concentration of drug at its receptor. The multiple factors that ultimately determine the concentration of a drug at its receptor are summarized in Figure 20-13.

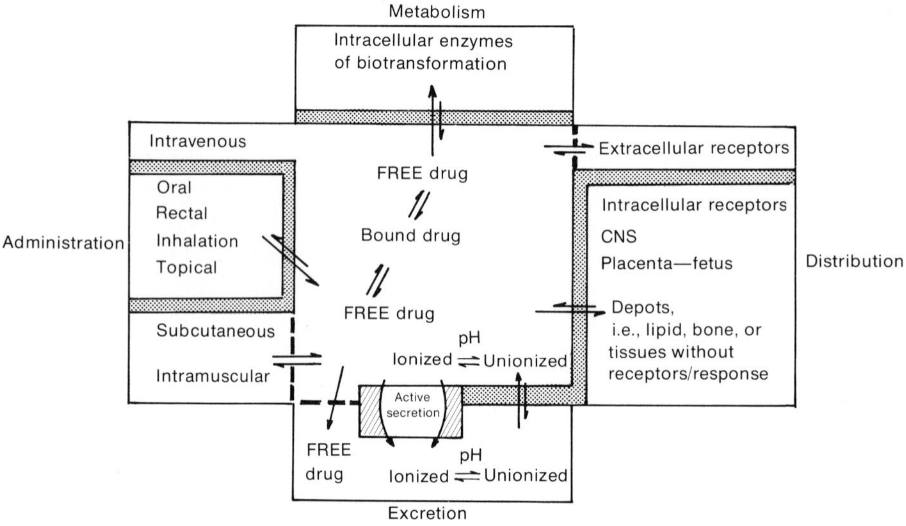

FIGURE 20–13 A diagrammatic summary of factors that affect the concentration of a drug at its receptors. The figure depicts the interrelationships of routes of administration, distribution, excretion, and biotransformation of a drug. From the discussion in the text, it should be obvious that drug absorption, distribution, excretion, and metabolism are influenced by lipid solubility, the pk_a of the compound, and the pH of all the environments into which the drug enters.

FURTHER READING

Ariëns, E. J. (ed.): Molecular Pharmacology. Vol. I. New York, Academic Press, 1964.
Bacq, Z. M. (ed.): Fundamentals of Biochemical Pharmacology. New York, Pergamon Press, 1971.
Csáky, T. Z. (ed.): Introduction to General Pharmacology. New York, Appleton-Century-Crofts, 1969.
Goldstein, A., Aronow, L., and Kalman, S. M. (eds.): Principles of Drug Action: The Basis of Pharmacology. 2nd ed. New York, John Wiley and Sons, 1974.
Goodman, L. S., and Gilman, A. (eds.): The Pharmacological Basis of Therapeutics. 5th ed. Ch. I. New York, Macmillan, 1975.
Gourley, D. R. H. (ed.): Interactions of Drugs with Cells. Springfield, Charles C Thomas, 1971.
Korolkovas, A. (ed.): Essentials of Molecular Pharmacology. New York, John Wiley and Sons, 1970.
Levine, R. R. (ed.): Pharmacology: Drug Actions and Reactions. Boston, Little, Brown, 1973.
McMahon, F. G. (ed.): Pharmacokinetics, Drug Metabolism and Drug Interactions. Mount Kisco, N.Y., Futura Publishing, 1974.
Notari, R. E. (ed.): Biopharmaceutics and Pharmacokinetics. 2nd ed. New York, Marcel Dekker, 1975.

APPENDIX A

STRUCTURE AND PROPERTIES OF α-AMINO ACIDS

The amino acids found in the hydrolysates of proteins are all α-amino acids, an α-amino acid being one in which the amino group is linked to the α-carbon, which is the carbon atom adjacent to the carboxyl group. Although the structure of an α-amino acid is often written as

$$R{-}\overset{\alpha}{C}H(NH_2){-}COOH$$

this structure is never predominant in an aqueous solution. Instead, the predominant structure in aqueous solution between pH 4 and pH 8 is the zwitterionic structure

$$R{-}CH(NH_3^+){-}COO^-$$

in which the amino group is protonated and thus is positively charged, while the carboxyl group is ionized to form a negatively charged carboxylate anion.

The α carbon employs sp^3 hybrid atomic orbitals in forming covalent bonds to the four atoms to which it is attached; therefore, the four bonds formed by the α carbon are directed toward the corners of a regular tetrahedron. If the substituent designated R in the amino acid stucture is any group other than —H, $—NH_3^+$, or $—COO^-$, the α carbon of an amino acid will have four different groups attached. A carbon to which four different atoms or groups are linked is known as an asymmetric carbon atom, because the four different atoms or groups may be attached to the carbon in two different spatial configurations that are mirror images of each other. Mirror-image forms of the same compound are known as *enantiomers*. Using the convention of representing bonds projecting above the plane of the page by wedges, and bonds projecting below the plane of the page by dashed lines, the two possible configurations for an α-amino acid are represented as shown in Figure A–1*A*. In these projection formulas it readily can be seen that the two structures are mirror images and are not superimposable. A more convenient convention is employed in Figure A–1*B*, in which vertical lines represent bonds projecting below the plane of the page, and horizontal lines represent bonds projecting above the plane of the page. Following these conventions, the enantiomer with the $—NH_3^+$ group to the left, when the carboxyl group is toward the top of the page, is known as an L-amino acid and its mirror image as a D-amino acid.

Enantiomers are also called *optical isomers* because they are optically active; that is, they possess the ability to rotate the plane of polarization of plane-polar-

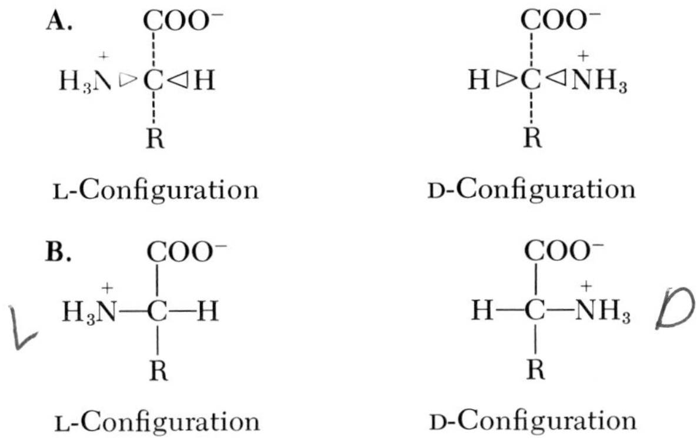

Figure A–1 Representation of the two possible configurations for an α-amino acid. *A*. Projection formulas with wedges representing bonds projecting below the plane of the page and dashed lines representing bonds projecting below the plane of the page. *B*. Projection formulas with horizontal lines representing bonds projecting above the plane of the page and vertical lines representing bonds projecting below the plane of the page.

ized light. If one of the isomers rotates the plane of polarization to the right (dextrorotatory), its mirror image, or enantiomer, will rotate the plane of polarization an equal amount to the left (levorotatory). Another way in which enantiomers may differ is in their rates of chemical reaction with other asymmetrical molecules. It is the latter difference that contributes to the great specificity of enzymes, the biological catalysts, which are themselves asymmetrical molecules (see Chapter 4).

APPENDIX B

STRUCTURE OF D-GLUCOSE

The open chain structure of D-glucose as well as the numbering system used to designate each carbon is shown in the following structure.

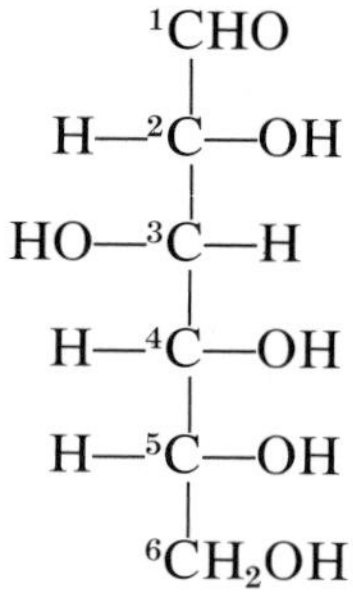

Glucose is a polyhydroxy aldehyde with four asymmetric carbon atoms, C-2, C-3, C-4 and C-5. A well-known reaction in organic chemistry is that of aldehydes with alcohols to form hemiacetals and acetals.

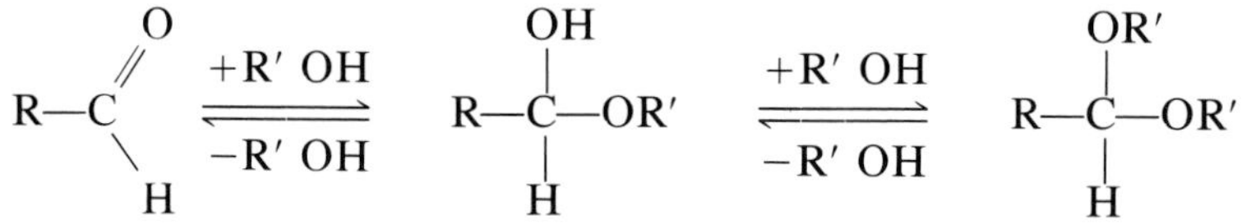

aldehyde hemiacetal acetal

In forming a hemiacetal, the aldehydic carbon atom is converted into an asymmetric carbon, provided R=H. Since glucose contains both an aldehyde group and hydroxyl groups, it can form an intramolecular cyclic hemiacetal in two ways, as shown in Figure B-1. The two cyclic hemiacetals formed are designated α-D-glucose and β-D-glucose. Both have the same configuration about C-2, C-3, C-4, and C-5, but they differ in the configuration of the newly formed asymmetric center at C-1. Isomers of this type are known as *anomers*, and C-1 is called an *anomeric carbon*. Hemiacetal formation is reversible in aqueous solution; therefore, an aqueous solution of D-glucose contains both α-D-glucose and β-D-glucose as well as a trace of the open chain form. At equilibrium, approximately 36 per cent of the glucose exists as the α-anomer and 64 per cent as the β-anomer.

The structures of α-D-glucose and β-D-glucose can also be represented with Haworth projection formulas. Since Haworth projection formulas more clearly indicate the cyclical nature of

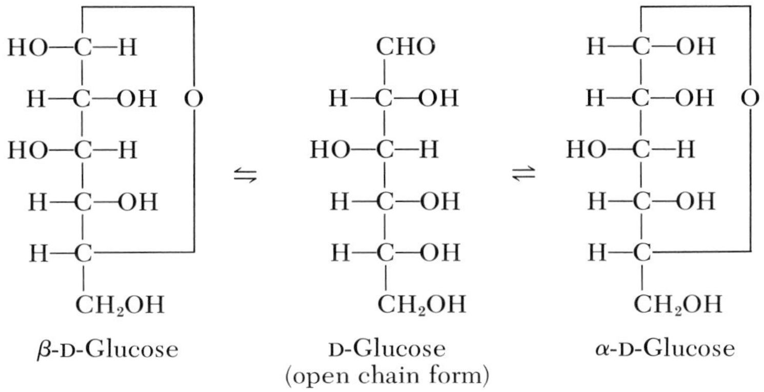

Figure B–1 Fischer projection formulas for D-glucose in its open chain form and in its two cyclic hemiacetal forms, α-D-glucose and β-D-glucose.

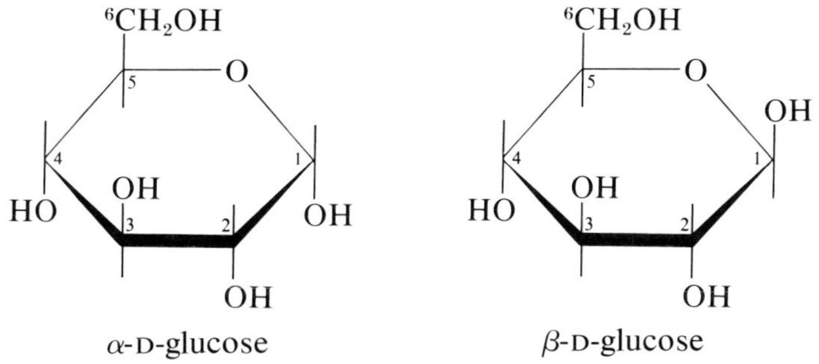

these hemiacetals and are in some ways more convenient, they are employed throughout this text.

One of the characteristic properties of hydroxy aldehydes, such as the open chain form of glucose, is that they are readily oxidized by mild oxidizing agents, such as Cu^{++}. For this reason glucose is said to be a reducing sugar. The cyclic hemiacetal forms of glucose are not directly oxidized by Cu^{++}, but since they are always in equilibrium with the open chain form, they too will lead to the reduction of mild oxidizing agents such as Cu^{++}.

APPENDIX C

EXAMPLE OF THE DETERMINATION OF THE AMINO ACID SEQUENCE OF A POLYPEPTIDE

The amino acid composition of a specific dodecapeptide was found to be Ala_2, Arg, Glu, Gly, Leu, Lys_2, Phe, Tyr_2, Val. It was determined that the N-terminal amino acid of the dodecapeptide was valine and the C-terminal amino acid, leucine. Hydrolysis of the dodecapeptide by trypsin yielded four peptides whose structures were determined and found to be those given in A to D.

Tyr-Glu-Lys	Phe-Gly-Arg	Val-Lys	Ala-Tyr-Ala-Leu
A	B	C	D

Since valine was the N-terminus and leucine the C-terminus of the dodecapeptide, it is apparent that peptide C must represent the amino acid sequence at the N-terminal end, and peptide D the amino acid sequence at the C-terminal end of the dodecapeptide. To establish the order of the A and B peptides in the interior of the dodecapeptide, another sample of the dodecapeptide was hydrolyzed by chrymotrypsin, the four peptides formed were sequenced, and their structures were found to be those given in E to H.

Ala-Leu	Glu-Lys-Ala-Tyr	Val-Lys-Phe	Gly-Arg-Tyr
E	F	G	H

The sequences Gly-Arg-Tyr in peptide H and Glu-Lys-Ala-Tyr in peptide F clearly establish that peptide B must precede peptide A in the dodecapeptide. Hence the structure of the dodecapeptide is unambiguously determined to be that shown in I.

Val-Lys-Phe-Gly-Arg-Tyr-Glu-Lys-Ala-Tyr-Ala-Leu
I

APPENDIX D

ACIDS AND BASES

THE IONIZATION OF ACIDS AND BASES

The Ionization of Monofunctional Acids

When a monofunctional acid, HA, is dissolved in water, the following equilibrium is rapidly established:

$$(1) \qquad \underset{\text{acid}}{HA} + \underset{\text{base}}{H_2O} \rightleftharpoons \underset{\text{conjugate acid}}{H_3O^+} + \underset{\text{conjugate base}}{A^-}$$

Water acts as a base by accepting the proton from the acid, HA, forming the hydronium ion, H_3O^+, and A^-. The H_3O^+ is referred to as the conjugate acid of water, and A^- is the conjugate base of the acid HA. From equation 1 it is apparent that the solvent plays a very important role in acid-base equilibria; in fact, the ionization of an acid is a property of the entire system, not just a property of the acid. For example, HCl is completely dissociated in H_2O to yield $H_3O^+ + Cl^-$, but in benzene HCl exists as dissolved HCl and does not dissociate.

The ionization constant or equilibrium constant K_a for the reaction represented by equation 1 can be written as

$$(2) \qquad K_a = \frac{a_{H_3O^+}\, a_{A^-}}{a_{HA}\, a_{H_2O}}$$

where a_x is the activity or "effective" concentration of species x at equilibrium. The activity of H_2O, the solvent, remains constant and is arbitrarily set at unity. hence, equation 2 becomes

$$K_a = \frac{a_{H_3O^+}\, a_{A^-}}{a_{HA}}$$

A more convenient but less exact expression for an equilibrium constant is obtained by substituting actual concentrations for activities. This constant will be designated K'_a

$$K'_a = \frac{[H_3O^+]\,[A^-]}{[HA]}$$

The ionization constant K_a is a measure of the extent of dissociation of an acid in aqueous solution. The extent of dissociation is also referred to as the strength of the acid. If an acid dissociates completely in aqueous solution, K_a is infinite, and the acid is classified as a strong acid. If only a fraction of the acid dissociates in aqueous solution, K_a is finite and the acid is designated a weak acid. The larger the value of K_a, the stronger the acid.

A further simplification commonly encountered is the use of the hydrogen ion or proton H^+, instead of H_3O^+, even though the hydrogen ion is always hydrated in aqueous solution. Hence, the dissociation of a weak acid and its K_a can be represented by

$$HA \rightleftharpoons H^+ + A^-$$

and

$$K'_a = \frac{[H^+]\,[A^-]}{[HA]}$$

Water itself is a weak acid as well as a weak base.

$$H_2O \rightleftharpoons H^+ + OH^-$$

and

$$K'_a = \frac{[H^+]\,[OH^-]}{[H_2O]}$$

Since $[H_2O]$ is large and does not change appreciably as a result of the very small amount that donates or accepts a proton, we can write

$$K'_a\,[H_2O] = [H^+]\,[OH^-] = K_w$$

where K_w is referred to as the ion product or autoprotolysis constant of H_2O. At 25° C, K_w is 1×10^{-14} M^2, In pure water $[H^+]$ must equal $[OH^-]$; therefore, $[H^+] = 1 \times 10^{-7}M$ in pure water at 25° C.

The Ionization of Weak Bases

A weak base, such as NH_3, reacts with H_2O as follows:

NH_3 +	H_2O	$\rightleftharpoons$	NH_4^+	+	OH^-
base	acid		conjugate acid		conjugate base

and

$$K'_b = \frac{[NH_4^+]\,[OH^-]}{[NH_3]}$$

where K'_b is the base ionization constant, a measure of the strength of the base,or the affinity of the base for a proton. The ionization of NH_4^+, the conjugate acid of NH_3, can be represented by

$$NH_4^+ \rightleftharpoons H^+ + NH_3$$

and

$$K'_a = \frac{[H^+][NH_3]}{[NH_4^+]}$$

Since the strength of an acid is determined by the affinity of its conjugate base for a proton, there must exist a simple relationship between the K'_a of an acid and the K'_b of its conjugate base. Multiplying the expressions for K'_a and K'_b, one obtains

$$K'_a K'_b = \frac{[H^+][NH_3]}{[NH_4^+]} \times \frac{[NH_4^+][OH^-]}{[NH_3]} = [H^+][OH^-] = K_w$$

or

$$K'_b = \frac{K_w}{K'_a}$$

Thus the larger the value of K'_a, the stronger the acid and the weaker its conjugate base. As is the general custom, we shall express the strength of a weak base in terms of the K'_a of its conjugate acid rather than the K'_b.

The pH and pK$_a$ Scales

The hydrogen ion concentration in aqueous solutions can vary over a very large range. In 1N HCl, for example, $[H^+] = 1M$, and in 1N NaOH, $[H^+] = 1 \times 10^{-14}M$. Since the range of hydrogen ion concentration is so large, it is convenient to express its concentration on a logarithmic scale, known as the pH scale, where pH is defined as

$$pH \equiv \log_{10}\frac{1}{[H^+]} = -\log_{10}[H^+]$$

Hence, when $[H^+] = 1M$, pH = 0; when $[H^+] = 1 \times 10^{-7}M$, pH = 7; when $[H^+] = 1 \times 10^{-14}$, pH = 14. The larger the pH, the lower the $[H^+]$. Two solutions that differ in pH by one unit differ in $[H^+]$ by a factor of 10.

The K'_a values for weak acids also cover a very broad range. Therefore, it is convenient to express K'_a values on a logarithmic scale, known as the pK'_a scale, where pK'_a is defined as

$$pK'_a \equiv \log_{10}\frac{1}{K'_a} = -\log_{10} K'_a$$

Hence, the larger the pK'_a, the weaker the acid and the stronger its conjugate base.

The Henderson-Hasselbalch Equation

Equation *a*, the expression for the ionization constant of an acid

(a) $$K'_a = \frac{[H^+][A^-]}{[HA]}$$

can be rearranged to give equation *b*

(b) $$\frac{1}{[H^+]} = \frac{1}{K'_a} \times \frac{[A^-]}{[HA]}$$

Taking the logarithm of both sides, equation *b* yields *c*,

(c) $$\log_{10}\frac{1}{[H^+]} = \log_{10}\frac{1}{K'_a} + \log_{10}\frac{[A^-]}{[HA]}$$

and from the definitions of pH and pK'_a, equation *c* is equivalent to *d*, which is known as the Henderson-Hasselbalch equation.

(d) $$pH = pK'_a + \log_{10}\frac{[A^-]}{[HA]}$$ Henderson-Hasselbalch equation

From the Henderson-Hasselbalch equation, it is apparent that the pH of a solution containing equimolar concentration of a weak acid and its conjugate base ($\frac{[A^-]}{[HA^-]} = 1$) will be equal to the pK'_a of the weak acid.

TITRATION AND BUFFERS

Titration of a Weak Acid

The reaction of a weak acid with a strong base such as OH^-, the hydroxide ion, produces H_2O and the conjugate base of the weak acid.

$$HA + OH^- \rightleftharpoons H_2O + A^-$$

Conversely, the titration of a weak base with a strong acid produces H_2O and the conjugate acid of the weak base.

$$A^- + H_3O^+ \rightleftharpoons HA + H_2O$$

These are both reversible reactions, but the equilibrium position of each is very far to the right. Therefore, in a titration experiment in which a strong base, such as OH^-, is added to a solution of weak acid HA, the number of equivalents of A^- in the solution will be approximately equal to the number of equivalents of OH^- added (provided the equivalents of OH^- added are less than the equivalents of HA present in the solution initially). Hence, when the weak acid is half titrated, i.e., the equivalents of OH^- added are equal to one half the equivalents of HA present initially, then $[A^-] = [HA]$, and the pH of the solution is equal to the pK'_a of HA. Plots of pH versus extent of titration with a strong base are depicted in Figure D–1. Inspection of these plots reveals that the shape of the titration curve is independent of the pK'_a of the acid, whereas the vertical displacement is dependent upon the pK'_a. The pK'_a of each acid can be determined by measuring pH at half titration (0.5 on the abscissa).

Buffers

Inspection of the titration curves of Figure D–1 also reveals that over a large central portion of each curve the addition of relatively large amounts of strong base (or strong acid) produces relatively small changes in the pH of the solution. This ability to resist a pH change upon addition of a strong base or a strong acid is known as buffer action, and it is observed in any solution containing a weak acid and its conjugate base. Therefore, a solution containing a weak acid and its conjugate base is referred to as a buffer, or buffer system. The maximum buffering power of a buffer occurs at a pH equal to the pK'_a of the weak acid, that is, at the pH at which $[HA] = [A^-]$. This is the point of least slope of the titration curve. The effective buffering range of a buffer composed of a weak acid and its conjugate base is roughly from 1 pH unit below to 1 pH unit above its pK'_a. Thus, the *range* of buffering power is independent of the buffer concentration. On the other hand, the buffering *capacity* of a buffer, the number of equivalents of strong base or strong acid required to produce a given pH change, is directly proportional to the concentration of the buffer.

Biological systems are protected against large fluctuations in the concentration of H^+ by the presence of intercellular and intracellular buffer systems. the most important inorganic buffer systems in human physiology are the phosphate and bicarbonate buffer systems.

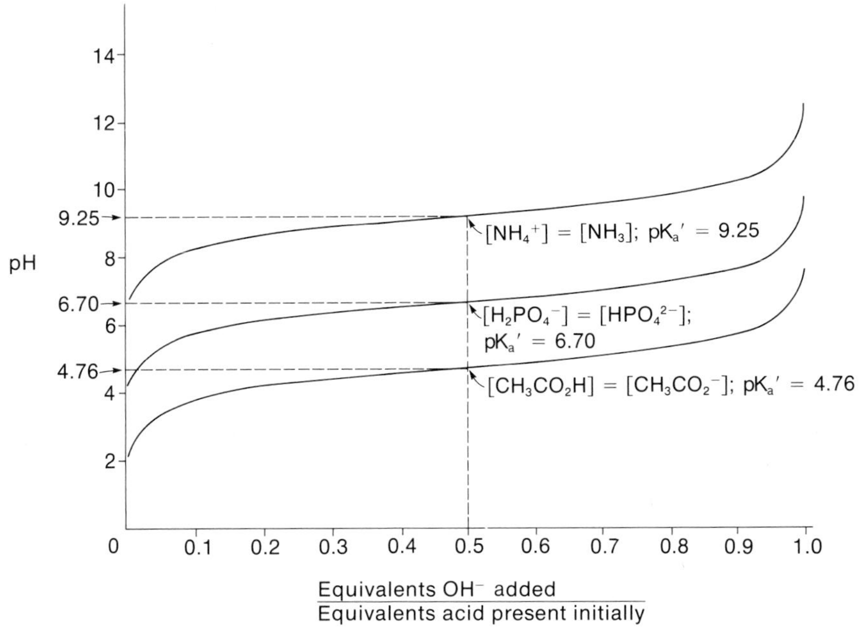

Figure D–1 Titration of three weak acids with a strong base (OH^-).

THE PHOSPHATE BUFFER SYSTEM

Phosphoric acid has three dissociable protons.

$$H_3PO_4 \overset{K'_{a_1}}{\rightleftharpoons} H_2PO_4^- + H^+ \quad pK'_{a_1} = 2.1$$

$$H_2PO_4^- \overset{K'_{a_2}}{\rightleftharpoons} HPO_4^{2-} + H^+ \quad pK'_{a_2} = 6.7$$

$$HPO_4^{2-} \overset{K'_{a_3}}{\rightleftharpoons} PO_4^{3-} + H^+ \quad pK'_{a_3} = 12.3$$

Since the second ionization of phosphoric acid has a pK'_a of 6.7, a solution containing $H_2PO_4^-$ and its conjugate base, HPO_4^{2-}, will constitute an effective buffer system near neutrality, the pH range of physiological interest. The Henderson-Hasselbalch equation for the biologically important phosphate buffer system is

$$pH = pK'_{a_2} + \log_{10}\frac{[HPO_4^{2-}]}{[H_2PO_4^-]} = 6.7 + \log_{10}\frac{[HPO_4^{2-}]}{[H_2PO_4^-]}$$

THE BICARBONATE BUFFER SYSTEM

The most important buffer system in the blood plasma of humans is the bicarbonate buffer system. This is a more complex buffer system than most, because carbonic acid, H_2CO_3, is in equilibrium with CO_2 dissolved in the plasma via a reaction catalyzed by carbonic anhydrase. Furthermore, the dissolved CO_2 equilibrates with gaseous CO_2 in the alveoli of the lung.

$$CO_{2(gas)} \rightleftharpoons CO_{2(dissolved)} + H_2O \underset{\text{carbonic anhydrase}}{\rightleftharpoons} H_2CO_3 \rightleftharpoons H^+ + HCO_3^-$$

The Henderson-Hasselbalch equation for the dissociation of H_2CO_3 is

$$pH = pK'_a + \log_{10}\frac{[HCO_3^-]}{[H_2CO_3]} = 3.88 + \log_{10}\frac{[HCO_3^-]}{[H_2CO_3]}$$

However, the concentration of H_2CO_3 is dependent upon the concentration of dissolved CO_2, and the concentration of dissolved CO_2 in turn is dependent upon the partial pressure of CO_2, P_{CO_2}, in the gas phase with which the solution is equilibrated. Therefore, if the entire coupled system is considered, the equation becomes

$$pH = 6.1 + \log_{10}\frac{[HCO_3^-]}{0.0301\ P_{CO_2}}$$

This shows that the P_{CO_2} in the alveolar air, which is regulated by the rate and depth of respiration, is an important factor in the regulation of plasma pH.

APPENDIX E

DERIVATION OF THE MICHAELIS-MENTEN EQUATION

The Michaelis-Menten theory is based on the following model for an enzyme-catalyzed reversible reaction

(1) $$E + S \underset{k_2}{\overset{k_1}{\rightleftharpoons}} ES \underset{k_4}{\overset{k_3}{\rightleftharpoons}} P + E$$

in which k_1, k_2, k_3, and k_4 are specific rate constants, S is substrate, P is product, E is free enzyme, and ES is an enzyme-substrate complex formed by the binding of substrate S at the active site of the enzyme. The velocity of the reaction is the net rate of appearance of P which is equal to the rate of appearance of P minus the rate of disappearance of P. This is given by

$$v = \frac{d[P]}{dt} = k_3[ES] - k_4[P][E]$$

If the initial reaction mixture contains only E and S (no P) and the initial velocity, v_0, is determined, the concentration of P will be very small when v_0 is measured. If [P] is sufficiently small, then $k_4\,[P][E] << k_3[ES]$ and the initial velocity will be approximated by equation 2

(2) $$v_0 = k_3[ES]$$

In addition to the approximation represented by equation 2, another assumption is made to simplify the analysis of enzyme kinetics. The second assumption is that the concentration of the enzyme-substrate complex is constant during the time in which the initial velocity is determined. This is known as the steady-state assumption. If the concentration of ES is constant, the rate of formation of ES must be equal to the rate of breakdown of ES. The rate of formation of ES can be represented by

(3) $$\text{rate of formation of ES} = k_1\,[E][S] + k_4\,[E][P]$$

However, since we are concerned with initial velocity conditions where $[P] << [S]$, we can substitute for equation 3 the approximation

(4) $$\text{rate of formation of ES} = k_1\,[E][S]$$

The rate of breakdown of ES is given by

(5) $$\text{rate of breakdown of ES} = k_2\,[ES] + k_3\,[ES]$$

Hence, from the steady-state assumption

$$\text{Rate of formation} = \text{Rate of breakdown}$$
$$k_1\,[E][S] = k_2\,[ES] + k_3\,[ES]$$

or

(6) $$\frac{[E][S]}{[ES]} = \frac{k_2 + k_3}{k_1}$$

Since k_2, k_3, and k_1 are constants, $\frac{k_2 + k_3}{k_1}$ is also a constant and is known as the Michaelis constant, K_m.

(7) $$K_m \equiv \frac{k_2 + k_3}{k_1}$$

Substituting K_m into equation 6, we obtain equation 8

(8) $$\frac{[E][S]}{[ES]} = K_m$$

The total enzyme concentration, $[E_T]$, is equal to the concentration of free enzyme plus the concentration of enzyme-substrate complex; therefore

$$[E_T] = [E] + [ES]$$

or

$$[E] = [E_T] - [ES]$$

Substituting this expression for E into equation 8 yields

(9) $$\frac{([E_T] - [ES])\,[S]}{[ES]} = K_m$$

Solving for [ES], we obtain

(10) $$[ES] = \frac{[E_T]\,[S]}{K_m + [S]}$$

Substitution of equation 10 into equation 2 yields

(11) $$v_0 = \frac{k_3\,[E_T]\,[S]}{K_m + [S]}$$

Since the approximation can be made that when v_0 is measured, the concentration of S is equal to $[S_0]$, the initial concentration of substrate, equation 11 can be approximated by

(12) $$v_0 = \frac{k_3[E_T][S_0]}{K_m + [S_0]}$$

Returning to a consideration of equation 2, $v_0 = k_3[ES]$, it is apparent that the maximum initial velocity, V_{max}, will be observed when the concentration of ES is maximum. The maximum concentration of ES will exist when $[ES] = [E_T]$, that is, when all of the enzyme present is in the form of the ES complex. Therefore,

(13) $$V_{max} = k_3[E_T]$$

Substituting 13 into 12 yields equation 14, the Michaelis-Menten equation.

(14) $$v_0 = \frac{V_{max}[S_0]}{K_m + [S_0]}$$ Michaelis-Menten equation

The binding of substrate by the enzyme, $E + S \underset{k_2}{\overset{k_1}{\rightleftharpoons}} ES$, is a reversible reaction; therefore, the dissociation constant, K_d, for the ES complex is

(15) $$K_d = \frac{[E]_{eq}[S]_{eq}}{[ES]_{eq}}$$

where $[\ ]_{eq}$ signifies concentrations at equilibrium. Although equation 15 is of the same form as the expression for K_m given in equation 8, the two equations are not identical, because the concentrations in equation 15 are equilibrium concentrations and those in equation 8 are steady-state concentrations. The significance of the difference in the two equations is more readily seen if the expression for K_d is derived in terms of the rate constants. At equilibrium, the velocity in the forward direction, v_f, must be equal to the velocity in the reverse direction, v_r. At equilibrium, $v_f = k_1[E]_{eq}[S]_{eq}$ and $v_r = k_2[ES]_{eq}$; therefore

$$k_1[E]_{eq}[S]_{eq} = k_2[ES]_{eq}$$

or

$$\frac{[E]_{eq}[S]_{eq}}{[ES]_{eq}} = \frac{k_2}{k_1} = K_d$$

Since K_d is equal to $\frac{k_2}{k_1}$ and K_m is equal to $\frac{k_2 + k_3}{k_1}$ (see equation 7), K_m is larger than K_d. If for a given enzyme-catalyzed reaction, k_3 is much less than k_2, then K_m will be approximately equal to K_d. Hence, the approximation that $K_m \approx K_d$ is only valid for those enzyme-catalyzed reaction in which $k_3 << k_2$.

APPENDIX F

CATALYTIC MECHANISMS

GENERAL ACID-BASE CATALYSIS

In aqueous solution, any acid other than H_3O^+ and any base other than OH^- is known respectively as a general acid and a general base. Some functional groups on proteins, which might function as general acids, are —COOH, 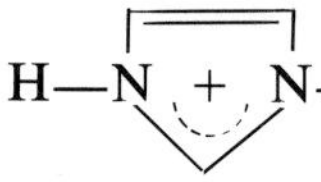, and $—NH_3^+$. Some of the functional groups, which might serve as general bases, are $—COO^-$, 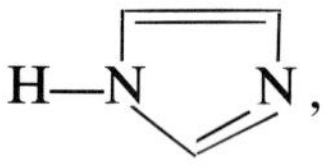, and $—NH_2$. In general acid catalysis, proton transfer by the general acid occurs at the transition state of the rate-determining step of the reaction. In general base catalysis, proton abstraction by the general base at the transition state of the rate-determining step of the reaction results in a stabilization of the transition state and a rate acceleration. Many simple organic reactions have been found to be subject to general acid and general base catalysis. For example, the hydrolysis of some esters is subject to general base catalysis. A proposed mechanism for the general base catalysis of ester hydrolysis involves the abstraction of a proton by a general base, B^-, from the water molecule as it forms a bond with the carbonyl carbon of the ester.

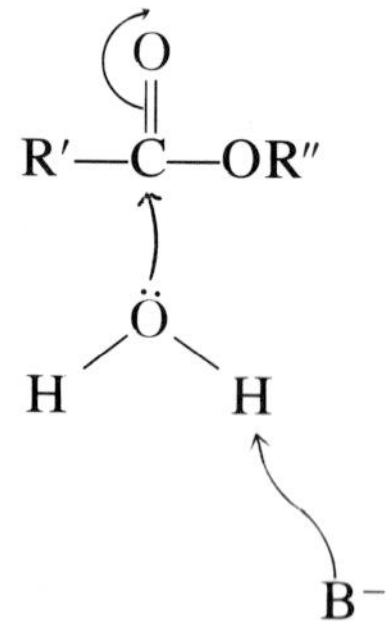

COVALENT CATALYSIS

In covalent catalysis, an intermediate containing a covalent linkage between the reactant and the catalyst is formed during the course of the reaction. One simple model for this type of catalysis is found in the nucleophilic catalysis of ester hydrolysis. For example, the hydrolysis of some esters is catalyzed by trimethylamine.

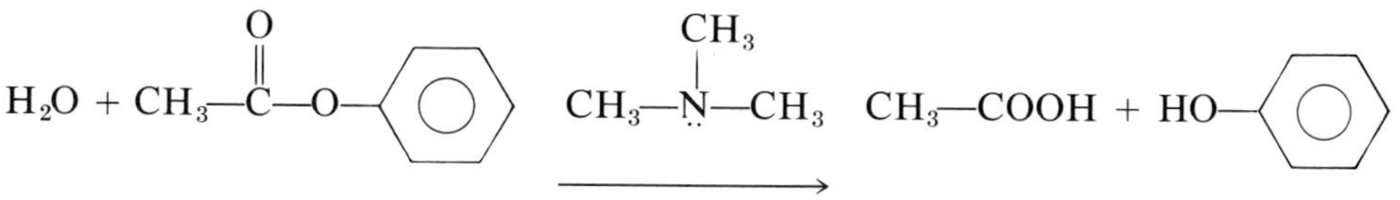

The mechanism of the trimethylamine catalysis involves a rapid nucleophilic attack of the amine on the carbonyl carbon of the ester, resulting in displacement of the phenolate ion and formation of a covalent intermediate between reactant and catalyst. The covalent intermediate is then rapidly attacked by H_2O and hydrolyzed.

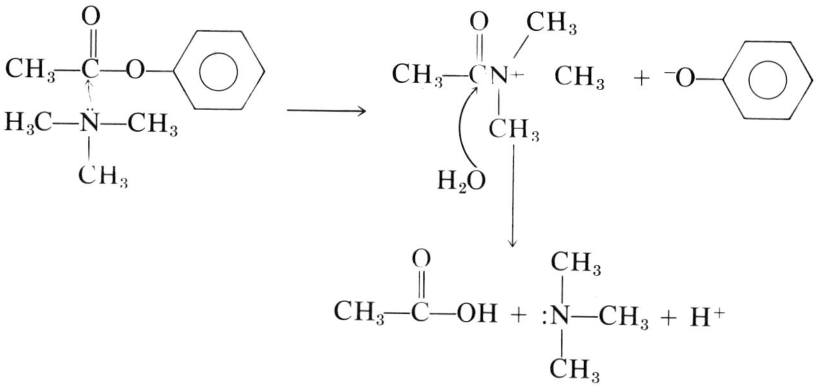

Some of the functional groups of a protein that could potentially function as nucleophiles are the hydroxyl group (—OH) of serine, the sulfhydryl group (—SH) of cysteine, the imidazole group (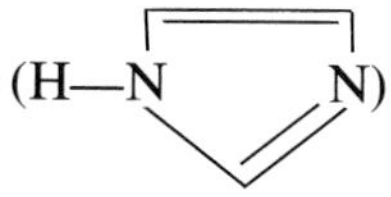) of histidine, and the amino group ($—NH_2$).

Approximation and Orientation of Reactants

The model reactions discussed in the foregoing sections are intermolecular reactions involving two or more molecules. Intermolecular reactions can occur only when the reactants come together in an orientation respective to one another that permits effective orbital overlap for bond formation. However, an enzyme-catalyzed reaction more closely resembles an intramolecular reaction, since the substrates and catalyst are brought very close together through the formation of the enzyme-substrate complex. Such an approximation of the reactants markedly accelerates a reaction. For example, converting the reaction discussed in the previous section to an intramolecular reaction (shown in the following equation) results in a rate enhancement of 10^3.

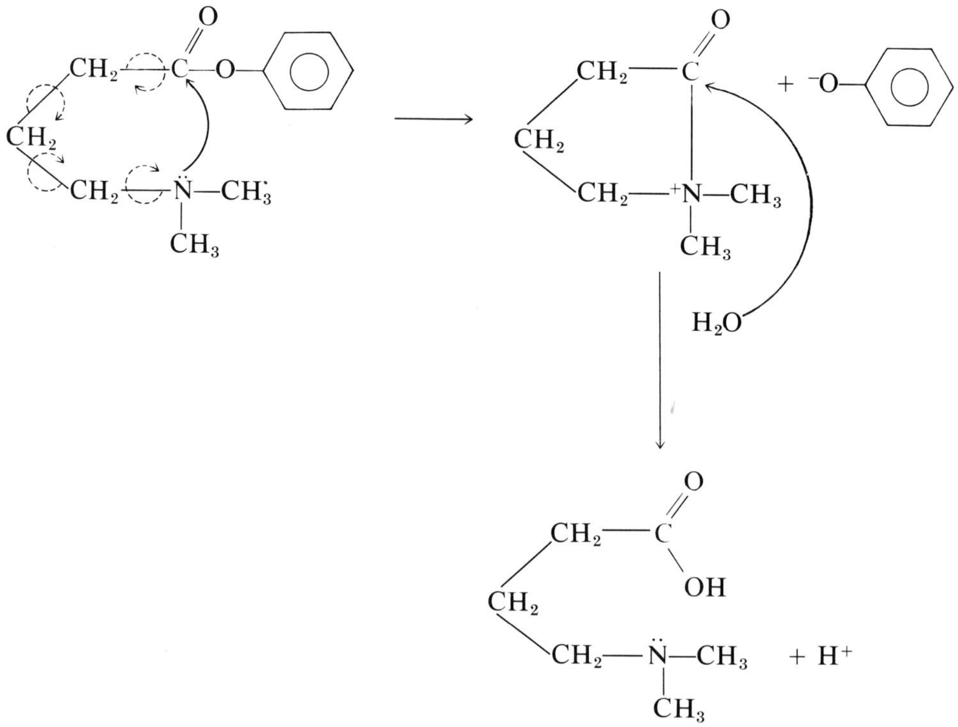

Even an intramolecular reaction of this type does not contain the rigid approximation and orientation that can be attained in the ES complex. Rotations about the C—C and C—N bonds, as indicated by the dashed arrows (⤻), lead to conformations of the molecule that do not have the reacting groups in close proximity with the proper orientation for orbital overlap and bond formation. From the results of studies with model systems in which the approximation and orientation are more rigid, it appears that the approximation and orientation of the reacting groups achievable through formation of the enzyme-substrate complex may account for rate enhancements of 10^6 to 10^8. Therefore, formation of the ES complex is a very important factor contributing to the catalytic effectiveness of all enzymes.

Induction of Strain or Distortion in the Substrate or Enzyme on Forming the ES Complex

Strain or distortion in the reactant molecule can result in large rate enhancements, provided the strain is at least partially relieved in the transition state. In the following simple example, the rate of thiol-disulfide exchange represented by reaction a is approximately 5×10^3 greater than that of reaction *b*.

a.

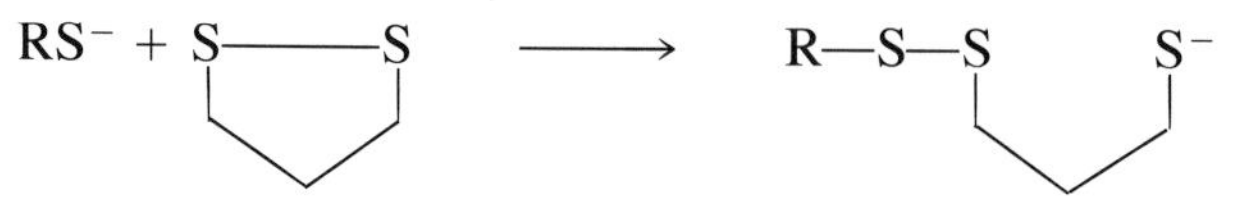

b.

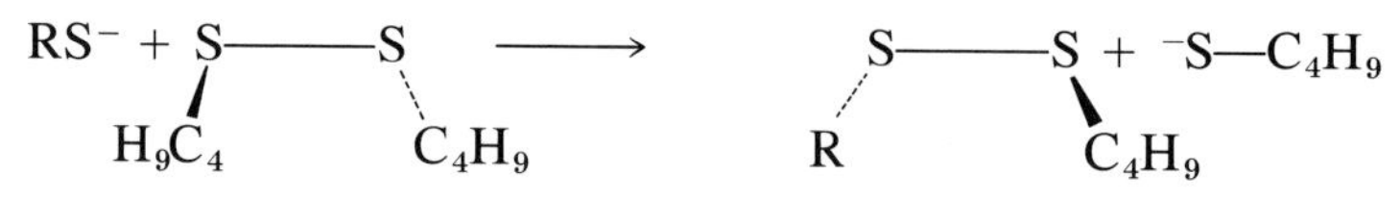

Due to the electrostatic repulsion of the unshared electrons on the two sulfur atoms in a disulfide, the most stable conformation for a disulfide, R—S—S—R, is one in which the dihedral angle is about 90° (looking down the S—S bond, 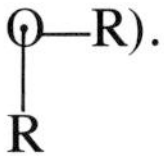). The cyclic disulfide in reaction *a* cannot assume this most stable conformation as a result of the constraints imposed by the ring; thus, the cyclic disulfide is more strained than the open chain disulfide of reaction *b*. Because some of this strain is relieved in the transition state as the two sulfur atoms move apart, the activation energy is less and the rate is greater for reaction *a* than for reaction *b*. In other model reactions, strain has been observed to produce rate enhancements of 10^8. Therefore, if the binding of substrate to enzyme in the formation of the ES complex were to induce a strain or distortion in the substrate or enzyme that was relieved in the transition state, very large rate enhancements might be expected.

APPENDIX G

FREE ENERGY CHANGES IN REDOX REACTIONS

The free energy changes in oxidation-reduction reactions can be obtained in the following way. Consider the redox couples:

$$A \rightleftharpoons A^{2+} + 2e^-$$
$$B \rightleftharpoons B^{2+} + 2e^-$$

If A and B^{2+} are mixed at pH 7.0 and 25°C, a net transfer of electrons from A to B^{2+} will occur until the reaction reaches equilibrium.

(1) $$A + B^{2+} \rightleftharpoons A^{2+} + B$$

At equilibrium, the redox potentials of the two redox couples must be equal. From the Nernst equation, the redox potentials at equilibrium are

$$E_A = E'_{0_A} + \frac{RT}{nF} \ln \frac{[A^{2+}]_{eq.}}{[A]_{eq.}}$$

and

$$E_B = E'_{0_B} + \frac{RT}{nF} \ln \frac{[B^{2+}]_{eq.}}{[B]_{eq.}}$$

Since $E_A = E_B$ at equilibrium,

$$E'_{0_A} + \frac{RT}{nF} \ln \frac{[A^{2+}]_{eq.}}{[A]_{eq.}} = E'_{0_B} + \frac{RT}{nF} \ln \frac{[B^{2+}]_{eq.}}{[B]_{eq.}}$$

or

(2) $$E'_{0_B} - E'_{0_A} = \frac{RT}{nF} \ln \frac{[A^{2+}]_{eq.}\,[B]_{eq.}}{[A]_{eq.}\,[B^{2+}]_{eq.}}$$

However, the equilibrium constant for reaction 1 is given by

(3) $$K = \frac{[A^{2+}]_{eq.}\,[B]_{eq.}}{[A]_{eq.}\,[B^{2+}]_{eq.}}$$

therefore, substituting equation 3 into 2 gives

$$E'_{0_B} - E'_{0_A} = \frac{RT}{nF} \ln K$$

or

$$\ln K = \frac{nF\Delta E_0'}{RT}$$

where $\Delta E_0' = E_{0_B}' - E_{0_A}'$, the difference in the standard redox potentials of the two couples. In Chapter 9 it was shown that $\Delta G^{0\prime}$, the standard free energy change for a reaction, is related to K, the equilibrium constant for the reaction, by the expression

(5) $$\Delta G^{0\prime} = -RT \ln K$$

Substituting equation 4 into 5 gives

(6) $$\Delta G^{0\prime} = -nF\Delta E_0'$$

Thus the standard free energy change of a redox reaction can be calculated from the difference in the standard redox potentials of the two couples using equation 6.

INDEX

Page numbers in *italic* indicate figures. Page numbers followed by t indicate tables.